Plant Metabolites: Methods, Applications and Prospects

Swapna Thacheril Sukumaran •
Shiburaj Sugathan • Sabu Abdulhameed
Editors

Plant Metabolites: Methods, Applications and Prospects

 Springer

Editors
Swapna Thacheril Sukumaran
Department of Botany
University of Kerala
Thiruvananthapuram, Kerala, India

Shiburaj Sugathan
Division of Microbiology
KSCSTE—Jawaharlal Nehru Tropical Botanic
Garden and Research Institute
Thiruvananthapuram, Kerala, India

Sabu Abdulhameed
Department of Biotechnology and
Microbiology
Kannur University
Kannur, Kerala, India

ISBN 978-981-15-5135-2   ISBN 978-981-15-5136-9 (eBook)
https://doi.org/10.1007/978-981-15-5136-9

© Springer Nature Singapore Pte Ltd. 2020
This work is subject to copyright. All rights are reserved by the Publisher, whether the whole or part of the material is concerned, specifically the rights of translation, reprinting, reuse of illustrations, recitation, broadcasting, reproduction on microfilms or in any other physical way, and transmission or information storage and retrieval, electronic adaptation, computer software, or by similar or dissimilar methodology now known or hereafter developed.
The use of general descriptive names, registered names, trademarks, service marks, etc. in this publication does not imply, even in the absence of a specific statement, that such names are exempt from the relevant protective laws and regulations and therefore free for general use.
The publisher, the authors, and the editors are safe to assume that the advice and information in this book are believed to be true and accurate at the date of publication. Neither the publisher nor the authors or the editors give a warranty, expressed or implied, with respect to the material contained herein or for any errors or omissions that may have been made. The publisher remains neutral with regard to jurisdictional claims in published maps and institutional affiliations.

This Springer imprint is published by the registered company Springer Nature Singapore Pte Ltd.
The registered company address is: 152 Beach Road, #21-01/04 Gateway East, Singapore 189721, Singapore

We proudly dedicate this book to Prof. (Dr.) M Haridas, professor emeritus and founder head of the Department of Biotechnology and Microbiology, Kannur University, Kerala, India. Prof. Haridas epitomises the triumph of inventiveness over all other settings, which make ones contribution less significant, a triumph of genius over circumstances. For decades, he has pursued and achieved excellence in both research and teaching. His achievements are powerful and lasting in their impact on a global scale. What makes them so inspiring to younger generations of scientists and teachers is that they are more necessary in the present world where science only could solve the problems of environment and human society.

Foreword

Despite the many triumphs of modern medicine, there remain many diseases for which effective treatments are still not available, or for which modern medicines are not affordable, especially for patients living in resource-poor areas. However, many parts of the world have century-old traditions of using herbal medicines to treat a variety of medical conditions, including infections, cancer, and immune disorders.

While many of these treatments have not been rigorously tested in controlled clinical trials, challenges in clinical development consist of incomplete knowledge on their active ingredients and mechanisms of action. In addition to containing compounds to treat specific diseases currently already targeted, plants can also harbor many more metabolites that, if properly identified, could have promise for novel medical applications, or can be the basis for further chemical modification to improve efficacy and reduce the risk of adverse effects.

Further success in this research area is contingent on progress in the isolation, purification, in vitro and in vivo characterization, and production of these plant metabolites. This book provides the latest updates from world experts, investigating a variety of plant metabolites aimed at preventing or treating medical conditions. As the reader will conclude, much progress is being made, and this exciting research opens the door to further explore unchartered territories in future years.

University of California, Davis Koen Van Rompay
Davis, CA, USA

Preface

Plants by way of their metabolites are a hot topic of research for all over the world. Both the types of metabolites—primary metabolites (produced for the existence) and secondary metabolites (produced as a result of metabolic errors or produced as non-essentials and accumulated as by-products)—find enormous relevance in pharmaceutical, food, agriculture and other sectors with commercial potential of billions of dollars. Plants ranging from the primitive to the highly evolved groups and those which survive in unique and diverse habitats are treasures of high-value molecules. Though there are concerted efforts taking place across the globe, the sector is yet to take a full bloom.

The book entitled "Plant Metabolites: Methods, Applications and Prospects" is compiled and edited by scientists from Kerala and most of the contributors of the different chapters are also from Kerala, India. Kerala is well known for its rich biodiversity and it has about 700 km long track of Western Ghats, the enlisted biodiversity hotspot. The state pioneers in sustainable utilization of biodiversity through novel practices, processes and products. Hence, the contribution to knowledge through this edited volume is directly from the players with expertise and experience evolved with a background of years. The background of herbal material utilization in Kerala goes remote in unrecorded history of folklore and in recorded history to the seventeenth century Latin classic, Hortus Malabaricus.

The book comprises twenty-two chapters with state-of-the-art knowledge on sources of plant metabolites ranging from small and primitive plant species to fully evolved trees, methods for extraction, purification and characterization of metabolites, applications of metabolites in various sectors, microbial biotransformation of metabolites for preparation of nutraceuticals, existing traditional knowledge-based practices for plant-based therapy and advanced techniques including that of gene technology for the enhanced and sustainable production of metabolites. While we have taken all the possible efforts to make the book a perfect one, there may be lapses which we are ready to accept and improve in future. We thank all who have contributed to this book and all the well-wishers for extending their support. We sincerely wish that the scientific fraternity across the globe will make use of the information presented in this book published under the banner of Springer-Nature.

Thiruvananthapuram, Kerala, India	Swapna Thacheril Sukumaran
Thiruvananthapuram, Kerala, India	Shiburaj Sugathan
Kannur, Kerala, India	Sabu Abdulhameed

Contents

1 **Plant Metabolomics: Current Status and Prospects** 1
   C. S. Sharanya, A. Sabu, and M. Haridas

2 **Plant Metabolites: Methods for Isolation, Purification, and Characterization** 23
   Shabeer Ali Hassan Mohammed, Renu Tripathi, and K. Sreejith

3 **Molecular Markers and Their Application in the Identification of Elite Germplasm** 57
   Karuna Surendran, R. Aswati Nair, and Padmesh P. Pillai

4 **Cell and Protoplast Culture for Production of Plant Metabolites** ... 71
   S. R. Saranya Krishnan, R. Sreelekshmi, E. A. Siril, and Swapna Thacheril Sukumaran

5 **Hairy Root Culture: Secondary Metabolite Production in a Biotechnological Perspective** 89
   Radhakrishnan Supriya, Radhadevi Gopikuttan Kala, and Arjunan Thulaseedharan

6 **Methods for Enhanced Production of Metabolites Under In Vitro Conditions** 111
   K. P. Rini Vijayan and A. V. Raghu

7 **Invasive Alien Plants: A Potential Source of Unique Metabolites** ... 141
   T. K. Hrideek, Suby, and M. Amruth

8 **Modified Plant Metabolites as Nutraceuticals** 167
   O. Nikhitha Surendran, M. Haridas, George Szakacs, and A. Sabu

9 **Ethnomedicine and Role of Plant Metabolites** 181
   Lekshmi Sathyaseelan, Riyas Chakkinga Thodi, and Swapna Thacheril Sukumaran

10 **Herbal Cosmeceuticals** 217
   Ramesh Surianarayanan and James Prabhanand Bhaskar

11	**Plant Secondary Metabolites as Nutraceuticals** . . . . Lini Nirmala, Zyju Damodharan Pillai Padmini Amma, and Anju V. Jalaj	239
12	**Bioactive Secondary Metabolites from Lichens** . . . . Sanjeeva Nayaka and Biju Haridas	255
13	**Algal Metabolites and Phyco-Medicine** . . . . Lakshmi Mangattukara Vidhyanandan, Suresh Manalilkutty Kumar, and Swapna Thacheril Sukumaran	291
14	**Bioactive Metabolites in Gymnosperms** . . . . Athira V. Anand, Arinchedathu Surendran Vivek, and Swapna Thacheril Sukumaran	317
15	**Flavonoids for Therapeutic Applications** . . . . Thirukannamangai Krishnan Swetha, Arumugam Priya, and Shunmugiah Karutha Pandian	347
16	**Plant-Based Pigments: Novel Extraction Technologies and Applications** . . . . Juan Roberto Benavente-Valdés, Lourdes Morales-Oyervides, and Julio Montañez	379
17	**Plant Lectins: Sugar-Binding Properties and Biotechnological Applications** . . . . P. H. Surya, M. Deepti, and K. K. Elyas	401
18	**Plant Metabolites as Immunomodulators** . . . . Sony Jayaraman and Jayadevi Variyar	441
19	**Polyphenols: An Overview of Food Sources and Associated Bioactivities** . . . . Alejandro Zugasti-Cruz, Raúl Rodríguez-Herrera, and Crystel Aleyvick Sierra-Rivera	465
20	**Plant Metabolites Against Enteropathogens** . . . . Praseetha Sarath, Swapna Thacheril Sukumaran, Resmi Ravindran, and Shiburaj Sugathan	497
21	**Molecular Chaperones and Their Applications** . . . . Gayathri Valsala, Shiburaj Sugathan, Hari Bharathan, and Tom H. MacRae	521
22	**Bioprospecting of Ethno-Medicinal Plants for Wound Healing** . . . . S. R. Suja, A. L. Aneeshkumar, and R. Prakashkumar	553

Editors and Contributors

About the Editors

Swapna Thacheril Sukumaran is a Professor at the Department of Botany, University of Kerala. She completed her PhD at the Department of Biotechnology, Cochin University of Science and Technology, India. Dr. Swapna has 25 years of teaching and research experience and has published nine books and over 100 research papers in journals and conference proceedings. Her research interests include *in vitro* secondary metabolite production, medicinal plant conservation and phytochemistry. Dr. Swapna is a recipient of research grants from the Department of Environment and Climate Change, Government of Kerala, Western Ghats Development Cell, Government of Kerala, University Grants Commission, Government of India and SERB, Department of Science and Technology, Government of India. She is currently serving as a Member of the Kerala State Biodiversity Board.

Shiburaj Sugathan has a background in Plant Sciences and has been working in Microbial Bioprospecting since 1995. He is currently a Senior Scientist and Head of the Division of Microbiology at Jawaharlal Nehru Tropical Botanic Garden and Research Institute, Thiruvananthapuram, India. He has over 25 years of experience in the area of microbial biotechnology and has published several papers in reputed national and international journals, three books, several book chapters and one Indian patent. Dr. Shiburaj received his PhD degree from the University of Kerala and completed his postdoctoral studies at Madurai Kamaraj University, India and Dalhousie

University, Canada. He is a recipient of a BOYCAST fellowship from the Department of Science & Technology, Government of India. Dr. Shiburaj is also a Visiting Professor at the University of Coahulia, Mexico. His current research interests include recombinant expression and scale up of industrial enzymes from Actinobacteria, and the characterization of antibiofilm molecules.

Sabu Abdulhameed is currently working as an Associate Professor at the Department of Biotechnology and Microbiology, Kannur University, India. He received his PhD from Cochin University of Science and Technology and joined the Biotechnology Division of CSIR-NIIST, Thiruvananthapuram as a Scientist Fellow. He has engaged in post-doctoral research work in microbial fermentation and biotransformation in the USA and France. His research interests are in microbial fermentation, therapeutic enzymes and the biotransformation of bioactive molecules. He has contributed extensively to fermentation and biotransformation in medicated wines and has three patents to his credit. Dr. Sabu has published five books and 70 research papers in international journals. He is a recipient of a postdoctoral fellowship from the Institute for Research and Development (IRD, France) and a Visiting Professorship from the Mexican Agency for International Cooperation and Development. He is also a member of many academic bodies and expert panels, and serves on the editorial boards of numerous research journals.

Contributors

M. Amruth Forestry and Human Dimension-Programme Division, KSCSTE—Kerala Forest Research Institute, Thrissur, Kerala, India

A. L. Aneeshkumar Ethnomedicine & Ethnopharmacology Division, KSCSTE—Jawaharlal Nehru Tropical Botanic Garden and Research Institute, Thiruvananthapuram, Kerala, India

R. Aswathi Nair Department of Biochemistry & Molecular Biology, Central University of Kerala, Kasaragod, Kerala, India

V. S. Athira Department of Botany, University of Kerala, Karyavattom Campus, Thiruvananthapuram, Kerala, India

Juan Roberto Banevente-Valdés Department of Chemical Engineering, Universidad Autónoma de Coahuila, Saltillo, Coahuila, México

Hari Bharathan Department of Zoology, Sree Narayana College, Kollam, Kerala, India

James P. Bhaskar ITC Life Sciences and Technology Centre, ITC Limited, Bangalore, Karnataka, India

M. Deepti Department of Biotechnology, University of Calicut, Malappuram, Kerala, India

K. K. Elyas Department of Biotechnology, University of Calicut, Malappuram, Kerala, India

Gayathri Valsala Division of Microbiology, KSCSTE—Jawaharlal Nehru Tropical Botanic Garden and Research Institute, Thiruvananthapuram, Kerala, India

Biju Haridas Division of Microbiology, KSCSTE—Jawaharlal Nehru Tropical Botanic Garden and Research Institute, Thiruvananthapuram, Kerala, India

M. Haridas Inter University Centre for Bioscience and Department of Biotechnology & Microbiology, Dr E K Janaki Ammal Campus, Kannur University, Kannur, Kerala, India

Shabeer Ali Hassan Mohammed Division of Molecular Parasitology & Immunology, CSIR-CDRI, Lucknow, Uttar Pradesh, India

T. K. Hrideek Department of Forest Genetics and Tree Breeding, KSCSTE—Kerala Forest Research Institute, Thrissur, Kerala, India

Anju V. Jalaj Department of Botany, University of Kerala, Thiruvananthapuram, Kerala, India

Sony Jayaraman Department of Biotechnology & Microbiology, Kannur University, Kannur, Kerala, India

R. G. Kala Advanced Centre for Molecular Biology and Biotechnology, Rubber Research Institute of India, Kottayam, Kerala, India

M. V. Lakshmi Department of Botany, University of Kerala, Thiruvananthapuram, Kerala, India

N. Lini Department of Biotechnology, Mar Ivanios College, Thiruvananthapuram, Kerala, India

Tom H. MacRae Department of Biology, Dalhousie University, Halifax, NS, Canada

Julio Montañez Department of Chemical Engineering, Universidad Autónoma de Coahuila, Saltillo, Coahuila, México

Lourdes Morales-Oyervides Department of Chemical Engineering, Universidad Autónoma de Coahuila, Saltillo, Coahuila, México

Sanjeeva Nayaka Lichenology Laboratory, CSIR—National Botanical Research Institute, Lucknow, Uttar Pradesh, India

O. Nikhita Surendran Department of Biotechnology & Microbiology, Inter University Centre for Bioscience, Kannur University, Dr E K Janaki Ammal Campus, Kannur, Kerala, India

Shunmugiah Karutha Pandian Department of Biotechnology, Science Campus, Alagappa University, Karaikudi, Tamil Nadu, India

Padmesh P. Pillai Department of Genomic Science, Central University of Kerala, Kasaragod, Kerala, India

R. Prakashkumar Ethnomedicine & Ethnopharmacology Division, KSCSTE—Jawaharlal Nehru Tropical Botanic Garden and Research Institute, Thiruvananthapuram, Kerala, India

Arumugam Priya Department of Biotechnology, Science Campus, Alagappa University, Karaikudi, Tamil Nadu, India

A. V. Raghu Kerala Forest Research Institute, Thrissur, Kerala, India

Resmi Ravindran Department of Pathology & Laboratory Medicine, University of California Davis Medical Center, Sacramento, CA, USA

K. P. Rini Vijayan Kerala Forest Research Institute, Thrissur, Kerala, India

Raúl Rodríguez-Herrera Food Research Department, Faculty of Chemistry, Autonomous University of Coahuila, Saltillo, Coahuila, Mexico

A. Sabu Inter University Centre for Bioscience, Department of Biotechnology and Microbiology, Dr E K Janaki Ammal Campus, Kannur University, Kannur, Kerala, India

S. R. Saranya Krishnan Department of Botany, University of Kerala, Thiruvananthapuram, Kerala, India

Praseetha Sarath Division of Microbiology, KSCSTE—Jawaharlal Nehru Tropical Botanic Garden and Research Institute, Thiruvananthapuram, Kerala, India

Lekshmi Sathyaseelan Department of Botany, University of Kerala, Thiruvananthapuram, Kerala, India

C. S. Sharanya Inter University Centre for Bioscience, Department of Biotechnology & Microbiology, Dr E K Janaki Ammal Campus, Kannur University, Kannur, Kerala, India

Crystel Aleyvick Sierra-Rivera Laboratory of Immunology and Toxicology, Faculty of Chemistry, Autonomous University of Coahuila, Saltillo, Coahuila, Mexico

E. A. Siril Department of Botany, University of Kerala, Thiruvananthapuram, Kerala, India

K. Sreejith Department of Biotechnology and Microbiology, Dr E K Janaki Ammal Campus, Kannur University, Kannur, Kerala, India

R. Sreelekshmi Department of Botany, University of Kerala, Thiruvananthapuram, Kerala, India

Suby Department of Forest Genetics and Tree Breeding, KSCSTE—Kerala Forest Research Institute, Thrissur, Kerala, India

Shiburaj Sugathan Division of Microbiology, KSCSTE—Jawaharlal Nehru Tropical Botanic Garden and Research Institute, Thiruvananthapuram, Kerala, India

S. R. Suja Ethnomedicine & Ethnopharmacology Division, KSCSTE—Jawaharlal Nehru Tropical Botanic Garden and Research Institute, Thiruvananthapuram, Kerala, India

Swapna Thacheril Sukumaran Department of Botany, University of Kerala, Thiruvananthapuram, Kerala, India

R. Supriya Advanced Centre for Molecular Biology and Biotechnology, Rubber Research Institute of India, Kottayam, Kerala, India

Karuna Surendran Department of Genomic Science, Central University of Kerala, Kasaragod, Kerala, India

M. K. Suresh Kumar Laboratory of Genetics and Genomics, National Cancer Institute—NIH, Bethesda, MD, USA

Ramesh Surianarayanan R&D Consultant, Chennai, Tamil Nadu, India

P. H. Surya Department of Biotechnology, University of Calicut, Malappuram, Kerala, India

Thirukannamangai Krishnan Swetha Department of Biotechnology, Science Campus, Alagappa University, Karaikudi, Tamil Nadu, India

George Szakacs Department of Applied Biotechnology and Food Science, Budapest University of Technology and Economics, Budapest, Hungary

Riyas Chakkinga Thodi Department of Botany, University of Kerala, Thiruvananthapuram, Kerala, India

A. Thulaseedharan Advanced Centre for Molecular Biology and Biotechnology, Rubber Research Institute of India, Kottayam, India

Renu Tripathi Division of Molecular Parasitology & Immunology, CSIR-CDRI, Lucknow, Uttar Pradesh, India

Jayadevi Variyar Department of Biotechnology & Microbiology, Kannur University, Dr. E K Janaki Ammal Campus, Kannur, Kerala, India

S. Vivek Department of Botany, University of Kerala, Thiruvananthapuram, Kerala, India

Alenjandro Zugasti-Cruz Laboratory of Immunology and Toxicology, Faculty of Chemistry, Autonomous University of Coahuila, Venustiano Carranza Blvd. and Jose Cardenas Valdes Street, Saltillo, Coahuila, Mexico

Zyju Damodharan Pillai Padmini Amma ThermoFisher Scientific, Dubai, UAE

Plant Metabolomics: Current Status and Prospects

C. S. Sharanya, A. Sabu, and M. Haridas

Abstract

Plant metabolomics deals with the interpretation of various metabolic pathways in contrast to other -omics technologies applied in systems biology. Metabolomics is a highly challenging field where the metabolite analysis is done by high-end technologies for proposing metabolic pathways. The high-throughput technologies utilized for these studies importantly include mass and nuclear magnetic resonance spectrometry. Both these techniques have their own specific features and it is usually difficult to interpret the data compared to genomics and transcriptomics data. The metabolic pathways in plants are changed during biotic and abiotic stresses. These changes can be noted at each step through metabolite analysis. Plants acclimatize to the changes during stress conditions by producing secondary metabolites by other mechanisms. We may analyse these changes by advanced techniques.

Keywords

Genomics · Transcriptomics · Proteomics · Metabolomics · GC MS · LC-MS

1.1 Introduction

In an era of developed high-throughput genomics (DNA sequencing), transcriptomics (gene expression analysis) and proteomics (protein analysis) metabolomics is the foremost of the '-omics' approaches emerging from the metabolic profiling. Thus, the significant results of all these technologies combine to become systems biology. The interactions among different organisms in the

C. S. Sharanya · A. Sabu · M. Haridas (✉)
Department of Biotechnology and Microbiology, Inter University Centre for Bioscience, Kannur University, Thalassery, Kerala, India

© Springer Nature Singapore Pte Ltd. 2020
S. T. Sukumaran et al. (eds.), *Plant Metabolites: Methods, Applications and Prospects*, https://doi.org/10.1007/978-981-15-5136-9_1

environment are through natural products. This is particularly important in the communication among the members of the same species. Though metabolomics and natural products discovery evolved independently, they have a great amount of structure-function in common and to share. Though these two fields have different origins historically, they have overlapping objectives. The convergence of metabolomics and natural product discovery has been greatly enhanced by the databases devoted to structural parameters, particularly small-molecule databases. Metabolomics is a highly challenging field for the essentially inclusive, nonbiased, high-throughput screening of complicated metabolites present in the plant extracts used as traditional medicine. Furthermore, we can consider it as a complex framework built of all these emerging technologies for a whole, systems biology (Fig. 1.1).

1.1.1 Genomics

The first -omics phase is genomics, which is of the entire genome sequencing of an organism. Thus genome can be defined as a complete set of genes inside a cell. Once the sequencing of entire genome is completed, this sequence can be used to analyse functions of the genes (functional genomics), compare the genome with another genome (comparative genomics) or to generate 3D structure of proteins in a protein family, thus hinting to their function (structural genomics). DNA is the basis of every genome and its further processing leads to the molecular machinery. Prior to the many post-translational modifications, DNA is transcribed first into RNA and then translated into protein. In the case of plant species, the genetic background ensures its production and resistance to unfavourable conditions. It also ensures the scope of improving production and the potential of resistance to environmental risks. We can repair the dysfunctions and can generate a healthy condition by examining the genetic information of any deleterious mutations, insertions or deletions from the normal state. There may be many genetic variants present in an organism of either benign or protective in nature, having an advantage against the condition of no variants. So, the variants are classified as simple, nucleotide variations and structural variations. Single nucleotide variations (SNv) have single insertions or deletions whereas structural variations have large copy number variation and inversions. Moreover, we can analyse these variations by the relatively recent, advanced methods in the area of microarray technology. Microarrays measure the differences in the DNA sequence of individual plants and analyse the expression of thousands of genes at a time. This reveals the abnormalities in chromosomes by a process called comparative genomic hybridization. In the case of agriculture, the significant genes are identified for the nutritional purpose, production and showing relevance in resistance against pathogens (Horgan and Kenny 2011).

The first plant genome sequenced was that of *Arabidopsis thaliana* (Arabidopsis Genome Initiative 2000) and the first crop genome was of rice (Eckardt 2000). The International Rice Genome Sequencing Project (IRGSP) was initiated in 1997 at a workshop held in Singapore. Many countries from all over the world participated in the endeavour and guidelines were proposed (Table 1.1).

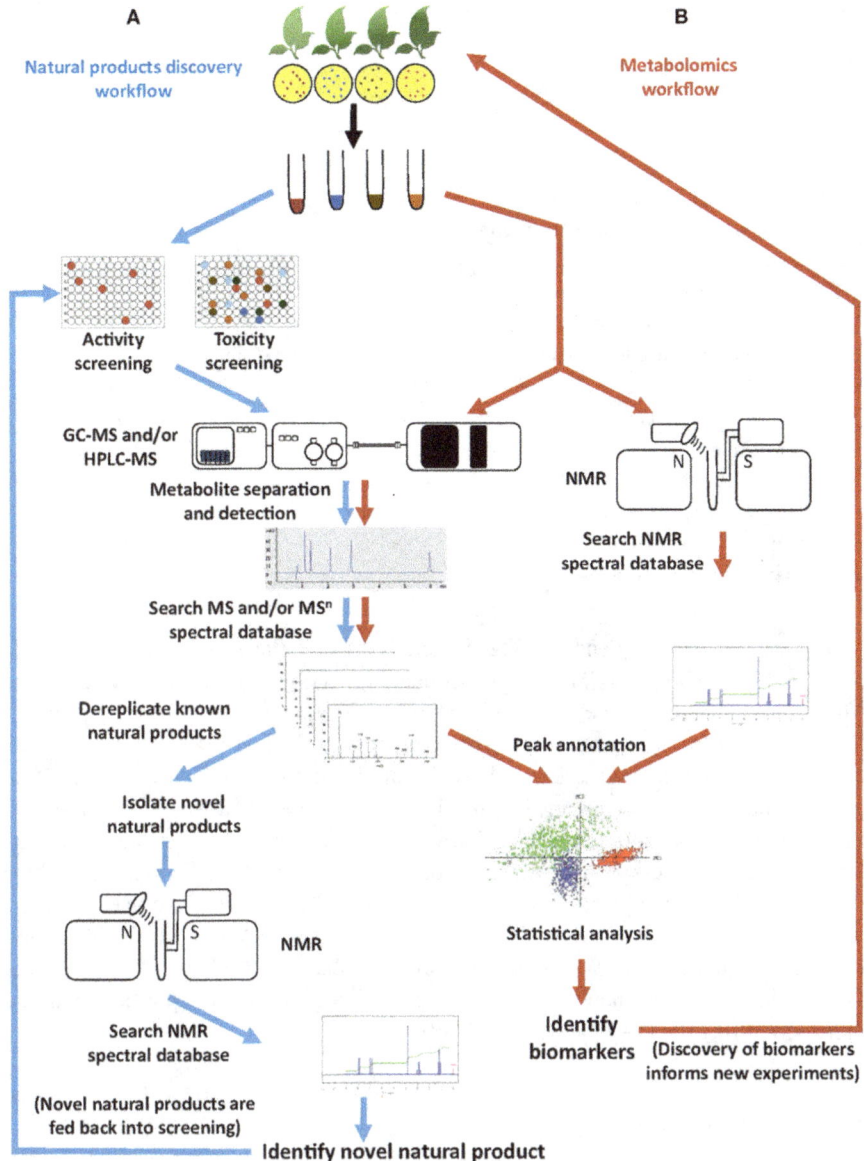

Fig. 1.1 Typical metabolomics work flow (Reproduced from Johnson and Lange 2015, under creative commons license https://doi.org/10.3389/fbioe.2015.00022)

Each country started working on different chromosomes, and finally the complete genome was published in the year 2000. Rice genome has a size of 400–430 million base pairs which is three times higher than *Arabidopsis thaliana* with a well-mapped genome, having 6000 markers and 40,000 expressed sequence tags, making rice a

Table 1.1 Countries involved in the International Rice Genome Sequencing Project

Research Institute	Chromosome
Rice Genome Research Program (RGP; Japan)	1, 6, 7, 8
Korea Rice Genome Research Program (Korea)	1
CCW (United States) CUGI (Clemson University) Cold Spring Harbor Laboratory Washington University Genome Sequencing Center	3, 10
TIGR (United States)	3, 10
PGIR (United States)	10
University of Wisconsin (United States)	11
National Center for Gene Research Chinese Academy of Sciences (China)	4
Indian Rice Genome Program (University of Delhi)	11
Academia Sinica Plant Genome Center (Taiwan)	5
Genoscope (France)	12
Universidad Federal de Pelotas (Brazil)	12
Kasetsart University (Thailand)	9
McGill University (Canada)	9
John Innes Centre (United Kingdom)	2

good candidate for sequencing. The IRGSP followed the strategy of map-based, clone-by-clone shotgun strategy for sequencing. This method was found to be efficient for exploiting the full potential of rice sequence and bacterial artificial chromosome libraries constructed from 'Nipponbare', a variety of *Oryza sativa* ssp. *japonica*. Molecular marker technology allows the identification of a gene that controls specific trait by genome having all genes, themselves acting as a marker and the process becomes more correct and effective. Hence, it becomes more significant while making allele-specific markers for the successive PCR reactions. Thus, end-sequencing, fingerprinting and marker-aided PCR screening are used to make sequence-ready contigs. After the completion of the project, all the data and annotated sequences are made publicly available at each IRGSP member's website. They maintain their sequence data with high quality and available publicly within a short period of time (Sasaki and Burr 2000).

1.1.2 Transcriptomics

Genome is the storage centre of all the biological data. However, it cannot transfer the data to cells directly, and for the transfer, a series of biochemical reactions are carried out with the help of enzymes and proteins. The entire process is referred to as gene expression. The sum of the results of genome expression is the transcriptome. The whole collection of transcripts in a species speaks to a key connection between the DNA and the phenotype whose biological data is required by the cell at a specific time. These RNA molecules are further translated into proteins, which dictate the nature of the biochemical reactions that the cell is able to carry out. Transcriptome profile is widely used for analysing the genetic regulation of specific cell types (Frith et al. 2005).

Hence, transcriptomics is the study of relative RNA transcript abundances, using microarray technologies and sequencing-based methods. Microarray technology has a library of known origin transcripts specialized for each purpose. These RNA sequences are then reverse transcribed into its complementary DNA from two different samples, labelled with different dyes. And after the cDNA synthesis, the amount of expressed genes can be quantified in which RNAs are up-regulated or down-regulated in a normal sample versus abnormal sample. Moreover, the results from these studies can be highly varying because transcription levels are highly influenced by many factors including disease phenotypes. To be of general use, transcriptomics data must be generated under a wide variety of conditions and compared, to eliminate the trivial sources of variation (Dunwell et al. 2001).

Transcriptome studies were earlier based on the gene predictions and the transcripts were reverse transcribed to cDNA to generate expressed sequence tags (ESTs). These sequences were analysed by the automated Sanger sequencing method, similar to the human genome project (Adams et al. 1991). Later on, initial studies of Serial Analysis of Gene Expression (SAGE) were carried out in the human pancreas (Velculescu et al. 1995). In plants, first studies were carried out in Arabidopsis in which expression of 45 genes were analysed through microarrays containing cDNA spotted on to glass slides (Schena et al. 1995). In the case of plant research, gene expression during different stress conditions are important. Furthermore, a high-throughput screening in proteomics has been developed where the analysis of proteins at different conditions can be studied. The analyses of the products of transcripts were more prevalent than transcriptome analysis. Alteration in gene expression can be monitored through transcriptome or proteome to differentiate between the two biological states of a cell.

1.1.3 Proteomics

Proteomics, the study of a complete set of proteins, the effectors of the biological system and their levels, are reliant on comparing mRNA levels as well as posttranslational mechanisms (Graves and Haystead 2002). In certain cases, mRNA level will be high and that cannot be measured by microarray chips in the transcriptome analysis. Therefore, the final product of the biological system, the proteins are profiled and this is considered to be the most significant data set to characterize a bio system. For characterizing proteins, we have to isolate the proteins initially using conventional chromatography techniques and different types of columns are available for this procedure. Columns used for isolation mainly include ion exchange, size exclusion, and affinity chromatography columns. For specific proteins, ELISA and western blotting can be used. But these are incapable of defining the protein expression levels. A complex protein mixture can be separated via SDS PAGE, 2D Gel electrophoresis, and 2D differential gel electrophoresis. In the cases of analysing protein expression, microarray chip-based high-throughput systems are available.

Functional analysis of a complete genome is challenging. So, mass spectrometry (MS) technique is considered better for scrutinizing complex protein mixtures with greater sensitivity. Amino acid sequencing of proteins can be achieved through the Edman degradation sequencing method and quantity of proteins analysed through stable isotope labelling with amino acids in cell culture (SILAC), isotope-coded affinity tag (ICAT) labelling and isobaric tag for relative and absolute quantitation (iTRAQ) techniques. The 3D structure of proteins can be analysed through X-ray crystallography and NMR spectroscopy. We can gain insights into their biological functions through structure data generated by these protocols. Furthermore, various bioinformatics tools are available for the 3D structure prediction, motif and domain analysis, protein-protein interactions and data analysis of MS. Evolutionary relationships can be studied through sequence and structure alignment tools. Thus, proteome investigation gives the total delineation of basic and structural functions of the cell, just as the reaction of the cell against different sorts of pressures/stresses and drugs, utilizing single or multiple proteomics strategies (Aslam et al. 2017).

1.1.4 Metabolomics

Current enrichment through the mass spectrometric study and by means of practical genomic research, the metabolomics area becomes strengthened in research. The foremost activities in this area entail setting up a multifaceted, wholly integrated strategy for the top line sample extraction, metabolite separation/detection/identification, computerized data procuring/handling/analysis, and ultimately quantification. Both analytical and computational trends are indispensable to acquire this goal (Hall et al. 2002).

The known profile of all compounds, synthesized by an organism, is known to be metabolome. The term was coined by Stephen Oliver (College of Manchester, UK) (Oliver et al. 1998). New definition of metabolomics considers it as all the metabolites of a living being gathered to be distinguished and quantified inside an organic framework under specific conditions (Fiehn 2002). A large variety of metabolites, with diverse physical properties, including ionic inorganic compounds, hydrophilic carbohydrates derived through biochemical means, and hydrophobic compounds, are coming under this metabolic pool. A wide variety of compounds are present in the plant kingdom. It appears that there is a complex network among these small molecules in plants, and by detecting the relation among these metabolites, metabolomic investigation contributes to the understanding of the connection essentially between genotype and metabolic outcome by handling key network components. Such a kind of metabolomic investigation, coordinating with transcriptome, has been effectively connected to explore the metabolite concentrations in plants (Hong et al. 2016).

Unlike transcriptome and proteome studies via genomic knowledge and central dogma, it is challenging to generate metabolomics data. In spite of the fact that metabolomics is downstream of the other genomics data (transcriptomics and proteomics), the metabolome of a species is not like transcriptome or proteome and cannot be accessed specifically by known genomic data using central dogma. Subsequently, metabolomics is used to get a relevant massive data set for the disclosure of genes and pathways through accurate and high-throughput technologies. Owing to this diversity in metabolites, the present metabolomics studies combine large varieties of intricate analytical tools to combine all the varying data from complex biological samples. Among these, a prominent analytic tool used for the analysis is NMR. However, its lower sensitivity and dynamic range make NMR less prominent. Hence, the most powerful technique used is a liquid or gas chromatography coupled with mass spectrometry (Hall et al. 2002).

1.2 Plant Metabolites

There are mainly two types of metabolites present in plants, primary and secondary. Primary metabolites are required for the growth and development and found to be present in every plant and secondary metabolites derive from primary metabolites, which are considered to be important in plant defence mechanisms, and their functions are specific. These are low-molecular-weight compounds that are often used as drugs, dyes, insecticides, flavours and fragrances with high commercial value. They are species, organ, tissue and cell-specific and might be produced in certain conditions like defence against pathogens and herbivores, certain stress conditions, attracting insects and animals for fertilization and/or seed dispersal or repellence of unwanted feeders. They are distributed in different combinations in different parts of the plants like leaves stem, bark, roots, shoots, etc. at different stages of development like seeds, seedlings and plantlets under different ecological conditions in different classes of plants. In normal prospects, they are produced for survival and reproduction (Guerriero et al. 2018).

These metabolites are formed mainly via three pathways in plants, isoprenoid pathway, shikimate pathway, and polyketide pathway. Shikimate pathway is the major pathway for the synthesis of aromatic compounds and further modifications are based on the plant species. In the case of carbon flux, the phenylpropanoid pathway is significant and about 20% of cell metabolism work is based on these and chorismate mutase, found to be the target enzyme in this pathway. Shikimate pathway is prominent because it leads to the production of lignin, flavonoids and anthocyanin. Furthermore, the next relevant pathway is the isoprenoid pathway for the production of terpenoids. There are mainly three classes of metabolites, terpenoids, phenols, and nitrogen- and sulphur-containing components (Thirumurugan et al. 2018).

1.2.1 Phenols

Plants possess a wide range of phenolic compounds derived from phenylalanine or tyrosine by-products of the shikimate pathway. These include flavonoids, lignins, stilbenes, tannins and lignans along with components of suberin and cutin having long-chain carboxylic acids. These components have been used as pesticides, antibiotics, UV protectants, for establishing symbiosis with rhizobia and so on.

Vanillin, the most important flavouring agent produced in vanilla orchid (*Vanilla planifolia*), constitutes about 1% only of the market share. Most of vanillin is produced by advanced synthetic biology technique for natural products, in laboratory by microorganisms. Hence, the microbial production of secondary metabolites ensures higher productivity and the metabolite can be analysed through mass spectrometry. Earlier, vanillin was produced from ferulic acid via biotransformation using *Streptomyces sannanensis, Escherichia coli* and *Pseudomonas fluorescens*. Later on, a combination of *Aspergillus niger* and *Pycnoporus cinnabarinus* is being used for the microbial catabolism of ferulic acid to vanillin (Pyne et al. 2019).

1.2.2 Terpenoids

Terpenoids, plant isoprenoids, form a major class of secondary metabolites derived from acetyl Co-A or glycolytic pathways and are commercially important components of aroma substance for cosmetics, food and beverages, vitamins (A, D, E) natural insecticides (pyrethrin), rubber and gutta-percha and solvents like turpentine. In the biosynthetic pathway of isoprenoid the starting precursor is acetyl CoA. Combination with two or three acetyl CoA molecules forms mevalonate. Mevlonic acid further gets transformed into isopentenyl pyrophosphate (IPP), the five-membered Skelton of terpenes via several reactions (pyrophosphorylation, decarboxylation, dehydration). IPP is also formed through glycolytic and photosynthetic reduction cycle via methylerythritol phosphate. Based on the number of five carbon units present terpenes are classified as monoterpene (2 C5 units), sesquiterpene (3 C5 units), di terpene (4 C5 units), triterpene (30 carbon), tetraterpene (40 carbon) and polyterpenes ([C5]n) (Kallscheuer et al. 2019).

Microbial production of terpenoids is also favoured globally. Linalool is a sesquiterpenoid produced via microbial system by expressing linalool synthase gene from *Lavandula angustifolia* to S*accharomyces cerevisiae* and significant amount of linalool is produced by the genetically modified organism. In addition to this, monoterpene α-pinene production is also favoured by expressing geranyl diphosphate synthase and a pinene synthase *in Escherichia coli* and *Corynebacterium glutamicum* (Huccetogullari et al. 2019).

1.2.3 Secondary Metabolite with Sulphur Group

Secondary metabolites produced during sulphur metabolism are not usual. But they have a significant role in plant defence mechanism. The main two categories of sulphur-containing compounds are glucosinolates and the alliins. These compounds are usually inactive in plants and in case of any external stimuli like herbivore attack, the enzyme producing these compounds is activated and the products are released. The hydrolytic enzymes activating these compounds are myrosinases and alliinases.

They are usually stored in high amount and released in unfavourable situation leading to the production of metabolites. Alliins are found in Alliaceae family that include onion, garlic and leeks. It expresses many pharmacological activities like antibacterial, antifungal, anticancer and antiviral properties (Thirumurugan et al. 2018).

1.2.4 Secondary Metabolites with Nitrogen Group

Secondary metabolites having nitrogen in their heterocyclic ring are found to be alkaloids and are derived from aromatic amino acids likes tyrosine, tryptophan, aspartic acid and lysine. They are classified into three major categories, one with nitrogen-containing heterocyclic group, the other with non-cyclic biological amines and finally steroidal alkaloids, which are not derived from amino acids. Furthermore, they can block ion channels, hinder proteins, and act on nervous system and membrane transport (Wink 2018). Most of the alkaloids are found to be in its salt form, basic in nature and are soluble in hydro-ethanolic media. There are about 15,000 alkaloids reported from 20% of plants species. Nicotine and atropine alkaloids are true alkaloids with nitrogen in their heterocyclic ring and found to be basic in nature, while caffeine and solanidine are pseudo alkaloids not derived from any amino acid and exhibits basicity. Common ring structure for the true alkaloids includes piperidines, pyrrolidines, isoquinolines, indoles, pyridines and pyrroles. Another kind of alkaloids which are basic in nature but do not have nitrogen in their heterocyclic ring are termed as proto-alkaloids. Mescaline, a phenyl ethyl amine derived alkaloid, is an example for proto-alkaloids (Chomel et al. 2016).

Furthermore, presence of a large variety of these plant secondary metabolites and their impact on communication with environment and in plant defence mechanisms make them a crucial target in the plant metabolome studies. Biotic and abiotic stresses including salinity, temperature, CO_2 availability, and water scarcity challenge the plants to thrive under these conditions. The current scenario with extreme climate changes affects plants' biochemical and physiological mechanisms, and can be identified and studied by modern technologies.

1.3 Methods in Metabolomics

Metabolomics has seen an exponential development in the most recent decade, driven by significant applications spreading over a wide range of domains in life. MS in combination with chromatography and nuclear magnetic resonance are the two noteworthy analytical tools for investigations of metabolites as a mixture. Inferable from its innate and significantly quick information delivery, MS assumes an inexorably prevailing strategy in the metabolomics field. Accordingly, the metabolomics field has moved far. Subjective strategies and basic examples acknowledge ways to deal with a scope of worldwide and focused quantitative methodologies. They are routinely utilized and give solid information, which bring in more prominent confidence in the derivations determined (Hong et al. 2016).

Metabolomics have effect on vast areas including pharmaceutics, drug development, toxicology, food and plant science, environmental science and medicine. Metabolic profiling studies were first started for analysing the human biological fluids to determine the variability in profiles across individuals. In 1940–1960s studies were carried out in paper chromatography and later on in 1970s, with the emergence of new techniques like GC/LC-MS, allowed the quantitative analysis of the metabolites. Earlier studies were done by Roger Williams and co-workers. Later on, Horning along with Pauling and Robinson teams started developing the metabolic profiling of body fluids by GC-MS (Lytovchenko et al. 2009).

Plants possess an invariant metabolism consisting of important metabolic pathways in various tissues, cells and subcellular organelles to produce metabolites essential for responses against infections, stress and the immediate environment. This coordinated action of all organelles helps in the synthesis of various chemical compounds with extraordinary roles, which make plants a tremendous model for metabolome analysis.

Plants are sessile life forms which cannot escape from the ecological conditions. They adjust themselves by changing metabolic conditions at disturbances in the system. The central metabolism in plants has changed with different developmental processes enabling plants to endure such natural dangers. The estimation of known essential metabolites reveals the plants' mechanism to adjust the fluctuating situations. These varying conditions can be assessed by MS-based technologies (Griffiths 2007; Sawada and Hirai 2013).

Metabolite analysis can be divided into numerous steps.

1. Experimental design.
2. Metabolites extraction, derivatization and analysis by GC/LC-MS.
3. MS files processing, interpretation and statistical analysis.

1.3.1 Designing an Experiment

For designing an experiment, the correct biological circumstances have to be chosen. One of the key factors in arranging metabolomics studies is to examine every conceivable source of variety that can apply credible effect on the theory (Jorge et al. 2016). The experimental conditions include environmental conditions under which the plant is developing, treatment undergone and the tissues used for further metabolomics analysis, and transfer the data into a biologically important information realm. It is required to consider the natural conditions and its impact on the metabolome when designing an experiment. The levels of multifaceted nature of those reactions are exposed to additional levels of troubles when the impacts of different abiotic or biotic stresses are examined. In regard to biotic stress, for instance, usually an outstanding arrangement of metabolites is shared between the plant and the connecting accomplice, impeding the distinguishing proof of changes focused to every life form (Jorge and António 2017).

Determining the size of the experiment is also important. So the experimental size is based on certain variables, including variation occurring during developmental stages, genotypes, growth conditions and tissues analysed. Furthermore, there are intra- and inter-species variations in the metabolites. For keeping sample integrity in small- and large-scale experiments, randomization is critical. Indeed, keeping the plants in similar levels of variation during the examination is also important. Compared to nursery or field tests, plants are exposed to more prominent variations under natural conditions, which are unavoidable, yet can promptly be perceived. Randomization ensures that treatment and control tests are similar, as far as the watched and unobservable features are concerned, and reducing the contrasts in soil and environment heterogeneity.

1.3.2 Sample Preparation

First of all, the plant tissues for the analysis has to be selected and the extraction procedure follows. The extraction process, either hot extraction or cold extraction, is chosen based on the hypothesis of the metabolites to be analysed. If thermolabile compounds are present, then cold extraction is more preferable than hot extraction. So, at the initial stage a hypothesis is created, and based on that plants are selected and extraction procedures followed. The observation intervals are similarly significant since numerous metabolites, including essential and optional mixes, may change during the diurnal cycle. Abrupt changes in the natural conditions, such as covering the plants during examining, can likewise alter the metabolic production (Jin et al. 2017; Henion et al. 1998). Since the establishment of Ayurvedic practices of Charaka and Susruta, different medicinal plant collection protocols are implemented for optimum results (Tavhare and Nishteswar 2014 and references there in).

A large number of techniques could be employed to preserve the integrity of plant metabolites during harvesting. Most popular technique is storing of plant tissues in liquid nitrogen. While using liquid nitrogen, there is a possibility of conversion of chemical constituents, which is not favourable. An alternative method is freeze clamping; but it is also challenging to use in large-scale investigations. The tissue must be frozen in every step of reaping and sample preparation, if fresh materials are used. Every sample should be properly labelled during the reaping and sample preparation. It may be advisable to utilize best quality plastic tubes that will persevere through outrageous temperatures, for example, liquid nitrogen. Use of screw capped cryogenic micro tubes can avoid the chance of explosion and loss of materials during the thawing procedure. As indicated by the planned examinations, tests can be put away for new material solidified at $-80\ °C$ or if solidified at $-20\ °C$ or room temperature. In this context, use of fresh materials are essential for analysis of volatile and semi-volatile compounds (Harborne 1984).

It may be essential to granulate, filtrate, fractionate, or pre-concentrate the tissues according to the nature of plant tissues. The extraction efficiency of the tissues can be improved by tissue crushing, which permits sample homogenization and increase in contact surface area. Moreover, for handling a large number of samples, there is a wide range of crushing instruments available in the market (e.g. ball processors and robotized cryogenic processors). However, a less difficult, conceivable decision is to utilize a precooled mortar and pestle with liquid nitrogen. The containers for extraction procedure must be contamination free. A large variety of metabolites are found in plants and their physicochemical properties are also different. This will create problems for selecting extraction protocol due to the variations in plant components. Extraction should be carried out with less degradation or modification of components.

A better extraction method must be followed for analysing a large number of small molecules. A few parameters must be considered all through the extraction system to remove the metabolites productively from cells or tissues of intrigue. More often than not, the examination of metabolites pursues two potential approaches, directed or untargeted. In focused metabolomics, a characterized gathering of known and biochemically commented on highlights is recognized, while untargeted examinations are fair to give a review of all the quantifiable analytes in a sample, including obscure compounds. The decision on the extraction technique will improve the part of metabolites in focused investigations, with comparable chemical properties. In spite of a few endeavours to improve the metabolite inclusion required in untargeted investigations, the choice of an extraction method will support a particular class of compounds, presenting some sort of specificity. Hence, a mix of extraction techniques and diagnostic advances is recommended to intensify the quantity of recognizable metabolic highlights (Stein 1999).

Suitable solvents for the extraction purpose may have to be chosen. Selection of solvents is based on their different properties like polarity, selectivity, inertia, sample solvent ratio and the interaction between metabolites and biological matrix. The disintegration rate in the dissolvable extraction might be moderate, depending on the associations among metabolites and biological matrix. Despite the fact that there may be a problem of compound degradation, the time lag in extraction and temperature may support metabolite extraction. Hence, we can utilize ultra-sonication for cell wall breaking as an efficient method in extraction procedure. It is likewise critical to store the concentrates properly, according to the strength of the metabolites in focus. Also, the solvent purity, pH and its vaporization are also considered because these may change the metabolite concentration and texture. For example, plasticizers or surfactants can ruin the analysis. Hence, we have to use the extraction buffer within a short period of time.

The actual mode of extraction is based on the water content and texture of the compound being isolated. It is advisable to wash initially the desired plant part in boiling ethanol for killing the hydrolytic enzymes. Alcohol is a good solvent for extracting most of the plant components. If green tissues are used, alcohol extraction leads to the removal of chlorophyll and many of the low-molecular-weight compounds. Further extraction would be completely free of green colour. The dried plant parts are usually extracted in organic solvents ranging from non-polar (ether, hexane, chloroform) to polar (ethyl acetate and alcohol) in a soxhlet apparatus. Other methods like sonication and refluxing are also used for extraction. Extractions with non-polar solvents extract lipids and terpenoids whereas polar solvents separate polar compounds. The physiological parameter, mainly temperature, has a profound effect on purifying the constituents from crude extract. The solubility of constituents is also an important factor, in which liquid–liquid extraction is most favoured. If non-polar constituents are needed, partitioning of crude extract with water would remove unwanted components. On the other hand, if polar compounds are needed, the crude extract may be extracted with suitable organic solvents after aqueous extraction. Further partition with acid or alkali would remove acidic and basic impurities (Jorge and António 2017). Acetonitrile is a better solvent for protein precipitation than methanol. In general, there is no 'generic protocol' for the extraction procedure. Any of the methods suitable to the component or metabolite that is being analysed need to be followed. By considering all these factors and after adopting suitable method for extraction, the concentrate should be perfect, with respect to the investigative procedure to be followed. Because of the highly sophisticated instruments (LC/MS) used for the further analysis of the analyte, an ideal concentrate of the sample is necessary. Otherwise, it will affect the ion suppression in instrument affecting the ionization efficiency of the compounds.

1.3.3 Metabolite Analysis via GC/LC-MS

Analysis of metabolites mainly depends upon the goal of the study. Mass spectrometry (MS) has been broadly utilized in plant metabolomics in combination with gas or liquid chromatography (GC and LC, respectively). The amount of metabolite measured is based on the specificity of the detection method and resolution of the chromatographic system. For the detection of volatile and semi-volatile compounds GC MS is more suitable (Miyagawa and Bamba 2019). It offers powerful measurement of a few metabolite classes in plants requiring a derivatization step for compounds with insufficient volatility in their normal state and thermal stability. There are a variety of methods for derivatization of compounds like silylation (adding silyl group by replacing hydrogen in the functional groups of metabolites), alkylation (using chloroformates, especially methyl chloroformate as the derivatizing agent), alkoxyamination and acylation. Alkylation cannot derivatize sugar and sugar alcohol. But silylation is broad and can generate variety of stable, less polar and volatile derivatives of parent compound. In plant metabolomics, the initial stage is the stabilization of carbonyl moiety in plant components by using methoxyamine in pyridine, thereby overwhelming the keto-enol tautomerism and the creation of multiple acetal or ketal structures. This process would aid to decrease the numbers of derivatives of reducing sugar and produce only two forms of – N=C< derivatives. In the next step, the functional groups like –OH, –COOH, –SH, and –NH groups are converted to TMS ethers, TMS esters, TMS sulphides or TMS amines, using a trimethyl silyl reagent usually BSTFA (N,O-bis(trimethylsilyl) trifluroacetamide) or MSTFA (N-methyl-N-(trimethylsilyl) trifluoroacetamide). TMS derivatization has been thoroughly investigated and found to be very efficient. But in the case of certain compounds, multiple derivatives are formed after derivatization and it would make analysis difficult. In such cases, LC-MS is very useful where derivatization procedure is not used (Griffiths 2007).

The TMS derivatization procedure is generally used in the case of steroids. For the hydroxyl group derivatization of steroids, TMS ethers are used. But considering the oxo group in steroids, there may be certain problems as they react under forceful conditions and the molecular ion of poly hydroxyl steroid TMS ethers would be absent due to the extensive loss of trimethyl silanol to give ions. This can be overcome by preparation of TMS alkyl dimethylsilyl derivatives, which would help in the interpretation of spectra. Sterically congested alkylsilyl ethers are more stable than TMS ethers and can be purified by conventional chromatographic approaches (Zarate et al. 2016).

Other than the derivatization procedure, micro chemical or enzymatic responses can help in structure resolution or can be utilized to give appropriate mass spectrometric properties. Simple reactions include conversion of vicinal hydroxyl group into acetonides or boronates, periodate oxidation of vicinal hydroxyl group, oxidation with sodium bismuthate, selective or complete oxidation of –OH group with chromic acid in acetone, reduction of carbonyl group with sodium borohydride or lithium aluminium hydride, etc. (Griffiths 2007).

The analyst must be aware that, the trimethylsilylation process, even under the mild conditions generally employed in plant metabolite profiling, leads to by-products conversion. For example, arginine in reaction with BSTFA is converted to ornithine. Hence for interpreting the results, the analyst should take care of the intermediates formed that are reactive or unstable in the metabolic processes. If any problem occurs during interpretation, the analyst can perform another derivatization process with respect to the specific functional groups (Lytovchenko et al. 2009).

The trimethylsilyl subsidiaries of many plant metabolites, including sugars (e.g. glucose and fructose), have fundamentally the same EI mass spectra. So for recognizing them, GC retention constraints are especially significant. Since retention times fluctuate with the column dimensions, stationary stage and temperature, relative retention times may be incorporated with appropriate parameters of correlation, called retention indices (RI) (Halket et al. 2005). Relative retention times are basically the proportions of analyte times to the hour of a selected standard compound. An increasingly widespread arrangement depends on purported Kováts indices (Kováts 1958) which relate to the retention times of the metabolites to the retention times of n-alkanes examined under similar conditions, even by co-infusion. For instance, a sugar subsidiary having a similar retention time as the n-alkane, C20 would be allotted an index of 2000 or a 'methylene unit' estimation of 20.00. One showing up somewhere between C20 and C21, would be doled out a list of 2050, etc. Most manufacturers' information frameworks don't deal with RI well. Nevertheless, the AMDIS (Stein 1999) from the NIST has astounding RI abilities and is promptly accessible for download (http://chemdata.nist.gov/mass-spc/amdis/). The software can peruse most makers' information documents and perform mass spectral deconvolution so as to 'clean up' the mass spectra preceding library looking. Likewise, client library creation is straightforward and spectra can be looked against the NIST database. The AMDIS programming has been applied to plant and urinary metabolites (Halket et al. 2005).

Building up a scientific MS-based strategy is an iterative procedure that can be very tedious to rely upon the strategy by necessities, specifically the quantity of target analytes, analyte ionization and matrix effects. So as to accumulate synthetic and structural data of the target analyte(s), technique advancement analyses ought to be consistently begun with an appropriate research planning. This is the beginning stage for every succeeding investigation.

1.3.4 LC-MS

LC-MS is an appropriate technique for the analysis of plant metabolites and it doesn't require derivatization of metabolites. The most regularly utilized ionization strategy is electrospray ionization (ESI), a delicate ionization method that presents minimal interior vitality and gives little data on structure since few fragments are created. Point-by-point disintegration and structural data from completely ionized species can be obtained by performing collision-induced dissociation (CID) tests more often than not, performed on a tandem MS instrument that permits two (MS/MS) or more (MSn) successive phases of mass spectrometric investigations (Sawada and Hirai 2013).

One significant factor when creating LC-ESI-MS/MS strategies is the determination of the column chemistry and its retention system. An incredible number of submissions have been accounted for the target examination of phytohormones and utilize reverse phase (RP) column, where stationary phase comprises non-polar C18 silica, with normal solvent system made out of organic or aqueous solvent mixtures (for example, water/acetonitrile or water/methanol) as mobile phase. As it is, polar analytes (for example, most essential metabolites) have no association with the non-polar C18 stationary phase, and in this way elute near the void volume without chromatographic retention. Subsequently, elective column chemistries (for example, hydrophilic cooperation chromatography (HILIC), permeable graphitic carbon (PGC) and anion exchange chromatography (AEC)) must be applied for the target examination of the wide scope of polar metabolites commonly found in the plant metabolome. In certain cases, normal phase columns are also used for the separation of analytes in urine using hydrophilic interaction liquid chromatography (HILIC)-MS and RP-HPLC-MS. These LCs are so useful for the analysis of highly polar compounds found in urine whereas RP columns are mostly conventional C18 columns that are predominantly used for plant metabolite analysis. Furthermore, the relatively newer UHPLC is more efficient with high pressure, increased chromatography resolution and peak intensity compared to HPLC (Ghosson et al. 2018).

Mass spectrometry is a prevailing scientific technique since it tends to be effectively applied regardless of whether one has just a modest amount accessible for investigation, as small as 10^{-12} g, 10^{-15} mol for a compound of mass 1000 Daltons

(Da). In a chemically complex mixture, it may be easy to identify compounds with small concentrations through mass spectrometry. The fundamental mass spectrometry procedures of instrumentation comprise (Adams et al. 1991) sample introduction, an example which can be a solid, liquid, or vapour stacked onto a mass spectrometry gadget and is vaporized (Arabidopsis Genome Initiative 2000); ionization, test compounds are ionized by one of a few accessible strategies to make ions (Aslam et al. 2017); sorting in analyser, the ions are sorted in an analyser as indicated by their m/z proportions using electromagnetic fields (Chomel et al. 2016); detector, the particles at that point go through a detector where the ion transition is changed over into a proportional electrical flow (Dunwell et al. 2001) and data transformation, the extent of the ion/electrical sign is changed over into a mass range (Sumner et al. 2003).

1.3.4.1 Sample Introduction

Sample can be introduced into the machine by two ways, either direct insertion or direct infusion. In the direct insertion method, sample is introduced in the probe and it directly enters into the ionization chamber through a vacuum interlock. If the machine is not using the vacuum interlock, allow the vacuum to maintain for a long time. After sample insertion, the sample is then exposed to a number of desorption forms, for example, laser desorption or direct warming, to encourage vaporization and ionization. In direct infusion method samples are introduced into capillary column in the form of gas or solution. This is efficient because a very small amount of sample can be introduced without compromising vacuum.

1.3.4.2 Ionization

The metabolite profiles substantially depend on the ionization strategy opted for LC-MS. In LC-MS, the ion-generating source is most significant and the main ionization sources are electron spray ionization suitable for semi-polar and polar compounds and atmospheric pressure chemical ionization (APCI), the essential method for ionizing less polar compounds, fast atom/ion bombardment (FAB), field desorption/field ionization (FD/FI), MALDI, and thermospray ionization. Along with this, electron impact (EI) and chemical ionization (CI) are also found to be sources of ionization. The distinctive ionization techniques work by either ionizing a neutral molecule through electron ejection, electron capture, protonation, cationization, or deprotonation, or by moving a charged atom from a condensed stage to the gas phase (Pitt 2009).

In the EI source, ionization is achieved through electron ejection and provides a net positive charge forming radical cations whereas electron ejection produces significant fragment ions. It is usually implemented for small molecules with non-polar nature. In the other strategy of electron capture ionization, a net negative charge of -1 is accomplished. This kind of ionization is well suited for compounds with a high electron affinity, for example, halogenated mixtures. It can give an insight of electron ionization and can give molecular mass just as fragmentation data. However, it regularly creates a lot of discontinuity and can be misty whether the most elevated mass particle is the atomic particle or a fragment (Lei et al. 2011).

In addition to electron capturing, proton addition is also another technique of ionization. Here, a proton is added to an atom, creating a net positive charge of +1 and this positive charge is found to be attached with basic residues of amines to form stable cations. MALDI, ESI and APCI utilize protonation techniques, which are suitable for the ionization of peptides. On the contrary, de-protonation is applied in some cases, where a net negative charge of -1 is accomplished through the expulsion of a proton from an atom. MALDI, ESI and APCI utilize de-protonation techniques and are exceptionally helpful for the ionization of acidic compounds including phenols, carboxylic acids and sulfonic acids. Like protonation, another technique, cationization adds a positively charged ion to a neutral molecule via non-covalent linkage, thus producing a charged complex mainly a cation adduct other than a proton (for example, soluble base, ammonium). MALDI, ESI and APCI utilize cationization for ionizing carbohydrates (Wang 2015).

Depending upon the proton affinity of samples, both positive and negative charged ions will be produced during the ionization procedure. So one must decide whether positive- or negative-charged ions need to be detected. Electron spray ionization is well suited for the ionization of small molecules and insists chemist to rely on these techniques. Among them, ESI and MALDI have unmistakably advanced to be the techniques for decision, with regard to bio-molecular investigation. ESI and MALDI ionization techniques offer a magnificent mass range and effectiveness which make these techniques more suitable for bio-molecular mass spectrometry (Fujimura and Miura 2014).

In the electro-ionization, a strong electric field is applied to the sample molecule, usually 70 eV energy would pass through the capillary tube and samples are bombarded with gas phase, resulting in the production of $[M]^+$ ions which are typically radical cations and thermal energy free electrons. In most of the cases, molecular ion is unstable and fragments are found to be stable. But there are both advantage and disadvantage in the electron ionization. The advantage is the ability to obtain structural information of compounds and their identification from the fragments produced. Nevertheless, the fragmentation is so extensive that molecular ion is not found and thus molecular weight cannot be determined. EI generate either positive or negative ions and the negative ion is produced by the capture of secondary low-energy electron from ionization of bath gas, mainly argon or nitrogen. Another essential condition for EI is that samples should be ionized in the gas phase (Shimizu et al. 2018).

The ionization technique used since 1980s is fast atom bombardment (FAB) for the analysis of metabolite in bile acid and steroids. Hence, FAB is significantly used for the polar or ionic biomolecules and a fast atom beam of neutral atom is found to be the source of ionization (6–8 keV KE, usually Xe/Ar). Protonated and deprotonated molecules which are stable and minute fragments are produced as a result of both positive and negative modes. This FAB system generally is found in liquid secondary ion mass spectrometry and negative ion FAB reduce the use of derivatization, hydrolysis, solvolysis necessary for GC-MS and make the analysis of bile acid and steroid samples easy (Jorge et al. 2016; Yang et al. 2019).

Similar to FAB, other ionization technique for polar and ionic biomolecule is electro spray ionization. In this, analytes are solubilized in suitable solvents like methanol, ethanol, mixture of acetonitrile and water or chloroform and alcohol. Solubilized analytes are sprayed from metal or fused silica capillary needle with a back pressure to the content of the capillary. Needle should be of 200–100 μm i.d. at a flow rate of 1–500 μL/min and electro spray achieved at approximately 4 kV both in positive and negative mode. The analyte sprays in the form of charged droplets is directed to an electrode with low potential, and during the travel time towards this electrode, the droplet loses its solvent, shrink and is broken into smaller droplets. These smaller droplets derived from the surface of their precursors encompass highest concentration of charge and hence the offspring droplets have an enhanced charge to mass ratio. Ultimately the droplets are converted to very small particles, where the charge density on the droplet goes beyond the surface tension and gas phase ions are desorbed. The ions are then transmitted to the vacuum chamber of mass spectrometer through a circular orifice in the counter electrode. Ions are transmitted to the high vacuum region of the mass spectrometer, and bypass through differentially pumped regions via skimmer lenses (Qi et al. 2014; Sumner et al. 2015).

Thermospray is also a technique like electrospray ionization in which thermal heating of the spray gives out vaporization and ionization and performs best with buffered aqueous ammonium acetate (0.1–0.01 M) as mobile phase. In the positive ion mode, the droplets become charged between the unevenly distributed cations and anions and as a result gaseous ($[M + H]^+$, $[M + NH_4]^+$, or $[M + Na]^+$) ions are produced, which are similar to that generated in ES. The analysis of acidic metabolites are mainly done in the negative ion mode, where $[M-H]^-$ ions are formed (Qi et al. 2014).

Neutral biomolecules like steroids, oxysterol, and bile acid acyl glucoside are ionized via APCI ionization technique and this process is similar to that of TS. The difference of APCI from ES is that a corona discharge electrode is additionally present. The analyte is solubilized in its corresponding solvent and is sprayed onto an atmospheric ion source and vaporization of solvent sample is carried out using heat. These vapours are then directed to the corona discharge electrode to produce ions. Other ionization techniques include electron capture atmospheric pressure chemical ionization, atmospheric pressure photoionization, desorption electrospray ionization mass spectrometry and matrix-assisted laser desorption/ionization (MALDI). Proteins, large molecules and peptides can be analysed by MALDI. The analytes to be ionized are first embedded into a matrix. The matrix is in 1% TFA in acetonitrile and spotted on a target plate, allowed to co-crystallize in air and admitted to high vacuum which absorbs energy at a wavelength of 337 nm, the wavelength of laser and the matrix becomes ionized. Both positive and negative ions are formed. These ions are analysed via TOF analyser (Qin et al. 2018).

1.3.4.3 Mass Analyser

After the ionization process, the molecules get separated by mass-to-charge ratio in the mass analyser. The time of flight, Quadrupole (Q) and Triple Quadrupole (QQQ) are the mass analysers in MS spectrometry. In the single quadrupole analyser 4 metal rods are located and the ions pass through the central axis of the metal rod. These two pair of rods are electrically connected and a radio frequency voltage is applied to these pairs and a direct current is superimposed to this RF voltage. Thus, a quadrupole field is generated between these rods and allows only a certain range of mass to pass through and the others are unstable and destroyed. Only ions with a particular m/z ratio is analysed and transferred to the detector. In a triple quadrupole, three quadrupole cells arranged linearly and allow product ion monitoring. The first quadrupole is similar to the classic form, which retrieve only specific mass of ions, whereas the second quadrupole functions as a collision cell, which allows gases to pass through the filtered ion molecules and their collision would produce fragments. The third quadrupole analyses the fragments produced via collision induced decomposition and thus signifies the MS/MS which is suited for the structure interpretation of a metabolite. Time-of-flight mass spectrometry is based on the fact that the ions, which are accelerated by high electric field in the drift, tube and linearly pass the ion to the detector. The main theory is that mass of ion is proportional to the flight velocity, and along with this, the length of the flight tube is also important. For a longer flight tube, the distance travelled by the ions are longer resulting in an increased space separation and hence an increased *m/z* (Ernst et al. 2014; Lee et al. 2010).

1.3.4.4 Detector

During the analysis, the ionized molecule pass through the mass analyser and finally reach the detector. These detectors are mainly of three types, which are used to transform −ve and +ve ions into electrons and then get amplified and detected. In certain cases, these ions hit on phosphor and emit photons that strikes a photomultiplier and signals are recorded. These technologies are based on Dynolyte photomultiplier, Electron multipliers and Micro channel plates in the Q-TOF MS instrument. Q-TOF has micro channel plate detectors in which a strong electronic field is created during the hitting of ions on surface and these get transferred to electronic signals. Subsequently, these signals pass through the orifice and a secondary electronic cascade is produced with higher amplification in a very low time response. These signals move next onto data management system and is once again amplified, while reaching to the time-to-digital converter (Ernst et al. 2014).

Mass spectra libraries for LC/MS investigation is less basic than GC/MS libraries mostly in light of the fact that coupled mass spectra change the contingent based on the instrument and method utilized. In this way, there is a need to gather further metabolite databases including information on the mixed mass spectra and structure together with other important data, to completely define the capability of LC/MS in plant metabolomics investigation.

1.4 Conclusions

Genomics gives an overview of the possible and complete set of genetic instructions embedded in DNA, while transcriptomics provides a profile of the RNAs, hence providing the gene expression patterns. Proteomics studies are on protein products made as per the genetic information and their interactions. Metabolomics is also an intermediate step like transcriptomics, which is a sandwich between DNA and proteins in understanding the entire metabolism of an organism. Metabolomics is one of the newest 'omics' sciences. The metabolome refers to the complete set of low-molecular-weight compounds in a sample. These compounds are the substrates and by-products of enzymatic reactions, and have a direct effect on the phenotype of the system. Metabolomics may be used to study the differences between a healthy and diseased plant as far as their secondary metabolites are taken into account.

In the domain of systems biology, metabolomics has received a far-reaching responsiveness, contributing to the field of life sciences. Especially, plant metabolomics reveals the change in metabolic pathways according to the environmental conditions. Plants produce primary metabolites during normal physiological functions, and from the primary metabolites secondary metabolites are synthesized through various pathways during varying conditions. So plant metabolomics relates to understanding the biochemical and regulatory mechanism during the biotic and abiotic stress conditions.

References

Adams MD, Kelley JM, Gocayne JD, Dubnick M, Polymeropoulos MH, Xiao H, Merril CR, Wu A, Olde B, Moreno RF (1991) Complementary DNA sequencing: expressed sequence tags and human genome project. Science 252:1651–1656. https://doi.org/10.1126/science.2047873

Arabidopsis Genome Initiative (2000) Analysis of the genome sequence of the flowering plant *Arabidopsis thaliana*. Nature 408:796–815. https://doi.org/10.1038/35048692

Aslam B, Basit M, Nisar MA, Khurshid M, Rasool MH (2017) Proteomics: technologies and their applications. J Chromatogr Sci 55:182–196. https://doi.org/10.1093/chromsci/bmw167

Chomel M, Guittonny-Larchevêque M, Fernandez C, Gallet C, DesRochers A, Paré D, Jackson BG, Baldy V (2016) Plant secondary metabolites: a key driver of litter decomposition and soil nutrient cycling. J Ecol 104:1527–1541. https://doi.org/10.1111/1365-2745.12644

Dunwell JM, Moya-León MA, Herrera R (2001) Transcriptome analysis and crop improvement (a review). Biol Res 34:153–164. https://doi.org/10.4067/s0716-97602001000300003

Eckardt NA (2000) Sequencing the rice genome. Plant Cell 12:2011–2017. https://doi.org/10.1105/tpc.12.11.2011

Ernst M, Silva DB, Silva RR, Vêncio RZN, Lopes NP (2014) Mass spectrometry in plant metabolomics strategies: from analytical platforms to data acquisition and processing. Nat Prod Rep 31:784–806. https://doi.org/10.1039/C3NP70086K

Fiehn O (2002) Metabolomics—the link between genotypes and phenotypes. Plant Mol Biol 48:155–171

Frith MC, Pheasant M, Mattick JS (2005) Genomics: the amazing complexity of the human transcriptome. Eur J Hum Genet 13:894–897. https://doi.org/10.1038/sj.ejhg.5201459

Fujimura Y, Miura D (2014) MALDI mass spectrometry imaging for visualizing *in situ* metabolism of endogenous metabolites and dietary phytochemicals. Meta 4:319–346. https://doi.org/10.3390/metabo4020319

Ghosson H, Schwarzenberg A, Jamois F, Yvin J-C (2018) Simultaneous untargeted and targeted metabolomics profiling of underivatized primary metabolites in sulfur-deficient barley by ultra-high performance liquid chromatography-quadrupole/time-of-flight mass spectrometry. Plant Methods 14:62. https://doi.org/10.1186/s13007-018-0329-0

Graves PR, Haystead TAJ (2002) Molecular Biologist's guide to proteomics. Microbiol Mol Biol Rev 66:39–63. https://doi.org/10.1128/MMBR.66.1.39-63.2002

Griffiths WJ (2007) Metabolomics, metabonomics and metabolite profiling. Royal Society of Chemistry, Cambridge

Guerriero G, Berni R, Muñoz-Sanchez JA, Apone F, Abdel-Salam EM, Qahtan AA, Alatar AA, Cantini C, Cai G, Hausman J-F, Siddiqui KS, Hernández-Sotomayor SMT, Faisal M (2018) Production of plant secondary metabolites: examples, tips and suggestions for biotechnologists. Genes (Basel) 9. https://doi.org/10.3390/genes9060309

Halket JM, Waterman D, Przyborowska AM, Patel RKP, Fraser PD, Bramley PM (2005) Chemical derivatization and mass spectral libraries in metabolic profiling by GC/MS and LC/MS/MS. J Exp Bot 56:219–243. https://doi.org/10.1093/jxb/eri069

Hall R, Beale M, Fiehn O, Hardy N, Sumner L, Bino R (2002) Plant metabolomics: the missing link in functional genomics strategies. Plant Cell 14:1437–1440. https://doi.org/10.1105/tpc.140720

Harborne JB (1984) Macromolecules. In: Harborne JB (ed) Phytochemical methods: a guide to modern techniques of plant analysis. Springer, Dordrecht, pp 243–276

Henion J, Brewer E, Rule G (1998) Peer reviewed: sample preparation for LC/MS/MS: analyzing biological and environmental samples. Anal Chem 70:650A–656A. https://doi.org/10.1021/ac981991q

Hong J, Yang L, Zhang D, Shi J (2016) Plant metabolomics: an indispensable system biology tool for plant science. Int J Mol Sci 17. https://doi.org/10.3390/ijms17060767

Horgan RP, Kenny LC (2011) 'Omic' technologies: genomics, transcriptomics, proteomics and metabolomics. Obstet Gynaecol 13:189–195. https://doi.org/10.1576/toag.13.3.189.27672

Huccetogullari D, Luo ZW, Lee SY (2019) Metabolic engineering of microorganisms for production of aromatic compounds. Microb Cell Factories 18:41. https://doi.org/10.1186/s12934-019-1090-4

Jin J, Zhang H, Zhang J et al (2017) Integrated transcriptomics and metabolomics analysis to characterize cold stress responses in Nicotiana tabacum. BMC Genomics 18. https://doi.org/10.1186/s12864-017-3871-71

Johnson SR, Lange BM (2015) Open-access metabolomics databases for natural product research: present capabilities and future potential. Front Bioeng Biotechnol 3. https://doi.org/10.3389/fbioe.2015.00022

Jorge TF, António C (2017) Plant metabolomics in a changing world: metabolite responses to abiotic stress combinations. In: Plant abiotic stress responses climate change. InTech. https://doi.org/10.5772/intechopen.71769

Jorge TF, Mata AT, António C (2016) Mass spectrometry as a quantitative tool in plant metabolomics. Philos Trans A Math Phys Eng Sci 374:20150370. https://doi.org/10.1098/rsta.2015.0370

Kallscheuer N, Classen T, Drepper T, Marienhagen J (2019) Production of plant metabolites with applications in the food industry using engineered microorganisms. Curr Opin Biotechnol 56:7–17. https://doi.org/10.1016/j.copbio.2018.07.008

Kováts E (1958) Gas-chromatographische charakterisierung organischer verbindungen. Teil 1: Retentions indices aliphatischer halogenide, alkohole, aldehyde und ketone. Helv Chim Acta 41:1915–1932. https://doi.org/10.1002/hlca.19580410703

Lee DY, Bowen BP, Northen TR (2010) Mass spectrometry—based metabolomics, analysis of metabolite-protein interactions, and imaging. BioTechniques 49:557–565. https://doi.org/10.2144/000113451

Lei Z, Huhman DV, Sumner LW (2011) Mass spectrometry strategies in metabolomics. J Biol Chem 286:25435–25442. https://doi.org/10.1074/jbc.R111.238691

Lytovchenko A, Beleggia R, Schauer N, Isaacson T, Leuendorf JE, Hellmann H, Rose JK, Fernie AR (2009) Application of GC-MS for the detection of lipophilic compounds in diverse plant tissues. Plant Methods 5:4. https://doi.org/10.1186/1746-4811-5-4

Miyagawa H, Bamba T (2019) Comparison of sequential derivatization with concurrent methods for GC/MS-based metabolomics. J Biosci Bioeng 127:160–168. https://doi.org/10.1016/j.jbiosc.2018.07.015

Oliver SG, Winson MK, Kell DB, Baganz F (1998) Systematic functional analysis of the yeast genome. Trends Biotechnol 16:373–378. https://doi.org/10.1016/S0167-7799(98)01214-1

Pitt JJ (2009) Principles and applications of liquid chromatography-mass spectrometry in clinical biochemistry. Clin Biochem Rev 30:19–34

Pyne ME, Narcross L, Martin VJJ (2019) Engineering plant secondary metabolism in microbial systems. Plant Physiol 179:844–861. https://doi.org/10.1104/pp.18.01291

Qi X, Chen X, Wang Y (2014) Plant metabolomics: methods and applications. Springer, Dordrecht

Qin L, Zhang Y, Liu Y, He H, Han M, Li Y, Zeng M, Wang X (2018) Recent advances in matrix-assisted laser desorption/ionisation mass spectrometry imaging (MALDI-MSI) for in situ analysis of endogenous molecules in plants. Phytochem Anal 29:351–364. https://doi.org/10.1002/pca.2759

Sasaki T, Burr B (2000) International rice genome sequencing project: the effort to completely sequence the rice genome. Curr Opin Plant Biol 3:138–114. https://doi.org/10.1016/S1369-5266(99)00047-3

Sawada Y, Hirai MY (2013) Integrated LC-MS/MS system for plant metabolomics. Comput Struct Biotechnol J 4:e201301011. https://doi.org/10.5936/csbj.201301011

Schena M, Shalon D, Davis RW, Brown PO (1995) Quantitative monitoring of gene expression patterns with a complementary DNA microarray. Science 270:467–470. https://doi.org/10.1126/science.270.5235.467

Shimizu T, Watanabe M, Fernie AR, Tohge T (2018) Targeted LC-MS analysis for plant secondary metabolites. Methods Mol Biol 1778:171–181. https://doi.org/10.1007/978-1-4939-7819-9_12

Stein SE (1999) An integrated method for spectrum extraction and compound identification from gas chromatography/mass spectrometry data. J Am Soc Mass Spectrom 10:770–781. https://doi.org/10.1016/S1044-0305(99)00047-1

Sumner LW, Mendes P, Dixon RA (2003) Plant metabolomics: large-scale phytochemistry in the functional genomics era. Phytochemistry 62:817–836. https://doi.org/10.1016/S0031-9422(02)00708-2

Sumner LW, Lei Z, Nikolau BJ, Saito K (2015) Modern plant metabolomics: advanced natural product gene discoveries, improved technologies, and future prospects. Nat Prod Rep 32:212–229. https://doi.org/10.1039/C4NP00072B

Tavhare S, Nishteswar K (2014) Role of Deepaneeya and Shwashara Dashemani in the management of Tamakashwasa w.s.r. to Bronchial Asthma: a review. Int J Ayurvedic Med 5

Thirumurugan D, Cholarajan A, Vijayakumar SSSR (2018) An introductory chapter: secondary metabolites. In: Secondary metabolites—sources and applications. https://doi.org/10.5772/intechopen.79766

Velculescu VE, Zhang L, Vogelstein B, Kinzler KW (1995) Serial analysis of gene expression. Science 270:484–487. https://doi.org/10.1126/science.270.5235.484

Wang G (2015) LC-MS in plant metabolomics. In: Qi X, Chen X, Wang Y (eds) Plant metabolomics: methods and applications. Springer, Dordrecht, pp 45–61

Wink M (2018) Plant secondary metabolites modulate insect behaviour-steps toward addiction? Front Physiol 9. https://doi.org/10.3389/fphys.2018.00364

Yang Y, Yin Y, Chen X, Chen C, Xia Y, Qi H, Baker PN, Zhang H, Han T-L (2019) Evaluating different extraction solvents for GC-MS based metabolomic analysis of the faecal metabolome of adult and baby giant pandas. Sci Rep 9:1–9. https://doi.org/10.1038/s41598-019-48453-1

Zarate E, Boyle V, Rupprecht U, Green S, Villas-Boas SG, Baker P, Pinu FR (2016) Fully automated trimethylsilyl (TMS) derivatisation protocol for metabolite profiling by GC-MS. Metabolites 7:1. https://doi.org/10.3390/metabo7010001

Plant Metabolites: Methods for Isolation, Purification, and Characterization

2

Shabeer Ali Hassan Mohammed, Renu Tripathi, and K. Sreejith

Abstract

Phytochemistry deals with primary and secondary metabolites/chemicals synthesized by the plants. The prime intention of this interdisciplinary branch is to make use of the phytochemicals for the well-being of human, animals, and agriculture. Efforts have been started since long years back to make use of the phytochemicals as medicines, cosmetics, bio-control agents, food, and food additives. Since phytochemistry is an interdisciplinary branch, its development depends also on the scientific and engineering advancements. As a result, the branch "phytochemistry" remained unexplored for a while and a related branch still exists under the label "traditional or folk medicine" even without knowing the active principles that contribute the medicinal effect. This chapter provides a complete guide for the researchers and beginners in the field of phytochemistry. The discussion starts with the initial steps such as plant collection and different processing methods to set up the raw material suitable for extraction. The different approaches to isolate a target molecule from the crude extract are also being discussed in further sections. The purification process requires expertize

---

S. A. Hassan Mohammed
Division of Molecular Parasitology and Immunology, CSIR-Central Drug Research Institute, Lucknow, Uttar Pradesh, India

Department of Biotechnology and Microbiology, Kannur University, Kannur, Kerala, India
e-mail: shabeer.m@cdri.res.in

R. Tripathi
Division of Molecular Parasitology and Immunology, CSIR-Central Drug Research Institute, Lucknow, Uttar Pradesh, India
e-mail: renu_tripathi@cdri.res.in

K. Sreejith (✉)
Department of Biotechnology and Microbiology, Kannur University, Kannur, Kerala, India
e-mail: sreejithk@kannuruniv.ac.in

© Springer Nature Singapore Pte Ltd. 2020
S. T. Sukumaran et al. (eds.), *Plant Metabolites: Methods, Applications and Prospects*, https://doi.org/10.1007/978-981-15-5136-9_2

and deep knowledge about the physicochemical properties of the compounds and the materials used for purification. In this aspect, we described the basics, advancement, and troubleshooting guides for several chromatographic techniques. DART, on the other hand, provides a direct analysis of raw plant materials on a real time basis which practically avoid the elaborated preprocessing and extraction procedures. The sophisticated analytical techniques like GCMS, LCMS, FTIR, NMR, and XRD together provide a strong basement for the structure analysis of the compound of interest.

Keywords

Plant metabolites · Phytochemistry · Isolation · Purification · Characterization · Chromatography

2.1 Introduction

The experimentation of medicinal as well as toxic effects of plants dates back to earlier days of human civilization. Due to the lack of proper methodologies, our ancient practitioners were not aware about the key constituents of plants that exert the desirable or undesirable effects. The evolution of "Phytochemistry" has passed different milestones. The very first use consisted of raw plant parts, which later progressed to the mechanical processing methods like crushing, grinding, and drying to increase the bio-availability of phyto-constituents. At this point they understood the role of some key components present in plants that are contributing to the desired biological effects. For the next centuries, much interest have been invested to improve the efficacy of the phytomedicine by trying different processing methods and use of water extracts to change the consistencies of the formulations that suites different routes of administration. Keeping such extracts for a while may have brought additional benefits, which thus initiated the impact of fermentation on phytomedicine. Heating water based extracts and further trial and error experiments may have given the role of volatile matters present in the extracts and the volatilization as a method to treat respiratory disturbances and also as a method of phytomedicine preparation. Further, discovery of different solvents through fermentation may have revolutionized the process of liquid extraction methods. Even though all these methods revolutionized the field of phytochemistry, the structural identity of the molecules present in the plants remained hidden for centuries due to the lack of proper detection methods.

Plants possess plenty of chemical classes with defined functional roles such as maintaining physiological role, metabolism, immunity, and defense. Among these, most of the compounds act as therapeutic agents, nutritional additives or supplements and toxic agents on several other species including human (Cos et al. 2006). Symbiotic associations between plants of same species or taxonomically distinct species such as fungi or bacteria are the classic examples for the positive effect of phytochemicals. Whereas, the allelopathic effect exerted by some plants to eradicate other microbes and plants stands the basis for the production of toxic chemicals by plants. Irrespective of the role of a phytochemical, proper planning and application can serve the humanity in different aspects including food, medicine, cosmetics, agriculture, etc.

This chapter deals with the methodologies developed so far in the field of phytochemical analysis. Isolation of the active principle from a raw plant material has to pass several hurdles and challenges, which require proper understanding of the different aspects of organic and analytical chemistry, without which the purification will be a difficult task.

2.2 Plant Material Collection and Processing

2.2.1 Collection

Different plant parts such as root, stem, bark, leaf, flower, and fruit may possess slightly different chemical profile. Thus, one should have a clear idea about the part of the plant that has to be collected for the desired experimental, medicinal, or other purposes. Care should be taken to ensure that the plant is disease free and there are no visible fungal infections or association with mosses (Paris et al. 1960; Harborne and Hora 1970). Though this step seems no need of extreme care, minute errors in this step may spoil the entire outcome or mislead to false positive conclusions. Collecting fresh plant parts are highly recommended for phytochemical analysis and aged, decayed, or fallen plant parts should be avoided.

The collected plant material's botanical identity should be authenticated by a reputed authority before starting the phytochemical analysis and the plant part used for authentication should be kept in herbarium for future references (Harborne 1973).

2.2.2 Common Contaminants

Contamination is a major trouble in every aspects of phytochemical analysis. The major contaminations that can occur during sample collection are the etiological agents of plant diseases, commensal microbes associated with the plant, soil, or dust particles, and so on (Paris et al. 1960; Harborne and Hora 1970; Harborne 1973). Collection of roots, stem, or bark near the soil may contain several soil fungi, bacteria, mosses, lichens, etc. The leaves of plants living in the moist and polluted area harbor a number of microbes and other chemical contaminants that may mislead to false interpretation. In most cases, the following sterilizing methods can help to reduce the amount of contaminants present in the sample.

2.2.2.1 Washing
Though washing itself is not a sole practice of sterilization, washing plant materials with sterile distilled water will reduce the burden to some extent.

2.2.2.2 Chemical Treatment

(a) *Sodium hypochlorite treatment*: In order to get rid of microbial contaminants, collected plant materials should be immersed in 10–20% solution of commercially available sodium hypochlorite solution (5.25%) for 10–20 min.
(b) *Ethyl alcohol*: Short-term exposure to 70% ethanol for few seconds is recommended for plant material surface sterilization. This treatment is not recommended for delicate plant parts.
(c) *Mercuric chloride*: Even though $HgCl_2$ was used for plant material surface sterilization, their routine use was not entertained due to their increased toxicity. Repeated rinses with sterile distilled water are highly recommended after its use.
(d) *Hydrogen peroxide*: Treatment with 30% H_2O_2 solution is one of the widely accepted plant material surface sterilization method.
(e) *Calcium hypochlorite*: 3.25% solution of calcium hypochlorite is considered as a safe sterilizing agent when compared to sodium hypochlorite.

Repeated washing with sterile distilled water after all the chemical treatment is a mandatory step; otherwise the chemical contaminants will be troublesome. Most of the treatment methods described above not only removes contaminants but also kills plant tissues which is necessary to prevent the enzymatic hydrolysis or oxidation that may damage the compound of interest.

2.2.3 Processing

Preprocessing the collected plant material is an important step to enhance the yield of desired phytochemicals. Preprocessing protocols break open the plant tissues and cells which releases the intracellular chemical depot. Use of intact plant materials for extraction may not give the desired phytochemicals in pleasing quantity.

2.2.3.1 Crushing/Grinding
Crushing the plant materials using mortar and pestle is a common method. Acid wash sand or glass powder can also be used to enhance the crushing process.

2.2.3.2 Chopping
Harder plant parts like matured stem, root, etc. require chopping using scissors. The main advantage of chopping is that it will not produce fine powders that may block column nozzle, solvent flow through column matrix, and so on.

2.2.3.3 Drying
Collected plant materials can be dried in several ways; drying under shade is recommended for most of the isolation protocols. But drying under shade may sometimes enhance fungal growth; such situations can be overcome by drying under moderate temperatures in an oven. Drying under shade in the presence of table fan will also enhance the drying process. Drying for long time and use of higher temperatures are not recommended for the extraction of volatile compounds.

2.3 Phytochemical Extraction Methods

Even though enormous extraction methods are available, the selection of appropriate method depends on the chemistry of the target molecule and the nature and consistency of the processed plant material. Separation of phytochemicals or any other molecules is based on the parameters such as acidity, alkalinity, molecular size, weight, polarity, and affinity towards certain substances or column matrix.

2.3.1 The Concept of Polarity

The concept of polarity possesses extreme importance in the isolation and purification of synthetic as well as natural compounds. The basic principle states that the solubility or mobility of compounds in fluids depends on the polarity of the compounds and the respective fluids (Altemimi et al. 2017). The polar compounds will only be dissolved in the polar solvents (sugar in water), and the nonpolar compounds will only be dissolved in nonpolar solvents (e.g., naphthalene in benzene). Similarly, mixing up of polar and nonpolar solvents will lead to the formation of two layers due to their immiscible nature. This method can also be used to separate a mixture of polar and nonpolar solvents (discussed later). A chart showing the polarity of commonly used solvents and their miscibility are presented in Fig. 2.1.

2.3.2 Cold Extraction Method

The powdered plant materials (dried at 50–60 °C) can be extracted with a series of solvents. Desired quantity of the sample can be treated with solvents in the order of nonpolar to polar. It is a prolonged process that may last up to 5–7 days with low speed agitation or stirring with magnetic beads in each solvent at least for 7–8 h. It should be ensured that the solvent level in the container is not dropped too much during the process. After the completion of the extraction process, the extract can be filtered through Whatman filter paper if necessary (Ingle et al. 2017).

2.3.3 Methods to Concentrate the Extracts

The extracts obtained from most of the extraction methods will be of large volume; such extracts can be concentrated using different methods such as rotary evaporator, vacuum drying, and manual drying.

2.3.3.1 Rotary Evaporator (Rotovap/Rotavap)
In this method, the solvent was evaporated under reduced pressure without affecting the stability of the chemical compounds dissolved in it. The system contains a motor to rotate the attached round-bottom flask, which contains the solvent with the

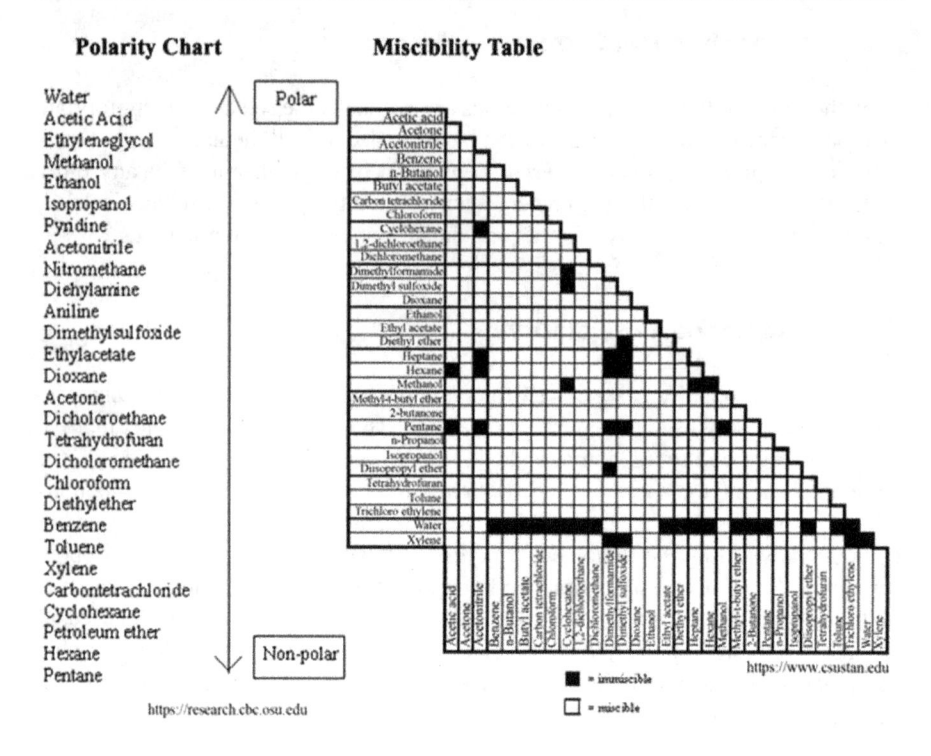

Fig. 2.1 Solvent polarity chart (adopted from https://research.cbc.osu.edu) and miscibility chart (adopted from https://www.csustan.edu)

molecule of interest. A vacuum pump connected with the vacuum tight channel reduces the pressure within the evaporator system. The rotating round-bottom (RB) flask is partially immersed (height of the RB flask can be adjusted) in a programmable water bath attached with the instrument. A condenser with circulating coils that carry the coolant will condense the evaporating solvents, which can then be collected at the collecting flask connected at the bottom of the condenser. As the round-bottom flask with the sample is rotating at a particular velocity, it will form a thin layer of sample-solvent mixture along the inner sides of rotating RB flask and provides large surface area to enhance the rate of evaporation. The rotation also provides uniform heat distribution to the sample and avoids overheating of the part immersed in water bath (JoVE Science Education Database 2019). Within few minutes, the sample/molecule of interest will get concentrated in the RB flask. The duration of operation depends on the amount of solvent present in the RB flask; large volume of solvent will prolong the process.

2.3.3.2 Soxhlet Extraction

Compounds with limited solubility in one solvent where the impurities are insoluble can be extracted with soxhlet extraction method. Though soxhlet extraction is carried out under continuous heating; separation of heat-sensitive compounds are

Fig. 2.2 Schematic diagram of soxhlet apparatus assembly (picture adopted from Kou and Mitra 2003)

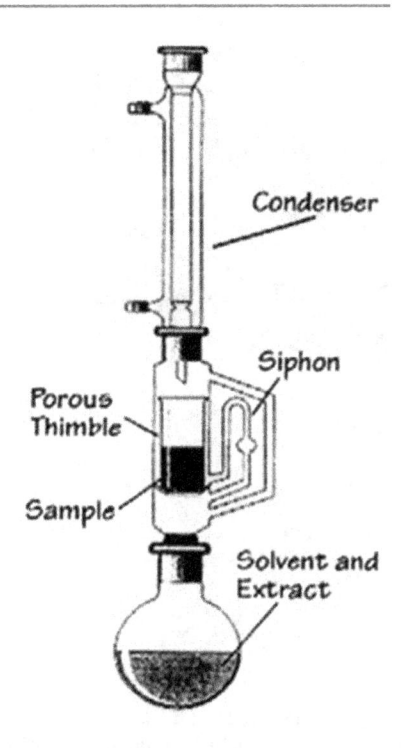

not recommended with this method and the stability of such compounds after extraction could not be assured (Nikhal and Dambe, 2010; Sutar et al. 2010). The extraction can be performed in a specially designed glass apparatus (Fig. 2.2) assembly. For starting an extraction, the sample should be placed in the thimble-holder; the solvents evaporated from the RB distillation flask was then condensed by the coolant circulating through the condenser and oozes into the thimble-holder. Once after the thimble-holder is filled with the condensed solvent, the overflow with the extracted content will be redirected to the distillation flask through "siphon" (Luque and Garcia 2000). The procedure should be repeated till the content in the thimble-holder is completely extracted; complete extraction can be ensured by observing the color and consistency (viscosity) of the condensate passing through the siphon. Care should be taken to apply gentle heat which is sufficient enough to evaporate the solvent in the distillation flask. Overheating and excessive boiling may cause undesirable effects.

2.3.4 Extraction of Volatile/Essential Oils

The essential oils present in plants can be extracted by two important methods. In the hydro-diffusion method (Fig. 2.3a), the steam circulating through the chopped/grinded plant material filled in the hydro-diffusion glass chamber will extract the essential oils, and condenses into the attached burette by the action of the condenser.

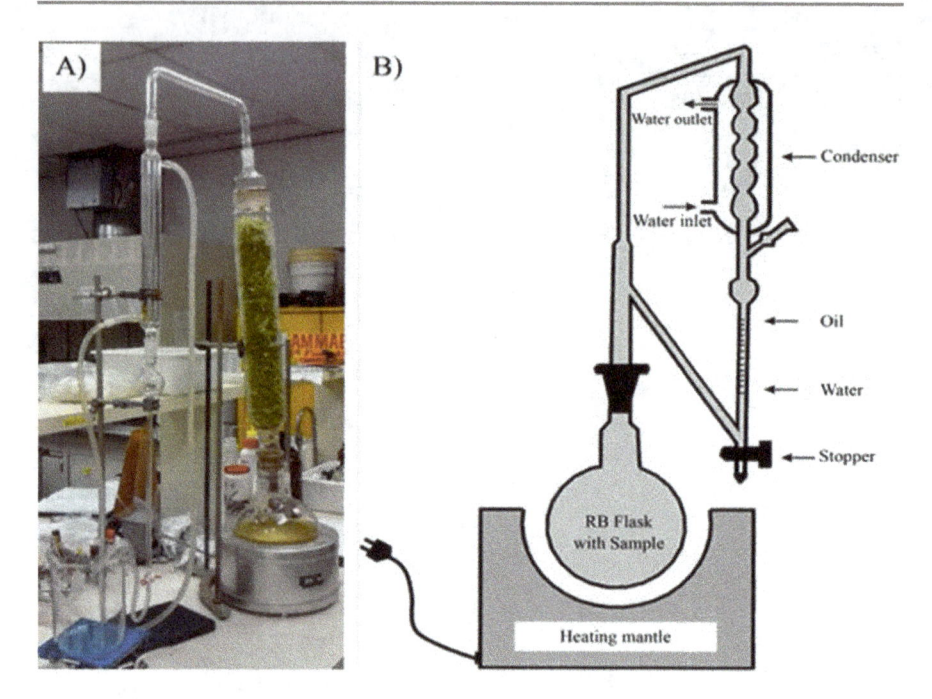

Fig. 2.3 (**a**) Hydro-diffusion unit (adapted from https://www.phytochemia.com/en/2014/08/04/extracting-essential-oils-in-the-lab/) and (**b**) Clevenger apparatus (adopted from Ismaiel et al. 2016)

In the second method, the round-bottom flask of Clevenger apparatus (Fig. 2.3b) (Clevenger 1928) containing the sample mixed with water is directly heated; the resulting vapor/steam carries the essential oils present in the sample which is then condensed and collected into the attached burette (Ismaiel et al. 2016). However, essential oil extraction using Clevenger apparatus is more efficient than hydro-diffusion because the pockets created during the packing of plant material in the hydro-diffusion column will only be poorly exposed to steam, which prevents proper extraction.

2.3.4.1 Microwave Hydro-Diffusion and Gravity Technology (MHG)

MHG is a new and green methodology for the extraction of essential oil which combines microwave heating and earth's gravity at atmospheric pressure. In this method, the flavors diffuse out from the plant material and dropping out of the microwave reactor by the influence of earth gravity. The cooling system fitted outside the microwave chamber will continuously condenses the extract. Since MHG requires less operation time (less than 30 min) when compare to the conventional methods (more than 3 h), they are considered as a time-saving method.

2.3.5 Liquid–Liquid Extraction (LLE)

Even though plethora of solvent extraction techniques is available till date, the most common and easy method that can be employed in phytochemistry is the two-phase solvent extraction. This method works on the basis of relative solubility of compounds in two entirely different immiscible solvents. Here, the extract obtained from any of the abovementioned processes can be mixed with equal volumes of immiscible solvent (e.g., mixing up of aqueous and organic solvents) in a separating funnel. As a result, the impurities with higher affinity towards the second phase will migrate to the other phase (Berk 2013).

2.3.6 Supercritical Fluid Extraction (SFE)

In order to better understand the SFE methodology, a basic understanding about supercritical fluids is necessary. By definition, all the substance in nature is characterized by a "critical point" attained at specific conditions of pressure and temperature (McHugh and Krukonis 1994). A fluid when exposed to a pressure and a temperature, which is higher than its critical point under which the physical state of the fluid is indistinguishable, is said to be "supercritical" fluid. The supercritical fluids possess wide application in a number of fields. From the last decades, much interest has been invested on the applications of supercritical CO_2. The important property that should be highlighted is its density and solvation characteristics equivalent to hexane at a pressure range of 200 bar. As a result, it acts as a nonpolar solvent at supercritical state and is sufficient to dissolve up to 1% (by mass) triglycerides. Small temperature drop or large pressure drop will lead to the precipitation of entire solutes. The main advantage of using supercritical CO_2 for extraction is its ability to evaporate completely to give products that are free from solvent residues (Sapkale et al. 2010).

The instrumentation setup for SCF extraction involves a cylinder which contains the material to be extracted. The super critical fluid is then pumped through this cylinder and carried to a separation chamber from where the extract and the gas are separated. Solvent properties of supercritical CO_2 can be altered by varying temperature and pressure and also by adding some entrainers such as ethanol (Patil and Shettigar 2010).

2.3.7 Microwave-Assisted Extraction (MAE)

MAE is a combined technology which makes use of the microwave energy for heating the solvent-sample mixture and the conventional solvent extraction method for extracting the phytochemicals. Enhanced extraction kinetics due to the rapid and uniform heating by microwaves is the major advantage of MAE. Typical MAE instrument consist of a microwave generator, the waveguide to propagate the generated microwaves to the microwave cavity, the applicator or resonance cavity where the sample is placed, and finally the circulator (Delazar et al. 2012).

2.3.7.1 Closed-Vessel Microwave Systems

The earlier versions of analytical MAE systems were of closed-vessel type with a multimode cavity that allows the processing of several samples at the same time under pressure and temperature feedback control. The instrumentation of closed-vessel MAE system includes microwave transparent closed vessel, an external body, and a transparent cap (e.g., polyetherimide) with inner Teflon liner. The focused high-pressure, high-temperature MAE system comprises an integrated closed vessel and a focused microwave system operating at a high pressure and temperature. Even though this system does not entertain simultaneous extraction, but the use of an auto-sampler will be useful to run an extraction sequences of up to 24 samples (Llompart et al. 2019).

2.3.7.2 Open-Vessel Microwave Systems

In contrast to the closed-vessel systems that are operating at high pressure, the open-vessel instruments are considered safe due to their operation at atmospheric pressure. Since it is an open-vessel instrument, reagents or solvents can be added at any stages of the process; it is also possible to connect a reflex system. The first open-vessel systems are of focused open-vessel MAE system in which the extraction is carried out at atmospheric pressure; hence, the boiling point of the solvents used for extraction determines the maximum temperature required for extraction. As the term "focused" indicates, the microwave irradiation focuses on the specific part of the extraction vessel which contains the sample. However, the open-vessel systems allow the use of only one flask at a time (Llompart et al. 2019).

2.3.7.3 Solvent-Free Microwave Extraction

The solvent-free microwave extraction involves extraction of sample using MAE without any solvents including water. The cooling system fitted outside the microwave chamber keep on condensing the vapors generated from the heated sample which can be collected in appropriate containers. The excess water is refluxed back to the extraction vessel to restore the water content of the sample. Since the extraction takes place without distillation or evaporation, it will not impact any negative effects on the final composition of the extracted compounds (Llompart et al. 2019).

2.3.8 Activity-Guided Purification

In most of the phytochemical analysis, the intention of the purification is to isolate a phytochemical with a desired biological activity such as antimicrobial, anticancer, and anti-inflammatory activity. Such a clear aim will be very useful to orient the experiment on the right track. For example, if a person screening a plant for potent antioxidants, it is better to check the antioxidant activity of different solvent extracts before moving on to the advanced purification techniques. If the user has observed a positive antioxidant activity at any one of the solvent extract, then it is not necessary to spend time on other extracts that were negative for the test. Similarly, screening for specific group of compounds such as sugars, phenolics, glycosides, and flavonoids can be detected by spraying specific detection reagents. In summary, go ahead with the desired fractions and discard the unwanted.

2.4 Working with Plant Extracts-Purification Methods

2.4.1 Column Chromatography

The sample for column chromatography can either be a solvent extract or a processed plant material (dried/crushed/powdered). The selection of column matrix of suitable mesh size and chemistry is an important step to achieve better purification. The column chromatography operates on the basis of a number of principles and is classified as size exclusion, affinity chromatography, ion exchange chromatography, etc. The selection of the suitable chromatographic technique lies in the nature and chemistry of the compounds to be separated.

2.4.1.1 Choice of Stationary Phase

Silica and alumina are polar adsorbents (Armarego and Chai 2003) suitable for the separation of polar components; these polar components are retained more strongly on the stationary phase and are eluted last when a series of mobile phase with gradually increasing polarity was used. Even though silica is recommended for most of the compounds, it is more preferable for the separation of basic components because of its slight acidic nature. Conversely, alumina is slightly basic which will retain acidic compounds more strongly. It is also good for the purification of amines and other weakly or moderately polar compounds.

Like polarity, acidity, and charge, the particle size of the column matrix material also plays significant role in the successful separation of the compounds. The common column matrixes such as silica or alumina are available in a variety of particle size that suits the purification of a number of compounds with different molecular size. The size is expressed by the mesh value, which implies that the matrix particles with graded mesh size act as a molecular mesh or filter. In short, silica particles with higher mesh size values such as "silica gel 230–400" indicates more number of holes per unit area; hence they are recommended for flash column. Similarly, a lower range of mesh size values (70–230 silica gel) indicates comparatively fewer number of holes per unit area and are highly recommended for gravity-based column (Millar 2012c).

Based on the water content of alumina, they are classified as type I, II, and III. In which type I has the least water content and type III contains the most. The amount of water in alumina particle determines their degree of polarity. Lower water content means there are more polar sites that are free to bind with polar compounds and will retain them on the column for long time. The particle size range for alumina ranges from 70–290 mesh (50–200 µm); however type II or III alumina with average mesh size of 150 is the most commonly used.

There are several other column matrixes such as cellulose-based (cellufine), dextran-based (sephadex, sephacryl), agarose-based (sepharose, Biogel A), polyacrylamide-based (Biogel P), and polystyrene-based (Dowex-50) materials that are currently in use. But most of these matrix materials are specially designed and highly recommended for peptide/protein purification practices and possess less significance in phytochemistry (http://technologyinscience.blogspot.com).

2.4.1.2 Preparing the Column

There are minute variations in the design of commercially available chromatographic glass column in which some of them contain a frit at the bottom, while others do not. Care should be taken to add a ball of cotton at the bottom of non-fritted column. A layer of acid washed sand can also be added optionally in both the columns to provide an even base and also to prevent concentration and streaking of the bands (Fig. 2.4).

In the dry-pack method shown in Fig. 2.5a, the column should be filled with suitable solvent and allow it to pass through the sand layer and cotton ball to remove the air bubbles if any. Add the stationery phase (Silica or Alumina) slowly and uniformly and allow it to settle; tap gently on the column for tight and uniform

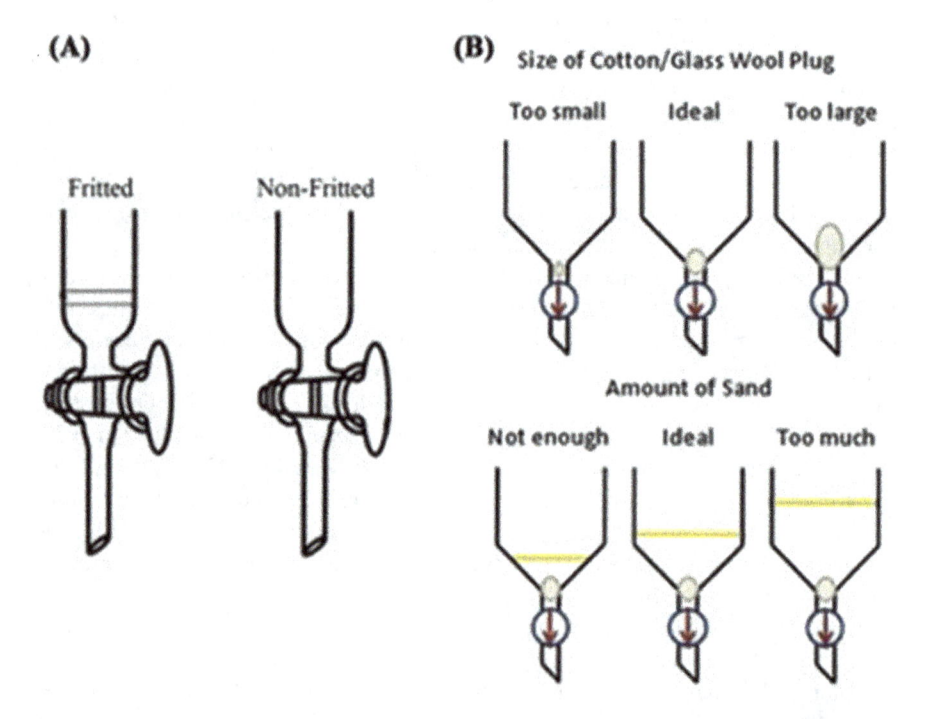

Fig. 2.4 (a) Differentiating fritted and non-fritted column. (b) Method of placing cotton ball and sand layer during column preparation (Millar 2017)

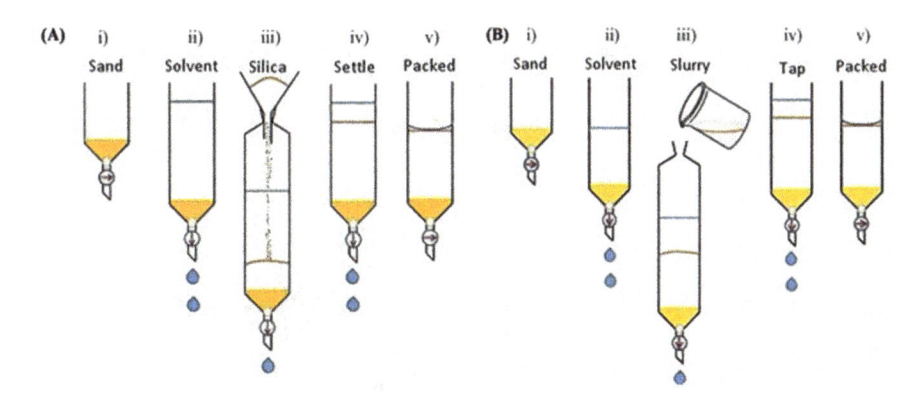

Fig. 2.5 Different steps of chromatography column preparation. (**a**) Dry-pack method, (**b**) slurry method (Figure adopted from Tips and Tricks for the Lab: Column Packing by Millar (2017), Copyright: Wiley-VCH Verlag GmbH & Co. KGaA, Weinheim)

packing. Allow the solvent to drain throughout the packing procedure to prevent overflow. At the same time it should not be allowed to drain completely. The major disadvantage of this method is the difficulty to get a well-packed column. Whereas, the slurry method in which the silica was made into a slurry with the solvent prior to adding into the column is a quick and easy method, but requires repeated washing which consumes large quantities of solvents. However, the slurry method is the widely accepted method for routine column chromatographic practices.

In the slurry method (Fig. 2.5b), fill the column with the solvent, simultaneously prepare a slurry of silica or alumina by mixing with the same solvent in a separate beaker and pour the slurry gently into the column. Keep the drain valve open throughout the packing process and should not allow the column to dry. In both the methods the column should be equilibrated by allowing the solvent to pass multiple times.

2.4.1.3 Sample Loading

Dissolve the sample in minimal volume of more polar solvent than the one that we are going to use for elution; load the sample onto the top of silica layer without disturbing the column. Alternatively, the sample dissolved in solvent can also be mixed with dry silica or alumina and can be loaded after drying. This method will be more useful if the sample is poorly soluble in the solvent. After loading the sample, add small amount of slurry or sand to form a thin layer over the sample to prevent column and sample disturbances while adding solvents.

2.4.1.4 Solvent System

As in the other purification methods, silica column can also be run with isocratic or gradient method. For the isocratic method, only one kind of solvent can be used for elution; the choice of solvent is based on the polarity of the compound. In the gradient method, combination of different solvents with varying proportions can be

Table 2.1 Commonly used mobile phase combinations in the order of increasing polarity for the purification of phytochemicals (Bajpai et al. 2016)

Fraction number	Solvent system	Ratio
1	Hexane	100
2	Hexane:ethyl acetate	10:01
3	Hexane:ethyl acetate	05:01
4	Hexane:ethyl acetate	01:01
5	Hexane:ethyl acetate	01:05
6	Hexane:ethyl acetate	01:10
7	Ethyl acetate	100
8	Ethyl acetate:methanol	10:01
9	Ethyl acetate:methanol	05:01
10	Ethyl acetate:methanol	01:01
11	Ethyl acetate:methanol	01:05
12	Ethyl acetate:methanol	01:10
13	Methanol	100

employed to make a gradient change in polarity of the solvent system which results in detailed elution of the compounds. An example for the sequence of solvent system used to elute plant secondary metabolites (Bajpai et al. 2016) is presented in Table 2.1.

Like this, several other combinations can also be tried which best suits the chemistry of the target molecule. It should always be kept in mind that the solvents that are mixing together should be completely miscible at standard conditions; if the mixture exists as two phases even after vigorous mixing, then it is not-recommended to use for elution in the column. The recommended solvent systems for normal phase silica are dichloromethane, hexane–acetone, and hexane–butyl acetate mixtures. Methanol or acetonitrile with water are the recommended mobile phase combinations for reversed-phase stationary phases. Hexane-dioxane mixtures are suitable for alumina-based columns (Pacáková 2000) (Table 2.2).

2.4.1.5 Normal Phase and Reverse Phase Chromatography

The methods that have been discussed so far are the normal phase chromatographic techniques in which the stationary phase is polar in nature which facilitates the separation of polar compounds. The hydroxyl groups (Silanol groups) on the surface of normal phase silica particle give the polar nature. Whereas, the stationary phase in the reversed-phase chromatography is hydrophobic in nature (nonpolar) that retains nonpolar compounds and helps in their separation. The silica without modification is highly polar and can be used for normal phase chromatography, whereas the modifications in C18 or other nonpolar chains or functional groups will make it hydrophobic and can be used as a stationary phase for reversed-phase chromatography. Some modifications to the silica will be discussed in the HPLC section.

Table 2.2 Commonly used solvent combinations for silica column (Millar 2012a)

No.	Isocratic (least polar to most polar)	Gradient combination (least polar to most polar)	Exotic solvents
1	• Pentane • Petroleum ether • Hexanes	Ether + petroleum ether or hexane or pentane	n-Butanol + acetic acid + water
2	• Ether or • Dichloromethane	Hexane + ethyl acetate or dichloromethane	2-Methyl-1-propanol + acetic acid + water
3	Ethyl acetate	Dichloromethane + methanol	1-Propanol + NH_4OH (aq) + water
		The ratio of the solvent combinations can be adjusted to give a complete spectrum of polarity. Greater than 10% methanol is not recommended for normal phase silica column since methanol will strip off silica from its –OH group	Exotic solvents are used only for extremely difficult separation or to avoid the use of expensive reverse phase matrix

Note: Hexane–Dioxane mixtures are used with alumina

2.4.1.6 Sample Collection After Column Chromatography

Even though sample collection is a simple step, which does not require much expertise, the concept of "dead volume" will always cause trouble in collecting the desired fraction at the right time. "Dead volume" theoretically describes the volume of solvent passing through the column in which no sample or analyte is present. For HPLC and UPLC columns, the manufactures will provide the estimated dead volume and the instruments were designed to cope with this. In the glass column, the users should calculate the dead volume so as to understand the point at which sample collection should start. In some cases, during elution a sample line (sometimes colored) will migrate with the solvent which will guide you to determine the starting of sample collection. If the sample line is absent, mark the maximum solvent level in the column, then measure the 2/3rd of the packed column matrix volume and make a second mark corresponding to the 2/3rd length below the first marking. This is the point at which sample collection should start (Fig. 2.6).

2.4.2 Thin Layer Chromatography (TLC)

TLC is an important technique which helps to determine which purification technique or solvent system can be employed at different stages of compound purification. Usually, TLC experiment is performed before the column chromatography; this is to determine which column matrix and solvent system is suitable for the separation of the compound of interest. TLC can be performed on a silica or alumina matrix, which is either pre-coated on aluminum sheets or can be coat manually on glass plates. The TLC plates usually contain a fluorescent indicator F 254 which enhances

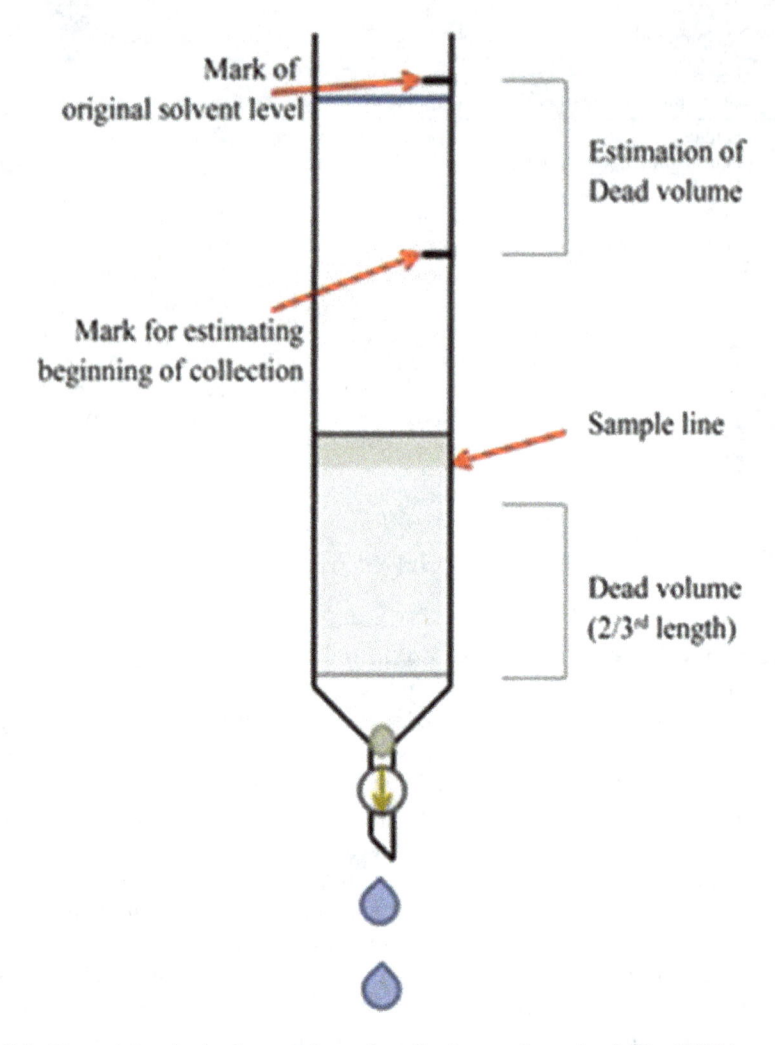

Fig. 2.6 Determining dead volume and sample collection starting point (Millar 2012b)

detection under UV light. TLC is easy to perform, but gives valuable information that helps to troubleshoot the complications in purification. Dip the capillary tube into the sample solution which will take up sample into it, spot it on to the TLC plate with proper care to avoid sample overloading (spot should not be more than 2 mm). Place the TLC plate in a container with the solvent system, close the top of the container with lid, allow it to run for few minutes, and remove the plate when the solvent front is 0.5–1 cm from the top of the plate. Visualize under UV lamp or apply detection reagents for visualization and identification. When visualizing under UV lamp, the compounds migrated along with the mobile phase can be seen as illuminated bands or spots. From their position, the R_f values of the corresponding

Fig. 2.7 Schematic diagram of a developed TLC plate showing the spotted sample, migrated analyte, and the solvent front

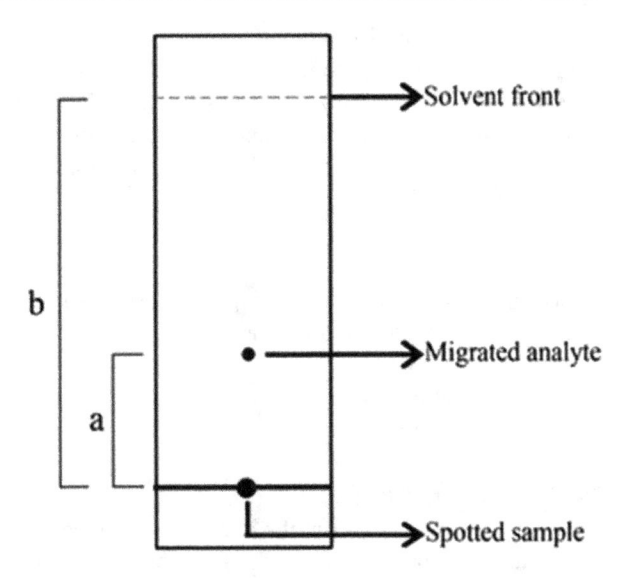

compounds can be ascertained (https://www.chemguide.co.uk/analysis/chromatography/thinlayer.html) (Fig. 2.7).

Calculation of R_f value (retardation factor)

$$= \frac{\text{Distance (cm) from baseline travelled by the analyte (a)}}{\text{Distance (cm) from baseline travelled by the solvent (b)}}.$$

2.4.2.1 Selection of Mobile Phases

The basics of selecting mobile phase depend up on the R_f value which ranges from 0 to 1. In the normal phase TLC, R_f value 0 denotes the compound is not moved. This indicates that, the polarity of the solvent is very low. Conversely, if the R_f is 1, the polarity of the mobile phase is too high. By noticing the R_f values, minor adjustments can be made to the existing mobile phase to attain better separation. The most common starting solvent system is hexane and ethyl acetate in the ratio 1:1; after determining the R_f values with this solvent system, further modifications in polarity of the mobile phase can be made. A mixture of highly polar compounds can be separated by using ethyl acetate, butanol, acetic acid, and water in the ratio 80:10:5:5, respectively. Ten percent or lesser amount of methanol in DCM is also recommended for the separation of polar compounds. Highly basic components in the mixture can be separated by making a mixture of 10% NH_4OH in methanol, and then make a 1–10% mixture of this in dichloromethane.

2.4.2.2 Anomalies in TLC

If the analyte is highly soluble in the mobile phase, it will travel further up the solvent front or plate. To resolve this, the polarity of the mobile phase should be reduced. Similarly, if the compound has higher affinity towards the stationery phase, then it will not move at all; under these circumstances, we have to increase the polarity of the mobile phase.

Streaks/Tailings

Acids, bases, and highly polar compounds produce streaks rather than spots in TLC plates when a neutral mobile phase is used. This can be resolved by adding few percent of acetic acid or formic acid (for acidic compounds) or few percent of triethylamine for basic compounds. Streaking with polar compounds can be resolved by adding methanol to the mobile phase (Fig. 2.8).

2.4.2.3 Preparative TLC

The compounds separated in the TLC can be recovered in the preparative scale by employing a preparative TLC. This can be performed on a glass plate of higher dimensions (~148 × 105 mm) coated with slurry of silica or alumina specific for TLC experiment. Since it is a higher dimension, it needs sufficient amount of mobile phase to complete the run and the run time is also longer than usual analytical methods. Once the run is completed, visualize the plates under UV and spot the exact location of the compound of interest or isolated bands, scrap off the separated spot (with silica of that area) from the plate and dissolve in a suitable solvent.

Fig. 2.8 Streaking or tailing of analytes in developed TLC plates

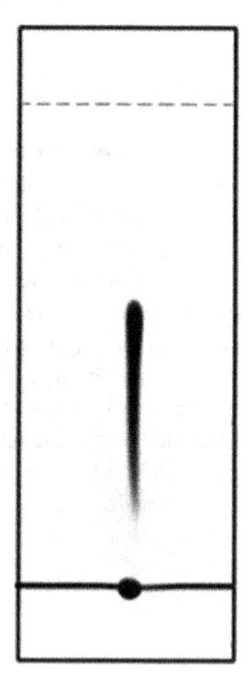

2.4.3 High-Performance Thin-Layer Chromatography (HPTLC)

High-performance thin-layer chromatography is the advanced instrumental form of conventional TLC. It uses specially designed small HPTLC plates with narrow particle size that provides better resolution, sensitivity, and ease of quantification. This sophisticated instrument, controlled by the integrated software, ensures highest possible degree of usefulness, reliability, and reproducibility of generated data.

2.4.3.1 Performing HPTLC Analysis

HPTLC analysis is more or less similar to that of the conventional TLC analysis, except some additional and advanced methodologies incorporated to improve the efficacy, sensitivity, and reproducibility of the analysis. HPTLC plates are usually pre-washed with methanol or mixtures such as methanol:chloroform (1:1), chloroform:methanol:ammonia (90:10:1), methylene chloride:methanol (1:1) or 1% ammonia solution. Plates exposed to humid environment require additional activation step by keeping in oven at 110–120 °C for 30 min. Plates taken from freshly opened box does not require activation. Long-term exposure to high temperatures should be avoided, as it may cause undesirable effects on the plates. HPTLC instruments contain semi-automatic/automatic sample applicators for spotting the samples. However, micro-syringes are highly recommended to apply samples in the absence of automatic sample applicators. The spotted plates can be developed in the chromatographic chamber using suitable mobile phase. Finally, visualization can be performed using UV illumination or by spraying special detection reagents that can distinguish different chemical groups. The software system of HPTLC instruments provides enormous applications such as densitometry for quantification and fingerprinting (Attimarad et al. 2011).

2.4.3.2 Application of HPTLC

Quantification of known samples with unknown concentration can be quantified by plotting a standard curve of varying concentration of the reference compound. The instrument will read the density of the isolated band of unknown concentration, and with the help of software programs the concentration will be calculated by correlating the band intensity with the reference compound.

2.4.3.3 Chromatographic Fingerprint Analysis

The chromatographic fingerprint analysis is useful to identify the chemical similarity between closely related plant species or different members of same groups. HPTLC is the most reliable method to perform the fingerprint analysis.

2.4.4 Identification Tests (Touchstone 1992; Jork et al. 1994)

2.4.4.1 Detection of Flavonoids
(a) Spray 1% ethanolic solution of aluminum chloride onto the TLC plate; appearance of yellow fluorescence in long wavelength UV light (360 nm) is positive for the test.
(b) Spray the TLC plates with 1% solution of ethanolamine diphenylborate in methanol and then with 5% ethanolic solution of polyethylene glycol for fluorescence stabilization. Irradiate for 2 min with intense 365 nm UV light and visualize under the same.

2.4.4.2 Detection of Phenols and Arylamines (Emerson Reagent)
Spray solution I (1 g aminoantipyrine (4-aminophenazone) in 100 mL 80% ethanol), allow to dry for 5 min, spray solution II (4 g potassium hexacyanoferrate (III) in 20 mL water, fill to 100 mL with ethanol), dry for 5 min, place the TLC plate in a chamber with vapor from 25% ammonium solution, make sure to avoid the contact between the sample layer and the liquid. Red-orange to salmon pink spot is positive for the reaction.

2.4.4.3 Test for Phenolic Compounds and Tannins
(a) Ferric chloride test
    The extract (50 mg) is dissolved in 5 mL of distilled water. To this few drops of neutral 5% ferric chloride solution are added. A dark green color indicates the presence of phenolic compound (Mace 1963).
(b) Gelatin test
    The extract (50 mg) is dissolved in 5 mL of distilled water and 2 mL of 1% solution of gelatin containing 10% NaCl is added to it. White precipitate indicates the presence of phenolic compounds (Evans 1997).
(c) Lead acetate test
    The extract (50 mg) is dissolved in distilled water and to this 3 mL of 10% lead acetate solution is added. A bulky white precipitate indicates the presence of phenolic compounds.
(d) Alkaline reagent test
    An aqueous solution of the extract is treated with 10% ammonium hydroxide solution. Yellow fluorescence indicates the presence of flavonoids.
(e) Magnesium and hydrochloric acid reduction
    The extract (50 mg) is dissolved in 5 mL of alcohol and few fragments of magnesium ribbon and concentrated hydrochloric acid (drop wise) are added. If any pink to crimson color develops, presence of flavonol glucosides is inferred (Harborne 1998).

2.4.4.4 Detection of Cannabinoids
Spray the freshly prepared Fast Blue B reagent (0.5 g Fast Blue B (tetraazotized di-o-anisidine) in acetone/water (9:1, v/v)) and then overspray with 0.1 M sodium

hydroxide solution. The migrated spots/bands containing cannabinoids will turn dark red/purple.

2.4.4.5 Detection of Phosphoric Acid Derivatives

Dry the developed TLC plate and treat at 60 °C, hydrolyze di- and triphosphates by spraying perchloric acid solution (1 M perchloric acid in water/acetone (1:1)-Solution I), spray the same reagent two times, and again dry the plates at 50 °C. Spray the warm plate with ammonium molybdate solution (Ammonium molybdate soln: 5 g $(NH_4)_6Mo_7O_{24} \cdot 4H_2O$ in 35 mL semi-conc. Nitric acid and 65 mL water-Solution II) and subsequently with solution III (Tin (II) chloride soln: 0.5 g $SnCl_2 \cdot 2H_2O$ in 100 mL 0.5 M hydrochloric acid). In this test, phosphates appear as blue to blue-green spots. Polyphosphates can also be detected by dipping the plates in a solution of ammonium molybdate (1 g) dissolved in water (8 mL) and perchloric acid (3 mL, 70%), filled up to 100 mL with acetone. Then phosphates appear as yellow-green spots on a blue background.

2.4.4.6 Detection of Carbohydrates/Sugars

(a) Spray with a freshly prepared solution containing 1 mL p-anisaldehyde, 1 mL 97% sulfuric acid in 18 mL ethanol and heat at 110 °C. Sugar phenylhydrazones produce green-yellow spots in 3 min. Sugars will produce blue, green, violet spots in 10 min. This test also detects digitalis glycosides.

(b) Mix a solution of 3% p-anisidine hydrochloride in *n*-butanol and spray on the TLC plate. Heat the treated plate at 100 °C for 2–10 min. Presence of aldohexoses can be confirmed by the appearance of green-brown spots. On the other hand, the appearance of yellow, green, and red spots indicates the presence of ketohexoses, aldopentoses, and uronic acids, respectively.

(c) Spray the plates with a solution containing 0.5 g thymol in 95 mL ethanol and 5 mL of 97% sulfuric acid and heat the plates for 15–20 min at 120 °C. Presence of sugars in the plate will form pink-colored spots.

(d) Spray the TLC plates with a solution containing 5 g urea in 20 mL of 2 M hydrochloric acid made up to 100 mL with ethanol, which is then heated at 100 °C. Ketoses and oligosaccharides containing ketoses will turn blue.

2.4.4.7 Detection of Carbohydrates and Reducing Sugars

(a) Spray the TLC plates with a solution containing 1.23 g p-anisidine and 1.66 g phthalic acid in 100 mL 95% ethanol. The inferences are as follows: hexoses—green, pentoses—red-violet, methylpentoses (sensitivity 0.5 µg)—yellow-green, uronic acids (sensitivity 0.1–0.2 µg)—brown.

(b) After drying the TLC plate, spray with the solution (0.93 g aniline and 1.66 g o-phthalic acid dissolved in 100 mL *n*-butanol saturated with water). Dry the plate with hot air, then heat to 105 °C for 10 min. Positive test results indicate the formation of different colors on colorless background and some spots give fluorescence at 365 nm.

(c) Spray the developed TLC plates with a solution containing 3.6 g m-phenylenediamine dihydrochloride in 100 mL 70% ethanol and heat briefly at

105 °C. Appearance of intensely fluorescence colors in UV light (254 and/or 366 nm).

2.4.4.8 Detection of Sweeteners, Saccharine and Cyclamate
Spray the TLC plate with a 0.2% solution of dichlorofluorescein in 96% ethanol, dry under warm air, spray with water, and view under UV light (360 nm). Appearance of fluorescent spots indicate the presence of sweeteners, saccharine and cyclamate

2.4.4.9 Detection of Aromatic Hydrocarbons
Spray the TLC plates with a solution containing 0.5–1.0 g tetracyanoethylene in dichloromethane or toluene. Aromatic hydrocarbons will show various colors, some of them last for a short time only.

2.4.4.10 Detection of Halogenated Hydrocarbons
Spray the TLC plates with a solution containing 0.1 g silver nitrate dissolved in 1 mL water, 10 mL 2-phenoxyethanol which is then made up to 200 mL with acetone and additional one drop hydrogen peroxide (30% solution). Irradiate the sprayed plates with unfiltered UV light until optimal contrast is achieved. Appearance of dark spots indicates the presence of halogenated hydrocarbons. Same combination of spraying reagent without hydrogen peroxide was also reported for the detection of pesticide ioxynil (3,5-diiodo-4-hydroxybenzonitrile).

2.4.4.11 Detection of Indoles and Thiazoles
Spray the plates with a freshly prepared solution of 1 g sodium nitrite in 100 mL hydrochloric acid and heat at 100 °C. Presence of indoles will form red color and thiazole derivatives will form light green.

2.4.4.12 For Detection of Organic Acids
Dip the TLC plate in a solution containing 0.1 g bromocresol green in 500 mL ethanol and 5 mL 0.1 M NaOH. Development of yellow spots on a blue background indicates the presence of acids.

2.4.4.13 Detection of Alpha-keto Acids
Spray the developed plates with a solution of 0.05 g 1,2-phenylenediamine in 100 mL of 10% aqueous trichloroacetic acid. Heat the plates at 100 °C for short term (not more than 2 min), appearance of green fluorescence spots in long wavelength UV light indicates the presence of alpha-keto acids.

2.4.4.14 Detection of Lipids/Phospholipids
(a) Spray 0.1% bromothymol blue in 10% aqueous ethanol (made alkaline with NH_4OH) on the dried TLC plates and observe for the development of blue-green colors.
(b) Spray the plates with a solution of 1 mg rhodamine 6 G in 100 mL acetone and observe under long wavelength UV.

(c) Spray the TLC plates with 0.1% gentian violet (crystal violet) dissolved in methanol and place in a container with bromine vapor. Presence of lipids will produce blue spots on yellow background.

2.4.4.15 Detection of Cholesterol and Its Esters
Spray the plates with 20% phosphotungstic acid in ethanol, heat at 110 °C for 5–15 min or until the spots become clearly visible. Presence of cholesterol and esters can be confirmed by the appearance of red spots.

2.4.4.16 Detection of Sterols, Steroids, and Bile Acids
Heavily spray the TLC plates with a solution containing 85% phosphoric acid in water (1:1, v/v) until the layer becomes transparent. Eventually, heat the plate for 10–15 min at 120 °C and observe for the development of various colors under visible and UV light.

2.4.4.17 Detection of Conjugated 3-Ketosteroids
Spray the plates with a solution containing 0.9 g p-phenylenediamine and 1.6 g phthalic acid in 100 mL 1-butanol saturated with water. Heat the plates at 100–110 °C and observe for the presence of yellow to orange spots.

2.4.4.18 Detection of Vitamins, Carotenoids, Steroids, Sapogenins, Steroid Glycosides, Terpenes
Spray the TLC plate with a solution containing 25 g antimony (III) chloride in 75 mL chloroform (generally a saturated solution of antimony (III) chloride in chloroform or carbon tetrachloride is used), heat for 10 min at 100 °C, and observe under long wavelength light (360 nm) for the presence of fluorescing spots.

2.4.4.19 Test for Phytosterols
Libermann-Burchard's test: The extract (50 mg) is dissolved in 2 mL acetic anhydride. To this, one or two drops of concentrated sulfuric acid are added slowly along the sides of the test tube. An array of color change shows the presence of phytosterols (Finar 1986).

2.4.4.20 Detection of Barbiturates
Freshly mix equal volumes of 2% ethanolic mercury (II) chloride and 0.2% ethanolic diphenlycarbazone and spray on to the TLC plates. Barbiturates will appear as pink spots on a violet background.

2.4.4.21 Detection of N-Substituted Barbiturates
Spray the TLC plate with 0.1% ethanolic solution of dichlorofluorescein and subsequently with 0.1% ethanolic solution of fluorescein sodium salt; eventually observe the plates for the appearance of fluorescent spots.

2.4.4.22 Detection of Aldehydes/Ketones

(a) Spray a solution containing 0.4 g 2,4-dinitrophenylhydrazine in 100 mL of 2 N hydrochloric acid onto the TLC plate and add 1 mL ethanol. Development of yellow-red spots indicates the presence of aldehydes/ketones.
(b) Spray the reagent containing 15 mg of 2,2′-diphenylpicrylhydrazyl (DPPH) dissolved in 25 mL chloroform and heat for 5–10 min at 110 °C. Appearance of yellow spots on a purple background is a positive indication for the test.

2.4.4.23 Detection of Glycosides and Glycolipids

Slightly spray the TLC plate with the reagent containing 10 mL of 10% diphenylamine in ethanol, 100 mL HCl and 80 mL glacial acetic acid. Cover the plate with a glass plate, heat 30–40 min at 110 °C until blue-colored spots develop as an indication for the presence of glycolipids.

2.4.4.24 Detection of Glycosides, Glycolipids

Spray the TLC plate with the solution (dissolve 0.1 g orcinol in 40.7 mL concentrated HCl, add 1 mL 1% ferric (III) chloride, and dilute to 10 mL) and heat at 80 °C for 90 min. Presence of glycolipids can be confirmed by the appearance of violet spots.

2.4.4.25 Detection of Alkaloids and Aromatic Hydrocarbons

Spray the TLC plate with a solution of 37% formaldehyde in concentrated sulfuric acid (1:10) immediately after taking the plate from the developing chamber without heat treatment. Appearance of spots with different color indicates a positive test result.

2.4.4.26 Detection of Amino Acids, Amines, Amino Sugars

Spray the TLC plates with a solution of 0.2 g ninhydrin in 100 mL ethanol and heat to 110 °C until reddish spots appear.

2.4.4.27 Detection of Amino Acids and Heterocyclic Amines

Dissolve 1 g ninhydrin and 2.5 g cadmium acetate in 10 mL glacial acetic acid and make up to 500 mL with ethanol. Spray the TLC plate with the solution and heat at 120 °C for 20 min. Observe the appearance of red, pink, or purple spots.

2.4.5 Other Chromatographic Techniques

Apart from the above mentioned techniques, there are several other chromatographic techniques of choice that are usually of less significant in the field of phytochemistry.

Ion-exchange chromatography: operates on the basis of electrostatic interaction between the stationary phase and the molecules to be eluted. The matrix material possesses a charge on its surface; as a result it binds and retains molecules with opposite charges which can be eluted by changing the pH, concentration, or ionic strength of the mobile phase. Such kind of mobile phase modifications are very easy in case of buffers; hence this method is more suitable for the separation of proteins. Based on the charge of the matrix material they are classified as anion-exchange (positively charged matrix) and cation-exchange (negatively charged) chromatography (Karlsson et al. 1998).

Gel-filtration or gel-permeation chromatography: uses dextran-based column matrixes (e.g., Sephadex, Sephacryl) to separate molecules based on their molecular size differences. This method is also highly recommended for peptide/protein applications and not for small molecules. For this method, several other column matrixes such as agarose-based (Sepharose, Biogel A) and polyacrylamide-based (Biogel-P) materials can also be used (Walls and Loughran 2011).

Affinity chromatography: A ligand with higher affinity towards the target protein will be attached with the column matrix (dextran, polyacrylamide, or cellulose matrix) which can form complexes with the target protein. Such complexes can then be separated by changing the ionic strength of the mobile phase (Wilchek and Chaiken 2000; Firer 2001).

Hydrophobic interaction chromatography: is also closely related with affinity chromatography in which the column matrix surface was attached with hydrophobic groups or ligands which facilitate the interaction of hydrophobic molecules present in the sample (Mahn and Asenjo 2005; Queiroz et al. 2001).

Dye-ligand chromatography: uses inexpensive synthetic dyes (azo, anthraquinone or phthalocyanine dyes) as affinity ligands for biomolecules. Here, the reactive group (e.g., chlorotriazine ring) of the dye is the key player to mediate the interaction with the target molecule (https://www.news-medical.net/life-sciences/Dye-ligand-Affinity-Chromatography-for-Protein-Purification.aspx).

2.4.6 HPLC (High-Performance Liquid Chromatography)

HPLC is a highly improved and advanced form of conventional column chromatography. In this method, the solvent or mobile phase is forced to pass through a chromatographic column packed with matrix of fine particle size and small diameter under high pressures of up to 400 atmospheres. Higher operating pressure and the use of a column packing material with smaller particle size (which provides larger surface area for interaction between the stationary phase and the molecules passing through it) allow better separation of the components present in the mixture. Purification of compounds in the preparative scale renders HPLC as a backbone of purification techniques. The major advantage of HPLC lies on its compatibility with other detection methods such as UV detector, fluorescence detector, refractive index detector, evaporative light scattering detector (ELSD), and mass detector.

There are two variants of HPLC that are commonly employed. One method depends on the relative polarity of the solvent and the other one on stationary phase. These methods were already discussed in brief in the column chromatography section.

2.4.6.1 Normal Phase HPLC

In the normal phase, the column is filled with silica particles of polar nature; as a result the polar compounds present in the sample will remain attached for longer time. On the other hand, the nonpolar compounds will move rapidly with the nonpolar mobile phase. As a result, compounds with least polarity will elute sooner and those with higher polarity will elute later.

2.4.6.2 Reversed-Phase HPLC

In the RP-HPLC, modifications to the column packing matrix (e.g., silica) by attaching long hydrocarbon chains (typically with 8 or 18 carbon atoms) to its surface will make it nonpolar. Hence, the stationary phase will retain the nonpolar compounds for longer time and the polar compounds will migrate faster with the polar mobile phase. Reversed-phase chromatography is the most widely used technique for routine use. However, other column chemistry should also be employed to separate compounds that are difficult to separate with conventional columns.

2.4.6.3 Selection of Suitable Chromatography Column

A chromatographic column that is more suitable for the desired purpose can be selected based on a number of factors. Figure 2.9 shows the selection of column

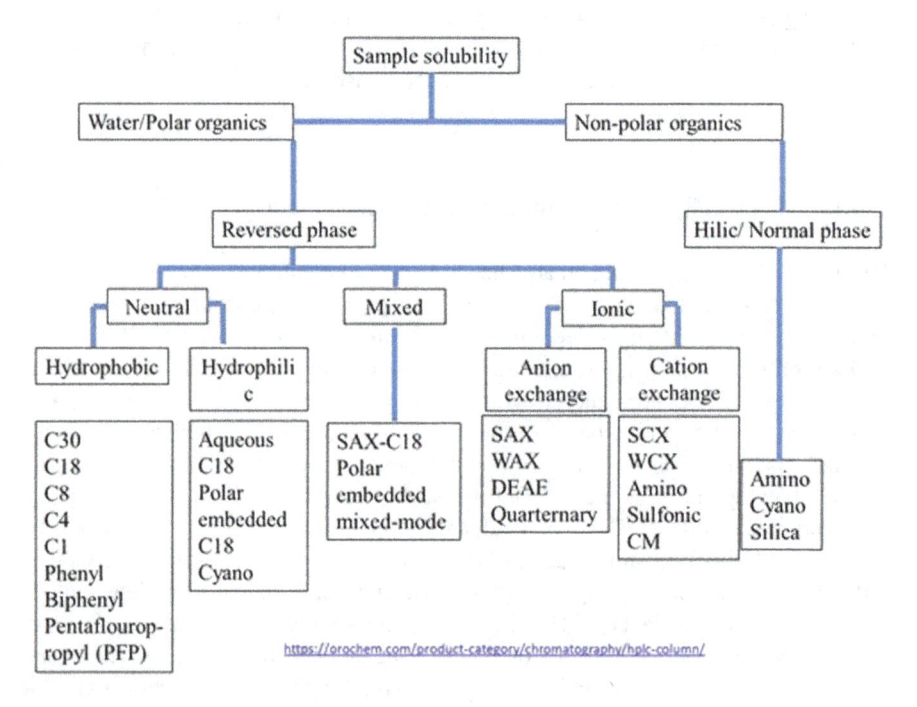

Fig. 2.9 Schematic representation of the selection of chromatographic stationary phases based on sample solubility

based on sample solubility. Since the column chemistry is an elaborate field, the columns presented in figure are not described in detail here.

2.4.6.4 Selection of Diluent and Mobile Phase

Sample preparation for liquid chromatography depends on the solubility of the sample and the compatibility of the column (Stationary phase) with the diluent. Similarly, compatibility of the mobile phases with the stationary phase should also be ensured before starting an analysis. For example, C18 column is not recommended to run with highly nonpolar organics such as hexane and toluene. Usually, the column manufacturers will provide a data sheet showing the solvent compatibility of the column. Same precautions should also be taken while choosing the diluent, preferably the diluent should be any of the mobile phase recommended for the column.

2.4.7 Structural Characterization of Metabolites

2.4.7.1 Liquid Chromatography-Mass Spectrometry

LCMS is a multi-unit instrument in which the working principles of the individual units are entirely different. The liquid chromatography unit (HPLC or UPLC) is coupled to the mass detector. The individual components in the mixture are first separated with in the LC column, which is then followed by ionization and separation of the ions on the basis of their mass/charge (m/z) ratio. The ion source is the important component in mass spectrometry, which generates ions for analysis. There are different types of ion sources specified for some compounds with unique chemical properties; such ionization methods includes APCI (atmospheric pressure chemical ionization), ESI (electrospray ionization), MALDI (matrix-assisted laser desorption ionization), etc.

The main advantage of this technology is that, the user can get an idea about the masses of components that are passing through the column. The advanced MSMS or triple quadrupole system can bring you to the next level of your compound identification. The fragmentation of the desired masses will generate a number of daughter ions with unique fragmentation pattern, which can then be interpreted with a thorough knowledge to identify the molecule. However, LCMS/MS is not sufficient to structurally identify a molecule; under these circumstances a combination of more advanced techniques such as FTIR, NMR, and XRD are mandatory for the identification and structure elucidation of the molecule.

2.4.7.2 Interpreting an MSMS Spectrum

Fragmentation of a (collision-induced dissociation, electron impact ionization, etc.) molecule in LC-MSMS or GCMS generates unique fragments that is called as fragmentation pattern or mass fingerprints. Such fragmentation pattern provides structural details of the molecule which can be interpreted with the help of McLafferty's rule (McLafferty et al. 1978). According to this rule, loss of certain chemical groups from a molecule during fragmentation follows a particular order

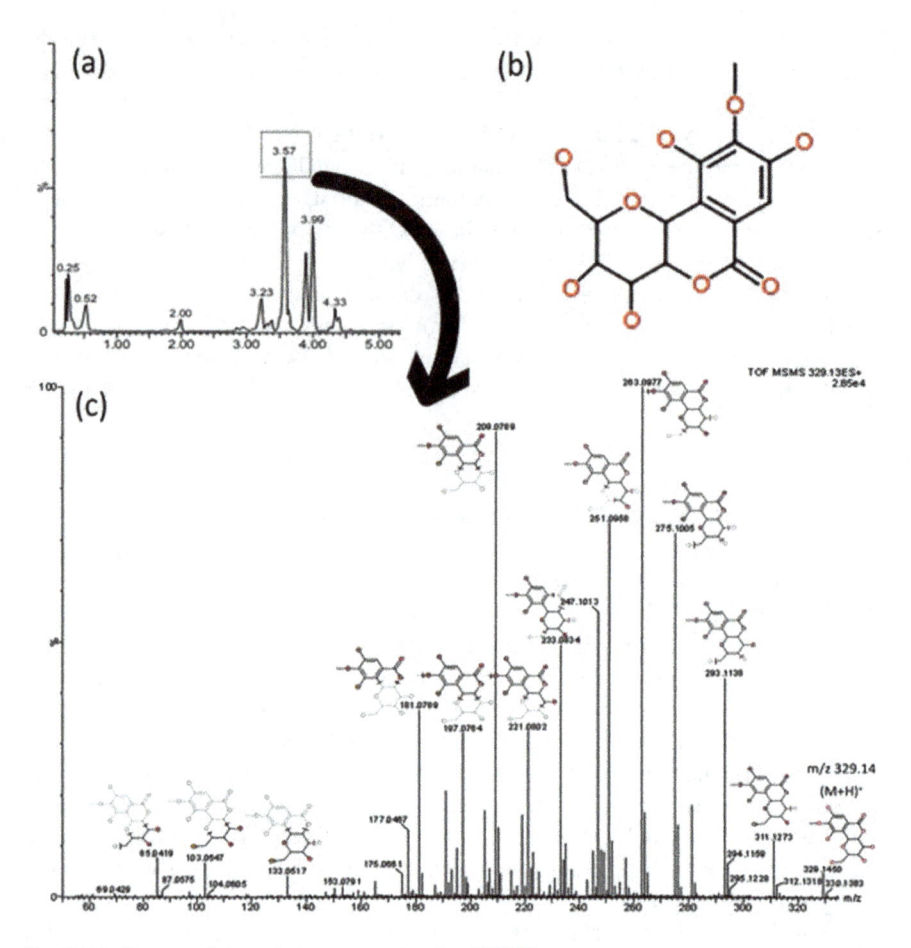

Fig. 2.10 Structure interpretation approach using MSMS spectrum

and each chemical group possesses different probability sequence of release. Some groups may get released immediately with mild collisions or rearrangements (–OH) whereas some other groups require stronger collisions and higher levels of structural rearrangements. Since chemical structure interpretation is a vast area, detailed chemistry of structure interpretation are not discussed in this chapter. In order to provide a basic knowledge to the readers, an example has been presented in Fig. 2.10c (done in our lab) which describes the structure prediction from the daughter ions generated after collision-induced dissociation of a phytochemical. The inset chemical structures placed near each daughter ions indicate the structure remained after each fragmentation and the groups released.

In most of the cases, mass spectrometry alone is not sufficient enough to accurately interpret the structure of an analyte. There are some advanced techniques/methodologies like ion mobility, hydrogen/deuterium (H/D) exchange, and chemical derivatization that can be coupled with MS for better results.

Ion mobility technique separates gas-phase ions based on their size/shape. This technique can be coupled with MS (IM-MS) to obtain additional information about unknown molecules (De Vijlder et al. 2018). The major advantage of IM-MS is its ability to distinguish isomeric forms of the analytes and their extended applications in the analysis of small molecules as well as biological macromolecules like peptides and proteins (Lapthorn et al. 2013). The ability to selectively remove background ion using IM-MS method also reduces complexity and confusion during data interpretation (Eckers et al. 2007).

Hydrogen/deuterium (H/D) exchange methodology is useful to interpret the structure of two closely related/different molecules that shows same elemental composition after preliminary MS analysis. H/D exchange works on the basis that different molecules contain different numbers of exchangeable hydrogen atoms. Heteroatom (oxygen, nitrogen, and sulfur) bound hydrogen atoms can readily exchange for deuterium, but those bound with carbon does not. Every H/D exchange will cause 1 Da mass difference which can be easily detected with conventional mass spectrometers (De Vijlder et al. 2018).

It is highly recommended to interpret the structure of phytochemicals or any other molecules with a combination of other analytical techniques such as NMR, FTIR, or XRD along with MSMS analysis. Because, MSMS analysis has several limitations

2.4.7.3 Detection of Molecules That Are Difficult to Identify with Routine Methods

Derivatization
Some pesticides (e.g., glyphosate) and phytochemicals especially with highly polar nature, ionic structure, absence of chromaphores, low molecular mass, and poor volatility are extremely difficult to identify with routine chromatographic analysis and detectors. Such compounds can be detected in the conventional chromatographic methods by derivatization with FMOC-Cl (9-fluorenylmethyl chloroformate). This reagent acts by increasing the molecular weight of the analyte to a detectable range and also by changing the polarity so as to retain them in the chromatographic column for longer time (Ibanez et al. 2006).

2.4.7.4 Gas Chromatography-Mass Spectrometry
GC/MS is a method of choice for the analysis of volatile small molecules such as benzenes, alcohols, aromatics, steroids, fatty acids, and hormones. Volatizing the sample is the first step of GC/MS, at which the sample is vaporized into the gas phase. The vaporized sample with the compound of interest will then move on to the capillary column packed with a stationary (solid) phase; this is the point at which separation of the compounds starts. The compounds are carried by an inert carrier gas such as argon, helium, or nitrogen. As the components become separated, they elute from the column at different times, which is generally referred to as their retention times. The compound to be analyzed by GC/MS must be thermally stable and volatile enough. In addition, functionalized compounds may require prior

chemical modification (derivatization) to eliminate undesirable adsorption effects that may possibly affect the result.

The components passing through the GC column are then ionized by the mass spectrometer using electron (Electron Impact (EI)) or chemical ionization (CI) sources. In the EI mode of ionization, a beam of electrons ionizes the molecules resulting in loss of one electron and is represented as M+ (a radical cation). In contrast to the ESI (soft ionization), EI is a hard ionization method which results in the fragmentation of the molecules and generate a "mass fingerprint" for each molecule. This information can then be used for the structure identification of the molecule.

CI, on the other hand, begins with the ionization of a suitable gas which creates a radical that will ionize the sample molecule to produce [M + H] + molecular ions. Since CI is a less energetic ionization mode, the molecules will fragment less than with EI. Even though CI yields less structural information, it is useful to detect the molecular ion that cannot be detected using EI. The further processes are similar to that of conventional mass spectrometers, in which the accelerated ions are separated based on their differences in mass-to-charge (m/z) ratios.

2.4.7.5 DART-MS (Direct Analysis in Real Time Mass Spectrometry) Analysis

This advanced technique allows the analysis of intact samples (e.g., intact plant leaf) by exposing them to a stream of heated gas. Since the technique uses intact objects as samples, it is highly useful to distinguish closely related plant varieties by comparing differences in their mass profile or marker molecules. DART generally makes use of the atmospheric pressure ionization (API)-related technique for the ionization of the sample components. Metastable states of rare gas atoms such as helium, nitrogen, and argon produced by the gas discharge source interact with water and produce ionized water clusters which react and ionize analytes on surfaces or exist as vapor in the ionizing chamber. Raw samples can be placed in the DART gas stream in front of the MS inlet, if the sample is an extract or liquid that can be placed on a sealed melting point capillary. DART is very useful for the analysis of extremely unstable marker molecules that can be lost during the processing.

2.4.7.6 FTIR (Fourier Transform-Infrared Spectroscopy)

FTIR measures the absorption of infrared radiation (IR) by the sample material versus wavelength, such IR absorption will be recorded as bands or peaks from which the user can identify molecular components and structures of the target molecule. When IR radiation is passed through the sample, some quantity of the radiation will be absorbed (absorbance) by the components in the sample and the remaining will passes through it (transmittance). As a result of this absorption, the molecule will get excited into a higher vibrational state. The resulting signal recorded at the detector is a spectrum representing the molecular fingerprint of the sample. Such molecular fingerprint will be unique for a specific molecule, and bit more clearly, the molecular fingerprint of two different molecules will not be the same at all. The signal obtained from the detector is an interferogram, which must be

Table 2.3 List of chemical groups and corresponding peaks formed by bending and stretching vibrations

Class	Absorbance range (cm^{-1})	
	Stretching vibrations	Bending vibrations
Alkanes	2850–3000	1350–1470, 1370–1390, 720–725
Alkenes	3020–3100, 1630–1680, 1900–2000	880–995, 780–850, 675–730
Alkynes	3300, 2100–2250	600–700
Arenes	3030, 1600 and 1500	690–900
Alcohols and phenols	3580–3650, 3200–3550, 970–1250	1330–1430, 650–770
Amines	3400–3500, 3300–3400, 1000–1250	1550–1650, 660–900
Aldehydes and ketones	2690–2840, 1720–1740, 1710–1720, 1690, 1675, 1745, 1780	1350–1360, 1400–1450, 1100
Carboxylic acids and derivatives	2500–3300 (acids), 1705–1720 (acids), 1210–1320 (acids) 1785–1815 (acyl halides) 1750 & 1820 (anhydrides) 1040–1100, 1735–1750 (esters) 1000–1300, 1630–1695 (amides)	1395–1440, 1590–1650, 1500–1560

analyzed with a computer using Fourier transforms to obtain an interpretable infrared spectrum. The FTIR spectra are usually presented as plots of intensity versus wavenumber (in cm^{-1}). Wavenumber is the reciprocal of the wavelength. The intensity can be plotted as the percentage of light transmittance or absorbance at each wavenumber.

Various bonded atoms and groups exhibit different degrees of infrared absorption at different wavelength. Table 2.3 represents the signature absorption peaks formed by bending and stretching vibrations of different chemical groups under different wavelength infrared radiation (https://www2.chemistry.msu.edu, Infrared spectroscopy). Such peaks observed with your samples indicate the presence of any of the chemical groups in your target molecule (Table 2.3).

2.4.7.7 Nuclear Magnetic Resonance (NMR) Spectroscopy

NMR spectroscopy measures the relative magnitude and direction (moment) of spin orientation of the nucleus of the individual atoms within a molecule in solution under high-intensity magnetic field. The degree of spin shift and the signal splitting characteristics of the individual atoms such as hydrogen or carbon of a molecule within the magnetic field is dependent on the nature and location of its surrounding groups.

The sample to be analyzed should be of high purity and dissolved in deuterated solvents (e.g., deuterochloroform (CDCl$_3$), deuterium oxide (D$_2$O), deuterated dimethyl sulfoxide (DMSO-d6), or deuterated benzene for hydrophilic analytes). From the NMR spectrum we will get the following information such as:

Chemical shift: Information about the atomic groups present in the molecule.
Spin-Spin coupling constant: Information about adjacent atoms.
Relaxation time: Information on molecular dynamics.
Signal intensity: Quantitative information, e.g., atomic ratios within a molecule that can be helpful in determining the molecular structure, and proportions of different compounds in a mixture (https://www.jeol.co.jp/en/products/nmr/basics.html).

2.4.7.8 XRD (X-Ray Diffraction)

Most of the phytochemicals will form crystals in their pure form without additional chemical practices. The three-dimensional chemical structure of such molecules can be easily determined by X-ray diffraction methods. In this method, a beam of X-rays is projected into the sample, the atoms in the sample will scatter some of the X-rays, and the scattered rays will interfere with each other. The interaction of the incident rays with the different atoms or atom groups in the sample produces constructive interference (and a diffracted ray) when the conditions of Bragg's law ($n\lambda = 2d\sin\theta$) is met. Bragg's law relates the wavelength of electromagnetic radiation to the diffraction angle and the lattice spacing in a crystalline sample. These diffracted X-rays are then detected, processed, and counted.

2.5 Conclusions

With the advance of modern technology, identification and structure interpretation of phytochemicals can be done instantly. Superficial knowledge and proper understanding about the available technologies and resources is a mandatory to make them useful for our studies. Several extraction and purification methods described earlier in this chapter were already replaced with modern technologies such as DART which eradicate the elaborate and hectic purification methods. However, it is not possible to completely exclude the traditional extraction methods especially while supplying samples for sophisticated and sensitive instruments like LCMS, NMR, and XRD, for which repeated processing of the sample is necessary. Detailed structure elucidation of the analytes requires information from different analytical techniques such as mass spectrometry, FTIR, NMR, and XRD.

References

Altemimi A, Lakhssassi N, Baharlouei A, Watson DG, Lightfoot DA (2017) Phytochemicals: extraction, isolation, and identification of bioactive compounds from plant extracts. Plan Theory 6(4). https://doi.org/10.3390/plants6040042

Armarego WLF, Chai CLL (2003) Chemical methods used in purification. In: Armarego WLF, Chai CLL (eds) Purification of laboratory chemicals, 5th edn. Butterworth-Heinemann, Burlington, pp 53–71. https://doi.org/10.1016/B978-075067571-0/50006-5

Attimarad M, Ahmed KKM, Aldhubaib BE, Harsha S (2011) High-performance thin layer chromatography: a powerful analytical technique in pharmaceutical drug discovery. Pharm Methods 2 (2):71–75. https://doi.org/10.4103/2229-4708.84436

Bajpai VK, Majumder R, Park J (2016) Isolation and purification of plant secondary metabolites using column-chromatographic technique. Bangladesh J Pharmacol 11:844. https://doi.org/10.3329/bjp.v11i4.28185

Berk Z (2013) Extraction. In: Berk Z (ed) Food process engineering and technology, 2nd edn. Academic Press, San Diego, pp 287–309. https://doi.org/10.1016/B978-0-12-415923-5.00011-3

Clevenger JF (1928) Apparatus for the determination of volatile oil. J Am Pharm Assoc 17(4):345–349. https://doi.org/10.1002/jps.3080170407

Cos P, Vlietinck AJ, Berghe DV, Maes L (2006) Anti-infective potential of natural products: how to develop a stronger in vitro 'proof-of-concept'. J Ethnopharmacol 106(3):290–302. https://doi.org/10.1016/j.jep.2006.04.003

De Vijlder T, Valkenborg D, Lemière F, Romijn EP, Laukens K, Cuyckens F (2018) A tutorial in small molecule identification via electrospray ionization-mass spectrometry: the practical art of structural elucidation. Mass Spectrom Rev 37(5):607–629. https://doi.org/10.1002/mas.21551

Delazar A, Nahar DL, Hamedeyazdan S, Sarker S (2012) Microwave-assisted extraction in natural products isolation. Methods Mol Biol (Clifton, NJ) 864:89–115. https://doi.org/10.1007/978-1-61779-624-1_5

Eckers C, Laures AM, Giles K, Major H, Pringle S (2007) Evaluating the utility of ion mobility separation in combination with high-pressure liquid chromatography/mass spectrometry to facilitate detection of trace impurities in formulated drug products. Rapid Commun Mass Spectrom 21(7):1255–1263. https://doi.org/10.1002/rcm.2938

Evans WC (1997) Treaseand evans pharmacognosy. Harcourt Brace and Company. Asia Pvt. Ltd, Singapore

Finar IL (1986) Stereo chemistry and the chemistry of natural products, vol 2. Longman, Singapore, p 518

Firer MA (2001) Efficient elution of functional proteins in affinity chromatography. J Biochem Biophys Methods 49(1–3):433–442. https://doi.org/10.1016/s0165-022x(01)00211-1

Harborne JB (1973) Methods of plant analysis. In: Phytochemical methods: a guide to modern techniques of plant analysis. Springer, Dordrecht, pp 1–32. https://doi.org/10.1007/978-94-009-5921-7_1

Harborne JB (1998) Textbook of phytochemical methods. A guide to modern techniques of plant analysis, 5th edn. Chapman and Hall Ltd, London, pp 21–72

Harborne JB, Hora FB (1970) Unpublished results

http://technologyinscience.blogspot.com/2011/12/matrix-or-media-used-in-column.html# (n.d.) XbRbZZIzbIU

https://www2.chemistry.msu.edu/faculty/reusch/VirtTxtJml/Spectrpy/InfraRed/infrared.htm (n.d.)

Ibanez M, Pozo OJ, Sancho JV, Lopez FJ, Hernandez F (2006) Re-evaluation of glyphosate determination in water by liquid chromatography coupled to electrospray tandem mass spectrometry. J Chromatogr A 1134(1–2):51–55. https://doi.org/10.1016/j.chroma.2006.07.093

Ingle KP, Deshmukh AG, Padole DA, Dudhare MS, Moharil MP, Khelurkar VC (2017) Phytochemicals: extraction methods, identification and detection of bioactive compounds from plant extracts. J Pharmacogn Phytochem 6(1):32–36

Ismaiel OA, Abdelghani E, Mousa H, Eldahmy SI, Bayoumy BE (2016) Determination of estragole in pharmaceutical products, herbal teas and herbal extracts using GC-FID. J Appl Pharm Sci 6(12):144–150. https://doi.org/10.7324/japs.2016.601220

Jork H, Funk W, Fischer W, Wimmer H (1994) Thin-layer chromatography: reagents and detection methods, vol Ib. Physical and chemical detection methods: activation reactions, reagent sequences, reagents II. Wiley, New York, pp 466

JoVE Science Education Database (2019) Organic chemistry. Rotary evaporation to remove solvent. JoVE, Cambridge. Source: Dr. Melanie Pribisko Yen and Grace Tang-California Institute of Technology. https://www.jove.com/science-education/5501/rotary-evaporation-to-remove-solvent

Karlsson E, Ryden L, Brewer J (1998) Protein purification. Principles, high resolution methods, and applications. Ion exchange chromatography, 2nd edn. Wiley, New York

Kou D, Mitra S (2003) Extraction of semivolatile organic compounds from solid matrices. In: Winefordner JD, Mitra S (eds) Sample preparation techniques in analytical chemistry. https://doi.org/10.1002/0471457817.ch3

Lapthorn C, Pullen F, Chowdhry BZ (2013) Ion mobility spectrometry-mass spectrometry (IMS-MS) of small molecules: separating and assigning structures to ions. Mass Spectrom Rev 32(1):43–71. https://doi.org/10.1002/mas.21349

Llompart M, Garcia-Jares C, Celeiro M, Dagnac T (2019) Extraction—microwave-assisted extraction. In: Worsfold P, Poole C, Townshend A, Miró M (eds) Encyclopedia of analytical science, 3rd edn. Academic Press, Oxford, pp 67–77. https://doi.org/10.1016/B978-0-12-409547-2.14442-7

Luque de Castro MD, García Ayuso LE (2000) Environmental applications—Soxhlet extraction. In: Wilson ID (ed) Encyclopedia of separation science. Academic Press, Oxford, pp 2701–2709. https://doi.org/10.1016/B0-12-226770-2/06681-3

Mace ME (1963) Histochemical localization of phenols in healthy and diseased Banana roots. Physiol Plant 16(4):915–925. https://doi.org/10.1111/j.1399-3054.1963.tb08367.x

Mahn A, Asenjo J (2005) Prediction of protein retention in hydrophobic interaction chromatography. Biotechnol Adv 23:359–368. https://doi.org/10.1016/j.biotechadv.2005.04.005

McHugh MA, Krukonis VJ (1994) Supercritical fluid extraction principles and practice. Butterworth-Heinemann, Boston. http://worldcat.org

McLafferty FW, Bockhoff FM (1978) Separation/identification system for complex mixtures using mass separation and mass spectral characterization. Anal Chem 50(1):69–76. https://doi.org/10.1021/ac50023a021

Millar S (2012a) Tips and tricks for the lab: column choices. ChemViews. https://doi.org/10.1002/chemv.201200074

Millar S (2012b) Tips and tricks for the lab: column troubleshooting and alternatives. ChemViews. https://doi.org/10.1002/chemv.201200084

Millar S (2012c) Tips and tricks for the lab: column packing. https://www.chemistryviews.org/details/education/2040151/Tips_and_Tricks_for_the_Lab_Column_Packing.html. https://doi.org/10.1002/chemv.20120005

Millar S (2017) Tips and tricks for the lab: column packing. ChemViews. https://doi.org/10.1002/chemv.201200053

Nikhal SB, Dambe PA (2010) Hydroalcoholic extraction of *Mangifera indica* (leaves) by soxhletion. Int J Pharm Sci Res 7(1):78–81. https://doi.org/10.13040/IJPSR.0975-8232.1

Pacáková V (2000) Herbicides—thin-layer (planar) chromatography. In: Wilson ID (ed) Encyclopedia of separation science. Academic Press, Oxford, pp 3006–3010. https://doi.org/10.1016/B0-12-226770-2/06391-2

Paris R, Durand M, Bonnet JL (1960) [Characterization of the chlorogenic acid in certain yeasts by paper chromatography]. Ann Pharm Fr. 18:769–772

Patil PS, Shettigar R (2010) An advancement of analytical techniques in herbal research. J Adv Sci Res 1(1):08–14

Queiroz JA, Tomaz CT, Cabral JM (2001) Hydrophobic interaction chromatography of proteins. J Biotechnol 87(2):143–159. https://doi.org/10.1016/s0168-1656(01)00237-1

Sapkale GN, Patil SM, Surwase US, Bhatbhage PK (2010) Supercritical fluid extraction: a review. Int J Chem Sci 8(2):729–743

Sutar N, Ranju G, Sharma U, Sharma U, Amit J (2010) Anthelmintic activity of *Platycladus orientalisleaves* extract. Int J Parasitol Res 2. https://doi.org/10.9735/0975-3702.2.2.1-3

Touchstone JC (1992) Practice of thin layer chromatography, 3rd edn. Wiley, New York

Walls D, Loughran ST (2011) Protein chromatography: methods and protocols. Methods in molecular biology. 2nd edn. pp 681. https://doi.org/10.1007/978-1-4939-6412-3

Wilchek M, Chaiken I (2000) An overview of affinity chromatography. Methods Mol Biol 147:1–6. https://doi.org/10.1007/978-1-60327-261-2_1

Molecular Markers and Their Application in the Identification of Elite Germplasm

3

Karuna Surendran, R. Aswati Nair, and Padmesh P. Pillai

Abstract

Genetic markers have attained great research interest due to their significant and wide application in plant and animal breeding programs and they find applications in identifying elite germ plasm for enhanced production of plant metabolites. In the diversified group of genetic markers, molecular markers are most preferred for genetic diversity assessment along with other molecular approaches including DNA barcoding, for the selection of superior chemotypes. The combined strategy of Genomics and Metabolomics comprising combination of molecular fingerprinting with quantitative analysis of metabolites is a powerful tool in the identification of elite plant materials with high therapeutic value. Furthermore, it is also used to address issues related to identification of various taxonomic varieties. This chapter is more focused on the usage and application of different molecular markers like RAPD, SSR, EST-SSR, and ISSR in the molecular characterization of chemotypes. Furthermore, the desirability of combination of molecular markers like ISSR with SCoT markers, ITS2, etc., to overcome limitations in the identification and assessment of genetic diversity among elite germplasm of different plant species, is also discussed. The authentication of plant materials having therapeutic efficiency, through such molecular tools, can greatly increase the efficacy of the herbal formulations and could yield tremendous benefits to the herbal industry and other stakeholders.

K. Surendran · P. P. Pillai (✉)
Department of Genomic Science, Central University of Kerala, Kasaragod, Kerala, India

R. Aswati Nair
Department of Biochemistry and Molecular Biology, Central University of Kerala, Kasaragod, Kerala, India

© Springer Nature Singapore Pte Ltd. 2020
S. T. Sukumaran et al. (eds.), *Plant Metabolites: Methods, Applications and Prospects*, https://doi.org/10.1007/978-981-15-5136-9_3

Keywords

DNA barcoding · Chemotype · Molecular fingerprinting · Genetic diversity · RAPD · SSR · ISSR · FISSR · EST-SSR · ITS2 · SNP · SCoT markers

3.1 Introduction

Genetic markers are variations in DNA sequences arising due to mutation or alteration in the genomic loci and can be used to identify individuals and/or species (Pierce 2012). Genetic markers have found wide applications in plant and animal breeding, genetics, evolutionary biology, and genetic resource conservation (Boopathi 2013). Broadly the genetic markers have been grouped as classical and DNA or molecular markers (Boopathi 2013), while morphological, cytological, and biochemical markers are categorized as classical markers (Al-Samarai and Al-Kazaz 2015). Accordingly, restriction fragment length polymorphism (RFLP), amplified fragment length polymorphism (AFLP), simple sequence repeats (SSRs), single-nucleotide polymorphism (SNP), and diversity arrays technology (DArT) markers are some types of molecular markers (Joshi et al. 1999). Technically, molecular marker assisted selection (MAS) strategies have been termed as "Smart Breeding" strategies and have a number of advantages over classical markers in genetic diversity assessment (Miah et al. 2013; Sharma et al. 2008).

3.2 Medicinal Plant Authentication and Need for Molecular Marker Development

Demand for herbal products is increasing in both developed and developing countries with more than 1000 companies involved in the production of herbal products (Nirmal et al. 2013). Annual revenue of the herbal industry has been estimated to be in excess of US$60 billion (Newmaster et al. 2013). Efficacy of herbal formulations depends on use of the correct chemotype of the plant having therapeutic efficiency in the preparation of formulations (Shinde et al. 2009; Ahmad et al. 2014). Generally, the herbal industry relies on traditional methods for identification of medicinal plants that include morphological characters and/or microstructures. However, such traditional identification methods are largely dependent on subjective experience and feelings, which easily bring confusion while identifying the plant material. Added to this, herbal preparation of rare and expensive medicinal plants are frequently adulterated or substituted with morphologically similar, less expensive, and easily available species (Snehal et al. 2018). There have been various case studies on substitution and adulteration of popular Ayurvedic drugs (Prakash et al. 2013; Rai et al. 2012). An example is the adulteration of *Swertia chirata* with *Andrographis paniculata* (Phoboo et al. 2011). In the absence of any systematic research combining genomics and metabolomics of medicinal plants, it becomes imperative that suitable methods be developed to identify and distinguish elite and/or superior chemotypes based on molecular markers.

3.3 Molecular Markers for Identification of Superior Chemotypes

Majority of studies using molecular markers have focused on identifying variations in ecotypes (Cortese et al. 2010) but not in relationship with their secondary metabolite content. Realizing the fact that molecular marker technique could be used to recognize elite varieties of plant species with therapeutic potential, many researchers have undertaken such experiments combining molecular fingerprinting with quantitative analysis of metabolites. Apparently, such experiments not only help in elite genotype identification, but can also be used for accurate taxonomic identification of plant material while developing formulations for the herbal industry, which is often fraught with problems of adulteration (Srivastava and Mishra 2009).

3.3.1 Random Amplified Polymorphic DNA

One of the widely used molecular marker for characterizing chemotypes of medicinal plants is RAPD (Sangwan et al. 1999) due to its high speed, low cost, and requirement of small amount of template DNA (Williams et al. 1990). In the medicinal plant, *Mucuna pruriens* commonly known as velvet bean or cow-itch, RAPD has been used for determining genetic diversity as well as its relation to metabolite content. The species is a store house of several secondary metabolites with principal metabolite being 3,4-dihydroxy-L-phenylalanine (L-DOPA) (Pras et al. 1993). Distribution of L-DOPA varies in different parts of the plant like 0.15% in dried leaves and pods, 0.49% in stem, and a very high concentration of 4.47–5.39% is reported in raw seeds (Bell and Janzen 1971). Three varieties of *M. pruriens* viz., *M. pruriens* (L.) DC, *M. prurient* (L.) DC var. *prurient*, and *M. pruriens* var. *utilis* are distributed across India with qualitative and quantitative morphological variations among the taxa which adds to its taxonomical ambiguity. Molecular marker analysis of *M. pruriens* (L.) DC belonging to varieties *prurient* and *utilis* using RAPD primers revealed high genetic variation in both varieties of *pruriens* and *utilis*. The total heterozygosity of *Mucuna* observed as 0.44 was greater than the values reported for other self-pollinating legumes of cosmopolitan distribution and also for many other tropical cross-pollinated tree species like *Gaultheria fragrantissima* (Padmesh et al. 2006). A uniform pattern of L-DOPA distribution was observed which contradicted earlier observations wherein higher L-DOPA synthesis was associated with plants grown at lower latitudes near the equator. Studies had clarified that L-DOPA synthesis is related as a function of irradiance that can be either stimulatory or inhibitory with L-DOPA synthesis only varying by 8% with latitude (Brain 1976; Wichers et al. 1983). In yet another example wherein RAPD had provided important insight into identification of elite medicinal plant germplasm is with the medium sized tree, *Poeciloneuron pauciflorum* Bedd. (Family: Clusiaceae). The plant has highly restricted distribution in Southern-Western Ghats of Kerala and Tamil Nadu States and is red listed as a critically endangered tree species by IUCN (*Poeciloneuron pauciflorum*. In: IUCN 2010. IUCN Red List of Threatened Species. Version 2010.4. www.iucnredlist.org as on 9 May 2011). The plant is a rich source of xanthones displaying various bioactivities like antimalarial (Ignatushchenko et al. 2000), effective against anti-methicillin-resistant *Staphylococcus aureus* (Rukachaisirikul et al. 2003, 2005; Sukpondma et al. 2005), selective cyclooxygenase-2 inhibition (Zou et al. 2005), tumor-promoting inhibition (Ito et al. 2003) and inhibitory effects on PAF-induced hypotension (Oku et al. 2005). RAPD profiling of *P. pauciflorum* collected from four different locations revealed high variability with cluster analysis demarcating the accessions collected from two different geographic regions representing two Southern states, viz., Tamil Nadu and Kerala. Despite the observed high genetic diversity, high genetic differentiation (Gst = 0.33), and low gene flow (Nm = 0.98), populations showed strong tendency to genetic drift and inbreeding depression. Authors have attributed the reasons for high genetic diversity and endemism of *P. pauciflorum* to factors that include recent speciation from a more widespread species, changes in distribution or habitat, breeding system, somatic mutations, multiple founder events, tropical forest fragmentation leading to decreased gene flow and/or increased inbreeding producing a high differentiation among remnant populations (Richter et al. 1994; Deshpande et al. 2001). The studies reinforce the importance of developing ex situ conservation strategies through conventional or non-conventional approach involving simple micropropagation for conservation of the species (Pillai et al. 2011).

3.3.2 Simple-Sequence Repeats (SSR)

Simple-sequence repeats (SSRs), also known as microsatellites, are short tandem repeated motifs that may vary in the number of repeats at a given locus (Tautz 1989). Thus, SSR markers are multi-allelic, relatively abundant, widely dispersed across the genome, co-dominant, and easily and automatically scored (Powell et al. 1996). They have been used for genetic diversity analysis of medicinal plants like *Punica granatum* (Hasnaoui et al. 2012) and *Chrysanthemum morifolium* (Feng et al. 2016). Use of SSR marker for elite germplasm identification has been undertaken in *Lagerstroemia speciosa* (Family: Lythraceae), commonly known as "Banaba," native of South–East Asia collectively named as "crape myrtles," "lilac of the south," or "Pride of India." Magnificent blooms of *L. speciosa* have attracted attention of many breeders resulting in the development of hundreds of hybrids through inter- and intraspecific hybridizations with diverse combination of flower color, growth habit, bark characteristics, and resistance traits (Egolf 1990). Besides its magnificent blooms, *L. speciosa* is also rich in various secondary metabolites with corosolic acid (CRA) being the principal metabolite. CRA (2α-hydroxyursolic acid) is reported to exhibit anticancer, antioxidative, antiobesity, and antiseptic properties in addition to its well-known antidiabetic property (Vijayan et al. 2015). SSR marker based genetic diversity analysis was carried out in 42 accessions representing 12 natural populations from Southern Western Ghats (SWG) to identify the elite genotypes by comparing the corosolic acid (CRA) content and genetic data. The 38 SSR markers used for the study were derived from microsatellites developed from *Lagerstroemia indica* and *L. caudate* (Pounders et al. 2007). The study revealed moderately high genetic diversity in *L. speciosa* with mean value of observed heterozygosity (HO) as 0.39, expected heterozygosity (HE) as 0.28, Shannon's index (I) as 0.30 at the inter-population level, and average Polymorphism Information Content (PIC) of 0.67. The observed genetic differentiation to be due to genetic drift based on values obtained for Nm as 1.40 (which was slightly >1) and coefficient of genetic differentiation (FST) in the range of 0.02–0.55. Analysis of molecular variance (AMOVA) revealed higher variance within population than among populations with 22% of the total variance distributed among regions while 50% of variance was distributed within population and only 28% among population. Data identified existence of two distinct groups with genetic diversity along a north–south geographical gradient which was confirmed by analysis of CRA content. Percentage distribution of CRA was found to be less in population from south of SWG (mean = 0.064%) compared to those from north of SWG (mean = 0.321%) (Fig. 3.1). The study successfully identified elite populations from north of SWG relatively rich in CRA and genetically more diverse and suggested further tissue culture propagation of the species (Jayakumar et al. 2014). Availability of various Expressed sequence tag libraries have opened up the possibility of identifying SSR within EST and some of these EST-derived SSR or EST-SSR may be linked to functional genes involved in secondary metabolite biosynthesis and hence are of particular relevance in genetic diversity analysis of elite chemotypes. EST-SSR has been used for genetic diversity analysis of medicinal plant, *Artemisia annua* L. (Family: *Asteraceae*), which is known for its principal metabolite artemisinin, an endoperoxide seco sesquiterpene lactone, that is effective against multidrug-resistant strains of *Plasmodium* species (Klayman 1993). The study, which estimated genetic diversity parameters of *A. annua* accessions collected from two geographic regions of Ladakh, India, suggested the importance of introducing elite genotypes into regions wherein accessions displayed lower genetic diversity for breeding programs and germplasm conservation (Kumar et al. 2014).

3.3.3 Inter Simple Sequence Repeat (ISSR)

Another important molecular marker used for genetic diversity analysis is Inter Simple Sequence Repeat (ISSR), which was developed by Zietkiewicz et al. (1994). ISSR retrieves variations in the microsatellite regions that are widely distributed throughout the genome, especially the nuclear genome. Thus, it is highly beneficial compared to other molecular approaches as it bypasses the challenge of characterizing individual loci (Ng and Tan 2015; Prashanth et al. 2015). ISSR primers are usually designed to any of the highly variable microsatellite sequences

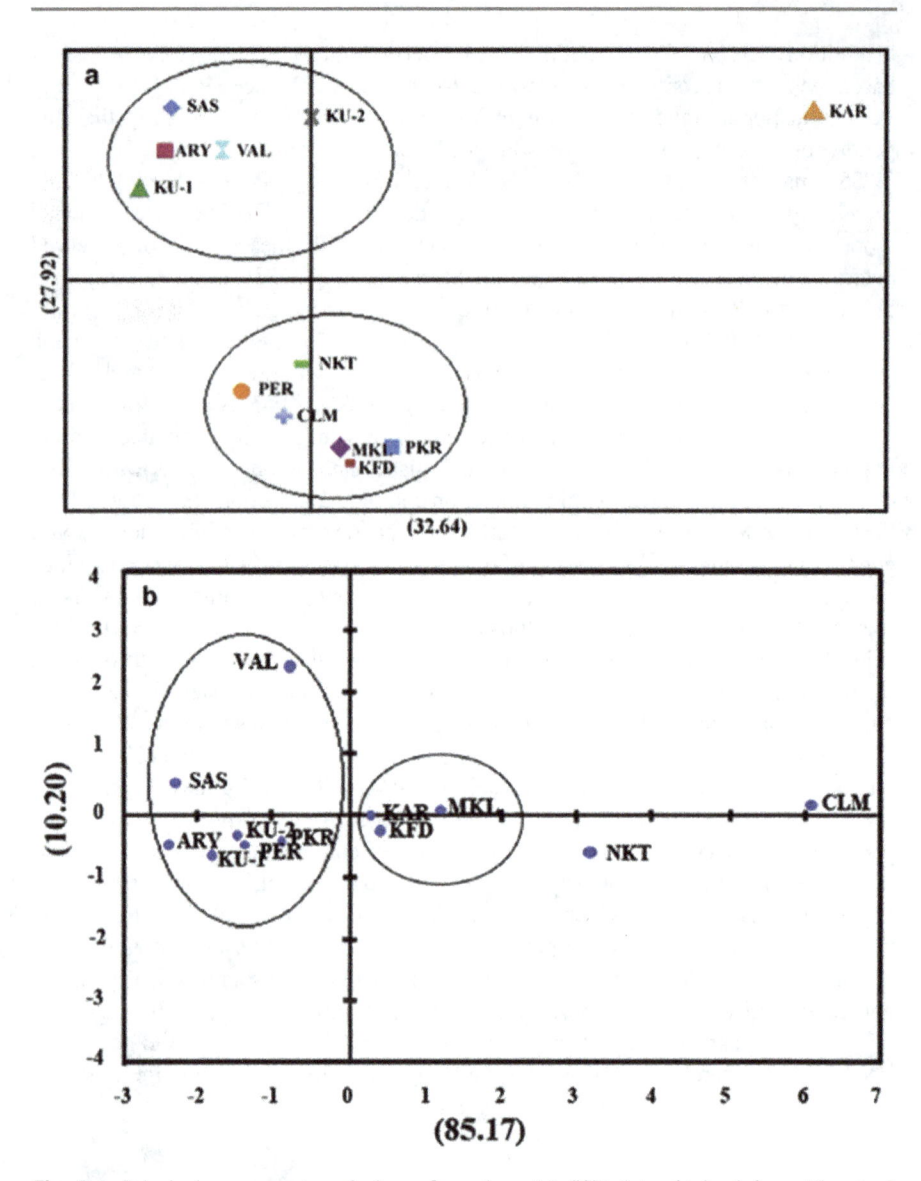

Fig. 3.1 Principal component analysis performed on (**a**) SSR data obtained from 12 natural populations of *L. speciosa*. (**b**) Estimate of corosolic acid content across these populations

with an arbitrary pair of bases at the 3' end. ISSR offers high degree of reproducibility, detection of high level of polymorphism, and is a relatively simple procedure (Pradeep Reddy et al. 2002). The medicinal herb, *Rauvolfia serpentina* L. (Family: Apocynaceae) is widely distributed in tropics like Himalayas, Indian peninsula and in other South-east Asian countries and across different parts of the world (Ghani 1998). Overexploitation of the plant for commercial purposes has led to

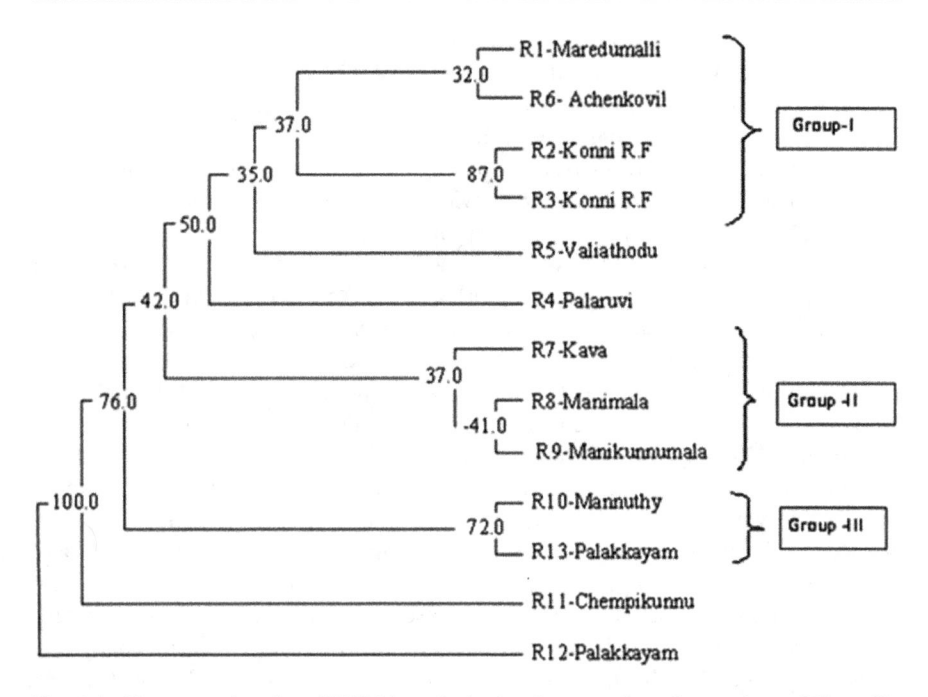

Fig. 3.2 Phenogram based on UPGMA analysis showing grouping of accessions of *Rauvolfia serpentina*. Numbers at the fork indicate bootstrap values

categorization of *R. serpentina* as a highly threatened species in India. Further challenges on seed-based propagation of *R. serpentina* are posed by poor seed viability and very low germination percentages (25–50%) attributed to the presence of cinnamic acid derivatives (Mitra 1976). Analysis of 13 accessions of *R. serpentine* collected from different locations in Western Ghats using 15 ISSR primers revealed moderate levels of polymorphism with an average of 6 polymorphic bands per primer and coefficient of genetic similarity in the range of 0.49–0.81. The high genetic variability of *R. serpentina* was attributed to its cosmopolitan distribution and high outcrossing rates (Pillai et al. 2012). The study also provided insight into such aspect as the genetic differentiation among *R. serpentina* accessions does not correlate with geographic distance as all samples from southern and central Kerala clustered together along with the lone accession from a distant location, viz., from Andhra Pradesh with few outliers (Fig. 3.2).

3.4 Combinations of Different Molecular Markers

Many of these molecular markers have been used in congruence to overcome the limitations of one for assessing and analyzing the nature and the extent of genetic diversity among elite germplasm of different plant species. RAPD and ISSR have been used for evaluating genetic diversity in elite germplasm of *Cymbopogon winterianus* Jowitt ex Bor (Family: Poaceae) across West Bengal, India (Bhattacharybr and Ghosh 2010). Aromatic essential oil of *C. winterianus* has considerable pharmaceutical importance and is widely used in cosmetics, soap, perfumery, and beverage industries (Mathur et al. 1988). Wide genetic variation was observed with RAPD analysis estimating 62.9% genetic diversity while ISSR data revealed 57.9% of genetic diversity within the population. Due to its polyallelic nature, RAPD displayed better resolving power over ISSR in estimating molecular diversity of different ecotypes of *C. winterianus*. A combination of ISSR with Fluorescent ISSR (FISSR) has been used for elite genotype identification in *Azadirachta indica* A. Juss commonly called Neem (Family: Meliaceae) (Chary 2011). It is a medicinal plant of significance with Charaka-Samhita bearing testimony to its importance in ancient medicine (Veerendra 1995; Tewari 1992). Neem is an important source of tetranortriterpenoid, Azadirachtin-A (Aza-A), a potent, environment friendly and biodegradable pesticide. DNA fingerprinting patterns obtained with ISSR and FISSR categorized the Neem accessions according to quantitative variations in Aza-A density. Accessions that exhibited high and low Aza-A clearly segregated into groups that distinctly overlapped with the clusters generated, both based on their Aza-content and also with respect to their habitat (Chary 2011). In another study, ISSR was used with SCoT based marker to distinguish *Senna obtusifolia* L from *Senna occidentalis* L, both belonging to Leguminosae family (Mao et al. 2017). *S. obtusifolia* is a famous traditional Chinese medicine with various pharmaceutical properties such as eyesight improvement (Yang et al. 2012), regulation of blood lipid level, hypertension (Li and Guo 2002) and hepatoprotective effect (Kim et al. 2009) attributed to the seeds. However, *S. occidentalis* is the common and indistinguishable adulterant and improper use of *S. occidentalis* seeds causes serious symptoms (Panigrahi et al. 2014; Teles et al. 2015). Cluster analysis based on ISSR and SCoT markers clearly distinguished the two species. The study also identified high level of genetic diversity in both *S. obtusifolia* and *S. occidentalis* with their genetic relationship showing a certain connection with geographical origin. Chemical profiling had identified the absence of aurantio-obtusin in *S. occidentalis*, a significant difference from *S. obtusifolia*. Also higher concentration of chrysophanol and several other anthraquinones like emodin, aloe-emodin, and physcion in *S. obtusifolia* seeds compared to *S. occidentalis* seeds signifies the importance of *S. obtusifolia* over *S. occidentalis* seeds for medicinal use. A similar distinction between the two species could also be made out by sequencing spacer region adjoining 5.8S locus of nuclear ribosomal cistron, viz., internal transcribed spacer 2 (ITS2) region. ITS2 (<300 bp) which has been proposed by Chen et al. (2010) as suitable for DNA barcoding of medicinal plants is the most commonly sequenced region successfully applied for identifying medicinal plants like *Hyoscyamus niger* (Xiong et al. 2016) and the traditional Chinese medicine officinal rhubarb (the dried root and rhizome of rhubarb source plants comprising three *Rheum* species; Zhou et al. 2017).

3.5 DNA Barcoding

DNA barcoding proposed by Hebert et al. (2003) sequences a small portion of the genome (<1000 bp) and DNA barcodes suggested for discriminating plants include candidate regions like matK, rbcL, trnH-psbA, ITS, trnL-F, 5S-rRNA, and 18S-rRNA. Realizing its usefulness in authenticating raw plant material in herbal formulations (Li et al. 2011), there has been an upsurge in number of barcoding studies in medicinal plants (Mishra et al. 2015). One such study using nuclear ITS2 and plastidial matK and rpoC1 has been undertaken in *Asparagus racemosus* Willd that possesses various pharmacological properties to distinguish from its adulterant, *A. gonoclados* Baker (Rai et al. 2012). Both species belonging to family Asparagaceae are closely related with no marked difference in flower and fruit structure except for differences in leaf structure and plant distribution. Greater intraspecific variations (0.232) differences among the two *Asparagus* species compared to interspecific variations (0.12) suggested the ineffectiveness of ITS2 gene as candidate barcode for Asparagaceae species. The same study also conducted DNA barcoding of *Hemidesmus indicus* R. Br., which possesses various pharmacologic properties and is used in various skin care formulations (Aminuddin and Girach 1991). *H. indicus* is often adulterated with *Decalepis hamiltonii* Wight and Arn., both belonging to family *Asclepiadaceae*, due to scarcity of raw material and elaborate extraction procedure. ITS2 estimated lesser intraspecific variations among *H. indicus* species compared to interspecific variations, suggesting ITS2 as a standard DNA barcode to identify medicinal plants of Asclepiadaceae family. Failure of plastidial marker, *rpoC1* and *matK1* to yield any amplification product in *Asparagus* and *Hemidesmus* species indicates that barcoding is not robust for some taxa and hence is not broadly applicable (Hebert and Gregory 2005). Furthermore, compared to nuclear genomic loci that are subject to rapid evolution, chloroplast regions lack sequence divergences within genera and hence are not suitable for the identification/ authentication of medicinal plants (Ruzicka et al. 2009). In the case of *Verbena officinalis* (Verbenaceae), a widely used drug, ITS2 revealed lower intraspecific variation within *V. officinalis* (Ruzicka et al. 2009). The same study also used RAPD and high-resolution melting analysis (HRM) for authentication of *Verbena* species. HRM evaluates differences in DNA melting curves of amplicons in the presence of an intercalating fluorescence dye (Wittwer et al. 2003). Depending on length, GC content and nucleotide sequence each amplicon has a characteristic melting temperature (Tm) and melting curve, which can be determined by a stepwise increase of temperature (Herrmann et al. 2006). HRM is efficient compared to earlier melting analyses in sensitivity and specificity as it genotypes only one SNP (single nucleotide polymorphisms) even in products as large as 544 bp (Liew et al. 2004a, b). In *Verbena* species, SCAR marker was observed to distinguish *V. officinalis* from other *Verbena* species except its closest relative *V. hastate*. However, HRM was identified as ideal for authentication of *V. officinalis*.

3.6 Conclusions

The past decades had witnessed successful development of molecular markers from RFLP to SNP and various combinations of DNA markers for different molecular approaches. The DNA markers like RAPD, ISSR, and SSR are powerful tools for the selection of superior chemotypes and in the analysis of genetic diversity of core germplasm collections. Development of molecular approaches using combination of various DNA markers has significantly assisted the process of characterization and authentication of medicinal plants, in a limited span of time using less resources. This will have tremendous impact on the drug industry as proper authentication of plant materials is often a challenge that undermine the popularity of herbal medicines.

References

Ahmad I, Khan MSA, Cameotra SS (2014) Quality assessment of herbal drugs and medicinal plant products. In: Encyclopaedia of analytical chemistry. American Cancer Society, pp 1–17. https://doi.org/10.1002/9780470027318.a9946

Al-Samarai FR, Al-Kazaz AA (2015) Molecular markers: an introduction and applications. Eur J Mol Biotechnol 9:118–130. https://doi.org/10.13187/ejmb.2015.9.118

Aminuddin Girach RD (1991) Pluralistic folk uses of *Hemidesmus indicus* (L.) R. Br. From south eastern India. J Econ Taxon Bot 15:715–718

Bhattacharya S, Ghosh T (2010) Efficiency of RAPD and ISSR markers in assessment of molecular diversity in elite germplasms of Cymbopogon winterianus across West Bengal, India. Emir J Food Agric 22(1):13

Bell EA, Janzen DH (1971) Medical and ecological considerations of L-Dopa and 5-HTP in seeds. Nature 229:136–137. https://doi.org/10.1038/229136a0

Boopathi NM (2013) Genetic mapping and marker assisted selection: basics, practice and benefits. Springer, New Delhi

Brain KR (1976) Accumulation of L-DOPA in cultures from mucunapruriens. Plant Sci Lett 7:157–161. https://doi.org/10.1016/0304-4211(76)90129-2

Chary P (2011) A comprehensive study on characterization of elite neem chemotypes through mycofloral, tissue-cultural, ecomorphological and molecular analyses using azadirachtin-A as a biomarker. Physiol Mol Biol Plants 17(1):49–64

Chen S, Yao H, Han J, Liu C, Song J, Shi L, Zhu Y, Ma X, Gao T, Pang X, Luo K, Li Y, Li X, Jia X, Lin Y, Christine L, Thomas M, Gilbert P (2010) Validation of the ITS2 region as a novel DNA barcode for identifying medicinal plant species. PLoS One 5(1):e8613

Cortese LM, Honig J, Miller C, Bonos SA (2010) Genetic diversity of twelve switchgrass populations using molecular and morphological markers. Bioenergy Res 3:262–271. https://doi.org/10.1007/s12155-010-9078-2

Deshpande AU, Apte GS, Bahulikar RA, Lagu MD, Kulkarni BG, Suresh HS, Singh NP, Rao MK, Gupta VS, Pant A, Ranjekar PK (2001) Genetic diversity across natural populations of three montane plant species from the Western Ghats, India revealed by inter simple sequence repeats. Mol Ecol 10:2397–2408. https://doi.org/10.1046/j.0962-1083.2001.01379.x

Egolf DR (1990) 'Caddo' and 'tonto' Lugerstroemia. Hort Sci 25:585–587. https://doi.org/10.21273/HORTSCI.25.5.585

Feng S, He R, Lu J, Jiang M, Shen X, Jiang Y, Wang Z, Wang H (2016) Development of SSR markers and assessment of genetic diversity in medicinal *Chrysanthemum morifolium* cultivars. Front Genet 7:Article 113

Ghani A (1998) Medicinal plants of Bangladesh: chemical constituents and uses. Asiatic Society of Bangladesh, Dhaka

Hasnaoui N, Buonamici A, Sebastiani F, Mars M, Zhang D, Vendramin GG (2012) Molecular genetic diversity of *Punica granatum* L. (pomegranate) as revealed by microsatellite DNA markers (SSR). Gene 493:105–112

Hebert PD, Cywinska A, Ball SL, Dewaard JR (2003) Biological identifications through DNA barcodes. Proc R Soc Lond Ser B Biol Sci 270(1512):313–321

Hebert PD, Gregory TR (2005) The promise of DNA barcoding for taxonomy. Syst Biol 54:852–859

Herrmann MG, Durtschi JD, Bromley LK, Wittwer CT, Voelkerding KV (2006) Amplicon DNA melting analysis for mutation scanning and genotyping: cross-platform comparison of instruments and dyes. Clin Chem 52:494–503. https://doi.org/10.1373/clinchem.2005.063438

Ignatushchenko MV, Winter RW, Riscoe M (2000) Xanthones as antimalarial agents: stage specificity. Am J Trop Med Hyg 62:77–81. https://doi.org/10.4269/ajtmh.2000.62.77

Ito C, Itoigawa M, Takakura T, Ruangrun H, Nishino H, Furukawa H (2003) Chemical constituents of Garcinia fusca: structure elucidation of eight new xanthones and their cancer chemopreventive activity. J Nat Prod 66:200–205. https://doi.org/10.1021/np020290s

Jayakumar KS, Sajan JS, Aswati Nair R, Padmesh Pillai P, Deepu S, Padmaja R, Agarwal A, Pandurangan AG (2014) Corosolic acid content and SSR markers in Lagerstroemia speciosa (L.) Pers.: a comparative analysis among populations across the Southern Western Ghats of India. Phytochemistry 106:94–103

Joshi SP, Ranjekar PK, Gupta VS (1999) Molecular markers in plant genome analysis. Curr Sci 77:230–240

Kim DH, Kim S, Jung WY, Park SJ, Park DH, Kim JM, Cheong JH, Ryu JH (2009) The neuroprotective effects of the seeds of Cassia obtusifolia on transient cerebral global ischemia in mice. Food Chem Toxicol 47(7):1473–1479

Klayman DL (1993) *Artemisia annua*: from weed to respectable antimalarial plant. In: Human medicinal agents from plants. ACS symp series: Washington, DC, USA, vol 534, pp 242–255

Kumar J, Bajaj P, Mishra GP, Singh SB, Singh H, Naik PK (2014) Utilization of EST-derived SSRs in the genetic characterization of *Artemisia annua* L genotypes from Ladakh. India Indian J Biotechnol 13:464–472

Li X-E, Guo B-J (2002) [Effect of protein and anthraquinone glucosides from cassia seed on serum lipid of hyperlipidemia rats]. Zhongguo Zhong Yao Za Zhi (China J Chin Mater Medica) 27:374–376

Li D-Z, Gao L-M, Li H-T, Wang H, Ge X-J, Liu J-Q, Chen Z-D, Zhou S-L, Chen S-L, Yang J-B, Fu C-X, Zeng C-X, Yan H-F, Zhu Y-J, Sun Y-S, Chen S-Y, Zhao L, Wang K, Yang T, Duan G-W (2011) Comparative analysis of a large dataset indicates that internal transcribed spacer (ITS) should be incorporated into the core barcode for seed plants. Proc Natl Acad Sci 108(49):19641–19646

Liew M, Pryor R, Palais R, Meadows C, Erali M, Lyon E, Wittwer C (2004a) Genotypingof single-nucleotide polymorphisms byhigh-resolution melting of small amplicons. Clin Chem 50:1156–1164

Liew M, Pryor R, Palais R, Meadows C, Erali M, Lyon E, Wittwer C (2004b) Genotyping of single-nucleotide polymorphisms by high-resolution melting of small amplicons. Clin Chem 50:1156–1164. https://doi.org/10.1373/clinchem.2004.032136

Mao R, Xia P, He Z, Liu Y, Liu F, Zhao H, Han R, Liang Z (2017) Identification of seeds based on molecular markers and secondary metabolites in Senna obtusifolia and Senna occidentalis. Bot Stud 58(1):43

Mathur AK, Ahuja PS, Pandey B, Kukreja AK, Mandal S (1988) Screening and evaluation of somaclonal variations for quantitative and qualitative traits in an aromatic grass, Cymbopogon winterianus Jowitt. Plant Breed 101:321–334. https://doi.org/10.1111/j.1439-0523.1988.tb00305

Miah G, Rafii M, Ismail M, Puteh A, Rahim H, Islam K, Latif M (2013) A review of microsatellite markers and their applications in rice breeding programs to improve blast disease resistance. Int J Mol Sci 14:22499–22528. https://doi.org/10.3390/ijms141122499

Mishra P, Kumar A, Nagireddy A, Mani DN, Shukla AK, Tiwari R, Sundaresan V (2015) DNA barcoding: an efficient tool to overcome authentication challenges in the herbal market. Plant Biotechnol J 14:8–21

Mitra GC (1976) Studies on the formation of viable & non-viable seeds in *Rauvolfia serpentine* Benth. Indian J Exp Biol 14:54–54

Newmaster SG, Grguric M, Shanmughanandhan D, Ramalingam S, Ragupathy S (2013) DNA barcoding detects contamination and substitution in North American herbal products. BMC Med 11:222. https://doi.org/10.1186/1741-7015-11-222

Ng WL, Tan SG (2015) Inter-simple sequence repeat (ISSR) markers: are we doing it right? ASM Sci J 9(1):30–39

Nirmal S, Pal S, Otimenyin S, Aye T, Mostafa E, Kundu S, Amirthalingam R, Subhash Mandal C (2013) Contribution of herbal products in global market. Pharma Rev. November – December, 95–104

Oku H, Ueda Y, Iinuma M, Ishiguro K (2005) Inhibitory effects of xanthones from guttiferae plants on PAF-induced hypotension in mice. Planta Med 71:90–92. https://doi.org/10.1055/s-2005-837760

Padmesh P, Reji JV, Jinish Dhar M, Seeni S (2006) Estimation of genetic diversity in varieties of *Mucuna pruriens* using RAPD. Biol Plant 50:367–372. https://doi.org/10.1007/s10535-006-0051-z

Panigrahi G, Tiwari S, Ansari KM, Chaturvedi RK, Khanna VK, Chaudhari BP, Vashistha VM, Raisuddin S, Das M (2014) Association between children death and consumption of Cassia occidentalis seeds: clinical and experimental investigations. Food Chem Toxicol 67:236–248

Phoboo S, Bhowmik PC, Jha PK, Shetty K (2011) Anti-diabetic potential of crude extracts of medicinal plants used as substitutes for *Swertia chirayita* using *in* vitro assays. J Plant Sci 7:48–55. https://doi.org/10.3126/botor.v7i0.4373

Pierce BA (2012) Genetics: a conceptual approach, 4th edn. W.H. Freeman, New York

Pillai PP, Sajan JS, Menon KM, Hemanthakumar AS, Pandurangan AG, Krishnan PN, Seeni S (2011) Analysis of genetic diversity in *Poeciloneuron pauciflorum* Bedd.—An endemic tree species from the Western Ghats of India. Am J Plant Sci 02:416–424. https://doi.org/10.4236/ajps.2011.23047

Pillai PP, Sajan JS, Menon KM, Jayakumar KSP, Subramoniam A (2012) ISSR analysis reveals high intraspecific variation in Rauvolfia serpentina L. – a high value medicinal plant. Biochem Syst Ecol 40:192–197

Pounders C, Rinehart T, Sakhanokho H (2007) Evaluation of interspecific hybrids between *Lagerstroemia indica* and *L. speciosa*. Hort Sci 42:1317–1322. https://doi.org/10.21273/HORTSCI.42.6.1317

Powell W, Machray GC, Provan J (1996) Polymorphism revealed by simple sequence repeats. Trends Plant Sci 1(7):215–222

Pradeep Reddy M, Sarla N, Siddiq EA (2002) Inter simple sequence repeat (ISSR) polymorphism and its application in plant breeding. Euphytica 128:9–17. https://doi.org/10.1023/A:1020691618797

Prakash O, Jyoti AK, Kumar P, Manna NK (2013) Adulteration and substitution in Indian medicinal plants: an overview. J Med Plants Stud 1:127–132

Pras N, Woerdenbag HJ, Batterman S, Visser JF, Van Uden W (1993) Mucunapruriens: improvement of the biotechnological production of the anti-Parkinson drug L-dopa by plant cell selection. Pharm World Sci 15:263–268. https://doi.org/10.1007/BF01871128

Prashanth NM, Yugander A, Bhavani NL, Nagar R (2015) DNA isolation and PCR amplification of turmeric varieties from Telangana State. Int J Curr Microbiol App Sci 4(5):485–490

Rai PS, Bellampalli R, Dobriyal RM, Agarwal A, Satyamoorthy K, Narayana DA (2012) DNA barcoding of authentic and substitute samples of herb of the family Asparagaceae and Asclepiadaceae based on the ITS2 region. J Ayurveda Integr Med 3:136–140

Richter T, Soltis P, Soltis D (1994) Genetic variation within and among populations of the narrow endemic, *Delphinium viridescens* (Ranunculaceae). Am J Bot 81:1070. https://doi.org/10.2307/2445302

Rukachaisirikul V, Kamkaew M, Sukavisit D, Phongpaichit S, Sawangchote P, Taylor WC (2003) Antibacterial Xanthones from the leaves of *Garcinia nigrolineata*. J Nat Prod 66:1531–1535. https://doi.org/10.1021/np0303254

Rukachaisirikul V, Tadpetch K, Watthanaphanit A, Saengsanae N, Phongpaichit S (2005) Benzopyran, biphenyl, and tetraoxygenated xanthone derivatives from the twigs of *Garcinia nigrolineata*. J Nat Prod 68:1218–1221. https://doi.org/10.1021/np058050a

Ruzicka J, Lukas B, Merza L, Göhler I, Abel G, Popp M, Novak J (2009) Identification of *Verbena officinalis* based on ITS sequence analysis and RAPD-derived molecular markers. Planta Med 75(11):1271–1276

Sangwan RS, Sangwan NS, Jain DC, Kumar S, Ranade SA (1999) RAPD profile based genetic characterization of chemotypic variants of *Artemisia annua* L. IUBMB Life 47:935–944. https://doi.org/10.1080/15216549900202053

Sharma A, Namdeo A, Mahadik K (2008) Molecular markers: new prospects in plant genome analysis. Pharmacogn Rev 2(3):23–34

Shinde VM, Dhalwal K, Potdar M, Mahadik KR (2009) Application of quality control principles to herbal drugs. Int J Phytomed 1(1):4–8. https://doi.org/10.5138/ijpm.2009.0975.0185.05786

Snehal S, Anand T, Mukund D, Kavita D (2018) Adulteration and need of substitution of raw materials—a review. Int J Ayurveda Pharma Res 6(7):92–95

Srivastava S, Mishra N (2009) Genetic markers—a cutting-edge technology in herbal drug research. J Chem Pharma Res 1(1):1–18

Sukpondma Y, Rukachaisirikul V, Phongpaichit S (2005) Xanthone and sesquiterpene derivatives from the fruits of *Garcinia scortechinii*. J Nat Prod 68:1010–1017. https://doi.org/10.1021/np0580098

Tautz D (1989) Hypervariability of simple sequences as a general source for polymorphic DNA markers. Nucleic Acids Res 17(16):6463–6471

Teles AVFF, Fock RA, Górniak SL (2015) Effects of long-term administration of Senna occidentalis seeds on the hematopoietic tissue of rats. Toxicon 108:73–79

Tewari DN (1992) Monograph on neem (Azadirachta indica A. Juss.). International Book Distributors, Dehra Dun, India, p 279

Veerendra HCS (1995) Variation studies in provenances of *Azadirachta Indica* (the neem tree). Indian Forester 121:1053–1056

Vijayan A, Padmesh Pillai P, Hemanthakumar AS, Krishnan PN (2015) Improved in vitro propagation, genetic stability and analysis of corosolic acid synthesis in regenerants of *Lagerstroemia speciosa* (L.) Pers. by HPLC and gene expression profiles. Plant cell tissue organ cult. PCTOC 120:1209–1214. https://doi.org/10.1007/s11240-014-0665-3

Wichers HJ, Malingr TM, Huizing HJ (1983) The effect of some environmental factors on the production of L-DOPA by alginate-entrapped cells of *Mucuna pruriens*. Planta 158:482–486. https://doi.org/10.1007/BF00397239

Williams JG, Kubelik AR, Livak KJ, Rafalski JA, Tingey SV (1990) DNA polymorphisms amplified by arbitrary primers are useful as genetic markers. Nucleic Acids Res 18:6531–6535. https://doi.org/10.1093/nar/18.22.6531

Wittwer CT, Reed GH, Gundry CN, Vandersteen JG, Pryor RJ (2003) High-resolution genotyping by amplicon melting analysis using LCGreen. Clin Chem 49:853–860

Xiong C, Hu ZG, Tu Y, Liu HG, Wang P, Zhao MM, SHIi YH, Wu L, Sun W, Chen SL (2016) ITS2 barcoding DNA region combined with high resolution melting (HRM) analysis of Hyoscyami Semen, the mature seed of Hyoscyamus niger. Chin J Nat Med 14(12):898–903

Yang J, Ye H, Lai H, Li S, He S, Zhong S, Chen L, Peng A (2012) Separation of anthraquinone compounds from the seed of Cassia obtusifolia L. using recycling countercurrent chromatography. J Sep Sci 35(2):256–262

Zhou Y, Xiao-Lei D, Zheng X, Huang M, Li Y, Wang X-M (2017) ITS2 barcode for identifying the officinal rhubarb source plants from its adulterants. Biochem Syst Ecol 70:177–185

Zietkiewicz E, Rafalski A, Labuda D (1994) Genome fingerprinting by simple sequence repeat (SSR)-anchored polymerase chain reaction amplification. Genomics 20(2):176–183

Zou J, Jin D, Chen W, Wang J, Liu Q, Zhu X, Zhao W (2005) Selective cyclooxygenase-2 inhibitors from *Calophyllum membranaceum*. J Nat Prod 68:1514–1518. https://doi.org/10.1021/np0502342

Cell and Protoplast Culture for Production of Plant Metabolites

4

S. R. Saranya Krishnan, R. Sreelekshmi, E. A. Siril, and Swapna Thacheril Sukumaran

Abstract

Plant secondary metabolites are biosynthetically derived from primary metabolites. Recent surveys have established that nearly 35% of molecules in the pharmaceutical industry are of plant origin. The production of plant secondary metabolites for a long period of time was relayed on natural stands of the source plant or cultivation in the field. This poses problems such as increasing pressure on naturally growing plant resources, consequent depletion of plants, the geographic variation that is causing inconsistency in the total content of metabolites, seasonal variation in metabolite content, and growth of plants. Also, a large area is to be maintained for raising resources for the industrial level extraction of metabolites. Scientists and biotechnologists have considered plant cell, tissue, and organ cultures as an alternative way to produce the corresponding secondary metabolites. Plant cell culture technology deals with the in vitro culture of plant cells for the production of specific metabolites. Different strategies, using in vitro systems, have been explored for improving the production of secondary metabolites. Alternate strategies such as in vitro mutagenesis and induction of polyploids can also be explored to achieve enhanced productivity of metabolites. It is now possible to manipulate the pathways that lead to enhanced production of secondary metabolites. Successful cloning of entire genes in a pathway to plasmid vectors and expression in microbial systems, makes simple and highly efficient metabolite production system in future.

Keywords

Elicitors . In vitro metabolite production . Metabolic engineering . Plant cell culture . Precursor molecules . Secondary metabolites

S. R. Saranya Krishnan · R. Sreelekshmi · E. A. Siril (✉) · S. T. Sukumaran
Department of Botany, University of Kerala, Thiruvananthapuram, Kerala, India

© Springer Nature Singapore Pte Ltd. 2020
S. T. Sukumaran et al. (eds.), *Plant Metabolites: Methods, Applications and Prospects*, https://doi.org/10.1007/978-981-15-5136-9_4

4.1 Introduction

Plant secondary metabolites always get attention among the scientific community due to its versatile uses in various industries like therapeutic, cosmetic, food, textile, and more. Kossel defined the concept of secondary metabolites in 1891. Later in 1921, Czapek concisely explained that secondary metabolites are the end product of nitrogen metabolism in plant cells (Bourgaud et al. 2001). These metabolites are low in occurrence as compared to primary metabolites and are stored in specialized organelles. Though the advancement in the field of phytochemistry revealed the physical, chemical, and biosynthetic characters of these compounds, there are many more compounds in the plant life that remain surreptitious to humanity.

The secondary metabolites largely contribute to the interaction of plants with their surroundings and make them fit to survive in the ecosystem they belong (Bourgaud et al. 2001). Most of them possess properties like antimicrobial, allelopathic, UV absorbing, and anti-feeding properties (Kabera et al. 2014).

Based on the metabolic pathways, these bioactive compounds in plants are mainly grouped into three large families: (a) alkaloids, (b) terpenes and steroids, and (c) phenolics (Taiz and Zeiger 2002; Goodwin and Mercer 2003). Nitrogen-containing secondary metabolites, which may or may not have a heterocyclic ring, fall under the alkaloid category. Alkaloids are derived from amino acids. Many of the alkaloids have therapeutic potential. Terpenoids are derived from five-carbon isoprene units. They are further classified on the basis of number of isoprene units present in it. Isoprene units, in turn, developed from isopentenyl pyrophosphate (IPP). Functionally, many of the terpenes have hormonal activity (e.g., cytokinins, gibberellins). Carotenoids are involved in photosynthesis. The economic importance of terpenes is also high. Phenolics or polyphenols are the most widespread compounds among plants. A general rule proposed by Quideau et al. (2011) suggests that the term polyphenol should be used to define plant secondary metabolites derived exclusively from the shikimate-derived phenyl propanoid and/or the polyketide pathway(s), featuring more than one phenolic ring and being devoid of any nitrogen-based functional group in their most basic structural expression (Cheynier et al. 2013).

4.2 In Vitro Production of Secondary Metabolites

Most of the secondary metabolites produced in plants are unique. So, the large-scale production of these compounds from natural stands will result in the depletion of natural resources. In vitro production of compounds through the application of various biotechnological techniques will be an effective alternative for the large-scale production of such compounds. We can use methods like callus, suspension, adventitious root, or hairy root cultures for the production of secondary metabolites. In this chapter, we discuss the possibility of cell and protoplast cultures for the production of bioactive compounds from plants.

4.3 Cell Culture

Plant cell cultures can be considered as a potentially renewable resource for active principles from plants. It is able to replace the traditional agricultural practices for the industrial production of valuable chemicals by cell culture. This technique was introduced in the 1960s for research and production of plant metabolites. Research in different aspects of plant cell cultures is still going on to improve the production of valuable chemicals. The major advantage of this technique is that it can act as a stable source of required compounds, it is independent of environmental conditions, large-scale production from limited space, and is very cost-effective if done in proper way (Bourgaud et al. 2001; Mulabagal and Tsay 2004; Vanisree & Tsay 2004).

The establishment of high yielding cell suspension culture requires skill and expertise. For the successful metabolite production, the establishment of cell culture from high yielding cell lines in optimal cultural conditions is important. Various methods can be applied for the stimulation of biosynthetic pathways of cells to get maximum productivity. Providing optimum growth conditions, using high yielding cell lines as inoculum, incorporation of elicitors or precursors, etc. are some of them. Chemical screening of plants to identify promising accession or elite individual contain the highest quantity of metabolite is often required to achieve high producing in vitro culture system (Pillai and Siril 2018). The development of callus is the primary step to the in vitro production of metabolites. A callus is an undifferentiated mass of loosely arranged thin-walled parenchyma cells developing from proliferating cells of the parent tissue (Dodds and Roberts 1985). The unique feature of callus is that the abnormal growth has the biological potential to develop normal roots, shoots, and embryoids, ultimately forming a plant. The callus is formed through three developmental stages: induction, cell division, and differentiation. During induction, the metabolic rate of cells is stimulated, the duration of which depends on physiological status and nutritional and environmental factors. Owing to increased metabolic rate cells accumulate high contents and finally divide to form a number of cells. Cellular differentiation and expression of certain metabolic pathways start in the third phase leading to synthesis and accumulation of secondary products. Many factors such as genotype, the composition of the nutrient medium, and physical growth factors such as light, temperature, humidity, and endogenous supply of growth regulators are important for callus induction (Kim and Kim 2002; Biswas et al. 2010). Callus cultures also serve as useful systems for investigating factors influencing the biosynthesis of natural products (Pretto and Santarem 2000). Callus culture raised from parent tissue can be maintained for an extended period by subculture and can, therefore, represent a convenient form of long-term maintenance of cell lines. These cultures can be used for initiating suspensions (Brown 1990). Many workers have reported the effectiveness of callus as an in vitro system for the production of secondary metabolites (Baumer et al. 1992; Anirudhan and Nair 2009; Janarthanam et al. 2010; Legha et al. 2012). Callus cultures of *Tinospora cordifolia* produced 3.19-fold enhanced berberine productivity (Pillai and Siril 2019) when cultures were fed with tyrosine as a precursor. Cell suspension culture systems could be used for large-scale culturing of plant cells from which secondary metabolites could be extracted. The advantage of this method is that it can ultimately provide a continuous, reliable source of natural products. The major advantages of a cell culture system over the conventional cultivation of whole plants are: (1) Useful compounds can be produced under controlled conditions, independent of climatic changes or soil conditions; (2) cultured cells would be free of microbes and insects; (3) the cells of any plants, tropical or alpine, could easily be multiplied to yield their specific metabolites; (4) automated control of cell growth and rational regulation of metabolite processes would reduce of labor costs and improve productivity; (5) organic substances are extractable from callus cultures (Vanisree et al. 2004).

To achieve enhanced productivity of metabolites through in vitro culture of plant cells, following aspects are considered in detail.

4.3.1 Providing Optimum Growth Condition Through Standardization

The development of an optimal culture medium with suitable environmental conditions for the effective production of bioactive compounds requires careful standardization. It is a tedious task that requires continuous trial-and-error workouts for days or even months to get the suitable media and environmental condition of desired compounds. But once standardized it can be used for the rest of the experiment. The important factors to be considered while standardization of culture conditions include media components (Singh and Dixit 2012), growth factors, pH, light, and aeration (Goleniowski and Trippi 1999; Lee and Shuler 2000). The productivity of certain compounds can be enhanced many times from their natural stands by providing optimal media conditions. Zenk et al. (1975) produced ten times higher the quantity of anthraquinones from cell suspension cultures of *Morinda citrifolia* L., than naturally occurring plants by providing optimal media conditions. Similarly, Zhong et al. (1996) produced ginsenoside from *Panax quinquefolium* by using suitable media formulations. In the case of anthraquinone production through callus cultures of *Oldenlandia umbellata*, color and nature of the callus varied with type and concentration of auxins used (Fig. 4.1a–l). Metabolite content in callus cultures varies with the variation in culture conditions. Most of the researchers point out that culture conditions play a major role in the accumulation of metabolites in plant cell suspension cultures.

4.3.2 Cell Line Selection

Cell cultures contain a population of cells in different physiological states, i.e., in heterogeneous conditions. The selection of cell lines should focus on developing highly productive and fast-growing cultures. The yield of bioactive compounds depends upon the genetic potential of the explants from which the callus is generated for stock culture. So the screening process should be done at the very beginning of the experiment.

Fig. 4.1 (**a–l**) Callus cultures raised in *Oldenlandia umbellata* using various auxins. (**a**) Cream-colored callus produced on 1 µM 2,4-D, (**b**) light yellow-colored callus produced in 2.5 µM 2,4-D, (**c**) pale white embryogenic calli from 5 µM 2,4-D, (**d**) compact cream callus from 10 µM 2,4-D, (**e**) compact brownish green callus from 1 µM picloram, (**f**) greenish white callus from 2.5 µM picloram, (**g, h**) brownish compact organogenic callus from 5 and 10 µM picloram, (**i**) shining yellow callus from 1 µM NAA, (**j**) golden yellow friable callus produced on 2.5 µM NAA, (**k**) bright yellow callus from 5 µM NAA, (**l**) yellowish friable callus from 10 µM NAA. (**m**) Stock cell suspension culture of *Morinda citrifolia* maintained in MS medium supplemented with 2.5 µM NAA (scale bar = 0.65 cm) producing anthraquinone (AQ). (**n**) AQ accumulation by using piroxicam, α-keto glutaric acid, and phenyl alanine, respectively, in *O. umbellata*. (**o**) AQ accumulated in adventitious root cultures by using pectin (50 mg L^{-1}) elicitation in *O. umbellata*. (**p**) Precursor α-ketoglutarate, (**q**) salicylic acid, (**r**) copper sulfate-treated suspension cultures of *Rubia cordifolia* producing anthraquinones

The secondary metabolite production from callus culture may vary for several subcultures due to somaclonal variations (Larkin and Scowcroft 1981). After a few subcultures, when genetic stability occurs, each callus might be an aggregate of homogenous cells and might show stable secondary metabolite production. Thus, homogeneous cell lines with higher productivity can be developed and further used for cell suspension cultures.

Along with the genetic profile of the mother plant, the accumulation of secondary metabolites in cell suspension cultures depends on the specific physical and chemical factors provided. Thus, callus from superior mother plant which are subjected to continuous selection along with optimal culture conditions can yield higher amounts of metabolites in cell cultures than natural stands (Fett-Neto et al. 1993; Goleniowski and Trippi 1999; Wang et al. 1999; Lee and Shuler 2000; Yogananth et al. 2009; Grover et al. 2012; Karwasara and Dixit 2012).

4.3.3 Elicitor and Precursor Feeding

Nowadays, elicitation and precursor feeding have been widely accepted as a popular approach for higher production of secondary metabolites from cell cultures. Elicitation is based on the concept that plants produce secondary metabolites as a part of the defense mechanism. Elicitors are the compounds that have a pathogenic origin, and plant cells show the same response to elicitors as and when pathogens challenge them. There are both biotic and abiotic elicitors, which can be used in cell cultures. Biotic elicitors have fungal, bacterial, or yeast origin (Akimoto et al. 1999). For example polysaccharides, glycoproteins, inactivated enzymes, purified curdlan, xanthan and chitosan, and salts of heavy metals (Rao and Ravishankar 2002). Generally, elicitors act as signal molecules which trigger defense mechanism in plants (Benhamou 1996; Marcos et al. 2003; Montesano et al. 2003; Sreeranjini and Siril 2019). In *Morinda citrifolia* use of pectin (20 mg/L) as an elicitor molecule in the culture medium containing 2.5 µM NAA produced deep yellow cell suspension (Fig. 4.1m) with twofold enhanced production of anthraquinone component alizarin (Sreeranjini and Siril 2015).

The general molecular regulatory mechanism for the stimulation of secondary metabolite pathways in plants is explained by Zhao et al. (2005). When the signal molecule is recognized by the surface receptor of the plasma membrane, it triggers a signal transduction network by the synthesis of transcription factors for the plant secondary metabolism, which activates the enzymes to catalyze specific metabolite production. So it is important to understand the exact molecular mechanism behind elicitor-mediated metabolite production in plants so that we can optimize commercial production of target compounds by using receptor-specific elicitors. The following table shows a list of generally used elicitors for the production of secondary metabolites (Table 4.1).

A precursor is any intermediate compound in the biosynthetic pathway of secondary metabolites. There were reports that the exogenous supply of precursors in cell cultures can increase the yield (Moreno et al. 1993). The use of inexpensive

Table 4.1 List of commonly used elicitors for secondary metabolite production

Sl No	Elicitor	Secondary metabolite	Plant species	Reference
1.	Agaropectin	Shikonin	*Lithospermum erythrorhizon*	Fukui et al. (1983)
2.	*Ascochyta rabiei*	Medicarpin, Maackiain	*Cicer arientium*	Daniel et al. (1990)
3.	Cellulase	Capsidol	*Capsicum annuum*	Chávez-Moctezuma et al. (1996)
4.	Chitosan	Rutacridone epoxide	*Ruta graveolens*	Eilert et al. (1984)
5.	*Erwinia carotovora*	Camalexin, indole glucosinolates	*Arabidopsis*	Brader et al. (2001)
6.	Fungal elicitor	Indole alkaloids	*Catharanthus roseus*	Zhao et al. (2001)
7.	Fungal glucan	Glyceollin	*Glycine max*	Ebel et al. (1985)
8.	Fungal mycelia	Steroid (Diosgenin)	*Dioscorea deltoidea*	Rokem et al. (1984)
9.	Fungal spores	Lubimin	*Datura stramonium*	Whitehead et al. (1990)
10.	*Pythium aphanidermatum*	N-acetyl-tryptamine	*Catharanthus roseus*	Eilert et al. (1986)
11.	*Trichoderma viride*	Ajmalicine	*Catharanthus roseus*	Namdeo et al. (2002)
12.	Yeast elicitor	Diterpenoid tanshinones	*Salvia miltiorrhiza*	Yan et al. (2005)
13.	Yeast extract	Silymarin	*Silybum marianum*	Sánchez-Sampedro et al. (2005)

precursors like amino acids in cell cultures is a cost-effective method for enhancing productivity.

4.3.4 Permeabilization

Generally, secondary metabolites are stored in vacuoles of plant cells. The products have to penetrate two membranes like plasma membrane and tonoplast to get released into the medium. The cell permeabilization techniques enable easy passage of molecules into and out of the cell. It is achieved either through the application of certain chemicals like isopropanol, dimethyl sulfoxide (DMSO), and polysaccharides like chitosan or by techniques like ultrasonication, electroporation, ionophoretic release, usage of high electric field pulses, and ultrahigh pressure (Knorr and Angersbach 1998). Firoozi et al. (2019) reported that sonication enhanced the production of metabolites in saffron.

4.3.5 Organ Culture Systems

In the recent past, various plant cell culture systems were exploited for the enhancement of high-value metabolites. Also, the comparison of chemical constituents in clonally propagated plants over plants propagated through conventional methods has been reported. In some cases, secondary metabolites are synthesized in certain cells, which also serve as storage cells for these metabolites. In callus, such organized structures are absent and thus unable to synthesize metabolites. In such cases, organized structures possessing the specialized cells are grown in vitro for isolation of biochemicals.

Shoot cultures have been established for many medicinal plants such as *Solidago chilensis* (Schmeda-Hirschmann et al. 2005), *Psoralea corylifolia* (Baskaran and Jayabalan 2008), *Hypericum hirsutum* and *H. maculatum* (Coste et al. 2011), *Teucrium polium* (Al-Qudah et al. 2011), *Merwilla plumbea* (Baskaran et al. 2012), *Ruta graveolens* (Bohidar et al. 2013), *Stevia rebaudiana* Bertoni (Singh and Dwivedi 2014), and *Glossogyne tenuifolia* (Chen et al. 2014), which have been shown to accumulate secondary metabolites to a greater extent than that by naturally raised plants.

The roots of several plants belonging to diverse families are reported to be the site of biosynthesis or accumulation of major secondary metabolites, including alkaloids, polyacetylene, sesquiterpenes, and quinones. Adventitious root culture is another alternative system for the production of secondary metabolites on a commercial scale. In the past, several workers have reported the efficiency of adventitious root tissues for biomass and metabolite productivity (Choi et al. 2000; Hahn et al. 2003; Kim et al. 2004; Yu et al. 2005). In vitro cultured roots grow vigorously and profusely in phytohormone-supplemented medium and have shown tremendous potential for the accumulation of valuable secondary metabolites (Murthy et al. 2008). On the other hand, for several metabolites of interest, the production is too low in the cultured cells, despite extensive studies on the optimization of growth and production media and cell line selection for the development of high producing strains. This low production might probably be because the metabolism is controlled in a tissue-specific manner and tissue de-differentiation, thus resulting in loss of production capacity. Therefore organized structures such as root, embryo, and shoot cultures have been focused as alternatives for the production of secondary metabolites (Verpoorte et al. 2002). Production of secondary metabolites from adventitious root cultures involves four discrete stages, namely, successful induction of adventitious roots from the desired explants, culturing of adventitious roots in liquid medium and establishing growth kinetics (developing suitable medium components and cultural environment for the biomass and metabolite accumulation), developing strategies for higher accumulation of metabolites including elicitation, precursor feeding and culturing of adventitious roots in large scale using bioreactors (Eibl and Eibl 2002; Murthy et al. 2008; Eibl et al. 2010; Murthy et al. 2014). Adventitious root induction is influenced by various factors, including genotype, plant growth regulators, other medium components, and cultural conditions (Liu et al. 2010; Zhu et al. 2010; Reis et al. 2011). Important phytochemicals have been produced by in vitro culture systems such as hepatoprotective diterpenoids from *Andrographis paniculata* (Jain et al. 2000), andrographolide from *A. paniculata* (Praveen et al. 2009), and chicoric acid, chlorogenic acid, and caftaric acid from *Echinacea purpurea* (Wu et al. 2007). Recent studies had shown that proper biochemical treatment and manipulation (addition of elicitors, precursors) can result in an increase in the production of some secondary compounds as well as accumulation of chemicals that would not usually be synthesized in the source plants (Dass and Ramawat 2009; Frankfater et al. 2009; Korsangruang et al. 2010; Lee et al. 2011; Coste et al. 2011). Anthraquinone accumulation in *O. umbellata* by using an inhibitor molecule (piroxicam) and precursor molecules (α-keto glutaric acid and phenylalanine) showed varying response (Fig. 4.1n). Cui et al. (2012) have reported the enhanced production of valtrate in the adventitious roots of *Valeriana amurensis* using elicitors and precursors. In *O. umbellata* adventitious root cultures (Fig. 4.1o) raised using IAA (1 μM) followed by elicitation by adding pectin (50 mg/L) facilitated 2.19-fold enhancement in the production of anthraquinone (Krishnan and Siril 2018). In *Rubia cordifolia*, addition of precursor molecule, α-ketoglutarate (Fig. 4.1p), elicitors salicylic acid (Fig. 4.1q), or copper sulfate (Fig. 4.1r) showed variations in AQ accumulation. Use of bioprocess technology for bioactive plant secondary metabolites production and metabolic engineering is of great interest (Georgiev et al. 2009). Bioreactor design and operations have to be optimized for enhanced production (Wang & Zhong 1996; Terrier et al. 2007; Miao et al. 2008). Adventitious roots of *Echinacea purpurea* were cultured in balloon-type airlift bioreactors, and optimized biomass accumulation and production of secondary metabolites (Hahn et al. 2009).

4.3.6 Hairy Root Cultures

Hairy root culture system in the production of metabolites is now applied to a wide variety of metabolites synthesized in roots. In many cases, metabolites synthesized by hairy roots are the same as those usually synthesized in intact parent roots, with similar or higher yields (Sevón and Oksman-Caldentey 2002). Hairy root cultures are raised by inoculating explant material in a suspension of *Agrobacterium rhizogenes*. The bacteria contain root-inducing (Ri) plasmid. The hairy root cultures are characterized by fast hormone-independent growth, lack of geotropism, lateral branching, and genetically stable nature (Hussain et al. 2012). Hormone-independent growth habits of hairy root coupled with a high degree of transformation stability of hairy root lines make them an attractive tool in the production of metabolites (Qiong et al. 2005). During the infection process, from Ri plasmid, transferred DNA (T-DNA) part and the genes contained in this segment are transferred to the host genome and expressed in the same way as the endogenous genes of the plant cells (Guivarch et al. 1999). Two sets of pRi genes are involved in the root induction process: the aux genes and the rol (root loci) (Jouanin 1984). The hairy roots are normally induced on aseptic, wounded parts of plants by inoculating them with *A. rhizogenes*. Transformation may be induced on aseptic plants grown from seed or on detached leaves, leaf discs, petioles, or stem segments. In many cases profusion of roots may appear directly from the site of inoculation, and in others callus will form initially and roots emerge subsequently from it (Tenea et al. 2008). The susceptibility of species to infection varies. The addition of acetosyringone, a phenolic compound produced during the wounding response of dicotyledonous plants, activates *vir* genes of *Agrobacterium* plasmid.

Transformed roots provide a promising alternative for the biotechnological exploitation of plant cells. *A. rhizogenes*-mediated transformation has also been used to produce transgenic hairy root cultures for the production of metabolite as demonstrated in *O. umbellata* (Krishnan and Siril 2016). The *A. rhizogenes*-mediated transformation can be used to transfer any foreign gene of interest placed in binary vector to the transformed hairy root clone, e.g., *Atropa belladonna*. Fivefold enhanced production of alkaloid scopolamine was achieved by using a binary vector system using *A. rhizogenes* (Hashimoto et al. 1993).

4.3.7 Protoplast Cultures

Protoplast culture is relatively a novel metabolite production method consisting of isolated protoplasts. A plant cell devoid of a cell wall constitutes protoplast. Protoplast represents the finest single system and offers diverse possibilities in the fields of hybrid production through the fusion of isolated somatic protoplasts under in vitro conditions and subsequent development of their product to a hybrid cell, or the whole plant is termed protoplast hybridization. Protoplasts can be employed as a material for genetic transformation, cell fusion, or somatic cell hybridization and induction of somatic mutations thus produce variant cell progenies characterized by novel qualities and improved productivity (Davey et al. 2005). DNA uptake into protoplasts has been especially important in transforming plants that are not amenable to other methods of gene delivery. Somatic hybridization by protoplast fusion enables nuclear and cytoplasmic genomes to be combined, fully or partially, at the interspecific and intergeneric levels to circumvent naturally occurring sexual incompatibility barriers (Aoyagi 2011). An advantage of using protoplasts for the production of useful metabolites is that the products are released readily into the culture medium, with the double benefits of increasing overall productivity and facilitating downstream processing in cases where the cell wall limits the secretion of useful products. Protoplasts are highly fragile, and therefore cannot be used for long-term production. Also, cell wall formation may occur in viable protoplasts if growth conditions are available. Therefore to achieve successful applications, protoplast immobilization in a suitable matrix and elicitor-mediated metabolite production is employed (Aoyagi 2011).

High-quality protoplasts are obtained through the following steps: enzymatic isolation of protoplast, purification of protoplast, and viability testing of protoplast. This was followed by protoplast culture on agar-embedded culture, liquid-over-agar, hanging drop culture, micro-culture chamber, multi-drop array technique, or feeder layer technique. These protoplasts were employed for immobilization in a suitable matrix. In general, commercially available cellulase and pectinase enzymes are routinely used for the isolation of protoplasts from various plant cells and tissues. In single-step procedure, a mixture of enzymes can be used (Power and Cocking 1970). Various kinds of pectinases mostly used in the fruit, paper, and textile industries are also useful in protoplast isolation. For the production of useful metabolites, large quantities of viable protoplasts are required. Protoplast isolation is an enzyme-catalyzed process; therefore, it is necessary to determine the optimum reaction conditions concerning the cell age, enzyme type, and concentration, reaction pH, and reaction time. In addition to these, there must have some mechanism to determine the density of viable isolated protoplasts accurately. The optimization of conditions for the isolation of protoplasts involves continuous monitoring and examinations.

Major applications of protoplast culture include development of novel hybrid plants through protoplast fusion, isolation of mutant lines developed spontaneously or through mutagenesis, single-cell cloning through isolated protoplast culture, and its regeneration to the whole plant, facilitating various genetic transformations through DNA or organelle uptake. However, in the past few attempts were made to produce metabolites through protoplast culture (Aoyagi 2011). In several cases protoplast when immobilized using suitable matrix-like alginate, leading to physiological changes to cells and consequent enhancement in the production of metabolites. Protoplast fusion products upon culture in vitro also found to produce increased metabolite compared to its normal counterparts.

Somatic hybrids were produced using protoplasts fusion of *Centella asiatica* and *Murraya koenigii* for improvement of secondary metabolites by Sadheeshna et al. (2010). Callus produced from the hybrid produced more active compounds and higher activity than parental cells. Protoplast-derived callus was derived from medicinal ginger *Boesenbergia rotunda*, for enhanced metabolite production (Tan et al. 2016). Yamada and Mino (1986) succeeded in producing high berberine-producing *Coptis* cells from protoplast culture. The fusion of protoplasts in microalgae for strain improvement with new or improved characteristics of industrial interest was also reported (Echeverri et al. 2019). Therefore this tool is applied in cell culture technology for the enhancement of productivity. But protoplast culture for the production of secondary metabolites has yet to be explored deeply.

4.3.8 Metabolic Engineering Approaches

To achieve significant enhancement in the production of drug molecules, novel tools based on molecular biology and rDNA technology are also developed. Metabolic engineering (ME) is such an approach envisaged to manipulate specific steps in a metabolic pathway. Metabolic engineering is defined as "improvement of cellular activities by manipulations of enzymatic transport and regulating the functions of the cell with the use of recombinant DNA technology." Metabolic engineering is a relatively new approach to understand and explain plant biochemical pathways. This discipline of applied life science was emerged due to the advancement of phytochemistry, biochemistry, molecular plant physiology, molecular biology, recombinant DNA technology, genomics, transcriptomics, metabolomics, etc.

The availability of gene transfer techniques has led to increased interest in using this technique to redirect metabolic fluxes in plants for industrial purposes (Hanson and Shanks 2002). However, complete mapping of a biosynthetic pathway is a prerequisite for any metabolic engineering program (Lau et al. 2014). Engineering secondary metabolite pathways can be accomplished by several strategies, including elicitor treatment to stimulate plant defense response system, enhancing the expression or activity of a rate-limiting enzyme, preventing feedback inhibition of a key enzyme, etc. To achieve the targeted alteration in the pathway, following steps are generally adopted: synthesis, i.e., construction of the recombinant strain with improved properties, analysis of the recombinant strain, and comparison with original strain and then design the next target for gene manipulation. By adopting metabolic engineering, following can be achieved: increased levels of fine chemicals, new compounds for screening for biological activity, new flower colors or food colors, new taste or fragrance of food, improved nutritional (or health-promoting) effect of food with lowered levels of unwanted compounds in food and fodder, and improved resistance against pest and disease. Therefore, to evolve a successful strategy for the large-scale production of specific drug or metabolite, a combination of plant cell culture technology and metabolic engineering can be used (Verpoorte et al. 1999). Advancement of such an approach may lead to the elegant construction of synthetic operons in a complex pathway and expressing it in a suitable microbial system (Chemler and Koffas, 2008).

4.3.9 Translational Research Approach in Plant Cell Culture Technology

Comprehensive translational research in plant cell culture technology is relevant to establish metabolite production in an economically viable state. In concept, Translational Plant Science (TPS) is the process through which knowledge from basic research on plant genetics, molecular biology, and genomics are provided to improve plants and their productivity. This field will be driven by scientists who combine deep knowledge in molecular plant science with an interdisciplinary mindset, global perspective of industry based on plant resources, broad awareness of economic, social, and political issues associated with applications of biotechnological tools. TPS is based on the ability to establish fruitful interaction with different disciplines in plant biology and other societal strata that play key roles in the bench-to-marketplace pipeline (academia-industry-government-consumers). Plant cell culture research emphasized on secondary metabolite production where the basic method for the metabolite production demonstrated lack of follow-up research which may hinder translation of such findings to societal needs. In this scenario, the translational research approach can be applied for further enhancement of productivity.

4.4 Conclusions

The outcome of cell culture-based metabolite production system is to achieve commercially feasible productivity. Then the technology will be ready to transfer to appropriate stakeholders for implementation. Production of drug molecules through cell culture involves important processes, and the processes developed by research can be protected by patents and can be transferred to end-users such as pharmaceutical entrepreneurs. Another benefit of the cell culture approach is that the method for phytochemical production evolved through research can help to reduce the cost of the phytochemical drug significantly and thus can be afforded by the common people. Also when cell culture-based approach is used for the large-scale production of phytochemicals or drug molecules, it reduces the pressure on our natural resources; in other words, it protects overexploitation of source plant for the extraction of the drug molecules.

References

Akimoto C, Aoyagi H, Tanaka H (1999) Endogenous elicitor-like effects of alginate on physiological activities of plant cells. Appl Microbiol Biotechnol 52:429–436
Al-Qudah TS, Shibli RA, Alali FQ (2011) In vitro propagation and secondary metabolites production in wild germander (*Teucrium polium* L.). In Vitro Cell Dev Biol Plant 47:496–505
Anirudhan K, Nair AS (2009) *In vitro* production of solasodine from *Solanum trilobatum*. Biol Plant 53:719–722
Aoyagi H (2011) Application of plant protoplasts for the production of useful metabolites. Biochem Eng J 56:1–8. https://doi.org/10.1016/j.bej.2010.05.004
Baskaran P, Jayabalan N (2008) Effect of growth regulators on rapid micropropagation and psoralen production in *Psoralea corylifolia* L. Acta Physiol Plant 30:345–351
Baskaran P, Ncube B, Van Staden J (2012) In vitro propagation and secondary product production by *Merwilla plumbea* (Lindl.) Speta. Plant Growth Regul 67:235–245
Baumer A, Groger D, Kuzovkina IN, Reisch J (1992) Secondary metabolites produced by callus cultures of various *Ruta* species. Plant Cell Tissue Org Cult 28:159–162
Benhamou N (1996) Elicitor-induced plant defence pathways. Trends Plant Sci 1:233–240
Biswas MK, Roy UK, Islam R, Hossain M (2010) Callus culture from leaf blade, nodal and runner segments of tree strawberry (*Fragaria* sp.) clones. Turkish J Biol 34:75–80
Bohidar S, Pattanaik S, Thiruvanoukkarasu M (2013) Improved furanocoumarin production in *Ruta graveolens* L. regenerated via in vitro stem internode cultures. Plant Biotechnol Rep 7:399–405
Bourgaud F, Gravot A, Milesi S, Gontier E (2001) Production of plant secondary metabolites: a historical perspective. Plant Sci 161:839–851
Brader G, Tas E, Palva ET (2001) Jasmonate-dependent induction of indole glucosinolates in *Arabidopsis* by culture filtrates of the nonspecific pathogen *Erwinia carotovora*. Plant Physiol 126:849–860
Brown JT (1990) The initiation and maintenance of callus culture. In: Pollard JW, Walker JM (eds) Methods in molecular biology, Plant cell tissue organ culture, vol vol 6. Humana Press, New Jersey, pp 57–65
Chemler JA, MAG K (2008) Metabolic engineering for plant natural product biosynthesis in microbes. Curr Opin Biotechnol 19:597–605
Chen CC, Chang HC, Kuo CL, Agrawal DC, Wu CR, Tsay H-S (2014) In vitro propagation and analysis of secondary metabolites in *Glossogyne tenuifolia* (Hsiang-Ju)—a medicinal plant native to Taiwan. Bot Stud 55:45–51

Cheynier V, Comte G, Davies KM, Lattanzio V, Martens S (2013) Plant phenolics: recent advances on their biosynthesis, genetics and ecophysiology. Plant Physiol Biochem 72:1–20

Choi SM, Son SH, Yun SR, Kwon OW, Seon JH, Park KY (2000) Pilot scale culture of adventitious roots of ginseng in a bioreactor system. Plant Cell Tissue Organ Cult 62:187–193

Coste A, Vlase L, Halmagyi A, Deliu C, Coldea G (2011) Effects of plant growth regulators and elicitors on production of secondary metabolites in shoot cultures of *Hypercium hirsutum* and *Hypericum maculatum*. Plant Cell Tissue Organ Cult 106:279–288

Cui L, Wang ZY, Zhou XH (2012) Optimization of elicitors and precursors to enhance valtrate production in adventitious roots of *Valeriana amurensis* Smir ex kom. Plant Cell Tissue Organ Cult 108:411–420

Daniel S et al (1990) Elicitor-induced metabolic changes in cell cultures of chickpea (*Cicer arietinum* L.) cultivars resistant and susceptible to *Ascochyta rabiei*. Planta 182:270–278

Dass S, Ramawat KG (2009) Elicitation of guggulsterone production in cell cultures of *Commiphora* wightii by plant gums. Plant Cell Tissue Organ Cult 96:334–353

Davey MR, Anthony P, Power JB, Lowe KC (2005) Plant protoplasts: status and biotechnological perspectives. Biotechnol Adv 23:131–171

Dodds JH, Roberts LW (1985) Experiments in plant tissue culture. Cambridge University Press, Cambridge, p 256

Ebel J, Stäb MR, Schmidt WE (1985) Induction of enzymes of phytoalexin synthesis in soybean cells by fungal elicitor. Primary and secondary metabolism of plant cell cultures. Springer, Berlin, pp 247–254

Echeverri D, Romo J, Giraldo N, Atehortua L (2019) Microalgae protoplasts isolation and fusion for biotechnology research. Rev Colomb. Biotecnol. XXI(1):71–82. https://doi.org/10.15446/rev.colomb.biote.v21n1.80248

Eibl R, Eibl D (2002) Bioreactors for plant cell and tissue cultures. In: Oksman-Caldentey KM, Barz WH (eds) Plant biotechnology and transgenic plants. Marcel Dekker, New York, pp 163–199

Eibl R, Werner S, Eibl D (2010) Bag bioreactor based on wave induced motion: characteristics and applications. Adv Biochem Eng Biotechnol 115:55–87

Eilert U, Ehmke A, Wolters B (1984) Elicitor-induced accumulation of acridone alkaloid epoxides in *Ruta graveolens* suspension cultures. Planta Med 50:508–512

Eilert UF, Constabel WG, Kurz W (1986) Elicitor-stimulation of monoterpene indole alkaloid formation in suspension cultures of *Catharanthus roseus*. J Plant Physiol 126:11–22

Fett-Neto AG et al (1993) Improved growth and taxol yield in developing calli of *Taxus cuspidata* by medium composition modification. Biotechnology (NY) 11:731

Firoozi B, Zare N, Sofalian O, Sheikhzade-mosadegh P (2019) In vitro indirect somatic embryogenesis and secondary metabolites production in the saffron: emphasis on ultrasound and plant growth regulators. J Agric Sci 25:1–10

Frankfater CR, Dowd MK, Triplett BA (2009) Effect of elicitors on the production of gossypol and methylated gossypol in cotton hairy roots. Plant Cell Tissue Organ Cult 98:341–349

Fukui H, Yoshikawa N, Tabata M (1983) Induction of shikonin formation by agar in *Lithospermum erythrorhizon* cell suspension cultures. Phytochemistry 22:2451–2453

Georgiev MI, Weber J, Maciuk A (2009) Bioprocessing of plant cell cultures for mass production of targeted compounds. Appl Microbiol Biotechnol 83:809–823

Goleniowski M, Trippi VS (1999) Effect of growth medium composition on psilostachyinolides and altamisine production. Plant Cell Tissue Organ Cult 56:215–218

Goodwin TW, Mercer EI (2003) Introduction to plant biochemistry. CBS Publishers and Distributers, New Delhi, p 567

Grover A et al (2012) Production of monoterpenoids and aroma compounds from cell suspension cultures of *Camellia sinensis*. Plant Cell Tissue Organ Cult 108:323–331

Guivarch A, Boccara M, Proteau M, Chriqui D (1999) Instability of phenotype and gene expression in long term cultures of carrot hairy root clones. Plant Cell Rep 19:43–50

Hahn EJ, Kim YS, Paek KY (2003) Adventitious root cultures of *Panax ginseng* C. A Meyer and ginsenoside production through large scale bioreactor system. J Plant Biotechnol 5:1–6

Hahn EJ, Wu CH, Paek Y (2009) Production of root biomass and secondary metabolites through adventitious root cultures of *Echinacea purpurea* in bioreactors. In: ISHS Acta Horticulturae 829: VI international symposium on in vitro culture and horticultural breeding. https://doi.org/10.17660/ActaHortic.2009.829.9

Hanson AD, Shanks JV (2002) Plant metabolic engineering—entering the S curve. Metab Eng 4:1–2. https://doi.org/10.1006/mben.2001.0213

Hashimoto T, Yun DJ, Yamada Y (1993) Production of tropane alkaloids in genetically engineered root cultures. Phytochemistry 32:713–718

Hussain MS, Fareed S, Ansari S, Rahman MA, Ahmad IZ, Saeed M (2012) Current approaches toward production of secondary plant metabolites. J Pharm Bioallied Sci 4:10–20. https://doi.org/10.4103/0975-7406.92725

Jain DC, Gupta MM, Saxena SK (2000) HPLC analysis and hepatoprotective diterpenoids from *Andrographis paniculata*. J Pharm Biomed Anal 22:705–709

Janarthanam B, Gopalakrishnan M, Sekar T (2010) Secondary metabolite production in callus cultures of *Stevia rebaudiana* Bertoni. Bangladesh J Sci Ind Res 45:243–248

Jouanin L (1984) Restriction map of an agropine-type Ri plasmid and its homologies with Ti plasmids. Plasmid 12:91–102

Kabera JN et al (2014) Plant secondary metabolites: biosynthesis, classification, function and pharmacological properties. J Pharm Pharmacol 2:377–392

Karwasara VS, Dixit VK (2012) Culture medium optimization for improved puerarin production by cell suspension cultures of *Pueraria tuberosa* (Roxb. ex Willd.) DC. In Vitro Cell Dev Biol Plant 48:189–199

Kim SH, Kim SK (2002) Effects of auxins and cytokinins on callus induction from leaf blade, petiole and stem segments of in vitro grown Sheridan grape shoots. Plant Biotechnol J 4:17–21

Kim SY, Hahn EJ, Murthy HN, Paek KY (2004) Adventitious root growth and ginsenoide accumulation in *Panax ginseng* cultures as affected by methyl jasmonate. Biotechnol Lett 26:1619–1622

Knorr D, Angersbach A (1998) Impact of high-intensity electric field pulses on plant membrane permeabilization. Trends Food Sci Technol 9:185–191. https://doi.org/10.1016/S0924-2244(98)00040-5

Korsangruang S, Soonthornchareonnon N, Chintapakorn Y, Saralamp P, Prathanturarug S (2010) Effects of abiotic and biotic elicitors on growth and isoflavonoid accumulation in *Pueraria candollei* var candollei and *P.candollei* var mirifica cell suspension cultures. Plant Cell Tissue Organ Culture 103:333–342

Krishnan SRS, Siril EA (2016) Induction of hairy roots and over production of anthraquinones in *Oldenlandia umbellata* L.: a dye yielding medicinal plant by using wild type *Agrobacterium rhizogenes* strain. Ind J Plant Physiol. https://doi.org/10.1007/s40502-016-0229-0

Krishnan SRS, Siril EA (2018) Elicitor mediated adventitious root culture for the large-scale production of anthraquinones from *Oldenlandia umbellata* L. Ind Crops Prod 114:173–179

Larkin PJ, Scowcroft WR (1981) Somaclonal variation—a novel source of variability from cell cultures for plant improvement. Theor Appl Genet 60:197–214

Lau W, Fischbach MA, Osbourn A, Sattely ES (2014) Key applications of plant metabolic engineering. PLoS Biol 12(6):e1001879. https://doi.org/10.1371/journal.pbio.1001879

Lee CWT, Shuler ML (2000) The effect of inoculum density and conditioned medium on the production of ajmalicine and catharanthine from immobilized *Catharanthus roseus* cells. Biotechnol Bioeng 67:61–71

Lee Y, Lee DE, Lee HS, Kim SK, Lee WS, Kim SH, Kim MW (2011) Influence of auxins, cytokinins and nitrogen on production of rutin from callus and adventitious roots of the white mulberry tree (*Morus alba* L.). Plant Cell Tissue Organ Cult 105:9–19

Legha MR, Prasad KV, Singh S, Kaur C, Arora A, Kumar S (2012) Induction of carotenoid pigments in callus cultures of *Calendula officianalis* L. in response to nitrogen and sucrose levels. In Vitro Cell Dev Biol Plant 48:99–106

Liu C, Callow P, Rowland LJ, Hancock JF, Song GQ (2010) Adventitious shoot regeneration from leaf explants of southern highbush blueberry cultivars. Plant Cell Tissue Organ Culture 103:137–144

Marcos M, Brader G, Palva TE (2003) Pathogen derived elicitors: searching for receptors in plants. Mol Plant Pathol 4:73–79

Miao Y, Ding Y, Sun QY, Xu ZF, Jiang L (2008) Plant bioreactors for pharmaceuticals. Biotechnol Genet Eng Rev 25:363–380

Moctezuma C, Patricia M, Lozoya-Gloria E (1996) Biosynthesis of the sesquiterpenic phytoalexin capsidiol in elicited root cultures of chili pepper (*Capsicum annuum*). Plant Cell Rep 15:360–366

Montesano M, Brader G, Palva ET (2003) Pathogen derived elicitors: searching for receptors in plants. Mol Plant Pathol 4(1):73–79. https://doi.org/10.1046/j.1364-3703.2003.00150.x

Moreno PRH, Van der Heijden R, Verpoorte R (1993) Effect of terpenoid precursor feeding and elicitation on formation of indole alkaloids in cell suspension cultures of *Catharanthus roseus*. Plant Cell Rep 12:702–705

Mulabagal V, Tsay HS (2004) Plant cell cultures - an alternative and efficient source for the production of biologically important secondary metabolites. Int J Appl Sci Eng 2(1):29–48

Murthy HN, Hahn EJ, Paek KY (2008) Adventitious roots and secondary metabolism. Chin J Biotechnol 24:711–716

Murthy HN, Lee EJ, Paek KY (2014) Production of secondary metabolites from cell and organ cultures: strategies and approaches for biomass improvement and metabolite accumulation. Plant Cell Tissue Organ Cult 118:1–16. https://doi.org/10.1007/s11240-014-0467-7

Namdeo A, Patil S, Fulzele DP (2002) Influence of fungal elicitors on production of ajmalicine by cell cultures of *Catharanthus roseus*. Biotechnol Prog 18:159–162

Pillai SK, Siril EA (2018) Elite screening and in vitro propagation of *Tinospora cordifolia* (Willd.) Miers ex Hook F. & Thoms. Proc Natl Acad Sci India B Biol Sci 89:551–557. https://doi.org/10.1007/s40011-018-0971-3

Pillai SK, Siril EA (2019) Enhanced production of berberine through callus culture of *Tinospora cordifolia* (Willd.) Miers ex Hook F. and Thoms. Proc Natl Acad Sci India B Biol Sci. https://doi.org/10.1007/s40011-019-01106-9

Power JB, Cocking EC (1970) Isolation of leaf protoplasts: macromolecular uptake and growth substance response. J Exp Bot 21:64–70

Praveen N, Manohar SH, Naik PM, Nageem A, Jeong J, Murthy HN (2009) Production of andrographolide from adventitious root cultures of *Andrographis paniculata*. Curr Sci 96:694–696

Pretto FR, Santarem ER (2000) Callus formation and plant regeneration from *Hypercium perforatum* leaves. Plant Cell Tissue Organ Culture 62:107–113

Qiong Y et al (2005) Efficient production and recovery of diterpenoid tanshinones in *Salvia miltiorrhiza* hairy root cultures with in situ adsorption, elicitation and semi-continuous operation. J Biotechnol 119:416–424

Quideau S et al (2011) Plant polyphenols: chemical properties, biological activities, and synthesis. Angew Chem Int Ed 50:586–621

Rao RS, Ravishankar GA (2002) Plant cell cultures: chemical factories of secondary metabolites. Biotechnol Adv 20:101–153

Reis RV, Borges APPL, Chierrito TPC (2011) Establishment of adventitious root culture of S*tevia rebaudiana* Bertoni. In a roller bottle system. Plant Cell Tissue Organ Cult 106:329–335

Rokem JS, Schwarzberg J, Goldberg I (1984) Autoclaved fungal mycelia increase diosgenin production in cell suspension cultures of *Dioscorea deltoidea*. Plant Cell Rep 3:159–160

Sadheeshna KS, Annie G, Aldous JH, Sasikala L (2010) Improvement of secondary metabolites by somatic hybridization of *Murraya koenigii* and *Centella asiatica*. Asian J Anim Sci 4(2):231–236

Sánchez S, Angeles M, Tárrago JF, Corchete P (2005) Yeast extract and methyl jasmonate-induced silymarin production in cell cultures of *Silybum marianum* (L.) Gaertn. J Biotechnol 119:60–69

Schmeda-Hirschmann G, Jordan M, Gerth A, Wilken D (2005) Secondary metabolite content in rhizomes, callus cultures and in vitro regenerated plantlets of *Solidago chilensis*. Z Naturforsch 60c:5–10

Sevón N, Oksman-Caldentey KM (2002) Agrobacterium rhizogenes-mediated transformation: root cultures as a source of alkaloids. Planta Med 68:859–868

Singh KV, Dixit VK (2012) Culture medium optimization for improved puerarin production by cell suspension cultures of *Pueraria tuberosa* (Roxb. ex Willd.) DC. In Vitro Cell Dev Biol Plant 48:189–199

Singh P, Dwivedi P (2014) Two-stage culture procedure using thidiazuron for efficient micropropagation of *Stevia rebaudiana*, an anti-diabetic medicinal herb. 3 Biotech 4:431–437

Sreeranjini S, Siril EA (2015) Optimising elicitors and precursors to enhance alizarin and purpurin production in adventitious roots of *Morinda citrifolia* L. Proc Natl Acad Sci India B Biol Sci 85:725–731. https://doi.org/10.1007/s40011-014-0395-7

Sreeranjini S, Siril EA (2019) Anthraquin0one production in cell cultures of *Morinda citrifolia* L. through co treatment with elicitors and precursors. J Cytol Genet 20:13–18

Taiz L, Zeiger E (2002) Plant physiology, 3rd edn. Sinauer Associates Inc., Sunderland, p 623

Tan H, Tan BC, Wong SM, Khalid N (2016) A medicinal ginger, *Boesenbergia rotunda*: from cell suspension cultures to protoplast derived callus. Sains Malaysiana 45(5):795–802

Tenea GN, Calin A, Gavrilla L, Cucu N (2008) Manipulation of root biomass and biosynthetic potential of *Glycyrrhiza glabra* L. plants by *Agrobacterium rhizogenes* mediated transformation. Romanian Biotechnol Lett 13:3922–3932

Terrier B, Courtois D, Henault N, Cuvier A, Bastin M, Aknin A, Dubreuil J, Petiard V (2007) Two new disposable bioreactors for plant cell culture: the wave and undertow bioreactor and the slug bubble bioreactor. Biotechnol Bioeng 96:914–923

Vanisree M, Tsay HS (2004) Plant cell cultures-an alternative and efficient source for the production of biologically important secondary metabolites. Int J Appl Sci Eng 2:29–48

Vanisree M, Chen YL, Shu-Fung L, Satish MN, Chien YL, Hsin Sheng T (2004) Studies on the production of some important secondary metabolites from medicinal plants by plant tissue cultures. Bot Bull Acad Sin 45:1–22

Verpoorte R, van der Heijden R, ten Hoopen HJG, Memelink J (1999) Metabolic engineering of plant secondary metabolite pathways for the production of fine chemicals. Biotechnol Lett 21:467–479

Verpoorte R, Contin A, Memelink J (2002) Biotechnology for the production of plant secondary metabolites. Phytochem Rev 1:13–25

Wang SJ, Zhong JJ (1996) A novel centrifugal impeller bioreactor. I. Fluid circulation, mixing, and liquid velocity profiles. Biotechnol Bioeng 51:511–519

Wang HQ, Yu JT, Zhong JJ (1999) Significant improvement of taxane production in suspension cultures of *Taxus chinensis* by sucrose feeding strategy. Process Biochem 35:479–483

Whitehead IM, Atkinson AL, Threlfall DR (1990) Studies on the biosynthesis and metabolism of the phytoalexin lubimin and related compounds in *Datura stramonium* L. Planta 182:81–88

Wu CH, Murthy HN, Hahn EJ, Park KY (2007) Large scale cultivation of adventitious roots of *Echinacea purpurea* in air lift bioreactors for the production of chichoric acid, chlorogenic acid and caftaric acid. Biotechnol Lett 29:1179–1182

Yamada Y, Mino M (1986) Instability of chromosomes and alkaloid content in cell lines derived from single protoplasts of cultured Coptis japonica cells. Curr Top Dev Biol 20:409–417. https://doi.org/10.1016/s0070-2153(08)60679-1

Yan Q, Hu Z, Tan RX, Wu JY (2005) Efficient production and recovery of diterpenoid tanshiniones in *Salvia miltiorrhiza* hairy root cultures with in situ adsorption, elicitation and semi-continuous operation. J Biotechnol 119:416–424

Yogananth N, Bhakyaraj R, Chanthuru A, Parvathi S, Palanivel S (2009) Comparative analysis of solasodine from in vitro and *in vivo* cultures of *Solanum nigrum* Linn. J Sci Eng Technol 5:99–103

Yu KW, Murthy HN, Jeong CS, Hahn EJ, Paek KY (2005) Organic germanium stimulates the growth of ginseng adventitious roots and ginsenoside production. Process Biochem 40:2959–2961

Zenk MH, El-Shagi H, Schulte U (1975) Anthraquinone production by cell suspension cultures of *Morinda citrifolia*. Planta Med 28:79–101

Zhao J, Zhu WH, Hu Q (2001) Selection of fungal elicitors to increase indole alkaloid accumulation in *Catharanthus roseus* suspension cell culture. Enzym Microb Technol 28:666–672

Zhao J, Davis LC, Verpoorte R (2005) Elicitor signal transduction leading to production of plant secondary metabolites. Biotechnol Adv 23:283–333

Zhong JJ, Bai Y, Wang SJ (1996) Effects of plant growth regulators on cell growth and ginsenoside saponin production by suspension cultures of *Panax quinquefolium*. J Biotechnol 45:227–234

Zhu XY, Chai SJ, Chen LP, Zhang MF, Yu JG (2010) Induction and origin of adventitious roots from chimeras of *Brassica juncea* and *Brassica oleraceae*. Plant Cell Tissue Organ Cult 101:287–294

Hairy Root Culture: Secondary Metabolite Production in a Biotechnological Perspective

Radhakrishnan Supriya, Radhadevi Gopikuttan Kala, and Arjunan Thulaseedharan

Abstract

Plants produce a variety of organic compounds that are not directly involved in the primary metabolic process, known as secondary metabolites, but essential for the proper functioning of the plant in relation to its environment. Plant secondary metabolites are valuable phytochemicals used in pharmaceuticals, like flavors, fragrance, dyes, food additives, insecticides etc. Secondary metabolites are directly extracted from plants through various techniques. The very low yield in wild plants and tedious extraction procedures are constraints for commercial production of secondary metabolites. The cell and tissue culture technology combined with recent developments in molecular biology offers a viable approach to overcome these difficulties. Hairy roots are highly branched ectopic roots produced in different plants as a result of infection with the gram-negative soil bacterium, *Agrobacterium rhizogenes*. During infection, *A. rhizogenes* transfers a DNA segment known as T-DNA from its root-inducing (Ri) plasmid to the plant and integrate with the plant genome. The oncogene present in the T-DNA stimulates the formation of highly proliferative hairy roots. Hairy roots produce the same biochemicals as their wild counterparts but at very high concentrations. The fast growth and genetic and biochemical stability of hairy roots over a long period made this a promising technology for plant secondary metabolite production. The large-scale continuous culture of hairy roots can be optimized in bioreactors and the production of the desired metabolite can be further enhanced by the addition of suitable elicitors. The present chapter

R. Supriya
Advanced Centre for Molecular Biology and Biotechnology, Rubber Research Institute of India, Kottayam, India - 686 009

Department of Biotechnology, University of Kerala, Thiruvananthapuram, Kerala, India - 695 581

R. G. Kala · A. Thulaseedharan (✉)
Advanced Centre for Molecular Biology and Biotechnology, Rubber Research Institute of India, Kottayam, India - 686 009

© Springer Nature Singapore Pte Ltd. 2020
S. T. Sukumaran et al. (eds.), *Plant Metabolites: Methods, Applications and Prospects*, https://doi.org/10.1007/978-981-15-5136-9_5

highlights the general perspectives and status of hairy root cultures for the production of valuable secondary metabolites.

Keywords

Agrobacterium rhizogenes · Bioreactors · Genetic transformation · Hairy root culture · Rol genes · Secondary metabolites

5.1　Introduction

Plants synthesize a variety of extremely diverse clusters of phytochemicals, known as secondary metabolites, which are not critical for living cells in their normal growth and development (Fraenkel 1959). They provide protection to the plant from many adverse environmental conditions (Stamp 2003) or attack from other species (Samuni-Blank et al. 2012). Because of their biological activities, many plant natural products have long been exploited by humans as pharmaceuticals, stimulants, and poisons (Petrovska 2012). Most of the higher plants produce commercially important organic compounds such as resins, tannins, oils, waxes, gums, dyes, natural rubber, aromas and fragrances, pharmaceuticals, pest repellents, pesticides etc. The extraction of secondary metabolites from the native plants possesses certain challenges due to their relatively low content and the complexity of extraction. The alternative to the traditional extraction technique was the production of plant cell suspension cultures. A major challenge for this technology, however, is that secondary metabolites are often synthesized at distinct developmental stages in specialized cells (Balandrin et al. 1985). In addition, certain compounds are not synthesized in undifferentiated cells (Berlin et al. 1985).

Hairy Root Culture (HRC) is an advanced technology adopted for the large-scale synthesis of essential secondary plant metabolites using the highly profuse hairy roots produced in plants as a result of infection with *Agrobacterium rhizogenes,* a gram-negative soil bacterium. During infection, *A. rhizogenes* transfers a DNA segment called transfer DNA (T-DNA) from its root-inducing (Ri) plasmid, carrying the genes for auxin biosynthesis and the plant oncogenes which help in the induction of hairy roots in hormone-free media. Hairy roots are characterized by their fast growth and profuse branching in the absence of external source of hormones. The long-lasting overproduction of secondary metabolites and the genetic stability of the hairy roots made this method more suitable for the production of many useful secondary metabolites (Flores et al. 1999; Shanks and Morgan 1999; Sevon and Oksman-Caldenty 2002). In addition to secondary metabolite production, the hairy root system can also be used for the production of valuable recombinant proteins, since the gene of interest can be incorporated into the plant genome through the T-DNA. For large-scale production the process can be optimized in a bioreactor level (Kim et al. 2002) and the production can be further enhanced by the addition of elicitors (Lu et al. 2008; Srivastava et al. 2006). The present chapter highlight the current status of hairy root culture for the production of essential plant secondary metabolites.

5.2 Plant Secondary Metabolites

Plants synthesize thousands of metabolites which directly and indirectly influence their growth and development, reproduction, defense against attack by different kinds of organisms, and survival in often adverse and changing environments. The plant-synthesized phytochemicals fall into two groups based on their role in fundamental metabolic processes, namely primary and secondary metabolites. By engaging in normal growth, development, and reproduction, primary metabolites perform important metabolic roles; therefore, they are more or less similar in all living cells. Secondary metabolites (SM) are compounds that are not important to a cell /organism in its normal metabolic process. But they play a critical role in the cell /organism's interaction with its environment. Such compounds also participate in plant defense from several biotic or abiotic stresses. Secondary metabolites (SMs) come from different families of metabolites and are highly inducible in response to various stresses in plants. Many SMs are used as medicines, flavors, fragrances, insecticides, and colorants and thus have tremendous economic potential. Plant secondary metabolites can be categorized into four main groups according to their chemical structure: phenolic compounds, terpenoids, alkaloids, and sulfur-containing compounds. Terpenoids are the major class of secondary metabolites composed of five carbon units synthesized through the acetate/mevalonate pathway or the glyceraldehyde/pyruvate pathway. Such phytochemicals can be antimicrobial, serve as attractants/repellents or as herbivore deterrents. Most of them have biological and pharmacological activities also and are therefore important in the field of medicine and biotechnology (Singh and Sharma 2015).

Around 25,000 of the secondary metabolites are thought to have good potential in a wide variety of industrial and pharmaceutical applications (Trémouillaux-Guiller 2013) and are often used in agricultural, food, pharmaceutical, cosmetic, and textile industries too. Such new applications would help to extend and enhance the continued utilization of the higher plants as renewable chemical sources, especially pharmaceutical compounds. Continued and intensive efforts in this area are important to the effective development of unique, valuable plant chemicals (Saurabh et al. 2015). The commercial importance of secondary metabolites is increasing day by day. This resulted in a renewed interest in research focused on developing alternative strategies for secondary metabolite production in cultures of plant cells and tissues.

5.3 *Agrobacterium rhizogenes* and Hairy Roots

Agrobacterium rhizogenes (updated scientific name: *Rhizobium rhizogenes*) belonging to *Rhizobiaceae* family is a gram-negative soil bacterium that produces hairy root disease in a wide range of dicotyledonous plants (Ranganath 2016). *Agrobacterium rhizogenes* is widely used in gene transfer studies due to its ease of use, economic viability, high transformation rate, and ability to induce rapid root growth in many dicotyledonous plants. They cause multi-branched adventitious roots at the site of infection termed "hairy roots" (HR) (Baranski 2008). Upon infecting the plant, the phytopathogen transfers a fragment of DNA from its large root-inducing (Ri) plasmid, called T-DNA (Transfer DNA), bringing a series of genes into the host plant genome. It encodes enzymes capable of altering plant hormonal metabolism such as auxin and cytokinin biosynthesis and induces neoplastic growth of roots characterized by high growth levels in hormone-free media and demonstrates genetic stability. Opines such as agropine and mannopine are also synthesized following the induction of the Ri plasmids of *A. rhizogenes*. It is found that in the same pattern of the corresponding wild-type organism, HRs too generate the phytochemicals. Hairy roots are swift-growing and plagiotropic (grow away from the vertical). The plagiotropic trait is beneficial, as the aeration in the liquid culture medium increases. To this end, the roots are grown in an aerated medium and have high biomass accumulation (Mishra and Ranjan 2008).

5.3.1 Gene Transfer and Hairy Root Induction

A. rhizogenes chemotactically react to certain signal molecules exuded by vulnerable wounded plant cells and get attached to them. Thus plant infection with *A. rhizogenes* cause HRs to grow at the site of infection. On infection, *A. rhizogenes* transfers T- DNA from the Ri plasmid to the host plant genome, which induce hairy roots to grow. HRs obtained by contact with specific strains of bacteria display numerous morphologies. The differences in virulence and morphology can be explained by the different plasmids harbored in different *A. rhizogenes* strains (Nguyen et al. 1992). One copy of a large Ri plasmid is contained in all *A. rhizogenes* strains usually.

The T-DNA consists of a left-hand T-DNA (TL-DNA) and a right-hand T-DNA (TR-DNA) region bounded by a 25 bp oligonucleotide repeat. The TR-DNA mostly contains two genes *iaaM* and *iaaH* for auxin biosynthesis, which is homologous to the Ti plasmids of *A. tumefaciens* (Klee and Romano 1994). The hairy root inducing four genes such as *rolA*, *rolB*, *rolC*, and *rolD* are located in the TL-DNA. Genes for agropine synthesis are found in the TR-DNA region in the Ri plasmids of agropine-producing strains, along with the tumor-inducing plasmid genes. T-DNA is transmitted stably to wounded plant cells and incorporated into the host genome (Chilton et al. 1982). To enable their expression in plant cells, they have eukaryotic regulatory sequences along with genes of bacterial origin. Even in the absence of TR-DNA-directed auxin synthesis of the mannopine type, which lacks tms (tumor shoot) loci, root induction occurs (Mishra and Ranjan 2008). The genes governing the synthesis of substances that induce the cells to differentiate into roots under the influence of endogenous auxin synthesis genes are residing in the Ri TL-DNA (Shen et al. 1988). Except for the 25 bp direct repeat border sequences, none of the other T-DNA sequences are required for the gene transfer to plants. Transmission of T-DNA is regulated by virulence genes residing in the Ri plasmid vir (virulence genes) and chv (chromosomal virulence) genes present in bacterial chromosomes. Various phenolic compounds released by wounded plant cells, such as acetosyringone and α-hydroxyacetosyringone, induce the transcription of the vir region. Acetosyringone and other sugars act synergistically to induce high levels of vir gene expression. The effect of growth medium and bacterial concentration on hairy root induction cannot be ruled out. For example, high-salt media favor HR formation in some plants (Giri et al. 1997) whereas low-salt media favors excessive bacterial multiplication in the medium. A high concentration of bacterial cells can decrease plant transformation by competitive inhibition, whereas suboptimal bacterial concentration results in lower bacterial availability for the same (Kumar et al. 2006).

5.3.2 General Perspectives of Hairy Root Culture

Hairy root cultures (HRC) have been studied for decades to produce useful metabolites present in the wild-type roots, because of their genetic stability and higher secondary metabolite accumulation. Moreover, genetic manipulations are much easier to be performed in its metabolic pathway of synthesis. The hairy roots can be cultured without the need for exogenous auxin source. Other features of HRC include incubation in dark and stable metabolite production over a long period. All these advantages allow many root-derived plant products to be produced through hairy root culture technology, which is thought not to be suitable for production via cell culture. This property of rapid root growth leads to high frequency of plantlet regeneration, which allows elite plant clonal propagation. In addition to the above uses the altered phenotype of hairy root regenerants (hairy root syndrome) is useful in ornamental plant breeding programs (Giri and Narasu 2000). In addition, the advent of key molecules to overcome the limiting culture parameters for regulating metabolic pathways has made it possible to increase the production of secondary metabolites by hairy roots (Guillon et al. 2006). The secretion and processing of such metabolites with the aid of trapping systems strengthen interest in these cultures. Interestingly, elicitation can be performed in HRC, and this elicitation helps to address various difficulties associated with the large-scale processing of the most commercially important bioactive secondary metabolites from wild and cultured plants, whether undifferentiated or differentiated cell cultures. HRs may use different elicitation methods to further increase their accumulation in both small- and large-scale outputs. But for performance through elicitation, HR cultivation has to be carried out in bioreactors, which require optimization. HRCs can also be exploited as biological farms for recombinant protein production. It thus provides enhanced potential for its industrial application. HR technology is now being strongly strengthened by the recent awareness on the molecular mechanisms underlying its profuse growth and secondary metabolite production (Fig. 5.1).

In addition to secondary metabolite overproduction, HRCs are also used in industries to produce recombinant animal proteins. Knowledge gained from the molecular mechanism of *A. rhizogenes* T-DNA transfer paves more opportunities

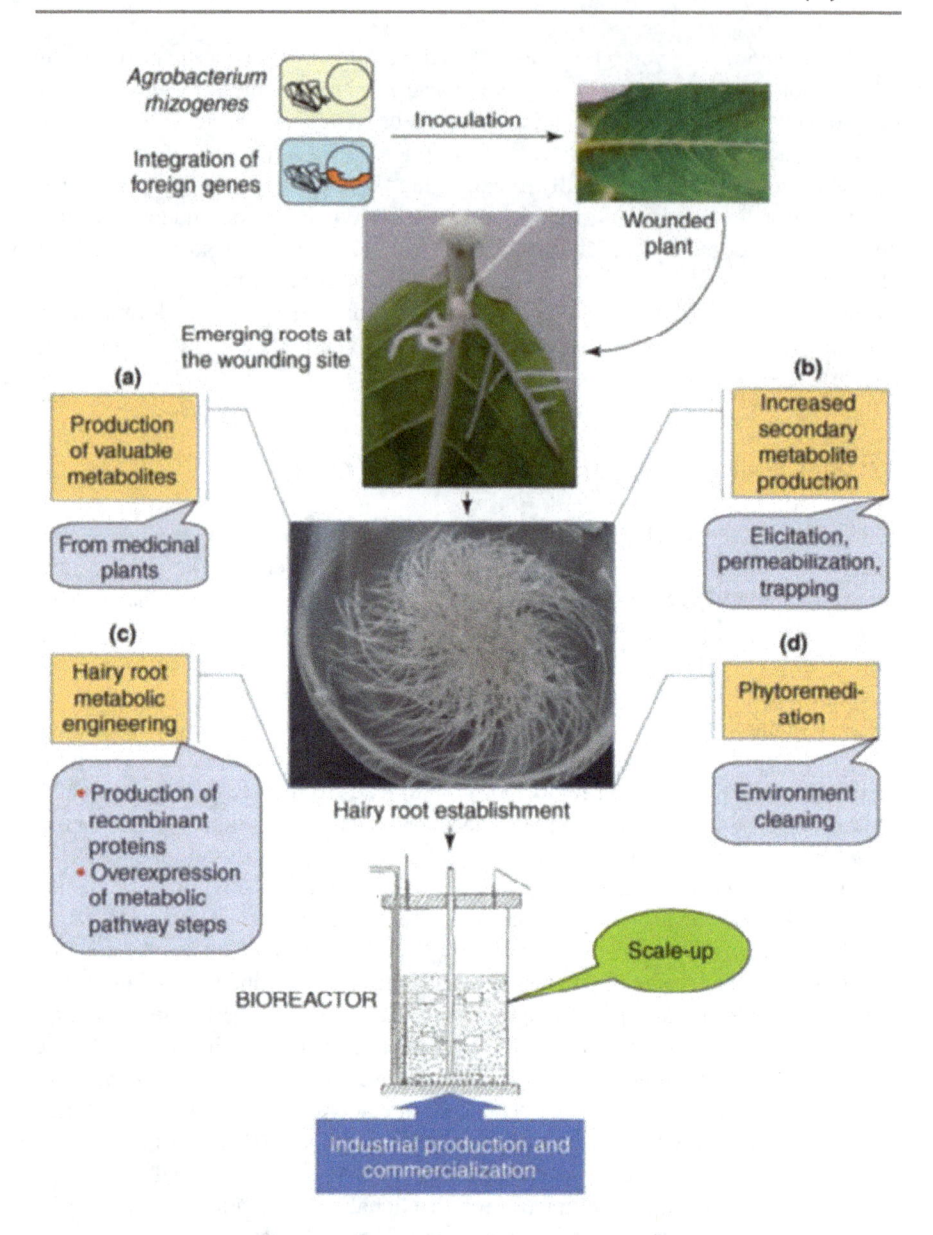

Fig. 5.1 Biotechnological prospects for hairy root research. *Agrobacterium rhizogenes* transfers the T-DNA segment from its large plasmid (Ri) into the plant genome after leaf infection. A few days later, roots emerge from the inoculation site of the leaf. Hairy roots develop in agitated liquid culture. To become an acceptable biotechnological process, hairy root cultures must be scaled up in a bioreactor in order (**a**) to produce valuable metabolites from medicinal plants, (**b**) to increase the production of secondary metabolites by chemical means, (**c**) to introduce foreign genes into plant genomes to produce recombinant proteins or to overexpress proteins that are otherwise limiting to metabolic pathways, and (**d**) to uptake heavy metals from phytoremediation systems. (Guillon et al. 2006; Reproduced with permission from Elsevier publications)

for creation of new metabolic engineering strategies. The methods of gene gain- or loss-of-function and transcriptome analyzes for the detection of new metabolic genes can also be demonstrated via hairy root systems. The hairy root system can be scaled up in bioreactors if they are to be used in industry for mass production of secondary metabolites (Guillon et al. 2008). The most general and prominent perspectives for hairy root culture depend on its genotype and phenotype stability. It is proved that HRCs exhibit high degree of chromosomal stability over a prolonged period of time. It is this stability that ensures its wide usage in research and industrial applications. By analyzing growth, DNA, expression of genes, and production of secondary metabolite, the stability of HRCs can be confirmed.

5.3.3 Effect of *A. rhizogenes* Infection on Plant Tissues

Transferring one or two fragments of T DNA from the root-inducing (Ri) plasmid to the host plant genome allows HR to grow in higher plants. More likely, HR formation from infected cells is controlled by *rolA*, *rolB*, and *rolC* genes. Among these, the most significant locus for HR induction is thought to be *rolB*. Roots were induced by a single gene *rolB*, while *rolC*, ORF13, and Ri plasmid TL DNA ORF14 independently promoted root induction by the rolB gene. The impact on rolB-mediated rooting of these genes were found to be in the order ORF13 > *rolC* < ORF14 (Seishiro and Kunihiko 1999).

The effects of gene expression on the secondary metabolism in the hairy root cultures have been studied. In *Artemisia carvifolia* Buch hairy root cultures, Dilshad et al. (2016) reported that flavonoid accumulation had been influenced by *rol* genes. Those harboring rolB were much more successful among the transgenic plants than the *rolC*-transformants in flavonoid biosynthesis. In *Artemisia duba rolA* transgenic hairy root cultures were reported to be viable and cultivable. This has also been an important source for artemisinin and its derivatives (Amanullah et al. 2016).

5.3.3.1 *RolA* Gene
The *rolA* gene in Ri-plasmid's TL-DNA in *A. rhizogenes* plays an important role in the development of hairy root syndrome in transgenic plants. A reduction in gibberellin A1 content was observed in transgenic tobacco plants expressing *rolA* gene. The transgenic plants were also characterized by dark green wrinkled leaves with a reduced length-width ratio, stunted growth, compact inflorescences, fewer flowers, delayed flowering, and shorter styles (Dehio et al. 1993). The *rolA* protein is also assumed to act as a transcription factor corresponding with gibberellin metabolism leading to a reduced gibberellin level (Meyer et al. 2000). The *rolA* gene was observed to induce rooting in transformed tobacco leaf discs (Serino et al. 1994). The *rolA* gene is also reported to confer tolerance to the fungal plant pathogen, *Fusarium oxysporum* in transgenic tomato plants (Bettini et al. 2016).

5.3.3.2 *RolB* Gene

In transformed plants, the *rol B* gene mainly participate in the induction of adventitious roots. The *rolB* gene codes for the protein beta-glucosidase that is able to hydrolyze indole-beta-glucosides. Estruch et al. (1991) proposed that physiological and developmental changes in transgenic plants expressing the *rolB* gene may be the result of increased intracellular indoleacetic acid (natural auxin) activity by the release of active auxins from inactive β-glucosides. Under in vitro conditions, the adventitious roots induced by the *rolB* gene produce abundant lateral roots (Pistelli et al. 2010). Such findings suggest the critical impact of the *rolB* protein on lateral root formation in plants. Reports indicate that the variations in growth potential between HR lines is due to differences in the degree of rolB gene expression (Tanaka et al. 2001). Earlier studies also indicate that *rolB* is not involved in the control of hormone metabolism. The observed auxin effects may be due to altered hormone stimuli. The *rolB* might function as a transcriptional coactivator/mediator as suggested by Moriuchi et al. (2004). For the production of secondary metabolites, the *rolB* gene-transformed *Vitis amurensis* calli were able to produce up to 3.15% dry weight of resveratrol compared to non-transformed calli. The enhanced production of secondary metabolites was found to be proportional to the abundance of *rolB* mRNA transcripts (Kiselev et al. 2007).

5.3.3.3 *RolC* Gene

The role of *rolC* gene on the modulation of plant growth and development is well documented. It is found that *rolC* directly affects plant size and architecture, such as reduced height and internode length, apical dominance, male fertility, increased number of flowers, and unusual changes in leaf size, shape, and color, thus raising the ornamental value of the plants (Casanova et al. 2005). The differences in color and height depend on various factors like the locus of integration which can be managed to a certain extent through specific promoters, mutation, copy number, changes in expression level of related genes, and somaclonal variations. The effect of *rolC* gene on plant morphology may be due to activity of cytokinin β-glucosidase which increases the cytokinin levels. Work with various classes of secondary metabolites, such as tropane alkaloids and pyridine alkaloids, has shown the *rolC* stimulatory impact on secondary metabolism (Matveeva and Sokornova 2018). Siva et al. (2015) have documented the strong stimulating effect of the *rolC* gene on the synthesis of shikimate-derived anthraquinone phytoalexins in the transgenic callus cultures of *Rubia cordifolia*.

5.3.3.4 *RolD* Gene

The reproductive phase transition in plants is altered by *A. rhizogenes* T-DNA oncogene, *rolD*. Among transgenic tobacco plants, the major traits imparted by the *rolD* gene are early flower setting and a strong enhancement in flowering potential. In *rolD* plants, early flowering accompanied by very fast growth of multiple lateral inflorescences is also observed (Mauro et al. 1996). Recent biochemical studies show that *rolD* encodes an ornithine cyclodeaminase (OCD) capable of catalyzing the NAD^+-dependent ornithine conversion to proline (Trovato et al. 2001). The accumulation of proline might have caused stress signaling to the plants and aided in proline mediated early flowering. An elevated proline level may affect the biosynthesis rate of hydroxyproline-rich glycoproteins (HRGs), which are structural components of the plant cell wall and typically play a key role in controlling cell division, cell extension, and cell wall self-assembly (Varner and Lin 1989). HRGs accumulation can also confer tolerance to a number of biotic and abiotic stresses.

5.3.4 Developments in Secondary Metabolite Production Through Hairy Root Culture

Hairy root culture (HRC) technology provides new possibilities for selected plant varieties to produce secondary plant metabolites in vitro. Compared to cell suspension cultures, hairy roots are genetically and biochemically stable over long periods in culture. The rapid growth of hairy roots is also an additional benefit, even in the absence of growth regulators in a sterile culture medium, to use it as a continuous source for valuable secondary metabolite production. In genetically transformed root cultures, genes encoding the desired secondary metabolites or the transcription factors were overexpressed for their enhanced production. Hence, genetically modified root cultures can be turned into green factories that have biotechnological potential to produce high-value plant-derived metabolites and pharmaceutically important recombinant proteins. Some of the secondary metabolites produced through HRCs are given in Table 5.1.

The extraction of secondary metabolites from plants cannot be economically carried out due to its complex structure and low level of production from the wild genotypes. Hence the synthesis of secondary metabolites in larger quantities is exploited through HRC. Some of the valuable secondary metabolites produced through HRC includes: shikonin, used as an anti-ulcer and anti-bacterial agent; L-DOPA, a precursor to dopamines, an essential neurotransmitter used in the treatment of Parkinson's disease; opiate alkaloids, mainly codeine and morphine alkaloids used for medical purposes; anthraquinone used for dyes and medicinal purposes; ginsenosides for medicinal purposes; berberine, an alkaloid used for cholera and bacterial dysentery; rosmarinic acid used for medicinal purposes; cardioactive or cardenolide glycosides used to cure heart diseases; valepotriates used as a sedative; and quinine used in the treatment of malaria (Korde et al. 2016).

Tropane alkaloids like (−) hyoscyamine, scopolamine (hyoscine), and atropine are widely used as anticholinergic and antispasmodic agents which affect the parasympathetic nervous system. The hairy roots of *Datura metel*, an important medicinal plant, is used for the production of atropine, a tropane alkaloid. For the overproduction of atropine, Shakeran et al. (2015) transformed *Datura metel* leaf segments with *A. rhizobium* A4, and to further increase the yield of atropine, biotic and abiotic elicitors were added to the hairy root cultures. The biomass accumulation and production of atropine was enhanced by all the elicitors. Among the elicitors

Table 5.1 Some of the valuable secondary metabolites produced through hairy root cultures

Secondary metabolite	Source (plant species and family)	Medicinal uses	Reference
Ajmalin Ajmalicine	*Rauvolfia micrantha* (Apocyanaceae)	Antihypertensive	Lorence et al. (2004)
Artimisinin	*Artemisia annua* (Asteraceae)	Treatment of fever, malaria	Patra and Srivastava (2014)
Campothecin	*Campotheca acuminate* (Nyssaceae)	Anticancer, antiviral	Lorence et al. (2004)
Ginsenosides	*Panax ginseng* (Araliaceae)	Anticancer, antioxidant, prevents cardiovascular risk	Jeong et al. (2005)
Ginkolides	*Ginko biloba* (Ginkgoaceae)	Ageing disorders	Ayadi and Tremouillaux (2003)
Jaceosidin	*Saussurea medusa* (Asteraceae)	Antitumorous	Zhao et al. (2004)
Morphine	*Papaver somniferum* (Papaveraceae)	Sedative, analgesic	Le Flem et al. (2004)
Phenolic acids	*Salvia miltiorrhiza* (Lamiaceae)	Anti-oxidant, anti-inflammatory, and anticancer	Kai et al. (2019)
Podophyllotoxin	*Linum flavum* (Linaceae)	Anticancer, inhibit the *Herpes simplex* type I virus replication	Malik et al. (2014)
Puerarin	*Pueraria phaseolides* (Fabaceae)	Hypothermic, spasmolytic, hypotensive, antiarrhythmic	Shi and Kintzois. (2003)
Saponin gypenoside	*Gynostemma pentaphyllum* (Cucurbitaceae)	Pharmacological activities	Cui et al. (1999)
Verbascoside	*Gmelina arborea* (Lamiaceae)	Fevers, skin problems and stomach disorders	Dhakulkar et al. (2005)
Withanolide	*Withania somnifera*	Anti-arthritis, antioxidant, anti-inflammatory	Murthy et al. (2008)

tested, nanosilver has been most successful in improving the atropine content of the hairy roots. Habibi et al. (2015) induced hairy root system in *Atropa belladonna* to increase the scopolamine content. The maximum amount of scopolamine (1.59 mg/ g^{-1} dry wt) was obtained in the bioreactor with an aeration of 1.25 vvm (volume per volume per minute) and agitation at 70 rpm.

Sanguinarine (SAN) is a quaternary benzylisoquinoline alkaloid which has been used in livestock production for many years as a natural growth promoter (NGP) and an alternative to antibiotics. It has recently been proven to have potential applications in the treatment of schistosomiasis and osteoarthritis. This chemical also reported to have antitumor, antimicrobial, and anti-inflammatory functions. The most important commercial source of SAN is *Macleaya cordata*, a traditional medicinal herb that belongs to the family *Papaveraceae*. Huang et al. (2018) stimulated hairy roots in *M. cordata* with *A. rhizogenes* to obtain a new source of sanguinarine. It was found that, contents of sanguinarine and dihydrosanguinarine produced by hairy roots were much higher when compared to wild-roots. Accordingly, this study showed that the hairy root system has additional potential for commercial bioengineering and sustainable sanguinarine production.

Owing to the wide therapeutic potential, resveratrol (3,5,4′-trihydroxy-trans-stilbene) (t-R) is among the most studied stilbenes. There have been detailed studies of its antioxidant, antiviral, anti-inflammatory, cardioprotective activity as well as anti-aggregation platelet and melanoma chemoprevention activities. The t-piceatannol derivative of hydroxylated t-R has a strong anticancer effect. Hairy root (HR) cells of tobacco have an inherent capacity to transform exogenous t-resveratrol (t-R) into t-piceatannol (t-Pn.). Hidalgo et al. (2017) developed a method for the bioconversion of exogenous t-resveratrol into piceatannol in amounts close to mg L^{-1} by transferring the human cytochrome P450 hydroxylase 1B1 (HsCYP1B1) gene to tobacco hairy roots. Similarly, the exogenous t-resveratrol was bioconverted into pterostilbene following heterologous expression of resveratrol O-methyl transferase from *Vitis vinifera* (VvROMT) in tobacco hairy roots.

Withania somnifera (L.) Dunal (*Solanaceae*) is an essential medicinal plant that produces pharmaceutically active compounds in its roots, called withanolides. It is one of the best known and most explored plants in traditional Ayurveda and holds its importance parallel to Ginseng in Chinese therapy (Abraham and Thomas 2017). Withaferin A is known for its anticancer properties and is reported to inhibit the growth of different lines of human cancer cells. It is a neotropic agent for recovery from nerve degeneration too. Therefore, it may be a promising compound for the treatment of diseases associated with neuronal degeneration such as Parkinson's disease and Alzheimer's. *W. somnifera* plants were transformed with *rolB* genes through *A. rhizogenes* strain R1601 mediated genetic transformation and hairy roots were induced from cotyledons and leaf explants (Murthy et al. 2008). The transformed hairy roots producing withanolideA has been reported and the concentration was 2.7-fold more than in untransformed cultured roots. Genetic transformation mediated by *A. rhizogenes* for hairy root induction has also been documented in *W. somnifera* by Saravanakumar et al. (2012). The accumulation of the phytochemical withaferin-A was quantified in the hairy roots and obtained 72.3 mg/g dry weight of the tissue.

Diosgenin is an important intermediate substance for the synthesis of many steroid hormones. For the optimization and overproduction of diosgenin, hairy root cultures of *Trigonella foenum-graecum* L. were established with *A. rhizogenes* strain A4 (Merkli et al. 1997). The highest concentration of diosgenin (0.040% dry weight) was observed when half-strength WP medium with 1% sucrose was used. The amount was almost twice that of detected in the 8-month-old un-transformed roots (0.024%). Almost 17 μg of diosgenin/g fresh weight was observed at optimum conditions. Adding 40 mg/L of chitosan as an elicitor increased the content of diosgenin to three times to that found in hairy roots that were non-elicited.

Sharafi et al. (2014) developed an efficient *A. rhizogenes*-mediated transformation systems for *Tribulus terrestris* L., an important medicinal plant, for the isolation of β-carboline alkaloids. From the cut edges of leaf explants, hairy roots were started appearing directly within 10–14 days after inoculation. The highest frequency transformation was 49%, which was accomplished using *A. rhizogenes* strain AR15834 in hormone-free MS medium after 28 days of inoculation. To detect ß-carboline alkaloids, isolated control and transgenic hairy roots grown in liquid media containing IBA were analyzed. At the end of 50 days of culture, harmine content was measured at 1.7 $\mu g/g^{-1}$ of the dried weight of transgenic hairy root cultures.

5.3.5 Effect of Elicitors on Secondary Metabolite Production in HRCs

Elicitors are substances that either initiate or enhance the biosynthesis of specific compounds when introduced in very small concentrations to a living cell system. The use of elicitors for improving the production of secondary metabolites in hairy root cultures has been carried out by several research groups. Factors responsible for enhancing the secondary metabolite production include elicitation with the right compound in the optimum concentration at the right culture stage. Selection of the right culture medium, concentration of the media components, and the culture conditions are also important factors. It is reported that the elicitor treatment results in higher yields of biomass as well as secondary metabolites if given in the late log phase, while in the early phase, this leads to an immediate increase in the accumulation of secondary metabolites but decreased biomass yield (Shilpa et al. 2010). It is also important to consider the signals involved in the elicitation process in order to select the appropriate elicitor.

In the last three decades, secondary metabolite accumulation in vitro or their efflux in the culture medium has been induced in the undifferentiated or differentiated tissue cultures of many plant species by applying a low concentration of biotic and/or abiotic elicitors. Some commonly used abiotic elicitors are jasmonic acid, methyl jasmonate, and chemicals like NaCl, $CaCl_2$, and KCl. Fungi like *Aspergillus niger*, *Alternaria* sp., and *Fusarium monoliformae*; bacterium like *Enterobacter sakazakii*; and Yeast extract are some biotic elicitors used for enhancing synthesis of secondary metabolites. Due to their synergistic effect, combined application of different elicitors, integration of precursor feeding or replenishment of medium or in situ product recovery from the roots/liquid medium has shown an improvement in secondary metabolite production (Halder et al. 2019). The application of elicitation in HRCs is highly valued due to the high growth rate in growth regulator-free medium, genetic and secondary metabolite production stability, and consistency in response to elicitor treatment (Table 5.2).

Table. 5.2 Metabolites produced in hairy roots with elicitors

Elicitors—abiotic/biotic	Plant species	Metabolites	References
Enterobacter sakazakii	*Ammi majus*	Scopoletin	Staniszewska et al. (2003)
Rhizoctonia bataticola	*Solanum tuberosum*	Sesquiterpene	Komaraiah et al. (2003)
Yeast extract, $AgNO_3$	*Salvia miltiorrhiza*	Tanshinones	Ge and Wu (2005)
Chitosan, Methyl jasmonate	*Panax ginseng*	Ginsenosides	Palazon et al. (2003)
Tween 80	*Beta vulgaris*	Betalaines	Thimmaraju et al. (2003)
Salicylic acid(SA), Methyl jasmonate	*Angelica gigas*	Decursin and Decursinol angelate	Rhee et al. (2010)
Phytopthora parasitica	*Cichorium intybus*	Esculin and Esculetin	Bais et al. (2000)
Yeast elicitor	*Salvia miltiorrhiza*	Rosmarinic acid	Chen et al. (2001)
Aspergillus niger, Alternaria sp., *Fusarium monoliformae*, yeast extract	*Datura metel* L.	Scopolamine	Ajungla et al. (2009)
Phenylalanine, chitosan, salicylic acid	*Psoralea corylifolia*	Daidzein, genistein	Lo et al. (2007)
Glycosphingolipids	*Artemisia annua*	Artimesinin	Wang et al. (2009)

Several studies have been reported using both biotic/abiotic elicitors to enhance production of valuable secondary metabolites in hairy roots. In *Centella asiatica* hairy roots, enhanced asiaticoside (7.12 mg g^{-1} DW) accumulation was recorded when 0.1 mM of methyl jasmonate (MJ) was added to the culture medium as an elicitor for 3 weeks (Kim et al. 2007). Pectic fragments released from the plant cell walls like oligogalacturonide (OGA) also exhibit specific elicitor activity to enhance secondary metabolite production (Lu et al. 2008). The isolated fraction OGA2 induced the accumulation of artemisinin in the hairy roots of *Artemisia annua*. A 55.2% increase in artemisinin over the control was observed when 16-day-old hairy root cultures were supplied with OGA elicitor for 4 days. According to Zhang et al. (2010), the reactive oxygen species (ROS) induced by OGA stimulated the biosynthesis of artemisinin in the hairy roots. In hairy root cultures of *Beta vulgaris* and *Tagetes patula* aqueous extracts of the green algae *Haematococcus pluvialis* and blue green alga *Spirulina platensis* increased the accumulation of betalains and thiophenes (Ramachandra Rao et al. 2001). In hairy root cultures treated with *Haematococcus pluvialis*, the accumulation of betalains showed a 2.28-fold increase on the 15th day compared with the control. Similarly, *Streptomyces platensis* extract treated hairy roots showed increased betalain development by 1.16-fold on 25th day over control. Likewise, on the 20th day an increased accumulation of thiophene by

1.2-fold was observed in cultures treated with *H. pluvialis* extract, over the untreated control. Putalun et al. (2007) found a sixfold increase in artemisinin production in the hairy roots of *Artemissia annua* by adding chitosan and yeast extract. Wang et al. (2009) observed an increased concentration of artemisinin stimulated by a fungal derived cerebroside, which is a glycosphingolipid, in the hairy roots. Zabetakis et al. (1999) researched the effect of elicitors like methyl jasmonate, fungal elicitor (yeast cell wall), and oligogalacturonides. Methyl jasmonate was found to be the most effective for the enhanced synthesis of tropane alkaloids (littorine, hyoscyamine, scopolamine). The role of biotic (yeast extract) and abiotic ($CaCl_2$, $AgNO_3$, $CdCl_2$) elicitors on accumulation and secretion of hyoscyamine and scopolamine in *Brugmansia candida* hairy root cultures was studied by Pitta-Alvarez et al. (2000). The release of scopolamine was preferentially increased when $AgNO_3$ and yeast extract was used in the medium. Singh et al. (1998) showed that when hairy roots of *Hyoscyamus muticus* were elicited with methyl jasmonate, it favored the production of sesquiterpenes early on the lubimin pathway. However, production of lubimin was favored as end product when fungal elicitors were used.

Srivastava et al. (2006) noticed a 14.8-fold increase of ajmalacine production after 1 week of treatment with 100 mM concentration of NaCl in the hairy root cultures of *Rauwolfia serpentina*. To induce hyoscyamine in elevated levels in *Datura*, potassium chloride and calcium chloride were used as elicitors (Harfi et al. 2016). They also identified that elicitors can also be used as single or in combination with polysaccharides (yeast extracts, chitosan), heavy metal ions (cobalt, silver, cadmium), and signal compounds (salicylic acid) for the enhanced production of secondary metabolites through HRC. Wang et al. (2016) observed that the maximum production of tanshinone reached to 4.9-fold increase over the control when 18-day-old *Salvia miltiorrhiza* hairy root cultures were exposed to methyl jasmonate and ultraviolet B irradiation.

5.3.6 Bioreactors for Large-Scale Production of Secondary Metabolites Through HRC

Industrial scale-up in the production of plant secondary metabolites can be done using bioreactors designed for the purpose, which meets the unstable productivity of plant cells, sensitivity, minimum oxygen requirements, and slow growth rate. The scale-up using bioreactors involves bioreactors of distinct sizes and features. The major attraction behind the use of hairy root technology for the production of important secondary metabolites is their genetic stability, higher biomass production, and enhanced product accumulation. However, a major challenge to produce it in large scale and utilization of this technology commercially is the lack of suitable bioreactor configurations by which the mass production of hairy roots and secondary metabolite accumulation can be easily scaled up and expedited. Bioreactors are self-contained units intended for optimization and monitoring as well as to provide homogeneous culture conditions like pH, air circulation, dissolved gasses and temperature for mass propagation of cells, tissues, organogenic propagules, or

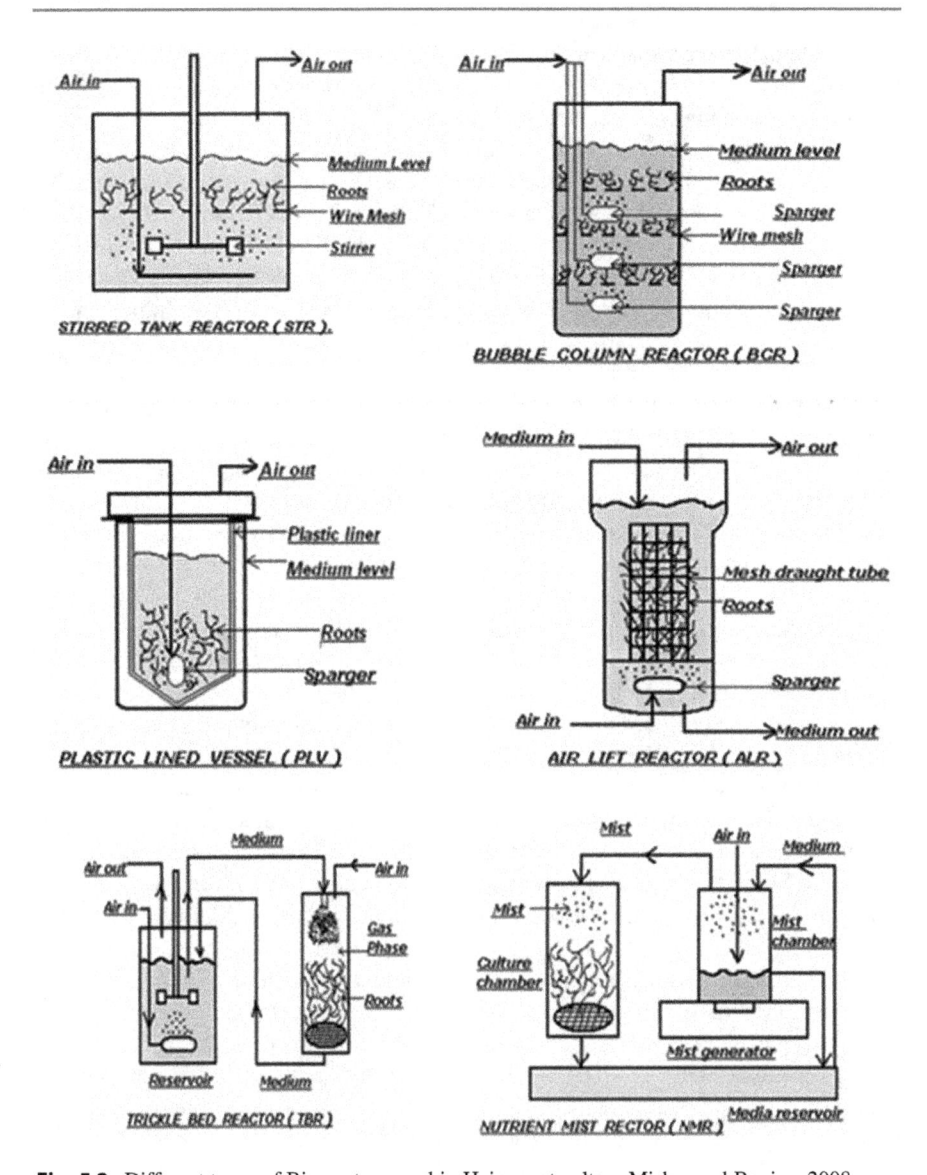

Fig. 5.2 Different types of Bioreactors used in Hairy root culture Mishra and Ranjan, 2008

somatic embryos in a sterile environment (Stiles and Liu 2013). In the event of bioreactor-mediated scale-up of secondary metabolites, various factors like the optimization of culture conditions, measurement of biomass (tissue and organ cultures), etc. should be considered (Ruffoni et al. 2010) (Fig. 5.2).

The HRCs are very delicate, fragile, and very sensitive in nature. These roots respond to even the slight changes in the culture conditions like temperature, pH, and shear stress. Variations in the morphology of hairy roots in its density, thickness, and root length also have a major impact on the accumulation of secondary metabolites (Srivastava and Srivastava 2012). The dynamic fibrous nature of hairy root growth and its fragile nature present unique limitations for scaling up in bioreactors. In culture, the hairy roots form an interlocked network that prevents mass transfer of nutrients and oxygen, resulting in a nonhomogeneous crop ecosystem from which a collection of senescent tissues get emerged (Vashishta and Sharma 2015).

For mass production of hairy roots, bioreactors in different configurations have been developed by various research groups worldwide. An optimized bioreactor for hairy root production should show the improved characteristics like: impose minimum hydrodynamic shear stress, maintenance of a sterile and homogeneous culture environment for steady and continuous production of the desired metabolite, efficient arrangements for light and CO_2 supply for photosynthesis if the cultures are mixotrophic or phototrophic, uniform distribution of oxygen and nutrients, and it should provide the need to act as a support matrix for roots. A major challenge is the designing of a reactor with all the necessary properties for the large-scale maintenance of hairy roots (Khan et al. 2018). Bioreactors for hairy roots are divided into three main groups, depending on the continuous phase used: liquid-phase reactors, gas-phase reactors, and hybrid reactors combining the two (Kim et al. 2002). The hairy root biomass is fully submerged in the liquid media in the liquid-phase reactors and hence the word "submerged reactors" is used as well. Mass transmission of gaseous media in liquid-phase reactors is rate-limiting. Some variants of liquid-phase reactors include stirred tank reactors, bubble column reactors, air-lift reactors, submerged convective flow reactors, etc. The impeller damaged the roots in early studies with stirred tank reactors, resulting in the development of callus and ultimately low biomass production (Hilton et al. 1988), and this problem was later solved by using a steel cage or mesh to separate impeller from roots (Hilton and Rhodes 1990). The hairy root biomass is not immersed in liquid in the gas-phase reactors, and the roots are exposed to air and a combination of fluid and air media. Some examples of gas-phase reactors are trickle bed reactor, droplet phase reactor, nutrient mist reactor, etc. In gas-phase reactors, liquid-phase mass transfer of oxygen and other nutrients is substantially increased (Khan et al. 2018). However, a major issue with gas-phase reactors is the uniform distribution of root tips in the reactor. This problem can be solved by initially growing the root tips in the liquid-phase method until the root tips are evenly distributed in the packing matrix and then running the gas-phase reactor (Ramakrishnan et al. 1994).

5.4 Conclusions

Secondary plant metabolites are generally defined as natural products that are synthesized by plants that are not necessary to sustain their growth and development. It is estimated that more than 200,000 distinct chemical compounds are formed as secondary metabolites by the plant kingdom, and most of them come from specialized metabolic activities. Most of these compounds play an important role in the competition and protection of interspecies among plants. Many plant-natural products are used as nutrients, medicines, fragrances, flavors, colorants, repellents etc. Since plants are a very rich source of these bioactive molecules, the consumption of herbal medicines and medicinal plants is widespread and increasing. The medicinal plants grow naturally, and the natural and wild fields form the key source of raw material. Habitat degradation and the loss of genetic diversity are threats for the availability of herbal medicines. Protection of environments overcomes cultivation problems and thus it enables plant engineering for the production of several bioactive compounds. Significant progress in the use of tissue culture and genetic transformation has been made in recent years to alter pathways for targeting metabolite biosynthesis. Hairy roots are unique in their regulation of genes and biosynthesis of metabolites. Their rapid growth provides further advantage to be used as a continuous source for a wide range of valuable secondary metabolite production. The effects of oncogenes of *A. rhizogenes* on plant morphogenesis and modulation of metabolic pathways have only recently been explained. In this chapter, HRC as a promising tool for secondary metabolites is explained. Current advances in hairy root culture techniques could enhance the production of secondary metabolites in vitro. Together with HRC, the use of bioreactors and genetic and metabolic engineering offers a promising area in improving the *in planta* production of secondary metabolites. In addition, HRC-followed heterologous in planta processing appears to be more cost-effective and environmentally sustainable than other emerging biotechnology platforms. Advances in multigene transformation, transcription factors, and targeting of cellular compartment techniques will allow for higher output levels in future engineered plants that will bring us closer to secondary metabolite development on an industrial scale.

References

Abraham J, Thomas D (2017) Hairy root culture for the production of useful secondary metabolites. In: Malik S (ed) Biotechnology and production of anti-cancer compounds. Springer International Publishing AG, Cham, pp 201–230

Ajungla L, Patil PP, Barmukh RB, Nikam TD (2009) Influence of biotic and abiotic elicitors on accumulation of hyoscyamine and scopolamine in root cultures of *Datura metel* L. Indian J Biotechnol 8:317–322

Amanullah BM, Rizvi Z, Zia M (2016) Production of artemisinin and its derivatives in hairy roots of *Artemisia dubia* induced by rolA gene transformation. Pak J Bot 48:699–706

Ayadi R, Tremouillaux-Guiller J (2003) Root formation from transgenic calli of Ginkgo biloba. Tree Physiol 23(10):713–718

Bais HP, Govindaswamy S, Ravishankar GA (2000) Enhancement of growth and coumarin production in hairy root cultures of Witloof chicory (*Cichorium intybus* L. cv. Lucknow local) under the influence of fungal elicitors. J Biosci Bioeng 90:648–653

Balandrin MF, Klocke JA, Wurtele ES, Bollinger WH (1985) Natural plant chemicals: sources of industrial and medicinal materials. Science 228:1154–1160

Baranski R (2008) Genetic transformation of carrot (*Daucus carota*) and other Apiaceae species. Transgenic Plant J 2(1):18–31

Berlin J, Beier H, Fecker L, Forche E, Noé W, Sasse F, Schiel O, Wray V (1985) Conventional and new approaches to increase the alkaloid production of plant cell cultures. In: Neumann KH,

Barz W, Reinhard E (eds) Primary and secondary metabolism of plant cell cultures. Springer, Berlin, pp 272–280

Bettini PP, Santangelo E, Baraldi R, Rapparini F, Mosconi P, Mauro ML (2016) *Agrobacterium rhizogenes rolA* gene promotes tolerance to *Fusarium oxysporum* f. sp. *lycopersici* in transgenic tomato plants (*Solanum lycopersicum* L.). J Plant Biochem Biotechnol 25:225–233

Bhartendu NM, Ritu R (2008) Growth of hairy-root cultures in various bioreactors for the production of secondary metabolites. Biotechnol Appl Biochem 49:1–10

Casanova E, Trillasa MI, Moysseta LR et al (2005) Influence of *rol* genes in floriculture. Biotechnol Adv 23:3–39

Chen H, Chen F, Chiu FCK, Lo CMY (2001) The effect of yeast elicitor on the growth and secondary metabolism of hairy root cultures of *Salvia miltiorrhiza*. Enzym Microb Technol 28:100–105

Chilton MD, Tepfer DA, Petit A, David C, Delbart F, Tempe J (1982) *Agrobacterium* rhizogenes inserts T-DNA into the genomes of the host plant root cells. Nature 295:432–434

Cui JF, Eneroth P, Bruhn J (1999) *Gynostemma pentaphyllum*: identification of major sapogenins and differentiation from Panax species. Eur J Pharm Sci 8:187–191

Dehio C, Grossmann K, Schell J, Schmulling T (1993) Phenotype and hormonal status of transgenic tobacco plants over- expressing the rolA gene of *Agrobacterium rhizogenes* T-DNA. Plant Mol Biol 23:1199–1210

Dhakulkar S, Ganapathi TR, Bhargava S, Bapat VA (2005) Induction of hairy roots in *Gmelina arborea* Roxb. and production of verbascoside in hairy roots. Plant Sci 169:812–818

Dilshad E, Ismail H, Haq I, Cusido RM, Palazon J, Estrada KM, Mirza B (2016) *Rol* genes enhance the biosynthesis of antioxidants in *Artemisia carvifolia* Buch. BMC Plant Biol 16:125–133

Estruch JJ, Schell J, Spena A (1991) The protein encoded by rolB plant oncogene hydrolyses indole glucosides. EMBO J 10:3125–3128

Flores HE, Vivanco JM, Loyola-Vargas VM (1999) Radicle biochemistry: the biology of root-specific metabolism. Trends Plant Sci 4:220–226

Fraenkel GS (1959) The raison d'Être of secondary plant substances these odd chemicals arose as a means of protecting plants from insects and now guide insects to food. Science 129 (3361):1466–1470

Ge XC, Wu JY (2005) Tanshinone production and isoprenoid pathways in Salvia miltiorrhiza hairy roots induced by Ag^+ and yeast elicitor. Plant Sci 168:487–491

Giri A, Narasu ML (2000) Transgenic hairy roots. Recent trends and application. Biotechnol Adv 18(1):1–22

Giri A, Banerjee S, Ahuja PS, Giri CC (1997) Production of hairy roots in *Acomtum heterophyllum* wall. Using *Agrobacterium rhizogenes*. In Vitro Cell Dev Biol Plant 33:280–284

Guillon S, Tre'mouillaux-Guiller J, Pati PK, Rideau M, Gantent P (2006) Hairy root research: recent scenario and exciting prospects. Curr Opin Plant Biol 9(3):341–346

Guillon S, Trémouillaux-Guiller J, Kumar PP, Gantet P (2008) Hairy roots: a powerful tool for plant biotechnological advances. In: Ramawat K, Merillon J (eds) Bioactive molecules and medicinal plants. Springer, Berlin

Habibi P, Piri K, Eljo A, Moghadam YA, Ghiasvand T (2015) Increasing scopolamine content in hairy roots of *Atropa belladonna* using bioreactor. Braz Arch Biol Technol 58:166–174

Halder M, Sarkar S, Jha S (2019) Elicitation: a biotechnological tool for enhanced production of secondary metabolites in hairy root cultures. Eng Life Sci 19(12):880–895

Harfi B, Khelifi-Slaoui M, Bekhouche M, Benyammi R, Hefferon K, Makhzoum A, Khelifi L (2016) Hyoscyamine production in hairy roots of three *Datura* species exposed to high-salt medium. In Vitro Cell Dev Biol Plant 52:92–98

Hidalgo D, Martínez-Márquez A, Moyano E, Bru-Martínez R, Corchete P, Palazon J (2017) Bioconversion of stilbenes in genetically engineered root and cell cultures of tobacco. Sci Rep 7:45331

Hilton MG, Rhodes MJC (1990) Growth and hyoscyamine production of hairy root cultures of Datura stramonium in a modified stirred tank reactor. Appl Microbiol Biotechnol 33:132–138

Hilton MG, Wilson PDG, Robins RJ, Rhodes MJC (1988) Transformed root cultures—fermentation aspects. In: Robins RJ, Rhodes MJC (eds) Manipulating secondary metabolism in culture. Cambridge University Press, Cambridge, pp 239–245

Huang P, Xia L, Liu W, Jiang R, Liu X, Tang Q, Xu M, Yu L, Tang Z, Zeng J (2018) Hairy root induction and benzylisoquinoline alkaloid production in *Macleaya cordata*. Sci Rep 8:11986. https://doi.org/10.1038/s41598-018-30560-0

Jeong GT, Park DH, Ryu HW et al (2005) Production of antioxidant compounds by culture of *Panax ginseng* C.A. Meyer hairy roots: I. Enhanced production of secondary metabolite in hairy root cultures by elicitation. Appl Biochem Biotechnol 121-124:1147–1157

Kai G, Liu S, Shi M, Han B, Hao X, Liu Z (2019) Biochemistry, biosynthesis, and medicinal properties of phenolic acids in *Salvia miltiorrhiza*. In: Lu S (ed) The *Salvia miltiorrhiza* genome. Compendium of plant genomes. Springer, Cham

Khan SA, Siddiqui MH, Osama K (2018) Bioreactors for hairy root culture: a review. Curr Biotechnol 7:417–427

Kim Y, Wyslouzil BE, Weathers P (2002) Secondary metabolism of hairy root cultures in bioreactors. In Vitro Cell Dev Biol Plant 38:1–10

Kim OT, Bang KH, Shin YS (2007) Enhanced production of asiaticoside from hairy root cultures of *Centella asiatica* (L.) urban elicited by methyl jasmonate. Plant Cell Rep 26:1941–1949

Kiselev KV, Dubrovina AS, Veselova MV et al (2007) The rolB gene-induced overproduction of resveratrol in *Vitis amurensis* transformed cells. J Biotechnol 123:681–692

Klee HJ, Romano CP (1994) The roles of phytohormones in development as studied in transgenic plants. Crit Rev Plant Sci 13(4):311–324

Komaraiah P, Reddy GV, Reddy PS, Raghavendra AS, Ramakrishna SV, Reddanna P (2003) Enhanced production of antimicrobial sesquiterpenes and lipoxygenase metabolites in elicitor-treated hairy root cultures of *Solanum tuberosum*. Biotechnol Lett 25:593–597

Korde N, Zhang Y, Loeliger K, Poon A, Simakova O, Zingone A, Costello R, Childs R, Noel P, Silver S, Mok M, Mo C, Young N, Landgren O, Soland E, Maric I (2016) Monoclonal gammopathy associated pure red cell aplasia. Br J Haematol 173:876–883

Kumar V, Sharma A, Prasad BCN, Gururaj HB, Ravishankar GA (2006) *Agrobacterium rhizogenes* mediated genetic transformation resulting in hairy root formation is enhanced by ultrasonication and acetosyringone treatment. Electron J Biotechnol 9(4):349–357

Le Flem-Bonhomme V, Laurain-Mattar D, Fliniaux MA (2004) Hairy root induction of *Papaver somniferum* var. album, a difficult-to transform plant by *A. rhizogenes* LBA 9402. Planta 218:890–893

Lo FH, Mak NK, Leung KN (2007) Studies on the anti-tumor activities of the soy isoflavone daidzein on murine neuroblastoma cells. Biomed Pharmacother 61:591–595

Lorence A, Medina-Bolivar F, Nessler CL (2004) Camptothecin and 10-hydroxycamptothecin from *Camptotheca acuminate* hairy roots. Plant Cell Rep 22:437–441

Lu H, Zhao XM, Bai XF, Du YG (2008) Primary study on resistance induced by oligogalacturonides to tobacco mosaic virus. Plant Prot 34:38–41

Malik S, Bilba O, Gruz J, Arroo RRJ, Strnad M (2014) Biotechnological approaches for producing aryltetralin lignans from *Linum* species. Phytochem Rev 13:893–913

Matveeva TV, Sokornova SV (2018) Agrobacterium rhizogenes-mediated transformation of plants for improvement of yields of secondary metabolites. In: Pavlov A, Bley T (eds) Bioprocessing of plant in vitro systems. Reference series in Phytochemistry. Springer, Cham

Mauro ML, Trovato M, De Paolis A et al (1996) The plant oncogene rolD stimulates flowering in transgenic tobacco plants. Dev Biol 180:693–700

Merkli A, Christen P, Kapetanidis I (1997) Production of disogenin by hairy root cultures by *Trigonella foenum graecum*. Plant Cell Rep 16:632–636

Meyer A, Tempe J, Costantino P (2000) Hairy root; a molecular overview. Functional analysis of *Agrobacterium rhizogenes* T-DNA genes. In: Stacey G, Keen NT (eds) Plant microbe interactions. APS Press, St. Paul, pp 93–139

Mishra BN, Ranjan R (2008) Growth of hairy-root cultures in various bioreactors for the production of secondary metabolites. Biotechnol Appl Biochem 49:1–10

Moriuchi H, Okamoto C, Nishihama R et al (2004) Nuclear localization and interaction of rolB with plant 14-3-3 proteins correlates with induction of adventitious roots by the oncogene rolB. Plant J 38:260–275

Murthy HN, Dijkstra C, Anthony P, White DA, Davey MR, Power JB, Hahn EJ, Paek KY (2008) Establishment of *Withania somnifera* hairy root cultures for the production of withanolide A. J Int Plant Biol 50:975–981

Nguyen C, Bourgaud F, Forlot P, Guckert A (1992) Establishment of hairy root cultures of *Psoralea* species. Plant Cell Rep 11:424–427

Palazon J, Mallol A, Eibl R, Lettenbauer C, Cusido RM, Pinol MT (2003) Growth and ginsenside production in hairy root cultures of *Panax ginseng* using a novel bioreactor. Planta Med 69:344–349

Patra N, Srivastava AK (2014) Enhanced production of artemisinin by hairy root cultivation of *Artemisia annua* in a modified stirred tank reactor. Appl Biochem Biotechnol 174:2209–2222

Petrovska BB (2012) Historical review of medicinal plants usage. Pharmacogn Rev 6:1–5

Pistelli L, Giovannini A, Ruffoni B, Bertoli A, Pistelli L (2010) Hairy root cultures for secondary metabolites production. In: Giardi MT, Rea G, Berra B (eds) Bio-farms for nutraceuticals. Advances in experimental medicine and biology, vol 698. Springer, Boston, MA, pp 167–184

Pitta-Alvarez SI, Spollansky TC, Giulietti AM (2000) The influence of different biotic and abiotic elicitors on the production and profile of tropane alkaloids in hairy root cultures of *Brugmansia candida*. Enzyme Microb Technol 26(2–4):252–258

Putalun W, Luealon W, De-Eknamkul W, Tanaka H, Shoyama Y (2007) Improvement of artemisinin production by chitosan in hairy root cultures of *Artemisia annua* L. Biotechnol Lett 29:1143–1146

Ramachandra Rao S, Tripathi U, Suresh B, Ravishankar GA (2001) Enhancement of secondary metabolite production in hairy root cultures of *Beta vulgaris* and *Tagetes patula* under the influence of micro algal elicitors. Food Biotechnol 15(1):35–46

Ramakrishnan D, Salim J, Curtis WR (1994) Inoculation and tissue distribution in pilot-scale plant root culture bioreactors. Biotechnol Tech 8(9)

Ranganath RR (2016) Hairy roots production through agrobacterium rhizogenes genetic transformation from Daucus carota explants. Int J Adv Res Biol Sci 3(8):23–27

Rhee HS, Cho HW, Son SY, Yoon SYH, Park JM (2010) Enhanced accumulation of decursin and decursinol angelate in root cultures and intact roots of *Angelica gigas* Nakai following elicitation. Plant Cell Tissue Organ Cult 101:295–302

Ruffoni B, Pistelli L, Bertoli A, Pistelli L (2010) Plant cell cultures: bioreactors for industrial production. In: Giardi MT, Rea G, Berra B (eds) Bio-farms for nutraceuticals, vol 698. Springer, Boston, MA, pp 203–221

Samuni-Blank M, Izhaki I, Dearing MD, Gerchman Y, Trabelcy B, Lotan A, Karasov WH, Arad Z (2012) Intraspecific directed deterrence by the mustard oil bomb in a desert plant. Curr Biol 22 (13):1218–1220

Saravanakumar A, Aslam A, Shajahan A (2012) Development and optimization of hairy root culture systems in *Withania somnifera* (L.) Dunal for withaferin-A production. Afr J Biotechnol 11:16412–16420

Saurabh P, Manila B, Niraj T, Sonal P, Bansal YK (2015) Secondary metabolites of plants and their role: overview. Curr Trends Biotechnol Pharm 9(3):293–304

Seishiro A, Kunihiko S (1999) Synergistic function of rolB, rolC, ORF13 and ORF14 of TL-DNA of agrobacterium rhizogenes in hairy root induction in Nicotiana tabacum. Plant Cell Physiol 40 (2):252–256

Serino G, Clerot D, Brevet J, Costantini P, Cardarelli M (1994) *rol* genes of *Agrobacterium rhizogenes* cucumopine strain: sequence, effects and pattern of expression. Plant Mol Biol 26:415–422

Sevon N, Oksman-Caldentey KM (2002) *Agrobacterium rhizogenes*-mediated transformation: root cultures as a source of alkaloids. Planta Med 68:859–868

Shakeran Z, Keyhanfar M, Asghari G, Ghanadian M (2015) Improvement of atropine production by different biotic and abiotic elicitors in hairy root cultures of *Datura metel*. Turk J Biol 39:111–118

Shanks JV, Morgan J (1999) Plant 'hairy root' culture. Curr Opin Biotechnol 10:151–155

Sharafi A, Sohi HH, Azadi P, Sharafi AA (2014) Hairy root induction and plant regeneration of medicinal plant *Dracocephalum kotschyi*. Physiol Mol Biol Plants 20:257–262

Shen WH, Petit A, Guern J, Tempe J (1988) Hairy roots are more sensitive to auxin than normal roots. Proc Natl Acad Sci U S A 85:3417–3421

Shi HP, Kintzois S (2003) Genetic transformation of *Pueraria phaseoloides* with Agrobacterium *rhizogenes* and puerarin production in hairy roots. Plant Cell Rep 21:1103

Shilpa K, Varun K, Lakshmi BS (2010) An alternate method of natural drug production: eliciting plant secondary metabolite using plant cell culture. J Plant Sci 5:222–247

Singh B, Sharma RA (2015) Plant terpenes: defense responses, phylogenetic analysis, regulation and clinical applications. 3 Biotech 5:129–151

Singh G, Gavrelli J, Oakey JS, Curtis WR (1998) Interaction of methyl jasmonate, wounding and fungal elicitation during sesquiterpene induction in *Hyoscyamus muticus* in root cultures. Plant Cell Rep 17:391–395

Siva R, Sairam V, Saleel Y, Promit B, Abhishek A, Sarang H, Abhishek K, Alifiya P, Anshul G, Siddharth D (2015) *Agrobacterium* mediated transformation in *Rubia cordifolia* for obtaining hairy root producing anthraquinone dye. Res Rev J Microbiol Biotechnol 4:16–17

Srivastava S, Srivastava AK (2012) *In vitro* Azadirachtin production by hairy root cultivation of *Azadirachta indica* in nutrient mist bioreactor. Appl Biochem Biotechnol 166:365–378

Srivastava A, Tripathi AK, Pandey R, Verma RK, Gupta MM (2006) Quantitative determination of reserpine, ajmaline, and ajmalicine in *Rauvolfia serpentina* by reversed-phase high-performance liquid chromatography. J Chromatogr Sci 44:557–560

Stamp N (2003) Out of the quagmire of plant defense hypotheses. Q Rev Biol 78(1):23–55

Staniszewska I, Krolicka A, Malinski E, Lojkowska E, Szafranek J (2003) Elicitation of secondary metabolites in in vitro cultures of *Ammi majus* L. Enzym Microb Technol 33:565–568

Stiles AR, Liu C (2013) Hairy root culture : bioreactor design and process intensification. In: Biotechnology of hairy root systems. Springer, Berlin, pp 91–114

Tanaka N, Fujikawa Y, Aly MAM, Saneoka H, Fujita K, Yamashita I (2001) Proliferation and rol gene expression in hairy root lines of Egyptian clover (*Trifolium alexandrinum* L.). Plant Cell Tissue Organ Cult 66:175–182

Thimmaraju R, Bhagyalakshmi N, Narayan MS, Ravishankar GA (2003) Food-grade chemical and biological agents permeabilize red beet hairy roots, assisting the release of betalaines. Biotechnol Prog 19:1274

Trémouillaux-Guiller J (2013) Hairy root culture: an alternative terpenoid expression platform. In: Ramawat K, Mérillon JM (eds) Natural products. Springer, Berlin

Trovato M, Maras B, Linhares F, Costantino P (2001) The plant oncogene *rolD* encodes a functional ornithine cyclodeaminase. Proc Natl Acad Sci U S A 98:13449–13453

Varner JE, Lin L-S (1989) Plant cell wall architecture. Cell 56:231–239

Vashishta M, Sharma N (2015) Nutrient mist reactor: a remarkable new approach. Int J Adv Sci Eng Technol 3:8–11

Wang JW, Zheng LP, Zhang B, Zou T (2009) Stimulation of artemisinin synthesis by combined cerebroside and nitric oxide elicitation in *Artemisia annua* hairy roots. Appl Microbiol Biotechnol 85:285–292

Wang CH, Zheng LP, Tian H, Wang JW (2016) Synergistic effects of ultraviolet-B and methyl jasmonate on tanshinone biosynthesis in *Salvia miltiorrhiza* hairy roots. J Photochem Photobio B Biol 159:93–100

Zabetakis A, Edwards R, O'Hagan D (1999) Elicitation of tropane alkaloid biosynthesis in transformed root cultures of *Datura stramonium*. Phytochemistry 50:53–56

Zhang B, Zou T, Yan Hua Lu YH, Wang JW (2010) Stimulation of artemisinin biosynthesis in *Artemisia annua* hairy roots by oligogalacturonides. Afr J Biotechnol 9:3437–3442

Zhao D, Fu C, Chen Y, Ma F (2004) Transformation of *Saussurea medusa* for hairy roots and jaceosidin production. Plant Cell Rep 23:468–474

Methods for Enhanced Production of Metabolites Under In Vitro Conditions

6

K. P. Rini Vijayan and A. V. Raghu

Abstract

In the present world health scenario, plant secondary metabolites are considered highly significant for its properties and its commercial importance. As the conventional manner for their production, which includes field cultivation was found unreliable to provide satisfactory results, the quest for an alternative source was inevitable. As a result, plant cell cultures emerged as an advanced and potential technique of plant secondary metabolites. Over the past decades, research followed different strategies to produce biomass and the subsequent extraction of bioactive compounds which serves as essences, aromas, pharmaceuticals, dyes, food supplements, etc. The overall process of in vitro management of secondary metabolites, starting from the selection of significant yielding up to scale-up of cultures, is versatile and receptive to modifications. The significant approaches for increasing production of secondary metabolites through in vitro technique include elicitation, supplementation of precursors, biotransformation, immobilization and the utilization of novel cambial meristematic cells, etc. Information about plant secondary metabolite pathways is too valid since each chemical compound retains its own specific pathway. Apart from the increased yield, the phase out of geological and periodic changes and interference of environmental causes for biosynthesis of these compounds further promote plant cell culture as a dependable technique. This provides an effective-defined production system which ensures the continuous yield of product with uniform condition. The present chapter summarizes the various methods adopted in the production of plant secondary metabolites and its enhancement using specific methods in in vitro.

K. P. Rini Vijayan · A. V. Raghu (✉)
Kerala Forest Research Institute, Peechi, Thrissur 680653, Kerala, India

© Springer Nature Singapore Pte Ltd. 2020
S. T. Sukumaran et al. (eds.), *Plant Metabolites: Methods, Applications and Prospects*, https://doi.org/10.1007/978-981-15-5136-9_6

Keywords

Secondary metabolites · Elicitation · Precursors · Cell culture · Bioreactor

6.1 Introduction

Metabolites are defined as low-molecular-weight organic and inorganic chemicals. They are the reactants, intermediates, or products of enzyme-mediated biochemical reactions (Dunn et al. 2011). In metabolomics, metabolites generally characterized as any molecule having less than 1 kDa in molecular weight (Bentley 1999). In 1891, Kossel introduced the term 'secondary metabolites' pointing towards the restricted occurrence of these compounds. He set forward the characteristic distinguishing aspect of secondary metabolites from primary metabolites and stated that secondary metabolites are present fortuitously and are not a prerequisite for plant life. But in consequence of several in-depth studies, the above definition subjected to an entire reconstruction and they are now known to be necessary to plant life, playing a crucial role in plant defence mechanism against bacterial, viral and fungal attack, in pollination as signal compounds and also in plant-animal interactions. It can be stated that Primary metabolites are those engaged in primary metabolic processes and are produced by citric acid (Krebs) cycle, glycolysis and photosynthesis and associated pathways. Whereas others, including many pathways derived from primary metabolic pathways, are considered 'Secondary' (Seigler 1998).

6.1.1 Classification of Secondary Metabolites

Plant secondary metabolites are considered as economically valuable products. They play the role of pharmaceuticals, flavours, fragrances, insecticides, dyes, etc. in our regular existence. Nitrogen-containing compounds, nitrogen and sulphur-containing compounds, terpenoids and phenolics are the main classification of secondary metabolites.

Compounds containing nitrogen include alkaloids. They are a group of diverse natural products that contain one or more basic nitrogen atoms in a heterocyclic ring (Da Silva et al. 2007; Michael 2008). They have a variety of pharmacological applications in modern drugs with analgesic (e.g. aspirin), anti-hyperglycaemic (e.g. piperine), anticancer (e.g. taxol), antiarrhythmic (e.g. lappaconitine) and antibacterial (e.g. sanguinarine) properties. The alkaloids like cocaine, caffeine, and nicotine show stimulant effects to CNS and psilocin to psychotropic implements (Ng et al. 2015).

Glucosinolates are the compounds that fall under the second category. They are biologically determined compounds which are the characteristic trait of Brassicaceae and related families within the order capparales. Recent studies have shown beneficial effects of glucosinolates, having regulatory functions in inflammation, stress response, phase I metabolism, and antioxidant activities as well as direct antimicrobial properties (Bischoff 2016). Terpenoids and steroids are gathered in nature from isoprenoid C5 units derived from isopentenyl (3-methyl but-3-en-1-ol) pyrophosphate. These C5 units are linked together in a head-to-tail manner. Terpenoids are composed of the most important group of active compounds in plants with over than 23,000 known structures. Steroids, carotenoids, and gibberellic acid are common among them (Kabera et al. 2014). They have a characteristic branched chain structure (Hanson 2001). About 60% of known natural products are terpenoids. Plant-derived terpenoids are employed for their aromatic qualities and perform a role in conventional herbal medicines. Well-known plant terpenoids include citral, menthol, camphor, etc. Phytosterols constitute a diverse group of natural products. Many of them have been described to possess different biological activities. Withanolides and phytoecdysteroids are excellent examples of biologically active phytosterols.

Phenolic compounds show anti-inflammatory, antioxidant, anti-carcinogenic and other biological properties, and may secure from oxidative stress and some diseases (Park et al. 2001). Plant phenolics are mainly categorized into phenolic acids, coumarins, flavonoids, tannins and lignins. The chemical structure having aromatic benzene ring with one or more hydroxyl groups are the chemical characteristics of phenolics. Plants produced this compounds mainly for protection from stress. Phenolics have significant roles in biosynthesis (e.g. lignin and pigment) and development in plants. They also provide structural stability and scaffolding shelter to plants (Bhattacharya et al. 2010).

6.1.2 Biosynthesis of Plant Secondary Metabolites

The most popular pathways in plants for the biosynthesis of secondary metabolites are shikimic acid pathway (for phenols, tannins, aromatic alkaloids), acetate-malonate pathway (for phenols and alkaloids) and mevalonic acid pathway (for terpenes, steroids and alkaloids) (Dewick 2002). The primary products of acetate malonate pathway are the fatty acids but this pathway further leads towards the production of several aliphatic and aromatic compounds, which are biosynthesized through the formation of polyketides. Acetyl-coenzyme A is the precursor of this pathway, which leads to the development of compounds including fatty acids, terpenoids, steroids, polyketides, aromatic compounds and acetyl esters and amides. The conversion of acetyl CoA to citrate and other tricarboxylic acids leads to the formation of the amino acids and their products, such as the nucleic acids and alkaloids (Vickery and Vickery 1981). The acetate mevalonate pathway (or isoprenoid pathway) is the key pathway for the synthesis of terpenoids and steroids in plants. The primary precursor of this pathway is mevalonic acid which is derived from acetyl CoA. Phenols, tannins and aromatic alkaloids are synthesized through shikimic acid pathway. In higher plants, numerous aromatic compounds are derived from the end-products (phenylalanine, tyrosine and tryptophan) of the shikimic acid pathway. Erythrose-4-phosphate and phosphoenol pyruvate (PEP) are precursors of shikimic acid pathway. The major pathways of secondary metabolism, such as those of terpenoid and phenylpropanoid metabolism, operate in essentially all higher plants. All secondary metabolic pathways derive from primary precursors and can be treated as extensions of these primary pathways (Rhodes 1994).

The human dependency on biological diversification, principally on plants to satisfy their regular demands, is as old as humankind. The role of plant secondary metabolites in our routine life is numerous as they provide pharmaceuticals, food supplements, flavours, fragrances, latex, tannins, gums, essential oils, etc. The method, field cultivation is traditionally used for the production of secondary metabolites from plants. But this procedure is not stable and encounters various disadvantages and difficulties such as overexploitation of plants from their natural habitat, poor yield, uncertainty of content because of geographical, seasonal and ecological diversity, use of a considerable amount of land and labour intensiveness. The structural and stereochemical complexity of specialized metabolites hinders most attempts to approach these compounds using chemical synthesis (Pyne et al. 2019). These drawbacks, and new findings of cell cultures capable of generating specific compounds, lead to the advancement of plant cell and organ culture as an alternative for field cultivation and chemical synthesis for the making of these compounds. Knowledge about the metabolic pathways and the enzymology of the product formation is highly required for in vitro secondary metabolite production and its manipulations (Naik et al. 2012). However, the production of compounds under in vitro conditions is not fully free from defects, and it has its disadvantages.

Research on plant cell cultures started in the late 1930s. In 1956, Pfizer Inc., an American multinational pharmaceutical corporation, filed the first patent for the production of plant metabolites through cell culture (Ratledge and Sasson 1992), and in 1983 the patent for the industrial scale production of 'shikonin' was filed by Mitsui petrochemical industries Ltd. (Fujita and Tabata 1987).

6.2 Plant Cell Cultures for Production of Metabolites

The manufacturing of plant metabolites in plant cell cultures may be qualitatively and quantitatively distinct from an intact plant (Tepe and Sokmen 2007). Examples like production of ajmalicine, Anthraquinones, Berberine, Ginsenoside and Shikonin from *Catharanthus roseus* (Lee-Parsons and Shuler 2002), *Morinda citrifolia* (Zenk 1977), *Coptis japonica* (Fujita and Tabata 1987), *Panax ginseng* (Matsubara et al. 1989) and *Lithospermum erythrorhizon* (Kim and Chang 1990), respectively, shows a remarkable variation in yield under in vitro culture conditions when matched with that of an intact plant.

Apart from the increased yield, the elimination of geographical and seasonal differences and interference of environmental features from the biosynthesis of these compounds also promote plant cell culture as a reliable system. It also offers a well-defined metabolite production system that ensures the regular supply of product with homogeneous quality. Production of unique compounds which are not found in parent plant and stereo- and regio-specific biotransformation for the production of new compounds from cheap precursors is also possible in in vitro condition. Plant cell culture provides an efficient downstream production system (Smetanska 2008). The improvement of compound yield by adding elicitors is again an interesting feature.

Above all these advantages, significant constraints in the production of metabolites in plant cell culture include limited production rate, physiological heterogeneity, genetic variability, low metabolite content, problems associated with product secretion, maintenance of aseptic condition, etc.

6.2.1 Enhanced Production of Secondary Metabolites

There are several reports on production of high amount of secondary metabolite in plant cell cultures than the parent plants. Anthocyanin production from *Perilla frutescens* (Zhong 2001), anthraquinone production from *Morinda citrifolia* (Zenk 1977), serpentine production from *Catharanthus roseus* (Moreno et al. 1995), shikonin from *Lithospermum erythrorhizon* (Kim and Chang 1990), sanguinarine from *Papaver somniferum* cultures (Dicosmo and Misawa 1995), etc. are well-known examples of established cell cultures.

Production of secondary metabolites (SMs) by introducing special media and manipulation of culture conditions either by applying stress or by adding precursors or elicitors facilitated stimulated, cause an increase of metabolite accumulation and establish the process controllable. In most plants, the production of secondary metabolites is not associated with cell growth. Secondary metabolites are usually formed under stress situations or at the end of the growth cycle, i.e. during the stationary phase (Dixon 2001; Verpoorte 2000). This realization contributes to implementing multistage culture systems. In a multistage culture system, the process of biomass accretion and the process of biosynthesis of metabolites are perhaps employed as two separate processes. In the first stage, the accumulation of biomass by cell division is encouraged. In the second or final stage, biosynthesis of compounds from accumulated cells is promoted. In such a system, manipulations can be done independently for each system. In the first stage, procedures for maximum production and multiplication of cells can be employed wherein the later stage methods for the stimulation and enhanced production of beneficial compounds can be incorporated. The two-stage culture system was first attempted by Zenk (1977) for the production of indole alkaloids in *Catharanthus roseus* cells. A similar procedure was adopted by Fujita et al. (1981, 1982) for commercial production of shikonin by cell cultures of *Lithospermum erythrorhizon* (Bhojwani and Razdan 1996).

6.2.2 Strategies for Establishing a Culture System

The process for establishing a culture system starts from the selection of the mother plant. The composition of the desired metabolite in the selected plant should be taken into consideration at this stage. Various reports indicate that there is a marked variation in the phytochemical profile or in the concentration of the desired metabolite in plants growing in different geographical and environmental conditions. It is well known that the chemical profile of different parts of a single plant shows variation. It is because in many plants the organ in which the biosynthesis occurs differs from the organ in which its accumulation is taking place. Nicotine is an excellent example for this. *Nicotiana tabacum* plants produce the pyridine alkaloid nicotine only in root tissues, and then nicotine gets translocated to the aerial parts through the xylem cells (Shoji and Hashimoto 2013). Since the production of nicotine has high metabolic costs, its production remains at basal levels under normal conditions, but increases to higher levels in plants attacked by insects or herbivores. This induction occurs through methyl jasmonate (MeJA) signalling (Baldwin 1998). Upon up-regulation of nicotine production, nicotine content in the xylem fluid increases, and finally nicotine concentration in leaves also increases (Baldwin 1989).

Theoretically, any part chosen from a plant can induce callus cultures, but younger and fresh explants are most preferable as starting material. It is challenging to obtain callus from a monocotyledonous plant when compared with a dicotyledonous one. Callus obtained from woody plants are difficult to grow and normally show slow growth. Callus functions as a meristem. Like primary and secondary meristem in plants, a callus is also devoid of phytochemicals, especially during active division or high mitotic activity. Any deceleration of meristematic activity will progressively permit product accumulation (Constabel and Vasil 1988). Normally, callus cultures are not regarded as a reliable system as suspension cultures for secondary metabolite production. It is employed mainly for screening. In particular instances, species-specific compounds are unavailable in callus cultures. This may be because of the absence of concomitant structural components, lack of precursors, absence of enzyme activity, and degradation of product (Constabel and Vasil 1988).

6.2.3 Selection of Cell Lines and Clones

The heterogeneity in the biochemical activity existing within a population of cells has been utilized to achieve highly productive cell lines (Ogino et al. 1978). The output of a heterogeneous culture would be an average of the productivity of its great and low-yielding cells. The heterogeneity of the cell population can also result from different responses of cells of different physiological states to environmental conditions (Berlin et al. 1988). Selection and cloning of high-yielding cells from a heterogeneous culture are, therefore, considered as an adequate method to enhance in vitro secondary metabolite production (Bhojwani and Razdan 1996). For this to happen, the culture should be initiated from a high-yielding genotype. From the initial cultures, cell lines capable of producing valuable compounds can be differentiated. Different techniques from visual screening to advanced techniques like HPLC and RIA were too adopted to establish high-yielding cell lines. Visual screening can be adopted if the expected metabolite is a pigment or a coloured compound. Production of shikonin, berberine, betanin, anthocyanin, etc. can be screened simply by visual identification. In such cases, the most coloured areas from the callus cultures can be isolated and cultured separately (Bhojwani and Razdan 1996). In many plant cell cultures, secondary metabolite production takes place only at the slow growth state or when the meristematic activity stops because of depletion of nutrients in culture media. This results in difficulty in isolation or selection of cells with maximum yield of the compound (Berlin et al. 1988). Next step is analytical screening which can be subdivided into direct and indirect methods. The direct analysis includes all techniques by which the isolated clone can be directly subcultured. When the selected clone can only be evaluated after analysis of cell extract and merely a part of the clone can be subcultured, it can be categorized as indirect analysis. However, the screening events which are not stable on the cellular level for at least a few passages can be stated as physiological states where the cell lines which maintain their notable characteristics over years can be stated as variants (Berlin et al. 1988).

Apart from callus cultures, cell suspension cultures can also be employed for screening purposes. This can be accomplished with a sharp, swiftly growing cell suspension consisting of microscopic aggregates of up to 50 cells. Cells with fluorescent activity can be singled out with the aid of UV light by the bare eye, or a fluorescent microscope can be used. For a considerable number of cell flow cytometry can be adopted (Aiken and Yeoman 1986; Hara et al. 1989; Adamse 1990). For colourless compounds, immunological tests or specific reactions of crushed cells can be used (Bhojwani and Razdan 1996). High-yielding cell lines can be selected by using certain selective forces as an alternative method. This method is based on the fact that improved knowledge of the mode of action of a toxic compound (antimetabolite) may benefit in the use of selection for resistance as a tool for establishing cell lines with higher levels of enzymes or specific compounds. In this process, a large population of cells is exposed to a lethal (or cytotoxic) inhibitor or environmental stress and only cells that are capable to resist the selection methods will grow. P-Fluorophenylalanine, an analogue of phenylalanine, was largely utilized to select high-yielding cell lines with respect to phenolics. Other selective agents consist of 5-methyltryptophan (5-MT); glyphosate, Para-fluorophenylalanine (PFP) and biotin have also been used to select high-yielding cell lines (Berlin et al. 1988).

6.2.4 Culture Medium Optimization

The growth of callus and organ culture and cyto-differentiation resulting in product accumulation depends on the composition and administration of media used in the culture (Zenk 1977). Production of secondary metabolites and culture growth are not a parallel process. Most probably, these metabolites are developed during the late static phase. Several physical and chemical factors influence the process of biomass accumulation and product formation. The culture media is the important basic factor influencing the growth and metabolism of cells in in vitro. The various critical factors associated with the culture media include (1) basal media and salt strength, (2) carbon source and concentration, (3) nitrogen source and concentration, (4) phosphate level and (5) plant growth regulators and concentration.

6.2.5 Influence of Basal Media and Salt Strength

The optimal media for plant cell cultures should comprise all the nutrients for the effective multiplication and growth of cells in culture. The principal components are nutrients (macro and micro), vitamins, amino acids (or nitrogen supplements), carbon source, growth regulators, organic supplements specific for certain cultures and solidifying agents (with the solid medium). Various commonly used media compositions are MS (Murashige and Skoog) media, B5 (Gamborg's) media, SH (Schenk and Hildebrandt) media, White's media, Nitsch media, N6 (Chu) media, WPM (Woody plant) media and LS (Linsmaier and Skoog) media.

The MS media was specifically formulated for *the* in vitro callus culture of *Nicotiana tabacum* in 1962. This media is characterized by a high concentration of nitrate, potassium and ammonium ions. Gamborg's B5 media was formulated for the callus and cell suspension culture of Glycine max. It has a higher percentage of nitrate and potassium but a nominal concentration of ammonia. SH media was formulated for the callus cultures of monocotyledonous and dicotyledonous plants. SH media is characterized by a high concentration of copper and myo-inositol concentration. White's media was earlier formulated for the root culture of the tomato. Low salt concentration with nitrate concentration 19% less than MS is the characteristic feature of this media. Nitsch media is specifically formulated for the anther culture of *Nicotiana tabacum* and it is characterized by the low concentration of micronutrients compared to MS media but a higher concentration of salt compared to White's media. Chu media was developed for the anther culture of rice. In this potassium nitrate serves as a source of nitrate. In 1981, Lioyd and McCrown developed woody plant media for *Kalmia latifolia*. WPM is a medium with less salt concentration to MS media and a high sulphate concentration.

Fujita et al. (1981) tested five basal media, viz., LS media, White's media, B5 media, B1 (Blaydes media) and Nitsch and Nitsch media in cell culture of *L. erythronhizon* for shikonin and biomass production. Among the tested media, the maximum biomass production was achieved in LS media, followed by B5 media. The other culture media did not show good results towards growth and multiplication. However, White's media (Bhojwani and Razdan 1996) showed promising results in the production of shikonin derivatives.

Apart from the basal media, when we consider the salt strength of culture medium, we can see a notable difference in the growth pattern and production of compounds regarding salt strength. For example, in ginseng adventitious root cultures, the strength of MS media influenced the results in ginsenoside production (maximum in 0.5 strength), biomass and growth rate (maximum in 0.75 strength) (Sivakumar et al. 2005a, b). In another example, full strength MS medium gave desirable results towards biomass production and the gymnemic acid accumulation in *Gymnema sylvestre* culture (Nagella and Murthy 2011).

6.2.6 Influence of Carbon Source and Concentration

In nature, CO_2 is assimilated to sucrose as the main trans locatable carbon source in plants. Around 10–25% of sucrose is present in the phloem and cell sap of plants and it is considered as higher than conventional concentrations used in tissue culture media (Cresswell et al. 1989). Plant cell cultures are grown heterotrophically in natural sugars like glucose, fructose, maltose, sucrose and their combinations (Murthy et al. 2014). Among these, sucrose is being used largely in culture media since it was found to be a perfect carbon source for biomass accumulation in most of the tested cultures. Sucrose can be considered as a substitute of glucose in cultures. However, the differences in the effects of sucrose and glucose upon culture growth and secondary metabolism (Fowler and Stephan-Sarkissian 1985) indicate that their method of uptake and use are controlled rather differently (Cresswell et al. 1989). While considering the two monosaccharides, viz., glucose and fructose, the uptake profiles are correlated but the former is always more rapidly removed from the medium than latter (Stephan-Sarkissian and Fowler 1986). There are very limited studies dealing with alternative carbon sources in culture media on secondary metabolism. As per the experience from the in vitro studies, sucrose is the most effective carbon source for secondary metabolite production in culture media (Mizukami et al. 1977; Stephan-Sarkissian and Fowler 1986), but exactly how the effect is mediated is unknown.

The effect of various carbon source on the production of the compound artemisinin in hairy root culture of *Artemisia annua* shows that the production of hairy roots at maximum level was found in medium supplemented with glucose. However, in the culture medium supplemented with fructose, the production of artemisinin was achieved twice that in the sucrose supplemented medium. Apart from the source material, the concentration of the carbon source also matters to secondary metabolite production. The sucrose concentration of 2.5% and 7.5% in the culture of *Coleus blumei* reported the production of rosmarinic acid with 0.8 g/L and 3.3 g/L, respectively (Misawa 1985). In *Catharanthus roseus* culture 8% (w/v) of sucrose concentration was found to be optimal for the production of indole alkaloid (Knobloch and Berlin 1980). In *Gymnema sylvestre* cell culture, 3 and 4% sucrose concentration were found to be ideal for the production of biomass and accumulation of the highest amount of gymnemic acid, respectively (Murthy et al. 2014). In *Solanum melongena* culture sucrose played dual role, as a carbon source and as an osmotic agent (Mukherjee et al. 1991). However, different carbon sources (sugars) have been recognized as a signalling molecule that impacts biomass and metabolism of cultured cells (Wang and Weathers 2007). Therefore, incorporating an appropriate carbon source at a proper concentration is a key factor for effective production of secondary metabolites in plant cell culture (Murthy et al. 2014).

6.2.7 The Source and Concentration of Nitrogen

The biomass production and metabolite accumulation in plant cell cultures are largely dependent on the appropriate nitrogen source and its concentration. It was found that the nitrogen concentration has an effect on the proteinaceous or amino acid products in cell suspension culture (Rao and Ravishankar 2002). Various culture media, such as MS, LS and B5, have both nitrate and ammonium as a source of nitrogen. The supply of nitrogen as nitrate executes a high energy demand on the anabolism of nitrogen whereas the supply of nitrogen as ammonia is more effective than supplying the more oxidized form (Cresswell et al. 1989). Urea, glutamine, glutamate and alanine (casein) are the other nitrogen sources used in culture media (Dougall 1980). In some cases, amino acids, for example, L-tryptophan and phenylalanine, are as nitrogen source. But in such cases, their effect on secondary product formation can be variable (Cresswell et al. 1989). It was found that the presence of L-tryptophan in the culture of *Catharanthus roseus* increased the yield of alkaloids (Doller 1978). In *Cinchona ledgeriana* cultures the addition of L-tryptophan induced the production of carboline alkaloid (Harkes et al. 1986). The incorporation of phenylalanine into the culture media of PFP resistant *Nicotiana tabacum* cell lines increased the accumulation of phenolics in culture (Berlin 1980). However, the presence of total nitrogen has a direct influence on growth kinetics and production of secondary compounds in culture. The complete removal of nitrate in *Chrysanthemum cinerariaefolium* cultures induced a twofold increase in the pyrethrin accumulation in the second phase of the culture (Rajasekaran et al. 1991). Studies were also reported that reduced levels of total nitrogen increased the accumulation of capsaicin (*Capsicum frutescens*), anthraquinone (*Morinda citrifolia*) and anthocyanin (*Vitis species*) (Yamakawa et al. 1983; Yeoman et al. 1990; Zenk et al. 1975).

6.2.8 Influence of Phosphate Level

Phosphorus is an essential part of nucleic acids and other structural compounds in plants. It reaches the plants in the form of primary or secondary orthophosphate anions $H_2PO_4^-$ and HPO_4^{2-} by a natural process, which requires respiratory energy. Unlike nitrate and sulphate, phosphate is not reduced in plants and used as the fully oxidized orthophosphate (PO_4^{3-}) form. Soluble potassium mono- and di-hydrogen phosphates are being used as phosphate source in culture media (George et al. 2008). The concentration of phosphate in the culture media influences the production of secondary metabolites. Higher levels of phosphate induces enhanced growth of cells, where it is having an undesirable effect on secondary product accumulation (Sasse et al. 1982). Reduced phosphate level induces both the product development and the level of key enzymes which leads to the formation of products (Rao and Ravishankar 2002). An increased production of caffeine was found under reduced phosphate level in the suspension culture of *Coffea arabica* (Bramble et al. 1991) and on the other side some other studies reported about the positive correlation of increased concentration of phosphate and higher production of metabolites. For example, a twofold phosphate level in the standard MS media showed a positive effect on the production of rosmarinic acid from *Lavendula vera* suspension culture (Ilieva and Pavlov 1996).

6.2.9 Influence of Plant Growth Regulators

Growth, multiplication and metabolite accumulation are significantly affected by plant growth regulators in culture media. Callus, organ and adventitious root cultures need an exogenous supply of plant growth regulators for proper growth and multiplication. The type of plant growth regulator and its concentration also influences metabolite accumulation. Major categories of plant growth regulators are auxins, cytokinins, gibberellins, ethylene, brassinosteroids, jasmonate and strigolactones (Jamwal et al. 2018). The response to plant growth regulators may vary according to the species, varieties and age of plant. Factors like environmental conditions, stage of development, physiological and nutritional status and endogenous hormonal balance also influence the response (Aftab et al. 2010; Idrees et al. 2010a, b, 2011; Naeem et al. 2009, 2010, 2011). The induction of plant cell and tissue cultures by external PGR signals may lead to the activation of a specific developmental pathway and lead to the regeneration of competent cells (Sriskandarajah et al. 2006). Several investigations have been conducted and confirmed in a number of plants to study the effects of plant growth regulators on secondary metabolite production (Weathers et al. 2005; Rojbayani 2007; Khan et al. 2008; Shilpashree and Rai 2009). The type and concentration of growth regulators (especially auxin or cytokinin) and its ratio dramatically alter both the growth and the metabolite formation in cultured plant cells (Mantell and Smith 1984). Among the auxins, naphthalene acetic acid (NAA) and indole acetic acid (IAA) have shown a positive effect on anthocyanin production in suspension culture of carrot, nicotine in tobacco, anthraquinone in morinda, etc. (Ravishankar et al. 1988; Seitz and Hinderer 1988; Sahai and Shuler 1984). In many cultures, presence of 2,4-dichlorophenoxyacetic acid (2,4-D) resulted in a negative effect on metabolite production. In such cases, the removal of 2,4-D or its replacement by NAA or IAA from the media showed a specific effect on metabolite production. In some other cases, 2,4-D showed a stimulatory effect on the production of carotenoids in carrot (Mok et al. 1976), anthocyanin in *Oxalis*, etc. (Meyer and Van Staden 1995). Reports show that cytokinins have an enhancing effect on secondary metabolite production in *Hypericum sampsonii* and *Hypericum perforatum* cultures (Liu et al. 2007). The application of Abscisic acid (ABA) in *Orthosiphon stamineus* cultures caused oxidative stress and stimulated the secondary metabolism. For example, *Cannabis sativa* produced desired level of terpenoid in cultures (Mansouri et al. 2009). The plant growth regulator, gibberellic acid (GA_3) has also demonstrated good results in the growth of hairy roots in *Cichorium intybus* culture (Bais et al. 2001). Hormones like salicylic acid (SA), methyl jasmonate and jasmonate (JA) also have a property to trigger plant secondary metabolite biosynthesis through distinct signalling pathways (Zhao et al. 2005) and they all function collectively with nitric oxide (NO) in enhancing secondary metabolite production in cultures (Jamwal et al. 2018).

Plant growth regulators do not respond with intermediary compounds of biosynthetic pathways but appear to shift cytological conditions for product formation to higher or lower levels. Also, hormone action is also not specific; i.e., the formation of all kinds of secondary metabolites has been found to be affected (Kurz and Constabel 1979). As both cell multiplication and differentiation have to occur in cell cultures when metabolite production is to be increased and exploited, a sensible approach would be a two-step strategy: promotion of growth of cells cultured in nutrient media with significant levels of auxins (2,4-D) followed by transmission of the cell mass to media with moderate levels or devoid of auxins (Zenk 1977).

6.2.10 Influence of Culture Conditions

Apart from the optimization of desired media, optimization of specific culture conditions is also relevant in secondary metabolites production in in vitro cultures. Various culture conditions such as light, temperature, medium pH and gases have been studied largely to understand their effect upon plant secondary metabolite production in various cell and organ cultures.

6.2.10.1 Light

Light may exert an important function on the accumulation of metabolites in cultures. When we consider the effect of light on secondary metabolite production in cultured cells, different sources of light, quality of light, the quantity of irradiation (intensity), photoperiod (duration), direction and quality (frequency or wavelength) should be taken into consideration. Light can be used as an energy source or just as an elicitor depending upon the cell culture and desired product. Pedrosal et al. (2017) in their recent study examined the influence of different wavelength (white, blue, green, red and yellow) in growth and production of plant metabolites in *Hyptis marrubioides* cultures and found that after 30 days of cultivation red light induced plant growth whereas blue and white light promoted rutin production. In another study carried out by Cioc et al. in 2018, to analyse the effect of light quality on growth and metabolite accumulation of *Myrtus communis* in in vitro condition, various treatments with light-emitting diodes (LEDs) were compared with a traditional fluorescent lamp and found that secondary metabolite production with accumulation of phenolics and flavonoids and major compound myricetin was found in higher concentration under red LEDs. The inhibitory effect of light in secondary metabolite production is also reported in the production of shikonin and monoterpenes in *Lithospermum erythrorhizon* (Tabata et al. 1974) and *Citrus limon* (Mulder-Krieger et al. 1988), respectively. For the production of anthraquinone in *Cassia tora* and alkaloids in *Scopolia parviflora* (Tabata et al. 1972), *Nicotiana tabacum* (Furuya et al. 1971) and *Catharanthus roseus* (Doller Von et al. 1976), they preferred a dark condition.

6.2.10.2 Temperature

The influence of temperature in cell growth and multiplication and in secondary metabolite product formation was noticed in several culture systems. In experiments, 17–25 °C temperature is commonly used for the formation of callus tissues and growth of cultured cells. But this conventional range of temperature may be varied according to the plant species used for study. The effect of temperature on secondary metabolism in *Eleutherococcus senticosus* somatic embryos was examined for 45 days in a bioreactor by Shohael et al. (2006). The somatic embryos were exposed to 12, 16, 24 and 30 °C. The effect of this treatment on growth, and production of eleutheroside B, E, E1, total phenolics, flavonoid and chlorogenic acid were noted and found that low (12 and 18 °C) and high (30 °C) temperature influenced a notable drop in fresh weight, dry weight, total phenolics, flavonoid content and eleutheroside production. In this the accumulation of increased level of eleutheroside E production was noted at low temperature. Toivonen et al. (1992) found that application of low temperature enhanced the overall fatty acid content per cell in dry weight. Studies conducted on *Digitalis lanata* cell cultures showed biotransformation of digitoxin to digoxin at a temperature of 19 °C whereas 32 °C favoured formation of purpurea glycoside A (Kreis and Reinhard 1992).

6.2.10.3 Medium pH

Usually, in tissue culture media the pH is adjusted between 5 and 6 before autoclaving. The concentration of hydrogen ions in the media changes because of nutrient uptake or due to the accumulation of metabolite in culture (Murthy et al. 2014). Earlier studies reported that pH of media decreases during ammonia assimilation and increases during nitrate uptake (McDonald and Jackman 1989). The release of alkaloids from root cultures *Datura stramonium* and *Catharanthus roseus* and thiophenes from root cultures of *Tagetes patula* was found to increase when the pH of the culture media (ranging from 4.8 to 7.0) was reduced to 3.5 (Saenz-Carbonell et al. 1993). The shoot cultures of *Stevia rebaudiana* was maintained in different pH levels (4.6, 5.8, 7.4) and it was found that the maximum level of total phenol and total flavonoid yield was at pH 4.6 (Radic 2016). In *Withania somnifera* hairy root cultures, pH 5.8 favoured biomass accumulation and pH 6 favoured the production of withanolide A in roots (Praveen and Murthy 2010). In *Panax ginseng*, pH 6 and 6.5 favoured both biomass production and accumulation of ginsenoside in hairy root cultures (Sivakumar et al. 2005a, b). In *Beta vulgaris*, the exposure of the root culture to the acidic medium showed better results in the release of betalain pigment while the culture kept the capability for regrowth and continued pigment accumulation. Here a 10 min exposure to pH 2 and subsequent transfer to standard growth medium resulted in the release of the pigment (Mukundan et al. 1998).

6.2.10.4 Agitation and Aeration of Culture Medium

The agitation and aeration of culture media promotes further growth by enhancing the mass transfer and uptake of nutrients from liquid and gaseous phases through cells/organs and the dispersion of an air bubble for effective oxygenation (Murthy et al. 2014). In culture, plant cells have low growth rates compared to microbial cells, which results in lower oxygen requirements for plant cells than the microbial ones. In some cases, the high oxygen concentration is indeed toxic to the metabolic activities of cells and may strip nutrients such as carbon dioxide from the culture broth (Chattopadhyay et al. 2002). The intensity of culture broth mixing, the degree of air bubble dispersion, the culture medium's capacity for oxygen, and the hydrodynamic stress inside the culture vessel affect proper aeration of the culture (Chattopadhyay et al. 2002). Normally, high agitation speed for effective mixing is restricted in plant cell cultures due to their shear sensitivity to hydrodynamic stress. The high shear rate and shear time associated with good mixing reduce the mean aggregate size, but this also involves a harmful effect on cell survival. Therefore, plant cells are generally grown in stirred tank bioreactors at a very low agitation speed (Murthy et al. 2014).

Aeration is another important aspect that should be taken into consideration in large-scale production of desired compounds. This facilitates three main functions in cell cultures such as supply of aerobic condition, desorption of volatile products and elimination of metabolic heat by mixing and air flow (Georgiev et al. 2009). Ambid and Fallot in 1981 studied the effect of the composition of the gaseous environment on the production of volatile compounds in fruit suspension culture. According to this study, the presence of carbon dioxide stimulated the synthesis of monoterpenes in muscat grape suspension and induced the formation of linalool in culture. In *Thalictrum minus* suspension cultures the supply of 2% of carbon dioxide was vital to avoid the cell browning and for the sustainable production of berberine (Kobayashi et al. 1991). However, it was noticed that the supply of carbon dioxide was not favourable for the accumulation of saponin (Thanh et al. 2006).

6.3 Enhanced Biosynthesis of Secondary Metabolites in Culture

In the recent past, various methods have been developed for the enhanced accumulation of plant secondary metabolites in cell culture. The major methods are elicitation, precursor feeding, nutrient feeding, permeabilization, immobilization, biotransformation, organ culture, bioreactor culture and metabolic engineering. The details are as follows.

6.3.1 Elicitation

Elicitation is the generated or improved biosynthesis of metabolite as a result of the supplementation of trace amount of elicitors (Radman et al. 2003). Elicitation is considered as the most proper and practically possible strategy for increasing production of secondary metabolites in cultures (Poulev et al. 2003). Elicitor is a substance which when applied in small quantity to a biological system induces or improves the biosynthesis of desired compounds which do have an important role in the acclimatization of the plant to a stressful condition. Many research findings show that elicitor can regulate numerous biochemical processes such as activate the expression of key genes and transcription factors and they have the capability to control an array of cellular activity at the biochemical and molecular level (Zhao et al. 2005).

On the basis of nature, elicitors can be classified into abiotic and biotic elicitors. Polysaccharides, yeast extract, fungal and bacterial preparations, etc. are biotic elicitors. The abiotic elicitors can be further divided into physical and chemical. Physical elicitors include UV radiation, osmotic stress, salinity, thermal stress, etc. and chemical elicitors include heavy metals, mineral salts, gaseous toxins, etc. The elicitor activity of ozone in *Melissa officinalis* shoot cultures, and cell suspension cultures of *Hypericum perforatum* and *Pueraria thomsnii* resulted in the accumulation of rosmarinic acid, hypericin and puerarin, respectively (Tonelli et al. 2015; Xu et al. 2011; Sun et al. 2012). Supplementation of sucrose as elicitor triggered the accumulation of bacoside A and withanolide A in shoot cultures of *Bacopa monnieri* (Naik et al. 2010) and *Withania somnifera* cell suspension cultures (Praveen and Murthy 2010). Proline and polyethylene glycol are two chemicals which induce osmotic stress in culture, elicited the accumulation of steviol glycoside in *Stevia rebaudiana* callus and suspension cultures. Apart from these, there are numerous reports for the elicitor action of jasmonic acid and methyl jasmonate (Munish et al. 2015; Silja et al. 2014; Gangopadhyay et al. 2011; Arehzoo et al. 2015; Xiaolong et al. 2015; Sharma et al. 2013; Izabela and Halina 2009; Firouzi et al. 2013). Among biological elicitors, chitin enhanced the accumulation of hypericin in shoot cultures of *Hypericum perforatum* (Sonja et al. 2014). It is reported that fungal preparations from *Fusarium oxysporum* and *Aspergillus niger* were found to enhance the production of phenyl propanoid in *Hypericum perforatum* (Sonja et al. 2014) and gymnemic acid in *Gymnema sylvestre* cell suspension culture (Chodisetti et al. 2013).

Elicitors have a capacity to trigger the intensity of the plant's response to biotic and abiotic stresses with the improved synthesis of signal compounds such as jasmonic acid (JA), salicylic acid (SA), ethylene (ET) and nitric oxide (NO) and its successive influence on metabolite production (Giri and Zaheer 2016). JA involves signal transduction pathways with the participation of jasmonate zim domain (JAZ) proteins, which are further subjected to proteasomal degradation via Skpl/Cullin/F-box (SCF) complex. JAZ protein degradation results in the release of MYB/MYC transcription factors, which ultimately stimulate JA-dependent gene expression in secondary metabolism (Wasternack and Hause 2013). Qian et al. in 2005 successfully modified the chemical structure of methyl jasmonate and established a couple of new elicitors. This novel hydroxyl containing jasmonate has given good results in inducing taxoid biosynthesis when compared to methyl jasmonate (Hu et al. 2006). Salicylic acid has also given promising results in biosynthesis and its complex interaction with other signal compounds such as JA, ET and NO has been studied in detail by various researchers (Qiao and Fan 2008; Boatwright and Pajerowska-Mukhtar 2013; Caarls et al. 2015). Giri and Zaheer (2016) reported that systemic acquired resistance (SAR) and localized acquired resistance (LAR) which helps in biosynthesis in cultures was initiated by

SA. Ethylene perception is accomplished by ET receptors. NO is a free radical that stimulates the mitogen-activated protein kinase (MAP kinase) signalling pathway, which results in altered gene expression (Giri and Zaheer 2016). The overall objective of elicitor treatment in plant cell culture is to misguide the cells for the possible biotic/abiotic attack, mimicking the external environment identical to natural conditions under stress-related damage in plants. The expression of elicitor treatment with altered genetic and biochemical activities in the cellular background are observed as an enhanced production of targeted chemicals, higher gene expression and exploration of new biomolecules (Caretto et al. 2010).

6.3.2 Precursor Feeding

A precursor is a compound which is transformed into a class of functional molecules within short steps. The yield of a specific compound can be increased by locating its key precursor, even though without knowledge of the entire synthetic pathway of that compound. The concept of the precursor is based upon the idea that any compound, which is an intermediate, in or at the beginning of a secondary metabolite biosynthetic route, stands a good chance of increasing the yield of the final product (Rao and Ravishankar 2002). The enzymatic reaction synthesizing the precursor often forms a bottleneck in its succeeding metabolism. Since metabolism is considered an equilibrium process, over-feeding of a precursor to cultured cells usually produce the increased amount of compounds of its downstream. Plant cell cultures convert the precursors into metabolites by using the enzyme system existing in them. The addition of loganin, tryptophan and tryptamine as precursors improved the production of secologanin (Contin et al. 1999) and indole alkaloids (Moreno et al. 1993) in *Catharanthus roseus* suspension culture. Seo et al. (2015) found that the inclusion of phenylalanine has increased the phenolic production in tartary buckwheat culture. In *Coleus blumei* cell cultures the supplementation of phenylalanine led to the increase in yield of rosmarinic acid (Ibrahim 1987). Phenylalanine is a precursor of the N-benzoylphenylisoserine side chain of taxol. Hence the addition of phenylalanine resulted in an enhanced yield of taxol in *Taxus cuspidate* cultures (Fett-Neto et al. 1994). Feeding of ferulic acid to the culture of *Vanilla planifolia* resulted in the increased accumulation of vanillin (Romagnoli and Knorr 1988). Feeding of cholesterol in *Holarrhena antidysenterica* cultures increased the accumulation of the metabolite, conessine (Panda et al. 1992).

6.3.3 Nutrient Feeding

The nutrient feeding method is another most advised technique for enhanced secondary metabolite production in plant cell culture. The best example of nutrient feeding is an increase in dry biomass along with the increase in ginsenoside content in ginseng adventitious root culture. In this after the nutrients in the culture medium showed a depletion (40 day old culture), the culture was replenished with 0.75 and

1.0 strength media after 10 and 20 days of cultivation. Such cultures showed increased levels of biomass and secondary metabolites (Jeong et al. 2008). Similar effect can be observed in *Echinacea purpurea* root cultures (Wu et al. 2007) and *Lithospermum erythrorhizon* cell cultures (Srinivasan and Ryu 1993).

6.3.4 Permeabilization

Secondary metabolites are generally accumulated in the vacuoles of cells. It is beneficial to release the product into the culture medium so that the purification process may become easier and continuous recovery and production may be performed. Removal of product from vacuoles would also lessen the possible product inhibition and thus enhancing the yield (Murthy et al. 2014). Cell permeabilization depends on the development of pores in one or more of the membrane systems of the plant cell, enabling the passage of various molecules into and out of the cell. Attempts have been made to permeabilize the plant cells only transiently, to maintain cell viability and have short time periods of increased mass transfer of substrate and metabolites to and from the cell (Brodelius and Nilsson 1983). Permeabilization of plant membranes for the release of secondary metabolites were often connected with the loss of viability of the plant cells treated with permeabilizing agents and methods. Various methods have been used to initiate product release from cultured plant cells (Dörnenburg and Knorr 1995). These methods consist of treatment with a solution of high ionic-strength (Tanaka et al. 1985), adjustment of external PH, transfer to media lacking phosphate (Berlin et al. 1988), permeabilization with dimethylsulphoxide (DMSO) (Brodelius and Nilsson 1983; Parr et al. 1984), chitosan (Knorr and Berlin 1987) and other chemicals (Felix 1982), electroporation (Brodelius et al. 1988, ultrasonic (Kilby and Hunter 1990) or ultra-high-pressure treatments, etc.

6.3.5 Immobilization

Immobilization is a technique which confines a catalytically active enzyme or cells on a fixed support and restricts its entry into the liquid state (Yeoman et al. 1990). Due to several specific characters of plant cells, such as slow growth, the low shear resistance and the tendency for cell aggregation, it is difficult to establish a large-scale fermentation process for plant cells. In such a condition, the immobilization has a number of advantages. They are (1) extended viability of cells in the stationary stage enabling maintenance of biomass over a delayed time period, (2) simplified downstream processing, (3) high cell density, (4) reduced shear stress, (5) increased product accumulation, (6) greater flow rate and (7) minimization of fluid viscosity (Murthy et al. 2014). In general, cell immobilization provides a continuous process operation, reuse of biocatalysts, separation of growth and production phases, and a simplified separation of biocatalysts from the culture medium, which allows product-orientated optimization of the medium and reduction of cultivation periods

(Klein et al. 1987). Factors such as high cell density and a high arresting potential of the matrices determines the suitable immobilization system in cultures (Dörnenburg and Knorr 1995). Two significant methods for immobilization are gel entrapment and surface immobilization. Because of the economical, simple and reproducible techniques, Gel entrapment is the most commonly used immobilization method (Dörnenburg and Knorr 1995). In the gel entrapment method, calcium alginate is the most preferred matrix. Other polymers used for immobilization include agar, agarose, carrageenan, chitosan, gelatin, gellan, polyacrylamide, polyacrylamide hydracide, polyphenylene oxide and combinations of agarose and gelatin, alginate and gelatin, alginate and chitosan, carrageenan and chitosan, and pectin and chitosan. The polymer should be cheap and non-toxic with high polymerization activity (Murthy et al. 2014). Brodelius et al. reported immobilization of *Catharanthus roseus* and *Morinda citrifolia* in 1979. In surface immobilization, the property of cultured plant cells to adhere to inert surfaces immersed in a liquid is exploited. In *Catharanthus roseus*, *Nicotiana tabacum* and *Glycin max*, the surface immobilization technique is reported for the secondary metabolite production (Archambault et al. 1989; Asada and Shuler 1989).

6.3.6 Selective Adsorption of Plant Products

In specific cases, synthesis and storage of secondary metabolites in plant cells take place in separated compartments (Dörnenburg and Knorr 1995). This is where a 'two phase system' with an artificial site for the accumulation of the secondary metabolite comes into the picture of in vitro production of plant secondary metabolites. The addition of an artificial site for the accumulation of compounds can be considered an adequate tool for increasing the in vitro secondary metabolism (Dörnenburg and Knorr 1995). In such systems, the elimination and sequestering of the compound in a non-biological compartment can be a solution for the low accumulation of compounds. This system also enhances the release of product from the culture, protection of the released product from degradation and evaporation of volatile compounds (Payne and Shuler 1988). There are a number of non-biological substances which were successfully used by many investigators for the accumulation of desired compounds in plant cell cultures. Stockigt and co-workers reported the use of the neutral polymeric resin XAD-4 for the recovery of indole alkaloids. However, adsorption by this resin proved to be relatively nonspecific. As an alternate for this polycarboxylic ester resin (XAD-7) was examined by Payne and Shuler (1988) for the selective adsorption of indole alkaloids. Apart from this activated charcoal, RP-8 (Lipophilic carrier), Zeolith, XAD-2, Polyethylene glycol, Polydimethylsiloxan and wofatite were also listed as efficient adsorption systems. The selective adsorption system facilitates the recovery and purification of the compound from the culture, thus reducing the production costs (Dörnenburg and Knorr 1995).

6.3.7 Biotransformation

Biotransformation is a process through which the functional groups of organic compounds are modified either stereo- or regio-specifically by living cultures, entrapped enzymes, or permeabilized cells to a chemically different product (Rao and Ravishankar 2002). Plant-cells can transform a large variety of substrate through reactions like hydroxylation, hydrolysis, glycosylation, oxidoreduction, glucosylation, hydrogenation, methylation, acetylation, esterification and isomerization of various substrates. The process has several advantages, such as (1) More than one reaction can be accomplished using cell cultures that express a series of enzyme activities. (2) In some instances, even non-producing cell culture can be used to synthesize the desired end product using an appropriate precursor. (3) The process of biotransformation may be simple where the process is mediated by one or more enzymes with many steps. (4) Single step biotransformation is comparatively efficient, as the yield decreased with an increase in steps and (5) Natural or synthetic substrate is used for biotransformation (Dave et al. 2014). Aromatic compounds, steroid, alkaloid, coumarin, terpenoid, lignin, etc. are the major chemical compounds used for biotransformation in plant cell cultures. It is not necessary for the compound to be a natural intermediate of metabolism and it can be of synthetic origin (Murthy et al. 2014). The ability of plant cultures to transform cheap and productive industrial by-products into rare and costly products had been utilized by many scientists for the production of relevant compounds. The conversion of nicotine into nornicotine by cell suspension culture of *Nicotiana tabacum* (Hobbs and Yeoman 1991), Codeinone to Codeine in *Papaver somniferum* plant cell culture (Furuya et al. 1984), Valencene to Nootkatone in *Citrus* spp. (Furusawa et al. 2005), the conversion of the proto catechuic aldehyde and caffeic acid to vanillin and capcinin in freely suspended and immobilized cells of *Capsicum fruitescens* (Rao and Ravishankar 2002) and bioconversion of paeonol into its glycosides by ginseng cultured cells and root culture (Li et al. 2005) are the example for biotransformation by plant cell cultures.

6.3.8 Organ Culture

As an alternative method for the production of secondary metabolites, organ cultures such as root, embryo and shoot culture have been developed in various plant species. Organ cultures are extensively investigated in the form of shoot and root cultures. In many medicinal plant species, shoot cultures that can accumulate higher concentrations of valuable compounds have been developed as a better alternative for exact plant material. Enhanced metabolite production in shoot cultures of *Bacopa monnieri* (bacoside production) and *Camptotheca acuminata* (Camptothecin production) were reported by Sharma et al. (2014) and Thomas (2009). Similar studies like increased crocin production from *Crocus sativus* organ cultures (Zeng et al. 2003), enhanced production of naphthoquinones in transgenic hairy root lines of *Lithospermum canescens* (Syklowska-Baranek et al. 2012), Withanolide A production from hairy roots of *Withania somnifera* (Sivanandhan et al. 2013), adventitious root cultures of *Morinda citrifolia* in bioreactors for the production of anthraquinone (Baque et al. 2012), etc. are excellent examples of well-established organ cultures for the production of secondary metabolites.

6.3.9 Cambial Meristematic Cells

In nature, cambial meristematic cells (CMCs) are responsible for the vascular cells like xylem and phloem. These cells are undifferentiated in nature and grow indefinitely to perform plant stem cell function. Typically, contemporary cell cultures are consisting of dedifferentiated cells (DDCs). These cells consist of a mixture of different cell types resulting in a significant level of heterogeneity, which significantly contributes to the variability routinely associated with suspension cultures (Roberts and Kolewe 2010). Lee et al. (2010) developed an innately undifferentiated cell line derived from cambium cells, which function as vascular stem cells. As CMCs are innately undifferentiated plant stem cells, it can bypass the negative effects related with the differentiation step and offer stability in product accumulation for longer periods (Lee et al. 2010). CMCs are characterized by superior growth when compared to DDCs (dedifferentiated cells). It has showed reduced aggregation in liquid media, which is very important, because cell cluster formation can inhibit oxygen and nutrient distribution, thereby decreasing product yield (Lee et al. 2010). The reduced aggregation size of CMCs also inhibits shear sensitivity where shear stress is a crucial limitation in the scale-up of plant cells, affecting cells by causing either lysis or death owing to fluid motion (Joshi et al. 1996). Apart from this, CMCs are highly responsive to elicitation (Lee et al. 2010). In *Catharanthus roseus* CMCs, the addition of elicitors increased the production of ajmalicine relative to the non-treated cells (Zhou et al. 2015). The use of CMCs also promotes cheaper downstream processing, which is especially important for the large (industrial)-scale production due to the direct release of a product into the media (Ochoa-Villarreal et al. 2016). The use of CMCs is a promising platform for the production of complex molecules. The utility of CMCs for industrial-scale production has already been established with the production of ginsenosides from *P. ginseng*. Further, numerous other products obtained from CMCs of different plant species are now in development (Ochoa-Villarreal et al. 2016).

6.4 Scale-Up of Plant Cell Cultures in Bioreactors

For the increased and continuous production of natural products, the entire process of culture should be standardized in a large-scale bioreactor. A large array of bioreactor designs have been tested and used for plant cell cultures. The stirred tank reactors, air lift reactors and bubble column reactors are some of them (Verpoorte et al. 2002). The historical background of large-scale culture in bioreactors especially for plant cells starts from the continuous culture of tobacco cells in 20 kL fermentor (Hashimoto et al. 1982; Azechi et al. 1983). In 1982, Fujita et al. standardized the first industrial production of shikonin by using 750 L bioreactor from *Lithospermum erythrorhizon*. Apart from this, the well-known examples of large-scale production of plant cells for the secondary metabolite production includes *Catharanthus roseus* culture in 85 L bioreactor for the production of serpentine (Smart and Fowler 1981), *Coleus blumei* culture in 450 L bioreactor for rosmarinic acid production (Ulbrich et al. 1985), *Panax ginseng* culture in 2000 L bioreactor for saponin production (Ushiama et al. 1986), etc. Cell growth and secondary metabolite production represent the main factors for selecting the suitable process mode such as a batch culture, fed batch culture, rapid fed batch culture, two-stage batch culture, and continuous culture for a particular species (Ruffoni et al. 2010). Depending upon the agitation system bioreactors can be mechanically agitated or pneumatically agitated. The important factors affecting the growth in bioreactors include supply of oxygen and carbon dioxide exchange, pH, minerals, carbohydrates, growth regulators, the liquid medium agitation and cell density (Heyerdahl et al. 1995; Kieran et al. 1997). The commercial usefulness of these processes is related to the value of the desired metabolite. During the culture, the products are either released into the medium or accumulated in the cells. After a suitable incubation period the biomass is harvested. After the outstanding production of shikonin from *Lithospermum erythrorhizon* by Mitsui petrochemicals industry Co. Ltd. in Japan, another initiative was taken by Nitto Denko Co. Ltd. for the mass production of *Panax ginseng* or Ginseng cells in 20 kL tanks and following their success large-scale plant cell culture has rapidly expanded as a new trend for the production of valuable secondary metabolites, and the investigations are still going on to extend the opportunity of this new technique to numerous species and also for the production of novel compounds.

6.5 Conclusions

Plant metabolites have always been a favourite topic with the world because of its varied uses. Since centuries back, plant secondary metabolites have been contributing to the area like medicine, cosmetics, food, etc. Secondary metabolites from natural herbs are inadequate for meeting the market demand, so we have to rely on other sources. Plant cell culture technique has emerged and evolved as an alternative strategy for the production of secondary metabolites which find application in various sections of industries like flavours, fragrances, pharmaceuticals, colouring agents, food additives, etc. The production of secondary metabolites through in vitro techniques has significantly improved through development of various approaches like application of elicitators, supplementation of precursors, and applying biotransformation, immobilization and cambial meristematic cell culture technologies, etc. One of the great advantages of this science is that by this method we can produce the desirable quantity of metabolites in lab, without exploiting nature.

References

Adamse P (1990) Selection of high-yielding cell lines of Tagetes using flow cytometry. In: Nijkamp HJ (ed) Progress in plant cellular and molecular biology. Kluwer, Dordrecht, pp 726–731

Aftab T, Khan MM, Idrees M, Naeem M, Singh M, Ram M (2010) Stimulation of crop productivity, photosynthesis and artemisinin production in *Artemisia annua* L. by triacontanol and gibberellic acid application. J Plant Interact 1:273–281

Aiken MM, Yeoman MM (1986) A rapid screening technique for the selection of high yielding capsaicin cell lines of *Capsicum frutescens* Mill. In: Morris P (ed) Secondary metabolism in plant cell cultures. Cambridge University Press, London, pp 224–229

Ambid C, Fallot J (1981) Role of the gaseous environment on volatile compound production by fruit cell suspension cultured *in vitro*. In: Schreier P (ed) Flavour '81. de Gruyter, Berlin, pp p529–p538

Archambault J, Volesky B, Kurz WGW (1989) Surface immobilization of plant cells. Biotechnol Bioeng 33:293–299

Arehzoo Z, Christina S, Florian G, Parvaneh A, Javad A, Seyed HM, Christoph W (2015) Effects of some elicitors on tanshinone production in adventitious root cultures of *Perovskia abrotanoides* Karel. Ind Crop Prod 67:97–102

Asada M, Shuler ML (1989) Stimulation of ajmalicine production from *Catharanthus roseus*: effects of adsorption *in situ*, elicitors, and alginate immobilization. Appl Microbiol Biotechnol 30:475–481

Azechi S, Hashimoto T, Yuyama T, Nagatuska S, Nakashizuka M, Nishiyama T, Murata A (1983) Continuous cultivation of tobacco plant cells in an industrial scale plant. Hakkokogaku 61:117–128

Bais HP, Sudha G, George J, Ravishankar GA (2001) Influence of exogenous hormones on growth and secondary metabolite production in hairy root cultures of *Cichorium intybus* L. cv. Lucknow local. In Vitro Cell Dev Biol Plant 37(2):293–299

Baldwin IT (1989) Mechanism of damage-induced alkaloid production in wild tobacco. J Chem Ecol 15:1661–1680

Baldwin IT (1998) Jasmonate-induced responses are costly but benefit plants under attack in native populations. Proc Natl Acad Sci 95:8113–8118

Baque MA, Moh SH, Lee EJ, Zhong JJ, Paek KY (2012) Production of biomass and useful compounds from adventitious roots of high-value-added medicinal plants in bioreactor. Biotechnol Adv 30:1255–1267

Bentley R (1999) Secondary metabolite biosynthesis: the first century. Crit Rev Biotechnol 19:1–40

Berlin J (1980) Para-flurophenylalanine resistant cell lines of tobacco. Z Pflanzenphysio 97(4):317–324

Berlin J, Mollenschott C, Wray V (1988) Triggered efflux of protoberberine alkaloids from cell suspension culture of *Thalictrum rugosum*. Biotechnol Lett 10:193–198

Bhattacharya A, Sood P, Citovsky V (2010) The roles of plant phenolics in defence and communication during *Agrobacterium* and *Rhizobium* infection. Mol Plant Pathol 11(5):705–719

Bhojwani SS, Razdan MK (1996) Plant tissue culture: theory and practice a revised edition. Elsevier, Dordrecht

Bischoff KL (2016) Glucosinolates. Nutraceuticals, pp 551–554

Boatwright JL, Pajerowska-Mukhtar K (2013) Salicylic acid: an old hormone up to new tricks. Mol Plant Pathol 114(6):623–634

Bramble JL, Graves DJ, Brodelius P (1991) Calcium and phosphate effects on growth and alkaloid production in *Coffea arabica*: experimental results and mathematical model. Biotechnol Bioeng 37:859–868

Brodelius P, Nilsson K (1983) Permeabilization of cultivated plant cells, resulting in release of intracellularly stored products with preserved cell viability. Eur J Appl Microb Biotechnol 17:275–280

Brodelius P, Deus B, Mosbach K, Zenk MH (1979) Immobilized plant cells for the production of natural products. FEBS Lett 103:93–97

Brodelius P, Funk C, Shillito RD (1988) Permeabilization of cultivated plant cells by electroporation for release of intracellularly stored secondary products. Plant Cell Rep 7:186–188

Caarls L, Pieterse CM, Van Wees SC (2015) How salicylic acid takes transcriptional control over jasmonic acid signalling. Front Plant Sci 6:170

Caretto S, Quarta A, Durante M, Nisi R, De Paolis A, Blando F, Mita G (2010) Methyl jasmonate and miconazole differently affect arteminisin production and gene expression in *Artemisia annua* suspension cultures. Plant Biol (Stuttg) 13(1):51–58

Chattopadhyay S, Farkya S, Srivastava AK, Bisaria VS (2002) Bioprocess considerations for production of secondary metabolites. Biotechnol Bioprocess Eng 7:138–149

Chodisetti B, Rao K, Gandi S, Giri A (2013) Improved gymnemic acid production in the suspension cultures of *Gymnema sylvestre* through biotic elicitation. Plant Biotechnol Rep 7:519–525

Cioc M, Szewczyk A, Zupnik M (2018) LED lighting affects plant growth, morphogenesis and phytochemical contents of *Myrtus communis* L. In Vitro Plant Cell Tissue Organ Cult 132:433

Constabel F, Vasil IK (1988) Cell culture and somatic cell genetics of plants: phytochemicals in plant cell culture. Academic Press Inc., London

Contin A, van der Heijden R, Verpoorte R (1999) Effects of alkaloid precursor feeding and elicitation on the accumulation of secologanin in a *Catharanthus roseus* cell suspension culture. Plant Cell Tissue Organ Cult 56:111–119

Cresswell RC, Fowler MW, Stafford A, Stephan-Sarkissian G (1989) Inputs and outputs: primary substrates and secondary metabolism. In: Kurz WGW (ed) Primary and secondary metabolism of plant cell culture. ll. Springer, Heidelberg

Da Silva MF, Soares MS, Fernandes JB, Vieria PC (2007) Alkyl, aryl, alkylarylquinoline, and related alkaloids. Alkaloids Chem Biol 64:139–214

Dave V, Khirwadkar P, Dashora K (2014) A review on biotransformation. Iindian J Res Pharm Biotechnol 2(2):1136–1140

Dewick PM (2002) Medicinal natural products. John Wiley & Sons Ltd, New York, p 495

Dicosmo F, Misawa M (1995) Plant cell and tissue culture: alternatives for metabolite production. Biotechnol Adv 13:425–435

Dixon RA (2001) Natural products and plant disease resistance. Nature 411:843–847

Doller G (1978) Influence of the medium on the production of serpentine by suspension culture of *Catharanthus roseus* (L.) G Don. In: Alfermann W, Reinhard E (eds) Gesellschaft flir Strahlen- und Umweltforschung, Minchen, pp 109–117

Doller Von G, Alfermann AN, Reinhard E (1976) Production of indole alkaloids in tissue culture of *Catharanthus roseus*. Planta Med 30:14–20

Dörnenburg H, Knorr D (1995) Strategies for the improvement of secondary metabolite production in plant cell cultures. Enzym Microb Technol 17(8):674–684

Dougall DK (1980) Nutrition and metabolism. In: Staba EJ (ed) Plant tissue culture as a source of biochemical. CRC Press, Boca Raton, pp 21–58

Dunn WB, Broadhurst DI, Atherton HJ, Goodacre R, Griffin JL (2011) Systems level studies of mammalian metabolomes: the roles of mass spectrometry and nuclear magnetic resonance spectroscopy. Chem Soc Rev 40:387–426

Felix H (1982) Permeabilized cells. Anal Biochem 120:21-1–21-34

Fett-Neto AGJM, Stewart SA, Nicholson JJ, Pennington DiCosmo F (1994) Improved taxol yield by aromatic carboxylic acid and amino acid feeding to cell cultures of *T. cuspidata*. Biotechnol Bioeng 44:967–971

Firouzi A, Mohammadi SA, Khosrowchahli M, Movafeghi A, Hasanloo T (2013) Enhancement of silymarin production in cell culture of *Silybum marianum* (L) Gaertn by elicitation and precursor feeding. Int J Geogr Inf Syst 19(3):262–274

Fowler MW, Stephan-Sarkissian G (1985) Carbohydrate source, biomass productivity and natural product yield in cell suspension culture. In: Neumann KH, Braz W, Reinhard E (eds) Primary and secondary metabolism of plant cell cultures. Springer, Berlin, pp 66–73

Fujita Y, Tabata M (1987) Secondary metabolites from plant cells: pharmaceutical applications and progress in commercial production. In: Green CE, Somers DA, Hackett WP, Biesboer DD (eds) Plant tissue and cell culture. Alan R. Liss, New York, pp 169–185

Fujita Y, Hara Y, Suga C, Morimoto T (1981) Production of shikonin derivatives by cell suspension culture of *Lithospermum erythrorhizon* ll. A new medium for the production of shikonin derivatives. Plant Cell Rep 1:61–63

Fujita Y, Tabata M, Nishi A, Yamada Y (1982) New medium and production of secondary compounds with the two staged culture method. In: Fujiwara A (ed) Plant tissue culture. Maruzen, Tokyo, pp 399–400

Furusawa M, Hashimoto T, Noma Y, Asakawa Y (2005) Biotransformation of *Citrus aromatics* Nootkatone and valencene by microorganisms. Chem Pharm Bull 53(11):1423–1429

Furuya T, Kojima H, Syono K (1971) Regulation of nicotine biosynthesis by auxins in tobacco callus tissues. Phytochemistry 10:1529–1532

Furuya T, Yoshikawa T, Taira M (1984) Biotransformation of codeinone to codeine by immobilized cells of *Papaver somniferum*. Phytochemistry 23(5):999–1001

Gangopadhyay M, Dewanjee S, Bhattacharya S (2011) Enhanced plumbagin production in elicited *Plumbago indica* hairy root cultures. J Biosci Bioeng 111:706–710

George EF, Hall MA, Klerk GJD (2008) The components of plant tissue culture media I: macro- and micro-nutrients. In: George EF, Hall MA, Klerk GJD (eds) Plant propagation by tissue culture. Springer, Dordrecht, pp 65–113

Georgiev MI, Weber J, Maciuk A (2009) Bioprocessing of plant cell cultures for mass production of targeted compounds. Appl Microbiol Biotechnol 83:809–823

Giri C, Zaheer M (2016) Chemical elicitors versus secondary metabolite production *in vitro* using plant cell, tissue and organ culture: recent trends and a sky view appraisal. Plant Cell Tissue Organ Cult 126(1):1

Hanson JR (2001) The development of strategies for terpenoid structure determination. Nat Prod Rep 18:607–617

Hara Y, Yamagata H, Morimoto T, Hiratsuka J, Yoshioka T, Fujita Y, Yamada Y (1989) Flow cytometric analysis of cellular berberine contents in high and low-producing cell lines of *Coptis japonica* obtained by repeated selection. Planta Med 55:151–154

Harkes PAA, De Jong PJ, Wiijnsma R, Verpoorte R, Van Der Leer T (1986) Influence of production media on *Cinchona* cell culture; spontaneous formation of B- carbolines from L-tryptophan. Plant Sci 47:71

Hashimoto T, Azechi S, Sugita S, Suzuki K (1982) Large-scale production of tobacco cells by continuous cultivation. In: Fujiwara A (ed) Plant tissue culture. Maruzen, Tokyo, pp 403–404

Heyerdahl PHO, Olsen AS, Hvoslef-Eide AK (1995) Engineering aspects of plant propagation in bioreactors. In: Aitken-Christie J, Kozai T, Smith MAL (eds) Automation and environment control in plant tissue culture. Springer, Dordrecht, pp 87–123

Hobbs MC, Yeoman MM (1991) Biotransformation of nicotine to nornicotine by cell suspension of *Nicotiana tabacum* L CV. Wisconsin-38. New Phytol 119:477–482

Hu FX, Huang J, Xu Y, Qian XH, Zhong JJ (2006) Responses of defence signals, biosynthetic gene transcription and taxoid biosynthesis to elicitation by a novel synthetic jasmonate in cell cultures of *Taxus chinensis*. Biotechnol Bioeng 94:1064–1071

Ibrahim RK (1987) Regulation of synthesis of phenolics. In: Constabel F, Vasil IK (eds) Cell cultures and somatic cell genetics of plants. Academic Press, San Diego, pp 77–95

Idrees M, Khan MM, Aftab T, Naeem M, Hashmi N (2010a) Salicylic acid-induced physiological and biochemical changes in lemongrass varieties under water stress. J Plant Interact 5(4):293–303

Idrees M, Khan MM, Aftab T, Naeem M (2010b) Synergistic effects of gibberellic acid and triacontanol on growth physiology, enzyme activities and essential oil content of *Coriandrum sativum* L. Asian Austalas J Plant Sci Biotechnol 4:24–29

Idrees M, Naeem M, Aftab T, Khan MM (2011) Salicylic acid mitigates salinity stress by improving antioxidant defence system and enhances vincristine and vinblastine alkaloids production in periwinkle [*Catharanthus roseus* (L) G. Don]. Acta Physiol Plant 33(3):987–999

Ilieva M, Pavlov A (1996) Rosmarinic acid by *Lavandula vera* MM cell suspension: phosphorus effect. Biotechnol Lett 50:189–229

Izabela G, Halina W (2009) The effect of methyl jasmonate on production of antioxidant compounds in shoot cultures of *Salvia officinalis* L. Herba Pol 55:238–243

Jamwal K, Bhattacharya S, Puri S (2018) Plant growth regulator mediated consequences of secondary metabolites in medicinal plants. J Appl Res Med Aromat Plants 9:26–38

Jeong CS, Murthy HN, Hahn EJ, Paek KY (2008) Improved production of ginsenosides in suspension cultures of ginseng by medium replenishment strategy. J Biosci Bioeng 105:288–291

Joshi JB, Elias CB, Patole MS (1996) Role of hydrodynamic shear in the cultivation of animal, plant and microbial cells. Chem Eng J Biochem Eng J 62:121–141

Kabera JN, Semana E, Mussa AR, He X (2014) Plant secondary metabolites: biosynthesis, classification, function and pharmacological properties. J Pharm Pharmacol 2:377–392

Khan T, Krupadanam D, Anwar SY (2008) The role of phytohormone on the production of berberine in the calli cultures of an endangered medicinal plant, turmeric (*Coscinium fenestratum* 1.). Afr J Biotechnol 7(18):3244–3246

Kieran P, MacLoughlin P, Malone D (1997) Plant cell suspension cultures: some engineering considerations. J Biotechnol 59(1–2):39–52

Kilby NJ, Hunter CS (1990) Repeated harvest of vacuole located secondary product from *in vitro* grown plant cells using 1.02 MHz ultrasound. Appl Microbiol Biotechnol 33:44–51

Kim DJ, Chang HN (1990) Enhanced shikonin production from *Lithospermum erythrorhizon* by *in situ* extraction and calcium alginate immobilization. Biotechnol Bioeng 36(5):460–466

Klein TM, Wolf ED, Wu R, Sanford JC (1987) High-velocity micro projectiles for delivering nucleic acids into living cells. Nature 327(6117):70–73

Knobloch KH, Berlin J (1980) Influence of medium composition on the formation of secondary compounds in cell suspension cultures of *Catharanthus roseus*. L. G. Don. Z Naturforsch 35C:551–556

Knorr D, Berlin J (1987) Effects of immobilization and permeabilization procedures on growth of *Chenopodium rubrum* cells. J Food Sci 52:1397–1400

Kobayashi Y, Fukui H, Tabata M (1991) Effect of carbon dioxide and ethylene on berberine production and cell browning in *Thalictrum minus* cell cultures. Plant Cell Rep 9:496–449

Kreis W, Reinhard E (1992) 12b-hydroxylation of digitoxin by suspension cultured *Digitalis lanata* cells: production of digoxin in 20 L and 300 L airlift reactors. J Biotechnol 26:257–273

Kurz WGW, Constabel F (1979) Plant cell cultures, a potential source of pharmaceuticals. Adv Appl Microbiol 25:209–240

Lee EK, Jin YW, Park JH (2010) Cultured cambial meristematic cells as a source of plant natural products. Nat Biotechnol 28:1213–1217

Lee-Parsons CWT, Shuler ML (2002) The effect of ajmalicine spiking and resin addition timing on the production of indole alkaloids from *Catharanthus roseus* cell cultures. Biotechnol Bioeng 79(4):408–415

Li W, Koike K, Asada Y, Yoshikawa T, Nikaido T (2005) Biotransformation of paeonol by *Panax ginseng* root and cell cultures. J Mol Catal B Enzym 35(4–6):117–121

Liu XN, Zhang XQ, Sun JS (2007) Effects of cytokinins and elicitors on the production of hypericins and hyperforin metabolites in *Hypericum sampsonii* and *Hypericum perforatum*. Plant Growth Regul 53:207–214

Mansouri H, Asrar Z, Szopa J (2009) Effects of ABA on primary terpenoids and tetrahydrocannabinol in *Cannabis sativa* L. at flowering stage. Plant Growth Regul 58:269–277

Mantell SH, Smith H (1984) Cultural factors that influence secondary metabolite accumulation in plant cell and tissue cultures. In: Mantell SH, Smith H (eds) Plant biotechnology. Cambridge University Press, Cambridge, pp 75–108

Matsubara K, Shigekazu K, Yoshioka T, Fujita Y, Yamada Y (1989) High density culture of *Coptis japonica* cells increases berberine production. J Chem Technol Biotechnol 46:61–69

McDonald KA, Jackman AP (1989) Bioreactor studies of growth and nutrient utilization in *Alfalfa* suspension cultures. Plant Cell Rep 8:455–458

Meyer HJ, Van Staden J (1995) The *in vitro* production of anthocyanin from callus cultures of *Oxalis linearis*. Plant Cell Tissue Organ Cult 40:55–58

Michael JP (2008) Quinoline, quinazoline and acridone alkaloids. Nat Prod Rep 25:166–187

Misawa M (1985) Production of useful plant metabolites. In: Fiechter A (ed) Plant cell culture, Advances in biochemical engineering/biotechnology. Springer, Berlin, pp 59–88

Mizukami H, Konoshima M, Tabata M (1977) Effect of nutritional factors on shikonin derivative formation in *Lithospermum* callus cultures. Phytochemistry 16:1183

Mok MC, Gabelman W, Skoog F (1976) Carotenoid synthesis in tissue cultures of *Daucus carota*. J Am Soc Hortic Sci 101:442–449

Moreno PRH, Van der Heijden R, Verpoorte R (1993) Effect of terpenoid precursor feeding and elicitation on formation of indole alkaloids in cell suspension cultures of *Catharanthus roseus*. Plant Cell Rep 12:702–705

Moreno PRH, van der Heijden R, Verpoorte R (1995) Cell and tissue cultures of *Catharanthus roseus*: a literature survey. Plant Cell Tissue Org Cult 42(1):1–25

Mukherjee SK, Sabapathi RB, Gupta N (1991) Low sugar and osmotic requirements for shoot regeneration from leaf pieces of *Solanum melangena* L. Plant Cell Tissue Organ Cult 25:13–16

Mukundan U, Bhide V, Singh G (1998) pH-mediated release of betalains from transformed root cultures of *Beta vulgaris* L. Appl Microbiol Biotechnol 50:241

Mulder-Krieger T, Verpoorte R, Svendse A, Cheffer J (1988) Production of essential oils and flavours in plant cell and tissue cultures. A review. Plant Cell Tissue Organ Cult 25:13–26

Munish S, Ashok A, Rajinder G, Sharada M (2015) Enhanced bacoside production in shoot cultures of *Bacopa monnieri* under the influence of abiotic elicitors. Nat Prod Res 29(8):745–749

Murthy HN, Dandin VS, Zhong JJ, Paek KY (2014) Strategies for enhanced production of plant secondary metabolites from cell and organ cultures. In: Paek KY, Murthy H, Zhong JJ (eds) Production of biomass and bioactive compounds using bioreactor technology. Springer, Dordrecht

Naeem M, Khan MM, Siddiqui MH (2009) Triacontanol stimulates nitrogen-fixation, enzyme activities, photosynthesis, crop productivity and quality of hyacinth bean (*Lablab purpureus* L.). Sci Hortic 121(4):389–396

Naeem M, Idrees M, Aftab T, Khan MM, Moinuddin MH (2010) Changes in photosynthesis, enzyme activities and production of anthraquinone and sennoside content of *Coffee senna* (*Senna occidentalisi* L.) by triacontanol. Int J Dev Biol 4:53–59

Naeem M, Khan MM, Idrees M, Aftab T (2011) Triacontanol mediated regulation of growth yield, physiological activities and active constituents of *Mentha arvensis* L. Plant Growth Regul 65:195–206

Nagella P, Murthy HN (2011) *In vitro* production of gymnemic acid from cell suspension cultures of *Gymnema sylvestre*. R. Br. Eng Life Sci 11:537–540

Naik PM, Manohar SH, Praveen N, Murthy HN (2010) Effects of sucrose and pH levels on *in vitro* shoot regeneration from leaf explants of *Bacopa monnieri* and accumulation of bacoside A in regenerated shoots. Plant Cell Tissue Organ Cult 100:235–239

Naik PM, Manohar SH, Praveen N, Upadhya V, Murthy HN (2012) Evaluation of bacoside A content in different accessions and various organ of *Bacopa monnieri* (L.) Wettst. Int J Geogr Inf Syst 18:387–395

Ng YP, Or TCT, Ip NY (2015) Plant alkaloids as drug leads for Alzheimer's disease. Neuro Chem Int 89:260–270

Ochoa-Villarreal M, Howat S, Hong S, Jang MO, Jin YW, Lee EK, Loake GJ (2016) Plant cell culture strategies for the production of natural products. BMB Rep 49(3):149–158

Ogino T, Hiraoka N, Tabata M (1978) Selection of high nicotine producing cell lines of tobacco callus by single cell cloning. Phytochemicals 22:2447–2450

Panda AK, Bisaria VS, Mishra S (1992) Alkaloid production by plant cell cultures of *Holarrhena antidysenterica*: II. Effect of precursor feeding and cultivation in stirred tank bioreactor. Biotechnol Bioeng 39:1052–1057

Park ES, Moon WS, Song MJ, Kim MN, Chung KH, Yoon JS (2001) Antimicrobial activity of phenol and benzoic acid derivatives. Int Biodeterior Biodegradation 47(4):209–214

Parr AJ, Robins RJ, Rhodes MCJ (1984) Permeabilization of *Cinchona ledgeriana* cells by dimethylsulfoxide. Effects on alkaloid release and long-term membrane integrity. Plant Cell Rep 3:262–266

Payne GF, Shuler ML (1988) Selective adsorption of plant products. Biotechnol Bioeng 31 (9):922–928

Pedrosal RCN, Branquinho NAA, Hara ACBAM, Costa AC, Silva FG, Pimenta LP et al (2017) Impact of light quality on flavonoid production and growth of *Hyptis marrubioides* seedlings cultivated *in vitro*. Rev Bras Farmacogn 27(4):466–470

Poulev A, O'Neal JM, Logendra S, Pouleva RB, Timeva V, Garvey AS, Gleba D, Jenkins IS, Halpern BT, Kneer R, Cragg GM, Raskin I (2003) Elicitation, a new window into plant chemodiversity and phytochemical drug discovery. J Med Chem 46:2542–2547

Praveen N, Murthy HN (2010) Establishment of cell suspension cultures of *Withania somnifera* for the production of withanolide A. Bioresour Technol 101:6735–6739

Pyne ME, Narcross L, Martin VJJ (2019) Engineering plant secondary metabolism in microbial systems. Plant Physiol 179(3):844–861

Qian ZG, Zhao ZJ, Xu YF, Qian XH, Zhong JJ (2005) Highly efficient strategy for enhancing taxoid production by repeated elicitation with a newly synthesized jasmonate in fed-batch cultivation of *Taxus chinensis* cells. Biotechnol Bioeng 90:516–521

Qiao W, Fan LM (2008) Nitric oxide signaling in plant responses to abiotic stresses. J Integr Plant Biol 50(10):1238–1246

Radic S (2016) Influence of pH and plant growth regulators on secondary metabolite production and antioxidant activity of *Stevia rebaudiana* (Bert). Period Boil 118(1):9–19

Radman R, Saez T, Bucke C, Keshavarz T (2003) Elicitation of plant and microbial systems. Biotechnol Appl Biochem 37:91–102

Rajasekaran T, Ravishankar GA, Venkataraman LV (1991) Influence of nutrient stress on pyrethrin production by cultured cells of pyrethrum (*Chrysanthemum cinerariaefolium*). Curr Sci 60:705–707

Rao SR, Ravishankar GA (2002) Plant cell cultures: chemical factories of secondary metabolites. Biotechnol Adv 20:101–153

Ratledge C, Sasson A (1992) Production of useful biochemicals by higher-plant cell cultures: biotechnological and economic aspects. Cambridge University Press, Cambridge, p 402

Ravishankar GA, Sharma KS, Venktaraman LV, Kodyan AK (1988) Effect of nutritional stress on capcinin production in immobilized cell cultures of *Capsicum annum*. Curr Sci 57:381–383

Rhodes MJC (1994) Physiological roles for secondary metabolites in plants: some progress, many outstanding problems. Plant Mol Biol 24(1):1–20

Roberts S, Kolewe M (2010) Plant natural products from cultured multipotent cells. Nat Biotechnol 28:1175–1176

Rojbayani SAK (2007) Induction and stimulation of some phytochemicals from *Hypericum perforatum* L. *in vitro*. PhD dissertation, Al Mustansiriya University

Romagnoli LG, Knorr D (1988) Effects of ferulic acid treatment on growth and flavour development of cultured *Vanilla planifolia* cells. Food Biotechnol 2(1):93–104

Ruffoni B, Pistelli L, Bertoli A, Pistelli L (2010) Plant cell cultures: bioreactors for industrial production. In: Giardi MT, Giuseppina R, Bruno B (eds) Bio-farms for nutraceuticals, pp 203–221

Saenz-Carbonell LA, Maldonado-Mendoza IE, Moreno-Valenzula O (1993) Effect of the medium pH on the release of secondary metabolites from roots of *Datura stramonium*, *Catharanthus roseus* and *Tagetes patula* cultured *in vitro*. Appl Biochem Biotechnol 38:257

Sahai OP, Shuler ML (1984) Environmental parameters influencing phenolics production by batch cultures of *Nicotiana tabacum*. Biotechnol Bioeng 26:111–120
Sasse F, Heckenberg U, Berlin J (1982) Accumulation of B-Carboline alkaloids and serotonin by cell cultures of *Peganum harmala*. Z Pflanzenphysiol l05:315
Seigler DS (1998) Plant secondary metabolism. Springer Science+Business media, New York
Seitz HU, Hinderer W (1988) Anthocyanins. In: Constabel F, Vasil I (eds) Cell culture and somatic cell genetics of plants, vol 5. Academic, San Diego, pp 49–76
Seo JM, Arasu MV, Kim YB, Park SU, Kim SJ (2015) Phenylalanine and LED lights enhance phenolic compound production in Tartary buckwheat sprouts. Food Chem 177:204–213
Sharma P, Yadav S, Srivastava A, Shrivastava N (2013) Methyl jasmonate mediates up-regulation of bacoside A production in shoot cultures of *Bacopa monnieri*. Biotechnol Lett 35:1121–1125
Sharma M, Ahuja A, Gupta R, Mallubhotla S (2014) Enhanced bacoside production in shoot cultures of *Bacopa monnieri* under the influence of abiotic elicitors. Nat Prod Res 29 (8):745–749
Shilpashree HP, Rai VR (2009) *In vitro* plant regeneration and accumulation of flavonoids in *Hypericum mysorense*. Int J Integr Biol 8(1):43–49
Shohael AM, Ali MB, Yu KW, Hahn EJ, Paek KY (2006) Effect of temperature on secondary metabolites production and antioxidant enzyme activities in *Eleutherococcus senticosus* somatic embryos. Plant Cell Tissue Organ Cult 85:219–228
Shoji T, Hashimoto T (2013) Smoking out the masters: transcriptional regulators for nicotine biosynthesis in tobacco. Plant Biotechnol 30:217–224
Silja PK, Gisha GP, Satheeshkumar K (2014) Enhanced plumbagin accumulation in embryogenic cell suspension cultures of *Plumbago rosea* L. following elicitation. Plant Cell Tissue Organ Cult 119:469–477
Sivakumar G, Yu KW, Hahn EJ, Paek KY (2005a) Optimization of organic nutrients for ginseng hairy roots production in large-scale bioreactors. Curr Sci 89:641–649
Sivakumar G, Yu KW, Paek KY (2005b) Production of biomass and ginsenosides from adventitious roots of *Panax ginseng* in bioreactor cultures. Eng Life Sci 5:333–342
Sivanandhan G, Kapil Dev G, Jeyaraj M (2013) Increase production of withanolide A, withannone and withaferin A in hairy root cultures of *Withania somnifera* (L.) Dunal elicited with methyl jasmonate and salicylic acid. Plant Cell Tissue Organ Cult 114:121–129
Smart NJ, Fowler MW (1981) Effect of aeration on large scale culture of plant cells. Biotechnol Lett 3:171–176
Smetanska I (2008) Production of secondary metabolites using plant cell cultures. Adv Biochem Eng Biotechnol 111:187–228
Sonja GS, Oliver T, Stéphane M, Alain D, Claude J, Daniel H (2014) Effects of polysaccharide elicitors on secondary metabolite production and antioxidant response in *Hypericum perforatum* L. shoot cultures. Sci World J 11:609–649
Srinivasan V, Ryu DD (1993) Improvement of shikonin productivity in *Lithospermum erythrorhizon* cell cultures by altering carbon and nitrogen feeding strategy. Biotechnol Bioeng 42:793–799
Sriskandarajah S, Prinsen E, Motyka V, Dobrev PI, Serek M (2006) Regenerative capacity of cacti *Schlumbergera* and *Rhipsalidopsis* in relation to endogenous phytohormones, cytokinin oxidase/ dehydrogenase, and peroxidase activities. J Plant Growth Regul 25:79–88
Stephan-Sarkissian G, Fowler MW (1986) The metabolism and utilization of carbohydrates by suspension cultures of plant cells. In: Morgan MJ (ed) Carbohydrate metabolism in cultured cells. Plenum, New York, pp 151–181
Sun L, Su H, Zhu Y, Xu M (2012) Involvement of abscisic acid in ozone-induced puerarin production of *Pueraria thomsnii* Benth. Suspension cell cultures. Plant Cell Rep 31:179–185
Syklowska-Baranek K, Pietrosiuk A, Gawron A (2012) Enhanced production of antitumour naphthoquinones in transgenic hairy root lines of *Lithospermum canescens*. Plant Cell Tissue Organ Cult 108:213–219

Tabata M, Yamamoto H, Hiraoka N, Konoshima M (1972) Organization and alkaloid production in tissue cultures of *Scopolia parviflora*. Phytochemistry 11:949–955

Tabata M, Mizukami H, Hiraoka N, Konoshima M (1974) Pigment formation in callus cultures of *Lithospermum erythrorhizon*. Phytochemistry 13:927–932

Tanaka H, Hirano C, Semba H, Towaza Y, Ohmomo Y (1985) Release of intracellularly stored 5′-phosphodiesterase with preserved plant cell viability. Biotechnol Bioeng 27:890–892

Tepe B, Sokmen A (2007) Production and optimisation of rosmarinic acid by *Satureja hortensis* L. callus cultures. Nat Prod Res 21:1133–1144

Thanh NT, Murthy HN, Pandey DM, Yu KW, Hahn EJ, Paek KY (2006) Effect of carbon dioxide on cell growth and saponin production in suspension cultures of *Panax ginseng*. Biol Plant 50:752–754

Thomas YDS (2009) *In vitro* culture of *Camptotheca acuminata* Decaisne in temporary immersion immersion system (TIS): growth, development and production of secondary metabolites. Dissertation, University of Hamburg

Toivonen L, Laakso S, Rosenqvist H (1992) The effect of temperature on growth, indole alkaloid accumulation and lipid composition of *Catharanthus roseus* cell suspension cultures. Plant Cell Rep 11:390–394

Tonelli M, Pellegrini E, D'Angiolillo F, Nali C, Pistelli L, Lorenzini G (2015) Ozone–elicited secondary metabolites in shoot cultures of *Melissa officinalis* L. Plant Cell Tissue Organ Cult 120:617–629

Ulbrich B, Wiesner W, Arens H (1985) Large scale production of rosmarinic acid from plant cell cultures of *Coleus blumei* Benth. In: Deus-Neumann B, Barz W, Reinhard E (eds) Secondary metabolism of plant cell culture. Springer, Berlin, pp 293–303

Ushiama K, Oda H, Westphal K (1986) Large scale tissue culture of *Panax ginseng* root. In: Somers B, Genbach D, Biesboer D, Hackett W, Green C (eds) Sixth international congress of plant tissue and cell culture. University of Minnesota, Minneapolis

Verpoorte R (2000) Secondary metabolism. In: Verpoorte R, Alfermann AW (eds) Metabolic engineering of plant secondary metabolism. Kluwer Academic Publishers, London, pp 1–29

Verpoorte R, Cotin A, Memelink J (2002) Biotechnology for the production of plants secondary metabolites. Phytochem Rev 1:13–25

Vickery ML, Vickery B (1981) Secondary plant metabolism. Macmillan Press, London

Wang Y, Weathers PJ (2007) Sugars proportionately affect artemisinin production. Plant Cell Rep 26:1073–1081

Wasternack C, Hause B (2013) Jasmonates: biosynthesis, perception, signal transduction and action in plant stress response, growth and development. Ann Bot 111(6):1021–1058

Weathers PJ, Bunk G, McCoy MC (2005) The effect of phytohormones on growth and artemisinin production in *Artemisia annaua* hairy roots. In Vitro Cell Dev Biol Plant 41:47–53

Wu CH, Murthy HN, Hahn EJ, Murthy HN (2007) Improved production of caffeic derivatives in suspension cultures of *Echinacea purpurea* by medium replenishment strategy. Arch Pharm Res 30:945–949

Xiaolong H, Min S, Lijie C, Chao X, Yanjie Z, Guoyin K (2015) Effects of methyl jasmonate and salicylic acid on tanshinone production and biosynthetic gene expression in transgenic *Salvia miltiorrhiza* hairy roots. Biotechnol Appl Biochem 62(1):24–31

Xu M, Yang B, Dong J, Lu D, Jin H, Sun L, Zhu Y, Xu X (2011) Enhancing hypericin production of *Hypericum perforatum* cell suspension culture by ozone exposure. Biotechnol Prog 27:1101–1106

Yamakawa T, Kato S, Ishida K, Kodama T, Minoda Y (1983) Production of anthocyanins by *Vitis* cells in suspension culture. Agric Biol Chem 47:2185–2191

Yeoman MM, Holden MA, Corchet P, Holden PR, Goy JG, Hobbs MC (1990) Exploitation of disorganized plant cultures for the production of secondary metabolites. In: Charlwood BV, Rhodes MJC (eds) Secondary products from plant tissue culture. Clarendon Press, Oxford, pp p139–p166

Zeng Y, Yan F, Tang L (2003) Increased crocin production and induction frequency of stigma like structure from floral organs of *Crocus sativus* L. by precursor feeding. Plant Cell Tissue Organ Cult 72:185–191

Zenk MH, El-Shagi H, Arens H, Stockigt J, Weiler EW Deus D (1977) Formation of the indole alkaloids serpentine and ajmalicine in cell suspension cultures of *Catharanthus roseus*. In: Barz WRE (ed) Plant tissue culture and its biotechnological application. Springer, Berlin, pp 27–44

Zenk M, El-Shagi H, Schulte U (1975) Anthraquinone production by cell suspension cultures of *Morinda citrifolia*. Planta Med 28(S 01):79–101

Zhao J, Davis LC, Verpoorte R (2005) Elicitor signal transduction leading to production of plant secondary metabolites. Biotechnol Adv 23(4):283–333

Zhong JJ (2001) Biochemical engineering of the production of plant-specific secondary metabolites by cell suspension cultures. Adv Biochem Eng Biotechnol 72:26

Zhou P, Yang J, Zhu J (2015) Effects of β-cyclodextrin and methyl jasmonate on the production of vindoline, catharanthine, and ajmalicine in *Catharanthus roseus* cambial meristematic cell cultures. Appl Microbiol Biotechnol 99:7035–7045

Invasive Alien Plants: A Potential Source of Unique Metabolites

T. K. Hrideek, Suby, and M. Amruth

Abstract

Most exotic invasive plants possess definite chemical advantage over their competitors, which account for their extraordinary success and domination over their host communities. Invasive plants deploy special metabolites to create unfavourable environment for growth and multiplication of competing native species. Though the chemical profile of allelo-metabolites in invasive species and their precise biochemical mechanism of action are diverse, their net effect and known action remains focussed more or less on growth retardation and inhibition of seed germination in target plants. While plant communities are affected by these allelo-metabolites resulting in poor species diversity, allelopathy also induces resistance towards certain allelochemicals in the affected populations as an adaptation and function as a stimulant for genetic variability in the populations. Knowledge on the mechanisms of detoxification of phytochemicals evolved as a defence mechanism by affected plants in the home range of invasives as well as in invaded locality will be helpful in combating exotic invasives as well as native weeds at minimal ecosystem damage. Genetic incorporation of such traits in native plants would also pay great dividends when they are deployed for reclaiming and restoring expensive ecosystem services that are lost due to the biological invasion. This chapter provides a broad overview of the process of plant invasions aided by allelo-metabolites and hint at a possible course of research for reversing such invasions.

T. K. Hrideek (✉) · Suby
Department of Forest Genetics and Tree Breeding, KSCSTE—Kerala Forest Research Institute, Peechi, Thrissur, Kerala, India

M. Amruth
Forestry and Human Dimension Programme Division, KSCSTE—Kerala Forest Research Institute, Peechi, Thrissur, Kerala, India

Keywords

Alien invasive plants · Invasion · Allelochemicals · Allelopathy · Detoxification

7.1 Introduction

Biological invasion is a total or near total domination of local plant or animal community by a single species or by a combination of species, which are called invasive species. Very often the invasive species are alien invasive, i.e. species that are introduced to a location (biotope or ecosystem) where they do not occur naturally. Most invasive species possess definite chemical advantage over their competitors, which account for their extraordinary success and domination in the biotic communities they prove to be invasive. The focus of discussion here is on invasive plant species, which are potential source of metabolites.

Invasive alien species (IAS) multiply and spread quickly; they monopolize ecosystem resources and thereby displace native plants; upset ecological dynamics resulting in large monoculture of invasive species in the landscapes. In other words, invasives succeed by altering the community structure and by appropriating the energy flow and material cycle of the system to own advantage. The process of biological invasion is enabled by a combination of ecological and socio-political factors operating in tandem with certain extraordinarily successful physiological and life history strategies of the invasive species in question. Thus, domination of the invasive species in the local plant community is often aided by factors such as high reproductive rate, fecundity, wide dispersal ability, absence of predation, extraordinary ecological amplitude of the species and their ability to competitively exclude the local species from the habitat. The traits that provide competitive advantage to IAS are part of unique survival strategies they evolved through natural selection and co-evolution. Some invasive species deploy special metabolites to create unfavourable environment for growth and multiplication of competing native species. Such metabolites are called allelochemicals and the process is called allelopathy. While control and prevention of biological invasion poses a serious challenge to the scientists, conservation professionals, park managers, agriculturists and policy makers worldwide, the scenario also opens up novel terrain for those who are in search of new metabolites and bioactive molecules. Before embarking on a detailed discussion on the allelopathic metabolites from IAS, it is important to have a look at the process of biological invasion and its impacts in detail.

7.2 Biological Invasion and Its Impact

Unlike the native species, which extend their territories by means of natural dispersal, IAS are mostly introduced outside their original home range by direct or indirect aid of human agency either intentionally or unintentionally as a part of the human enterprise. Invasive species could be flora, fauna or microbes (Fig. 7.1) and

7 Invasive Alien Plants: A Potential Source of Unique Metabolites

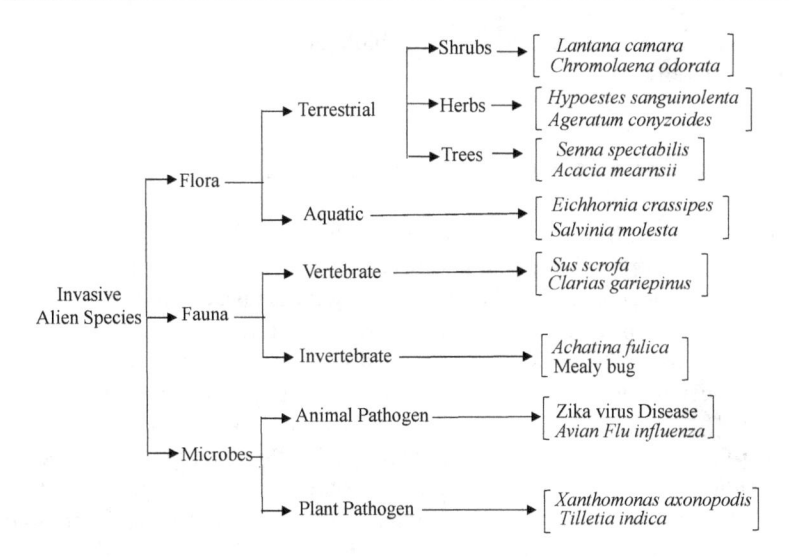

```
                                          ┌─ Lantana camara       ┐
                              ┌→ Shrubs → │  Chromolaena odorata  │
                              │           └                       ┘
                              │           ┌─ Hypoestes sanguinolenta ┐
                ┌→ Terrestrial├→ Herbs  → │  Ageratum conyzoides     │
                │             │           └                          ┘
                │             │           ┌─ Senna spectabilis ┐
                │             └→ Trees  → │  Acacia mearnsii   │
       ┌→ Flora ┤                         └                    ┘
       │        │                         ┌─ Eichhornia crassipes ┐
       │        └→ Aquatic ──────────────→│  Salvinia molesta     │
       │                                  └                       ┘
       │        ┌→ Vertebrate ───────────→┌─ Sus scrofa        ┐
Invasive│        │                         │  Clarias gariepinus│
Alien Species ├→ Fauna ┤                   └                    ┘
       │        │                         ┌─ Achatina fulica ┐
       │        └→ Invertebrate ─────────→│  Mealy bug       │
       │                                  └                  ┘
       │          ┌→ Animal Pathogen ────→┌─ Zika virus Disease ┐
       └→ Microbes┤                        │  Avian Flu influenza│
                  │                        └                     ┘
                  └→ Plant Pathogen ──────→┌─ Xanthomonas axonopodis┐
                                           │  Tilletia indica       │
                                           └                        ┘
```

Fig. 7.1 Diversity of invasive lifeforms and some examples of invasive species

some of them could prove to be economically benign. However, an alien species is designated as invasive only when it inflicts an adverse impact on the environment, economy or human health. Over the past 200 years, there is an upsurge of incidence involving the introduction of non-native species outside their original home range due to the increased movement of humans across the world.

There exist numerous theories to explain the success of plant invasions in terms of biotic and abiotic factors. 'Novel Weapon Hypothesis' put forth by Callaway and Ridenour (2004) proposes that in an exotic habitat, the invasive species have a competitive advantage over the native species because they possess a novel weapon, i.e. a trait that is new to the host community of native species. This enables the exotic species to emerge as dominant and invasive in a community of native species. Such physiological and life history strategies are traits that allow certain species to establish earlier, faster, and in greater numbers. These traits include adaptations to reproduce both asexually and sexually, accelerated growth and rapid multiplication, high dispersal ability, tolerance to a wide range of environmental conditions, phenotypic plasticity, ability to survive on wide range of nutritional modes and food types, association with humans, etc. This also helps to explain why certain alien invasive species remain non-invasive in their original home range where its native biotic community possess competitive and adaptive strategies to put population of the species under check. When introduced in a new community, which lacks such specific checks and adaptive strategies, the invasive potential of the species is unleashed and culminate in total domination of the habitat by them.

Complementing to these strategies is the allelopathy exerted by some of the invasive plant species, which deploy allelochemicals into the habitat, thereby inflicting a variety of detrimental and benign effects on the growth and survival of their host plant community. The invasive species are widely accepted to have adverse impacts on the ecosystem and biological diversity, animal health and national economy.

7.2.1 Impact on Ecosystem and Biodiversity

Ecological impacts are those that affect quality, structure and function of ecosystems, often referring to the loss of unique habitats and erosion of biological diversity. Immediate and direct impacts of biological invasion are on native biota, leading to decline in their population and species extinctions in extreme events of biological invasion. Species with narrow distributional limits, such as endemic species, succumb to extinction easily. Invasive plants disrupt or capture the established flow of energy and cycling of nutrients in the ecosystems, thereby reducing the availability of resources for the native flora and fauna. Consequences of invasion are far-reaching. For instance, disruption of competitive or mutualistic relationships among the species in a community leads to changes in community compositions and vegetation structure, which alter the qualitative aspects of landscapes including the aesthetic value and hydrological services (Lorenzo et al. 2013). Thus, the biological invasion is akin to large-scale habitat degradation with an immediate and dramatic impact on the biodiversity. Such changes can pose major challenges to the livelihood and well-being of the local communities owing to decline in food availability, medicinal plant diversity and raw materials, which can in turn lead to malnutrition, famine, diseases and loss of local knowledge associated with the bio-resources, especially in the developing economies of tropical countries.

7.2.2 Impact on Health

A majority of IAS contain toxic compounds that can harm humans when exposed to these plants. Impacts on human health can be diverse, including physiological effects, discomfort, congestion in respiratory tracts, skin irritations, allergies, poisoning, disease and even death (Ageliki and Helen 2018). Some of the impacts of IAS on human well-being are indirect. Direct negative impacts are often similar in both the native and invaded range, while indirect impacts are often unpredictable, as they constitute a novel element in existing ecological networks. One such instance is the presence of harmful pyrrolizide alkaloids in honey that is sourced from localities infested by the invasive *Echium plantagineum*. These alkaloids are a proven threat to health especially for infants and foetuses (Pimentel 2011).

7.2.3 Impact on Economy

The estimated economic damage due to invasive species worldwide totals more than US $1.4 trillion, which is roughly 5% of the global economy in the year 2012. The annual cost incurred due to invasives in the USA alone is estimated to be $120 billion, with more than 100 million acres affected (Anonymous 2012). Estimated damage due to 79 harmful invasive species totals up to US $185 billion, which corresponds to 1.4% of gross domestic product (GDP) of the USA in 1993. George et al. (2014) also provide an estimate of the economic cost of biological invasion in eight countries including India, Brazil and South Africa. The estimated economic loss incurred due to biological invasion in India is found to tune of 12% of the GDP, which is the highest among the countries studied.

7.3 Allelochemicals: Nature and Mode of Action

Allelopathy, in simple terms, is the process by which organisms influence the growth, survival, development and reproduction of other organisms using the biochemicals produced by them. While this is a well-researched area in crop sciences, the role of allelopathy in the natural and human modified ecosystems are yet to be understood well. Allelopathic effects of plants were noticed as far back as 300 BC in the context of agriculture. The term *allelopathy* is derived from the Greek word *allelo* meaning 'each-other' or 'mutual' and *pathy* meaning 'harm' or 'suffering' and was first used in 1937 by Austrian scientist Hans Molisch (Willis 2010). Though originally allelopathy was taken for a non-benign relation among organisms, the recent definitions of the term is expanded to include relations across a wide variety of organisms where both beneficial and detrimental effects of the allelochemicals are noticed. According to the working definition for allelopathy agreed upon by the International Allelopathy Society in 1998, it is 'any process involving secondary metabolites produced by plants, algae, bacteria and viruses that influence the growth and development of agricultural and biological systems' and any 'study of the function of secondary metabolites, their significance in biological organization, their evolutionary origin and elucidation of the mechanisms involving plant-plant, plant-microorganisms, plant-virus, plant-insect, plant-soil-plant interactions' fall within the purview of allelopathy research (Mallik and Inderjit 2002).

According to Fang and Cheng (2015), plant allelopathy is akin to the interaction between 'receptor' and 'donor' plants and may exert either desirable effects (e.g., for agricultural management, such as weed control, crop protection, or crop reestablishment) or undesirable effects (e.g., autotoxicity, soil sickness, or biological invasion). Allelochemicals, which are either non-nutritive secondary metabolites or decomposition products, are the active agents of allelopathy. These compounds belong to various chemical families and can be grouped into various classes based on their chemical similarities. One such early attempt to group the known allelochemicals identified 14 classes (Rice 1974) and a subsequent attempt (Putnam 1985) evolved 11 groups. A much more recent attempt (Li et al. 2010) to group the compounds based on their structural variability and properties grouped these into 10 broad categories, which are listed below. The list provides an idea of the diversity of biochemical groups that help exerting allelopathic effects.

1. Water-soluble organic acids, straight-chain alcohols, aliphatic aldehydes and ketones.
2. Simple unsaturated lactones.
3. Long-chain fatty acids and polyacetylenes.
4. Quinines (benzoquinone, anthraquinone and complex quinines).
5. Phenolics.
6. Cinnamic acid and its derivatives.
7. Coumarins.
8. Flavonoids.
9. Tannins.
10. Steroids and terpenoids (sesquiterpene lactones, diterpenes and triterpenoids).

These allelochemicals are synthesized in plant body through metabolism of carbohydrates, fats and amino acids in the different biosynthetic pathways such as methylerythritol phosphate, shikimic acid or mevalonic acid pathways. Allelochemicals thus produced are usually present in all plant parts like leaves, roots, stem, inflorescence and pollen grains at varying concentrations. These are released into the rhizospheric soil by a variety of mechanisms, such as leaching, decomposition of plant residues, volatilization and root exudation. In the soil, these compounds either may degrade or are transformed to other compounds that may continue to exert allelopathic action. Quality and quantity of allelochemicals released vary across plant species, cultivar, age, plant organ, season, physiological or pathological stress, etc. (Burgos et al. 1999). Severity of inhibition and toxicity by the allelochemicals of invasive species are determined by many factors such as form of chemicals, concentration, flux rate, age and metabolic state of the plant, and prevailing climate and environmental conditions.

Being secondary metabolites, a majority of allelochemicals do not have any important role to play in primary metabolic processes that are essential for plant survival (Corcuera 1993). However, some of the allelo-compounds, which are intermediary products of lignification, are capable of activating plant defence mechanisms when plants are exposed to pathogens, and thereby playing both structural and physiological roles in the plant body (Einhellig 1995).

7.3.1 Mode of Action of Allelochemicals

Allelopathic compounds act on target species by directly influencing their physiological processes such as mineral uptake, stomatal opening and closure, pigment synthesis, photosynthesis, respiration, protein synthesis, leghemoglobin biosynthesis, nitrogen fixation, plant-water relations, cell division, cellular expansion, cell wall construction and phyto-hormonal balance or by inhibiting specific enzymes (e.g. indole acetic acid oxidase). They may also inhibit pollen, spore and seed germination, or may result in DNA or RNA modifications (Rice 1984; Wink and Twardowski 1992). Yet another mode of action of allelochemical compounds is by inflicting oxidative damage by inhibiting cellular antioxidant mechanisms in the target species.

Allelochemicals also indirectly affect many physiological processes and trigger phenotypic responses to specific compounds, though these responses need not necessarily reflect their primary mode of action (Lotina et al. 2006). A generalized summary of major physiological and biochemical mechanisms of action of various allelochemicals in affected plants are provided in Table 7.1.

7.4 Impact of Allellopathic Substances to Environment

7.4.1 Impact on Soil Fertility and Soil Microbiota

Allelochemicals can persist in soil, affecting both the neighbouring plants and those germinating subsequently. Their fate in the soil in biochemical degradative sequences is yet to be fully elucidated and interpretation of the results of field studies on allelochemicals are extremely difficult. Nevertheless, it is possible to make some generalized statements on the impacts of allelochemicals on soil fertility.

Soils are important source of allelochemicals from which these leach into the aquatic systems, thereby inflicting allelopathic effect on the aquatic communities also. The quantities and chemical properties of edaphic allelochemicals are determined by the type and composition of the above-ground plant community—which is the primary source of secondary metabolites in the soil, especially in the root zone. Concentrations of allelochemicals in the soil solution are influenced by physicochemical factors of the soil such as nutrient status, organic matter content, pH, cation and anions exchange capacity, water content and soil microbial activity (Barazani and Friedman 1999). Among these factors, organic matter and its dynamics assume significance in regulation of allelochemical concentrations in the soil solution, because organic contents determine adsorption and desorption of the allelochemicals in the soil solid phase and their transportation across the soil profile (Kobayashi 2004).

Microorganisms play diverse roles in making the soil fertile and are very crucial in decomposition of organic matter, breakdown of metabolic by-products and agrochemicals, nitrogen fixation, enhancing the bioavailability of nitrates, sulphates, phosphates and essential metals. Soil microbes play key roles in ecosystems too, as they influence a large number of important ecosystem processes such as nutrient acquisition (Sprent and Platzmann 2001), carbon cycling (HoEgberg et al. 2001) and soil formation (Rilling and Mummey 2006). The allelopathic effect may be induced directly through the release of allelochemicals or indirectly through the suppression of an essential symbiont, which in turn affect the nutritional and other edaphic environmental conditions. Some plant species control the presence of allelochemicals in soil indirectly through an intermediary microbial medium (Barazani and Friedman 1999).

Microbial community structure in the rhizosphere and its functions are significantly influenced by soil property and the plant species. The soil microbial flora such

Table 7.1 A summary of major physiological and biochemical mechanisms of action of allelochemicals in affected plants (modified after Fang and Cheng 2015)

Mode of action	Cellular level action	Reference
Inducing changes in cell structure	Acts by widening or shortening of the root cells; inducing nuclear abnormalities; increasing vacuole numbers, etc. Allelochemical: Volatile monoterpenes, eucalyptol, camphor	Bakkali et al. (2008), Pawlowski et al. (2012)
Inhibiting cell division and cell elongation	Acts by inhibition of cell proliferation and DNA synthesis in plant meristems Allelochemical: monoterpenoids (camphor, 1,8-cineole, betapinene, alpha-pinene, and camphene)	Nishida et al. (2005)
	Inhibition of the mitotic process, as in the case of lettuce where action is noticed at the G2-M checkpoint Allelochemical: 2(3H)-benzoxazolinone (BOA)	Sanchez et al. (2008)
	Resulting in the reduced number of cells in each cell division and damaging the tubulins and resulting in polyploid nuclei. Allelochemical: Sorgoleone	Hallak et al. (1999)
Inducing imbalances in the antioxidant system	Inducing the production of reactive oxygen species (ROS) in the contact area in recipient plants	Bais et al. (2003)
	Altering the activity of antioxidant enzymes such as superoxide dismutase (SOD, peroxidase (POD)	Zeng et al. (2001), Yu et al. (2003)
	Alter the activity of antioxidant enzyme—ascorbic acid peroxidase, etc.	Zuo et al. (2012)
Augmenting cell membrane permeability	Inhibits the activity of antioxidant enzymes leading to increase free radical levels. This can result in accelerated membrane lipid peroxidation and membrane potential alteration, which diminish the scavenging effect on activated oxygen, damaging the entire membrane systems in affected plants	Lin et al. (2000), Zeng et al. (2001)
Creating imbalances in the plant growth regulator system	Endogenous hormone levels are altered by stimulating IAA oxidase activity and inhibiting the reaction of POD with IAA, bound GA or IAA e.g.: Phenolic allelochemicals	Yang et al. (2005)
	Altering the contents of plant growth regulators or by inducing imbalances in various phytohormones, especially those connected with growth and development of plants as in seed germination and seedling growth	Yang et al. (2005)

(continued)

Table 7.1 (continued)

Mode of action	Cellular level action	Reference
Influencing the functions and activities of various enzymes	By exerting effects on the synthesis, functions, contents and activities of various enzymes. Instance-Tannic acid can suppress the action of POD, CAT, and cellulase and reduce the synthesis of amylase and acid phosphatase in the endosperm Chlorogenic acid, caffeic acid and catechol Can inhibit the key enzyme λ-phosphorylase involved in seed germination	Rice (1984), Einhellig (1995)
Influencing respiration	Plant growth is inhibited by influencing different stages in respiration, such as: – Electron transfer in the mitochondria – Oxidative phosphorylation – CO_2 generation – ATP enzyme activity By reducing oxygen intake: – NADH oxidation is retarded – ATP synthesis enzyme activity is inhibited – ATP formation in mitochondria is reduced – Disturbs plant oxidative phosphorylation – Ultimately inhibiting respiration	Fang and Cheng (2015)
Reducing the photosynthetic efficiency	Allelochemicals act by: – Inhibiting or damaging the photosynthesis system and – Accelerating the decomposition of photosynthetic pigments leading to: – Decreased photosynthetic pigment contents – Reduced ATP synthesis enzyme activity – Reduced stomatal conductance and transpiration, all inhibiting the photosynthetic process	Yu et al. (2003), Wu et al. (2004)
	Allelochemicals affect photosynthesis by influencing the function of PS II	Weir et al. (2004)
Altering water and nutrient uptake	Allelochemicals affect nutrient absorption in plant roots or induce water stress through long-term inhibition of water utilization. Allelochemicals can inhibit the activities of Na+/K+-ATPase involved in the absorption and transport of ions at the cell plasma membrane, which suppresses the cellular absorption of K+, Na+, or other ions	Fang and Cheng (2015)

(continued)

Table 7.1 (continued)

Mode of action	Cellular level action	Reference
Influencing protein and nucleic acid synthesis and metabolism	Alkaloids closely integrate with DNA and increase the temperature of DNA cleavage, inhibit DNA polymerase I and prevent the transcription and translation of DNA, or inhibit protein biosynthesis	Wink and Latzbruning (1995)
	Interfering with protein synthesis and cell growth by inhibiting amino acid absorption and transporting	Abenavoli et al. (2003)
Acting on microorganisms and the ecological environment	Population of soil microbes are put under stress due to allelochemicals and could trigger positive and negative feedbacks for inducing allelopathy	Stinson et al. (2006), Inderjit et al. (2011), Wu et al. (2015)

as bacteria, fungi and actinomycetes isolated from rhizosphere soil of the invasives were found to vary from that of the native plant's rhizosphere (Ehrenfeld 2003). The invasion in turn alters the microbial community of the soil along with the structure, function and properties of the soil. Invasives affect the soil physicochemical properties such as pH and moisture content, thereby affecting the community structure and metabolic activity of the soil microorganisms (Congyan et al. 2015). Plant invasions may also alter level of carbon dioxide and oxygen and thereby affecting relationships of soil microbial mutualists on plants (Andrew and Lilleskov 2009).

It has been suggested that some of the bacterial and fungal association in the rhizosphere of invaded plants may act antagonistically with other soil microbial pathogens, thereby providing protection to the plant (Berg and Smalla 2009). For instance, certain fungi, which are sources of antimicrobials, were proven efficient against plant microbial pathogens and were used as biocontrol agents against such pathogens.

There are three primary modes by which allelochemicals become useful for an invasive plant to establish:

1. By inhibiting the growth of the competing plants:
   For example, the invasive grass *Holcus lanatus* succeed in dominating the habitat by inhibiting the growth of native species with the aid of allelopathic compounds that alter the composition of soil microbial flora in favour of the grass (Wardle et al. 2011).
2. By protecting invasives from soil pathogens and herbivores:
   Some plant invaders use allelopathic chemicals to annihilate pathogenic soil microbes and herbivores (Weidenhamer and Callaway 2010).
3. By promoting growth of invasive plants by acting as growth regulatory factors.

As reported by Hierro et al. (2014) the plant invasion could be aided by release of regulatory factors into the soil, which promote abundant growth of the invaded species. The success of invasion by *Acacia dealbata* is due to the release of allelopathic chemicals to the soil, which inhibit the native plant growth and affect soil microbiota, especially the bacterial population compared to that of fungi (Lorenzo et al. 2013). Barazani and Friedman (1999) have reported that allelopathic bacteria are found to evolve in the soil in areas where a single crop is grown successively. This results in yield reduction and the same cannot be restored by application of minerals. While most growth-inhibiting allelopathic bacterial species affecting higher plants are generalists, some do exhibit specificity; for instance, dicotyledonous plants are more susceptible to *Pseudomonas putida* than monocotyledons. The release of allelochemicals from plant residues in plots of 'continuous crop cultivation' or from allelopathic living plants may induce the development of specific allelopathic bacteria. Interaction between soil organisms and plants is an important aspect in allelopathy, which warrants deeper investigation.

7.4.2 Impact on Flora and Fauna

The allelopathic activity may reduce both diversity and population size of other species by impeding their competitive ability or by affecting diverse aspects of ecology, including occurrence, growth, community succession, composition of communities, dominance, physiognomy, diversity and productivity. Some allelochemicals cause oxidative stress in the target plants by inhibiting antioxidant mechanism of the target species, using this an allelopathic plant can inflict phytotoxicity on the target species. Allelopathy is also reported to induce autotoxicity where the allelochemicals inhibit germination and growth of same species. Allelopathy is observed in several tree species. For instance, the *Leucaena leucocephala* contains a toxic nonprotein amino acid called mimosine in its leaves which inhibits growth of other species while leaving its own seedlings unaffected (Patrick et al. 2002). The allelochemicals reduce the density and diversity of native species in a locality through drastic inhibition of their seed germination. Studies on allelochemicals revealed that *Leucaena* inter-planted with crops in an alley cropping system causes the reduction of wheat and turmeric yield. Invasive plants use allelopathy to guard their space by using roots to draw more water out of the soil. The Sorghum genus includes plants whose roots remove sorgoleone, a poisonous substance that blocks the respiration and photosynthesis of the other plants that come into contact with (Ramona 2018).

Allelochemicals are stored in different locations in plant tissues as inactive form and the environmental factors determine their concentration in the plant body. Many of these chemicals are acute or chronic toxins or deterrents for herbivorous insects. Long-period interaction and co-evolution has helped evolution of specialist herbivorous insects, which rely heavily on ingested plant allelochemicals for their own chemical defence from their predators and pathogens. Some herbivorous insects are found to be synthesizing pheromone required to regulate their aggregation, attracting, alarm and mating from the allelochemicals acquired from their host plants (Elsayed 2011).

Large number of allelochemicals are known to interfere with the reproductive system of animals, affecting their fitness as a species, population size and species composition in the community. Some of the allelochemicals are known to have antihormonal effects in animals as they mimic the structure of sexual hormones such as coumarins which dimerize (combine with a similar molecule) to form dicoumarols, or isoflavones. A large number of secondary metabolites are mutagenic and lead to either malformation of the offspring, death of the embryo or the premature abortion of the embryo (Wink 1999). A number of allelochemicals including many mono- and sesquiterpenes and alkaloids belong to this category.

7.5 Role of Allelopathy in Plant Invasion

Allelopathy plays a crucial role in plant succession by shaping the species composition of plant communities, and allelochemicals are one of the agents by which dominance is achieved by invasive species in a plant community. It is speculated that plants, specifically the noxious weeds, exude chemicals from their roots that are detrimental to the growth of other plants. Though researchers have suggested allelopathy as mechanism for impressive success of invasive plants (Hierro and Callaway 2003), the difficulty of distinguishing the effect of allelopathy from that of resource competition has hindered investigations on the role of phytotoxic allelochemicals in plant communities. The results of an investigation on the potential interacting influences of allelopathy and resource competition on plant growth–density suggested that plant–plant interference is a combined effect of allelopathy and resource competition with many other factors (Uddin and Randall 2017). Nevertheless, the experimental design, target-neighbour mixed-culture in combination of plant grown at varying densities with varying level of phytotoxins and monoculture proved to be successful in separating allelopathic effects from competition.

Some of the best known plant invasives in the world are also proven to be allelopathic, which includes species such as *Elytrigia repens* (Korhammer and Haslinger 1994), several Centaurea species (Ridenour and Callaway 2003), *Cyperus rotundus* (Agarwal et al. 2002) and *Parthenium hysterophorus* (Kanchan and Jayachandra 1980). El-Ghareeb (1991) tested the allelopathic impacts of the invasive plant *Tribulus terrestris* on surrounding annual vegetation in an abandoned field of Kuwait. It has been observed that the shoot leachates of *T. terrestris* inhibited the germination and radicle elongation in majority of the target species, which included both native and exotic annuals. Allelopathic effects of the IAS *Kochia scoparia* on the native grass *Bouteloua gracilis* were also evaluated in Petri dish bioassays (Karachi and Pieper 1987). While the water extracts from ground tissues of the invader had no effects on germination, the growth of seedling of the native grass was significantly suppressed by the same treatment (Hierro and Callaway 2003).

7.6 Invasive Plants as Source of Allelochemicals

The allelochemicals, most of them being metabolites, are generally stored in plant cells in bound form, and released into the environment through special glands on the stems and leaves (Putnam and Duke 1978) or are released in the form of vapour, leachates from the foliage, exudates from roots and decomposition products of dead and worn out plant tissues. Table 7.2 summarizes information on allelochemicals borne by 15 well-known IAS. Out of 35 metabolites listed, which are known to possess allelopathic effect, most are growth suppressants and seed germination inhibitors.

7.7 Detoxification of Allelochemicals

Some plants are adapted to nullify and survive the allelopathic effect caused by other plants with which it shares habitat. The same is valid in case of vast majority of plants that co-habit with IAS in their original habitats. They, in addition to surviving the noxious neighbour, manage to keep population of the allelopathic plants under check in their community. In the absence of such adaptive bulwarks, native plants in the habitats where the IAS are introduced succumb to allelochemicals by exposing the habitat for biological invasion. This adds an additional dimension to the poorly understood and already complex allelopathic interactions. Allelopathy and plant defence are two closely related aspects of phytotoxins. Notwithstanding the agronomic and ecologic significance of the phytotoxins, their target sites, mechanisms of adaptations and detoxification are inadequately understood.

The process of detoxification of the allelochemicals in the environment may involve their oxidation, hydroxylation, glycosylation, conjugation with sulphate, phosphate, etc. and influenced by both biotic and abiotic factors such as microbes besides the characteristics of soil and chemicals of neighbouring plants. For instance, detoxification may involve abiotic reactions such as oxidation or biotic mineralization of the allelochemicals and the microorganisms or enzymes may polymerize allelochemicals such as phenolic acids to humus-like materials (Duxbury 1981). These chemicals could also be adsorbed or oxidized by the clay minerals (Okamura and Kuwatsuka 1988). A wide variety of secondary plant metabolites undergoes detoxification during glycosylation and conversion to glucoside whereby their chemical reactivity is reduced (Sicker et al. 2000). Reactivity of cyanogenic glycosides, glucosinolates, volatile terpenoids and phenolics is reduced essentially by means of glycosylation.

Large number of plant species demonstrate tolerance to benzoxazinoids by rapidly metabolizing them to less phytotoxic glucoside or glucoside carbamate derivatives. According to Schulz and Wieland (1999), this is an adaptive capability acquired by these plants while they co-evolved with allelopathic species within the same communities. The mechanism evolved by plant system for detoxifying the allelochemicals opens up an opportunity for incorporating similar capability for

Table 7.2 Known effects of allelochemicals produced by some IAS on the other plant species

	Plant name	Allelochemicals	Action on other plant species	Reference
1	*Leucaena leucocephala* (Lam.) de Wit	Phenolic acid, quercetin, Mimosine	Inhibits seed germination	Chou and Kuo (1986)
2	*Lantana camara* L.	Lantadene A and B and salicylic acid	Growth retardant; repellent; inhibitory effect on germination and shoot elongation	Yi et al. (2005), Arpana (2015)
3	*Mikania micrantha* Kunth	Caffeic acid, p-hydroxybenzaldehyde, resorcinol and vanillic acid	Inhibits seed germination; growth suppression	Ismail and Chong (2002)
4	*Chromolaena odorata* L.	Phenolics, alkaloids and amino acids	Inhibits seed germination	Ambika and Jayachandra (1984)
5	*Ageratum conyzoides* L.	p-Coumaric acid, gallic acid, ferulic acid, hydroxybenzoic acid, anisic acid and syringic acid	Growth inhibitory effects	Smita (2008)
6	*Amaranthus spinosus*	Phenolic acids, alkaloids and sesquiterpene lactone, quercetin, linoleic acid	Inhibits seed germination	Suma (1998), Eva and Chakraborty (2015)
7	*Parthenium hysterophorus* L.	Caffeic acid, vanillic acid, ferulic acid, chlorogenic acid, p-coumaric acid and p-hydroxybenzoic acid and among the organic acids, fumaric acid	Inhibits seed germination; growth suppression	Kanchan and Jayachandra (1980)
8	*Eichhornia crassipes* (Mart.) Solms	Linoleic acid, glycerol-1,9-12 (ZZ)-octa decadienoic acid and N-phenyl-2-napthylamine	Inhibits growth in several aquatic flora; increases the protein content and decreases superoxide dismutase activity in some aquatic flora	Yang et al. (1992)
9	*Argemone mexicana*	Salicylic acid, p-hydroxybenzoic acid, vanillic acid, cinnamic acid	Growth inhibitor	Burhan and Shaukat (1999)
10	*Pistia stratiotes* L.	Linoleic acid, gamma-linoleic acid, alpha asarone	Growth inhibitor	Alliota et al. (1991)
11	*Datura stramonium* L.	Tropane alkaloids, scopolamine, hyoscyamine	Yield reduction	Lovett et al. (1981)

(continued)

Table 7.2 (continued)

	Plant name	Allelochemicals	Action on other plant species	Reference
12	*Melitous alba* Desr	Coumarin	Inhibits seed germination	Schnute (1984)
13	*Pteridium aquilinnum* (L.) Kuhn	Selligueain A	Inhibiting root and stem growth and root metaxylem cell size	de Jesus et al. (2016)
14	*Radicula sylvestris* (L.) Druce	Salicylic, p-hydroxybenzoic, vanillic, syringic acids, hirsutin and pyrocatechole isothiocyanates	Inhibits seed germination and seedling growth	Yamane et al. (1992)
15	*Helianthus tuberosus* L.	Salicylic acid, p hydroxybenzaldehyde, cinnamic acid, o-coumarinic acid, p-coumarin acid and coumarin	Inhibits seedling growth	Tesio et al. (2011)

plant defence against these chemicals by a variety of means including the recombinant DNA technologies.

Figure 7.2 is a generalized schematic diagram, which summarizes the fate of allelochemicals released into the soil environment and their possible impacts on the native plant community. Following the terminologies of allelopathy, an allelochemical produced in the 'source' plant is 'conveyed' to 'receiver', which are either 'impaired' or are 'assisted' (Coder 1999). Allelochemicals belonging to one or more classes of metabolites is present in most parts of the source plant at variable concentrations. It is released into the soil as root exudate or indirectly during the decomposition of plant parts. The allelochemical released into the soil may alter the pH of the soil and nutrient availability in soil. Once in the soil medium, the allelochemicals may: (1) directly inhibit the growth of receiver plant or the germination of seeds, (2) undergo microbial decomposition and transformation to release more toxic allelochemicals that can fatally impact upon the receiver plant's life or (3) undergo detoxification either through microbial mediation or by undergoing mineralization. The plants that are fatally affected by the allelochemicals or by its more toxic variants released by its microbial decomposition would display signs of severe physiological stress and yellowing or falling of immature leaves. Some of the plants, which escaped the effect of allelochemicals by detoxifying them, may not display any signs of stress. Yet another set of plants, which are partially affected by the primary impact of allelochemicals, such as the altered soil nutrients and soil pH, may appear stunted. This is a simplified scheme of a much more complex and detailed process of the production, release, decomposition, detoxification and transformation of the allelochemicals in the soil environment as elaborated in the chapter.

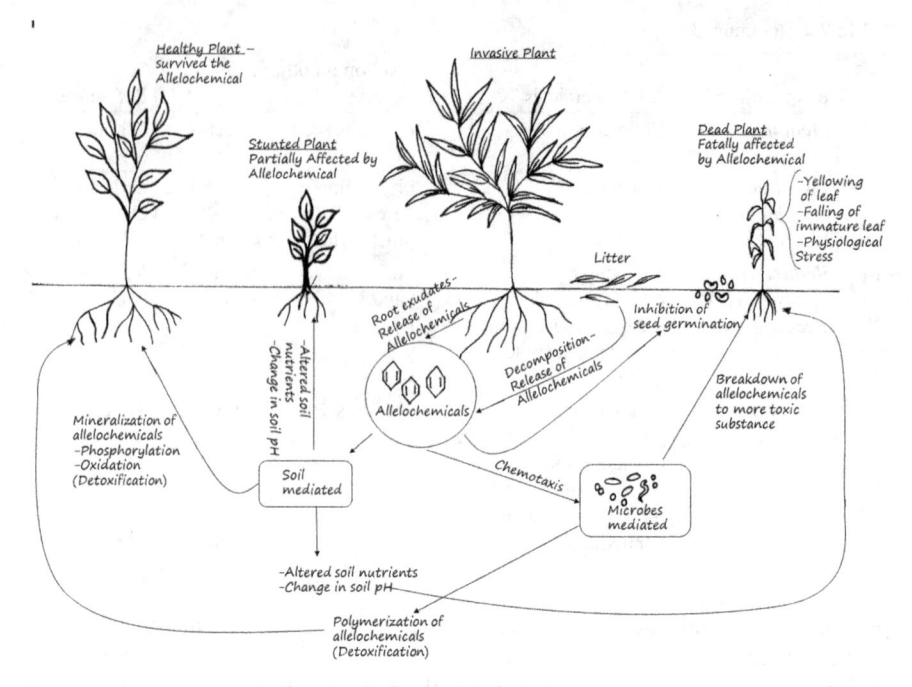

Fig. 7.2 A generalized scheme showing the fate of allelochemicals in soil and their impact on the native plant community

7.8 Biological Invasion: A Case Study from India

To illustrate the complexity of the allelopathic interactions in nature and their often-unimagined consequences, let us present here a real-life story from a far-flung locality from the Western Ghats mountain ranges in the peninsular India. The case pertains to biological invasion in the Wayanad Wildlife Sanctuary in Kerala. With an extent of 344.44 sq. km, the sanctuary is mostly tracts of tropical forests which are contiguous with the larger protected area network of Nagarhole and Bandipur National Parks in Karnataka state and Mudumalai National park of Tamil Nadu state making it a valuable and most extensive habitat of tiger and Asiatic elephant in south India. The sanctuary is integral part of Nilgiri Biosphere Reserve. Having located along the Western Ghats mountain ranges, one of the eight 'hottest hot-spots' of global biological diversity and a UNESCO world heritage site, the sanctuary is biologically rich with high degree of floristic endemism and rarity. The protected area network also functions as the major forested catchment of the largest and most prominent River Kaveri in the peninsular India. Forest area of these protected areas are contiguous across the windward and leeward slopes of the Western Ghats, so it has a spectrum of forest types which includes dry deciduous forests, moist deciduous forests, semi-evergreen forests and wet evergreen forests mixed with bamboo brakes

Fig. 7.3 Invasion of *Senna spectabilis* at Wayanad wildlife sanctuary, India

and shola grasslands. Eastern tracts of the landscape on the leeward side is relatively dry with a brief wet and a lengthy dry season, while the western windward slopes are more or less wet with less pronounced and brief dry season. This climatic gradient and diversity of habitats facilitate migration of herbivore population during the dry summer months. Presence of large grazers and predators and their seasonal movements to the wetter western tracts are heightened during the summer season. This increases the chances of wild animals wandering into the human habitations along the fringes of the sanctuary triggering human-wildlife conflicts and causalities. The persistent issue of human-wildlife conflicts and its management are proving increasingly expensive from the point of biodiversity conservation and ensuring benefit flows such as ecosystem services and livelihood opportunities for the local communities. Any additional stress in the habitats such as biological invasions would aggravate human-wildlife conflicts besides endangering the survival of hundreds of species of endemic flora and fauna.

A survey conducted in the year 2019 revealed the presence of 67 known invasive species within the territories of Wayanad Wildlife Sanctuary alone. Prominent among these are *Senna spectabilis, Maesopsis eminii, Lantana camara, Chromolaena odorata* and *Mikania micrantha*. Two invasive tree species among these, *Maesopsis eminii* and *Senna spectabilis,* are found to be exerting severe stress on the forest habitat by altering vegetation structure and reducing the habitat viability. These are medium-sized trees and native of Tropical America and Central Africa, respectively, which grow up to 20–25 m in height with a canopy spread of 15–20 diameter. While *S. spectabilis* belongs to the family Fabaceae (sub family Caesalpiniaceae) *M. eminii* belongs to the family Rhamnaceae (Figs. 7.3 and 7.4).

Among the special characteristics possessed by these species which enable them to proliferate are the early reproductive maturity, high fecundity, production of large quantities of seeds, wide dispersal ability and the ability to germinate under a wide range of physical conditions. They also have high growth rates and capability of vegetative reproduction. Seeds of *M. eminii* and *S. spectabilis* are dispersed from

Fig. 7.4 Invasion of *Maesopsis eminii* at Wayanad wildlife sanctuary, India

their fruits by explosive mechanism which makes them reach out to large areas. *S. spectabilis* regenerates both from seeds and by means of coppicing when cut. Coppices can develop from both stumps and roots. It is reported that their seeds can stay viable in the soil seed banks for a period of up to 3 years (Irwin and Barneby 1982). These characteristics provide them an edge over their native counterparts in proliferating the habitat. The invasive species reduce regeneration and establishment of native species by competing with them for energy and resources. This helps these invasive species to occupy and establish in an area in high density and thus out-compete the native plants. The overall biological richness when compared between two, one-hectare plots (one infested with IASs and the other with no presence of IAS) showed that the uninfested natural forest had high degree of species richness which was almost three time richer than the infested plot. This indicates that both *M. eminii* and *S. spectabilis* are homogenizing the flora and reducing the native biodiversity in the infested patches.

7.8.1 Impact of *S. spectabilis* and *M. eminii* on Rhizosphere Microbiota

A comparison of the magnitude of fungus, actinomycetes and bacteria in the rhizosphere soil in natural forest and the forested locality colonized by the *M. eminii* and *S. spectabilis* in Wayanad Wildlife Sanctuary revealed that the presence of IAS exert a considerable degree of impact on the soil microbial community. It was observed that the number of fungal isolates and species richness of fungi were higher in Invaded Forest Soil (IFS) as compared to that of Un-invaded Forest Soil (UFS). Nearly 17 clearly distinguishable fungal isolates were obtained from the IFS sample. The isolates are less in number in the case of UFS. Among the species identified, a few were common in both the soil samples. The soil samples from invaded forest had double the number of Actinomycete population than that in the

UFS. Apparently, the presence of invasive plants has opened up more niche to Actinomycetes to establish. The morphological analysis suggested that IFS isolates are richer in terms of actinomycetes (Anjusha et al. 2019; Athira et al. 2019). Interestingly, the load of pathogenic bacteria in the IFS is lesser than that isolated from UFS. Further morphological examination revealed that both gram-positive and gram-negative bacilli are equally distributed in the UFS and IFS samples, in which the numbers of 'cocci' were comparatively lower than that of bacilli in both the samples. It is assumed that an unknown stimulus driven by the invasive species have attracted a vast array of distantly related microbial species to the rhizosphere and contributed to microbial diversity. The allelopathic effects are attributed for the difference in composition of bacterial flora. Most of the IFS isolates of fungus, bacteria and actinomycetes obtained from the rhizosphere of the invasive species were known to control major plant pathogenic fungi such as *Fusarium oxysporum, Sclerotium rolfsii* and *Pestalotiopsis maculans* (Prajna et al. 2019; Aneesha and Hrideek 2018), which could be providing a competitive advantage for the invasives over the native plant species. The actual mechanisms and metabolites involved in the process are yet to be investigated in detail.

7.8.2 Impact of Invasive Plants on Seed Germination in Native Species

A study on the allelopathic effect of the leaf extracts of *S. spectabilis* is on seed germination and seedling growth of three native and naturally occurring forest species *Bauhinia variegate, Syzygium cumini* and *Shorea roxburghii* were carried out under the laboratory conditions. The results of the study suggest that the extracts from both old and fresh leaves are capable of inhibiting seed germination of native species and can retard the seedling growth at significant degree. Analysis of germination parameters revealed drastic reduction in germination rate of the seeds when irrigated with the leaf extracts. For instance, in *Bauhinia variegate, Syzygium cumini* and *Shorea roxburghii*, while the highest germination percentage in control sets of seeds were 24%, 32% and 5%, respectively, the same has declined to 1%, 9% and 2% on treating with 100% concentration of leaf extract. This indicates allelopathic effect of the *S. spectabilis* is on seed germination in native trees. As in the case of germination parameters, a drastic reduction in the seedling characters of the three tree species were also observed. A phytochemical profiling of the leaf extracts is expected to help identifying the growth-inhibiting compounds in the leaf extracts of *S. spectabilis* (Krishnapriya et al. 2018).

A few observations from the field study that are listed below are also suggestive of the presence of allelochemical(s) in *S. spectabilis*:

1. Ginger rhizomes meant to be planted in the forthcoming season are often stored during the fallow period by the farmers of Wayanad by leaving them covered with soil under a mulch of the foliage from native tree species. It has been reported by farmers that bulk of the rhizomes stored under the mulch from the *S. spectabilis*

has failed to sprout, while the majority of the rhizomes obtained from same field and stored under the similar conditions but mulched with leaves of plants other than the *S. spectabilis* have succeeded in sprouting.
2. The saplings of some of the fast-growing native species planted as part of a restoration experiment on a land that was infested with these invasive trees species reported poor growth.
3. It has been noted that *M. eminii* and *S. spectabilis* were least preferred by wild herbivores as fodder.
4. Suffocation was reported by the labourers engaged in the cutting and burning of *S. spectabilis*.

7.9 Conclusions

The study of the allelo-metabolites points towards two emerging areas of practical application of the allelochemicals in biological control of weeds and crop productivity improvement:

1. Use of Allelochemicals as natural growth and germination suppressor.
2. Use of recombinant DNA technology to genetically equip cultivars to produce allelochemicals.

Invasive plants assume special importance in this as allelochemicals of the invasives can be deployed as natural herbicides and pesticides with relatively less impact on ecology than their synthetic counterparts. The efforts to develop weedicides and pesticides from these natural metabolites would facilitate augmentation of food and industrial raw materials production with low environmental externalities. Complementing to these efforts are the initiatives for incorporating the allelopathic traits from wild or cultivated plants into the crop plants through traditional breeding or through genetic engineering. Genetic basis of allelopathy has now been demonstrated in wheat and rice. Specific cultivars with increased allelopathic potential are known in both these crops.

Allelo-metabolites may have wide-ranging impact on ecosystems. Most of their indirect impacts on soil properties, nutrient status and soil biota are not yet sufficiently understood. While plant populations are affected by allelometabolites resulting in poor species diversity and genetic variation in the plant community, allelopathy may also induce resistance towards certain allelochemicals in the affected populations as an adaptation and may function as a stimulant for genetic variability in the populations. This means allelopathy could be a driving factor in evolution of beneficial traits and symbiotic association as in the case of plants with ability to detoxify the allelochemicals. The mechanisms of detoxification of phytochemicals evolved by plants during their interaction with invasive plants in its home range as well as in invaded locality may help develop tools to combat invasives as well as other weeds more effectively by minimizing the ecosystem damage. Genetic incorporation of such traits in native plants would also pay great dividends when they are deployed for restoration of natural ecosystems affected by invasives. Even a partial success in these lines would prove to be a great boon as the costs of combating biological invasion due to IAS were found to be as high as 12% of GDP of some of the countries.

In this chapter, attempts were made to show how invasive plants deploy allelochemicals as a successful growth and germination suppressor for domination of plant communities, which they eventually colonize. While the chemical profile of allelo-metabolites in invasive species and their precise biochemical mechanism of action are diverse, their net effect and known action remains focussed more or less on growth retardation and inhibition of seed germination in target plants. Allelopathy is only one among the many strategies simultaneously deployed by the invasive species; nevertheless, allelochemicals have a significant role in the process. Understanding their mode of action could prove to be crucial in controlling biological invasion by deploying strategies that nullify the effect of allelopathy induced by them.

Biological invasion in the natural environment, especially in the complex tropical ecosystems, has to be viewed differently from the simplified weed-crop system, as understanding the strategies of domination and responses in the former would be more rich and rewarding in reclaiming and restoring expensive ecosystem services lost due to the biological invasion. Since the invasion induces ecological stress and adaptive responses at various levels (from individual genetic level to the community level), a complex biochemical process is set on the roll by providing infinite opportunities in fashioning and deploying of allelo-metabolites for suppression of the native biota. The adaptive defence strategy of the affected species also could prove to be a valuable source of genetic information and bioactive metabolite. Thus, an in-depth understanding of both mode of action of allelochemicals and the responses they generate may lead to the discovery of promising and unique bioactive metabolites.

References

Abenavoli MR, Sorgoná A, Sidari M, Badiani M, Fuggi A (2003) Coumarin inhibits the growth of carrot (*Daucus carota* L. cv. Saint Valery) cells in suspension culture. J Plant Physiol 160 (3):227–237

Agarwal A, Gahlot A, Verma R, Rao P (2002) Effects of weed extracts on seedling growth of some varieties of wheat. J Environ Biol 23:19–23

Alliotta G, Monaco P, Pinto G, Pollio A, Previtera L (1991) Potential allelochemicals from *Pistia stratiotes* L. J Chem Ecol 17(11):2223–2234

Ambika SR, Jayachandra (1984) *Eupatorium odoratum* L. in plantations—an allelopathy or a growth promoter. In: Placrosym-V. Proceedings of the fifth annual symposium on plantation crops, pp 247–259

Andrew C, Lilleskov EA (2009) Productivity and community structure of ectomycorrhizal fungal sporocarps under increased atmospheric CO_2 and O_3. Ecol Lett 12:813–822. https://doi.org/10.1111/j.1461-0248.2009.01334.x

Aneesha KA, Hrideek TK (2018) Impact of invasion of *Senna spectabilis* on soil microbes. Post graduate thesis submitted Department of Environmental Science, University of Calicut, p 47

Angeliki F. Martinou, Helen E. Roy (2018) In: Mazza G, Tricarico E (eds) Introduction from local strategy to global frameworks: effects of invasive Alien species on health and wellbeing. CABI International. ISBN 13-9781 78639 0981

Anjusha A, Suby, Hrideek TK, Sabu A, Shiburaj S (2019) Analysis of soil fungi associated with the invasive plant species *Maesopsis eminii* Engl. in Wayanad Wildlife Sanctuary of Nilgiri biosphere. In: International conference on exploring the scope of plant genetics resources. 22–24 May 2019. Department of Botany, University of Kerala, pp 285–287. ISBN 978-81-940888-0-6

Anonymous (2012) Economic impacts of invasive species in the Pacific Northwest economic region a report prepared by the PNWER Invasive Species Working Group January, p 5

Arpana M (2015) Allelopathic properties of *Lantana camara*. Int Res J Basic Clin Stud 3(1):13–28. https://doi.org/10.14303/irjbcs.2014.048

Athira KP, Shabeer Ali, Hrideek TK, Sabu A, Shiburaj S (2019) Rhizosphere actinomycetes population promotes the invasion and establishment of *Maesopsis eminii* Engl. in Wayanad Wildlife Sanctuary of Peninsular India. In: International conference on "exploring the scope of plant genetics resources, 22–24 May 2019. Dept. of Botany, University of Kerala, pp 291–293. ISBN 978-81-940888-0-6

Bais HP, Vepechedu R, Gilroy S, Callaway R, Vivanco JM (2003) Allelopathy and exotic plant invasion: from molecules and genes to species interactions. Science 301:1377–1380

Bakkali F, Averbeck S, Averbeck D, Idaomar M (2008) Biological effects of essential oils—a review. Food Chem Toxicol 46:446–475. https://doi.org/10.1016/j.fct.2007.09.106

Barazani O, Friedman J (1999) Allelopathic bacteria and their impact on higher plants. Crit Rev Microbiol 27:741–755

Berg G, Smalla K (2009) Plant species and soil type cooperatively shape the structure and function of microbial communities in the rhizosphere. FEMS Microbiol Ecol 68(1):1–13. https://doi.org/10.1111/j.1574-6941.2009.00654.x

Burgos NR, Talbert RE, Mattice JD (1999) Cultivar and age differences in the production of allelochemicals by *Secale cereale*. Weed Sci 47:481–485

Burhan N, Shaukat SS (1999) Allelopathic potential of *Argemone mexicana* L.—a tropical weed. Pak J Biol Sci 2:1268–1273

Callaway RM, Ridenour WM (2004) Novel weapons: invasive success and the evolution of increased competitive ability. Front Ecol Environ 2:436–443

Chou CK, Kuo YL (1986) Allelopathic research of subtropical vegetation in Taiwan. III. Allelopathic exclusion of understory by *Leucaena leucocephala* (Lam.) de Wit. J Chem Ecol 12:1431–1448

Coder Kim D (1999) Potential allelopathy in different tree species. University of Georgia Daniel B. Warnell School of Forest Resources Extension publication FOR 99-003, p 5

Congyan W, Xiao H, Liu J, Wang L, Daolin D (2015) Insights into ecological effects of invasive plants on soil nitrogen cycling. Am J Plant Sci 6:34–46

Corcuera LJ (1993) Biochemical basis for the resistance of barley to aphids. Phytochemistry 33(4):741–747

Duxbury T (1981) Toxicity of heavy metals to soil bacteria. FEMS Microbiol Lett 11:217–220

Ehrenfeld JG (2003) Effects of exotic plant invasions on soil nutrient cycling processes. Ecosystems 6:503–523

Einhellig FA (1995) Allelopathy—current status and future goals. In: Inderjit KM, Dakshini M, Einhellig FA (eds) Allelopathy: organisms, processes and applications. American Chemical Society, Washington, DC, pp 1–24

El-Ghareeb RM (1991) Suppression of annuals by *Tribulus terrestris* in an abandoned field in the sandy desert of Kuwait. J Veg Sci 2(2):147–154

Elsayed G (2011) Plant secondary substance and insect behavior. Arch Phytopathol Plant Protect 44(16):1534–1549

Eva S, Chakraborty P (2015) Allelopathic effect of *Amaranthus spinosus* Linn. on growth of rice and mustard. J Trop Agric 5(2):139–148

Fang C, Cheng Z (2015) Research Progress on the use of plant allelopathy in agriculture and the physiological and ecological mechanisms of allelopathy. Front Plant Sci 6. https://doi.org/10.3389/fpls.2015.01020

George M, Gren I-M, McKie B (2014) Economics of harmful invasive species: a review. Diversity 6:500–523. https://doi.org/10.3390/d6030500

Hallak AMG, Davide LC, Souza IF (1999) Effects of sorghum (*Sorghum bicolor* L.) root exudates on the cell cycle of the bean plant (*Phaseolus vulgaris* L.) root. Genet Mol Biol 22:95–99. https://doi.org/10.1590/S1415-47571999000100018

Hierro JL, Callaway RM (2003) Allelopathy and exotic plant invasion. Plant Soil 256:29–39

Hierro LJ, Maron JL, Callaway RM (2014) The biogeographical approach to plant invasion: the importance of studying exotics in their introduced and native range. J Ecol 93(1):5–15. https://doi.org/10.1111/j.0022-0477.2004.00953.x

HoÈgberg P, Nordgren A, Buchmann N, Taylor AF, Ekblad A, HoÈgberg MN, Nyberg G, Ottosson-LoÈfvenius M, Read DJ (2001) Large-scale forest girdling shows that current photosynthesis drives soil respiration. Nature 411:789

Inderjit WDA, Karban R, Callaway RM (2011) The ecosystem and evolutionary contexts of allelopathy. Trends Ecol Evol 26:655–662. https://doi.org/10.1016/j.tree.2011.08.003

Irwin HS, Barneby RC (1982) The American Cassiinae—a synoptical Revision of Leguminosae tribe Cassiae subtribe Cassiinae in the New World. Mem N Y Bot Gard 35:1–918

Ismail BS, Chong TV (2002) Effect of aqueous extracts and decomposition of *Mikania micrantha* H.B.K. debris on selected agronomic crops. Weed Biol Manag 2:31–38

de Jesus Jatoba L, Varela RM, Molinillo JMG, Ud Din Z, Juliano Gualtieri SC, Rodrigues-Filho E (2016) Allelopathy of bracken Fern (*Pteridium arachnoideum*): new evidence from green fronds, litter, and soil. PLoS One 11(8):e0161670. https://doi.org/10.1371/journal.pone.0161670

Kanchan SD, Jayachandra (1980) Allelopathic effects of *Parthenium hysterophorus* L. II. Leaching of inhibitors from aerial vegetative parts. Plant Soil 55:61–66

Karachi M, Pieper RD (1987) Allelopathic effects of kochia on blue grama. J Range Manag 40:380–381

Kobayashi K (2004) Factors affecting phytotoxic activity of allelochemicals in soil. Weed Biol Manag 4:1–7

Korhammer SA, Haslinger E (1994) Isolation of a biologically active substance from rhizomes of Quackgrass (*Elymus repens* (L.) Gould). J Agric Food Chem 42:2048–2050

Krishnapriya J, Vinod A, Muraleekrishanan K, Hrideek TK (2018) A study on the allelopathic effect of *Senna spectabilis* (dc.) HS Irwin & Barneby on germination and growth of native species. In: Mohanan KV, Radhakrishnan VV, Suhara Beevy S, Yusuf A, Gangaprasad A (eds) Modern trends in conservation, utilization and improvement of plant genetic resources. Gregor Mendel Foundation, Calicut University, Kerala, India, pp 92–98. isbn:978-81-935133-1-6

Li ZH, Wang Q, Ruan X, Pan CD, Jiang DA (2010) Phenolics and plant allelopathy. Molecules 15:8933–8952

Lin WX, Kim KU, Shin DH (2000) Rice allelopathic potential and its modes of action on Barnyard grass (*Echinochloa crus*-galli). Allelopathy J 7:215–224

Lorenzo P, Pereira CS, Echeveria SR (2013) Differential impact on soil microbes of allelopathic compounds released by the invasive *Acacia dealbata* Link. Soil Biol Biochem 57:156–163

Lotina-Hennsen B, King-Diaz B, Aguilar MI, Terrones MH (2006) Plant secondary metabolites. Targets and mechanisms of allelopathy. In: Reigosa MJ, Pedrol N, González L (eds) Allelopathy. Springer, Dordrecht, pp 229–265

Lovette JV, Levitt JV, Duuffield SMG (1981) Alleopathic potential of *Datura stramonium* (thornapples). Weed Res 21:165–170

Mallik AU, Inderjit (2002) Problems and prospects in the study of plant allelochemicals: a brief introduction. In: Mallik AU, Inderjit (eds) Chemical ecology of plants: allelopathy in aquatic and terrestrial ecosystems. Birkhäuser, Basal

Nishida N, Tamotsu S, Nagata N, Saito C, Sakai A (2005) Allelopathic effects of volatile monoterpenoids produced by *Salvia leucophylla*: inhibition of cell proliferation and DNA synthesis in the root apical meristem of *Brassica campestris* seedlings. J Chem Ecol 31:1187–1203. https://doi.org/10.1007/s10886-005-4256-y

Okamura Yokamura Y, Kuwatsuka S (1988) pH and concentration dependency of adsorption of phenolic acids on clay minerals. Clay Sci 7:139–150

Patrick KKY, Wong FTW, Wong JTY (2002) Mimosine, the Allelochemical from the leguminous tree *Leucaena leucocephala*, selectively enhances cell proliferation in dinoflagellates. Appl Environ Microbiol 68(10):5160–5163

Pawlowski A, Kaltchuk-Santos E, Zini CA, Caramao EB, Soares GLG (2012) Essential oils of *Schinus terebinthifolius* and *S. molle* (Anacardiaceae): mitodepressive and aneugenic inducers in onion and lettuce root meristems. South Afr J Bot 80:96–103. https://doi.org/10.1016/j.sajb.2012.03.003

Pimentel D (2011) Biological invasions: economic and environmental costs of alien plant, animal and microbe species. CRC Press, Boca Raton

Prajna KP, Mallikarjuna S, Hrideek TK, Sabu A, Shiburaj S (2019) Characterization of bacterial flora associated with the rhizosphere soil of *Maesopsis eminii* Engl. in Wayanad Wildlife Sanctuary of Western Ghats. In: International conference on exploring the scope of plant genetics resources, 22–24 May. Department of Botany, University of Kerala, pp 323–325. ISBN 978-81-940888-0-6

Putnam AR (1985) Weed allelopathy. In: Putnam AR, Duke SO (eds) Weed physiology: reproduction and ecophysiology, vol 1. CRC Press, Boca Raton, pp 131–155

Putnam AR, Duke WB (1978) Allelopathy in agroecosystems. Annu Rev Phytopathol 16:431–451

Ramona C (2018) Allelopathy and allelochemical interactions among plants. Scientific papers. Series A. Agronomy Vol. LXI, No. 1

Rice EL (1974) Allelopathy. Academic Press, New York

Rice EL (1984) Allelopathy, 2nd edn. Academic Press, New York

Ridenour WM, Callaway RM (2003) Root herbivores, pathogenic fungi, and competition between *Centaurea maculosa* and *Festuca idahoensis*. Plant Ecol 169:161–170

Rillig MC, Mummey DL (2006) Mycorrhizas and soil structure. New Phytol 171:41–53

Sanchez-Moreiras AM, de la Peña TC, Reigosa MJ (2008) The natural compound benzoxazolin-2 (3H)-one selectively retards cell cycle in lettuce root meristems. Phytochemistry 69:2172–2179

Schnute M (1984) The allelopathic aspects of *Melilotus alba* through coumarin. J Washington Acad Sci 74(4):117–120. http://www.jstor.org/stable/24537454

Schulz M, Wieland I (1999) Variation in metabolism of BOA among species in various field communities—biochemical evidence for co-evolutionary processes in plant communities? Chemoecology 9:133–141. https://doi.org/10.1007/s000490050044

Sicker D, Frey M, Schulz M, Gierl A (2000) Role of natural benzoxazinones in the survival strategy of plants. Int Rev Cytol 198:319–346

Smita L (2008) Allelopathic impact of *Ageratum conyzoides* Linn. towards some crop and weed plants. Thesis submitted Department of Botany, Aligarh Muslim University, Aligarh (India), pp 273

Sprent Janet I, Platzmann J (2001) Nodulation in legumes (Ed facsimile). Royal Botanic Gardens Kew, London

Stinson KA, Campbell SA, Powell JR, Wolfe BE, Callaway RM, Thelen GC (2006) Invasive plant suppresses the growth of native tree seedlings by disrupting below ground mutualisms. PLoS Biol 4:e140. https://doi.org/10.1371/journal.pbio.0040140

Suma S (1998) A brief study on the environmental physiology of *Amaranthus spinosus*. L. PhD thesis, Department of Botany, Bangalore, India, Bangalore University, pp 112

Tesio F, Weston LA, Ferrero A (2011) Allelochemicals identified from Jerusalem artichoke (*Helianthus tuberosus* L.) residues and their potential inhibitory activity in the field, laboratory. Sci Hortic 129(3):361–368

Uddin MN, Randall WR (2017) Allelopathy and resource competition: the effects of *Phragmites australis* invasion in plant communities. Bot Sci 58(1):29. https://doi.org/10.1186/s40529-017-0183-9

Wardle DA, Bardgett D, Callaway RM, Van der Putten W (2011) Terrestrial ecosystem responses to species gain and losses. Science 332:1273–1277

Weidenhamer, Callaway RM (2010) Direct and indirect effects of invasive plants on soil chemistry and ecosystem function. J Chem Ecol 36(1):59–69

Weir TL, Park SW, Vivanco JM (2004) Biochemical and physiological mechanisms mediated by allelochemicals. Curr Opin Plant Biol 7:472–479

Willis RJ (2010) The history of allelopathy. Springer, Dordrecht

Wink M (1999) Functions of plant secondary metabolites and their exploitation in biotechnology. In: Annual plant reviews, vol 3. CRC Press, Boca Raton

Wink M, Latzbruning B (1995) Allelopathic properties of alkaloids and other natural-products. In: Inderjit A, Dakshini KMM, Einhellig FA (eds) Allelopathy: organisms, processes, and applications. American Chemical Society Press, Washington, DC, pp 117–126

Wink M, Twardowski T (1992) Allelochemical properties of alkaloids. Effects on plants, bacteria and protein biosynthesis. In: Rizvi SJH, Rizvi V (eds), Allelopathy. Basic and applied aspects. Chapman and Hall, London, pp 129–150

Wu FZ, Pan K, Ma FM, Wang XD (2004) Effects of ciunamic acid on photosynthesis and cell ultrastructure of cucumber seedlings. Acta Hort Sin 31:183–188

Wu ZJ, Yang L, Wang RY, Zhang YB, Shang QH, Wang L, Ren Q, Xie ZK (2015) In vitro study of the growth, development and pathogenicity responses of *Fusarium oxysporum* to phthalic acid, an autotoxin from *Lanzhou lily*. World J Microbiol Biotechnol 31:1227–1234

Yamane A, Fujikura J, Ogawa H, Mizutani J (1992) Isothiocyanates as allelopathic compounds form *Rorippa indica* Heirn. (Cruciferae) roots. J Chem Ecol 18:1941–1954. https://doi.org/10.1007/BF00981918

Yang SY, Yu ZW, Sun WH, Zhao BW, Yu SW, Wu HM, Huang SY, Zhou HQ, Ma K, Lao XF (1992) Isolation and identification of anti-algal compounds from root system of water hyacinth. Acta Phytophysiol Sin 18:399–402

Yang QH, Ye WH, Liao FL, Yin XJ (2005) Effects of allelochemicals on seed germination. Chin J Ecol 24:1459–1465

Yi Z, Zhang M, Ling B, Xu D, Ye J (2005) Inhibitory effects of *Lantana camara* and its contained phenolic compounds in *Eichhornia crassipes* growth. J Appl Ecol 17:1637–1640

Yu JQ, Ye SF, Zhang MF, Hu WH (2003) Effects of root exudates and aqueous root extracts of cucumber (*Cucumis sativus*) and allelochemicals, on photosynthesis and antioxidant enzymes in cucumber. Biochem Syst Ecol 31:129–139. https://doi.org/10.1016/S0305-1978(02)00150-3

Zeng RS, Luo SM, Shi YH, Shi MB, Tu CY (2001) Physiological and biochemical mechanism of allelopathy of secalonic acid Fonhigher plants. Agron J 93(1):72–79. https://doi.org/10.2134/agronj2001.93172x

Zuo SP, Ma YQ, Ye LT (2012) In vitro assessment of allelopathic effects of wheat on potato. Allelopathy J 30(1):1–10

Modified Plant Metabolites as Nutraceuticals

8

O. Nikhitha Surendran, M. Haridas, George Szakacs, and A. Sabu

Abstract

Plants harbor a wide range of organic compounds, and based on their direct participation in growth, development, and well-being of host, these organic compounds are mainly referred to as primary or secondary metabolites. The primary metabolites, distributed among all plants, perform essential metabolic roles. Phyto sterols, acyl lipids, nucleotides, amino acids, and organic acids are present in all plants and these compounds are directly involved in their metabolism forming obvious examples of primary metabolites. In contrast, secondary metabolites do not appear to participate directly in growth and development, but they play important roles in the adaptation of plants to their environment. These substances are often differentially distributed among limited taxonomic groups within the plant kingdom and have wide range of chemical structures. In traditional medicine, secondary metabolites have been used for centuries due to their significant biological activities and they find applications in fine chemicals, pharmaceutics, flavors, pesticides, cosmetics, fragrances, and more recently in nutraceuticals (functional foods). Nutraceuticals are food or part of food that perform important roles in providing health benefits, including the prevention and treatment of various diseases. The basic principle of herbal nutraceuticals is to prevent nutritionally induced acute and chronic diseases, thereby promoting optimal health, longevity, and quality of life. Fermentation is an effective protocol of enhancing the value of nutraceuticals. Ayurvedic system of medicine had adopted methods for preparing fermented plant extracts several millennia ago, that could be equated to present-day nutraceuticals.

O. Nikhitha Surendran · M. Haridas · A. Sabu (✉)
Department of Biotechnology and Microbiology, Inter University Centre for Bioscience, Kannur University, Kannur, Kerala, India

G. Szakacs
Department of Applied Biotechnology and Food Science, Budapest University of Technology and Economics, Budapest, Hungary

Keywords

Primary metabolites · Secondary metabolites · Nutraceuticals · Fermentation · Herbal nutraceuticals

8.1 Introduction

Metabolites can be defined as the chemical substances which are essential to the metabolism of a particular organism or to a particular metabolic process. A plant cell produces two types of metabolites: primary and secondary metabolites. The primary metabolites (e.g., carbohydrates, amino acids, and lipids) of plants are widely distributed in nature and directly involved in growth and development, while the secondary metabolites have no apparent function in the growth, development, or reproduction of plants. Secondary metabolites of plants are derived by unique biosynthetic pathways from primary metabolites and act as pollinator attractants, defense chemicals against microorganisms, insects and higher predators, and even other plants (allelochemics) and also represent chemical adaptations to environmental stresses. However, the absence of secondary metabolites does not cause negative effects in the plants but they may play important roles in plant well-being by interacting with the ecosystems and also represent an important source of active pharmaceuticals (Irchhaiya et al. 2015; Bourgaud et al. 2001; Balandrin et al. 1985).

Secondary metabolites present in the plants determine the nutritional quality of food, color, taste, or smell and they also have antioxidative, anticarcinogenic, antihypertension, anti-inflammatory, antimicrobial, immunostimulating, and cholesterol-lowering properties (Irchhaiya et al. 2015). Plant secondary metabolites are commercially used as biologically active compounds like pharmaceuticals, flavors, pesticides, fragrances, etc. Examples of commonly used plant secondary metabolites are nicotine, pyrethrins, and rotenone, which have pesticidal activity and are used in limited quantities (Balandrin et al. 1985). Due to their important biological activities, extracts containing plant secondary metabolites have been used for centuries in traditional medicine for the treatment or prevention of diseases. Currently, they find applications in cosmetics, fine chemicals, and more recently in nutraceuticals or functional foods (Lavecchia et al. 2013). Almost 80% of world's population depends on traditional medicine for their fundamental health care needs and most of these drugs are derived from herbs (Sasidharan et al. 2011).

Recently, the demand for food products containing bioactive compounds as well as nonfood products like dietetics and pharmaceuticals has increased. Preparations from food materials, often fortified, that are regularly consumed to fill deficiency of essential nutrients and to maintain the general health are known as food supplements. It may be seen that there is no universally accepted definition for nutraceuticals. Nutraceuticals/functional foods are semi-purified plant products not consumed as regular foods, finding more value than foods but less than pharmaceuticals (Ranzato et al. 2014). For healthy survival and longevity, the required amount of nutrient compounds like vitamins, fats, proteins, carbohydrates, etc. should be consumed.

When a functional food helps to provide nutrients and also aids in the prevention and treatment of diseases other than anemia, then it can be termed as nutraceutical (Cencic and Chingwaru 2010; Pandey et al. 2010). Green tea is considered as a nutraceutical because it has been used for weight loss and cancer treatment. Another example is *Ginkgo biloba*. It is also used widely as nutraceutical due to its beneficial effect in improving cognitive function (Shinde et al. 2014).

Currently, nutraceuticals have received significant interest due to their potential nutritional, safety, and therapeutic effects. The term nutraceutical is mainly applied to products that are derived from herbal products, dietary supplements (nutrients), specific diets, and processed foods like cereals, soups, and beverages. These products are used as medicine other than nutrition. However, both pharmaceutical and nutraceutical compounds are used to treat or prevent diseases, but only pharmaceutical compounds have governmental sanction (Nasri et al. 2014).

8.2 Secondary Metabolites of Plants

Based on the involvement in basic metabolic processes, phytochemical constituents of plants are categorized as primary and secondary metabolites. Primary metabolites are present in all plants and they are more or less similar in all living cells and participate in basic physiological functions. Some common examples of primary metabolites include sugar, amino acids, tricarboxylic acids or Krebs cycle intermediates, proteins, nucleic acids, and polysaccharides. Conversely, secondary metabolites are products of subsidiary pathways as the shikimic acid pathway (Hussein and El-Anssary 2018).

Secondary metabolites produced from each plant family, genus, and species can sometimes be used as taxonomic features in classifying plants. Secondary compounds are derived from plant cells in smaller quantities than primary metabolites. The extraction and purification of secondary metabolites are difficult, as the synthesis of these metabolites is carried out by specialized cells at particular developmental stages. Therefore, the available secondary metabolites are high value–low volume products. Some of these compounds are used as medicines, flavorings, or recreational drugs (Kabera et al. 2014; Irchhaiya et al. 2015). The presence of secondary metabolites in the plant as a mixture (rather than a single compound) has many beneficial medicinal effects. The combination of secondary products in a particular plant may be taxonomically distinct. Due to this phenomenon, each plant may exhibit unique medicinal actions (Saranraj and Sivasakthi 2014).

Secondary metabolites can be subdivided into distinct groups mainly based on their chemical structure and biosynthetic pathways. These subgroups can again be broadly separated in terms of the nature of their ecological roles, ultimate effects, and comparative toxicity in consumption. Therefore, the phytochemicals are largely and most prevalently grouped by the chemical nature and synthetic aspects. In this manner, plant secondary metabolites are of three classes which include terpenoids, phenolic metabolites, and alkaloids (Kennedy and Wightman 2011). Each family of

Table 8.1 Classes of secondary metabolites

Terpenoids	Phenolic compounds	Alkaloids
Monoterpenoids, iridoids, sesquiterpenoids, sesquiterpene lactones, diterpenoids, triterpenoid saponins, steroid saponins, cardenolides and bufadienolides, phytosterols, cucurbitacins, nortriterpenoids, other triterpenoids and carotenoids	Anthocyanins, anthochlors, benzofurans, chromones, coumarins, minor flavonoids, flavonones and flavonols, isoflavonoids, lignans, phenols and phenolic acids, phenolic ketones, phenylpropanoids, quinonoids, stilbenoids, tannins and xanthones	Amaryllidaceae, betalain, diterpenoid, indole, isoquinoline, lycopodium, monoterpene, sesquiterpene, peptide, pyrrolidine and piperidine, pyrrolizidine, quinoline, quinolizidine, steroidal, and tropane compounds

compounds comprises several other classes of metabolites (Table 8.1) (Dillard and German 2000).

Nonprotein amino acids, amines, cyanogenic glycosides, glucosinolates, and purines and pyrimidines are considered as other nitrogen-containing constituents (Dillard and German 2000). There is no clear-cut boundary between alkaloids and other nitrogen-containing natural compounds. Compounds like amino acids, proteins, peptides, nucleotides, nucleic acid, and amines are not usually called alkaloids and they are coming under nitrogen-containing metabolites (Kabera et al. 2014).

8.2.1 Terpenoids

The terpenoids or isoprenoids including both primary metabolites and secondary metabolites are derived from the five-carbon precursor isopentenyl diphosphate (IPP) (Hussain et al. 2012). These are the largest group of phytochemicals present in green foods, soy plants, and grains. Terpenes play an important role in the photosynthetic reactions of plants and relates to their necessity to fix carbon by using photosensitizing pigments. By consuming these compounds via food helps the animal for hormonal and growth regulatory functions (vitamin A), and the presence of these molecules in animal tissues also provides protection from chronic damage and growth dysregulation. The combining of terpenes and free radicals by partitioning them into fatty membranes by virtue of their long carbon side chain exhibits antioxidant activity. The most studied of the terpenes are the tocotrienols and tocopherols which are antioxidants in nature (Dillard and German 2000).

Terpenes have been investigated for a range of medicinal uses such as antibacterial, antifungal and anticancer compounds. Sesquiterpenoids are the largest family of terpenoids and have a distinct range of biological activities including plant growth regulators, antioxidants, anti-feedant, toxic and antibiotic substances, insect juvenile hormone mimics, and phytoalexins (antifungal) (Singer et al. 2003).

8.2.2 Phenolic Compounds

Phenolic acids, polyphenols, and flavonoids are the most important dietary phenolics. Plant phenolics are formed through the shikimic acid pathway producing a group of phenolics called phenylpropanoids. Hydroxyl-cinnamic acids and coumarins are important members of this group (Singer et al. 2003).

Structurally, phenolic compounds share at least one aromatic hydrocarbon ring with one or more hydroxyl groups attached and these are ranging from simple low-molecular weight compounds, such as the simple phenylpropanoids, coumarins, and benzoic acid derivatives, to more complex structures such as flavanoids, tannins, and stilbenes. Among these, the largest and most diverse group is of flavonoids, that comprises about 6000 compounds. All flavonoids have a common underlying structure of two 6-carbon rings, with a 3-carbon bridge, that normally forms a third ring. Based on the modifications of this basic structure, flavoids can be subdivided into chalcones, flavones, flavonols, flavanones, isoflavones, flavan-3-ols, and anthocyanins (Kennedy and Wightman 2011). Flavonoids exhibit important biological activities and they act against inflammation, free radicals, free radical mediated cellular signaling, allergies, platelet aggregation, etc. They also act against microbes causing ulcers, viruses, tumors, and hepatotoxins. Flavonoids have been shown to possess many other biological effects like inhibiting the angiotensin-converting enzyme that raises blood pressure, inhibiting cyclooxygenase which forms prostaglandins, and inhibiting enzymes present in estrogenic pathway. The in vitro inhibitory actions of certain flavonoids could also play important roles in prevention of platelet aggregation reducing the cause of cardiac diseases and thrombosis. They may also bind estrogen receptors in several tissues to function as phytoestrogens, reducing the risk of estrogen-related cancers due to its deficiency. The most common flavonoid present in higher plants is quercetin, which has the ability to inhibit a number of enzymes and smooth muscle contraction and also it exhibits mutagenic activity and allergenic properties. Quercetin has also been demonstrated to have activities in the proliferation of rat lymphocytes and it possess anti-inflammatory, antibacterial, antiviral, and anti-hepatotoxic activities (Dillard and German 2000). The phenolic compounds (around 8000 or so) are produced by way of either the shikimic acid pathway or the malonate/ acetate pathway (Hussain et al. 2012).

8.2.3 Alkaloids

Organic compounds which have at least one nitrogen atom in a heterocyclic ring are known as alkaloids which are biosynthesized mainly from amino acids (Hussain et al. 2012). Over 20% plant species contain alkaloids, and are structurally diverse group of over 12,000 or so known cyclic nitrogen-containing compounds (Kennedy and Wightman 2011). Acridones, aromatics, carbolines, ephedras, ergots, imidazoles, indoles, bisindoles, indolizidines, manzamines, oxindoles, quinolines, quinozolines, phenylisoquinolines, phenylethylamines, piperidines, purines, pyrrolidines, pyrrolizidines, pyrroloindoles, pyridines, and simple tetrahydroisoquinolines are some of the examples of basic types of alkaloids. Alkaloids have key applications in pharmacological context including analgesia, local anesthesia, vasoconstriction, cardiac stimulation, respiratory stimulation and relaxation, muscle relaxation, and toxicity. They also demonstrate antineoplastic, antibacterial, antifungal, antiviral and allelopathic, hypertensive and hypotensive properties. Mutagenic or carcinogenic activity and cytotoxic activity are also reported. Nicotine and anabasine are examples for alkaloids which are used as insecticides (Hussein and El-Anssary 2018).

8.3 Nutraceuticals

The term nutraceutical is a combination of two words, nutrition and pharmaceuticals, and it was coined by Dr. Stephen DeFelice, Chairman of the Foundation for Innovation in Medicine in 1989. Nutraceuticals are any substances which may be considered a food or part of food, that perform important role in providing health benefits including the prevention and treatment of various ailments ranging from heart diseases to cancer. Presently over 470 nutraceuticals are available and are used to improve life expectancy, support the structure or function of the body, delay the aging process, and also prevent chronic diseases. Nutraceuticals act as antidiabetics, anticancer agents, cardiovascular agents, anti-obese agents, immune boosters, etc. (Rajasekaran et al. 2008; Nasri et al. 2014). Nutraceutical has advantage over the medicine because they may not have side effects under normal use, and are natural dietary supplements, easily available, often at low cost, helping to improve medical condition and health. It is believed that healing systems using nutraceuticals are better and conforms to the old perception that foods can also provide medicinal benefits rather than being the sources of nutrients and energy. Nutraceuticals can be obtained from plants, animals, minerals, or microbial sources (Chauhan et al. 2013). Let food be your medicine and medicine be your food—the philosophy by Hippocrates is key fact behind thoughts of nutraceuticals. Functional food provides the required amount of vitamins, fats, proteins, and carbohydrates necessary for human body for healthy survival (Das et al. 2012). Changing food may cause problems in supply of sufficient quantities of nutrients and warrants the necessity to provide the essential nutrients separately. Modern diet and human health have undergone drastic changes in recent years. So the nutraceuticals are recognized to play an important role in the public health promotion regime. This has increased the worldwide awareness of the significance of nutraceuticals. Nutraceuticals are a group of products which include isolated nutrients, dietary supplements, herbal products, and processed foods (Pandey et al. 2010).

There is no exact definition of nutraceuticals or functional foods or of similar terms, like health foods, or terms related to herbal products (Siddiqui et al. 2018). Nutraceuticals are also known by various terms like functional foods, dietary supplements, genetically engineered designer foods, medical foods, pharmaceutical foods, phyto-nutrients, etc. They are derived from herbal or botanical raw materials that provide additional physiological benefit and protection against chronic diseases. Essentially, the burden of correcting faulty food habits would fall on the secondary metabolites present in plants used as food sources. Also, it could be adopted from alien sources provided such source materials are established to have sufficient and supporting nutrient values. The recent surge of research activities in the realm of developing nutraceuticals has shown the right interest of the operators in health and wellness business. The worldwide nutraceutical market is ever expanding because the nutraceuticals possess multiple therapeutic properties such as anti-obesity effects, cardiovascular effects, antidiabetic effects, immune enhancement, natural antioxidant activity, and anti-inflammatory effects. All these phenomena are on increase since the change in lifestyles are the main uncontrollable, negatively influencing parameter, relevant in industry. Currently, the consumers worldwide use nutraceutical products more than the processed foods but less than pharmaceuticals to improve health, delay the aging process, and prevent chronic diseases. So it can be considered as a rapidly growing industry at 7–12% per year (Shinde et al. 2014; Kuppusamy et al. 2014).

The nutraceutical market worldwide has grown because of the public's view of modern diet, and human health has undergone drastic changes in recent years. Nutraceuticals are a group of products that can be categorized as dietary fiber, prebiotics, probiotics, polyunsaturated fatty acids, antioxidants, and other different types of herbal/natural foods. Some of the most common age-associated chronic diseases such as obesity, cardiovascular diseases, cancer, osteoporosis, arthritis, diabetes, and cholesterol could be tackled to certain extent with the help of nutraceuticals. Nutraceuticals containing beneficial phytochemicals of high nutritional value have been linked with several health benefits as a result of their medicinal properties conceptualized partly from the age-old practices and partly by the contemporary evidence-based research outcomes. Long-term use of nutraceuticals as a food has a nutritional role and impact may arise in the health in due course (Rajasekaran et al. 2008; Das et al. 2012).

8.3.1 Categories of Nutraceuticals

Nutraceuticals are nonspecific biological therapies that have beyond their nutritional value, used to promote wellness, prevent malignant processes and control disease symptoms. These can be grouped into three categories (Fig. 8.1) (Patil 2011; Chintale Ashwini et al. 2013).

Substances with established nutritional functions are categorized as nutrients. Herbals include herbs or botanical products as concentrates and extracts while the reagents derived from other sources showing specific functions, such as sports nutrition, weight-loss supplements, and meal replacements, are grouped as dietary supplements (Patil 2011; Chintale Ashwini et al. 2013). The dietary supplements are to be provided for the enhanced needs of performance of people engaged in specific activities, often sports events, which would demand higher than normal biological functions. And, they are to be of dietary in nature. In other words, they should not be

```
                    ┌─────────────────────────────────┐
                    │          Nutrients              │
              ┌─────│ Vitamins, minerals, amino acids │
              │     │      and fatty acids.           │
              │     └─────────────────────────────────┘
      ┌───────────────┐  ┌─────────────────────────────────┐
      │               │  │           Herbals               │
      │ Nutraceuticals│──│   Herbs or botanical products   │
      │               │  └─────────────────────────────────┘
      └───────────────┘
              │     ┌─────────────────────────────────┐
              │     │       Dietary supplements       │
              └─────│  Pyruvate, chondroitin sulphate,│
                    │   steroid hormone precursors etc.│
                    └─────────────────────────────────┘
```

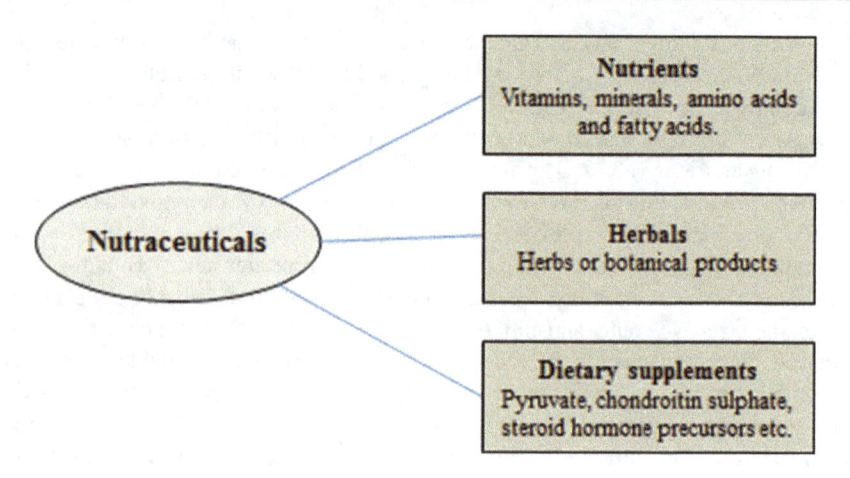

Fig. 8.1 Categories of nutraceuticals

enhancers of bio-functions like anabolic steroids, often illegally used by sports persons for enhanced performance.

8.3.2 Role of Nutraceuticals in Diseases

Nutraceuticals are medicinal foods that are formulated for safety and potential nutritional and therapeutic effects. Majority of the nutraceuticals are claimed to be maintaining well-being, improving health, and harmonizing immunity. Prevention and treatment with nutraceuticals are considered by public health authorities as a powerful instrument in maintaining health and also to act against nutritionally induced acute and chronic diseases (Patil 2011). Kumar and Kumar (2015) reviewed that almost all diseases are due to the deviant and exaggerated oxidative stress. Therefore, antioxidants are very much essential in the treatment of almost all diseases. Phytochemicals have different pharmaceutical properties and can be used variously. For example, flavonoids have profound effects through possession of anticancerous property, since they act as antioxidants, antimicrobials, stimulant to hormonal action, stimulation of biosynthesis or inhibition of enzymes, interfering with DNA replication, etc. Anticancer activity expressed by phytochemicals is more attractive than expressed by any other currently described class of chemicals. PUFAs (Omega-3 polyunsaturated fatty acids) which play an important role in the regulation of inflammation are another class of secondary metabolites possessing therapeutic effects. It has been shown to decrease the production of inflammatory eicosanoids, cytokines, and reactive oxygen species, possess immunomodulatory effects, and attenuate inflammatory disorders. Antioxidants and water- and fat-soluble vitamins are the most commonly known nutrients. Antioxidants are mainly used in the form of dietary intake or supplementation. They are attributed with many potential benefits. In general, they may be useful also in the prevention of cancer and cerebrovascular diseases. Parkinson's disease may be a well-known example that is prevented by high dietary intake of vitamin E (Patil 2011).

Some example of nutraceuticals possessing potent anticancer activities are curcumin (turmeric), green tea (catechins), silymarin (artichoke), capsaicin (red chili), and genistein (El-sherbiny et al. 2016). Isoflavones, polyphenols, lycopenes, resveratrol, etc. are some other examples for nutraceuticals which have shown evidences to prevent, reverse, or delay the carcinogenic processes. The major side effects and toxicity of cancer treatment due to radiation and chemotherapy are found to be reduced by nutraceuticals. It also helps cancer patients to lead a better life. The constituents of green tea mainly inhibit angiogenesis and neovascularization and its intensely researched active constituent is epigallocatechin-3-gallate (EGCG). EGCG prevents proliferation of various types of tumor cell such as colon, breast, and head and neck cancer cell lines. Capsaicin, the active constitutes of *Capsicum* sp., lags the multiplication of cancer cells, regulates apoptosis, and hinders inflammatory response. So many reports are revealing the promising role of capsaicin in cancer management. The active principles of saffron are crocetin and crocin, known for their antineoplastic activity, through inhibition of cell proliferation and initiation of apoptosis. Curcumin (active constitute of turmeric), probably the most widely studied nutraceutical, is well known for its anti-inflammatory activity and it also possess potential effects on broad spectrum of cancers, metabolic syndrome, diabetes, obesity, and atherosclerosis. Estrogen receptor (ER) is a key marker diagnosed with breast cancer. The isoflavones have the ability to control the ER and play an important role in the cancer treatment (Siddiqui et al. 2018).

Flavonoid, phenolic acids, stilbenes, and curcuminoids are the most prominent groups with the capacity to hinder the induction of carcinogenic process and to suppress cancer progression. One of the most important stilbenes is resveratrol, which has multiple bioactivities including anticancer, anti-carcinogenesis, and anti-inflammatory effects. It also possesses a natural antiproliferative activity as a result of its role as a phyto alexin (plant antibiotic). Quercetin, one of the active constitutes possessing anti-inflammatory potential, is obtained mainly from various fruits and vegetables, and come under the flavonoid class. Quercetin plays an important role in the induction of cell apoptosis but the low bioavailability (like resveratrol) is the main challenge for these active ingredients of nutraceuticals (Ranzato et al. 2014).

8.4 Fermented Nutraceuticals

Plant-based formulations have been used for long as remedial dosage forms against various diseases. Traditional medicinal formulae hold much of the prospects. This has led to the use of a number of fermented phyto-metabolites as anticarcinogenic and cardioprotective agents. Perceptions on the plant secondary metabolites, especially fermented ones, and their therapeutic potentials are changing. The knowledge about the biological activity of the fermented secondary metabolites of plants gained at the cellular and molecular level would be useful for development of nutraceuticals possessing therapeutic potential, including anticancer trials. Herbal-based or plant-originated cell cycle regulators might form a new set of drug leads. This would reduce the burden of diseases affecting humans and animals by simple dietary intake of modified/fermented plant-derived secondary metabolites.

Optimizing of advanced biological conversion processes such as enzymatic, microbial, and physicochemical processes including fermentation is an important part of the bioprocessing and nutraceutical development research agenda. Fermentation has been showed to be a methodology for making food last longer and better. It is an incredibly powerful, traditional way to increase health benefits and used by many cultures around the world to enhance quality of foods and to prevent spoilage. Fermentation predigests foods and enables their better absorption and the secondary metabolites may get biotransformed enabling them often to be more active than the originals. A potential application of a fermented nutraceutical (fermented papaya preparation) in acute respiratory illnesses has been demonstrated. An in vivo placebo-controlled, cross-over clinical study in different age groups of healthy subjects has shown better activity of biotransformed secondary metabolites (Marotta et al. 2012a, b, c). As the microbiome expands its frontiers by research, scientists are learning more about how bacteria in fermented foods and drinks can aid digestion, immunity, and overall health. Hence, the aptitude/market for fermented health products has grown significantly in recent years.

The benefits of bacteria together with the desire for naturally processed products make fermentation an attractive option for consumers. Furthermore, interest in traditional food processing and preparation has gone up. In south Asia, yogurt is a household name for millennia. Far East also has a long history of fermented foods. It is also implemented to significant proportions in Europe and America. The market research giant Mintel pointed to Kombucha's unique features and gut-friendly microbes, repositioning it as a successful alternative to soft drinks with functional ingredients, such as vitamins, minerals, and protein. This shows that a strong and greater consumer movement is growing towards healthier food norm. It may be seen that fermentation closely aligns with a clean practice across many culinary traditions to avoid additives and preservatives. Recently more and more manufacturers are promoting the use of fermentation. Hence, dairy products top the sector in food and beverages, with applications of *Streptococcus thermophilus* and *Lactobacillus bulgaricus*.

A noteworthy trend seen among dietary supplement manufacturers now is the use of fermentation in product lines targeting more digestibility, absorption, bioavailability, and overall gut health. Also, significantly higher bioavailability for the fermented form of the vitamins and minerals has been shown by experimental studies. Fermented supplements shall offer a more natural way for the body to get the important nutrients it needs in sufficient amount. They may not generate many of the problems that the synthetic products may cause, such as nausea or digestive upsets, problems in absorption, and unleash their full nourishing potential.

8.5 Ayurveda: Nutraceuticals and Fermented Nutraceuticals

Human intake has wide varieties of cooked or non-cooked plant products as foods, drugs, and other dietary forms as supplements which modify the functioning of systems of the body, especially the central nervous system (CNS). The CNS active properties of the materials consumed are attributable to the secondary metabolites of plants. In many cases, the phytochemicals affecting the human CNS might be linked either to the ecological roles in the plant life or to the physiological and molecular similarities in the biology of two disparate forms manifesting the functions of living system. There are evidences for the efficacy of a range of readily available fresh or processed, single or several plant-based extracts and chemicals that may positively modify the brain functioning. Many researchers have already been attached to the contemporary research theme of enhancers of brain performances and compounds/ extracts that significantly reduce the velocity of brain function deterioration. Many of these candidate phytochemicals/extracts can be grouped by the chemical nature of their active secondary metabolites into different categories discussed in the beginning of this chapter as alkaloids, terpenes, and phenolic compounds. They include much researched and discussed compounds like curcumin, piperine, resveratrol, etc. upon which the entire humanity has showed great interest.

Approximately one-half of all licensed drugs that were registered worldwide in the two decades period prior to 2000s were natural products or their synthetic derivatives. However, only a few of a total of nearly a hundred psychotropic drugs in this period fell within this class (Newman and Cragg 2007). Although the contemporary psychiatric drugs include multitude of synthetic psychotropic medications designed to modify aspects of brain function, until now there are only few mainstream options for improving brain function for different groups of cognitively normal people. These groups include the enlarging segments of the aging societies that suffer from natural, age-related declines of brain functioning. Even the people who suffer from dementia are offered few treatment options and the available options are generally potentially toxic acetyl cholinesterase (AChE) inhibitors. The progenitors of such AChE inhibitors were initially derived from phyto alkaloid (Mukherjee et al. 2007). These chemicals generally have a less than favorable effect/side effect profile (National Institute for Health and Clinical Excellence 2006). However, it may be seen that, as a contrast to the side-effect loaded regulated drug regime, the multitude of herbal supplements that purport to improve brain function aspects are becoming increasingly popular as commonly used items even in developed societies (Bent 2008). A huge amount of scientific literature focuses on the psychoactive herbal extracts and their phytochemicals from traditional and folk, ethnic medicines. The vast majority of these papers describe in vitro investigations of the potential mechanisms of action of phytochemicals in a reverse pharmacology mode. Also, there is a comparatively small amount of literature found on their efficacy in humans. The heritage of Ayurveda perfectly suits to this field.

The Ayurvedic nutraceuticals provides nutrition on one hand and medical or health benefits on the other, including the prevention and treatment of diseases. The science of food and nutritionals in Ayurveda was well developed and that had been categorized into different classes. One such category included food specifically suitable and advised for the newborns, children and diseased people and advocated in summer, shows that the concept of nutraceuticals could be found indirectly. Food in Ayurveda has been resorted to avert the degenerative changes caused by aging, convalescence after an illness, enhancing the defense system and maintaining the vigor and vitality. Apart from the applicability of the food types, they have been classified based on the nature/structure of the food materials, in such a way that the classification indirectly showed the primary and secondary nature of the metabolites. To sum up, it showed the secondary metabolites of the plant materials as well (Tripathi Reprint 2010). Food in childhood promotes growth and development, maintains a higher metabolic rate and ensures increased performance in adulthood, and leads to elevated catabolism in old age wear and tear leading to degenerative changes. For this precise reason, it is important to take these factors in consideration while advocating any nutraceutical.

Other functions of nutraceuticals are recognized to offer specific benefits in certain physiological conditions such as lactation and involution of uterus after child birth. Yet another role of nutraceuticals is to serve at organ level such as addressing the problems of respiratory system and gastrointestinal tract and preventing abdominal distension. A great repertoire of nutraceuticals has entered the market following ethno pharmacological findings from traditional, Ayurvedic practices (Subhose et al. 2005; Larsen and Berry 2003; Kalia 2005; Kokate et al. 2002; Pandey et al. 2010). It has been shown experimentally that fermentation may transform/modify plant secondary metabolites into better bioactive compounds that are suitable as biopharmaceuticals/nutraceuticals and often such medicated wines are used in Ayurveda. Berberine, a secondary metabolite found in many herbs used in Ayurvedic preparations, gets derivatized into better inhibitors implicated in inflammation (Chandra et al. 2012). The berberine derivatives reduced experimentally produced paw edema of animals. In another experiment, the derivatives of piperine formed by fermentative biotransformation were found to be better inhibitors of lipoxygenase than piperine (Sharanya et al. 2019). This shows that fermentation is an excellent production protocol for better nutraceuticals.

8.6 Conclusions

Nutraceuticals are natural bioactive products having medicinal or health benefits. Plants have been considered as a rich source of biologically active compounds, widely used in traditional medicine to prevent/treatment of chronic diseases. Secondary metabolites derived from plants play a tremendous role in medicinal system to develop pharmaceutics, flavors, pesticides, cosmetics, fragrances, and more currently nutraceuticals. Currently, plant-derived nutraceuticals possess great importance in medicinal and health care system due to their potential nutritional, safety, and therapeutic effects.

References

Balandrin MF, Klocke JA, Wurtele ES et al (1985) Natural plant chemicals: sources of industrial and medicinal materials. Science 228(4704):1154–1160

Bent S (2008) Herbal medicine in the United States: review of efficacy, safety, and regulation. J Gen Intern Med 23(6):854–859

Bourgaud F, Gravot A, Milesi S et al (2001) Production of plant secondary metabolites: a historical perspective. Plant Sci 161(5):839–851

Cencic A, Chingwaru W (2010) The role of functional foods, nutraceuticals, and food supplements in intestinal health. Nutrients 2(6):611–625

Chandra DN, Abhilash J, Prasanth GK et al (2012) Inverted binding due to a minor structural change in berberine enhances its phospholipase A2 inhibitory effect. Int J Biol Macromol 50 (3):578–585

Chauhan B, Kumar G, Kalam N et al (2013) Current concepts and prospects of herbal nutraceutical: a review. J Adv Pharm Technol Res 4(1):4–8

Chintale Ashwini G, Kadam Vaishali S, Sakhare Ram S et al (2013) Role of nutraceuticals in various diseases: a comprehensive review. Int J Res Pharm Chem 3(2):290–299

Das L, Bhaumik E, Raychaudhuri U et al (2012) Role of nutraceuticals in human health. J Food Sci Technol 49(2):173–183

Dillard CJ, German JB (2000) Phytochemicals: nutraceuticals and human health. J Sci Food Agric 80(12):1744–1756

El-sherbiny IM, El-baz NM, Hefnawy A (2016) Potential of nanotechnology in nutraceuticals delivery for the prevention and treatment of cancer. Forum Nutr 4:117–152

Hussain MS, Fareed S, Saba Ansari M et al (2012) Current approaches toward production of secondary plant metabolites. J Pharm Bioallied Sci 4(1):10–20

Hussein RA, El-Anssary AA (2018) Plants secondary metabolites: the key drivers of the pharmacological actions of medicinal plants. Herbal medicine. IntechOpen, London, pp 11–30

Irchhaiya R, Kumar A, Yadav A et al (2015) Metabolites in plants and its classification. World J Pharm Pharm Sci 4(1):287–305

Kabera JN, Semana E, Mussa AR et al (2014) Plant secondary metabolites: biosynthesis, classification, function and pharmacological properties. J Pharm Pharmacol 2:377–392

Kalia AN (2005) Text book of industrial pharmacognocy. CBS Publisher and Distributor, New Delhi, pp 204–208

Kennedy DO, Wightman EL (2011) Herbal extracts and phytochemicals: plant secondary metabolites and the enhancement of human brain function. Adv Nutr 2(1):32–50

Kokate CK, Purohit AP, Gokhale SB (2002) Nutraceutical and cosmaceutical. Pharmacognosy, 21st edn. Nirali Prakashan, Pune, pp 542–549

Kumar K, Kumar S (2015) Role of nutraceuticals in health and disease prevention: a review. South Asian J Food Technol Environ 1(2):116–121

Kuppusamy P, Yusoff MM, Maniam GP et al (2014) Nutraceuticals as potential therapeutic agents for colon cancer: a review. Acta Pharm Sin 4(3):173–181

Larsen LL, Berry JA (2003) The regulation of dietary supplements. J Am Acad Nurse Pract 15 (9):410–414

Lavecchia T, Rea G, Antonacci A et al (2013) Healthy and adverse effects of plant-derived functional metabolites: the need of revealing their content and bioactivity in a complex food matrix. Crit Rev Food Sci Nutr 53(2):198–213

Marotta F, Naito Y, Jain S et al (2012a) Is there a potential application of a fermented nutraceutical in acute respiratory illnesses? An in-vivo placebo-controlled, cross-over clinical study in different age groups of healthy subjects. J Biol Regul Homeost Agents 26(2):283–292

Marotta F, Yadav H, Kumari A et al (2012b) Cardioprotective effect of a biofermented nutraceutical on endothelial function in healthy middle-aged subjects. Rejuvenation Res 15(2):178–181

Marotta F, Catanzaro R, Yadav H et al (2012c) Functional foods in genomic medicine: a review of fermented papaya preparation research progress. Acta Biomed 83(1):21–29

Mukherjee PK, Kumar V, Mal M et al (2007) Acetylcholinesterase inhibitors from plants. Phytomedicine 14(4):289–300

Nasri H, Baradaran A, Shirzad H et al (2014) New concepts in nutraceuticals as alternative for pharmaceuticals. Int J Prev Med 5(12):1487–1499

National Institute for Health and Clinical Excellence (2006) Donepezil, galantamine, rivastigmine and memantine for the treatment of Alzheimer's disease (amended). NICE, London

Newman DJ, Cragg GM (2007) Natural products as sources of new drugs over the last 25 years. J Nat Prod 70(3):461–477

Pandey M, Verma RK, Saraf SA (2010) Nutraceuticals: new era of medicine and health. Asian J Pharm Clin Res 3(1):11–15

Patil CS (2011) Current trends and future prospective of nutraceuticals in health promotion. Bioinfo Pharm Biotechnol 1(1):1–7

Rajasekaran A, Sivagnanam G, Xavier R (2008) Nutraceuticals as therapeutic agents: a review. Res J Pharm Technol 1(4):328–340

Ranzato E, Martinotti S, Calabrese CM et al (2014) Role of nutraceuticals in cancer therapy. J Food Res 3(4):18–25

Saranraj P, Sivasakthi S (2014) Medicinal plants and its antimicrobial properties: a review. Global J Pharmacol 8(3):316–327

Sasidharan S, Chen Y, Saravanan D et al (2011) Extraction, isolation and characterization of bioactive compounds from plants' extracts. Afr J Tradit Complement Altern Med 8(1):1–10

Sharanya CS, Shabeer Ali H, Sabu A et al (2019) Fermentation of poly herbal preparations as in Ayurveda: a novel protocol for drug lead discovery. J Nat Ayurvedic Med 3(3):1–8

Shinde N, Bangar B, Deshmukh S et al (2014) Nutraceuticals: a review on current status. Res J Pharm Technol 7(1):110–113

Siddiqui ZH, Hareramdas B, Abbas ZK et al (2018) Use of plant secondary metabolites as nutraceuticals for treatment and management of cancer: approaches and challenges. In: Akhtar MS, Swamy MK (eds) Anticancer plants: properties and application. Springer, Singapore, pp 395–413

Singer AC, Crowley DE, Thompson IP (2003) Secondary plant metabolites in phytoremediation and biotransformation. Trends Biotechnol 21(3):123–130

Subhose V, Srinivas P, Narayana A (2005) Basic principles of pharmaceutical science in Ayurveda. Bull Indian Inst Hist Med (Hyderabad) 35(2):83–92

Tripathi B, Samhita S, Prakashan CS (2010) Reprint. Varanasi, Purvakhand, Chapter 3, Verse1-4

Ethnomedicine and Role of Plant Metabolites

9

Lekshmi Sathyaseelan, Riyas Chakkinga Thodi, and Swapna Thacheril Sukumaran

Abstract

Plants have been used for the treatment of various human ailments since ancient times. Bioactive compounds of plant origin are of great interest in modern medicine due to the lack of harmful side effects and are less expensive compared to synthetic medicines. So nowadays, traditional medicinal plants grab more attention in the field of medicine. However, the knowledge about these plant wealth is confined only to certain tribal groups and are orally passed through their generations. Here arises the need for documentation of such traditional uses of plants for treating various ailments and metabolites related to the medicinal properties. This chapter summarizes 94 significant ethnomedicinal plant species belonging to 41 families that are used by different tribal communities, worldwide and reported secondary metabolites responsible for the bioactivity.

Keywords

Bioactive compounds · Synthetic medicines · Ethnomedicine · Tribal community

9.1 Introduction

Ethnomedicine is defined as those principles or beliefs and practices involved in curing diseases, which are the products of native cultural developments and are not derived clearly from the abstract framework of modern medicine (Bannerman et al. 1983). Ethnomedicinal practice has been a multi-disciplinary system consisting of the use of plants or plant products, spirituality, and the natural environment as a source of remedy for people since time immemorial (Lowe et al. 2001). Various

---

L. Sathyaseelan · R. C. Thodi · S. T. Sukumaran (✉)
Department of Botany, University of Kerala, Thiruvananthapuram, Kerala, India

Fig. 9.1 Relative percentage of three major groups of bioactive compounds present in selected plants

- Phenols 59%
- Terpenes 34%
- Alkaloids 7%

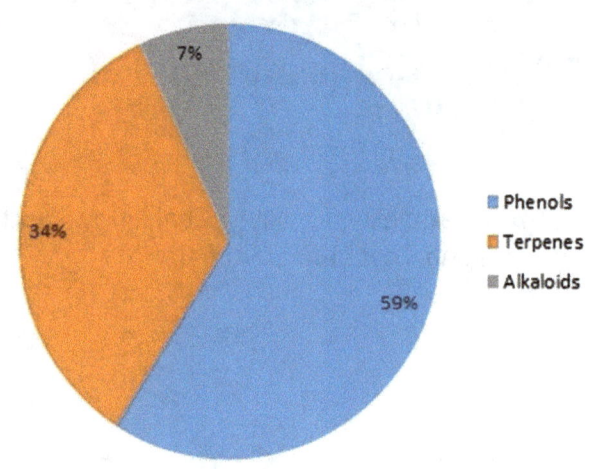

chemical constituents present in the plant sources are broadly categorized as primary and secondary metabolites based on their chemical structure and biosynthetic derivation. Secondary metabolites are compounds that have no direct role in the growth and development of plants but exhibit different kinds of pharmacological properties, which can be again classified based on their chemical structure and functional groups present in it. The major classes of secondary metabolites include alkaloids, terpenoids, phenolics, flavonoids, and glycosides, which may act as lead molecules for the development of new drugs in modern medicines (Velu et al. 2018). The secondary metabolites of medicinal plants often act individually, additively, or synergistically in the improvement of health (Schütz et al. 2006). Among a total of 88 different phytoconstituents isolated from the commonly used ethnomedicinal plants in India, 52 belonged to phenolic compounds followed by terpenes (30) and alkaloids (6). Among the phenolic compounds, the majority are flavonoids. The relative proportion of the three major groups of secondary metabolites are represented in Fig. 9.1.

According to records released by the World Health Organization (WHO), ethnomedicine has retained its popularity in all regions of the developing world, and its usage is rapidly expanding in industrialized countries (Thong et al. 1993). For example, the traditional herbal system of China accounts for 30–50% of the total medicinal consumption. In Mali, Ghana, Zambia, and Nigeria, the primary treatment for 60% of kids with malaria is with the use of herbal medicine, whereas in London, San Francisco, and South Africa, 70% of individuals living with HIV/AIDS use traditional medicine. At present-day world market, herbal medicine stands at over US $60 billion. Approximately 25% of medicine contains active constituents that are derived from higher plants (World Health Organization 2003).

Human civilization started using plants and plant products during ancient times and perhaps ethnobiology is the first science that was initiated with the evolution or existence of humans on this planet (Barukial and Sarmah 2011). Several initiatives have been taken at a global level for the protection and development of medicinal plants. Such initiatives are evident in the initiatives of WHO for the improvement of health and community-based upkeep activities by international organizations, including the World Bank, the International Development Research Centre (IDRC), and UNDP. The exertion by the WHO to know and encourage the use of local medicinal plant knowledge systems in the health segment, principally in developing countries, is prominent (Shukla and Gardener 2006). Today, ethnomedical practices and principles are part of a total trust system that transcends class, ethnicity, and religious belief in such a manner that the terms "folk or traditional" can be used to describe truly universal practices. In Europe and the Caribbean, the coming back to the traditional (ethnomedicinal) system of healthcare is not constrained to the poor but outspreads to all social classes (Lowe et al. 2001).

India is an ancient reservoir of traditional medicine, and the people of India were familiar with a far larger number of medicinal plants than the native of any other nation in earth. India is one of the 12 mega-biodiversity countries of the World, having rich vegetation with a wide variety of plants with medicinal value (Erah 2002). The traditional medicine in India is created on diverse systems such as Ayurveda, Unani, and Siddha. Shepherds and Hermits, who exist in nearby forests, are the best resourceful persons to recognize and utilize plants (Sharma et al. 2012). Indian traditional medicinal system and its knowledge provide a low cost and alternative source of primary health. The history of ethnomedicinal knowledge against various ailments in India can be recorded from the local people in different areas. In India, around 16,000 species of higher plants are identified; out of these 7500 species are used for medicinal purposes and healthcare purposes by different ethnic groups (Bhuyan 2015). Some ancient tribes till now acquire their food and medicines for particular diseases collected by dwelling from remote unreachable forest areas; unfortunately, they don't know the scientific explanation of ethnomedicine used for curing diseases. But the civilized people have a good amount of knowledge about the scientific background of modern medicines but lacks knowledge about traditional medicine. Thus, the knowledge of plants from ethnomedicine could be the best clues for modern drug development (Brahmam 2000). Therefore, the need for documentation of traditional knowledge on the ethnomedicinal use of plants has been considered as a high priority to support the discoveries of drugs for benefiting humanity (Cox and Balick 1994; Dutta and Dutta 2005). Hence, the present chapter focusses on the documentation of ethnomedicinally important plants, especially flowering plants, emphasizing the scientific validation of individual phytoconstituent along with its biological property based on available literature.

9.2 Sources of Ethnomedicine

Angiosperms or flowering plants are the most diverse group of plants with 64 orders and 416 families, according to the APG IV system of plant classification (Christenhusz and Byng 2016). In the present study, a total of 94 species of significant flowering plants have been reported as being used by local tribes to cure various diseases. The plants included in the study belonged to 41 families, among which the most dominant family was Asteraceae, followed by Rutaceae. For convenience, various families with ethnomedicinal plants are explained in alphabetical order.

9.2.1 Acanthaceae

Andrographis paniculata (Burm.f) Wall. ex Nees is a common ethnomedicinally important plant belonging to the family Acanthaceae. It is commonly known as "Kiriyath" or "Sirata" in local languages. Local tribes (Apatani) of Assam used the cold extracted dried leaves and bark against dysentery. *A. paniculata* is a famous traditional medicine, widely used to treat sore throat, flu, and upper respiratory tract infections (Lee et al. 2008). The Teli tribe of Bangladesh use the leaves of this plant to treat lung infections and liver disorders (Rahmatullah et al. 2012). A class of neoandrographolide isolated from *A. paniculata* possesses anti-inflammatory property. The studies regarding this property revealed that neoandrographolide inhibits NO production by macrophages and thereby inducing activation of MAPKinase signaling pathway, which eventually leads to platelet aggregation (Fabricant and Farnsworth 2001). The plant is hepatoprotective by preventing liver toxicity, with anticancer and anti-diarrhea effect, but there is no specific constituent reported (Jarukamjorn and Nemoto 2008). The diterpenoid compound isolated from *A. paniculata* (Andrographolide) showed cytotoxicity against breast cancer, leukemia, lung cancer, and human epidermoid carcinoma. Traditionally it is used in Indian, Thai, and Chinese systems of medicines to treat cancer (Kumar et al. 2004; Parveen et al. 2019).

The leaves of *Adhatoda vasica* are mostly used in the treatment of respiratory disorders in Ayurveda in the Indian system of medicine. The alkaloids, vasicine, and vasicinone present in the leaves possess respiratory stimulant activity (Baquar 1997). Vasicinone, the auto-oxidation product of vasicine, has been reported to cause bronchodilatory effects both in vitro and in vivo (Shinawie 2002). Kaempferol, quercetin, and vitexin are also reported from this plant, which has hepatoprotective potential (Maurya and Singh 2010).

9.2.2 Amaranthaceae

Seeds and leaves of *Amaranthus tricolor* are used by tribals of North Sikkim of India to treat gastric problems. The seeds are ground into powder, mixed with water, and taken as an infusion to cure general gastric problems. Also, the curry prepared from the green leaves of *A. tricolor* is effective in treating diarrhea (Pradhan and Badola 2008). Cox 1 and 2 enzymes were inhibited by compounds isolated from the species of *A. tricolor* (Jayaprakasam et al. 2004*).*

9.2.3 Anacardiaceae

The stem bark of *Lannea coromandelica* is made into a paste, mixed with molasses, and a pill is prepared from it. One pill is taken orally twice a day for 3 days against elephantiasis (Rahaman and Karmakar 2015). Quercetin, β-sitosterol, palmitate, myricadiol, protocatechuic acid, etc. are the major compounds that have been reported from the stem bark (Yun et al. 2014). Quercetin is reported to have antiparasitic activity (Mead and McNair 2006). The leaves of *Mangifera indica* (commonly known as "Aam") are immersed in cold water for about 30 min, and the extract is taken daily to check diabetes (Bandyopadhyay 2017). Mangiferin is the major antidiabetic compound reported in the leaves of *M. indica* (Telang et al. 2013).

9.2.4 Annonaceae

The plant juice of *Annona squamosa* along with paste form of *Datura* leaves is used against snakebite (Verma et al. 2010). Beta-sitosterol is isolated from the plant (Pandey and Barve 2011) and this compound is reported to possess antivenom property (Achika et al. 2014). *A. crassiflora* Mart. is traditionally used as an antidote for snake venom, and acetogenins were also obtained from this plant (Mesquita et al. 2005). Stigmasterol and beta sitosterol isolated from the plant (Luzia and Jorge 2013) are reported as snake antivenom (Achika et al. 2014). The Paniya tribes of Wayanad district, Kerala, use the whole plant of *Annona muricata* (*Mullatha*) against cancer (Marjana et al. 2018). Among the different phytochemicals present in *A. muricata*, acetogenins are found to be the major group of compounds (Coria-Téllez et al. 2018). These annonaceous acetogenins were reported to be capable of blocking ATP production in mitochondria. This mechanism of action was shown to be effective against cancer cells that produce higher amounts of ATP in comparison to normal cells, thus limiting the ability of cancer cells to grow (Waechter et al. 1997).

9.2.5 Apiaceae

The Apatani tribes of Arunachal Pradesh, India, used the plant *Centella asiatica* against gastric disorders by consuming the whole plants (Kala 2005). Juice of shoots of this plant is used to treat gastritis and constipation. *C. asiatica* possesses numerous pharmacological activities such as antibacterial, antidepressant, antiemetic, antineoplastic, antioxidant, antithrombotic, anxiolytic, gastroprotective, immunomodulatory, antigenotoxic, nerve regenerative, reproductive, and wound healing due to the presence of several saponin constituents, including asiaticoside, asiatic acid, madecassic acid, and some other bioactive compounds (Roy et al. 2013). Whole plant parts of *Centella asiatica* are crushed and are used to cure tuberculosis by the Lushai tribe of North East India (Sajem and Gosai 2006). The compounds responsible for the antimycobacterial effect are octadectrienoic acid and n-hexadecanoic acid (Suresh et al. 2010).

The shoot of *Coriandrum sativum* is helpful in digestion, and the shoot mixed with fenugreek and thyme took along with tea relieves stomach pain (Pradhan and Badola 2008). Freshly prepared leaf juice of *C. sativum* commonly known as "dhoney," is taken in an empty stomach every morning by the tribal people of West Bengal to cure diabetes (Bandyopadhyay 2017). Compounds like camphor, eugenol, trans-β-ocimene, geraniol, α-pinene, limonene, p-cymene, 1, 8-cineole, and thujone, which help in pancreatic β-cell restoration and insulin secretion, were reported in *C. sativum* (Broadhurst et al. 2000). Root juice of *Daucus carota* is used in the tribal areas of North Maharashtra for treating jaundice (Badgujar and Patil 2008). Kaempferol is the active therapeutic phytoconstituent present in *D. carota* that is capable of ameliorating the effect of hepatotoxic agents (Jain et al. 2012).

The roots of *Angelica sinensis* are used in traditional Chinese medicine to treat women's reproductive problems such as dysmenorrhea, amenorrhea, and blood deficiency (Wei et al. 2016). Ferulic acid, Z-ligustilide, and E-ligustilide isolated from this plant possess nephroprotective property (Bunel et al. 2015). Lhoba tribe of Tibet uses the plant *Angelica apaensis* as a hypotensive drug (Li et al. 2015). Oxypeucedanin, oxypeucedanin hydrate, isoimperatorin, byakangelicin, and byakangelicol are the major compounds reported in the plant which have anti-HIV activity (Qiong et al. 2008).

9.2.6 Apocynaceae

Kurichya tribes of Kerala use *Alstonia scholaris* (Analivega) against bites of venomous snakes. Pentacyclic triterpenes reported in the plant possess antivenom property (Meenatchisundaram 2008). Rhizome juice of *Rauvolfia serpentina* of the family Apocynaceae is used internally for the treatment of snake poison (Marjana et al. 2018). Various alkaloids identified in *Rauvolfia* include ajmaline, ajmalimine, ajmalicine, deserpidine, indobine, indobinine, reserpine, reserpiline, rescinnamine, rescinnamidine, serpentine, serpentinine, and yohimbine, which are responsible for the various pharmacological activities (Srivastava et al. 2006; Goel et al. 2009). The alkaloid "serpentine" neutralize snake venom (Gupta and Peshin 2012). *Holarrhena antidysenterica* is another plant of the family Apocynaceae, effective against dysentery, as indicated by its name. The juice of the bark of this plant is taken as a remedy for dysentery. A major alkaloid conessine isolated from this plant has anti-amoeboid property (Stephenson 1948). The juice of stem bark of *Wrightia tomentosa* is also reported to be effective for stomach ailment and chronic dysentery (Barukial and Sarmah 2011). Rutin was isolated from *W. tomentosa* (Muruganandam et al. 2000), which is reported to have a gastroprotective property (Abdel-Raheem 2010).

Various plant parts of *Nerium oleander* ("Kaner") are used traditionally in Pakistan for the treatment of swellings, leprosy, eye, and skin diseases. The leaves also possess cardiotonic, antibacterial, anticancer, and antiplatelet aggregation activity and depress the central nervous system. Four CNS depressant cardenolides including neridiginoside, nerizoside, neritaloside, and odoroside-H with CNS depressant activity have been isolated from the plant (Begum et al. 1999).

9.2.7 Araceae

Small piece of the dried rhizome of *Acorus calamus* is taken by the Lepcha tribe of North Sikkim, India, for curing distressing cough (Pradhan and Badola 2008). Mucirin is a bioactive fraction isolated from *A. calamus* which can be used as a potential drug candidate to treat diseases related to mucus hypersecretion, cough, bronchitis, etc. (Berlian et al. 2016). The juice made from the rhizome of *Acorus calamus* Linn. (Sweet flag) is effective against menstrual cycle irregularity and excessive uterine bleeding (Taid et al. 2014). In Chinese traditional medicine, dried rootstock of *A. gramineus* is used as a digestant and expectorant, and in the treatment of diarrhea and epilepsy. Beta asarone is one of the major bioactive compounds reported in *A. gramineus* (Tang and Eisenbrand 2013).

9.2.8 Asclepiadaceae

Hemidesmus indicus, commonly known as "Indian sarsaparilla," is used in polyherbal preparations against cancer (Turrini et al. 2018). Hemidesminine, a coumarin isolated from this plant, shows anticancer activity (Mandal et al. 1991). The plant roots are also used as an antipyretic, antidiarrheal, and blood purifier. They are used for the treatment of blood diseases, biliousness, dysentery and diarrhea, respiratory disorders, skin diseases, leprosy, leucorrhoea, leukoderma, itching, syphilis, bronchitis, asthma, eye diseases, epileptic fits in children, lack of appetite, burning sensation, rheumatism, and kidney and urinary disorders (Swathi et al. 2019).

9.2.9 Asteraceae

Artemisia vestita is a common traditional medicinal plant that has been used in Tibet and China, for treating various inflammatory diseases. Flavones isolated from *Artemisia vestita* exhibit anti-inflammatory and immunosuppressive properties (Yin et al. 2008). An investigation of different species of *Artemisia* showed a range of biological activities, including antimalarial, antibacterial, antifungal, cytotoxic, and antihepatotoxic and antioxidant activity. Artemisinin is a well-known antimalarial compound isolated from the Chinese herb *Artemisia annua*. Terpenoids, flavonoids, coumarins, caffeoylquinic acids, etc. are the major classes of phytoconstituents reported in the genus (Bora and Sharma 2011).

Erigeron breviscapus is an important plant used by the Lhoba tribes of China for the treatment of cardiovascular and cerebral vessel diseases. The main active components identified in the herb include flavonoids, coumarins, lignins, hydroxycinnamic acids, pyromeconic acids, scutellarin, and erigesides. Scutellarin is also reported to have anticancer potential (Zhang et al. 2007). The decoction of the herb *Ageratum conyzoides* is also given to cure stomach ailments such as diarrhea, dysentery, and intestinal colic with flatulence. Eugenol and beta-pinene isolated from the plant (Chauhan and Rijhwani 2015) have gastroprotective and anti-ulcer activity (Morsy and Fouad 2008; Rozza et al. 2011).

Water mixed with crushed leaves of *Artemisia vulgaris* is used for taking bath to prevent and cure allergy by the Lepcha tribe of North Sikkim, India (Pradhan and Badola 2008). Eucalyptol isolated from the plant is reported to be the major anti-inflammatory agent (Jiang et al. 2019). Leaves of *Helianthus annuus*, belonging to the family Asteraceae, are crushed and mixed with water and used for bath to cure allergy and skin diseases (Pradhan and Badola 2008). Grandiflorolic, kaurenoic, and trachylobanoic acids are the three diterpene acids found in *Helianthus annus* which possess anti-inflammatory property (Díaz-Viciedo et al. 2008).

Jaintia tribes of North East India use the crushed flowers of *Spilanthes paniculata* to cure toothache and cavity formation (Sajem and Gosai 2006), which is also a common practice in the state of Kerala. Spilanthol isolated from this plant is reported to have anti-inflammatory activity (Wu et al. 2008). *Pseudelephantopus spicatus* is commonly used by Chayahuita, an ethnic group of Peruvian Amazonia, against leishmaniasis. Antileishmanial sesquiterpene lactones were isolated from this plant, which has strong bioactivity towards *Leishmania amazonensis* (Odonne et al. 2011). The plant juice of *Cichorium intybus* ("Tareezax") is used in the Federally Administered Tribal Areas (FATA) of Pakistan against jaundice, hepatitis, enlarged spleen, and diarrhea (Aziz et al. 2018). The phenolic compound AB-IV obtained from the plant is reported to possess antihepatotoxic activity (Ahmed et al. 2003).

9.2.10 Bignoniaceae

The plant bark, leaves, and pods decoction of *Oroxylum indicum* (L.) is exploited to treat malarial infection and is used by Naga tribe in Assam (Jamir and Lanusunep 2012). Chrysin (5,7-dihydroxyflavone) is a major compound isolated from *O. indicum,* which is reported to have antibacterial (Babu et al. 2006), antioxidant, antihemolytic (Chaudhuri et al. 2007), and anti-inflammatory activities (Ali et al. 1998). The antihemolytic activity of this compound may be beneficial in the treatment of malaria-induced hemolysis. Bark and seeds of *Oroxylum indicum* is taken orally to relieve from dysentery and diarrhea by local people belonging to the Chutia tribe (Borah et al. 2012). To recover from snake bite, Bodo Kachari tribe of North East India use the bark and seeds of *Oroxylum indicum* (Basumatary et al. 2014).

9.2.11 Caesalpiniaceae

The water extracts of bark, leaves, and roots of *Bauhinia racemosa* are taken against jaundice (Badgujar and Patil 2008). Kaempferol, one of the important constituents, was isolated from the plant and reported to have antidiabetic and hepatoprotective activity (Filho 2009). The dried bark of *Cassia didymobotrya* is used as antihemorrhagic by the Marakwet community in Kenya (Kipkore et al. 2014). Tender leaves of *Cassia fistula* are ground with turmeric and the paste is applied for skin disease (Satyavathi et al. 2014). Srividhya et al. (2017) reported the presence of amentoflavone in the leaves of *C. fistula*. Amentoflavone is a potential bioactive compound and is an irreversible inhibitor of lymphocyte proliferation (Guruvayoorappan and Kuttan 2008), an inhibitor of phospholipase, and an inhibitor of nitric oxide synthase in macrophages (Woo et al. 2005).

9.2.12 Calophyllaceae

The latex from the bark of *Calophyllum inophyllum* is rubbed on the skin for treating psoriasis by the Kurichiya and Kuruma tribes of Kerala (Marjana et al. 2018). Amentoflavone is an active compound isolated from the plant (Li et al. 2007). Amentoflavone is effective against psoriasis since it suppresses NF-κB-mediated inflammation and keratinocyte proliferation (An et al. 2016). The seeds of *C. inophyllum* are used against rheumatism and leprosy in Australia (Cock 2011).

9.2.13 Caricaceae

Tribal people of West Bengal, India, use the latex of *Carica papaya* to cure dysentery. A mixture of fresh latex of papaya and lime water and a cup of cow's milk is taken in an empty stomach daily (Bandyopadhyay 2017). In African countries like Gambia, papaya fruits have been used to treat skin burns (Gurung and Škalko-Basnet 2009). The compounds namely hydroxyanthraquinones, tannins, alkaloids, saponins, and a well-known proteolytic enzyme papain were isolated from this plant (Oloyede 2005; Nwofia et al. 2012). In Australia, the leaves of papaya are used against cancer (Otsuki et al. 2010). The enzyme papain is effective against cancer (Fauziya and Krishnamurthy 2013).

9.2.14 Combretaceae

Bark and seeds of *Terminalia chebula* Retz. is taken orally to relieve from dysentery and diarrhea by the local people belonging to the chutia tribe of Assam, India (Borah et al. 2012). In the case of gastric and stomach pain, the powder of the dried fruit of *Terminalia chebula* is orally taken by the Bodo-Kachari tribes of North East India (Basumatary et al. 2014). Nepali and bhutia tribes consumed the fruits and bark of

T. chebula for the treatment of diarrhea and indigestion (Sajem and Gosai 2006). In the case of indigestion and diarrhea, powder of crushed bark and fruits of *Terminalia chebula* are consumed by the Lepcha and Bhutia tribes (Idrisi et al. 2010). Chebulagic acid is a major constituent of *T. chebula*, and this compound is reported to have gastroprotective effects (Liu et al. 2017). Gallotannins are a group of compounds present in *T. chebula* which possess antimicrobial (Engels et al. 2009) and antioxidant properties (Zhao et al. 2005). The fruits of *Combretum apiculatum* are used against cancer by the Marakwet community of Kenya (Kipkore et al. 2014).

9.2.15 Cucurbitaceae

The ripen fruits of *Cucurbita pepo* are the source of Curcurbitiacin B that have antihepatotoxic effect (Valan et al. 2010). The Lepcha tribes of North Sikkim used the plant against jaundice. Fresh leaf paste of *Cucurbita pepo* acts as a soothing agent and is applied on the burnt portion (Pradhan and Badola 2008). The seeds of *Momordica charantia* are used in China to treat infections and immune disorders (Fong et al. 1996).

In the tribal regions of North Maharashtra, the fruit and leaf juice of *M. charantia* is used in treating jaundice (Badgujar and Patil 2008). Polyphenols such as ferulic acid, cinnamic acid, and ascorbic acid isolated from the plant are reported to have liver protective action (Haque et al. 2011). Cucurbitane triterpenoids present in *M. charantia* are compounds responsible for the antidiabetic activity of this plant (Murakami et al. 2001). Another compound called Charantin, which is a steroidal glycoside present in *M. charantia*, has got blood sugar lowering property equivalent to insulin (Pitipanapong et al. 2007). This plant also contains cytotoxic proteins such as momorcharin and momordin (Ortigao and Better 1992).

9.2.16 Ebenaceae

Diospyros melanoxylon gum is effective against jaundice. The tribal people of the Indian state Andhra Pradesh mix the gum in water and consume orally once a day for 7 days (Rahaman and Karmakar 2015). A pentacyclic triterpene (lupeole) responsible for the medicinal property was isolated from *D. melanoxylon* (Sunitha et al. 2001). An early report suggested that lupeole has hepatoprotective action (Mallavadhani et al. 2001). Traditionally *D. chloroxylon* is used as a remedy for treating diabetes (Gupta et al. 2017). Studies reported that a novel compound, lup-20 (29)-ene-3α,6 β-diol, has prominent antihyperglycemic activity (Sharma et al. 2018). Tender twigs of *Diospyros scabra* are used as toothbrushes as a remedy for dental problems by Marakwet tribal community, Kenya (Kipkore et al. 2014).

9.2.17 Euphorbiaceae

Acalypha fruticosa Forssk. ("Paruvathazhi") belonging to the family Euphorbiaceae is used to treat stomach ache and infections of different tribes such as paniyars, kattunaikkans, and Paliyars of Kerala, India (Sandhya et al. 2006). The crushed leaves of this plant are used against scorpion sting by the Marakwet tribal community of Kenya (Kipkore et al. 2014). Four types of compounds like 5-O-β-D-glucopyranoside **1,** 2-methyl-5,7-dihydroxychromone, acalyphin**,** apigenin, and kaempferol 3-O-rutinoside isolated from the plant possess anti-inflammatory activity (Rajkumar et al. 2010).

According to the report published by Karuppusamy (2007), *Phyllanthus amarus* has been used by the tribes in Tamil Nadu (Paliyan) to treat jaundice. Gallic acid is one of the major compounds in *P. amarus*, which is reported to have hepatoprotective activity (Lee et al. 2006) and gastroprotective effect (Nanjundaiah et al. 2011). Here decoction of the whole plant is administered orally. *P. urinaria* is used in China to treat jaundice, hepatitis B, neprolithiasis, and painful disorders and Corilagin, geraniin, and gallic acid are the major reported metabolites (Li et al. 2004).

Kumaran and Karunakaran (2006) reported different kinds of phenolic compounds from *Emblica officinalis* L. (Nelli) such as geraniin, gallic acid, furosin, corilagin, and methyl gallate, all these compounds are significant nitric oxide scavengers than standard curcumin. The fruit is rich in quercetin, phyllaemblic compounds, gallic acid, tannins, flavonoids, pectin, and vitamin C. The fruits, leaves, and bark are rich in tannins. The root contains ellagic acid and lupeol, and bark contains leucodelphinidin (Arora et al. 2003; Kim et al. 2005).

The other investigations on amla berry include brain nourishment property, strengthening nail, hair, and teeth due to the presence of high minerals, especially calcium content, and use of decoction of the leaves for the treatment of diabetes (Treadway 1994). The higher amounts of vitamin C (Sri et al. 2013) and ellagic acid (Fatima et al. 2017) found in amla is reported to be responsible for its antidiabetic property. The berries are also medicinally used for the treatment of diarrhea (Mehmood et al. 2011). Gastroprotective effect of gallic acid and quercetin was reported by Dykes and Rooney (2007) and de la Lastra et al. (1994). *Euphorbia hirta* leaf extract is mixed with sugar and taken for dysentery in the tribal regions of Andhra Pradesh (Satyavathi et al. 2014). Quercetin isolated from *E. hirta* have antidiuretic potential (de la Lastra et al. 1994).

9.2.18 Fabaceae

Cajanus cajan has been used traditionally as a laxative and was identified as an antimalarial remedy (Ajaiyeoba et al. 2005). In the folk medicine of China, the plant is used for internal parasites (Zu et al. 2006). The leaf extract of this plant has been reported to contain a chalcone called as Cajachalcone, which is an antimalarial compound (Ajaiyeoba et al. 2013). It is believed that chalcone derivatives having antimalarial activity interact with the parasite *P. falciparum* enzyme cysteine protease, which is a key enzyme involved in hemoglobin degradation within the acidic food vacuole of the intraerythrocytic parasite (Shenai et al. 2000). Inhibition of this enzyme hampers digestion of hemoglobin within the food vacuole and proves fatal for the parasite.

The leaves of *C. cajan* are made into a paste and taken orally against jaundice by the Santal tribe of West Bengal. An unknown protein (43 KD) was purified from *C. cajan* with hepatoprotective action (Sarkar et al. 2006). The seed oil of *Millettia pinnata* is applied on the affected area to get relief from itching of the skin (Rahaman and Karmakar 2015). Anti-inflammatory pterocarpanoids have been reported in this plant (Wen et al. 2019).

The leaf extract of *Butea monosperma* is used against pimples and the flower paste is applied externally in skin diseases (Barukial and Sarmah 2011). Kattunaika tribes in Wayanad use *Butea monosperma* against stomachache. The isoflavonoid buteaspermin isolated from the stem bark of *B. monosperma* is reported to have osteogenic property. The stem bark of *B. monosperma* displays antifungal as well as anti-inflammatory property, which is due to the presence of an active constituent medicarpin (Bandara et al. 1989). The flower extract of this plant is used in India for the treatment of liver disorders and two antihepatotoxic flavonoids namely isobutrin and butrin have been isolated from it (Mengi and Deshpande 1995). The flower extract of this plant is also reported to have antiestrogenic (Shah and Baxi 1990) and antifertility activities (Razdan et al. 1970). Butin isolated from the flowers showed both male and female contraceptive properties (Bhargava 1989).

Glycyrrhiza glabra is used internally for Addison's disease, asthma, bronchitis, peptic ulcer, arthritis, allergic complaints, and steroid therapy (Cooper et al. 2007). The isoflavonoid glabridin isolated from root of *Glycyrrhiza glabra* has been reported to possess anti-inflammatory potential which inhibits cyclooxygenase enzyme. The plant which is also known as licorice is widely used to treat cough because it contains active component glycyrrhizin and glabridin (Yokota et al. 1998).

9.2.19 Juglandaceae

Fresh bark juice of *Juglans regia* is taken by the Lepcha tribes of Sikkim, India, to remove worms from the stomach (Pradhan and Badola 2008). The seeds of *J. regia* are used traditionally in Chinese medicine to treat cancer and tannins are reported to be responsible for the anticancer property of this plant (Cai et al. 2004).

9.2.20 Lamiaceae

Traditionally *Ocimum sanctum* is the well-known plant used by tribes against inflammation and fever and the flavonoids including cirsilineol, cirsimaritin, isothymusin, isothymonin, apigenin, rosmarinic acid, and appreciable quantities of eugenol isolated from *O. sanctum*, are reported to have antioxidant and high inflammatory potential (Kelm et al. 2000).

9.2.21 Lauraceae

The leaves of *Cinnamomum zeylanica* are also reported to be effective against cough (Barukial and Sarmah 2011). The flavonoids isolated from *C. zeylanica*, namely protocatechuic acid, cinnamtannin B-1, urolignoside, rutin, and quercetin-3-O-R-L-rhamnopyranoside, are reported to have antioxidant activities (Jayaprakasha et al. 2006). Cinnamaldehyde present in the bark oil of *C. zeylanica* possesses antibacterial property (Al-Bayati and Mohammed 2009).

9.2.22 Liliaceae

The Bodo-Kachari tribes of North East India use the powdered roots of *Asparagus racemosus* for treating jaundice (Basumatary et al. 2014). Leaves of this plant are also used to cure urinary disorders and stomach ache by orally consuming the dried powder of leaves (Sajem and Gosai 2006). Racemofuran and racemosol are two compounds isolated from *A. racemosus* which are reported to have gastroprotective and antihepatotoxic activities (Muruganadan et al. 2000). Traditionally *Sansevieria roxburghiana* is used as a cardiotonic, expectorant, febrifuge, purgative, tonic in glandular enlargement and rheumatism. Alkaloid sanseveirine, isoflavonoid (cambodianol), and a triterpene (lupeol) were isolated from *S. roxburghiana*. Lupeol has been reported to have anti-inflammatory potential (Saleem 2009). Studies on the plant have also reported analgesic, antidiabetic, and anticancer activities (Nesamani 2005).

The juice extracted from the bulb of *Allium cepa* is used to cure gastric disorders in federally administered tribal areas of Pakistan (Aziz et al. 2018). Diallylthiosulfinate (allicin), methyl allylthiosulfinate, and allyl methyl thiosulfinate isolated from this plant showed antibacterial and antifungal activities (Hughes and Lawson 1991). Kuruma tribes of Wayanadu use the leaves of *Aloe vera* to treat cuts and wounds (Thomas et al. 2014). Aloe emodin is an anthraquinone present in Aloe latex which has anti-inflammatory properties since it competitively inhibits thromboxane (Robson et al. 1982).

9.2.23 Lythraceae

In the tribal regions of Andhra Pradesh, the crushed and mildly heated leaves of *Woodfordia fruticosa* are used to get relief from rheumatic pain (Satyavathi et al. 2014). Flavonoids, anthraquinone glycosides, and polyphenols have been isolated from this species (Das et al. 2007). A large number of flavonoids and other phenolics have proved their noteworthy effects on immune system function and inflammatory processes (Middleton Jr and Kandaswami 1992).

9.2.24 Malvaceae

Abelmoschus moschatus is used in traditional Chinese medicine to treat depression and anxiety (Liu et al. 2006). The bark and fruits of *Abelmoschus esculentus* are useful against the diabetic condition. Reports suggest that oleanolic acid, beta sistostenol, myricetin, and kaempferol are the four main compounds which play a role in exhibiting antidiabetic effect brought by *Abelmoschus esculentus*. The bioflavonoids myricetin and kaempferol are reported to have anticancer properties also (Prabhune et al. 2017). The bark and fruit extracts of *Abelmoschus esculentus* (Common name: "Dherash") are also considered useful for blood cancer patients in the tribal regions of West Bengal (Bandyopadhyay 2017).

9.2.25 Mimosaceae

In the tribal regions of North Maharashtra, the bark of *Acacia catechu* is used to cure jaundice (Badgujar and Patil 2008). A number of phenolic components were isolated from species *A. catechu* such *as* aromadendrin, 4-hydroxybenzoic acid, kaempferol quercetin, and catechin (Li et al. 2010), which possessed hepatoprotective property against carbon tetra chloride induced toxicity (Sobhy et al. 2011; Singh et al. 2001a, b; Zhang et al. 2014). Juice or paste of crushed bark of *Entada pursaetha* ssp. *Sinohimalensis* is applied externally to cure skin diseases (Pradhan and Badola 2008). Among the various phytochemicals reported from *E. pursaetha*, tetradecanoic acid, n- hexadecanoic acid, and vitamin E have a role in anti-inflammatory and antioxidant activities (Kalpana et al. 2012).

9.2.26 Myrtaceae

The juice made from the bark of *Syzygium cumini* is considered as effective against stomachache and gastric problems (Rahaman and Karmakar 2015). Gallic acid, ellagic acid, corilagin and related ellagitannins, 3,6-hexahydroxydiphenoyl-glucose were isolated and gallic acid have gastroprotective potential (Bhatia and Bajaj 1975). Tribals in Tamil Nadu use the seeds of *Eugenia jambolana* which have been reported to be a rich source of flavonoids, which accounts for their radical scavenging activity (Ravi et al. 2004). *Eucalyptus globulus* leaves have been used in China to treat influenza, headache, cough, eczema, and dermatomycosis. The compound named macrocarpal C isolated from the leaves of the plant is reported to possess strong antifungal activity (Wong et al. 2015).

9.2.27 Piperaceae

Ingestion of raw leaves of *Piper betle* is considered as effective against malaria in tribal regions of West Bengal (Bandyopadhyay 2017). Phenanthrene alkaloid isolated from the root of *P. betle* has been shown to possess antifertility effects, traditionally the mixture component of black pepper and root of *P. betle* is taken as a contraceptive by women (Pakrashi et al. 1976). Certain diaza- analogs of phenanthrene possess antimalarial activity (Yapi et al. 2000).

9.2.28 Poaceae

In the tribal regions of North Sikkim, fermented seeds of *Eleusine coracana* are taken with traditional drink as medicine to the gastric patients. Imbalance of oxidative molecules is believed to initiate digestive system diseases including stomach ulcer and gastric carcinoma. Specifically, gastric damage has been suggested to be mediated by the generation of free radicals (Lobo et al. 2010). Ethanol is metabolized in the body and releases superoxide anion and hydroperoxy free radicals. Gallic and cinnamic acid have potential gastrointestinal protectivity (Dykes and Rooney 2007). Antioxidants can prompt gastroprotective and healing effects by increasing the amount of gastric mucus glycoprotein and inhibition of prostaglandin production (Nartey et al. 2012). ROS are important factors in ethanol-induced and NSAIDs-related mucosal damage and a report revealed that the flavonoid rutin prevents ethanol-induced gastrointestinal abnormalities (La Casa et al. 2000). Antioxidants can scavenge ROS and are expected to heal or prevent gastric ulcers (Park et al. 2019). The powdered grains are also baked into chapattis (Bread) and given to treat diarrhea. Gruel is made by the powdered grains of *Hordeum vulgare* and given in case of painful indigestion. Barley water with honey is used in bronchial coughs in the tribal areas of North Sikkim (Pradhan and Badola 2008). 3,4-Dihydroxybenzaldehyde and 4-hydroxycinnamic acid are the major phenolic compounds in barley grains that has antioxidant activity (Omwamba and Hu 2010).

9.2.29 Portulacaceae

Portulaca oleracea is used in Chinese folk medicine as a diuretic, antiseptic, antiscorbutic, febrifuge, antispasmodic, and vermifuge. Reports suggest that this plant has a wide variety of pharmacological properties such as anti-inflammatory, antibacterial, skeletal muscle relaxant, wound-healing, and in vitro antitumor activities. Two alkaloids named $(3R)$-3,5-bis(3-methoxy-4-hydroxyphenyl)-2,3-dihydro-2($1H$)-pyridinone and 1,5-dimethyl-6-phenyl-1,2-dihydro-1,2,4-triazin-3 ($2H$)-one, together with two known compounds $(7'R)$-*N*-feruloylnormetanephrine and *N-trans*-feruloyltyramine, isolated from *P. oleracea* were reported to have cytotoxic activity against A549 cancer cell lines (Tian et al. 2014).

9.2.30 Rubiaceae

Apatani tribes in the Arunachal Pradesh state of India use the stem paste of *Paederia foetida* (Rubiaceae) against gastritis, diarrhea, and stomach disorders (Kala 2005). Beta-sterol has been reported to have gastroprotective action (Navarrete et al. 2002). *Rubia cordifolia* is used all over the world to treat skin diseases such as eczema, dermatitis, skin ulcers, and cancer (Karodi et al. 2009). A hepatoprotective compound rubiadin is isolated from the plant (Rao et al. 2006). A compound named hydroxytectoquinone isolated from *R. cordifolia* showed anti-inflammatory and anticancer properties (Ghosh et al. 2010)

9.2.31 Rutaceae

Irula tribes in Tamil Nadu uses the Rutaceae member, *Feronia elephantum* for curing urinary problems, dysentery, and diarrhea (Sundararajan et al. 2006). The leaf juice of this plant is taken by the Santal tribes of Bangladesh as a diuretic and to control vomiting (Rahman 2015). The roots and leaves of *F. elephantum* were reported to contain a flavonoid characterized as 5-hydroxy-2-(4-hydroxyphenyl)-7-methoxy-6-(3-methylbut-2-enyl) chroman-4-one which along with three furanocoumarins named imperatorin, bergapten, and xanthotoxin were effective for ethanol-induced gastric problem (Birdane et al. 2007; Intekhab and Aslam 2009). *Zanthoxylum bungeanum* is used in traditional Chinese medicine against digestive disorders, toothache, stomach ache, and diarrhea (Zhang et al. 2017). About eight flavonoids were isolated from the species, including vanillic acid, rutin, quinic acid, chlorogenic acid, epicatechin 5-feruloyquinic acid, syringetin-3-glucoside, rutin, hyperoside, quercetin-3-arabinoside, quercitrin, and isorhamnetin-3-glucoside (Yang et al. 2013).

Glycosmis pentaphylla is another plant whose roots are effective against jaundice. The decoction of roots of this plant is given with milk for treating jaundice (Barukial and Sarmah 2011). An acridone alkaloid isolated from *Glycosmis pentaphylla* showed antimalarial activity (Ito et al. 1999). The leaf juice and fruits of *Aegle marmelos* are used for dysentery by the tribals of Assam (Barukial and Sarmah 2011) and also used for diabetes (Ruhil et al. 2011). In the tribal regions of Andhra Pradesh, India, the leaf juice of *A. marmelos* (Maredu) is given with pepper seeds twice a day to control diabetes (Satyavathi et al. 2014). Studies revealed that umbelliferone β-D-galactopyranoside present in *A. marmelos* has significant antidiabetic, antihyperlipidemic, and antioxidant activities. Therefore, umbelliferone β-D-galactopyranoside may be regarded as one of the major factors for the antidiabetic potential of *Aegle marmelos* (Kumar et al. 2013). Lupeole showed activity against inflammation (Geetha and Varalakshmi 2001). Fruits of *Citrus medica* are reported to be good for indigestion. Chewing dried fruit skin of *C. medica* as well as *C. reticulate* helps in preventing dysentery (Pradhan and Badola 2008). This maybe due to the presence of limonene and β pinene having gastro intestinal property by enhancing the peptide enzymes found in the intestine (Rozza et al. 2011). *Ruta graveolens* is traditionally used against rheumatic pain by the tribal people of China. The flavonoid Rutin isolated from the plant have the capacity to decrease capillary fragility (Parray et al. 2012).

9.2.32 Sapindaceae

The leaf paste of *Cardiospermum halicacabum*, belonging to the family Sapindaceae, is mixed with castor oil and applied to the affected area to cure skin burns (Satyavathi et al. 2014). Quercetin-3-O-α-L-rhamnoside, kaempferol-3-O-α-L-rhamnoside, apigenin-7-O-β-D-glucuronide, apigenin 7-O-β-D-glucuronide methyl ester, apigenin 7-O-β-D-glucuronide ethyl ester, chrysoeriol, apigenin, kaempferol, luteolin, quercetin, methyl 3,4-dihydroxybenzoate, p-coumaric acid, 4-hydroxybenzoic acid, hydroquinone, protocathehuic acid, gallic acid, and indole-3-carboxylic acid are pharmacologically important compounds isolated from *C. halicacabum* showing anti-inflammatory and antioxidant property (Huang et al. 2011).

9.2.33 Saururaceae

The fresh juice of leaves of *Houttuynia cordata* Thunb. is applied in the treatment of stomach pain and amoebic dysentery (Laloo and Hemalatha 2011). The flavonoids quercetin and isoquercetin isolated from *H. cordata* shows diuretic action (Nakamura et al. 1936; Kimura 1953). Barnaulov et al. (1982) reported the gastroprotective effects of the flavonoid quercetin. The aerial parts of *H. cordata* are traditionally used in China to treat cancer. The flavonols like quercetin, rutin, and phenolic acids such as chlorogenic acid are the major components responsible for the anticancer activity (Cai et al. 2004).

9.2.34 Scrophulariaceae

Santal tribes of Bangladesh use the plant *Scoparia dulcis* as a traditional remedy for diabetes (Rahman 2015). Scopadulin is a novel tetracyclic diterpene that has been reported in *S. dulcis* (Latha et al. 2004). Terpenoids are a class of compounds that have been suggested to be responsible for the antidiabetic potential of many plants.

9.2.35 Simaroubaceae

Dried bark of *A. altissima* is used in Chinese herbal medicine as an astringent, antidiarrheic, and hemostatic agent. The major constituents of the bark are quassinoids which showed amebicidal and antimalarial activity (Tang and Eisenbrand 2013; Veena et al. 2019). The bark of *Quassia amara* ("Guavo") is traditionally used against malaria in Northern Brazil (Barbetti et al. 1987). Quassinoids are a class of triterpenoids possessing anticancer, antimalarial, and anti-inflammatory properties in the family Simaroubaceae (Dou et al. 1996).

9.2.36 Solanaceae

For knee arthritis, the roots of *Datura metel* L. are ground into a paste and taken orally by the tribal people of West Bengal. The five known sesquiterpenes were isolated from the flower of *D. metel* and investigations showed that the plant crude extract showed anti-inflammatory, anaphylaxis and used against psoriasis treatment (Kuang et al. 2008). The leaves of this plant are used by the Santal tribes of Bangladesh to cure asthma and rheumatism (Rahman 2015). The seeds of *Solanum surattense* are active against malaria. Certain tribes in West Bengal (Santal) boil the seeds in water and taken orally to treat malaria (Rahaman and Karmakar 2015). Certain glycoalkaloids have been isolated from this plant (Tekuri et al. 2019), which are reported to have antimalarial activity (Chen et al. 2010).

9.2.37 Thymeliaceae

Oil extracted from the wood of *Aquilaria malaccensis* of the family Thymeliaceae is effective in managing skin diseases (Barukial and Sarmah 2011). Linalool and its corresponding acetate derivative from the oil have a role in the anti-inflammatory activity (Wang et al. 2018).

9.2.38 Trichopodiaceae

A glycoprotein fraction isolated from the leaves and fruits of *Trichopus zeylanicus* Gaertn. ("Arogyappacha") was observed to possess good anti-fatigue property. These studies were first investigated by Pushpangadan et al. (1988), with the basic knowledge of Kani tribes in Agasthyamala of Kerala who uses this for getting instant stamina and better health (Evans et al. 2002). Clinical studies further proved that this glycoprotein fraction is a good immunomodulant (Singh et al. 2001a, b).

9.2.39 Verbenaceae

The tribal people of Wayanad use the leaf and root extract of *Vitex negundo* against malaria (Marjana et al. 2018). *Vitex negundo* L. is the promising candidate in the traditional scenario, and the root methanolic extract was observed to have antagonistic activity against venom toxicity of *Viper russelii* (Alam and Gomes 2003). Vitexin and isovitexin were fractioned from the root of *V. negundo* (Srinivas et al. 2001). Ravishankar (Ravishankar and Shukla 2007) reported immunomodulatory effect of the plant and a flavonoid and a triterpene saponin (oleanolic acid) are reported from the leaves (Surveswaran et al. 2007). The leaves of *V. negundo* are used against arthritis and also as memory enhancers by the Teli tribe of Bangladesh (Rahmatullah et al. 2012). A vitedoin (a lignan alkaloid) was isolated from the leaves of *V. negundo* (Ono et al. 2004). Some unknown flavonoid fractions of seed interrupt some stages of spermiogenesis which may reduce fertility (Bhargava 1989). Leaf extract also showed anti-inflammatory activity by suppressing the activity of prostaglandin synthetase and thereby also reducing the pain of the body (Telang et al. 1999). Several plant secondary metabolites like flavonoids, quinonoids, polyphenols, and terpenoids are observed to have protein-binding and enzyme-inhibiting properties (Havsteen 1983; Selvanayagam et al. 1996) which also limit snake venom phospholipase A2 (PLA2) actions (Alcaraz and Hoult 1985), of both Viper and Cobra. Triterpenoid saponins present in *V. negundo* may also contain venom inactivation processes (Alam and Gomes 2003).

9.2.40 Vitaceae

The Santal tribal people of West Bengal use the root paste of *Ampelocissus latifolia* (Roxb.) to treat elephantiasis. The root is made into a paste, warmed, and applied topically on the affected area. Among the various compounds identified in *A. latifolia*, hexadecanoic acid has the property of antioxidant, 5-alpha-reductase inhibitor, antifibrinolytic activities (Kala et al. 2011). It also possesses anti-inflammatory activity (Aparna et al. 2012).

9.2.41 Zingiberaceae

The juice from the rhizome of *Curcuma longa* has been taken orally against gastric disorders (Basumatary et al. 2014). The "apatani" tribes consumed the whole plant for relieving cholera and constipation (Kala 2005). Ethnomedicinally it is used against various stomach disorders and skin diseases. The clinical studies about *Curcuma longa* are well established (Ammon and Wahl 1991). Current traditional Indian medicine claims the use of turmeric against biliary disorders, anorexia, coryza, cough, diabetes, wounds, hepatic disorders, rheumatism, and sinusitis (Kurup 1977). The rhizome paste of *C. longa* is mixed with cow milk and used in treating jaundice (Badgujar and Patil 2008). Curcumin is used as a drug for increasing the bile secretion from the liver for proper digestion and absorption (Fabricant and Farnsworth 2001); this major constituent from *C. longa* is also reported to have a wound-healing property by Gujral et al. (1953).

The anti-inflammatory activity of curcumin was reported by Ghatak and Basu (1972). *Curcuma* powder has been used to increase the mucin content of gastric juice in the stomach. It may be, therefore, beneficial in protecting the gastric mucosa against irritants (Ammon and Wahl 1991). A juice made from the rhizome of *Curcuma longa* L. is taken orally by the Bodo tribe's against gastric and stomach disorders. Juba and Kusoom tribes of Manipur extracted nectar from the flower of *C. longa* and was used to cure ulcer in mouth and stomach (Yumnam and Tripathi 2012). Curcumin is a nontoxic gastroprotective agent (Yashavanth et al. 2018). The rhizome of *Curcuma zedoaria* when eaten raw cures diarrhea and colic, and helps in digestion. Curcumin was also reported from *C. zedoaria* by Lobo et al. (2009). In North East India, Bodo-Kachari tribes use the paste made of the fruit of *Amomum aromaticum* to cure cough. To get relief from stomach disorder, a paste of the rhizome of *Zingiber officinale* (Ginger) is consumed among the Bodo-Kachari tribes (Basumatary et al. 2014). 6-Gingerol is a biologically active compound isolated from ginger which possesses antioxidant, anti-inflammatory, analgesic, and antipyretic properties (Dugasani et al. 2010). Tribes in Kerala use *Z. officinale* against pain, inflammation, arthritis, urinary infections, and gastrointestinal disorders. Gallic acid and cinnamic acid isolated from Ginger have ability for gastroprotection (Nanjundaiah et al. 2011). Gargling with seed decoction of *Amomum subulatum* and with water is used to treat teeth and gum infection (Pradhan and Badola 2008). Secondary metabolites such as tannins, flavonoids, and alkaloids of this plant showed antimicrobial activity (Gopal et al. 2012). *Amomum compactum* is used in traditional Chinese medicine to cure stomach disorders. Cineole is the major compound identified from this plant (Tang and Eisenbrand 2013).

9.2.42 Ethnomedicinal Plants Used by Tribal Communities in India

A huge number of tribal communities are existing in various states of India and they have their own traditional medicinal practices which involve the use of many significant flowering plant species. Some of the major plant species used by the tribal communities of India for treating various disease conditions and the corresponding bioactive compounds that have been reported in the available literature are mentioned in Tables 9.1, 9.2, 9.3, 9.4, 9.5.

9.3 Conclusions

About 41 families are mentioned here, which are used by various tribes across India for treating different diseases. Among these plants, some are scientifically validated by the discovery of the corresponding bioactive compounds. Phenolic compounds are the dominating and diverse group with versatile pharmacological actions among the different bioactive components isolated from the selected plants. The pharmacological properties of the various compounds isolated from plants have justified the traditional use of these plants for treating different disease conditions. However, the majority of the plants are yet to be studied for providing a scientific basis for their use as medicines. The search for traditional knowledge about medicinal plants has great significance in the current scenario since people prefer herbal medicines over synthetic drugs due to the lack of side effects. So more research works are needed in this field of medicinal plants, leading to the discovery of novel bioactive

Table 9.1 Ethnomedicinal plants used by tribal communities in India against dysentery and gastric complaints

Sl No	Name of the plant	Family	Bioactive compounds	Pharmacological action	References
1.	*Andrographis paniculata*	Acanthaceae	Neoandrographolide	Anti-inflammatory activity	Fabricant and Farnsworth (2001)
2.	*Centella asiatica*	Apiaceae	Asiaticoside, asiatic acid, madecassic acid (Saponins)	Antithrombotic, gastro protective, immunomodulatory, antigenotoxic, regenerative, reproductive and wound healing	Roy et al. (2013)
3.	*Holarrhena antidysentrica*	Apocynaceae	Conessine (alkaloid)	Bark juice effective against dysentery	Stephenson (1948)
4.	*Wrightia tomentosa*	Apocynaceae	Rutin (flavonoid)	Gastroprotective	Abdel-Raheem (2010)
5.	*Ageratum conyzoides*	Asteraceae	Eugenol, beta pinene and Caffeic acid	Gastroprotective and anti-ulcer action	Morsy and Fouad (2008)
6.	*Oroxylum indicum*	Bignoniaceae	5,7-Dihydroxyflavone (flavone)	Antibacterial	Babu et al. (2006)
7.	*Terminalia chebula*	Combretaceae	Chebulagic acid Gallotannins	Gastroprotective Antioxidant Antimicrobial	Liu et al. (2017), Zhao et al. (2005), Engels et al. (2009)
8.	*Acalypha fruticosa*	Euphorbiaceae	2-Methyl-5, 7-dihydroxychromone 5-O-β-D-glucopyranoside **1**, acalyphin, apigenin and kaempferol 3-O-rutinoside	Anti-inflammatory	Rajkumar et al. (2010)
9.	*Phyllanthus emblica*	Euphorbiaceae	Gallic acid (phenol) Quercetin (flavonoid)	Gastroprotective	Dykes and Rooney (2007)
10.	*Euphorbia hirta*	Euphorbiaceae	Quercetin (flavonoid)	Gastroprotective	de la Lastra et al. (1994)
11.	*Syzygium cumini*	Myrtaceae	Gallic acid, ellagic acid, corilagin and related ellagitannins, 3,6-hexahydroxydiphenoyl-glucose	Effective against stomachache and gastric problem	Bhatia and Bajaj (1975)

(continued)

Table 9.1 (continued)

Sl No	Name of the plant	Family	Bioactive compounds	Pharmacological action	References
12.	*Eleusine coracana*	Poaceae	Gallic acid Cinnamic acid	Gastro protective.	Dykes and Rooney (2007)
13.	*Paedaria foetida*	Rubiaceae	β-Sterol	Gastroprotective	Navarrete et al. (2002)
14.	*Feronia elephantum*	Rutaceae	Bergapten, imperatorin, xanthotoxin and marmesin	Gastroprotective and anti-ulcer	Birdane et al. (2007)
15.	*Citrus medica*	Rutaceae	Limonene and β pinene	Gastroprotective	Rozza et al. (2011)
16.	*Houttuynia cordata*	Saururaceae	Quercetin (flavonoid)	Gastroprotectant	Cai et al. (2004)
17.	*Ailanthus altissima*	Simaroubaceae	Quassinoid (terpene)	Amebicidal	Tang and Eisenbrand 2013, Veena et al. (2019)
18.	*Curcuma longa*	Zingiberaceae	Curcumin (flavonoid)	Gastroprotectant and against mouth ulcer	Yashavanth et al. 2018, Yunnam and Tripathi (2012)
19.	*Curcuma zedoaria*	Zingiberaceae	Curcumin (flavonoid)	Gastroprotectant and against mouth ulcer	Yashavanth et al. 2018 Yunnam and Tripathi (2012)
20.	*Zingiber officinale*	Zingiberaceae	Gallic acid and cinnamic acid 6-Gingerol	Gastroprotectant	Basumatary et al. (2014), Nanjundaiah et al. (2011)

Table 9.2 Ethnomedicinal plants used against diabetes

Sl No	Name of plant	Family	Bioactive compounds	Pharmacological action	References
1.	*Mangifera indica*	Anacardiaceae	Mangiferin	Antidiabetic	Telang et al. (2013)
2.	*Coriandrum sativum*	Apiaceae	Camphor, eugenol, trans-β-ocimene, geraniol, α-pinene, limonene, p-cymene, 1,8-cineole and Thujone	Pancreatic β-cell restoration and insulin secretion	Broadhurst et al. (2000)
3.	*Emblica officinalis*	Euphorbiaceae	Vitamin C, Ellagic acid	Antidiabetic	Sri et al. (2013) Fatima et al. (2017)
4.	*Abelmoschus esculentus*	Malvaceae	Oleanolic acid, β-sistostenol (triterpene), myricetin, and kaempferol (flavone)	Antidiabetic	Prabhune et al. (2017)
5.	*Aegle marmelos*	Rutaceae	Umbelliferone β-D-Galactopyranoside	Antidiabetic and Antihyperlipidemic	Kumar et al. (2013)
6	*Diospyros chloroxylon*	Ebenaceae	Lup-20(29)-ene-3α,6 β-diol	Antihyperglycemic	Sharma et al. (2018)

Table 9.3 Ethnomedicinal plants used against cancer

Sl No	Name of plant	Family	Bioactive compounds	Pharmacological action	References
1.	*Annona muricata*	Annonaceae	Acetogenins	Capable of blocking ATP production in mitochondria	Waechter et al. (1997)
2.	*Hemidesmus indicus*	Asclepiadaceae	Hemidesminine	Anticancer property	Mandal et al. (1991)
3.	*Abelmoschus esculentus*	Malvaceae	Myricetin and Kaempferol	Anticancer property	Prabhune et al. (2017)
4.	*Carica papaya*	Caricaceae	Papain (enzyme)	Anticancer activity	Fauziya and Krishnamurthy (2013)
5.	*Juglans regia*	Juglandaceae	Tannins	Anticancer activity	Cai et al. 2004

Table 9.4 Ethnomedicinal plants used against snake bite or insect poisoning

Sl No	Name of plant	Family	Bioactive compounds	Pharmacological action	References
1.	*Annona squamosa*	Annonaceae	Stigmasterol	Anti-snake venom	Achika et al. (2014)
2.	*Annona crassiflora*	Annonaceae	Stigmasterol, Beta-sitosterol	Anti-snake venom	Achika et al. (2014) Luzia and Jorge (2013)
4.	*Alstonia scholaris*	Apocynaceae	Pentacyclic triterpenes	Anti-snake venom	Meenatchisundaram (2008)
5.	*Rauvolfia serpentina*	Apocynaceae	Alkaloid "serpentine"	Neutralizes snake venom	Gopal et al. (2012)
1.	*Annona squamosa*	Annonaceae	Stigmasterol	Anti-snake venom	Achika et al. (2014)
2.	*Annona crassiflora*	Annonaceae	Stigmasterol, β-sitosterol	Anti-snake venom.	Achika et al. (20140 Luzia and Jorge (2013)
4.	*Alstonia scholaris*	Apocynaceae	Pentacyclic triterpenes	Anti-snake venom	Meenatchisundaram (2008)
5.	*Rauvolfia serpentina*	Apocynaceae	Serpentine (alkaloid)	Neutralizes snake venom	Gupta and Peshin (2012)

Table 9.5 Ethnomedicinal plants used against respiratory problems

Sl. No	Name of plant	Family	Bioactive compounds	Pharmacological action	References
1	*Centella asiatica*	Apiaceae	Octadectrienoic acid and *n*-hexadecanoic acid	Inhibitory activity against *Mycobacterium tuberculosis*	Suresh et al. (2010)
2	*Acorus calamus*	Araceae	Mucirin	Effective against mucus hypersecretion, cough, bronchitis, etc.	Berlian et al. (2016)
3	*Glycyrrhiza glabra*	Fabaceae	Glycyrrhizin and Glabridin	Anti-inflammatory	Yokota et al. (1998)
4	*Cinnamomum zeylanicum*	Lauraceae	Protocatechuic acid, Cinnamtannin B-1, Urolignoside, Rutin, and Quercetin-3-O-R-L-Rhamnopyranoside	Possess antioxidant activity	Jayaprakasha et al. (2006)
5	*Hordeum vulgare*	Poaceae	Dihydroxybenzaldehyde and 4-hydroxycinnamic acid	Possess antioxidant activity	Omwamba and Hu (2010)
6	*Adhatoda vasica*	Acanthaceae	Vasicine and Vasicinone	Respiratory stimulant activity	Baquar (1997)

compounds that may act as potential lead molecules for the development of new and less expensive drugs in the future.

References

Abdel-Raheem IT (2010) Gastroprotective effect of rutin against indomethacin-induced ulcers in rats. Basic Clin Pharmacol Toxicol 107(3):742–750

Achika JI, Arthur DE, Gerald I, Adedayo A (2014) A review on the phytoconstituents and related medicinal properties of plants in the Asteraceae family. Int Organ Sci Res J Appl Chem 7(8):1–8

Ahmed B, Al-Howiriny TA, Siddiqui AB (2003) Antihepatotoxic activity of seeds of *Cichorium intybus*. J Ethnopharmacol 87(2–3):237–240

Ajaiyeoba EO, Bolaji OM, Akinboye DO, Falade CO, Gbotosho GO, Ashidi JS, Okpako LC, Oduola OO, Falade MO, Itiola OA, Houghton PJ (2005) In vitro anti-plasmodial and cytotoxic activities of plants used as antimalarial agents in the southwest Nigerian ethnomedicine. J Nat Remed 5(1):1–6

Ajaiyeoba EO, Ogbole OO, Abiodun OO, Ashidi JS, Houghton PJ, Wright CW (2013) Cajachalcone. An antimalarial compound from *Cajanus cajan* leaf extract. J Parasitol Res 2013:703781

Alam MI, Gomes A (2003) Snake venom neutralization by Indian medicinal plants (*Vitex negundo* and *Emblica officinalis*) root extracts. J Ethnopharmacol 86(1):75–80

Al-Bayati FA, Mohammed MJ (2009) Isolation, identification, and purification of cinnamaldehyde from *Cinnamomum zeylanicum* bark oil. An antibacterial study. Pharm Biol 47(1):61–66

Alcaraz MJ, Hoult JR (1985) Effects of hypolaetin-8-glucoside and related flavonoids on soybean lipoxygenase and snake venom phospholipase A2. Arch Int Pharmacodyn Ther 278(1):4–12

Ali RM, Houghton PJ, Raman A, Hoult JRS (1998) Antimicrobial and antiinflammatory activities of extracts and constituents of *Oroxylum indicum* (L.) Vent. Phytomedicine 5(5):375–381

Ammon HP, Wahl MA (1991) Pharmacology of *Curcumalonga*. Plantamedica 57(01):1–7

An J, Li Z, Dong Y, Ren J, Huo J (2016) Amentoflavone protects against psoriasis-like skin lesion through suppression of NF-κB-mediated inflammation and keratinocyte proliferation. Mol Cell Biochem 413(1–2):87–95

Aparna V, Dileep KV, Mandal PK, Karthe P, Sadasivan C, Haridas M (2012) Anti-inflammatory property of n-hexadecanoic acid: structural evidence and kinetic assessment. Chem Biol Drug Des 80(3):434–439

Arora S, Kaur K, Kaur S (2003) Indian medicinal plants as a reservoir of protective phytochemicals. Teratog Carcinog Mutagen 23(S1):295–300

Aziz MA, Adnan M, Khan AH, Shahat AA, Al-Said MS, Ullah R (2018) Traditional uses of medicinal plants practiced by the indigenous communities at Mohmand Agency, FATA, Pakistan. J Ethnobiol Ethnomed 14(1):2

Babu KS, Babu TH, Srinivas PV, Kishore KH, Murthy USN, Rao JM (2006) Synthesis and biological evaluation of novel C (7) modified chrysin analogues as antibacterial agents. Bioorg Med Chem Lett 16(1):221–224

Badgujar SB, Patil MB (2008) Ethnomedicine for jaundice from tribal areas in North Maharashtra. Nat Prod Radiance 7(1):79–81

Bandara BR, Kumar NS, Samaranayake KS (1989) An antifungal constituent from the stem bark of *Butea monosperma*. J Ethnopharmacol 25(1):73–75

Bandyopadhyay D (2017) Herbal folk remidies and ethno medicine of Bankura District of West Bengal. Am J Ethnomed 16(4):1–4. https://doi.org/10.21767/2348-9502.100016

Bannerman RH, Burton J, Wen-Chieh C (eds) (1983) Traditional medicine and health care coverage: a reader for health administrators and practitioners. World Health Organization, Geneva

Baquar SR (1997) Medicinal and poisonous plants of Pakistan. Rosette Printas, Karachi, pp 95–96

Barbetti P, Grandolini G, Fardella G, Chiappini I (1987) Indole alkaloids from *Quassia amara*. Planta Med 53(03):289–290

Barnaulov OD, Manicheva OA, Zapesochnaya GG, Shelyuto VL, Glyzin VI (1982) Effects of certain flavonoids on the ulcerogenic action of reserpine in mice. Pharm Chem J 16(3):199–202

Barukial J, Sarmah JN (2011) Ethnomedicinal plants used by the people of Golaghat district, Assam, India. Int J Med Arom Plants 1(3):203–211

Basumatary N, Teron R, Saikia M (2014) Ethnomedicinal practices of the Bodo-Kachari tribe of Karbi Anglong district of Assam. Int J Life Sci Biotechnol Pharm Res 3(1):161–167

Begum S, Siddiqui BS, Sultana R, Zia A, Suria A (1999) Bio-active cardenolides from the leaves of *Nerium oleander*. Phytochemistry 50(3):435–438

Berlian G, Tandrasasmita OM, Tjandrawinata RR (2016) Effect of mucirin, a bioactive fraction of *Acorus calamus* l, as mucin regulator in human lung epithelial cultured cells. J Chem Pharm Res 8(8):24–31

Bhargava SK (1989) Antiandrogenic effects of a flavonoid-rich fraction of *Vitex negundo* seeds: a histological and biochemical study in dogs. J Ethnopharmacol 27(3):327–339

Bhatia IS, Bajaj KL (1975) Chemical constituents of the seeds and bark of *Syzygium cumini*. Planta Med 28(08):346–352

Bhuyan M (2015) Comparative study of ethnomedicine among the tribes of North East India. Int Res J Soc Sci 4(2):27–32

Birdane FM, Cemek M, Birdane YO, Gülçin İ, Büyükokuroğlu ME (2007) Beneficial effects of *Foeniculum vulgare* on ethanol-induced acute gastric mucosal injury in rats. World J Gastroenterol *13*(4):607

Bora KS, Sharma A (2011) The genus *Artemisia*: a comprehensive review. Pharm Biol 49 (1):101–109

Borah SM, Borah L, Nath SC (2012) Ethnomedicinal plants from Disoi valley reserve forest of Jorhat district, Assam. Plant Sci Feed 2(4):59–63

Brahmam M (2000) Indigenous medicinal plants for modern drug development Programme: revitalisation of native health traditions. Adv Plant Sci 13(1):1–10

Broadhurst CL, Polansky MM, Anderson RA (2000) Insulin-like biological activity of culinary and medicinal plant aqueous extracts in vitro. J Agric Food Chem 48(3):849–852

Bunel V, Antoine MH, Nortier J, Duez P, Stévigny C (2015) Nephroprotective effects of ferulic acid, Z-ligustilide and E-ligustilide isolated from *Angelica sinensis* against cisplatin toxicity in vitro. Toxicol In Vitro 29(3):458–467

Cai Y, Luo Q, Sun M, Corke H (2004) Antioxidant activity and phenolic compounds of 112 traditional Chinese medicinal plants associated with anticancer. Life Sci 74(17):2157–2184

Chaudhuri S, Banerjee A, Basu K, Sengupta B, Sengupta PK (2007) Interaction of flavonoids with red blood cell membrane lipids and proteins: antioxidant and antihemolytic effects. Int J Biol Macromol 41(1):42–48

Chauhan A, Rijhwani S (2015) A comprehensive review on phytochemistry of *Ageratum conyzoides* Linn.(Goat weed). Int J Eng Technol Manag Appl Sci 3 www.ijetmas.com

Chen Y, Li S, Sun F, Han H, Zhang X, Fan Y, Tai G, Zhou Y (2010) In vivo antimalarial activities of glycoalkaloids isolated from Solanaceae plants. Pharm Biol 48(9):1018–1024

Christenhusz MJ, Byng JW (2016) The number of known plants species in the world and its annual increase. Phytotaxa 261(3):201–217

Cock IE (2011) Medicinal and aromatic plants—Australia. Ethnopharmacology. Encyclopedia of Life Support Systems (EOLSS)

Cooper H, Bhattacharya B, Verma V, McCulloch AJ, Smellie WSA, Heald AH (2007) Liquorice and soy sauce, a life-saving concoction in a patient with Addison's disease. Ann Clin Biochem 44(4):397–399

Coria-Téllez AV, Montalvo-Gónzalez E, Yahia EM, Obledo-Vázquez EN (2018) *Annona muricata*: a comprehensive review on its traditional medicinal uses, phytochemicals, pharmacological activities, mechanisms of action and toxicity. Arab J Chem 11(5):662–691

Cox PA, Balick MJ (1994) The ethnobotanical approach to drug discovery. Sci Am 270(6):82–87

Das PK, Goswami S, Chinniah A, Panda N, Banerjee S, Sahu NP, Achari B (2007) *Woodfordia fruticosa*: traditional uses and recent findings. J Ethnopharmacol 110(2):189–199

Díaz-Viciedo R, Hortelano S, Girón N, Massó JM, Rodriguez B, Villar A, de las Heras B (2008) Modulation of inflammatory responses by diterpene acids from *Helianthus annuus* L. Biochem Biophys Res Commun 369(2):761–766

Dou J, Khan IA, McChesney JD, Burandt CL Jr (1996) Qualitative and quantitative high performance liquid chromatographic analysis of quassinoids in Simaroubaceae plants. Phytochem Anal 7(4):192–200

Dugasani S, Pichika MR, Nadarajah VD, Balijepalli MK, Tandra S, Korlakunta JN (2010) Comparative antioxidant and anti-inflammatory effects of [6]-gingerol,[8]-gingerol, [10]-gingerol and [6]-shogaol. J Ethnopharmacol 127(2):515–520

Dutta BK, Dutta PK (2005) Potential of ethnobotanical studies in north East India: an overview. Indian J Traditi Knowl 4(1):7–14

Dykes L, Rooney LW (2007) Phenolic compounds in cereal grains and their health benefits. Cereal Foods World 52(3):105–111

Engels C, Knodler M, Zhao YY, Carle R, Ganzle MG, Schieber A (2009) Antimicrobial activity of gallotannins isolated from mango (*Mangifera indica* L.) kernels. J Agric Food Chem 57 (17):7712–7718

Erah PO (2002) Herbal medicines: challenges. Trop J Pharm Res 1(2):53–54

Evans DA, Subramoniam A, Rajasekharan S, Pushpangadan P (2002) Effect of *Trichopus zeylanicus* leaf extract on the energy metabolism in mice during exercise and at rest. Ind J Pharmacol 34(1):32–37

Fabricant DS, Farnsworth NR (2001) The value of plants used in traditional medicine for drug discovery. Environ Health Perspect 109(Suppl 1):69–75

Fatima N, Hafizur RM, Hameed A, Ahmed S, Nisar M, Kabir N (2017) Ellagic acid in *Emblica officinalis* exerts anti-diabetic activity through the action on β-cells of pancreas. Eur J Nutr 56 (2):591–601

Fauziya S, Krishnamurthy R (2013) Papaya (*Carica papaya*): source material for anticancer. CIBTech J Pharm Sci 2(1):25–34

Filho VC (2009) Chemical composition and biological potential of plants from the genus *Bauhinia*. Phytother Res 23(10):1347–1354

Fong WP, Poon YT, Wong TM, Mock JWY, Ng TB, Wong RNS, Yao QZ, Yeung HW (1996) A highly efficient procedure for purifying the ribosome-inactivating proteins α-and β-momorcharins from *Momordica charantia* seeds. N-terminal sequence comparison and establishment of their n-glycosidase activity. Life Sci 59(11):901–909

Geetha T, Varalakshmi P (2001) Anti-inflammatory activity of lupeol and lupeol linoleate in rats. J Ethnopharmacol 76(1):77–80

Ghatak N, Basu N (1972) Sodium curcuminate as an effective anti-inflammatory agent. Indian J Exp Biol 10(3):235

Ghosh S, Das Sarma M, Patra A, Hazra B (2010) Anti-inflammatory and anticancer compounds isolated from *Ventilago madraspatana* Gaertn., *Rubia cordifolia* Linn. and *Lantana camara* Linn. J Pharm Pharmacol 62(9):1158–1166

Goel MK, Mehrotra S, Kukreja AK, Shanker K, Khanuja SPS (2009) In vitro propagation of *Rauwolfia serpentina* using liquid medium, assessment of genetic fidelity of micropropagated plants, and simultaneous quantitation of reserpine, ajmaline, and ajmalicine. In: Protocols for in vitro cultures and secondary metabolite analysis of aromatic and medicinal plants. Humana Press, Totowa, pp 17–33

Gopal K, Baby C, Mohammed A (2012) *Amomum subulatum* Roxb: an overview in all aspects. Int Res J Pharm 3(7):96–99

Gu Q, Zhang X-M, Wang R-R, Liu Q-M, Zheng Y-T, Zhou J, Chen J-J (2008) Anti-HIV active constituents from *Angelica apaensis*. Nat Prod Res Dev 20(2)

Gujral ML, Chowdhury NK, Saxena PN (1953) The effect of certain indigenous remedies on the healing of wounds and ulcers. J Indian Med Assoc 22(7):273–276

Gupta YK, Peshin SS (2012) Do herbal medicines have potential for managing snake bite envenomation? Toxicol Int 19(2):89

Gupta S, Sidhu MC, Ahluwalia AS (2017) Plant-based remedies for the management of diabetes. Curr Bot 8:34–40

Gurung S, Škalko-Basnet N (2009) Wound healing properties of *Carica papaya* latex: in vivo evaluation in mice burn model. J Ethnopharmacol 121(2):338–341

Guruvayoorappan C, Kuttan G (2008) Inhibition of tumor specific angiogenesis by amentoflavone. Biochem Mosc 73(2):209–218

Haque ME, Alam MB, Hossain MS (2011) The efficacy of cucurbitane type triterpenoids, glycosides and phenolic compounds isolated from *Momordica charantia*: a review. Int J Pharm Sci Res 2(5):1135

Havsteen B (1983) Flavonoids, a class of natural products of high pharmacological potency. Biochem Pharmacol 32(7):1141–1148

Huang MH, Huang SS, Wang BS, Wu CH, Sheu MJ, Hou WC, Lin SS, Huang GJ (2011) Antioxidant and anti-inflammatory properties of *Cardiospermum halicacabum* and its reference compounds ex vivo and in vivo. J Ethnopharmacol 133(2):743–750

Hughes BG, Lawson LD (1991) Antimicrobial effects of *Allium sativum* L.(garlic), *Allium ampeloprasum* L.(elephant garlic), and *Allium cepa* L.(onion), garlic compounds and commercial garlic supplement products. Phytother Res 5(4):154–158

Idrisi MS, Badola HK, Singh R (2010) Indigenous knowledge and medicinal use of plants by local communities in Rangit Valley, South Sikkim, India. NeBIO 1:34–45

Intekhab J, Aslam M (2009) Isolation of a flavonoid from *Feronia limonia*. J Saudi Chem Soc 13 (3):295–298

Ito C, Kondo Y, Rao KS, Tokuda H, Nishino H, Furukawa H (1999) Chemical constituents of *Glycosmis pentaphylla*. Isolation of a novel naphthoquinone and a new acridone alkaloid. Chem Pharm Bull 47(11):1579–1581

Jain PK, Khurana N, Pounikar Y, Patil S, Gajbhiye A (2012) Hepatoprotective effect of carrot (*Daucus carota* L.) on paracetamol intoxicated rats. Int J Pharm Pharm Technol 1(2):17–22

Jamir NS, Lanusunep PN (2012) Medico-herbal medicine practiced by the Naga tribes in the state of Nagaland (India). Indian J Fund Appl Life Sci 2:328–333

Jarukamjorn K, Nemoto N (2008) Pharmacological aspects of *Andrographis paniculata* on health and its major diterpenoid constituent andrographolide. J Health Sci 54(4):370–381

Jayaprakasam B, Zhang Y, Nair MG (2004) Tumor cell proliferation and cyclooxygenase enzyme inhibitory compounds in *Amaranthus tricolor*. J Agric Food Chem 52(23):6939–6943

Jayaprakasha GK, Ohnishi-Kameyama M, Ono H, Yoshida M, Jaganmohan Rao L (2006) Phenolic constituents in the fruits of *Cinnamomum zeylanicum* and their antioxidant activity. J Agric Food Chem 54(5):1672–1679

Jiang Z, Guo X, Zhang K, Sekaran G, Cao B, Zhao Q, Zhang S, Kirby GM, Zhang XY (2019) The essential oils and eucalyptol from *Artemisia vulgaris* L. prevent acetaminophen-induced liver injury by activating Nrf2-Keap1 and enhancing APAP clearance through non-toxic metabolic pathway. Frontiers Pharmacol 10:782

Kala CP (2005) Ethnomedicinal botany of the Apatani in the Eastern Himalayan region of India. J Ethnobiol Ethnomed 1(1):11

Kala SMJ, Balasubramanian T, Soris PT, Mohan VR (2011) GC-MS determination of bioactive components of Eugenia singampattiana Bedd. Int J ChemTech Res 3(3):1534–1537

Kalpana Devi V, Shanmugasundaram R, Mohan VR (2012) GC-MS analysis of ethanol extract of *Entada pursaetha* dc seed. Biosci Discov 3(1):30–33

Karodi R, Jadhav M, Rub R, Bafna A (2009) Evaluation of the wound healing activity of a crude extract of *Rubia cordifolia* L.(Indian madder) in mice. Int J Appl Res Nat Prod 2(2):12–18

Karuppusamy S (2007) Medicinal plants used by Palyan tribe of Sirumalai Hills of southern India. Nat Prod Radiance 6:436–442

Kelm MA, Nair MG, Strasburg GM, DeWitt DL (2000) Antioxidant and cyclooxygenase inhibitory phenolic compounds from *Ocimum sanctum* Linn. Phytomedicine 7(1):7–13

Kim HJ, Yokozawa T, Kim HY, Tohda C, Rao TP, Juneja LR (2005) Influence of amla (*Emblica officinalis* Gaertn.) on hypercholesterolemia and lipid peroxidation in cholesterol-fed rats. J Nutr Sci Vitaminol 51(6):413–418

Kimura Y (1953) Studies on the standardization of crude drugs. III. On the production conditions and component of crude drugs (2). On the component of *Houttuynia cordata*. Yakugaku Zasshi 73:196–197

Kipkore W, Wanjohi B, Rono H, Kigen G (2014) A study of the medicinal plants used by the Marakwet Community in Kenya. J Ethnobiol Ethnomed 10(1):24

Kuang HX, Yang BY, Xia YG, Feng WS (2008) Chemical constituents from the flower of *Datura metel* L. Arch Pharm Res 31(9):1094–1097

Kumar RA, Sridevi K, Kumar NV, Nanduri S, Rajagopal S (2004) Anticancer and immunostimulatory compounds from *Andrographis paniculata*. J Ethnopharmacol 92 (2–3):291–295

Kumar V, Ahmed D, Verma A, Anwar F, Ali M, Mujeeb M (2013) Umbelliferone β-D-galactopyranoside from A*egle marmelos* (L.) corr. An ethnomedicinal plant with antidiabetic, antihyperlipidemic and antioxidative activity. BMC Complement Altern Med 13(1):273

Kumaran A, Karunakaran RJ (2006) Nitric oxide radical scavenging active components from *Phyllanthus emblica* L. Plant Foods Hum Nutr 61(1):1

Kurup PNV (1977) Handbook of medicinal plants, vol I. Central Council for Research in Indian medicine and Homoeophathy (CCRIMH), New Delhi

La Casa C, Villegas I, De La Lastra CA, Motilva V, Calero MM (2000) Evidence for protective and antioxidant properties of rutin, a natural flavone, against ethanol induced gastric lesions. J Ethnopharmacol 71(1–2):45–53

Laloo D, Hemalatha S (2011) Ethnomedicinal plants used for diarrhea by tribals of Meghalaya, Northeast India. Pharmacogn Rev 5(10):147

de la Lastra A, Martin MJ, Motilva V (1994) Antiulcer and gastroprotective effects of quercetin: a gross and histologic study. Pharmacology 48(1):56–62

Latha M, Pari L, Sitasawad S, Bhonde R (2004) Insulin-secretagogue activity and cytoprotective role of the traditional antidiabetic plant *Scoparia dulcis* (sweet Broomweed). Life Sci 75 (16):2003–2014

Lee CY, Peng WH, Cheng HY, Chen FN, Lai MT, Chiu TH (2006) Hepatoprotective effect of Phyllanthus in Taiwan on acute liver damage induced by carbon tetrachloride. Am J Chin Med 34(03):471–482

Lee SA, Tsai HT, Ou HC, Han CP, Tee YT, Chen YC, Wu MT, Chou MC, Wang PH, Yang SF (2008) Plasma interleukin-1β,-6,-8 and tumor necrosis factor-α as highly informative markers of pelvic inflammatory disease. Clin Chem Lab Med 46(7):997–1003

Li Y, Ooi LS, Wang H, But PP, Ooi VE (2004) Antiviral activities of medicinal herbs traditionally used in southern mainland China. Phytother Res 18(9):718–722

Li YZ, Li ZL, Hua HM, Li ZG, Liu MS (2007) Studies on flavonoids from stems and leaves of *Calophyllum inophyllum*. Zhongguo Zhong Yao Za Zhi 32(8):692–694

Li X, Wang H, Liu C, Chen R (2010) Chemical constituents of *Acacia catechu*. Zhongguo Zhong Yao Za Zhi 35(11):1425–1427

Li F, Zhuo J, Liu B, Jarvis D, Long C (2015) Ethnobotanical study on wild plants used by Lhoba people in Milin County, Tibet. J Ethnobiol Ethnomed 11(1):23

Liu IM, Liou SS, Cheng JT (2006) Mediation of β-endorphin by myricetin to lower plasma glucose in streptozotocin-induced diabetic rats. J Ethnopharmacol 104(1–2):199–206

Liu W, Shang P, Liu T, Xu H, Ren D, Zhou W, Wen A, Ding Y (2017) Gastroprotective effects of chebulagic acid against ethanol-induced gastric injury in rats. Chem Biol Interact 278:1–8

Lobo R, Prabhu KS, Shirwaikar A, Shirwaikar A (2009) *Curcuma zedoaria* Rosc.(white turmeric): a review of its chemical, pharmacological and ethnomedicinal properties. J Pharm Pharmacol 61 (1):13–21

Lobo V, Patil A, Phatak A, Chandra N (2010) Free radicals, antioxidants and functional foods: impact on human health. Pharmacogn Rev 4(8):118

Lowe H, Payne-Jackson A, Beckstrom-Sternberg SM, Duke JA (2001) Jamaica's Ethnomedicine: its potential in the healthcare system. Pelican Publishers, Gretna

Luzia DM, Jorge N (2013) Bioactive substance contents and antioxidant capacity of the lipid fraction of *Annona crassiflora* Mart. seeds. Ind Crop Prod 42:231–235

Mallavadhani UV, Panda AK, Rao YR (2001) *Diospyros melanoxylon* leaves: a rich source of pentacyclic triterpenes. Pharm Biol 39(1):20–24

Mandal S, Das PC, Joshi PC, Das A, Chatterjee A (1991) Hemidesminine—A new coumarino-lignoid from *Hemidesmus indicus* R. Br. ChemInform 22(37):712–713

Marjana MP, Remyakrishnan CR, Baiju EC (2018) Ethnomedicinal flowering plants used by Kurumas, Kurichiyas and Paniyas tribes of Wayanad district of Kerala, India. Int J Biol Res 3:01–08

Maurya S, Singh D (2010) Quantitative analysis of total phenolic content in *Adhatoda vasica* Nees extracts. Int J PharmTech Res 2(4):2403–2406

Mead JR, McNair N (2006) Antiparasitic activity of flavonoids and isoflavones against *Cryptosporidium parvum* and *Encephalitozoon intestinalis*. FEMS Microbiol Lett 259(1):153–157

Meenatchisundaram S (2008) Anti-venom activity of medicinal plants—a mini review. Ethnobot Leaflets 2008(1):162

Mehmood MH, Siddiqi HS, Gilani AH (2011) The antidiarrheal and spasmolytic activities of *Phyllanthus emblica* are mediated through dual blockade of muscarinic receptors and Ca^{2+} channels. J Ethnopharmacol 133(2):856–865

Mengi SA, Deshpande SG (1995) Evaluation of ocular anti-inflammatory activity of *Butea frondosa*. Indian J Pharmacol 27(2):116

Mesquita MLD, Desrivot J, Fournet A, Paula JED, Grellier P, Espindola LS (2005) Antileishmanial and trypanocidal activity of Brazilian Cerrado plants. Mem Inst Oswaldo Cruz 100(7):783–787

Middleton E Jr, Kandaswami C (1992) Effects of flavonoids on immune and inflammatory cell functions. Biochem Pharmacol 43(6):1167–1179

Morsy MA, Fouad AA (2008) Mechanisms of gastroprotective effect of eugenol in indomethacin-induced ulcer in rats. Phytother Res 22(10):1361–1366

Murakami T, Emoto A, Matsuda H, Yoshikawa M (2001) Medicinal foodstuffs. XXI. Structures of new cucurbitane-type triterpene glycosides, goyaglycosides-a,-b,-c,-d,-e,-f,-g, and-h, and new oleanane-type triterpene saponins, goyasaponins I, II, and III, from the fresh fruit of Japanese *Momordica charantia* L. Chem Pharm Bull 49(1):54–63

Muruganadan S, Garg H, Lal J, Chandra S, Kumar D (2000) Studies on the immunostimulant and antihepatotoxic activities of Asparagus racemosus root extract. J Med Arom Pl Sci 22:49–52

Muruganandam AV, Bhattacharya SK, Ghosal S (2000) Indole and flavanoid constituents of *Wrightia tinctoria, W. tomentosa* and *W. coccinea*. Indian J Chem Sect BOrg Chem Incl Med Chem 39:125–131

Nakamura H, Ota T, Fukuchi G (1936) The constituents of diuretic drugs (II). The flavonal glucoside of *Houttuynia cordata* Thunb. J Pharm Soc Japan 56:68

Nanjundaiah SM, Annaiah HNM, Dharmesh SM (2011) Gastroprotective effect of ginger rhizome (*Zingiber officinale*) extract: role of gallic acid and cinnamic acid in H^+, K^+-ATPase/*H. pylori* inhibition and anti-oxidative mechanism. Evid-Based Complem Altern Med 2011:249487

Nartey ET, Ofosuhene M, Kudzi W, Agbale CM (2012) Antioxidant and gastric cytoprotective prostaglandins properties of *Cassia sieberiana* roots bark extract as an anti-ulcerogenic agent. BMC Complem Altern Med 12(1):65

Navarrete A, Trejo-Miranda JL, Reyes-Trejo L (2002) Principles of root bark of *Hippocratea excelsa* (Hippocrataceae) with gastroprotective activity. J Ethnopharmacol 79(3):383–388

Nesamani S (2005) Oushadha Sasyangal. The State Institute of Languages Kerala, Thiruvananthapuram

Nwofia GE, Ojimelukwe P, Eji C (2012) Chemical composition of leaves, fruit pulp and seeds in some *Carica papaya* (L.) morphotypes. Int J Med Arom Pl 2(1):200–206

Odonne G, Herbette G, Eparvier V, Bourdy G, Rojas R, Sauvain M, Stien D (2011) Antileishmanial sesquiterpene lactones from *Pseudelephantopus spicatus*, a traditional remedy from the Chayahuita Amerindians (Peru). Part III. J Ethnopharmacol 137(1):875–879

Oloyede OI (2005) Chemical profile of unripe pulp of *Carica papaya*. Pak J Nutr 4(6):379–381

Omwamba M, Hu Q (2010) Antioxidant activity in barley (*Hordeum vulgare* L.) grains roasted in a microwave oven under conditions optimized using response surface methodology. J Food Sci 75(1):C66–C73

Ono M, Nishida Y, Masuoka C, Li JC, Okawa M, Ikeda T, Nohara T (2004) Lignan derivatives and a norditerpene from the seeds of *Vitex negundo*. J Nat Prod 67(12):2073–2075

Ortigao M, Better M (1992) Momordin II, a ribosome inactivating protein from *Momordica balsamina*, is homologous to other plant proteins. Nucleic Acids Res 20(17):4662

Otsuki N, Dang NH, Kumagai E, Kondo A, Iwata S, Morimoto C (2010) Aqueous extract of *Carica papaya* leaves exhibits anti-tumor activity and immunomodulatory effects. J Ethnopharmacol 127(3):760–767

Pakrashi A, Chakrabarty B, Dasgupta A (1976) Effect of the extracts from *Aristolochia indica* Linn. on interception in female mice. Experientia 32(3):394–395

Pandey N, Barve D (2011) Phytochemical and pharmacological review on *Annona squamosa* Linn. Int J Res Pharmaceut Biomed Sci 2(4):1404–1412

Park JU, Kang JH, Rahman MAA, Hussain A, Cho JS, Lee YI (2019) Gastroprotective effects of plants extracts on gastric mucosal injury in experimental Sprague-Dawley rats. BioMed Res Int 2019:8759708

Parray SA, Bhat JU, Ahmad G, Jahan N, Sofi G, IFS M (2012) *Ruta graveolens*: from traditional system of medicine to modern pharmacology: an overview. Am J PharmTech Res 2(2):239–252

Parveen R, Parveen B, Parveen A, Ahmad S (2019) *Andrographis paniculata*: from traditional to nano drug for cancer therapy. In: Nanomaterials and plant potential. Springer, Cham, pp 317–345

Pitipanapong J, Chitprasert S, Goto M, Jiratchariyakul W, Sasaki M, Shotipruk A (2007) New approach for extraction of charantin from *Momordica charantia* with pressurized liquid extraction. Sep Purif Technol 52(3):416–422

Prabhune A, Sharma M, Ojha B (2017) *Abelmoschus esculentus* (Okra) potential natural compound for prevention and management of diabetes and diabetic induced hyperglycemia. Int J Herb Med 5(2):66–68

Pradhan BK, Badola HK (2008) Ethnomedicinal plant use by Lepcha tribe of Dzongu valley, bordering Khangchendzonga Biosphere Reserve, in North Sikkim, India. J Ethnobiol Ethnomed 4(1):22

Pushpangadan P, Rajasekharan S, Ratheshkumar PK, Jawahar CR, Nair VV, Lakshmi N, Amma LS (1988) 'Arogyappacha'(*Trichopus zeylanicus* Gaerin), The 'Ginseng'of Kani Tribes Of Agashyar Hills (Kerala) for ever green healh and vitality. Ancient Sci Life 8(1):13

Rahaman CH, Karmakar S (2015) Ethnomedicine of Santal tribe living around Susunia hill of Bankura district, West Bengal, India: the quantitative approach. J Appl Pharm Sci 5(2):127–136

Rahman AHMM (2015) Ethnomedicinal survey of angiosperm plants used by Santal tribe of Joypurhat District, Bangladesh. Int J Adv Res 3(5):990–1001

Rahmatullah M, Khatun Z, Hasan A, Parvin W, Moniruzzaman M, Khatun A, Mahal MJ, Bhuiyan SA, Mou SM, Jahan R (2012) Survey and scientific evaluation of medicinal plants used by the Pahan and Teli tribal communities of Natore district, Bangladesh. Afr J Tradit Complement Altern Med 9(3):366–373

Rajkumar V, Guha G, Kumar RA (2010) Therapeutic potential of Acalypha fruticosa. Food Chem Toxicol 48(6):1709–1713

Rao GMM, Rao CV, Pushpangadan P, Shirwaikar A (2006) Hepatoprotective effects of rubiadin, a major constituent of Rubia cordifolia Linn. J Ethnopharmacol 103(3):484–490

Ravi K, Ramachandran B, Subramanian S (2004) Effect of *Eugenia jambolana* seed kernel on antioxidant defense system in streptozotocin-induced diabetes in rats. Life Sci 75(22):2717–2731

Ravishankar B, Shukla VJ (2007) Indian systems of medicine: a brief profile. Afr J Tradit Complement Altern Med 4(3):319–337

Razdan MK, Kapila K, Bhide NK (1970) Study of the antioestrogenic activity of alcoholic extracts of petals and seeds of Butea frondosa. Indian J Physiol Pharmacol 14(1):57–60

Robson MC, Heggers JP, Hagstrom WJ (1982) Myth, magic, witchcraft, or fact? *Aloe vera* revisited. J Burn Care Rehabil 3(3):157–163

Roy DC, Barman SK, Shaik MM (2013) Current updates on *Centella asiatica*: phytochemistry, pharmacology and traditional uses. Med Plant Res 3(4):70–77

Rozza AL, de Mello Moraes T, Kushima H, Tanimoto A, Marques MOM, Bauab TM, Hiruma-Lima CA, Pellizzon CH (2011) Gastroprotective mechanisms of *Citrus* lemon (Rutaceae) essential oil and its majority compounds limonene and β-pinene: involvement of heat-shock protein-70, vasoactive intestinal peptide, glutathione, sulfhydryl compounds, nitric oxide and prostaglandin E2. Chem Biol Interact 189(1–2):82–89

Ruhil S, Balhara M, Dhankhar S, Chhillar AK (2011) *Aegle marmelos* (Linn.) Correa: a potential source of phytomedicine. J Med Plants Res 5(9):1497–1507

Sajem AL, Gosai K (2006) Traditional use of medicinal plants by the Jaintia tribes in North Cachar Hills district of Assam, Northeast India. J Ethnobiol Ethnomed 2(1):33

Saleem M (2009) Lupeol, a novel anti-inflammatory and anti-cancer dietary triterpene. Cancer Lett 285(2):109–115

Sandhya B, Thomas S, Isabel W, Shenbagarathai R (2006) Ethnomedicinal plants used by the Valaiyan community of Piranmalai hills (Reserved forest), Tamilnadu, India—a pilot study. Afr J Tradit Complement Altern Med 3(1):101–114

Sarkar K, Ghosh A, Kinter M, Mazumder B, Sil PC (2006) Purification and characterization of a 43 kD hepatoprotective protein from the herb *Cajanus indicus* L. Protein J 25(6):411–421

Satyavathi K, Satyavani S, Padal TSN, Padal SB (2014) Ethnomedicinal plants used by primitive tribal of Pedabayalu Mandalam, Visakhapatnam District, AP, India. Int J Ethnobiol Ethnomed 1:1–7

Schütz K, Carle R, Schieber A (2006) Taraxacum—a review on its phytochemical and pharmacological profile. J Ethnopharmacol 107(3):313–323

Selvanayagam ZE, Gnanavendhan SG, Balakrishna K, Rao RB, Sivaraman J, Subramanian K, Puri RK (1996) Ehretianone, a novel quinonoid xanthene from *Ehretia buxifolia* with antisnake venom activity. J Nat Prod 59(7):664–667

Shah KG, Baxi AJ (1990) Phytochemical studies and antioestrogenic activity of *Butea frondosa* flowers. Indian J Pharm Sci 52(6):272

Sharma PV, Sharma G, Dhanvantari N (2012) Varanasi Chaukhambha Orientalia 2012(1):3

Sharma AK, Sharma TC, Singh R, Payal P, Gupta R, Sharma MC (2018) A novel compound Lup-20 (29)-ene-3α, 6β-diol identified in petroleum ether extract of *Diospyros melanoxylon* Roxb. leaves and to reveal its antidiabetic activity in rats. Pharmacogn Mag 14(55):245

Shenai BR, Sijwali PS, Singh A, Rosenthal PJ (2000) Characterization of native and recombinant falcipain-2, a principal trophozoite cysteine protease and essential hemoglobinase of *Plasmodium falciparum*. J Biol Chem 275(37):29000–29010

Shinawie A (2002) Wonder drugs of medicinal plants. Ethnobotany. Mol Cell Biochem 213 (1–2):99–109

Shukla S, Gardner J (2006) Local knowledge in community-based approaches to medicinal plant conservation: lessons from India. J Ethnobiol Ethnomed 2(1):20

Singh B, Gupta DK, Chandan BK (2001a) Adaptogenic activity of a glyco-peptido-lipid fraction from the alcoholic extract of *Trichopus zeylanicus* Gaertn. Phytomedicine 8(4):283–291

Singh B, Saxena AK, Chandan BK, Agarwal SG, Anand KK (2001b) In vivo hepatoprotective activity of active fraction from ethanolic extract of *Eclipta alba* leaves. Indian J Physiol Pharmacol 45(4):435–441

Sobhy EA, Tailang M, Benyounes S, Gauthaman K (2011) Antimalarial and hepatoprotective effects of entire plants of *Anastatica hierochuntica*. Int J Res Phytochem Pharmacol 1:24–27

Sri KS, Kumari DJ, Sivannarayana G (2013) Effect of Amla, an approach towards the control of diabetes mellitus. Int J Curr Microbiol Appl Sci 2(9):103–108

Srinivas S, Rao SS, Rao MB, Raju MV (2001) Chemical constituents of the roots of *Vitex Negundio*. Indian J Pharm Sci 63(5):422

Srivastava A, Tripathi AK, Pandey R, Verma RK, Gupta MM (2006) Quantitative determination of reserpine, ajmaline, and ajmalicine in *Rauvolfia serpentina* by reversed-phase high-performance liquid chromatography. J Chromatogr Sci 44(9):557–560

Srividhya M, Hridya H, Shanthi V, Ramanathan K (2017) Bioactive Amento flavone isolated from *Cassia fistula* L. leaves exhibits therapeutic efficacy. 3 Biotech 7(1):33

Stephenson RP (1948) The pharmacological properties of conessine, isoconessine, and neoconessine. Br J Pharmacol Chemother 3(3):237

Sundararajan P, Dey A, Smith A, Doss AG, Rajappan M, Natarajan S (2006) Studies of anticancer and antipyretic activity of *Bidens pilosa* whole plant. Afr Health Sci 6(1):27–30

Sunitha S, Nagaraj M, Varalakshmi P (2001) Hepatoprotective effect of lupeol and lupeol linoleate on tissue antioxidant defence system in cadmium-induced hepatotoxicity in rats. Fitoterapia 72(5):516–523

Suresh M, Rath PK, Panneerselvam A, Dhanasekaran D, Thajuddin N (2010) Anti-mycobacterial effect of leaf extract of *Centella asiatica* (Mackinlayaceae). Res J Pharm Technol 3(3):872–876

Surveswaran S, Cai YZ, Corke H, Sun M (2007) Systematic evaluation of natural phenolic antioxidants from 133 Indian medicinal plants. Food Chem 102(3):938–953

Swathi S, Amareshwari P, Venkatesh Kand Roja Rani A (2019) Phytochemical and pharmacological benefits of *Hemidesmus indicus*: an updated review. J Pharmacogn Phytochem 8(1):256–262

Taid TC, Rajkhowa RC, Kalita JC (2014) A study on the medicinal plants used by the local traditional healers of Dhemaji district, Assam, India for curing reproductive health related disorders. Adv Appl Sci Res 5(1):296–301

Tang W, Eisenbrand G (2013) Chinese drugs of plant origin: chemistry, pharmacology, and use in traditional and modern medicine. Springer Science & Business Media, Berlin

Tekuri SK, Pasupuleti SK, Konidala KK, Amuru SR, Bassaiahgari P, Pabbaraju N (2019) Phytochemical and pharmacological activities of *Solanum surattense* Burm. f—a review. J Appl Pharm Sci 9(03):126–136

Telang RS, Chatterjee S, Varshneya C (1999) Study on analgesic and anti-inflammatory activities of *Vitex negundo* Linn. Indian J Pharmacol 31(5):363

Telang M, Dhulap S, Mandhare A, Hirwani R (2013) Therapeutic and cosmetic applications of mangiferin: a patent review. Expert Opin Ther Pat 23(12):1561–1580

Thomas B, Arumugam R, Veerasamy A, Ramamoorthy S (2014) Ethnomedicinal plants used for the treatment of cuts and wounds by Kuruma tribes, Wayanadu districts of Kerala, India. Asian Pac J Trop Biomed 4:S488–S491

Thong D, Carpenter B, Krippner S (1993) A psychiatrist in paradise: treating mental illness in Bali. White Lotus, Bangkok

Tian JL, Liang X, Gao PY, Li DQ, Sun Q, Li LZ, Song SJ (2014) Two new alkaloids from *Portulaca oleracea* and their cytotoxic activities. J Asian Nat Prod Res 16(3):259–264

Treadway L (1994) Amla: traditional food and medicine. J Am Bot Coun 31:26

Turrini E, Catanzaro E, Muraro MG, Governa V, Trella E, Mele V, Calcabrini C, Morroni F, Sita G, Hrelia P, Tacchini M (2018) *Hemidesmus indicus* induces immunogenic death in human colorectal cancer cells. Oncotarget 9(36):24443

Valan MF, Britto A, Venkataraman R (2010) Phytoconstituents with hepatoprotective activity. Int J Chem Sci 8(3):1421–1432

Veena H, Navya N, Sandesh GK, Thippeswamy NB (2019) Antibacterial and antidiarrheal activity of *Simarouba amara* (Aubl.) bark. J Appl Pharm Sci 9(05):088–096

Velu G, Palanichamy V, Rajan AP (2018) Phytochemical and pharmacological importance of plant secondary metabolites in modern medicine. In: Bioorganic phase in natural food: an overview. Springer, Cham, pp 135–156

Verma C, Bhatia S, Srivastava S (2010) Traditional medicine of the Nicobarese. Ind J Trad Knowl 9(4):779–785

Waechter AI, Hocquemiller R, Laurens A, Cavé A (1997) Glaucafilin, an acetogenin from *Annona glauca*. Phytochemistry 44(8):1537–1540

Wang S, Yu Z, Wang C, Wu C, Guo P, Wei J (2018) Chemical constituents and pharmacological activity of agarwood and *Aquilaria* plants. Molecules 23(2):342

Wei WL, Zeng R, Gu CM, Qu Y, Huang LF (2016) Angelica sinensis in China—a review of botanical profile, ethnopharmacology, phytochemistry and chemical analysis. J Ethnopharmacol 190:116–141

Wen R, Lv HN, Jiang Y, Tu PF (2019) Anti-inflammatory pterocarpanoids from the roots of *Pongamia pinnata*. J Asian Nat Prod Res 21(9):1–8

Wong JH, Lau KM, Wu YO, Cheng L, Wong CW, Yew DTW, Leung PC, Fung KP, Hui M, Ng TB, Bik-San Lau C (2015) Antifungal mode of action of macrocarpal C extracted from *Eucalyptus globulus* Labill (Lan An) towards the dermatophyte *Trichophyton mentagrophytes*. Chinese Med 10(1):34

Woo ER, Lee JY, Cho IJ, Kim SG, Kang KW (2005) Amentoflavone inhibits the induction of nitric oxide synthase by inhibiting NF-κB activation in macrophages. Pharmacol Res 51(6):539–546

World Health Organization. Traditional Medicine (2003) Fact sheet No 134. http://www.who.int/mediacentre/factsheets/fs134/en/

Wu LC, Fan NC, Lin MH, Chu IR, Huang SJ, Hu CY, Han SY (2008) Anti-inflammatory effect of spilanthol from *Spilanthes acmella* on murine macrophage by down-regulating LPS-induced inflammatory mediators. J Agric Food Chem 56(7):2341–2349

Yang LC, Li R, Tan J, Jiang ZT (2013) Polyphenolics composition of the leaves of Zanthoxylum bungeanum Maxim. Grown in Hebei, China, and their radical scavenging activities. J Agric Food Chem 61(8):1772–1778

Yapi AD, Mustofa M, Valentin A, Chavignon O, Teulade JC, Mallie M, Chapat JP, Blache Y (2000) New potential antimalarial agents: synthesis and biological activities of original diaza-analogs of phenanthrene. Chem Pharm Bull 48(12):1886–1889

Yashavanth HS, Haniadka R, Rao S, Rao P, Alva A, Palatty PL, Baliga MS (2018) Turmeric and its principal polyphenol curcumin as a nontoxic gastroprotective agent: recent update. In: Polyphenols: prevention and treatment of human disease. Academic Press, New York, pp 319–325

Yin Y, Gong FY, Wu XX, Sun Y, Li YH, Chen T, Xu Q (2008) Anti-inflammatory and immunosuppressive effect of flavones isolated from Artemisia vestita. J Ethnopharmacol 120(1):1–6

Yokota T, Nishio H, Kubota Y, Mizoguchi M (1998) The inhibitory effect of glabridin from licorice extracts on melanogenesis and inflammation. Pigment Cell Res 11(6):355–361

Yumnam JY, Tripathi OP (2012) Traditional knowledge of eating raw plants by the Meitei of Manipur as medicine/nutrient supplement in their diet

Yun XJ, Shu HM, Chen GY, Ji MH, Ding JY (2014) Chemical constituents from barks of *Lannea coromandelica*. Chinese Herb Med 6(1):65–69

Zhang Y, Shi P, Qu H, Cheng Y (2007) Characterization of phenolic compounds in *Erigeron breviscapus* by liquid chromatography coupled to electrospray ionization mass spectrometry. Rapid Commun Mass Spectrom 21(18):2971–2984

Zhang X, Wu Z, Weng P (2014) Antioxidant and hepatoprotective effect of (−)-epigallocatechin 3-O-(3-O-methyl) gallate (EGCG3″Me) from Chinese oolong tea. J Agric Food Chem 62(41):10046–10054

Zhang M, Wang J, Zhu L, Li T, Jiang W, Zhou J, Peng W, Wu C (2017) *Zanthoxylum bungeanum* Maxim.(Rutaceae): a systematic review of its traditional uses, botany, phytochemistry, pharmacology, pharmacokinetics, and toxicology. Int J Mol Sci 18(10):2172

Zhao X, Sun H, Hou A, Zhao Q, Wei T, Xin W (2005) Antioxidant properties of two gallotannins isolated from the leaves of *Pistacia weinmannifolia*. Biochim Biophys Acta 1725(1):103–110

Zu YG, Fu YJ, Liu W, Hou CL, Kong Y (2006) Simultaneous determination of four flavonoids in Pigeonpea [*Cajanus cajan* (L.) Millsp.] leaves using RP-LC-DAD. Chromatographia 63(9–10):499–505

Herbal Cosmeceuticals

10

Ramesh Surianarayanan and James Prabhanand Bhaskar

Abstract

Herbal cosmeceutical is a rapidly growing category in the field of cosmetics and personal care products. The safety perception is the major contributing factor for the growth of this category. Therefore, the design of these products requires introspection from seed to shelf stage of the product. From good agricultural practices, through selection of ingredients for the base formulation, herbal research to eco-friendly packaging are the aspects that need to be considered to design herbal cosmeceutical products that are safe to environment and safer to consumers. This chapter talks about ways of making a cosmeceutical formulation base green, current herbal knowledge and future research perspective, regulatory considerations and industry approach to herbal products in terms of product claims.

Keywords

Cosmeceuticals · Herbal · Natural · Green · Organic · Cosmetics

10.1 Introduction

Why Herbal Cosmeceuticals? It is closer to nature, safer to health and environment.

R. Surianarayanan
R&D Consultant, Chennai, Tamil Nadu, India

J. P. Bhaskar (✉)
ITC Life Sciences and Technology Centre, ITC Limited, Bangalore, India
e-mail: james.bhaskar@itc.in

When the word 'Cosmetics' is mentioned, the first thing that comes to a consumer's mind is colour cosmetic products like lipsticks, nail polish, blushes, eye shadow and concealers. These are the products that follow original definition of cosmetics 'masking and enhancing appearance through temporary effect'. The field of Cosmetic Science has evolved in the twentieth century into areas like personal wash and personal care. These cosmetic products extended into areas like soaps, body wash, moisturizers, shampoo, toothpaste, moisturizing creams and lotions. Still these areas revolve around new definition of cosmetics—masking and maintaining. The innovation and growth in the sector have not stopped here. It is moving further and progressing rapidly into new areas that have been defined by scientists as cosmeceuticals, though this term still does not have a regulatory standing. The market is flooded with products that deliver skin benefits from masking, to maintaining to correct and enhance. Multiple products exist, to address skin pigmentation, acne, ageing skin, sun protection, hair fall, dandruff, sensitive teeth, bleeding gums cavities, and the list is growing. The other areas developing in this field are hygiene and feel-good products—from hand sanitizers to deodorants and antiperspirants.

The global cosmeceuticals market was USD 45.47 billion in 2017 and is estimated to reach to USD 72.99 billion by 2023 at a CAGR (Compound annual growth rate) of 8.21% during the forecasted period. The global cosmeceuticals market is outpacing all other product segments in the personal care products and cosmetic industry (Research and Markets 2018). While product innovations in these areas and explorations of new areas have been done effectively by cosmetic companies, the cosmetic ingredient suppliers have made significant contribution to the rapid growth of cosmeceutical industry bringing in innovations in all areas of cosmetic science like surfactants, preservatives, emollients, rheology modifiers, and active ingredients.

Cosmetic and cosmeceutical products are unique consumer products. Products should have good sensory, look, feel and smell both as a product and after application experience on skin, hair and teeth. The products are used by consumers regularly for many years without medical supervision. Therefore it is a challenge to design a cosmeceutical product that is efficacious and at the same time safe for long-term use. The category of herbal cosmeceuticals is growing rapidly. This is because of the safety perception of this category of product, i.e., safe to environment and safer to consumers. These products are formulated considering green ingredients both in the base formulation and actives and also extended to environmentally friendly packaging.

Cosmeceuticals deliver actives through a base like cream, lotion, shampoo, gels, oils, paste and the safety of the base and the actives needs to be considered. There are some controversies on the safety of excipients currently used in the base formula of cosmeceutical products. At the same time, the safety of some chemical actives like triclosan used in toothpaste as antibacterial, hydroquinone as skin fairness ingredient have also been studied and restricted for use in cosmeceuticals. Cosmetic raw material suppliers are constantly innovating and have launched many excipients from renewable sources. Industry is conducting research on herbal actives used by various traditions, bringing in a scientific approach in terms of identifying the active principles and standardizing the extracts.

Therefore, herbal cosmeceuticals involve intense research in the area of herbal ingredients used by various traditions like Ayurveda, traditional Chinese medicine, etc. This is because these traditional medicine records tell us the herb and its benefits, but does not explain the set of phyto-principles of the herb that deliver the benefits and this knowledge is important to standardize the herbal active extracts for consistent benefit delivery. Research should also be carried out on herbs of African and south American origin that may not be listed in any of other traditional medicine list.

Herbal cosmetics and cosmeceutical do not have any specific regulation. They fall under Cosmetics regulatory guidelines. In India, herbal cosmetics are registered under the regulatory body 'Ayush'. There are private bodies that have laid down guidelines for herbal cosmetics. Companies who wish to market their products emphasizing its herbal nature get approval from these private bodies and label the approvals appropriately on the pack. Companies evolve different strategies to market herbal cosmeceuticals, emphasizing more on the safety, both to consumers and environment.

Therefore, the design of herbal cosmeceutical product needs to focus on the following aspects:

1. Selection of ingredients for the base formulation.
2. Studying herbal actives to generate safe, efficacious and standardized herbal extracts.
3. Evolve a suitable strategy for getting product approval under a herbal platform.
4. Marketing claims.

10.2 Selection of Ingredients for the Base Formulation

The building block of a cosmeceutical formulation base generally comprises surfactants (as foaming agents, emulsifiers, solubilizers, antibacterial agents, conditioning agent), emollients, humectants, rheology-modifying agents, preservatives and perfume. There are no official guidelines for herbal cosmetic/cosmeceuticals. However private bodies have laid down exclusion of some of the traditionally used cosmetic ingredients. This includes petroleum-based oils, ethoxylates, coal tar colours and certain preservatives. There are evidences that indicate certain chemicals used in cosmetics may have long-term health effects. European Union (EU)'s precautionary approach to consumer safety and the Registration, Evaluation, Authorisation and Restriction of Chemicals (REACH), chemicals legislation has triggered research in the field of green, safer ingredients for cosmetics.

The ingredients generally selected for the herbal cosmeceutical base formulation are of herbal origin, mineral origin, molecules synthesized and extracted using green chemistry and free from ingredients of petroleum source.

Cosmetic ingredients can be generally classified as follows:

(a) *Natural*: any material, which has been harvested, mined or collected, and which subsequently may have been processed, without chemical reaction, to yield a chemical or chemicals that are identifiable in the original source material. For the purpose of this statement, 'without chemical reaction' would permit washing, decolourizing, distilling, grinding/milling, separation and/or concentration of the material by physical means.
(b) *Nature identical*: material produced by commercial synthesis which produces an end product chemically identical to that found in nature. For example, vitamin B3, vitamin E and menthol are natural ingredients, but currently the molecules are synthesized. Such materials are usually synthesized due to factors such as difficulty in supply (seasonal variation in quality and/or quantity), harm to the original source material (particularly animals and also plants), inability to produce required commercial quantity by non-synthetic means, cost effectiveness, etc.
(c) *Naturally derived*: There is widespread use of the term 'naturally derived' without any indication of what this means, ideal for the marketer but much too vague to be informative. It can reasonably be argued that everything can be claimed to be 'naturally derived' if you care to back track far enough. According to Australian Society of Cosmetic Chemists (ASCC) Technical Committee, 'naturally derived' are the materials where the majority of the molecule (not less than 50% by weight) is derived from natural materials (Warby 2011). There may be room for debate on the minimum quantum of natural source material, but it is clear that this figure must be substantial to give any scientific (rather than marketing) credibility to this term.

In an ideal world there can be only two types of ingredients, natural and synthetic. While there still may be some debate over the need for all these definitions, specifically nature identical and naturally derived, these are included as they are in common use and are presently needed to inform the consumer. Whatever may be the approach, nature identical, naturally derived and natural ingredients can be further substantiated for the safety to humans and environment if green chemistry is followed during refining or synthetic processes.

Green chemistry is an approach of chemical synthesis and extraction procedures considering environment. Anastas and Warner (1998) has originally published 12 principles of green chemistry, which is considered as a road map for chemists to implement. One of the key principles to green chemistry is elimination of solvents in chemical processes or replacement of hazardous solvent with environmentally cleaner solvents. Subsequently, there are several new extraction processes such as ultrasonic extraction (Reverchon 1997), instant controlled pressure drop process (Rezzoug et al. 2005; Allaf et al. 2013), accelerated solvent extraction (Brachet et al. 2001), sub-critical water extraction (Ozel et al. 2003), and solvent-free micro wave extraction (Lucchesi et al. 2004). These techniques are applied in the extraction of essential oil of finer quality for use in cosmeceuticals. The traditional methods of extraction and synthesis of some of the cosmeceutical ingredients like sunscreens, emollient esters and perfume compounds have been modified to suit the green chemistry (Mariani et al. 2000; Genta et al. 2002a, b; Hamelin et al. 2002; Villa et al. 2003; Gambaro et al. 2006; Villa et al. 2008).

There are various building blocks and examples of green ingredients that can be incorporated in herbal cosmeceuticals. The actual suppliers of the materials and the grade approved by private green cosmetic regulatory bodies as 'green/natural' should be consulted and appropriate grades of the ingredients should be used.

10.2.1 Emollients

Petroleum-based emollients are widely used in cosmetic cream and oils, e.g., liquid paraffin of various viscosities, petroleum jelly and petroleum waxes. These emollients are highly stable, and therefore the shelf life of products formulated with petroleum-based emollients is high. Replacement of petroleum waxes and oils in a green cosmeceutical base formulation is a challenge. Substitutes are esters, natural waxes, butters and vegetable oils. These ingredients vary highly in polarity and also have poor shelf life because rancidity can set in and cause olfactory issues in the formulation. So, choice of these emollients and preserving them require the knowledge of skilled cosmetic scientists. Ingredients like cocoa butter, shea butter, vegetable oils such as olive oil and coconut oil, jojoba oil, sunflower oil, and traditional esters like isopropyl myristate and isopropyl palmitate are used in herbal cosmeceutical base. Innovations in this area have enabled the launch of new products; examples have been provided in Table 10.1.

Table 10.1 Examples of green emollients

S. No.	INCI name	Application
1	Coco glycerides	Derived from coconut oil. It is a medium spreading emollient with good emulsifying properties. Coco glycerides find applications in sunscreen preparations to solubilize crystalline UV filters, and dispersing pigments
2	Sucrose cocoate	Sucrose cocoate is a blend of sucrose esters and fatty acid esters of coconut oil. It is a hydrophilic emollient used in surfactant systems to increase foam density, creaminess and viscosity. Also used in skin care cream base to enhance the moisturizing effect
3	Cetyl ricinoleate	Cetyl ricinoleate is used in cream base to impart dry silky feel. It optimizes the viscosity of creams and lotions
4	Coco caprylate	Coco caprylate is a clear, slightly yellow, medium polar oil used in cream formulations to improve the spreadability of cream base formulation. The ingredient is being evaluated as a green alternative to volatile silicones
5	C15–19 alkane	High purity alkane from vegetable origin, biodegradable, inert and nonpolar. It can substitute volatile silicones and impart gliding sensation and dry powdery after feel
6	C10–C18 triglycerides	Emollient naturally derived

10.2.2 Emulsifiers

Traditional cream bases are vanishing cream and cold cream. Potassium soap (potassium stearate/palmitate) is the primary emulsifier for vanishing cream. This emulsifier has limitation of being stable only in alkaline pH. Bees wax—borax is the emulsifier of cold cream and it is a weak emulsifier. However, there is a restriction of using Borax in cosmetics. With the advent of petroleum industry, ethoxylated emulsifiers have come into play. They are non-ionic, high HLB (Hydrophile-Lipophile Balance), effective emulsifiers, stable over a wide range of pH and are cost effective. Currently ethoxylates are restricted in green cosmeceuticals and substitution is a great challenge.

Lecithin is a natural emulsifier but is expensive and is challenging to formulate a stable emulsion with lecithin alone. There is innovation in naturally derived, non-ionic emulsifiers, examples being sugar esters, sorbitan esters, glyceryl esters, alkyl glucosides and olive oil esters.

Naturally derived anionic emulsifiers like glyceryl stearate citrates, acyl glutamates, acyl lactylates and acyl phosphates were also developed as green alternate emulsifiers. These emulsifiers are derived by esterification of natural hydrophilic acids like citric, lactic and glutamic acids with fatty alcohols and in some cases fatty acids and glycerine. Examples of emulsifiers which are considered green are shown in Table 10.2.

10.2.3 Surfactants

Classical surfactant systems used in shampoos and cleansers are a combination of sodium or ammonium lauryl ether sulphate (SLES, ALES) and cocamidopropyl betaine. They are cost effective, work synergistically and formulations with good foam and viscosity can be obtained. For green formulation use of ethoxylated surfactants is restricted. Therefore, sodium or ammonium lauryl ether sulphate has to be excluded from the formulation. This is a challenge to the formulators. Some green body regulations allow sodium and ammonium lauryl sulphate, the non-ethoxylated counterpart of SLES/ALES.

Natural surfactants, including saponins from plant sources, such as Yucca (*Yucca glauca*) or Quillaia (*Quillaja saponaria*) are used in some of the green cosmetics. Saponin-rich extracts are, however, often highly coloured and difficult to formulate into an aesthetically acceptable foaming product.

Alkyl Polyglucosides (APG): Non-ionic, biodegradable, plant derived surfactant is a classical naturally derived surfactant widely used in green cosmetics and cosmeceuticals (Rybinski and Hill 1998). APG is synthesized using two natural ingredients, glucose and fatty alcohols (Joshi and Sawant 2007), with variable length of fatty alcohol chain (Fig. 10.1). Based on the length of fatty alcohol chain, APG can be used as a cleansing agent or as an emulsifier (Hill et al. 2008). APG is available commercially with International Nomenclature of Cosmetic Ingredients (INCI) names like lauryl glucoside, decyl glucoside and cetearyl glucoside.

10 Herbal Cosmeceuticals

Table 10.2 Examples for green emulsifiers

S. No.	INCI name	Application
1	Glyceryl stearate citrate	Glyceryl stearate citrate is a fatty acid monoglyceride and it is used in cosmetics as an emulsifier, stabilizing ingredient and also as an emollient ingredient. Glyceryl stearate citrate helps skin and hair to retain moisture
2	Polyglyceryl-3 ricinoleate	Polyglyceryl-3 ricinoleate is condensed fatty acids from castor oil. It is a non-ionic, polyglycerol-based emulsifier, which has good liquid crystal promoting properties. It can be used without cosmetic creams and lotions, as an anhydrous system for lip-care
3	Glyceryl stearate	A naturally derived fatty acid, most often used as an emulsifier and is derived from palm kernel, vegetable or soy oil. It is useful for making water-in-oil emulsions. It acts as a lubricant on the skin's surface, which gives the skin a soft and smooth appearance
4	Sorbitan stearate	The sorbitan esters are produced by reacting the polyol, sorbitol, with a fatty acid (stearic acid, lauric acid, oleic acid, palmitic acid). Both sorbitol and the fatty acids are naturally occurring and are used in a variety of products including skin care products, skin cleansing products, moisturizers, eye makeup and other makeup
5	Cetearyl glucoside	Cetearyl glucoside is formed by the condensation of cetearyl alcohol (fatty acid) with glucose. Can be naturally derived (from coconut/corn oil) or chemically synthesized. It serves as an emulsifier in oil in water formulations in a variety of cosmetic products such as creams, moisturizer, sunscreen, anti-ageing treatment, conditioner and cleanser
6	Sucrose polystearate	Sucrose stearate is a mixture of the sugar sucrose and a natural substance, stearic acid. It helps to enhance the effectiveness of other ingredients and serves as an emulsifier, emollient, surfactant and skin conditioning agent
7	Olive oil polyglyceryl-6- esters	Olive oil polyglyceryl-6- esters is a waxy material, is an O/W emulsifier power, with high versatility, easy to use and stable at acidic and alkaline pH, and serves as a good emollient and skin conditioning agent
8	Glyceryl olivate	Glyceryl Olivate is a monoester of glycerin and the fatty acids derived from olive oil. It serves as a surfactant and emulsifying agent, and used for skin conditioning

There are many naturally derived surfactants allowed in green cosmetics; examples of green surfactants are shown in Table 10.3.

10.2.4 Rheology-Modifying Agents

Traditionally, carbomers are used in cosmetic formulation bases. Carbomers are effective, inert and relatively resistant to microbial contamination. Clear gels can be formulated using carbomers. Various grades of carbomers have been developed for

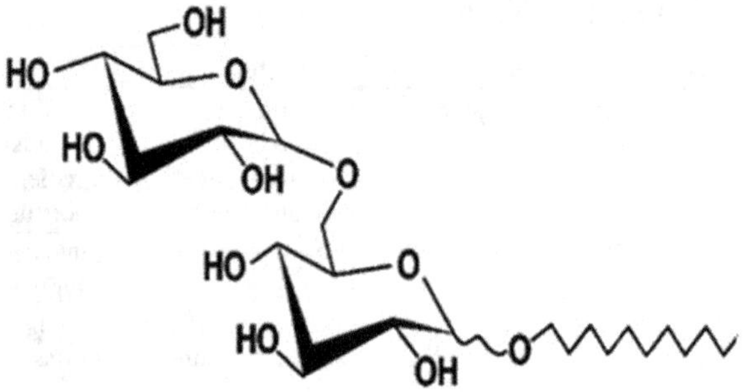

Fig. 10.1 Molecular structure of alkyl polyglucosides

Table 10.3 Examples for green surfactants

S. No.	INCI name	Application
1	Coco glucoside	Coco glucoside is a non-ionic surfactant that can be used as a foaming, cleansing, conditioning, and viscosity building agent to liquid cleansers and shampoos. It is derived from renewable raw materials such as coconut oil and corn and fruit sugars, and is completely biodegradable
2	Sodium lauroyl glutamate	Sodium lauroyl glutamate is derived from glutamic acid (an amino acid), lauric acid (a fatty acid derived from renewable plant resources) and sodium. It is used as a gentle cleanser more suitable for sensitive skin
3	Cocamidopropyl betaine	Cocamidopropyl betaine is mainly used as surfactant in cosmetic products and helps to clean skin and hair by helping water to mix with oil and dirt so that they can be rinsed away
4	Disodium cocoamphodiacetate	Disodium cocoamphodiacetate is an amber liquid with a faint fruity odour. In cosmetics and personal care products, this ingredient is used in the formulation of shampoos and other hair and skin cleansing products
5	Potassium Cocoate (coconut oil soap)	Potassium cocoate soap is a coconut oil derived potassium salt. Potassium cocoate contains glycerin, which serves as a natural moisturizer in personal care formulations
6	Disodium cocoyl glutamate	Sodium cocoyl glutamate is a mild surfactant with very good detergency and foaming power. It is produced from vegetable oils and readily biodegradable. It is also especially well-suited for cleansing sensitive skin
7	Coco betaine	Cocamidopropyl betaine is used as a foam booster in shampoos. It is a medium strength surfactant also used in bath products like hand soaps. It is also used in cosmetics as an emulsifying agent and thickener, and to reduce irritation purely ionic surfactants would cause
8	Sodium coco sulphate	Sodium dodecyl sulphate is an organic sodium salt that is the sodium salt of dodecyl hydrogen sulphate. It has a role as a detergent and a protein denaturant

application in shampoos, face washes, creams and lotions. Silicone-based polymeric emulsifiers are also widely used that impart silky feel to skin and hair, the sensories that cannot be matched by alternate ingredients evaluated till date. Biodegradability and sustainability are concerns for carbomers and silicone polymers, and therefore not recommended for green cosmetics. While a number of natural and naturally derived rheology-modifying agents are available (Table 10.4), but these do not serve as an effective substitute for synthetic polymers. This is again a challenge while formulating green cosmetics and cosmeceuticals.

10.2.5 Solvents/Humectants

Propylene glycol and butylene glycol are widely used in cosmetics and are restricted in green cosmetics since the ingredients are of petroleum source. Corn-based propane diol and biotechnology-based butylene and pentylene glycols have been introduced in the market and are approved for use in green cosmetics. Glycerin, derived from vegetable oils, is also allowed in green cosmetics.

Table 10.4 Examples for green rheology-modifying agents

Herbal source	
Cellulose gum	1.0% in shampoos and lotions, 0.5% in liquid makeup, 0.5–1.0% in toothpaste, 1% in lineaments, 30–50% in denture adhesives
Guar gum	<1% in hair conditioners, <0.5% in conditioning shampoos, <0.3% in aerosol
Mineral source	
Bentonite	0.5–5% in emulsions and suspensions
Hectorite	2% in creams and lotions; 2% in shampoos
Magnesium aluminium silicate	0.2–2% in lotions and suspensions
Marine source	
Alginic acid	0.2–2% in creams and gels; 0.1–1% in toothpastes
Carrageenan	0.4% in skin creams and lotions, 0.6–1.2% in toothpaste, 0.25–0.75% in gels/lotions
Biotechnology source	
Xanthan gum	0.1–3% in creams and lotions; 0.1–3% in shampoos and conditioners, 0.1% In mascaras, 0.1–3% in toothpaste, 0.1–1% in suspensions
Sclerotium gum	1% in gel, 0.2% in emulsion as a thickener

Naturally derived rheology-modifying agents (modified through synthetic process)
Hydroxypropyl guar, cetyl hydroxyethyl cellulose, hydroxypropyl methyl cellulose, Hydroxypropyl cellulose, methylcellulose, hydroxyethyl cellulose, quaternium-18 bentonite, stearalkonium bentonite, quaternium-18 hectorite, stearalkonium hectorite

10.2.6 Preservatives

Selection of preservative for green cosmetics and cosmeceuticals is a big challenge. Strong and proven preservatives approved by European Union is restricted for use in green cosmetics. Nature identical preservatives like benzoates and sorbates are approved preservatives for green cosmetics. However, these preservatives are only effective at low pH, i.e., below 5. Formulators have evaluated plant extracts to boost the preservative efficacy, some examples being grape seed extract and honey suckle extract. Some moisturizing and perfumery ingredients like glyceryl caprylate, glyceryl caprate, caprylyl glycol, glyceryl undecylenate, phenethyl alcohol, sodium anisate/sodium levulinate and benzyl alcohol are also evaluated to boost the preservative efficacy.

10.2.7 Antioxidants

Natural vitamin E is allowed as an antioxidant in green cosmetics. Antioxidants like butylated hydroxytoluene and butylated hydroxylanisole (BHT, BHA) are restricted for use in green cosmetics.

10.2.8 Chelating Agents

Ethylene diamine tetra acetic acid (EDTA) disodium salt is widely used in cosmetics as chelating agent. However, it is not natural and has some environmental biodegradability issues. Phytic acid, a natural plant-derived chelating agent, and its salt are approved for use in green cosmetics as chelating agents. They are effective chelating agents but the cost is considerably higher. Tetrasodium glutamate diacetate that has relatively good biodegradability profile has been approved as a chelating agent in soaps.

10.2.9 Fragrances

Synthetic fragrances are restricted for use in natural cosmetics. A wide range of fragrances are prepared from plants and are stabilized before adding to fragrance formulations.

10.2.10 Colours

Coal tar colours that are widely used in cosmetics are not approved for use in green cosmetics. It is a challenge to stabilize the colour using natural colourants.

With a restriction in use of some of the ingredients as mentioned above, it is indeed a challenge for a chemist to develop a green cosmeceutical formulation base, with excellent sensorial benefits and stability delivered at an affordable cost.

10.3 Herbal Actives

Herbal product technology has transformed over years. This has two aspects, (a) good agricultural and collection practice and (b) research.

Good Agricultural and Collection Practices (GACP) is essential to control the following factors that can affect the phyto-principle composition of the herb.

- Genetic variants (gene level).
- Geographical and nutritional factors—altitude, soil composition, microbial load/association, climate, temperature, etc.
- Seasonal changes (rainfall, drought, water stress, etc.)
- Seasonal variations, e.g., Alkaloid composition in the leaves of *Adhatoda vasica* is reported to vary according to seasonality, the highest being in the months of August and December to January (Patil et al. 2013).
- Association patterns including animals and insects.
- Lunar period.
- Collection and storage.

The World Health Organisation (WHO) has laid out the guidelines on good agricultural and collection practices.

10.3.1 Research

Research and data on herbs currently published are at herbal level and active molecule level. Research work should step to characterization of phyto-principles in herbs.

10.3.1.1 Herbal Level

According to traditional system of medicine, Ayurveda, all the activities of mind and body are governed by three energy forms vata, pitta and kapha. For good health, these three principles should be in perfect balance that will enable healthy mind and body a reflection of all features of good appearance. Ayurveda teaches the way of life leading to good health, stressing ways for internal detoxification and proper diet. Ayurvedic treatment include herbal formulations and has preparation for skin and hair care. Ayurveda has mentioned seven layers of skin and doshas (aspects that can affect the health and appearance of skin) associated with seven layers. Ayurveda has prescribed various internal and external herbal treatments for enhancing the health of skin and hair (Pallavi 2015). Examples of external herbal treatments given in

Table 10.5 Examples of herbal cosmeceuticals from Ayurveda

1. *Roma shatana (depilatory)*: Kshara (alkaline substances), Haridra (Curcumalonga), Haratala (arsenic tri sulphide), Hingu (*Ferula narthex*), Takra (butter-milk), Samudralavana (common salt)
2. *Padadarihara (crack healers)*: Lepa prepared of Madhuchishta (bees wax) with Vasa (animal fat), Sarjachurna (exudates of Shorearobusta) mixed with ghrutha (ghee), Madanaphaladilepa, Upodakaditaila, Saindhavadilepa, Katutaila
3. *Khalityahara (for baldness)*: Lepa prepared of Maricha (*Piper nigrum*), Manashila (arsenic dioxide), Kasisa (Iron sulphate), Tuttha (copper sulphate), Tagara (Vallerianawalici), Gunjakalka siddha taila
4. *Darunakahara (dandruff)*: Gunjaditaila, Bhringarajataila, Chitrakadyataila, Dhatturataila, triphaladyataila, Nilotpaladitaila
5. *Palityahara (grey hairs):* Ketakyaditaila, Nimbabeejatailanasya, Kashmaryaditaila
6. *Yavanapidaka (pimples and acne):* Dhanyaka (Criandrumsativum), Vacha (Acoruscalamus), Lodhra (Sympocusreacemosus), Saindhava (potassium salt), Sarshapa (*Brassica nigra*), Manjishtaditaila, Kumkumaditaila, Shalmalikantakalepa
7. *Nyaccha and Vyanga (black spots and moles):* Bala (Sidacordifolia), Atibala (Sidarhombifolium), Haridralepa, Agaru Krishna chandanalepa, Kumkumadilepa

Table 10.6 List of herbs used for various cosmeceutical preparations

S. No.	Benefit	Herbs
1	Skin barrier function	Aloe, coconut oil, Macadamia oil, sunflower oil, Marula seed oil, Moringa seed oil, Murumuru seed butter, olive oil, Sacha inchi oil, wheat germ oil, Avacodo oil, borage seed oil, jojoba oil, cocoa butter, Shea butter, almond oil, Chalumoogra oil
2	Acne	Argan seed oil, burdock, neem, wild pansy, Amaranth, Arnica, birch, Calendula, Celandine exudes, rhubarb, Orange peel, willow bark, lavender oil
3	Skin pigmentation	Bearberry, grape, Liquorice, soy, cranberry, cucumber, Ginkgo, turmeric, citrus fruit, mulberry
4	Wrinkles/anti-ageing	Bosewillia, Gingko, grape, green tea, Gotu kola, pomegranate, Rosa mosqueta, carrot
5	Anti-inflammatory	Chamomile, witch hazel, Liquorice, turmeric, Indian birthwort, chaulmoogra oil, Arnica, cranberry, dandelion, evening primrose oil, avocado oil, Gingko, ginseng, Gotu kola, Calendula extract, Boswellia
6	Skin tanning	Chaste berry fruit extract (8)
7	Skin firming	Cola seed extract, Gingko extract, guarana, Coleus, green tea extract, red sandalwood, sage, witch hazel, ginger, cinnamon, birch, rosemary
8	Wound healing	Elder berry, green tea extract, St John's wort
9	Hair growth	Liquorice, soy, Amla, Bhringraj, Brahmi

Ayurveda text are shown in Table 10.5. List of herbs for various cosmeceutical benefits from other literature sources are also shown in Table 10.6.

Traditional knowledge of Ayurveda, Chinese or other traditional medicines is on herbal level. Most of the other literature also discuss at herbal level.

10.3.1.2 Molecule Level

Advent of advanced analytical techniques and high-throughput screening techniques has transformed herbal research. Now the research is based on the active principles of the herb and studying the benefits of the active principle isolated from the herb for therapeutic benefits. The active moieties from the herbs can be broadly classified into various class of compounds like lipids, terpenoids, flavonoids, phenols, alkaloids, carbohydrates, glycosides and hydroxyacids.

10.3.1.3 Lipids

Lipids are hydrophobic molecules and are emollients that form occlusive film on skin and reduce transepidermal water loss, and also correct the barrier function of the skin. Lipids can also act as antifungal, and heal wound and ulcers (Alvarez and Rodríguez 2000). Lipids can be simple fatty acids like stearic acid, palmitic acid, lauric acid, or as fatty alcohols like cetyl acohol and stearyl alcohol. Herbal lipids are also present in complex form as triglycerides of fatty acid in vegetable oils, or long chain fatty acid-alcohol ester in waxes, or with no fatty moieties like phosphate, amino acids or sugars (phospholipid, ceramides, glycolipids), or as butters with various fatty acid, fatty alcohol and ester combination. Saturated fatty acids impart thicker consistency than the unsaturated fatty acids.

Vegetable oils, e.g., coconut oil, sunflower oil, olive oil and borage oil, are all triglycerides of fatty acids. The fatty acid composition is different in different oils. Coconut oil is a triglyceride of saturated fatty acid and therefore is resistant to rancidity. Coconut oil is a good moisturizer and has anti-eczema benefit owing to the lauric acid content. Oils like sunflower oil and borage seed oil, because of polyunsaturated fatty acid (PUFA), linoleic and linolenic acid content, aid in enhancing skin barrier properties and are used to treat psoriasis.

Hydrolysis of vegetable oil yields fatty acid and glycerine. Glycerin is a humectant and is widely used in cosmetics. Fatty acids are emollients and are also widely used in creams and lotion base. Fatty acids are neutralized with alkali to form soap. Lecithin is a phospholipid from soy, is a natural emulsifier and has good skin moisturizing properties. Soy lecithin is used to formulate liposome for enhancing dermal drug delivery.

10.3.1.4 Phenols

Phenols are characterized by having at least one aromatic ring bearing one or more hydroxyl groups having astringent and antioxidant properties. Herbal phenols include gallic acid, ellagic acid, lawsone, curcumin, ferulic acid, rosmarinic acid, coumarin, vanillin, arbutin, usnic acid, epicatechin and epigalocatechin.

In cosmeceuticals gallic acid is used for its astringent properties in astringent lotions. Coumarins are phenylpropanoids derived from the cyclization of *ortho*-hydroxy-cinnamic acid. Umbelliferone, esculetin and scopoletin are coumarins used in cosmeceuticals as fragrances and skin-whitening products (Burlando et al. 2010). Vanillin is a flavouring agent and ellagic and ferulic acids are evaluated in skin-whitening and anti-ageing products. Arbutin is also used in skin-whitening product. Curcumin is used in cosmeceutical for anti-inflammatory and skin-whitening properties. Lawsone is the colouring principle of henna. In fresh henna leaves, lawsone is in glycoside form, and in the hair shaft it gets hydrolyzed by enzymes and lawsone is released. Lawsone used as glycoside has the ability to colour skin and hair. Lawsone as such is insoluble in water. Rosmarinic acid is used in cosmeceuticals for its antioxidant, anti-inflammatory and antibacterial properties. Usnic acid isolated from lichen is an antimicrobial agent used in deodorants and antidandruff products. Epicatechin and epigalocatechin and their gallates are potent antioxidants and work synergistically with sunscreens to boost sun protection factor (Karamcheti et al. 2016).

10.3.1.5 Terpenoids

Terpenes found in nature occur as hydrocarbons, alcohols and their glycosides, ethers, aldehydes, ketones, carboxylic acids, and esters. Depending on the number of 2 methylbutane subunits terpenes are classified in mono (2 Isoprene units), sesqui, di, ses, tri, tetra and polyterpenes if more than 8 units. Among many volatile constituents, terpenoids play important role in essential oils. Terpenoids in essential oils are generally present as mono and sesqui terpenoids including cyclic and acyclic form such as hydrocarbons, alcohols and their glycosides, ethers, aldehydes, ketones, carboxylic acids, and esters.

Examples of terpenoids in essential oil are geraniol, menthol, limonene and pinene. Other examples of terpennoids are carotenoids, which are potent antioxidant used in anti-ageing products, and retinoic acid used in the treatment of acne. Alpha Bisabolol, a perfumery compound, is used for its skin-soothing properties. It has also anti-inflammatory action; ursolic acid is used for its antioxidant, anti-inflammatory, antibacterial and antifungal properties.

10.3.1.6 Alkaloids

Examples of alkaloids used in cosmeceuticals are capsaicin, caffeine, piperine, spilanthol, hydrocotyline, etc. (Lautenschläger 2014). Capsaicin is a counterirritant used in pain balms; it improves the microcirculation and reduces inflammation. Caffeine improves microcirculation, has lipolytic properties, and is used to treat cellulite. Piperine internally improves the bioavailability of curcumin. On dermal application it is found to initiate melanization and therefore evaluated for treatment of vitiligo. Spilanthol reduces muscle contraction of the expression lines and relaxes them resulting in fast and visible smoothening of wrinkles in the skin. The alkaloid hydrocotyline has wound-healing benefits.

10.3.1.7 Carbohydrates

Carbohydrates are highly hydroxylated compounds occurring as simple units, dimers, and polymeric forms of various lengths, termed polysaccharides.

Polysaccharides are used in cosmetics for their rheology-modifying properties (Maia Campos et al. 2014). One of the most important skin cosmeceutical ingredient is aloe gel. Most of its moisturizing and soothing and healing properties are due to polysaccharides in the gel. The gel contains monosaccharides (glucose and fructose) and polysaccharides (glucomannans/polymannose). These are derived from the mucilage layer of the plant and are known as mucopolysaccharide. The most prominent monosaccharide is mannose-6-phosphate, and the most common polysaccharides are called glucomannans [beta-(1,4)-acetylated mannan]. Acemannan, a prominent glucomannan, has also been found (Amar et al. 2008). Recently, a glycoprotein with anti-allergic properties, called alprogen and novel anti-inflammatory compound, C-glucosyl chromone, has been isolated from *Aloe vera* gel.

10.3.1.8 Flavonoids
Some of the flavonoids used in cosmeceuticals are beta carotene, lutein, zeaxanthin and lycopene. Flavonoids are used for its potent antioxidant benefit for photo protection, anti-ageing and skin-whitening products. Beta carotene is a precursor of vitamin A, and is widely used in anti-ageing products (Igielska-Kalwat et al. 2012). Oral supplementation of flavonoids has proven to benefit skin appearance and texture (Juturu et al. 2016). Table 10.7 shows various flavonoids and their herbal source.

10.3.1.9 Glycosides
Glycoside is a molecule in which sugar is bound to another functional group via glycosidic bond. Example of glycoside for cosmeceutical use is arbutin. Arbutin is a potent tyrosinase inhibitor and used in skin-whitening products (Lim et al. 2009).

10.3.1.10 Hydroxyl Acids
Examples of alpha hydroxy acids are glycolic, lactic, citric and malic acid. Glycolic and lactic acid are widely used in anti-ageing products. They are exfoliants and also enhance the cell turnover rate. They help in collagen synthesis that leads to reduction

Table 10.7 Flavonoids from natural sources

S. No.	Herb	Flavonoid
1	Apple	Apigenin
2	Broccoli	Kaempferol
3	Celery	Luteolin
4	Lettuce, olives, onions	Quercetin
5	Citrus	Fisetin, Hesperetin
6	Grape fruit	Naringin
7	Berries	Anthocyanins, Cyanidin
8	Cherries	Delphinidin
9	Red grapes	Malvidin, Peonidin
10	Green tea	Epicatechin, Epicatechin gallate

in wrinkles (Moghimipour 2012). Beta hydroxyl acids, e.g., salicylic acid, also possess similar benefit. Owing to poor solubility in water and more skin irritation potential, beta hydroxyl acids are generally used at less than 2% in the formulation. INCI Dictionary of Natural Ingredients (published by Aubrey cosmetics) available online, and Cosmos standard data on the actives for green and organic cosmetics give a detailed list of herbal actives currently available in the market for use in cosmetic products (Aubrey 2014).

Herbal active knowledge therefore stands on herbal level and the active principle level. For cosmeceutical formulation where aesthetics—look, feel and smell of the product is critical, herbal actives at molecular level can play an important role. However herbal research has to step further. At herbal level, the herbal extract is standardized based on a marker; therefore other phyto-principles in extracts are not standardized. At molecule level we may end up in using a natural chemical.

10.3.2 Finger Printing and Evaluation Phyto-Principles

At the next level, it would be important to look at the synergy that might be brought forth based on the interactions between the various phyto-principles present in the herb. For example, curcumin is the active principle of turmeric and is known, but how the other phyto-principles of turmeric synergize to enhance the benefits of curcumin is an important study and that is unknown. The data can enable to use the herb holistically. In this case we have to identify the molecules (Phyto-principles) of herb that work synergistically and the dosage. This is important to standardize the extract. This study will enable to discover more potent cosmeceutical active extract that are also safe and standardized. There are not many publications in this area, rather publications talk only about synergy between herbs and at herbal level. In a multi-herbal formulation, the study is much more complex, considering possibility of interaction of some of the phyto-principles of the herbs ending up in a new molecule in the final extract. Taking herbal research to higher levels and using cosmeceutical base excipients with skill and knowledge can take herbal cosmeceutical products to greater levels of efficacy and standard.

10.4 Regulatory Aspects of Herbal Cosmeceuticals

Following points are to be considered while discussing regulatory aspects of herbal cosmeceuticals:

- Cosmeceutical is the term developed by scientists and not recognized globally by regulatory bodies.
- Product claims allowed for cosmetics should also be used for cosmeceuticals. Words like cure and therapy are not allowed in product claims: Regular use of the product *will help to prevent acne* and **not** regular use of product *will cure acne*.

- There are cosmetic products like high SPF sunscreens that are sold as OTC drug in the USA market. In Japan, skin-whitening products are classified as quasi drugs.
- In India, herbal cosmeceuticals can be registered as Ayurvedic product. 'Ayush' is the approving authority for Ayurvedic products and is independent of Ministry of health that approves cosmetic products. Ayurvedic product can have therapeutic product claims, efficacy data supported through clinical studies, and the efficacy should be derived from the herbs listed in Ayurvedic text. Formulation base can use cosmetic excipients approved by European Union and Bureau of Indian Standards.
- Ayurvedic cosmeceutical need not list, on the label, all the ingredients in the formula. Only active ingredients and key excipients have to be indicated on the label. Cosmetic regulation stipulates listing, on the label, all ingredients used in the formula.
- Ayurvedic cosmeceuticals sold in India will be treated as cosmetics when exported out of India. So product efficacy claim on label should be modified accordingly when Ayurvedic cosmeceuticals are exported from India to other countries.
- There are private bodies that have stipulated guidelines for natural cosmetics and cosmeceuticals. As discussed in the earlier section, these private bodies restrict use of many excipients traditionally used in regular cosmetic products and approved by European guidelines for cosmetics.
- Therefore Ayurvedic cosmeceutical formulations have to be modified for certification, as natural cosmeceutical, from these private bodies.
- Certified natural/organic cosmetics and cosmeceuticals are also launched in the market. Private bodies classify herbal cosmetics as natural and organic.
- Organic cosmeceuticals use organically certified ingredients in the formulation. Level of organic ingredients, in the formulation, for organic certification is stipulated in the regulation. Private bodies that certify natural herbal cosmeceutical have also laid down regulation for organic cosmetics/cosmeceuticals.
- It is very important to note that all products that are registered as natural/organic with private bodies should also have registration, with ministry of health/drug control authorities, as cosmetics.

10.4.1 Background of Natural/Organic Cosmetics Certification Bodies

Organic certification is mainly for foods and agricultural practice. Most of the countries have their own guidelines for organic certification. United States of America, Canada, Czech Republic, France, Germany, Greece, Ireland, Switzerland, Sweden, United Kingdom, Australia, China, India, Japan, Singapore and Cambodia are the main countries having regulation for organic foods and farming. There are many private bodies authorized for inspection and certification.

When organic cosmetic ingredients (e.g., organic vegetable butters, vegetable oil, essential oils) are procured for natural/organic cosmeceuticals, it is important to know the country and type of organic certification that has been obtained for the ingredient. Slowly from foods, cosmetic and cosmeceutical companies attempted to get organic certification for their products. Some products achieved in getting 100% organic approval for a cosmetic/cosmeceutical. For example, a lip balm or cold balm can be formulated using organic vegetable butter, organic vegetable oils and organic essential oils combination. However, this could not be extended to other products since most of the cosmetic base excipients, like surfactants, are not from organic agriculture. Therefore, private bodies developed regulations for herbal and organic certifications for cosmetic products and the regulations were not harmonized. Subsequently, five leading private bodies, having independently herbal and organic body regulation, from Europe, came together to have broader perspective in setting up regulation for herbal and organic cosmetics. The five private bodies are BDIH (Germany), Cosmebio (France), Ecocert (France), ICEA (Italy) and Soil Association Great Britain. From these efforts COSMOS standards (COSMetic Organic and Natural Standard) has set regulation for natural and organic cosmetic products. The standards are globally recognized by the cosmetic industry.

10.4.2 COSMOS Guidelines

By adhering to the COSMOS guidelines, cosmetic products can use COSMOS signatures (Fig. 10.2) on packaging to confirm that the cosmetic product meets the minimum requirement to be considered as organic or natural. Various private bodies are authorized to give certification for cosmetic products based on COSMOS guidelines. The entire regulation on natural and organic cosmetics from COSMOS can be seen in: https://cosmos-standard.org. The details have been divided into different sections.

Fig. 10.2 The label of COSMOS on approved products

- For Consumers
- For Cosmetic companies and manufacturers

For Certifiers in the *consumers section*, brief information is given on the background on herbal and organic cosmetics, what to see on the label and more importantly list of products approved by COSMOS (Fig. 10.2).

For *Cosmetic companies*, information on authorized certifying bodies, certified ingredient data base, approved raw materials data base, and forms and documents links are given. For *certifiers*, links on certified cosmetic products, certified ingredients, approved raw materials, list of certifying bodies and forms and documents are given.

In the COSMOS standards, detailed guidelines are given on the following aspects:

1. Introduction covering objectives and documents.
2. Regulations.
3. Scope.
4. Definitions.
5. General aspects—precautionary principles, animal testing and sustainability.
6. Origin and processing of ingredients covering ingredient categories and rules for organic percentage calculation.
7. Composition of total product—rules for cosmetic products under natural and organic certification, calculation rules for natural origin percentage.
8. Regulation on storage, manufacturing and packaging including regulation for fabrics that can be used products like facial wipes, face masks, etc.
9. Environmental management plan.
10. Details on labelling and communication.
11. Certification.
12. Mentions implementation of this regulation Version 3, from 1st January 2019.

This is followed by appendix on physical process allowed, chemical processes allowed for processing agro-ingredients, examples of processes not allowed, ingredients of mineral origin allowed, other ingredients allowed, physically processed agro-ingredients that must be organic, chemically processed agro-ingredients that must be made from organic origin agro-ingredients, exceptions regarding toxicity and biodegradability data and packaging materials.

There are efforts to bring in guidelines for herbal/organic cosmeceuticals and it has reached to commendable levels. Ecovia Intelligence (formerly Organic Monitor), UK, continuously monitors the challenges and advances in the area of green cosmetics. It is always a challenge to a formulation chemist to adopt these regulations and deliver products with great sensory, good stability, and at an affordable cost. Representations have been made on this and have been discussed. The regulation is being updated periodically. Herbal cosmetics/cosmeceuticals can be marketed just with registration from ministry of health as cosmetics. However, private natural/organic body certification can be effectively used as a marketing tool for herbal products, as a proof for the claim, natural/organic/herbal.

10.5 Herbal Cosmetics/Cosmeceuticals Marketing Claims

Industry, currently adopts various strategies to sell herbal cosmeceuticals, and evolve marketing claims based on traditional herb route, natural route, technology claim, purity claim and proof of natural/organic herbal content through approval from private body regulators.

Traditional herb route: Herbal products are sold under traditional herbal medicine routes, like Ayurveda. In India, Ayurvedic cosmeceuticals can be registered under Ayush. In India, products registered under Ayush do not require cosmetic product registration under Ministry of Health. However, products registered as cosmetics in many countries claim the use of traditional herbal knowledge in their products.

Natural route: Claim on use of herbal ingredients, as actives that deliver specific function, in the product. The herbal ingredients used are not necessarily from traditional medicine text.

Technology claim: This is based on any breakthrough in herbal technology that is incorporated in the product.

All the above claims focus on herbal actives incorporated in regular cosmeceutical base and not green cosmeceutical base.

Purity claim: Claiming exclusion of controversial ingredients in the product formulation base, for example—paraben free, sodium lauryl ether sulphate free. Conceptually this claim adds to the safety of herbal product.

Proof of natural/organic herbal content through approval from private body regulators: Through appropriate natural/organic approval logos on the label, safety of the products can be communicated. Also, the approval authorities through their websites list the approved products for consumers to revalidate. This additionally gives product exposure to consumers. In this approach there is restriction on ingredients that can be used in the formulation base and guidelines laid out, by private bodies, for natural/organic cosmetics have to be followed. Companies that are adopting green approach to their products are also seen by consumers, to add value for corporate social responsibility (CSR). CSR is how the companies manage business processes to produce an overall positive impact on society. It covers sustainability, social impact and ethics (Skye 2019).

10.6 Conclusions

Herbal cosmeceuticals require careful design in all areas of technology from seed to shelf to cater to consumer need and environmental requirements. Herbs are capsules of safe and effective medicine. Appropriately designed herbal research can effectively tap the benefits. Current knowledge, at herbal level and active molecule level, has to be upgraded to understand how to use herbs holistically and also deliver consistent benefits. At the same time base formulation of a product has to be re-designed to be environmentally friendly and safer to consumers. Using only green excipients, the required sensory benefits from the product, like look, feel and smell, have to be delivered to consumers. The product shelf life cannot be compromised. Supply chain of herbs has to be managed effectively. Simultaneously research on environmentally friendly packaging has to be initiated and effectively delivered. A challenge and positive results of the efforts can benefit all. Herbal cosmeceuticals are considered as safe to use, environmentally friendly and free from side effects. Due to this, herbal cosmeceuticals are gaining more and more acceptance among users and they have immense potential to grow significantly in the years to come.

References

Allaf T, Tomao V, Ruiz K, Chemat F (2013) Instant controlled pressure drop technology and ultrasound assisted extraction for sequential extraction of essential oil and antioxidants. Ultrason Sonochem 20(1):239–246

Alvarez AMR, Rodríguez MLG (2000) Lipids in pharmaceutical and cosmetic preparations. Grasasy Aceites 51(1–2):74–96

Amar S, Resham V, Saple DG (2008) Aloe vera: a short review. Indian J Dermatol 53(4):163–166

Anastas PT, Warner JC (1998) Green chemistry: theory and practice. Oxford University Press, New York

Aubrey H (2014) INCI dictionary of natural ingredients with an index of common names. Copyright © 1994–2013 by Aubrey Organics, Inc., Organica Press, Tampa, FL

Brachet A, Rudaz S, Mateus L, Christen P, Veuthey JL (2001) Optimisation of accelerated solvent extraction of cocaine and benzoylecgonine from coca leaves. J Sep Sci 24(10/11):865–873

Burlando B, Verotta L, Cornara L, Massa EB (2010) Herbal principles in cosmetics: properties and mechanisms of action, 1st edn. CRC Press, Boca Raton

COSMOS natural and organic standards. https://cosmos-standard.org

Gambaro R, Villa C, Baldassari S, Mariani E, Parodi A, Bassi AM (2006) 3,3,5-Trimethylcyclohexanols and derived esters: green synthetic procedures, odour evaluation and in vitro skin cytotoxicity assays. Int J Cosmet Sci 28(6):439–446

Genta MT, Villa C, Mariani E, Longobardi M, Loupy A (2002a) Green chemistry procedure for the synthesis of cyclic ketals from 2-adamantanone as potential cosmetic odourants. Int J Cosmet Sci 24(5):257–262

Genta MT, Villa C, Mariani E, Loupy A, Petit A, Rizzetto R, Mascarotti A, Morini F, Ferro M (2002b) Microwave-assisted preparation of cyclic ketals from a cineole ketone as potential cosmetic ingredients: solvent-free synthesis, odour evaluation, in vitro cytotoxicity and antimicrobial assays. Int J Pharm 231(1):11–20

Hamelin J, Bazureau JP, Texier-Boullet F (2002) In: Loupy A (ed) Microwaves in organic synthesis. Wiley-VCH Verlag GmbH, Weinheim, pp 253–293

Hill K, Rybinski WV, Stoll G (2008) Alkyl polyglycosides: technology, properties, and applications. ISBN: 978-3-527-61468-4, September 2008, 251 pages

Igielska-Kalwat J, Wawrzyńczak A, Nowak I (2012) β-Carotene as an exemplary carotenoid and its application in cosmetic industry. CHEMIK 66(2):140–144

Joshi VY, Sawant MR (2007) Novel stereo controlled glycosylation of 1,2,3,4,6-penta-o-acetyl-b-D-glucopyranoside using MgO-ZrO2 as an environmentally benign catalyst. Catal Commun 8:1910–1916

Juturu V, Bowman JP, Deshpande J (2016) Overall skin tone and skin-lightening-improving effects with oral supplementation of lutein and zeaxanthin isomers: a double-blind, placebo-controlled clinical trial. Clin Cosmet Investig Dermatol 9:325–332

Karamcheti SK, Surianarayanan R, Shivakumar HG (2016) Formulation and evaluation of a sunscreen cream containing avobenzone photostabilised by green tea polyphenols. International Federation of Societies of Cosmetic Chemists (IFSCC Magazine) 19(2):73–85

Lautenschläger H (2014) Alkaloids in cosmetic applications. Kosmetik Int 12:22–25

Lim YJ, Lee EH, Kang TH, Ha SK, Oh MS, Kim SM, Yoon TJ, Kang C, Park JH, Kim SY (2009) Inhibitory effects of arbutin on melanin biosynthesis of α-melanocyte stimulating hormone-induced hyperpigmentation in cultured brownish guinea pig skin tissues. Arch Pharm Res 32:367

Lucchesi M, Chemat F, Smadja J (2004) An original solvent free microwave extraction of essential oil from spices. Flavour Fragr J 19(2):134–138

Maia Campos PMBG, de Melo MO, de Camargo Junior FB (2014) Effects of polysaccharide-based formulations on human skin. In: Ramawat K, Mérillon JM (eds) Polysaccharides. Springer, Cham

Mariani, E, Genta MT, Bargagna A, Neuhoff C, Loupy A, Petit A (2000) In: Acierno, D, Leonelli C, Pellacani GC, Mucchi (eds) Application of the microwave technology to synthesis and materials processing. Modena

Moghimipour E (2012) Hydroxy acids, the most widely used anti-aging agents. Jundishapur J Nat Pharm Prod 7(1):9–10

Ozel MZ, Gogus F, Lewis AC (2003) Subcritical water extraction of essential oils from Thymbra spicata. J Agric Food Chem 82(3):381–386

Pallavi K (2015) Concept of cosmeceuticals in ayurveda for skin, beauty & body care. https://doi.org/10.13140/RG.2.1.1288.5201

Patil S, Ojha R, Kaur G, Nepali K, Aggarwal S, Dhar KL (2013) Estimation of seasonal variation of two major pyrrolo [2,1-b] quinazoline alkaloids of *Adhatoda vasica* by HPLC. Nat Prod J 3:30–34

Research and Markets (2018) Global cosmeceuticals market 2017–2023: cosmeceuticals market is outpacing all other product segments in personal care, Dublin, June 11, 2018 (GLOBE NEWS-WIRE)—The "global cosmeceuticals market—by product type, active ingredients type, distribution channel, region—market size, company profiles, industry trends (2017–2023)". ResearchAndMarkets.com

Reverchon E (1997) Supercritical fluid extraction and fractionation of essential oils and related products. J Supercrit Fluids 10(1):1–37

Rezzoug SA, Boutekedjiret C, Allaf K (2005) Optimization of operating conditions of rosemary essential oil extraction by a fast controlled pressure drop process using response surface methodology. J Food Eng 71(1):9–17

Rybinski WV, Hill K (1998) Alkyl polyglycosides—properties and applications of a new class of surfactants. Angew Chem Int Ed 37(10):1328–1345

Skye S (April 22, 2019) What is corporate social responsibility? Business News

Villa C, Mariani E, Loupy A, Grippo C, Grossi GC, Bargagna A (2003) Solvent-free reactions as green chemistry procedures for the synthesis of cosmetic fatty esters. Green Chem 5:623

Villa C, Trucchi B, Gambaro R, Baldassari S (2008) Green procedure for the preparation of scented alcohols from carbonyl compounds. Int J Cosmet Sci 30(2):139–144

Warby JR (2011) Position paper natural. https://ascc.com.au/natural/Natural. Accessed 14 Jan 2011

Plant Secondary Metabolites as Nutraceuticals

11

Lini Nirmala, Zyju Damodharan Pillai Padmini Amma, and Anju V. Jalaj

Abstract

The term "Nutraceutical" refers to a group of substances which are either food or parts of food that provide medical or health benefits. The period of emergence of nutrients as medicines in the pharmaceutical world is of great importance and that draws attention of scientists and researchers towards the substantial benefits of nutraceuticals. During the last century, the interest towards plant-derived products decreased among human community as a result of the shift towards "fast foods." As the changed lifestyle resulted in various noncommunicable diseases and health complications, now the mankind is searching for plants or plant-derived products for healthy life, which is evident from the growth of nutraceutical market as a million-dollar industry at a global level. Plants produce enormous variety of phytochemicals called *secondary metabolites* which are valuable source of pharmaceuticals and nutraceuticals. The secondary metabolites are species specific or tribe specific and are endowed with healing potential and other benefits. This chapter intends to provide an account of various plant-derived secondary metabolites used as nutraceuticals for the betterment of human life.

Keywords

Plant secondary metabolites · Nutraceuticals

L. Nirmala (✉)
Department of Biotechnology, Mar Ivanios College, Thiruvananthapuram, Kerala, India

Z. D. P. Padmini Amma
ThermoFisher Scientific, Dubai, UAE

A. V. Jalaj
Department of Botany, University of Kerala, Kariavattom, Thiruvananthapuram, Kerala, India

© Springer Nature Singapore Pte Ltd. 2020
S. T. Sukumaran et al. (eds.), *Plant Metabolites: Methods, Applications and Prospects*, https://doi.org/10.1007/978-981-15-5136-9_11

11.1 Introduction

"*Let food be thy medicine and medicine be thy food*" is a famous quote of Hippocrates (460–377 BC), the well-recognized father of modern medicine, who conceptualized the relationship between the use of appropriate foods for health and their therapeutic benefits. What is food? And how can food become medicine? Search for answers to these questions advanced the nutritional research and thereby laid down a new therapeutic as well as industrial area called nutraceuticals.

Any substance that provides us with nutrition is called *food*. If the food that we consume provides extra physiological benefits apart from mere nutrition, it is referred as *a functional food.* Food is an inevitable part of every human's day-to-day life. The British Dietetic Association (BDA) reported the influence of food in determining one's mood, behavior, and learning process. So a person must choose the right food. As per the Global Nutrition Report-2018, released by World Health Organization, an unimaginable proportion of the world population suffers from the burden of malnutrition (The Burden of Malnutrition 2018). The wrong choice of food, as well as malnutrition, makes one unhealthy. So the need for a healthy diet is necessary for establishing a peaceful and productive world. The great concern of the scientific community on this issue brings about tremendous progress in the field of nutritional science. It has changed the concept of nutritive food to functional food—a food that meets the additional health requirements in addition to the normal nutritional supply. The term functional food was proposed in Japan during 1980s and the country has a unique regulatory approval process for the commercialization of functional food products. However, there is a slight difference between functional foods and nutraceuticals. "When food is being cooked or prepared using scientific intelligence with or without knowledge of how or why it is being used, the food is called functional food." When functional food aid in the prevention and or treatment of disease or disorder other than anemia, it is called nutraceutical (Cencic and Chingwaru 2010).

Nutraceuticals are substances that are derived from food or functional food and provide additional health benefits to the consumer other than nutrition. Kalra (2003) defined the nutraceutical product as "a substance, which has a physiological benefit or provides protection against chronic diseases." The hybrid term nutraceutical is derived from two words: "*nutrition*" and "*pharmaceutics*." Stephen Defelice (founder and chairman of the foundation for innovation in medicine (FIM), Cranford, New Jersey) holds the credit of coining the term "nutraceutical" (Brower 1998). The term is applied to products that are isolated from herbal products, dietary supplements (nutrients), specific diets, and processed foods such as cereals, soups, and beverages that are used not only for nutrition but also as medicine. Nutraceuticals, in contrast to pharmaceuticals, are substances, which are not usually patent-protected. Both pharmaceutical and nutraceutical compounds might be used to cure or prevent diseases, but only pharmaceutical compounds are under strict governmental sanctions and regulations (Ziesel 1999; Kalra 2003). Unlike conventional food, the nutraceutical is mainly consumed for its medicinal properties. Nutraceuticals may be used to improve health, delay the aging process, prevent chronic diseases, increase life expectancy, or support the structure and/or function of the body (Zhao 2007).

Even though there are medicines for most of the chronic diseases, interest in nutraceuticals is increasing nowadays because of their safety aspects. The therapeutic effects of nutraceuticals are utilized in the healing system called "nutritional therapy" and where the nutraceuticals are employed to detoxify the body, manage vitamin and mineral deficiencies, and promote healthy digestion and dietary habit (Nasri et al. 2014a, b). The search for new natural bioactive ingredients led to the utilization of microbial, animal, and plant species to a larger extent. Extensive research on the realm of functional food started during the 1990s which contributed to the development of thousands of natural bioactive ingredients. All nutraceuticals are broadly divided into two categories (De Felice 2002):

(a) *Potential nutraceuticals*: Products that hold the promise of a particular health or medical benefit. Most of the present-day nutraceuticals fall in this category.
(b) *Established nutraceuticals*: Potential nutraceuticals which were succeeded to demonstrate their benefits in a sufficient number of clinical trials are called established nutraceuticals.

Nutraceuticals derived from food, based on their established function, pharmacological action, chemical constitution, and source, are categorized into different types. One such category is called *probiotics*. This group comprises beneficial living microorganisms such as bacteria and yeast (Hill et al. 2014). Yogurt, kefir, buttermilk, pickles, etc. are the major sources of probiotics. The second category of nutraceuticals is referred to as *prebiotics*. These are the dietary fibers that are essential for the growth of beneficial bacteria in the gut. *Nutrients* and *herbals* are the other two important categories of nutraceuticals. *Nutrients* comprise vitamins, minerals, polyunsaturated fatty acids, omega 3-fatty acids, and different enzymes that are required for maintaining healthy systems in human beings. *Herbals* are concentrates, extracts, or purified products of plant origin.

11.2 Plant Secondary Metabolites

Plants were found to be the most promising source of health-enhancing substances. These substances are collectively called *phytochemicals*. These are nothing but the metabolites produced within the plants and are having specific biological actions. *Plant metabolites* are a group of intermediate compounds or molecules that are part of the anabolic and catabolic pathways in plants. A diverse array of molecules is identified as plant metabolites and they are broadly grouped into *primary metabolites* and *secondary metabolites*. Primary metabolites are the compounds/molecules which are required for the vital growth and development of plants. They may be produced within plants at the end of anabolic pathways such as photosynthesis (sugars, polysaccharides, lipids, proteins, etc.) or may serve as the substrates to meet primary energy requirement of the plant (carbohydrates, lipids, proteins,

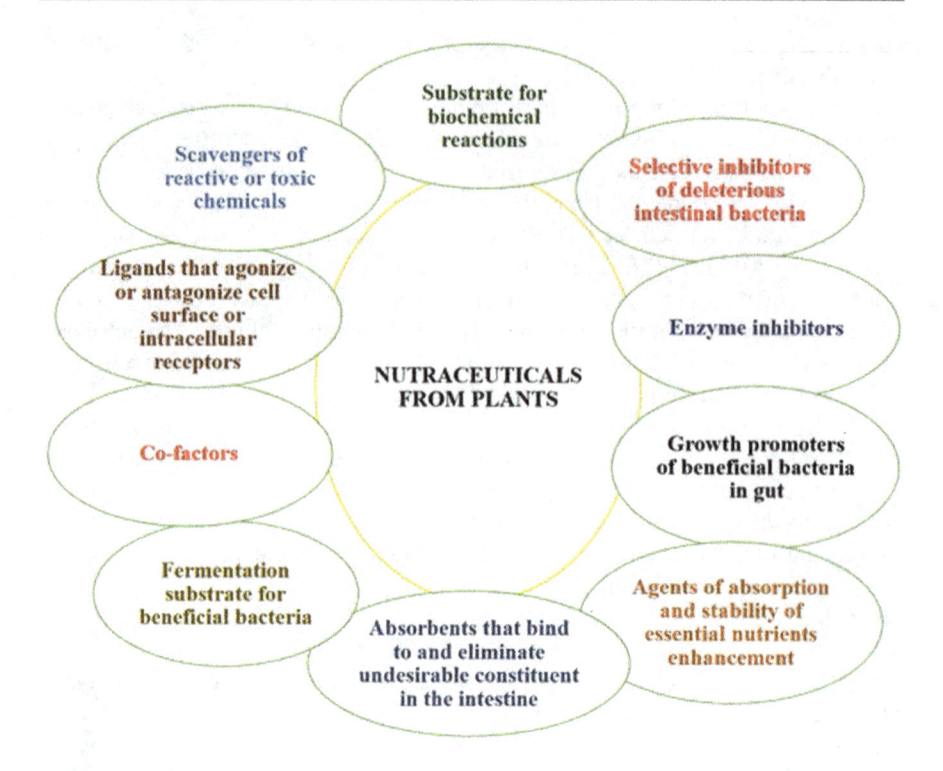

Fig. 11.1 Functions of plant-derived nutraceuticals within human body

etc.). In contrast to the primary metabolites, secondary metabolites are not required for the growth and development of plant but they are believed to play some unknown functions either for the plant or for other organisms connected with the plant (alkaloids, terpenoids, and phenolics). Dillard and German (2000) described the specific actions of plant-derived metabolites within our bodies. An outline of various roles of plant metabolites is given in Fig. 11.1.

Harborne (1973) identified the three major classes of plant chemicals as terpenoids, phenolics, alkaloids and other nitrogen-containing plant constituents. Classification of major plant secondary metabolites which can be utilized as nutraceuticals are shown in Fig. 11.2.

11.2.1 Terpenoids

Terpenoids are the largest group of secondary metabolites with diverse existence as sterols, hormones, and pigments. Terpenoids are synthesized by the mevalonic acid pathway from the precursor compound acetyl CoA. Isoprene units (C_5H_8) are the building units of terpenes, which combine in different proportion to form various classes of terpenes. Terpenes could commonly be found in oils and they may acquire

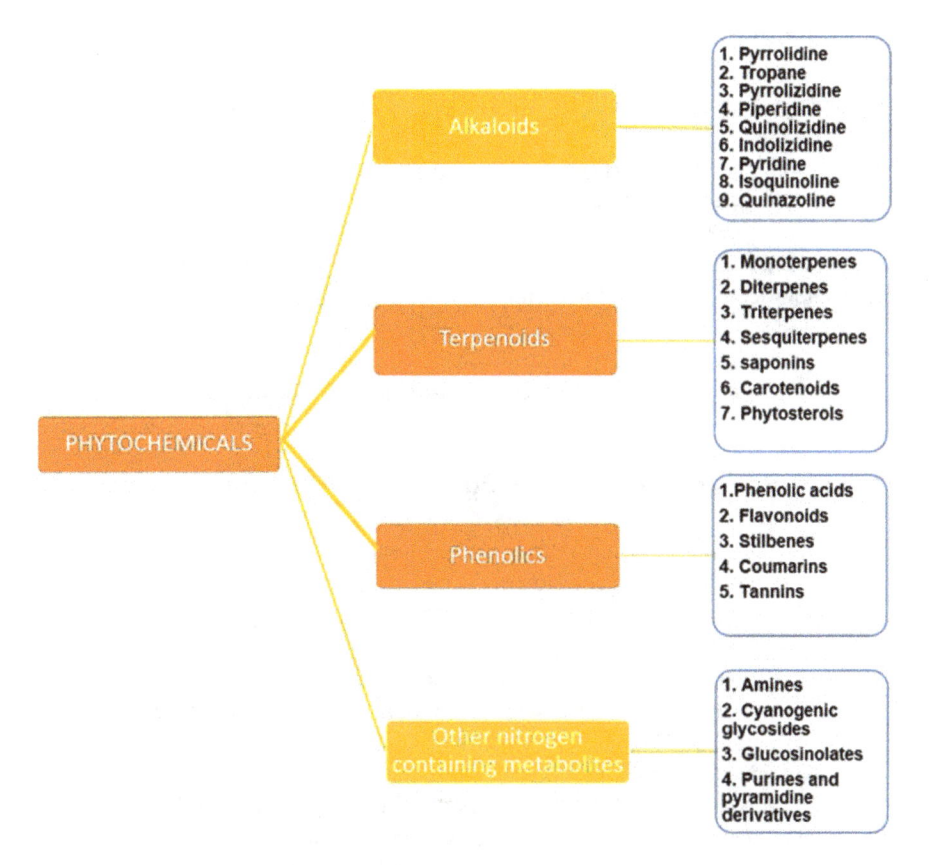

Fig. 11.2 Classification of plant secondary metabolites

cyclic or acyclic structure. As delineated by Harborne, the terpenoids include monoterpenoids, iridoids, sesquiterpenoids, sesquiterpene lactones, diterpenoids, triterpenoid saponins, steroid saponins, cardenolides and bufadienolides, phytosterols, cucurbitacins, nortriterpenoids, other triterpenoids, and carotenoids.

11.2.2 Alkaloids

Alkaloids are nitrogen-containing secondary metabolites present in plants. They are synthesized via either acetate-malonate or mevalonic acid pathway with the amino acid tyrosine as the precursor compound. The alkaloids include indole, isoquinoline, pyrrolidine and piperidine, pyrrolizidine, quinoline, quinolizidine, steroidal, and tropane compounds. Alkaloids from plants are generally recovered by acid-base extraction method. Various toxic reactions are reported for many alkaloids in organisms but they are also reported to have immense pharmacological importance.

11.2.3 Phenolics

Phenolics are the most widely distributed group of natural products in the plant kingdom. They have a hydroxyl group (–OH) covalently linked to an aromatic hydrocarbon ring. Phenol or carbolic acid is the simplest phenolic compound in this group. In general, plant phenolics are derived from phenyl propanoid pathway (PPP), but in species with a greater concentration and diversity of phenolics, compounds use alternative pathways and mechanisms for phenylpropanoid synthesis. Phenolic compounds hold a prime position in the life of human beings due to their widespread applications and are powerful antioxidants due to their higher efficiency of chelation. Plant phenolics are reported as potential sources of new natural drugs, antibiotics, insecticides, and herbicides (Cozier et al. 1997).

11.2.4 Other Nitrogen-Containing Plant Constituents

Other nitrogen-containing constituents include nonprotein amino acids, amines and purines, and pyrimidines. In addition to the above said major secondary metabolites, cyanogenic glycosides and glucosinolates are also formed in plants when the parts are crushed. The presence of cyanogenic glycosides is well known in plant families such as Leguminosae, Poaceae, and Rosaceae. The enzymes glucosidase and hydroxy nitrile initiate the formation of various types of cyanogenic glycosides such as amygdalin, linamarin, and lotaustralin. These compounds are mainly associated with the defensive response of plants.

Several bioactive secondary metabolites derived from plants have the potential or have already been established as pharmaceuticals or nutraceuticals (Rea et al. 2010). The demand for new alternatives for control of chronic diseases has increased considerably in recent decades. Increasing uptake of functional foods with a high content of phytochemicals (bioactive compounds) along with physical exercise is an alternate option to prevent or correct chronic degenerative diseases (Cameron et al. 2005). A large number of compounds from various natural sources such as crop, roots, leaves, and some other plants have been reported. These metabolites include different types of economically important compounds, among which are the antibiotics, pigments, toxins, pheromones, enzyme inhibitors, immunomodulatory agents, receptor antagonists and agonists, pesticides, antitumor agents, and elicitors in animals and plants (Evangelista and Moreno 2007).

11.3 Plant Secondary Metabolites as Nutraceuticals

Nutraceuticals provide an excellent option for humans to stay natural with a high quality of life. They are also a boon to some patients who are reluctant to go through conventional chemical therapy. Nutraceuticals are known to be effective against age-related and chronic diseases. Although the mechanisms of action of different nutraceuticals are still under investigation, many nutraceuticals have been proved to be associated with different types of biological activities. A large number of potential nutraceuticals are under different phases of research and development. Some of the therapeutic applications of plant-derived nutraceuticals are described below.

11.3.1 Cancer and Allied Conditions

Nutraceuticals derived from plants have gained a lot of attention in the field of cancer research. The sense of well-being of patients undergoing cancer therapy is enhanced by the use of nutraceuticals (Grimble 2003). Many of these nutraceuticals are rich sources of antioxidants and they would target signaling pathways related to redox-mediated transcription factors. They are also known to moderate the endocrine system, immunological cascade, and enzymes related to inflammation. DNA repair and the cleavage process are also found to be mediated by such compounds (Ranzato et al. 2014). Many nutraceuticals like green tea, curcumin, isoflavones, polyphenols, lycopene, resveratrol, etc. are found to reverse, prevent, or delay the carcinogenic process. Chemopreventive components found in fruits and vegetables also have potential anticarcinogenic and antimutagenic activities. A list of potent nutraceuticals that can be utilized as therapeutic agents in cancer is shown in Table 11.1.

Phytoestrogens are another type of phytopharmaceuticals recommended for the prevention of prostate and breast cancers (Limer and Speirs 2004). The importance of vegetables and fruits in human diet is justified by the presence of proven anticancer compounds such as carotenoids, lycopenes, daidzen, biochanin, isoflavanones, and genistein in them (Kruger et al. 2002). Another important class of anticancer agents is the Flavonoids. They protect against cancer by acting as antioxidants. Citrus fruits are rich sources of flavonoids (Elisa et al. 2007). Soya foods are rich in isoflavones, which protect against different types of cancers including breast, uterine, lung, colorectal, and prostate cancers (Oboh 2006). An antiproliferative and apoptotic compound found in many plants, called *Betulinic acid*—a pentacyclic triterpenoid—is found effective against human melanoma (Tan et al. 2003).

11.3.2 Cardiac Diseases

Nutraceuticals along with physical exercise are recommended for the prevention and treatment of cardiovascular diseases. Some of the phytochemicals are reported to reduce arterial disease by altering cellular metabolism and signaling. Flavonoids act as cardioprotective agents either by blocking the angiotensin-converting enzyme, blocking the cyclooxygenase enzymes that break down prostaglandins, or preventing platelet aggregation. They could also protect the vascular system that carries oxygen and nutrients to cells (Nasri et al. 2014a, b). Promising nutraceuticals that can be used in the treatment of cardiovascular diseases are shown in Table 11.2. The citrus biflavonoid *hesperidin* is used for the treatment of venous insufficiency and hemorrhoids (Garg et al. 2001). Armolipid Plus (containing red yeast rice,

Table 11.1 Nutraceuticals with anticancer potential

Nutraceutical	Source	Action	Reference
Lycopene	Tomato, guava, watermelon papaya	Decrease oxidative stress and damage to DNA	Shirzad et al. (2013)
β Carotene	Vegetables and fruits	Scavenge free radicals	Stahl and Sies (2005)
Resveratrol	Red grapes	Suppress COX 2 expression	Yang et al. (2001)
Epicatechin-3 gallate (ECGC)	Tea plant (*Camellia sinensis*)	Inhibit tumor cell proliferation	Yang et al. (2001)
Glucosinolates	Cruciferous vegetables	Block enzymes that promote liver, lung, colon, esophageal, breast and stomach cancer	Hidgon et al. (2007)
Ellagic acid	Walnut, strawberry, pomegranate	Prevent cancer cell proliferation	Li et al. (2003)
Tannins	Raspberry, blueberry, cranberry, tea	Prevent cancer cell proliferation	Li et al. (2003)
Saponins	Peas, spinach, alfalfa, tomato, potato	Prevent cancer cell proliferation	Li et al. (2003)
Crocin and crocetin	Saffron (*Crocus sativus*)	Inhibit cancer cell proliferation	Zheng et al. (2016)
Curcumin	Turmeric	Controls expression of TNF-α, NF-κβ, COX-2, and interleukins	Kuo et al. (2011)
Piperine	Black pepper	Inhibit cell proliferation	Zheng et al. (2016)
Diosgenin	Fenugreek	Inhibit COX enzymes and NF-κβ activity	Raju and Mehta (2009)

policosanol, berberine, folic acid, astaxanthin, and coenzyme Q10) is a commercial nutraceutical with proven cardioprotective effect. It could reduce hypertension and cholesterol levels and thereby prevent cardiovascular diseases (Mazza et al. 2015).

11.3.3 Diabetes

Diabetes mellitus is a very serious health issue that shows an alarming increase in worldwide occurrence. Global Diabetes Report says that more than 422 million in the world are suffering from diabetes (WHO 2016). As diabetes leads to several complications such as blindness, kidney failure, stroke, lower limb amputation, and

Table 11.2 Potential nutraceuticals for cardiac care

Nutraceutical	Source	Action	Reference
Sterols/stanols	Fruits, vegetables, seeds, nuts	Reduce LDL-C in blood, reduce intestinal absorption of cholesterol	de Jong et al. (2003)
Octacosanol	Fruits and whole grains	Lowers lipid level	Heidarian et al. (2013)
Lycopene	Tomato, papaya, watermelon	Increases HDL-C and reduces the risk of cardiovascular diseases	Agarwal and Rao (2000)
Allicin	Garlic	Inhibits platelet aggregation and enhances vasodilation	Apitz-Castro et al. (1983)
Quercetin	Fruits, leaves, and seeds of many plants	Lowers blood pressure and protects heart	Ried and Fakler (2014)
Bromelain	Pineapple	Improves immune system and protects heart	Rabelo et al. (2004)

Table 11.3 List of plant nutraceuticals as potential antidiabetic agents

Nutraceuticals	Source	Action	Reference
Lophenol, 24-methylenecycloartanol	*Aloe barbadensis*	Reduction in the fasting and random blood glucose levels and HbA1c levels in mice, hypoglycemia	Tanaka et al. (2006), Misawa et al. (2008)
Mangiferin, stigmasterol, and β-sitosterol	*Urena lobata* (leaves)	Inhibition of dipeptidyl peptidase IV activity	Agarwal and Rao (2000)
Oleanolic acid	*Erythrina indica* (stem bark)	Hypoglycemic effect	Apitz-Castro et al. (1983)
Kaempferol 3-Ogentiobioside	*Cassia alata* (leaves and flowers)	Antidiabetic	Varghese et al. (2013)
Phytosterols, tannins, phenolic compounds	*Platycladus orientalis* (leaves)	Reduced TC, TG, LDL, VLDL; increased HDL levels. Antioxidant property	Dash et al. (2014)

heart attacks, it is regarded as a serious health issue. Many of the antidiabetic drugs such as Pioglitazone were reported to have side effects. So search for natural agents to treat diabetes has given worldwide attention these days. Many plant-derived preparations such as diasulin, pancreatic tonic 180 cp, chakrapani, diabecon, bitter gourd powder, dia-car, diabetes-daily care, gurmar powder, epinsulin, diabecure, syndrex, and diabeta are now sold as antidiabetic agents after successful clinical studies (Modak et al. 2007). A list of common nutraceuticals against diabetes are given in Table 11.3.

Table 11.4 Plant-derived nutraceuticals as therapeutics in other ailments

Property	Nutraceuticals and source	Reference
Anti-obese	Capsaicin conjugated linoleic acid	Rubin and Levin (1994)
Anti-inflammatory	Gamma linolenic acid from nuts, vegetable oils, seed oil of hemp, etc.	Rouhi-Broujeni et al. (2013)
	Gentianin from gentian root	Nasri et al. (2014a, b)
	Curcumin from turmeric	Nasri et al. (2014a, b)
	Aloins and aloesin from *Aloe*	Joanne et al. (2012)
Osteoarthritis	Glucosamine and chondroitin sulfate	Gharipour et al. (2013)
Eye disorders	Lutein and zeaxanthin from cornflower, marigold flower, broccoli, green peas, etc.	Brookmeyer et al. (2007)
Antioxidants	Butylated hydroxy toluene (BHT)	Joanne et al. (2012)
Expectorant	Glycyrrhizin and liquirtin from liquorice	Joanne et al. (2012)
Purgative	Sennoside from senna	Joanne et al. (2012)
Antimalarial	Phloretin from apple leaves	MacDonald and Bishop (1952)

11.3.4 Plant-Derived Nutraceuticals in Managing Other Diseases

Nutraceuticals are used against many disorders such as arthritis, malaria, metabolic abnormalities, and asthma. A list of plant secondary metabolites which are reported to have therapeutic potential in such cases is given in Table 11.4.

11.4 Nutraceuticals as Food Additives

Public awareness on the risk of using synthetic food additives has increased and the search for safer alternatives took the food manufacturers to plant nutraceuticals as they are safer. Several plant secondary metabolites are used as potent food additives. Some of these bioactive compounds have antimicrobial activities widely proven by in vitro tests. Essential oils generated by aromatic plants are one such kind. The antimicrobial activities of essential oils make them good candidates for use as natural additives in foods and food products as they can be added as bioactive components in packaging materials. Currently, more than 3000 essential oils are known, with 300 of them having a commercial interest in food, pharmaceutical, sanitary, or cosmetic industries (Juana and Manuel 2018).

Table 11.5 Fruit and vegetable waste as rich source of nutraceuticals

Plant	Part	Nutraceuticals	Reference
Apple	Peel and pomace	Epicatechin, catechins, anthocyanins, quercitin glyco-sides, chlorogenic acid, hydroxycinnamates, phloretin glycosides, procyanidins	Wolfe and Liu (2003)
Banana	Peel	Gallocatechin, anthocyanins, delphindin, cyaniding, catecholamine	Someya et al. (2002), Kanazawa and Sakakibara (2000)
Citrus fruits	Peel	Hesperidin, naringin, eriocitrin, narirutin	Coll et al. (1998)
Guava	Skin and seeds	Catechin, cyanidin 3-glucoside, galangin, gallic acid, homogentisic acid, kaempferol	Deng et al. (2012)
Grapes	Seed and skin	Coumaric acid, cafeic acid, ferulic acid, chlorogenic acid, cinnamic acid, neochlorogenic acid, p-hydroxy-benzoic acid, protocatechuic acid, vanillic acid, gallic acid, proanthocyanidins, quercetin 3-o-gluuronide, quercetin, resvaratrol	Negro et al. (2003), Maier et al. (2009)
Pomegranate	Peel and pericarp	Gallic acid, cyanidin-3,5-diglucoside, cyanidin-3-diglucoside, delphinidin-3,5-diglucoside	Noda et al. (2002), Gil et al. (2000)
Carrot	Peel	Phenols, beta-carotene	Chantaro et al. (2008)
Cucumber	Peel	Pheophytin, phellandrene, caryophyllene	Zeyada et al. (2008)
Potato	Peel	Gallic acid, cafeic acid, vanillic acid	Zeyada et al. (2008)

11.5 Fruits and Vegetable Waste as Nutraceuticals

A potential source of nutraceuticals is the wastes such as peels and seeds derived during the fruits and vegetable processing. These materials are rich in bioactive compounds, such as phenolic compounds, carotenoids, and vitamins. A list of such compounds is presented in Table 11.5. The compounds and materials isolated from the fruit and vegetable waste such as sugars, minerals, organic acid, dietary fiber, and phenolics have antitumoral, antiviral, antibacterial, cardioprotective, and antimutagenic activities. Efforts to increase the awareness on the nutritional and economic value of such by-products will ensure their effective utilization.

11.6 Global Market Value of Nutraceuticals

Demand for nutraceuticals and functional foods is increasing year by year. The global nutraceuticals market reached a value of US$ 266 Billion in 2018 and expected to grow at a CAGR of 7.6% to reach $734.60 billion by 2026. According to a recent report, the total market for nutraceuticals in India is growing at 21% per annum. It is currently valued at INR 44bn (€621 m), but could be worth more than

INR 95bn in 4 years. The Asia Pacific is the largest market of nutraceuticals. China, India, Japan, and Australia are the main countries coming under this area. Amway Corporation, Abbott Laboratories Inc., Advanced Orthomolecular Research Inc., Ajinomoto Co., Baxter International Inc., Beneo-Orafti S.A., Boehringer Ingelheim, Cargill Inc., Cosucra Groupe Warcoing S.A., Croda International Plc, Danisco Als, Groupe Danone S.A., Icu Medical, Herbalife Ltd., Natrol Inc., and Matsun Nutrition are the most leading manufacturers of nutraceuticals around the world (Dublin 2019).

11.6.1 Areas of Concern in Using Nutraceuticals

As the market of herbal medicinal products is on the rise globally, their safety becomes a major concern. The lack of quality control, ambiguity in the laws governing the production/processing of botanical drugs, etc. may provide room for adulteration or contamination of nutraceuticals that would pose a threat to the health of consumers. The general public—consumers of the products—must be made aware of the risks associated with the use of these products to ensure that all medicines they use are safe and having required quality. Another major concern is the lack of adequate knowledge about the materials themselves. Most of the medicinal plants have not been characterized well and the efficacy and mechanism of action of many phytochemicals are not validated in vivo. The metabolic pathways of the biosynthesis of different phytochemicals and the effect of environmental modulation in the biosynthesis are also not studied well (Patil 2011). The toxicity of nutraceuticals—which are purer forms compared to the natural ones—also has to be studied well and toxicity/safety data should be made available after proper pharmacological/toxicological studies. The safety and quality of some nutraceuticals can be compromised via contamination with toxic plants, metals, mycotoxins, pesticides, fertilizers, drugs of abuse, etc. Safety and toxicity assessment of nutraceuticals need to be done with the knowledge of pharmacokinetic/toxicokinetic studies. Interaction studies are essential to determine efficacy, safety, and toxicity when nutraceuticals are used along with therapeutic drugs (Martin et al. 2018).

11.7 Conclusions

Nutraceuticals increase longevity and quality of life. The use of plant secondary metabolites as phytonutrients or nutraceuticals is rapidly increasing and many people choose these products for the treatment of various health challenges globally. In the developing world, herbal medicine is considered the primary source of healthcare. The demand for nutraceuticals is associated with the prevalence of diseases such as cancer, diabetes, and cardiovascular diseases. India being a biodiversity-rich country stand high among developing countries such as Brazil, Mexico, and South Africa in nutraceutical consumption. If proper platforms are developed for plant product research, India can become the highly competent in nutraceutical production zone in the world. Since nutraceuticals provide physiological or pharmacological effect, they may cause an adverse effect in susceptible individuals. Hence extensive research and proper law making on the use of them are required for their effective and safe use.

References

Agarwal S, Rao AV (2000) Tomato lycopene and its role in human health and chronic diseases. Can Med Assoc J 163:739–744

Apitz-Castro R, Cabrera S, Cruz MR, Ledezma E, Jain MK (1983) Effects of garlic extract and of three pure components isolated from it on human platelet aggregation, arachidonate metabolism, release reaction and platelet ultrastructure. Thromb Res 32:155–169

Brookmeyer R, Johnson E, Ziegler-Graham K, Arrighi HM (2007) Forecasting the global burden of Alzheimer's disease. Alzheimers Dement 3:186–191

Brower V (1998) Nutraceuticals: poised for a healthy slice of the healthcare market? Nat Biotechnol 16:728–731

Cameron SI, Smith RF, Kierstead KE (2005) Linking medicinal/nutraceutical products research with commercialization. Pharm Biol 43:425–433

Cencic A, Chingwaru W (2010) The role of functional foods, nutraceuticals, and food supplements in intestinal health. Nutrients 2(6):11–25

Chantaro P, Devahastin S, Chiewchan N (2008) Production of antioxidant high dietary fiber powder from carrot peels. LWT Food Sci Technol 41:1987–1994

Coll M, Coll L, Laencina J, Tomas-Barberan F (1998) Recovery of favanones from wastes of industrially processed lemons. Eur Food Res Technol 206:404–407

Cozier A, Jensen E, Lean MEJ, McDonald MS (1997) Quantitative analysis of flavonoids by reversed-phase high performance liquid chromatography. J Chromatogr 761:315–321

Dash AK et al (2014) Antidiabetic along with antihyperlipidemic and antioxidant activity of aqueous extract of *Platycladus orientalis* in streptozotocin-induced diabetic rats. Curr Med Res Pract 4(2014):255–262

De Felice SL (2002) FIM rationale and proposed guidelines for the Nutraceutical Research & Education act—NREA. Foundation for Innovation in Medicine

De Jong A, Plat J, Mensink RP (2003) Metabolic effects of plant sterols and stanols. J Nutr Biochem 14:362–369

Deng GF, Shen C, Xu XR, Kuang RD, Guo YJ, Zeng LS, Gao LL, Lin X, Xie JF, Xia EQ (2012) Potential of fruit wastes as natural resources of bioactive compounds. Int J Mol Sci 13:8308–8323

Dillard CJ, German JB (2000) Phytochemicals: nutraceuticals and human health. J Sci Food Agric 80:1744–1756

Dublin (2019) Nutraceuticals: global market outlook (2017–2026). Research and markets

Elisa T, Maurizio LG, Santo G, Danila DM, Marco G (2007) *Citrus* flavonoids: molecular structure, biological activity and nutritional properties: a review. Food Chem 104(2):466–479

Evangelista ZM, Moreno AE (2007) Metabolitos secundarios de importancia farmacéutica producidos por actinomicetos. Biotecnologia 11:37–50

Garg A, Garg S, Zaneveld LJ, Singla AK (2001) Chemistry and pharmacology of the *Citrus* bioflavonoid hesperidin. Phytother Res 15:655–669

Gharipour M, Ramezani MA, Sadeghi M, Khosravi A et al (2013) Sex based levels of C-reactive protein and white blood cell count in subjects with metabolic syndrome: Isfahan healthy heart program. J Res Med Sci 18:467–472

Gil MI, Tomás-Barberán FA, Hess-Pierce B, Holcroft DM, Kader AA (2000) Antioxidant activity of pomegranate juice and its relationship with phenolic composition and processing. J Agric Food Chem 48:4581–4589

Grimble RF (2003) Nutritional therapy for cancer cachexia. Gut 52:1391–1392

Harbone JB (1973) Phytochemical methods: a guide to modern technique of plant analysis. Chapman and Hall, London

Heidarian E, Rafieian-Kopaei M, Ashrafi K (2013) The effect of hydroalcoholic extract of *Allium latifolium* on the liver phosphatidate phosphatase and serum lipid profile in hyperlipidemic rats. J Babol Univ Med Sci 15:37–46

Higdon JV, Delage B, Williams DE, Dashwood RH (2007) Cruciferous vegetables and human cancer risk: epidemiologic evidence and mechanistic Bais. Pharmacol Res 55:224–236

Hill C, Guarner F, Reid G, Gibson GR, Merenstein DJ, Pot B, Morelli L, Canani RB, Flint HJ, Salminen S, Calder PC, Sanders ME (2014) Expert consensus document. The international scientific Association for Probiotics and Prebiotics consensus statement on the scope and appropriate use of the term probiotic. Nat Rev Gasteroenterol Hepatol 11(8):506–514

Joanne B, Linda AA, Phillipson DJ (2012) Herbal medicines. RPS Publishing, London

Juana F-L, Manuel V-M (2018) Introduction to the special issue: application of essential oils in food systems. Foods 7(56):1–4

Kalra E (2003) Nutraceutical—definition and introduction. AAPS Pharm Sci 5(3):25

Kanazawa K, Sakakibara H (2000) High content of dopamine, a strong antioxidant in cavendish banana. J Agric Food Chem 48:844–848

Kruger CL, Murphy M, DeFrietas Z, Pfannkuch F, Hiembach J (2002) An innovative approach to the determination of safety for a dietary ingredient derived from a new source: case study using a crystalline lutein product. Food Chem Toxicol 40:1535–1549

Li H, Wang Z, Liu Y (2003) Reviews in the studies on tannin activity of cancer prevention and anticancer. Zhong Yao Cai 26:444–448

Limer JL, Speirs V (2004) Phyto-estrogens and breast cancer chemoprevention. Breast Cancer Res 6:119–124

MacDonald RE, Bishop CJ (1952) Phloretin: an antibacterial substance obtained from apple leaves. Can J Bot 30:486–489

Maier T, Schieber A, Kammerer DR, Carle R (2009) Residues of grape (*Vitis vinifera* L.) seed oil production as a valuable source of phenolic antioxidants. Food Chem 112:551–559

Martin JR, Kim BP, Watt J (2018) Adverse effects of nutraceuticals and dietary supplements. Annu Rev Pharmacol Toxicol 58:583–601

Mazza A, Lenti S, Schiavon L et al (2015) Nutraceuticals for serum lipid and blood pressure control in hypertensive and Hypercholesterolemic subjects at low cardiovascular risk. Adv Ther 32:680–690

Misawa E et al (2008) Administration of phytosterols isolated from *Aloe vera* gel reduce visceral fat mass and improve hyperglycemia in Zucker diabetic fatty (ZDF) rats. Obes Res Clin Pract 2 (2008):239–245

Modak M, Dixit P, Londhe J, Ghaskadbi S, Paul A, Devasagayam T (2007) Indian herbs and herbal drugs used for the treatment of diabetes. J Clin Biochem Nutr 3:163–173

Nasri H, Baradaran A, Shirzad H, Rafieian-Kopaei M (2014a) New concepts in nutraceuticals as alternative for pharmaceuticals. Int J Prev Med 5(12):1487–1499

Nasri H, Motamedi P, Dehghani N, Nasri P, Taheri Z, Kinani F et al (2014b) Vitamin D and immune system. J Renal Endocrinol 1:5–7

Negro C, Tommasi L, Miceli A (2003) Phenolic compounds and antioxidant activity from red grape marc extracts. Bioresour Technol 87:41–44

Noda Y, Kaneyuki T, Mori A, Packer L (2002) Antioxidant activities of pomegranate fruit extract and its anthocyanidins: delphinidin, cyanidin, and pelargonidin. J Agric Food Chem 50:166–171

Oboh G (2006) Antioxidant properties of some commonly consumed and underutilized tropical legumes. Eur Food Res Technol 224:61–65

Patil CS (2011) Current trends and future prospective of nutraceuticals in health promotion. Bio Info Phram Biotechnol 1:1–7

Rabelo APB, Tambourgi EB, Pessoa A Jr (2004) Bromelain partioning in two phase aqueous systems containing PEO-PPO-PEO block copolymers. J Chromatogr B 807:61–68

Raju J, Mehta R (2009) Cancer chemopreventive and therapeutic effects of diosgenin, a food saponin. Nutr Cancer 61:27–35

Ranzato E, Simona M, Cinzia MC, Giorgio C (2014) Role of nutraceuticals in cancer therapy. J Food Res 3(4):18–25

Rea G, Antonacci A, Lambreva M, Margonelli A, Ambrosi C, Giardi MT (2010) The Nutrasnacks Project: basic research and biotechnological programs on nutraceutical. In: Giardi MT, Rea G, Berra B (eds) Bio-farms for nutraceuticals: functional food and safety control by biosensors, vol 698. Springer US, pp 1–16

Ried K, Fakler P (2014) Potential of garlic (*Allium sativum*) in lowering high blood pressure: mechanisms of action and clinical relevance. Integr Blood Press Control 7:71

Rouhi-Broujeni A, Heidarian E, Darvishzadeh-Boroojeni P, Rafieian-Kopaei M, Gharipour M (2013) Lipid lowering activity of *Moringa pergerina* seeds in rat: a comparison between the extract and atorvastatin. Res J Biol Sci 8:150–154

Shirzad H, Kiani M, Shirzad M (2013) Impact of tomato extract on the mice fibrosarcoma cells. J Herb Med Pharmacol 2:13–16

Someya S, Yoshiki Y, Okubo K (2002) Antioxidant compounds from bananas (*Musa cavendish*). Food Chem 79:351–354

Stahl W, Sies H (2005) Bioactivity and protective effects of natural carotenoids. Biochim Biophys Acta 1740:101–107

Tan Y, Yu R, Pezzuto JM (2003) Betulinic acid-induced programmed cell death in human melanoma cells involves mitogen-activated protein kinase activation. Clin Cancer Res 9(7):2866–2875

Tanaka M et al (2006) Identification of five phytosterols from *Aloe vera* gel as anti-diabetic compounds. Biol Pharm Bull 29(2006):1418–1422

The Burden of Malnutrition (2018) In: Global nutrition report, World Health Organization

Varghese GK et al (2013) Antidiabetic components of *Cassia alata* leaves: identification through α-glucosidase inhibition studies. Pharm Biol 51(2013):345–349

WHO (2016) Global report on diabetes. WHO, Geneva

Wolfe KL, Liu RH (2003) Apple peels as a value-added food ingredient. J Agric Food Chem 51:1676–1683

Yang CS, Landau JM, Huang MT, Newmark HL (2001) Inhibition of carcinogenesis by dietary polyphenolic compounds. Annu Rev Nutr 21:381–406

Zeisel SH (1999) Regulation of "nutraceuticals". Science 285:185–186

Zhao J (2007) Nutraceuticals, nutritional therapy, phytonutrients, and phytotherapy for improvement of human health: a perspective on plant biotechnology application. Bentham Science Publishers, Sharjah

Zheng J, Zhou Y, Li Y, Xu DP, Li S, Li HB (2016) Spices for prevention and treatment of cancers. Forum Nutr 8:1–35

Bioactive Secondary Metabolites from Lichens

12

Sanjeeva Nayaka and Biju Haridas

Abstract

Lichens are traditionally used as medicine since the medieval period. More than 1000 secondary metabolites have been identified in lichens. It is still unknown why the lichens produce such a plethora of secondary metabolites. However, scientists have successfully utilized them for taxonomy and bioprospecting. The extracts of lichens have exhibited wide range of biological activities, such as antimicrobial (antibacterial, antifungal, antiviral, anti-HIV), antioxidant, anti-inflammatory, antipyretic, analgesic, anti-ulcer, and anticancer activities. The lichen metabolites are also being assayed and found useful as hepatoprotective, cardiovascular protective, gastrointestinal protective, antidiabetic, and probiotic, which are considered as the lifestyle diseases of modern days. Polyketides are one of the major groups of secondary metabolites produced by lichens involving polyketide synthase genes. Interestingly among the 1000 secondary metabolites known from lichens only a few are isolated and tested for their biological activities, while in all remaining cases, activity is indirectly attributed to the presence of various metabolites. In the present chapter, a total of 35 secondary metabolites that are isolated from lichens were tested for biological activities and are listed along with their structure, substance class, and occurrence. Further scope for bioprospecting studies is also discussed.

S. Nayaka (✉)
Lichenology Laboratory, CSIR—National Botanical Research Institute, Lucknow, Uttar Pradesh, India

B. Haridas
Microbiology Division, KSCSTE—Jawaharlal Nehru Tropical Botanic Garden and Research Institute, Thiruvananthapuram, Kerala, India

© Springer Nature Singapore Pte Ltd. 2020
S. T. Sukumaran et al. (eds.), *Plant Metabolites: Methods, Applications and Prospects*, https://doi.org/10.1007/978-981-15-5136-9_12

Keywords

Lichenized fungi · Biodiversity · Biological activity · Lichen substances · Lichen chemistry

12.1 Introduction

Lichens by definition are symbiotic plant-like organisms, usually composed of a fungal partner (mycobiont) and one or more photosynthetic partners (photobiont), most often either a green alga or cyanobacterium. They are an outstandingly successful group, exploiting a wide range of habitats throughout the world and dominating about 8% of terrestrial ecosystems (Nash 2008). In the world, about 20,000 species of lichens are known at present (Lücking et al. 2016) and among them India represents 2750 species (Sinha 2018; Mao and Dash 2019). The dual nature of the lichens is now widely recognized and the lichen products are widely used in traditional medicine for centuries (Nash 2008). During the medieval period, lichens figured prominently in the herbals used by medicinal practitioners (Hale 1983). Lichens have been used in traditional medicine since the time of the first Chinese and Egyptian civilizations. *Pseudevernia furfuracea* (L.) Zopf. found in an Egyptian antique vase from the 18th Dynasty (1700–1600 BC) is considered as a clear evidence in support of this (Llano 1948).

In India, reference to lichens as medicine can be traced back to Rigveda (6000–4000 BC) and Atharva veda (1500 BC) where it is called as "Shipal." Later, Sanskrit literature referred lichens as "Shailaya" and "Shilapushp" in Sushruta Samhita (1000 BC), Charaka Samhita (300–200 BC), and several Nighantu (ancient dictionaries) (1100–1800 AD) (Kumar and Upreti 2001). About 150 lichens occurring in India are known to have medicinal value either in traditional medicine or in biological assays. Nayaka et al. (2010) reviewed in detail the ethnolichenological and traditional uses of lichens in India and listed their biological activities. It is revealed that a total of 36 species are used in traditional medicine either in India or elsewhere, 55 have been screened for antimicrobial activity, 57 for antioxidant property, while about 37 for anticancer and cytotoxicity. Some of the common macrolichens utilized as medicine are *Cetraria islandica* (L.) Ach.; *Cladonia rangiferina* (L.) Weber; *Dolichousnea longissima* (Ach.) Articus; *Evernia prunastri* (L.) Ach.; *Hypogymnia physodes* (L.) Ach.; *Hypotrachyna cirrhata* (Fr.) Divakar, A. Crespo, Sipman, Elix, and Lumbsch; *Peltigera canina* (L.) Willd. The medicinal values of various lichens are attributed to the presence of unique metabolites, which are discussed in this chapter.

12.2 Lichen Chemistry

Lichen produces a great number of organic compounds that attracted the researchers as early as 1830s. These compounds are usually referred as lichen substances. Vulpinic acid (Bebert 1831), picrolichenic acid (Alms 1832), and usnic acid (Knop 1844) are few of the lichen substances isolated from lichens in the early ages of lichen chemistry research. Gmelin (1858) published the first review on lichen substances. Zopf (1907) was one of the pioneers who first described over 150 lichen compounds. However, the structural elucidation of many of these compounds came from the painstaking work of Asahina and co-workers in Japan during 1930s (Asahina and Shibata 1954). The lichen substances can be categorized as primary and secondary metabolites. The primary metabolites are intracellular in nature and belong to classes of proteins, amino acids, polyols, carotenoids, polysaccharides, and vitamins. Both mycobiont and photobiont contribute to synthesizing these primary metabolites. Whereas, secondary metabolites are extracellular, stored either on the cortex or in medulla and mostly synthesized by mycobiont. Lichens are reported to produce more than 1000 secondary metabolites (Elix 2014) among them except for 50–60, remaining all are unique to lichens. The secondary metabolites in lichens are derived from the polyketide pathway (Elix 1996) through three pathways, shikimic acid pathway, mevalonic acid pathway, and acetate–malonate pathway. Polyketides are built primarily from combinations of acetate (acetyl-CoA) and malonate (malonyl-CoA). The shikimic acid pathway provides an alternative route to aromatic compounds, particularly the aromatic amino acids L-phenylalanine, L-tyrosine, and L-tryptophan. Mevalonic acid pathways, on the other hand, produce mainly terpenoids, which are derived from C5 isoprene units (Dewick 2002). Depsides, depsidones, and dibenzofurans are formed by the acetate–malonate pathway. The most important of these are the esters and the oxidative coupling products of simple phenolic units related to orcinol and p-orcinol. Most depsides and depsidones are colorless compounds that occur in the medulla of the lichen, however, usnic acids, yellow cortical compounds formed by the oxidative coupling of methylphloroacetophenone units are found in the cortex of many lichen species. Anthraquinones, xanthones, and chromones are all pigmented compounds which occur in the cortex, which are also produced by the acetate–malonate pathway by intramolecular condensation of long-folded polyketide units rather than the coupling of phenolic units. The shikimic acid biosynthetic pathway produces two major groups of pigmented compounds, which occur in the cortex, pulvinic acid derivatives, and terphenyl quinones. Terpenes and steroids are produced by the mevalonic acid pathway. Lichens can produce secondary metabolites as much as 20% of their thallus dry weight, but generally amount varies between 5 and 10%.

12.3 Lichen Chemistry in Taxonomy

It is not yet clear why lichens produce such a large number of secondary metabolites. But, scientists have wisely utilized these metabolites for taxonomy and bioprospecting. The application of chemical discriminators to lichen taxonomy began inadvertently when thallus color was accepted as a valid generic or specific character (Elix 2014). Generally, yellowish lichens have usnic acid, orange ones have anthroquinones, while grayish one would have atranorin. However, most lichen substances are colorless and can be detected only by indirect means. Nylander (1866) was the first lichenologist to conduct chemical tests on lichen thalli for taxonomic purposes. He detected the presence of various lichen substances by spotting reagents such as aqueous solution of potassium hydroxide, iodine solution, and calcium hypochlorite directly on the lichen thallus or medulla to produce characteristic color changes. Asahina developed an additional spot test reagent P or PD, an alcoholic solution of p-phenylenediamine. Further, he also invented a microcrystallization technique for more definitive recognition of individual lichen acids on a routine basis (Asahina and Shibata 1954). Subsequently, the techniques for identification of lichen substances for taxonomic purpose improvised. Wachtmeister (1952) introduced paper chromatography which is superseded by thin layer chromatography (TLC) introduced and improvised by Culberson and coworkers (Culberson 1972; Culberson and Ahmadjian, 1979; Culberson et al. 1981; Culberson and Johnson, 1982). In recent times, several more techniques are introduced such as high-performance thin layer chromatography, gas chromatography coupled with mass spectroscopy, and high-performance liquid chromatography. The major disadvantage of these techniques mostly is the expense of the equipment and purified solvents. Therefore, they are used only on special occasions and for routine lichen taxonomic work TLC still remains as the best technique. It can be noted that among hundreds of metabolites that lichen produces only about 200 are utilized for taxonomy purpose.

12.4 Lichen Chemistry in Bioprospecting

Production of secondary metabolites is costly to the organisms in terms of nutrient and energy, so one would expect that the plethora of metabolites produced by lichens would have biological significance to the organisms. Recent field and laboratory studies have shown that many of these compounds are indeed involved in important ecological roles (Bjerkea et al. 2005; Lawrey 1995; Rikkinen 1995). It is also now established that most of the lichens utilized in the traditional medicine have potential lichen substances, acting either independently or in combination with others. Some of the possible biological significance of lichen metabolites in nature can be summarized as antibiotic activities—provide protection against microorganisms; photoprotective activities—aromatic substances absorb UV light to protect algae (photobionts) against intensive irradiation; promote symbiotic equilibrium by affecting the cell wall permeability of photobionts; chelating agents—capture and supply important minerals from the substrate; antifeedant/antiherbivory activities—protect the lichens from insect and animal feedings; hydrophobic properties—prevent saturation of the medulla with water and allow continuous gas exchange; stress metabolites—metabolites secreted under extreme conditions. Biological assay has proven that lichen crude extracts or isolated compounds have antimicrobial (antibacterial, antifungal, antiviral, anti-HIV), antioxidant, anti-inflammatory, antipyretic, analgesic, anti-ulcer, and anticancer activities. The lichen metabolites are also being assayed and found useful as hepatoprotective, cardiovascular protective, gastrointestinal protective, antidiabetic, and probiotic, which are considered as the lifestyle disease of modern days. It is also becoming a trend in the recent days to produce silver and gold nanoparticles using lichen extract for improved applications under the flag of "green chemistry." Therefore, it can be said that "sky is the limit" for bioprospecting lichens and their metabolites.

Burkholder and coworkers (Burkholder et al. 1944; Burkholder and Evans 1945) initially discovered that extracts of 52 species of lichens collected from eastern North America inhibited the growth of several bacteria. Their study resulted in a mad race among scientists throughout the world for testing lichens for antimicrobial activities. Hence, antimicrobial activity is the most common study that is attempted using lichens so far. However, initial studies were limited to experimenting with crude extracts and only in the recent times, identification and isolation of active metabolites are attempted. Several reviews are available summarizing the bioactive or pharmacological potential of lichens (Boustie and Grube, 2005; Molnár and Farkas 2010; Shukla et al. 2010; Mitrović et al. 2011; Zambare and Christopher, 2012; Shrestha and St. Clair 2013). These reviews are mostly categorized according to various biological activities exhibited by lichens. It can be noted that about 95 % of the studies involved crude extract and indirect inference to the secondary metabolites they are possessing. In this communication, we are listing the secondary metabolites that are isolated from lichens for bioactivity testing.

12.5 Secondary Metabolites Isolated from Lichens

12.5.1 β-Alectoronic Acid (8′-O-Ethyl-β-Alectoronic Acid)

Substance class: Diphenyl ether, **Molecular formula:** $C_{30}H_{36}O_9$.

Occurrence: *Asahinea chrysantha* (Tuck.) W.L. Culb. and C.F. Culb.; *Parmelia birulae* Elenkin (Elix 2014) etc.

β-alectoronic acid (Fig. 12.1) isolated from *Alectoria sarmentosa* (Ach.) Ach. showed fairly good antimicrobial activity against *Staphylococcus aureus, Mycobacterium smegmatis,* and *Candida albicans* (Gollapudi et al. 1994).

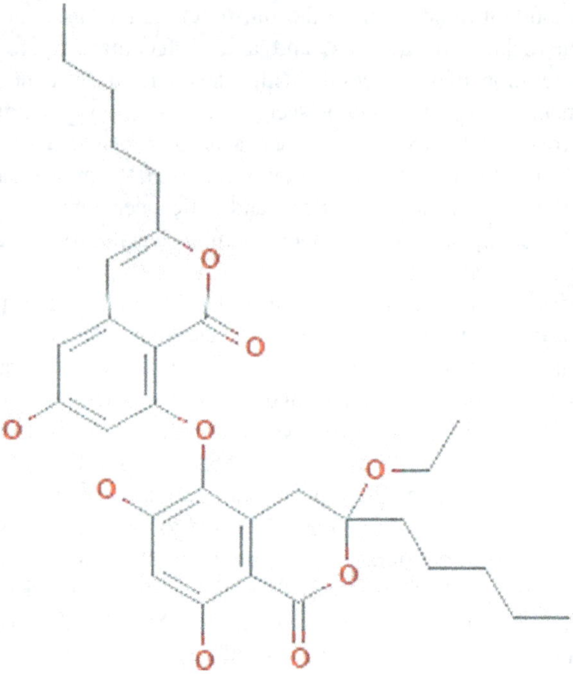

Fig. 12.1 Structure of
β-alectoronic acid

12.5.2 16-O-Acetylleucotylic Acid (16β-Acetoxy-22-Hydroxyhopane-4α-Oic Acid)

Substance Class: Terpenoids, **Molecular formula:** $C_{32}H_{52}O_5$.

Occurrence: *Myelochroa entotheiochroa* (Hue) Eilx and Hale (Elix 2014), *M. aurulenta* (Tuck.) Elix and Hale etc.

The acid isolated from *M. aurulenta* exhibited strong antiproliferative activity against human leukemia cell lines HL-60 with an EC_{50} value of 21 μM, whereas leucotylic acid, a derivative of 16-O-acetly-leucotylic acid had a higher EC_{50} value (72 μM) (Tokiwano et al. 2009) (Fig. 12.2).

12.5.3 Argopsin (1′-Chloropannarin)

Substance class: β-Orcinol Depsidones, **Molecular formula:** $C_{18}H_{14}C_{12}O_6$.

Occurrence: *Argopsis* spp., *Micarea lignaria* (Ach.) Heldl., *M. leprosula* (Th. Fr.) Coppins & A. Fletcher (Elix 2014) etc.

Argopsin was discovered from the lichen *Argopsis friesiana* Müll. Arg. by Huneck and Lamb (1975) (Fig.12.3). Argopsin isolated from continental Antarctic lichens showed excellent concentration and time-dependent cytotoxicity in terms of intracellular lactate dehydrogenase release in rat hepatocytes. The IC_{50} was >50 μg/mL, after incubation of 8 h and 27 μg/mL after 24 h (Correché et al. 2004). In another

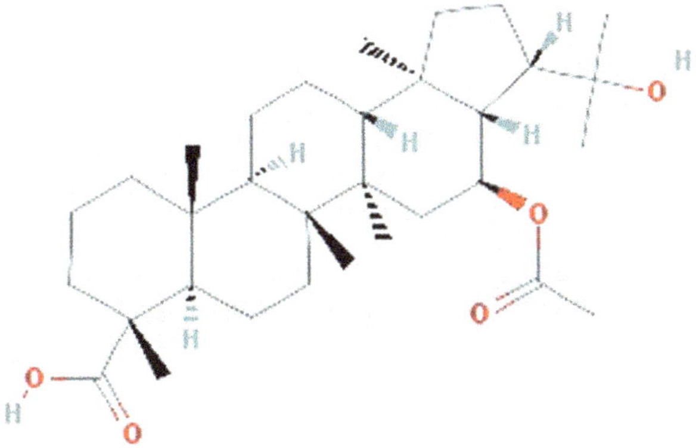

Fig. 12.2 Structure of 16-*O*-acetly-leucotylic acid

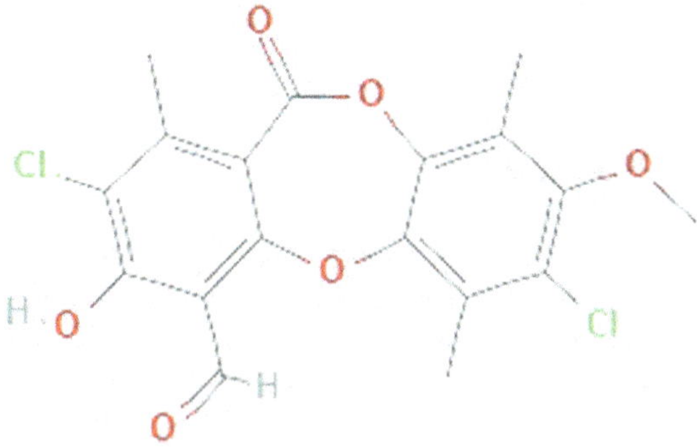

Fig. 12.3 Structure of argopsin

experiment, same compound from same lichen showed a stronger cytotoxicity in comparison to reference material colchicine against lymphocytes cell culture (Correche et al. 2002). Hidalgo et al. (1994) reported the antioxidant activity (AA) of Argopsin isolated from *Erioderma chilense* Mont., which inhibited rat brain homogenate auto-oxidation at concentration of 0.58 µM (AA 27) and β-carotene oxidation at 0.8 µM (AA 23). Fernández et al. (1998) reported the photoprotection properties of argopsin which inhibited photobinding to 8-Methoxypsoralen –human serum albumin (HAS) by 31.7% at a concentration of 10 µM and irradiated at 360 nm.

12.5.4 Alectosarmentin

Substance class: Dibenzofurans, **Molecular formula:** $C_{15}H_{10}O_6$.

Occurrence: *Alectoria sarmentosa* (Ach.) Ach., *Cladonia strepsilis* (Ach.) Grognot (Elix 2014) etc.

Alectosarmentin is a novel lichen metabolite isolated from *A. sarmentosa* by Gollapudi et al. (1994), which exhibited good antimicrobial activity against *S. aureus, M. smegmatis,* and *C. albicans* (Fig. 12.4).

Fig. 12.4 Structure of alectosarmentin

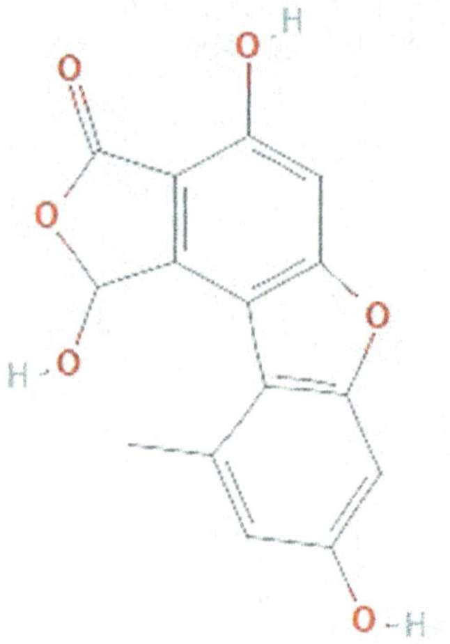

12.5.5 Atranorin

Substance class: β-Orcinol depsides, **Molecular formula:** $C_{19}H_{18}O_8$.

Occurrence: Very common in most of the lichens (Elix 2014).

Atranorin isolated from *Cladonia foliacea* (Huds.) Willd. and *Physcia aipolia* (Ehrh.) Fürnr. demonstrated a strong antimicrobial activity against pathogenic bacteria (Yılmaz et al. 2004; Ranković et al. 2008). Whereas, atranorin extracted from *Parmotrema dilatatum* (Vain.) Hale and *Parmotrema tinctorum* (Despr.) Hale exhibited weak inhibitory activity against *Mycobacterium tuberculosis* (Honda et al. 2010). Similarly, atranorin extracted from *Lepraria lobificans* Nyl. was found to be a weak antimicrobial agent (Kokubun et al. 2007). Commercially available atranorin exhibited effective anticancer activity, but at higher concentration (200 μm) it is capable of inducing a massive loss in mitochondrial membrane potential, along with caspase-3 activation in cell line HT-29 (human colon adenocarcinoma) and phosphatidylserine externalization in cell lines A2780 (human ovarian carcinoma) and HT-29 (Bačkorová et al. 2012). Atranorin isolated from *Bacidia stipata* I.M. Lamb showed a lower activity inhibiting the prostate androgen-sensitive (LNCaP) and androgen-insensitive (DU-145) human prostate cancer cells only at high concentrations (25 and 50 μM) (Russo 2012). Commercially available atranorin showed lowest treatment level of 50 μM as effective against cancer cell lines HL-60 cells 24 h after administration (Bačkorová et al. 2011). Atranorin isolated from continental Antarctic lichens showed poor, but concentration and time-dependent cytotoxicity in terms of intracellular lactate dehydrogenase release in rat hepatocytes. The IC_{50} was 111 μg/mL after incubation of 8 h and 82 μg/mL after 24 h (Correché et al. 2004). Hidalgo et al. (1994) reported the antioxidant activity of atranorin isolated from *Placopsis* sp., which inhibited rat brain homogenate auto-oxidation at a concentration of 5 μM (AA 7.3) and β-carotene oxidation at 0.84 μM (AA 6.5). Fernández et al. (1998) reported the photoprotection properties of

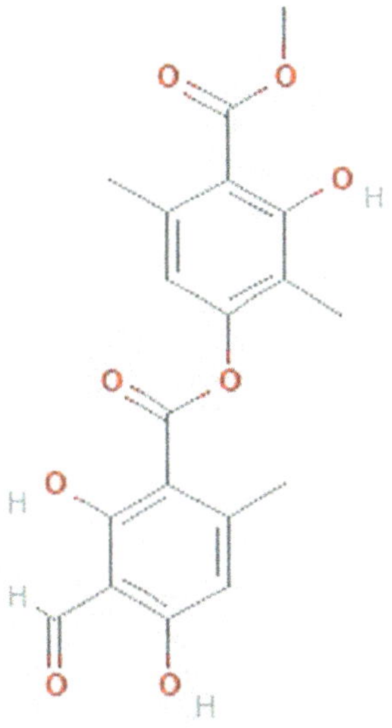

Fig. 12.5 Structure of atranorin

atranorin, which inhibited photobinding to 8-Methoxypsoralen–human serum albumin (HAS) by 20.1% at concentration 10 µM and irradiated at 360 nm (Fig. 12.5).

12.5.6 Barbatic Acid

Substance class: β-Orcinol depsides, **Molecular formula:** $C_{19}H_{20}O_7$.

Occurrence: Widespread in lichens species of *Cladia, Cladonia, Usnea* etc.

Martins et al. (2010) reported barbatic acid in the extract of *Cladia aggregata* (Sw.) Nyl. and proved that the crude or purified form of the same inhibited the growth of four multi-strain resistant strains of *S. aureus* (Fig. 12.6).

12.5.7 Epiphorellic Acid-1

Substance class: Diphenyl ethers, **Molecular formula:** $C_{26}H_{34}O_8$.

Occurrence: *Coelopogon abraxas* Brusse, *C. epiphorellus* (Nyl.) Brusse and Kärnefelt (Elix 2014) etc.

Epiphorellic acid-1 isolated from *C. epiphorellus* showed significant ($P < 0.001$) inhibitory effect on human prostate carcinoma DU-145 cells at the concentration (6–50 µmol/l) that is nontoxic to normal human prostatic epithelial cells (Russo 2006) (Fig. 12.7).

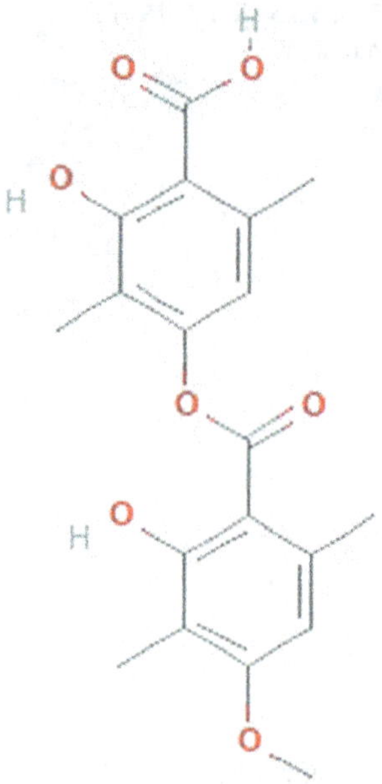

Fig. 12.6 Structure of barbatic acid

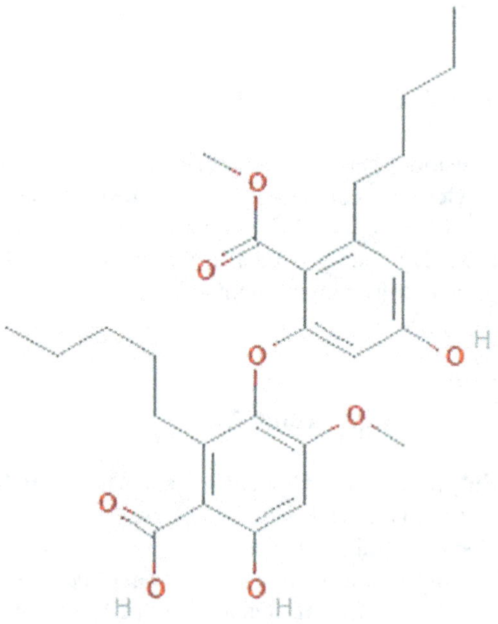

Fig. 12.7 Structure of epiphorellic acid-1

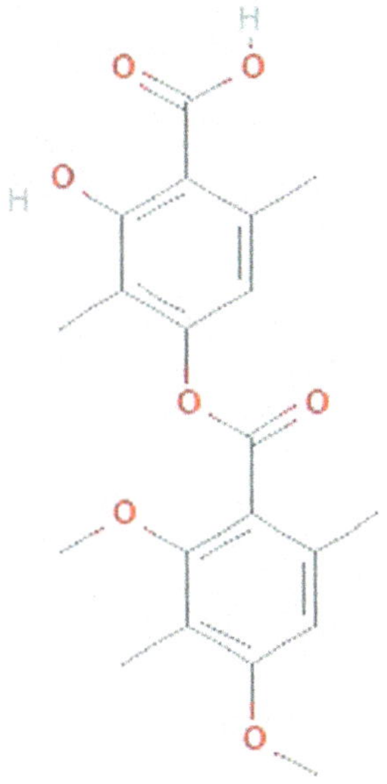

Fig. 12.8 Structure of diffractaic acid

12.5.8 Diffractaic Acid

Substance class: β-Orcinol depsides, **Molecular formula:** $C_{20}H_{22}O_7$.

Occurrence: *Cladia muelleri* (Hampe) Parnmen, Elix and Lumbsch, *Usnea subcavata* Motyka, *Dolichousnea diffracta* (Vain.) Articus (Elix 2014) etc.

Diffractaic acid isolated from *U. subcavata* Motyka was found to be the most active compound with MIC value 15.6 µg/mL, 41.7 µM against tuberculosis bacteria *M. tuberculosis* (Honda et al. 2010). Diffractaic acid isolated from *Protousnea magellanica* (Mont.) Krog showed a good antiproliferative activity against HCT-116 (colon carcinoma) cells from 25 µM, with IC50 value of 42.2 µM, whereas a reduction of viability in MCF-7 (breast adenocarcinoma) and HeLa (cervix adenocarcinoma) cells was observed at concentrations higher than 50 µM (Brisdelli et al. 2012). Diffractaic acid isolated from *P. magellanica* showed a lower activity inhibiting the prostate androgen-sensitive (LNCaP) and androgen-insensitive DU-145 (human prostate) cancer cells only at high concentrations (25 and 50 µM) (Russo 2012). Diffractaic acid isolated from continental Antarctic lichens showed poor, concentration and time-dependent cytotoxicity in terms of intracellular lactate dehydrogenase release in rat hepatocytes. The IC_{50} was 149 µg/mL after incubation of 8 h and 119 µg/mL after 24 h (Correché et al. 2004). Same acid isolated from parmelioid lichens reported as potent antiproliferative agent against leukotrienes-mediated inflammation with IC_{50} value of 2.6 µM (Kumar and Müller 1999). Diffractaic acid isolated from *D. diffracta* (Vain.) Articus showed analgesic effect in mice in vitro (Okuyama 1995) (Fig. 12.8).

12.5.9 Divaric Acid (2,4-Dihydoxy-6-Propylbezoic Acid; Olivetol Carboxylic Acid)

Substance class: Monocyclic aromatic compounds, **Molecular formula:** $C_{10}H_{12}O_4$.

Occurrence: *Cladonia macaronesica* Ahti (Elix 2014), *Evernia divaricata* (L.) Ach and other lichens.

Divaric acid is quite a rare secondary metabolite in lichens. Yuan et al. (2010) for the first time reported the antibacterial properties of Divaric acid from *E. divaricata* (L.) Ach., which showed very potent inhibitory activity against bacteria *B. subtilis*, *S. aureus*, *E. coli*, and *P. aeruginosa*. Divaricatic acid isolated from *Protousnea malacea* (Stirt.) Krog showed a lower activity inhibiting the prostate androgen-sensitive (LNCaP) and androgen-insensitive (DU-145) human prostate cancer cells only at high concentrations (25 and 50 µM) (Russo 2012) (Fig. 12.9).

12.5.10 Divaricatic Acid

Substance class: Orcinol depsides, **Molecular formula:** $C_{21}H_{24}O_7$.

Occurrence: *Canoparmelia texana* (Tuck.) Elix and Hale, *Evernia divaricata* (L.) Ach. (Elix 2014) etc.

Divaricatic acid isolated from continental Antarctic lichens showed moderate, but concentration and time-dependent cytotoxicity in terms of intracellular lactate dehydrogenase release in rat hepatocytes. The IC_{50} was ≥ 50 µg/mL after incubation of 8 h and 32 µg/mL after 24 h (Correché et al. 2004). Hidalgo et al. (1994) reported the antioxidant activity of divaricatic acid isolated from *P. malacea*, which inhibited rat brain homogenate auto-oxidation at concentration of 5 µM (10) and β-carotene oxidation concentration of 0.8 µM (AA 8.6) (Fig. 12.10).

12.5.11 Evernic Acid

Substance class: Orcinol depsides, **Molecular formula:** $C_{17}H_{16}O_7$.

Fig. 12.9 Structure of divaric acid

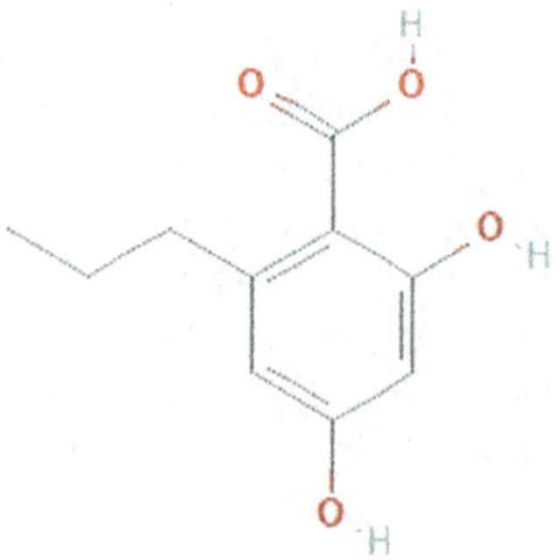

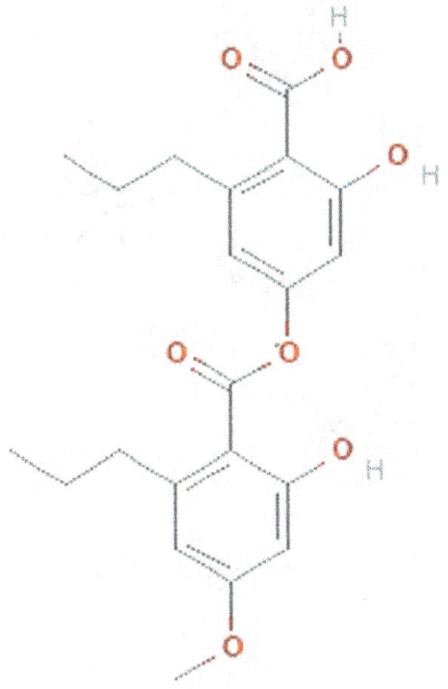

Fig. 12.10 Structure of divaricatic acid

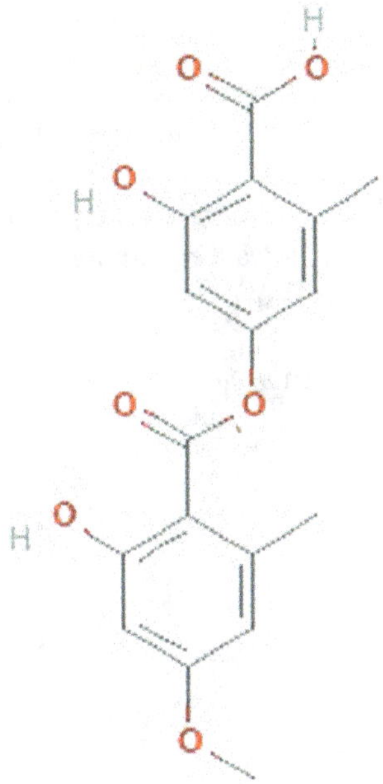

Fig. 12.11 Structure of evernic acid

Occurrence: *Evernia prunastri* (L.) Ach.

Evernic acid extracted from the lichen *E. prunastri* exhibited weak inhibitory action against multidrug-resistant bacteria *S. aureus* (Kokubun et al. 2007) (Fig. 12.11).

12.5.12 Fumarprotocetraric Acid

Substance class: β-Orcinol depsidones, **Molecular formula:** $C_{22}H_{16}O_{12}$.

Occurrence: *Cladonia phyllophora* Ehrh. *C. foliacea,* and *C. furcata* (Elix 2014) and other lichens.

Fumarprotocetraric acid isolated from *C. foliacea* and *C. furcata* demonstrated potential antimicrobial activity against several bacterial strains (Ranković and Misić 2008; Yılmaz et al. 2004), whereas fumarprotocetraric acid isolated from continental Antarctic lichens showed poor, but concentration and time-dependent cytotoxicity in terms of intracellular lactate dehydrogenase release in rat hepatocytes. The IC_{50} was >150 µg/mL after incubation of 8 h and > 150 µg/mL after 24 h (Correché et al. 2004) (Fig. 12.12).

12.5.13 Gyrophoric Acid

Substance class: Orcinol tridepsides, **Molecular formula:** $C_{24}H_{20}O_{10}$.

Occurrence: *Punctelia borreri* (Turner) Krog, *Umbilicaria polyphylla* (L.) Baumg. and *Xanthoparmelia pokornyi* (Körb.) O. Blanco, A. Crespo, Elix, D. Hawksw. and Lumbsch, (Elix 2014), *Umbilicaria hirsuta* (Sw.) Ach and other lichens.

Gyrophoric acid extracted from *U. polyphylla* (L.) Baumg. and *X. pokornyi* (Körb.) O. Blanco, A. Crespo, Elix, D. Hawksw. and Lumbsch demonstrated a strong antimicrobial activity against pathogenic bacteria (Ranković et al 2008; Candan et al. 2006). Gyrophoric acid isolated from *U. hirsuta* (Sw.) Ach. exhibited good anticancer activity against cell lines A2780 and HT-29 at higher concentration (200 µM), however, its effect is much lesser than the usnic acid and atranorin

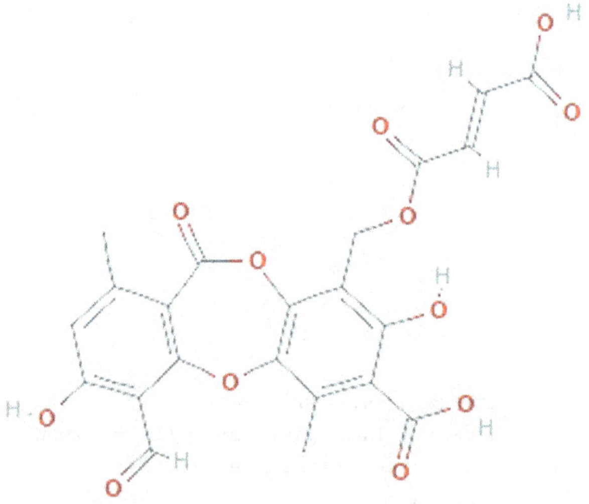

Fig. 12.12 Structure of fumarprotocetraric acid

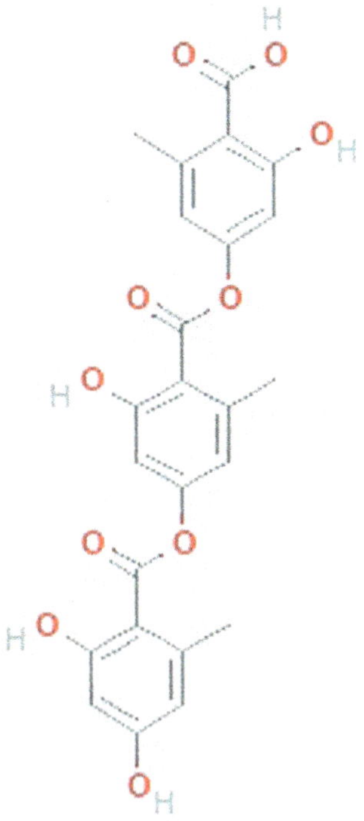

Fig. 12.13 Structure of gyrophoric acid

(Bačkorová et al. 2012). Though it is ineffective at lowest concentrations, concentrations at 100 µM gyrophoric acid induced a strong effect in HL-60 cells as quickly as only 24 h after incubation (Bačkorová et al. 2011). In another experiment, gyrophoric acid isolated from *Lasallia pustulata* (L.) Mérat in combination with usnic acid showed strong wound closure effects on HaCaT cells (Burlando et al. 2009). Gyrophoric acid isolated from continental Antarctic lichens showed poor, but concentration and time-dependent cytotoxicity in terms of intracellular lactate dehydrogenase release in rat hepatocytes. The IC_{50} was >150 µg/mL after incubation of 8 h and 61 µg/mL after 24 h (Correché et al. 2004). Gyrophoric acid isolated from parmelioid lichens was reported as potent antiproliferative agent against leukotrienes-mediated inflammation with IC_{50} value of 1.7 µM (Kumar and Müller, 1999) (Fig. 12.13).

12.5.14 Hypostictic Acid

Substance class: β-Orcinol depsidones, **Molecular formula:** $C_{19}H_{16}O_8$.

Occurrence: *Xanthoparmelia quintaria* (Hale) Hale (Elix 2014) and other lichens.

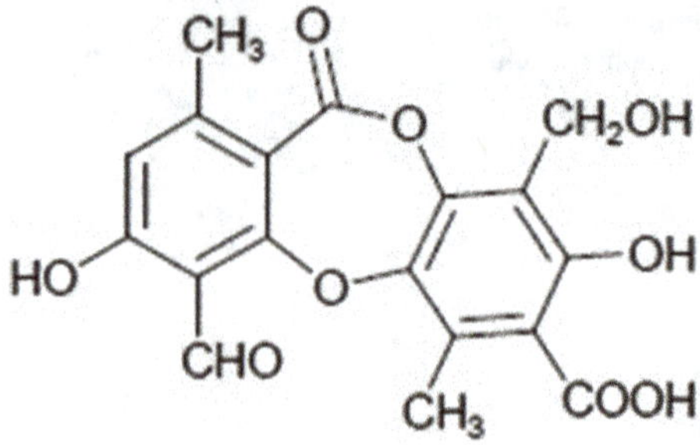

Fig. 12.14 Structure of hypostictic acid

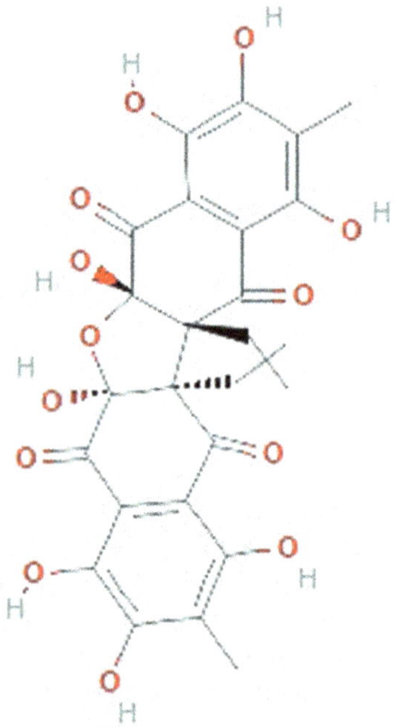

Fig. 12.15 Structure of hybocarpone

Hypostictic acid (Fig. 12.14) extracted from *Psuedoparmelia sphaerospora* (Nyl.) Hale exhibited middle range of antimicrobial activity against *M. tuberculosis* (MIC value 94.0 µg/ml, 251 µM) (Honda et al. 2010).

12.5.15 Hybocarpone

Substance class: Naphthaquinones, **Molecular formula:** $C_{26}H_{24}O_{13}$.

Occurrence: *Heterodermia hybocarponica* Elix, *Lecanora conizaeoides* Nyl. (Elix 2014) and other lichens.

Hybocarpone (Fig. 12.15) isolated from *Lecanora conizaeoides* Nyl. exhibited the strongest activity of all strains of *S. aureus* tested with MIC ranging from 4 to 8 µ/mL (813–16.3 µM). This is an important finding as it suggests that these compounds are opaque to the efflux mechanisms expressed by the strains *S. aureus* tested (Kokubun et al. 2007).

12.5.16 Lecanoric Acid

Substance class: Orcinol depsides, **Molecular formula:** $C_{16}H_{14}O_7$.

Occurrence: *Parmotrema tinctorum* (Despr.) Hale (Elix 2014) and other lichens.

Lecanoric acid (Fig. 12.16) isolated from *Ochrolechia androgyan* (Hoffm.) Arnold showed relatively strong antimicrobial activity against several bacterial strains (Ranković and Misić 2008). Cytotoxicity assay was carried out in in vitro for lecanoric acid isolated from *P. tinctorum* and its orsellinates derivatives obtained by structural modification with sulforhodamine B (SRB) using HEp-2 larynx carcinoma, MCF7 breast carcinoma, 786-0 kidney carcinoma, and B16-F10 murine melanoma cell lines, in addition to a normal (Vero) cell line. It is found that *n*-butyl orsellinate was the most active compound, with IC_{50} values ranging from 7.2 to 14.0 μg/mL, against all the cell lines tested. The compound was more active (IC_{50} = 11.4 μg/mL) against B16-F10 cells. However, lecanoric acid and methyl orsellinate were less active against all cell lines, having an IC_{50} value higher than 50 μg/mL (Bogo et al. 2010).

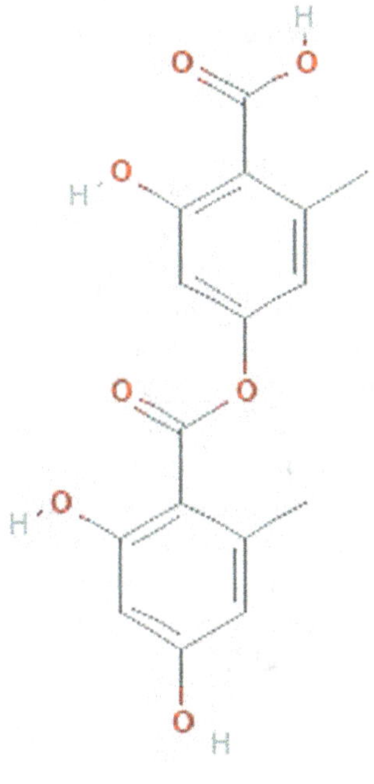

Fig. 12.16 Structure of lecanoric acid

Fig. 12.17 Structure of lobaric acid

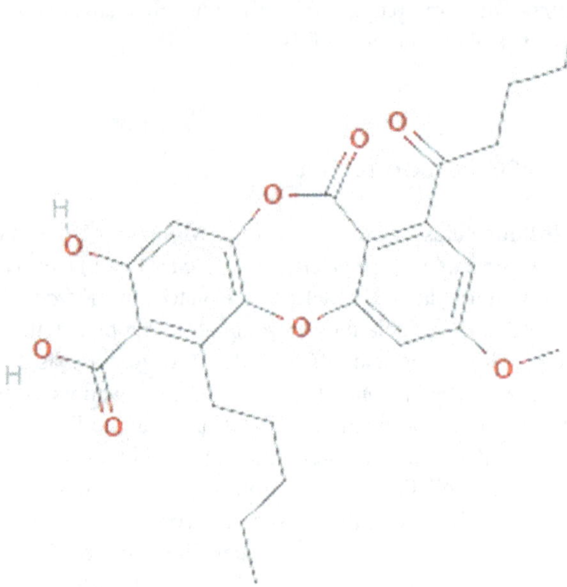

12.5.17 Lobaric Acid

Substance class: Orcinol depsidones, **Molecular formula:** $C_{25}H_{28}O_8$.
Occurrence: *Protoparmelia badia* (Hoffm.) Hafellner (Elix 2014).

Lobaric acid extracted from *Sterocaulon dactylophyllum* Flörke found to be an encouraging metabolite which displayed MIC of 8 µg/mL (17.5 µM) against *S. aureus* (Kokubun et al. 2007). Lobaric acid isolated from *Streocaulong alpinum* Laurer ex Funck exhibited antiproliferative activity at the higher concentrations only on HeLa (cervix adenocarcinoma) and HCT-116 cells (colon carcinoma) (Brisdelli et al. 2012). Lobaric acid isolated from continental Antarctic lichens showed moderate, but concentration and time-dependent cytotoxicity in terms of intracellular lactate dehydrogenase release in rat hepatocytes. The IC_{50} was >100 µg/mL, after incubation of 8 h and 43 µg/mL after 24 h (Correché et al. 2004). Lobaric acid isolated from *S. alpinum* caused a significant reduction in DNA synthesis as measured by thymidine uptake, in three malignant cell lines (T-47D and ZR-75-1—breast carcinoma; K-562—erythroleukemia), the dose inducing 50% of maximum inhibition (ED_{50}) was between 14.5 and 44.7 µg/mL. The proliferative response of mitogen-stimulated lymphocytes was inhibited with mean ED_{50} of 24.5 µg/mL. Significant cell death occurred in the three malignant cell lines at concentrations above 30 µg/mL and up to 38% cell death was observed at 15 mg/mL lobaric acid in mitogen-stimulated lymphocytes (Ögmundsdóttir et al. 1998) (Fig. 12.17).

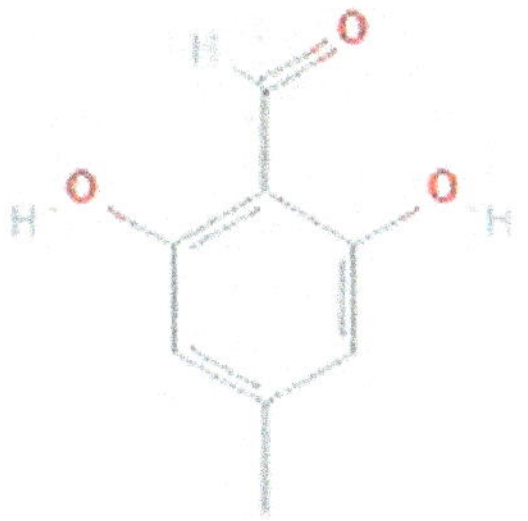

Fig. 12.18 Structure of methyl β-orsellinate (atranol)

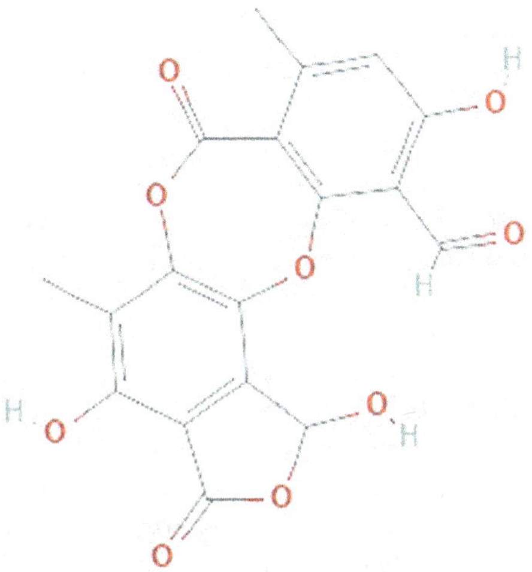

Fig. 12.19 Structure of norstictic acid

12.5.18 Methyl β-Orsellinate (Methyl β-Orcinolcarboxylate; Atranol)

Substance class: Monocyclic aromatic derivatives, **Molecular formula:** $C_8H_8O_3$.

Occurrence: *E. prunastri* (Elix 2014), *H. cirrhata, P. furfuracea,* and other lichens.

Methyl β-orsellinate isolated from *H. cirrhata* found to be excellent antimicrobial agent against azole-resistant strains of *C. albicans* and *S. cerevisiae* in the range of 10–400 µg/mL. Further, it also exhibited anticancer activity against liver, colon, ova, or mouth (oral) cancer cells of humans in the range 1–10 µg/mL (Khanuja et al. 2007). Earlier, atranol was found to be a major bioactive molecule in the crude extract of *P. furfuracea* which was highly active against bacteria like *B. subtilis, E. coli, P. digiratum,* and *S. cerevisiae* (Caccamese et al. 1985) (Fig. 12.18).

12.5.19 Norstictic Acid

Substance class: β-Orcinol depsidones, **Molecular formula:** $C_{18}H_{12}O_9$.

Occurrence: *Xanthoparmelia substrigosa* (Hale) Hale (Elix 2014) and other lichens.

Norstictic acid extracted from *Ramalina* sp. showed significant inhibition of growth of *M. tuberculosis* with MIC value of value 62.5 µg/mL, 168 µM) (Honda et al. 2010) (Fig. 12.19).

12.5.20 Olivetoric Acid

Substance class: Orcinol depsides, **Molecular formula:** $C_{26}H_{32}O_8$.
Occurrence: *Cetrelia olivetorum* (Elix 2014), *P. furfuracea* etc.

Olivetoric acid isolated from *P. furfuracea* displayed dose-dependent antiangiogenic activities, inhibited cell proliferation, and disrupted endothelial tube formation in adipose tissue. Olivetoric acid also inhibited the formation of actin stress fibers in a dose-dependent manner, which may be due to the decrease in tube formation (Koparal and Ulus 2010) (Fig. 12.20).

12.5.21 Pannarin

Substance class: β-Orcinol depsidones, **Molecular formula:** $C_{18}H_{15}ClO_6$.
Occurrence: *Pannaria conoplea* (Ach.) Bory (Elix 2014), *Psoroma pallidum* Nyl, *Psoroma* spp. etc.

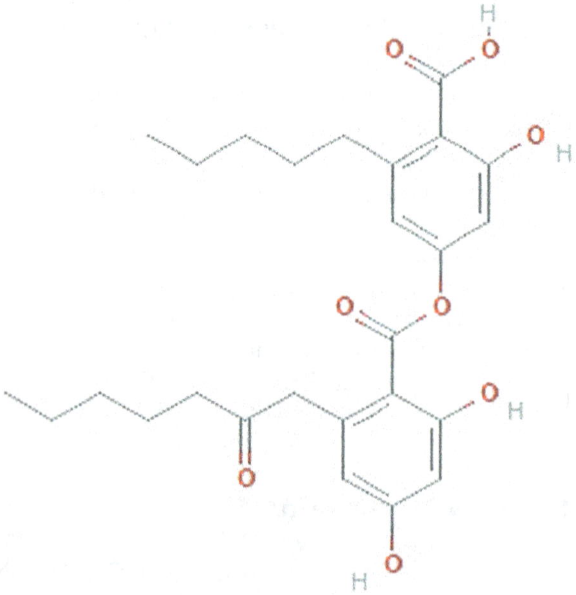

Fig. 12.20 Structure of olivertoric acid

Fig. 12.21 Structure of pannarin

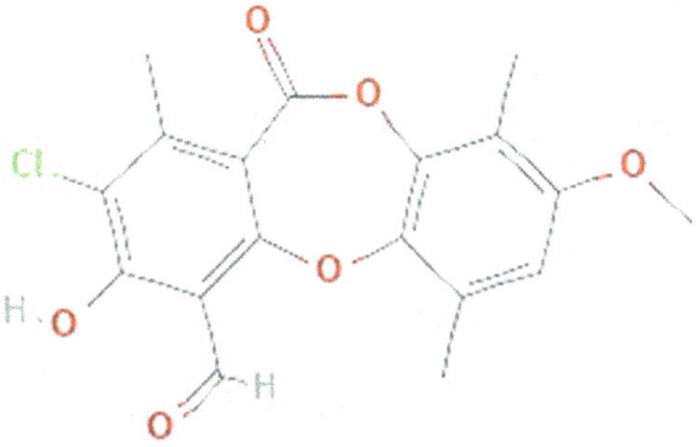

Fig. 12.22 Structure of parietin

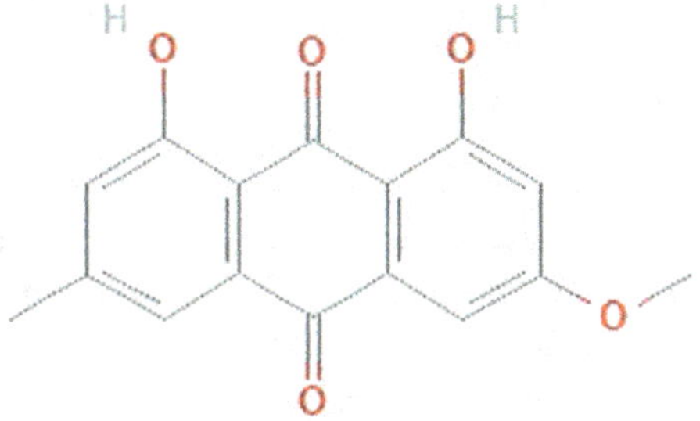

Pannarin isolated from *Psoroma* spp. showed a significant inhibitory effect on M14 (human melanoma) cells at a concentration of 12.5–50 µM. It also induced apoptotic cell death substantiated by DNA fragmentation and increased caspase-3 activity. It also inhibited superoxide anion formation (Russo 2008). In another experiment, pannarin isolated from same lichen at concentration 6–50 µmol/L, which is nontoxic to normal human prostatic epithelial cells showed significant ($P < 0.001$) inhibitory effect on human prostate carcinoma DU-145 cells (Russo 2006). Pannarin isolated from continental Antarctic lichens showed moderate, but concentration and time-dependent cytotoxicity in terms of intracellular lactate dehydrogenase release in rat hepatocytes. The IC_{50} was 86 µg/mL after incubation of 8 h and 12 µg/mL after 24 h (Correché et al. 2004). In another experiment, same compound from same lichen showed a stronger cytotoxicity in comparison to reference material colchicine against lymphocytes cell culture (Correché et al. 2002). Hidalgo et al. (1994) reported the antioxidant activity of pannarin isolated from *P. pallidum* Nyl., which inhibited rat brain homogenate auto-oxidation at concentration of 0.57 µM (AA 13) and β-carotene oxidation at 0.88 µM (AA 23). Fernández et al. (1998) reported the photo protection properties of pannarin which inhibited photobinding to 8-Methoxypsoralen–human serum albumin (HAS) by 40.4% at concentrations of 10 µM when irradiated at 360 nm (Fig. 12.21).

12.5.22 Parietin (Physcoin)

Substance class: Anthraquinones, **Molecular formula:** $C_{16}H_{12}O_5$.

Occurrence: *Xanthoria parietina* (L.) Th. Fr. (Elix 2014).

Parietin isolated from *X. parietina* proved as a significant anticancer agent in some cell lines (A2780, Jurkat, or HT-29) at concentrations of 50 µM (Bačkorová et al. 2011) (Fig. 12.22).

12.5.23 Physodic Acid

Substance class: Orcinol depsidones, **Molecular formula:** $C_{26}H_{30}O_8$.

Occurrence: *H. physodes* (L.) Nyl. (Elix 2014) and other lichens.

Physodic acid isolated from *H. physodes* is a weak (in comparison to usnic acid and streptomycin) antimicrobial agent inhibited the tested microorganisms but at high concentrations (1 µg/mL) (Ranković et al 2008). Türk et al. (2006) also showed similar results with physodic acid from *P. furfuracea*. However, Physodic acid isolated by Kokubun et al. (2007) from *H. physodes* showed good antibacterial activity against multidrug-resistant bacteria *S. aureus* with MIC of 32 µg/mL (68.0 µM). Similarly, physodic acid isolated from *A. sarmentosa* showed fairly good antimicrobial activity against *S. aureus, M. smegmatis,* and *C. albicans* (Gollapudi et al. 1994) (Figs. 12.22 and 12.23).

12.5.24 Protocetraric Acid

Substance class: β-Orcinol depsidones, **Molecular formula:** $C_{18}H_{14}O_9$.

Fig. 12.23 Structure of physodic acid

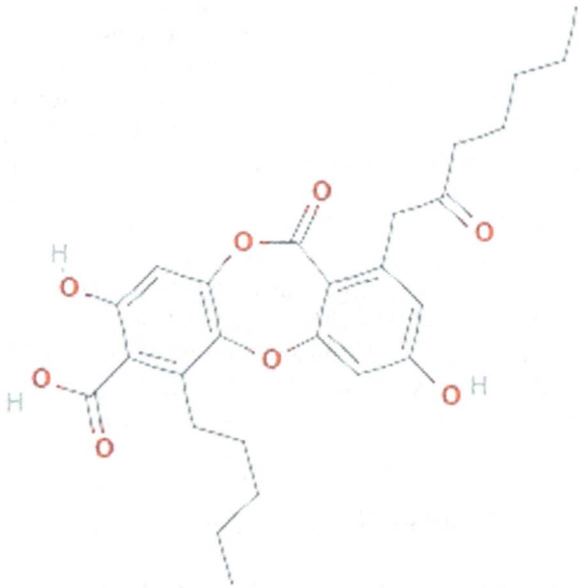

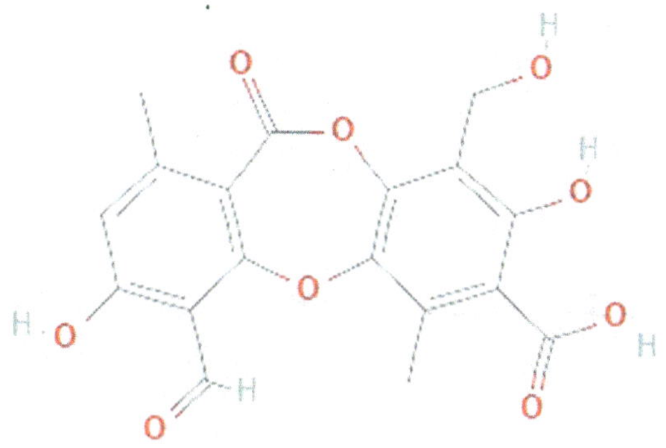

Fig. 12.24 Structure of protocetraric acid

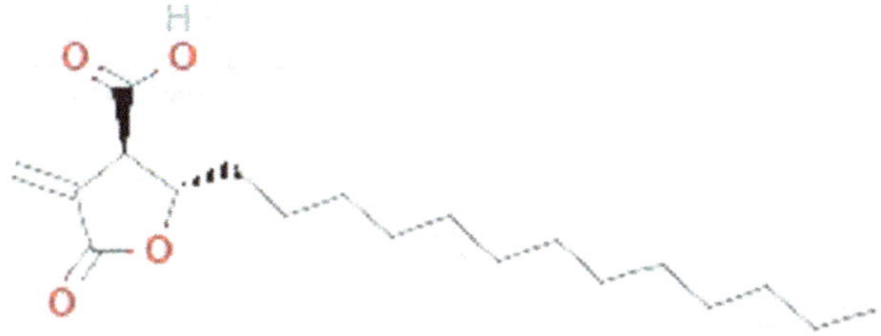

Fig. 12.25 Structure of protolichesterinic acid

Occurrence: *Flavoparmelia caperata* (Elix 2014).

Procetraric acid extracted from *P. dilatatum* showed middle range of antimicrobial activity against *M. tuberculosis* (MIC value 125 µg/mL, 334 µM) (Honda et al. 2010). Protocetraric acid isolated from *F. caperata* showed strong antimicrobial activity against several bacterial strains (Ranković and Misić 2008) (Fig. 12.24).

12.5.25 Protolichesterinic Acid

Substance class: Aliphatic acids, **Molecular formula:** $C_{19}H_{32}O_4$.

Occurrence: *C. islandica* (Elix 2014).

Protolichesterinic acid isolated from *Cornicularia aculeata* (Schreb.) Ach. exhibited as stronger cytotoxic effect against three human cancer cell lines, MCF-7 (breast adenocarcinoma), HeLa (cervix adenocarcinoma), and HCT-116 (colon carcinoma). The cell survival dropped significantly after treatment with concentrations higher than 25 µM (Brisdelli et al. 2012). Protolichesterinic acid isolated from *R. melanophthalma* (DC.) Leuckert showed dose-dependent relationship in the range of 6.25–50 µM concentrations and inhibited activity of androgen-sensitive (LNCaP) and androgen-insensitive (DU-145) human prostate cancer cells (Russo et al. 2012). Protolichesterinic acid isolated from *C. islandica* caused a significant reduction in DNA synthesis as measured by thymidine uptake, in three malignant cell lines (T-47D and ZR-75-1—breast carcinoma; K-562—erythroleukemia), the dose inducing 50% of maximum inhibition (ED_{50}) was between 1.1 and 24.6 µg/mL. The proliferative response of mitogen-stimulated lymphocytes was inhibited with mean ED_{50} of 8.4 µg/mL. Significant cell death occurred in the three malignant cell-lines at concentrations above 20 µg/mL and up to 38% cell death was observed at 20 µg/mL lobaric acid in mitogen-stimulated lymphocytes (Ögmundsdóttir et al. 1998) (Fig. 12.25).

12.5.26 Psoromic Acid

Substance class: β-Orcinol depsidones, **Molecular formula:** $C_{18}H_{14}O_8$.
Occurrence: *Usnea inermis* Motyka (Elix 2014).

Psoromic acid isolated from continental Antarctic lichens showed moderate, but concentration and time-dependent cytotoxicity in terms of intracellular lactate dehydrogenase release in rat hepatocytes. The IC_{50} was 76 µg/mL after incubation for 8 h and 11 µg/mL after 24 h (Correché et al. 2004) (Fig. 12.26).

Fig. 12.26 Structure of psoromic acid

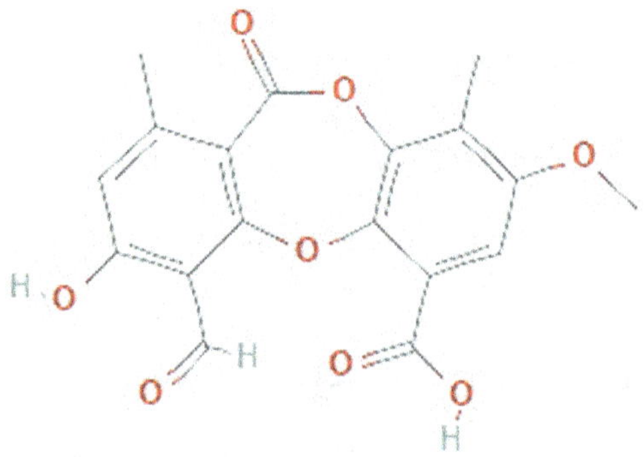

Fig. 12.27 Structure of retigeric acid

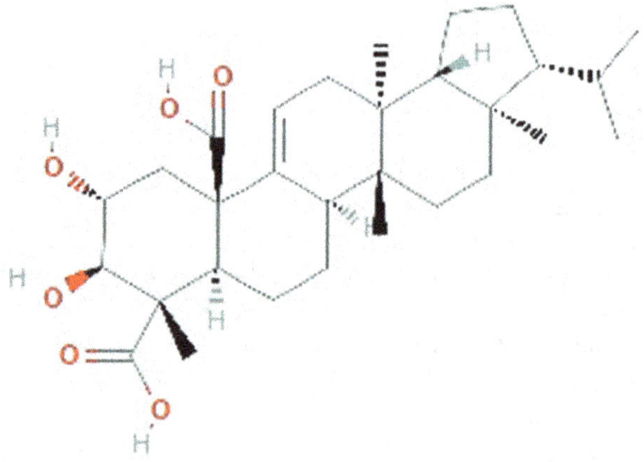

12.5.27 Retigeric Acid B

Substance class: Terpenoids, **Molecular formula:** $C_{30}H_{46}O_6$.

Occurrence: *Lobaria retigera* (Borry) Trevis. (Elix 2014).

Retigeric acid B, a naturally occurring pentacyclic triterpenic acid isolated from *Lobaria kurokawae,* inhibited prostate cancer cell proliferation and induced cell death in a dose-dependent manner, but exerted very little inhibitory effect on noncancerous prostate epithelial cell viability (Liu et al. 2010) (Fig. 12.27).

12.5.28 Rhizocarpic Acid

Substance class: Pulvinic acid derivatives, **Molecular formula:** $C_{28}H_{23}NO_6$.

Occurrence: *Rhizocarpon geographicum* (L.) DC. (Elix 2014) and other lichens.

Rhizocarpic acid extracted from *Psilolechia lucida* (Ach.) M. Choisy was reported to be a promising antimicrobial agent with MICs 32 to 64 μg/mL at 68.2 to 136 μM. (Kokubun et al. 2007) (Fig. 12.28).

12.5.29 Salazinic Acid

Substance class: β-Orcinol depsidones, **Molecular formula:** $C_{18}H_{12}O_{10}$.

Occurrence: *Xanthoparmelia tasmanica* (Hook. f. Taylor) Hale (Elix 2014) and other lichens.

Salazinic acid extracted from *Parmotrema lichexanthonicum* Eliasaro & Adler exhibited weak antimicrobial activity against *M. tuberculosis* (MIC value 4250 μg/mL, 643 μM) (Honda et al. 2010). Salazinic acid isolated from continental Antarctic lichens showed poor, but concentration and time-dependent cytotoxicity in terms of intracellular lactate dehydrogenase release in rat hepatocytes. The IC_{50} was 154 μg/

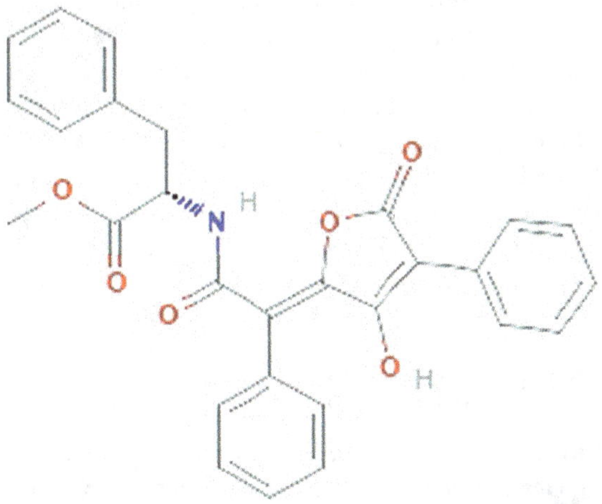

Fig. 12.28 Structure of rhizocarpic acid

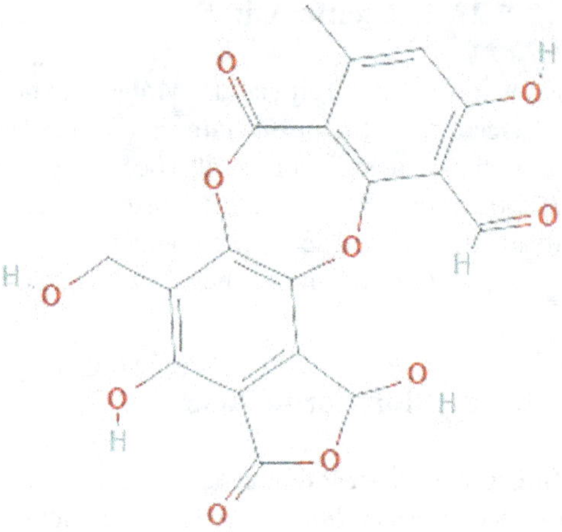

Fig. 12.29 Structure of salazinic acid

mL after incubation of 8 h and 18 µg/mL after 24 h (Correché et al. 2004) (Fig. 12.29).

12.5.30 Sphaerophorin

Substance class: Orcinol depsides, **Molecular formula:** $C_{23}H_{28}O_7$.
Occurrence: *Sphaerophorus fragilis* (L.) Pers. (Elix 2014) and other lichens.

Sphaerophorin isolated from *S. globosus* (Huds.) Vain. showed a significant inhibitory effect on M14 (human melanoma) cells at concentration 12.5–50 µM. It also induced apoptotic cell death substantiated by DNA fragmentation and increased caspase-3 activity. It also inhibited superoxide anion formation (Russo 2008). In another experiment, phaerophorin isolated from same lichen showed at concentrations ranging from 6 to 50 µmol/L that is nontoxic to normal human prostatic epithelial cells showed significant ($P < 0.001$) inhibitory effect on human prostate carcinoma DU-145 cells (Russo 2006). Sphaerophorin isolated from continental Antarctic lichens showed excellent concentration and time-dependent cytotoxicity in terms of intracellular lactate dehydrogenase release in rat hepatocytes. The IC_{50} was 30 µg/mL after incubation of 8 h and 3 µg/mL after 24 h (Correché et al. 2004). In another experiment, same compound from same lichen showed a stronger cytotoxicity in comparison to reference material colchicine against lymphocytes cell culture (Correché et al. 2002) (Fig. 12.30).

12.5.31 Stictic Acid

Substance class: β-Orcinol depsidones, **Molecular formula:** $C_{19}H_{14}O_9$.
Occurrence: *Xanthoparmelia conspersa* (Ehrh.) Hale (Elix 2014).

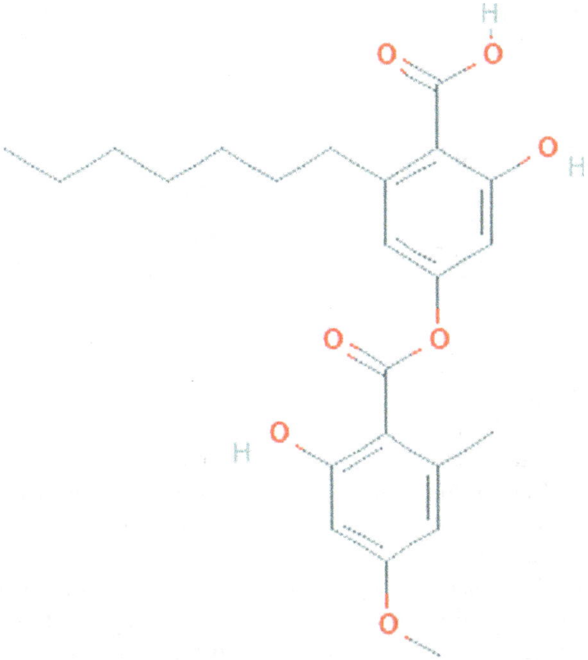

Fig. 12.30 Structure of sphaerophorin

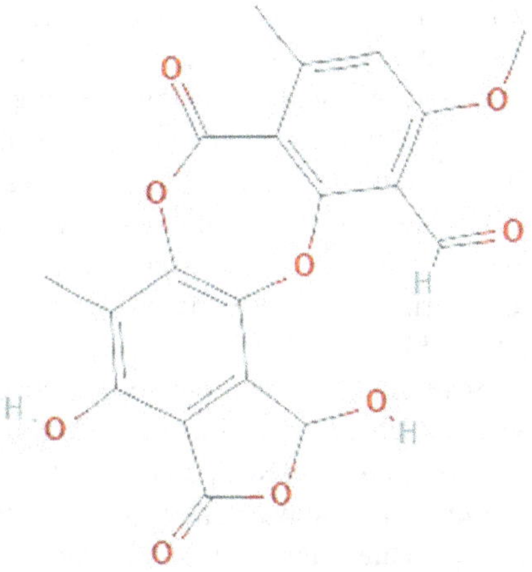

Fig. 12.31 Structure of stictic acid

Stictic acid isolated from *X. conspersa* showed weak (in comparison to usnic acid and streptomycin) antimicrobial activity against several strains of bacteria (Ranković and Misić 2008). Stictic acid isolated from continental Antarctic lichens showed moderate, but concentration and time-dependent cytotoxicity in terms of intracellular lactate dehydrogenase release in rat hepatocytes. The IC_{50} was 88 µg/mL after incubation of 8 h and 5 µg/mL after 24 h (Correché et al. 2004) (Fig. 12.31).

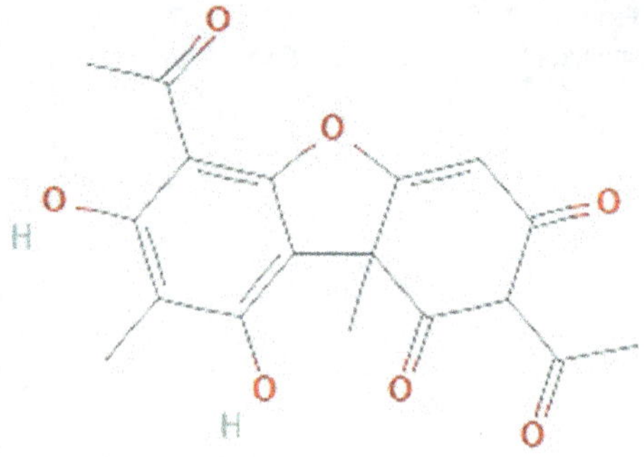

Fig. 12.32 Structure of usnic acid

12.5.32 Usnic Acid

Substance class: Usnic acid derivatives, **Molecular formula:** $C_{18}H_{16}O_7$.
Occurrence: *Usnea* spp. (Elix 2014) and other lichens.

It is one of the common and abundant secondary metabolite found in large amount in lichens such as *Cladonia, Evernia, Ramalina,* and *Usnea*. Lichens produce usnic acid up to 8% of their dry weight of the thalli, but undergo seasonal variation, reaching maximum in late spring and early summer, and low levels in autumn and winter. Also, usnic acid contents depend on geographic locality, insolation, and other ecological conditions (Bjerkea et al. 2005). As mentioned earlier, it was first isolated in 1844 by Knop. It is known to be present in three forms, i.e. (+)-usnic acid, (−)-usnic acid, and isousnic acid (Shibata and Taguchi 1967). Both (+) and (−)-usnic acid are biologically important. Probably it is the only lichen metabolite commercialized as pharmaceutical. It is used in antiseptic creams including "Usno" and "Evosin" and is sometimes believed to be more effective than penicillin salves in the treatment of external wounds and burns. It is also used in the treatment of tuberculosis (Nash 2008) and is a very effective drug used in antifeedant products, mouth rinses and dentifrices (Durazo et al. 2004), and cosmetics (Najdenova et al. 2001) (Fig. 12.32).

Several reviews are already available describing the biological potential of usnic acid (Cocchietto et al. 2002; Ingolfsdottir, 2002; Luzina and Salakhutdinov 2018). A large portion of research on usnic acid mostly deals with antimicrobial (antibacterial, antiviral, fungicidal, antiprotozoal) activities and hence can be employed as antitubercular, antimalarial, and anti-influenza agents. These studies prove beyond doubt that usnic acid is a potential antimicrobial agent. The antibiotic action of usnic acid is due to the inhibition of oxidative phosphorylation, an effect similar to that shown by dinitrophenol (Abo-Khatwa et al. 1996). Usnic acid also acts as a potential antioxidant, insecticidal, larvicidal, analgesic, and anti-inflammatory agent. Usnic acid has also shown anticancer properties (Mayer et al. 2005) and it is antineoplastic (Takai et al. 1979). The antioxidant action of usnic acid is related to the phenolic fragment, responsible for quenching free radicals, while antitumor activity is due to apoptosis (Luzina and Salakhutdinov 2018). In one of the studies, usnic acid was reported to inhibit the osteoclast differentiation (Lee 2015), which can lead to hypogenesis, fracture, osteoporosis, rheumatoid arthritis, periodontal disease, Paget's disease, transformative bone cancer, etc. Usnic acid can inhibit cellular differentiation of follicular cells, thereby inhibiting hair growth (Chao et al. 2015). Commercially available usnic acid exhibited effective anticancer activity, capable of inducing a massive loss in mitochondrial membrane potential, along with caspase-3 activation in cell line HT-29 and phosphatidylserine externalization in cell lines A2780 and HT29 (Bačkorová et al. 2012). Nanoparticles of usnic acid displayed very good water solubility, cell permeability, and bioavailability (Qu et al. 2015). The ecological role of usnic acid can be attributed to screening of excess of light (Rundel 1978) as it is a cortical substance. Therefore, lichens growing on high altitudes could be potential source of usnic acid for light screening product development.

Usnic acid isolated from continental Antarctic lichens showed excellent, but concentration and time-dependent cytotoxicity in terms of intracellular lactate dehydrogenase release in rat hepatocytes. The IC_{50} was 25 µg/mL after incubation of 8 h and 21 µg/mL after 24 h (Correché et al. 2004). (+)-usnic acid isolated from parmelioid lichens reported as potent antiproliferative agent against leukotrienes-mediated inflammation with IC_{50} value of 2.1 µM (Kumar and Müller 1999). Usnic acid also a gastro-protective compound, since it reduces oxidative damage and inhibits neutrophil infiltration in indomethacin-induced gastric ulcers in rats (Odabasoglu et al. 2006). It also exhibited strong larvicidal activity against the third and fourth instar larvae of the house mosquito (*Culex pipiens*), and larval mortality was dose-dependent (Cetin et al. 2008).

Few studies have shown that extensive uses of usnic acid as food supplement caused intoxication, severe hepatotoxicity leading to liver failure, weight loss, and allergy (Guo et al. 2008). These studies warn usage of usnic acid in pharmacological agent. On the other hand, *Cladonia* spp., lichens containing usnic acid are eaten in large quantity by wild animals, such as reindeers, indicating it does not cause toxicity. Further, pharmacokinetic studies in rabbits (Krishna and Venkataramana 1992) have shown that (+)-usnic acid is absorbed very well after oral administration. Thus, if usnic acid is used for therapy, it can be administered orally (Elo et al. 2007). Usnic acid being a wonder molecule for pharmacopeia, there is a need for derivatization to create a new biologically active compound with reduced toxicity (Fig. 12.32).

12.5.33 Variolaric Acid

Substance class: Orcinol depsidones, **Molecular formula:** $C_{16}H_{10}O_7$.

Occurrence: *Ochrolechia parella* (L.) A. Massal. (Elix 2014).

Variolaric acid isolated from continental Antarctic lichens showed moderate, but concentration and time-dependent cytotoxicity in terms of intracellular lactate dehydrogenase release in rat hepatocytes. The IC_{50} was 95 µg/mL after incubation of 8 h and 64 µg/mL after 24 h (Correché et al. 2004) (Fig. 12.33).

Fig. 12.33 Structure of variolaric acid

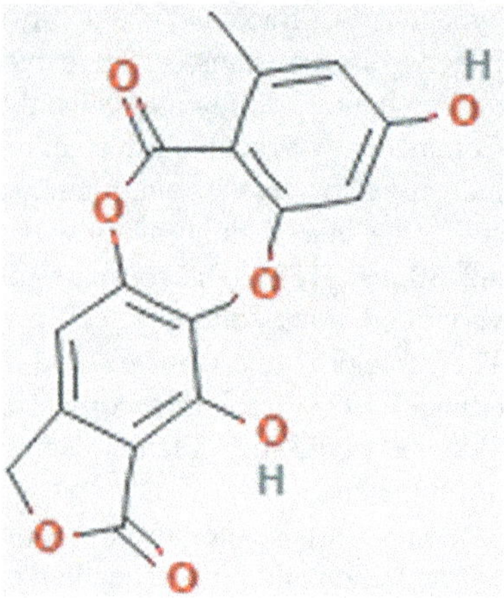

Fig. 12.34 Structure of vicanicin

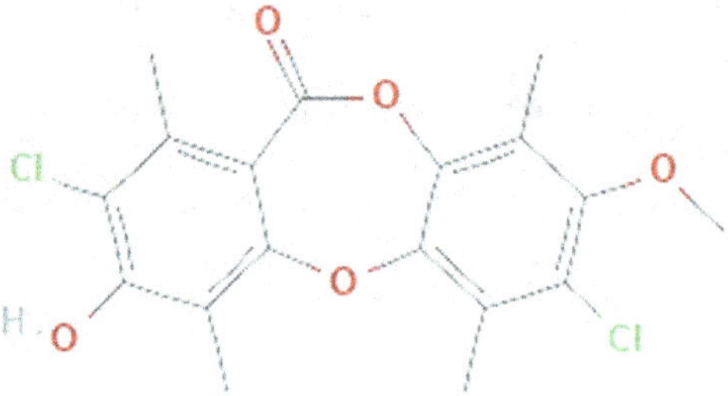

12.5.34 Vicanicin

Substance class: β-Orcinol depsidones, **Molecular formula:** $C_{18}H_{16}Cl_2O_5$.

Occurrence: *Teloschistes flavicans* (Sw.) Norman (Elix 2014) and other lichens.

Vicanicin (Fig. 12.34) isolated from *P. pallidum* Nyl., *P. pulchrum* Malme induced a significant loss of viability in a dose-dependent manner in HeLa (cervix adenocarcinoma) and HCT-116 colon carcinoma) cells with IC_{50} values of 67 μM and 40.5 μM, respectively, but did not show any cytotoxic effect on MCF-7 cells (breast adenocarcinoma) (Brisdelli et al. 2012). Vicanicin isolated from *Psoroma dimorphum* Malme showed a clear dose–response relationship in the range of 6.25–50 μM concentrations and inhibited activity of androgen-sensitive (LNCaP) and androgen-insensitive (DU-145) human prostate cancer cells (Russo 2012). Vicanicin isolated from continental Antarctic lichens showed poor, but concentration and time-dependent cytotoxicity in terms of intracellular lactate dehydrogenase release in rat hepatocytes. The IC_{50} was >150 μg/mL after incubation of 8 h and > 75 μg/mL after 24 h (Correché et al. 2004).

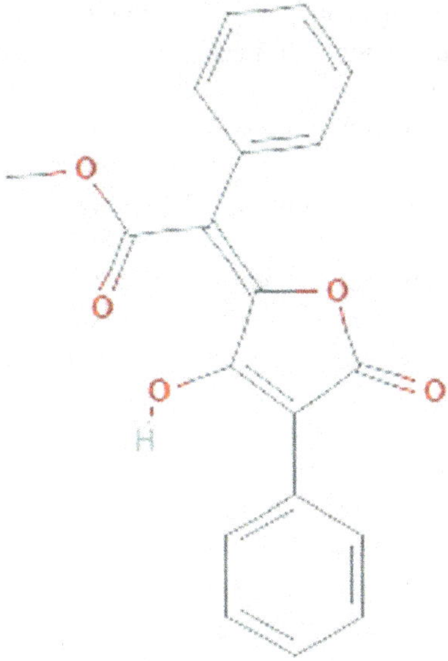

Fig. 12.35 Structure of vulpinic acid

12.5.35 Vulpinic Acid

Substance class: Pulvinic acid derivatives, **Molecular formula:** $C_{19}H_{14}O_5$.
  Occurrence: *Letharia vulpina* (L.) Hue (Elix 2014).

Vulpinic acid (Fig. 12.35) isolated from *L. vulpina* exhibited fairly good antimicrobial activity against several microbes with MIC ranging between 16 and 32 μg/mL, but it is less when compared to usnic acid in the same study (2–16 μg/mL) (Lauterwein et al. 1995). Vulpinic acid isolated from *L. vulpina* induced a stimulation of cell proliferation at lower concentration of (0.1–1.0 μM), however, EC_{50} values were significantly higher. It was more toxic on HaCaT than on tumor cells, except for the Neutral Red assay on A431 (Burlando et al. 2009).

12.6 Conclusions

Lichens are not always beneficial. Woodcutter's eczema among forestry and horticultural workers (and in lichen collectors) is a good example of dermatitis allergy caused due to lichen spores, which contain lichen compounds (Aalto-Korte et al. 2005). Along with usnic acid, Molnár and Farkas (2010) listed 11 other compounds that were reported to be allergic in various literature. Interestingly the list includes atranorin, diffractaic acid, and physodic acid.

It can be noted that among 1000 secondary metabolites known from the lichens relatively few substances have been screened in detail for biological activity and therapeutic potential. Sometimes the activities of crude extracts were indirectly attributed to the secondary metabolites present in them. Further, the exact molecular mechanisms of the action of lichen secondary metabolites are almost entirely unknown. Limited bioactive experimentation in lichens is mostly owing to their slow growth and nonavailability in bulk quantity. Overexploitation of the useful lichen would lead to loss of biodiversity. Sometimes the quantity of the purified compound is insufficient for structural elucidation and pharmacological testing. Therefore, culture of lichens in bioreactors and mycobiont culture seems to be ideal alternative methods. In lichen tissue culture, the mycobiont synthesizes required secondary metabolites only if suitable conditions are provided. These conditions include carbon sources similar to the carbohydrate supplied by the photobiont; precursors of the targeted end products and mimicking the conditions to which lichens are exposed in the wild. However, mycobiont in culture can yield novel products than from their original thallus, which is proven to be bioactive. For example, axenic cultures of *Bunodophoron patagonicum* isolated either from spores or from thallus fragments formed two chemosyndromes (suites of chemotypes) of depsides and dibenzofurans synchronically. *Physconia distorta* grown on nutrient-rich media produced mainly oleic acid, linoleic acid, stearic acid, and their triglyceride derivatives. An area that needs to be explored in future for lichenology research is the whole genome sequencing, identification, and characterization of genes responsible for the synthesis of lichen metabolites. Further prospects include transferring these genes in fast-growing microfungi and producing required metabolite at industrial scale. So far, no such successful microfungi system has been identified for mass production of lichen metabolites and hence there is a lot of scope in this area. Another possible area that needs to be explored is derivatization of lichen compounds. According to a study among all 1562 newly approved drugs during the period 1981–2014, the portion comprised of unaltered natural product was only 4%, whereas portions of natural product derivatives, synthetic drugs with natural product pharmacophores, or mimics of natural products accounted for 21% and 31%, respectively. This fact confirms the high significance of the derivatization of natural compounds as a tool for searching for more active agents or compounds with newly discovered biological properties.

Acknowledgment We thank Director, CSIR-NBRI and KSCSTE-JNTBGRI for providing infrastructural facilities, Dr. Siljo Joseph and members of Lichenology laboratory for their cooperation during the study.

References

Aalto-Korte K, Lauerma A, Alanko K (2005) Occupational allergic contact dermatitis from lichens in present-day Finland. Contact Dermatitis 52:36–38

Abo-Khatwa AN, Al-Robai AA, Al-Jawhari DA (1996) Lichen acids as uncouplers of oxidative phosphorylation of mouse-liver mitochondria. Nat Toxins 4:96–102

Alms I (1832) Über einen neuen Stoff in der Variolaria amara Ach. Ann Pharm 1:61–68

Asahina Y, Shibata S (1954) Chemistry of lichen substances. Japan Society for the Promotion of Science, Tokyo. vi + 240 pp.

Bačkorová M, Bačkor M, Mikeš J, Jendželovský R, Fedoročko P (2011) Variable responses of different human cancer cells to the lichen compounds parietin, atranorin, usnic acid and gyrophoric acid. Toxicol In Vitro 25(1):37–44. https://doi.org/10.1016/j.tiv.2010.09.004

Bačkorová M, Jendželovský R, Kello M, Bačkor M, Mikeš J, Fedoročko P (2012) Lichen secondary metabolites are responsible for induction of apoptosis in HT-29 and A2780 human cancer cell lines. Toxicol In Vitro 26(3):462–468. https://doi.org/10.1016/j.tiv.2012.01.017

Bebert (1831) Sur une nouvelle substance decouverte dans Ie lichen vulpinus. Rapporteurs: Robiquet et Blondeau. J Pharm Sci Access 17:696–700

Bjerkea JW, Elvebakka A, Domínguezb E, Dahlback A (2005) Seasonal trends in usnic acid concentrations of Arctic, alpine and Patagonian populations of the lichen Flavocetraria nivalis. Phytochemistry 66:337–344

Bogo D, de Matos MF, Honda NK, Pontes EC, Oguma PM, da Santos EC, de Carvalho JE, Nomizo A (2010) In vitro antitumour activity of Orsellinates. Z Naturforsch C J Biosci 65(1–2):43–48. https://doi.org/10.1515/znc-2010-1-208

Boustie J, Grube M (2005) Lichens—a promising source of bioactive secondary metabolites. Plant Genet Resour 3(2):273–287. https://doi.org/10.1079/PGR200572

Brisdelli F, Perilli M, Sellitri D, Piovano M, Garbarino JA, Nicoletti M, Celenza G (2012) Cytotoxic activity and antioxidant capacity of purified lichen metabolites: an in vitro study. Phytother Res 27(3):431–437. https://doi.org/10.1002/ptr.4739

Burkholder PR, Evans AW (1945) Further studies on the antibiotic activity of lichens. Bull Torrey Bot Club 72:157–164

Burkholder PR, Evans AW, McVeigh I, Thornton HK (1944) Antibiotic activity of lichens. Proc Natl Acad Sci U S A 30:250–255

Burlando B, Ranzato E, Volante A, Appendino G, Pollastro F, Verotta L (2009) Antiproliferative effects on tumour cells and promotion of keratinocyte wound healing by different lichen compounds. Planta Med 75(06):607–613. https://doi.org/10.1055/s-0029-1185329

Caccamese S, Toscano RM, Bellesia F, Pinetti A (1985) Methyl-b-orcinolcarboxylate and depsides from *Parmelia furfuracea*. J Nat Prod 48:157–158

Candan M, Yılmaz M, Tay T, Kıvança M, Türk H (2006) Antimicrobial activity of extracts of the lichen Xanthoparmelia pokornyi and its gyrophoric and stenosporic acid constituents. Zeitschrift Für Naturforschung C 61(5–6):319–323. https://doi.org/10.1515/znc-2006-5-603

Cetin H, Tufan O, Turk AO, Tay T, Candan M, Yanikoglu A, Sumbul HH (2008) Insecticidal activity of major lichen compounds. (−)-and (+)- usnic acid against the larvae of house mosquito, Culex pipiens L. Parasitol Res 2:1277–1279

Chao HS, Lee SK, Lee SH (2015) Composition for inhibiting growth of body hair comprising Usnic acid as effective ingredient. KR101511446 (Patent)

Cocchietto M, Skert N, Nimis P et al (2002) A review on usnic acid, an interesting natural compound. Naturwissenschaften 89:137–146. https://doi.org/10.1007/s00114-002-0305-3

Correché ER, Carrasco M, Giannini F, Piovano M, Garbarino J, Enriz D (2002) Cytotoxic screening activity of secondary lichen metabolites. Acta Farm Bonaer 21:273–278

Correché ER, Enríz RD, Piovano M, Garbarino J, Gómez-Lechón MJ (2004) Cytotoxic and apoptotic effects on hepatocytes of secondary metabolites obtained from lichens. Altern Lab Anim 32(6):605–615. https://doi.org/10.1177/026119290403200611

Culberson CF (1972) Improved conditions and new data for the identification of lichen products by a standardized thin-layer chromatographic method. J Chromatogr 72:113–125

Culberson CF, Ahmadjian K (1979) Standard method zur Dunnschicht chromatographi von Flechten substanzen. Herz 5:1–24

Culberson CF, Johnson A (1982) Substitution of methyl tert.-butyl ether for diethyl ether in standardized thin-layer chromatographic method for lichen products. J Chromatogr 238:483–487

Culberson CF, Culberson WL, Johnson A (1981) A standardized TLC analysis of B-orcinol depsidones. Bryologist 84:16–29

Dewick PM (2002) Medicinal natural products: A biosynthetic approach, 2nd edn. Wiley, Chichester

Durazo FA, Lassman C, Han S et al (2004) Fulminant liver failure due to usnic acid for weight loss. Am J Gastroenterol 99(5):950–952. https://doi.org/10.1111/j.1572-0241.2004.04165.x

Elo H, Matikainen J, Pelttari E (2007) Potent activity of the lichen antibiotic (+)-usnic acid against clinical isolates of vancomycin-resistant enterococci and methicillinresistant Staphylococcus aureus. Naturwissenschaften 94(6):465–468. https://doi.org/10.1007/s00114-006-0208-9

Elix JA (1996) Biochemistry and secondary metabolites. In: Nash T III (ed) Lichen biology. Cambridge University Press, pp 154–180

Elix JA (2014) A catalogue of standardized chromatographic data and biosynthetic relationships for lichen substances, 3rd edn. Author, Canberra

Fernández E, Reyes A, Hidalgo ME, Quilhot W (1998) Photoprotector capacity of lichen metabolites assessed through the inhibition of the 8-methoxypsoralen photobinding to protein. J Photochem Photobiol B 42(3):195–201. https://doi.org/10.1016/s1011-1344(98)00070-0

Gmelin L (1858) Handbuch der organischen Chemie. Bd V, pp 94–97

Gollapudi SR, Telikepalli H, Jampani HB, Mirhom YW, Drake SD, Bhattiprolu KR, Mitscher LA (1994) Alectosarmentin, a new antimicrobial Dibenzofuranoid Lactol from the lichen, Alectoria sarmentosa. J Nat Prod 57(7):934–938. https://doi.org/10.1021/np50109a009

Guo L, Shi Q, Fang JL (2008) Review of usnic acid and *Usnea barbata* toxicity. J Environ Sci Health C Environ Carcinog Ecotoxicol Rev 26(4):317–338

Hale ME (1983) The biology of lichens, 3rd edn. Edward Arnold Ltd, London

Hidalgo ME, Fernández E, Quilhot W, Lissi E (1994) Antioxidant activity of depsides and depsidones. Phytochemistry 37(6):1585–1587. https://doi.org/10.1016/s0031-9422(00)89571-0

Honda NK, Pavan FR, Coelho RG, de Andrade LSR, Micheletti AC, Lopes TIB (2010) Antimycobacterial activity of lichen substances. Phytomedicine 17:328–332

Huneck S, Lamb IM (1975) 1′-Chloropannarin, a new depsidone from *Argopsis friesiana*: notes on the structure of pannarin and on the chemistry of the lichen genus Argopsis. Phytochemistry 14 (7):1625–1628. https://doi.org/10.1016/0031-9422(75)85363-5

Ingolfsdottir K (2002) Usnic acid. Phytochemistry 61:729–736

Khanuja SPS, Tiruppadiripuliyur RSK, Gupta VK, Srivastava SK, Verma SC, Saikia D, Darokar MP, Shasany AK, Pal A (2007). Antimicrobial and anticancer properties of methyl-betaorcinolcarboxylate from lichen (*Everniastrum cirrhatum*). US Patent No. 0099993A1

Knop W (1844) Chemisch-physiologische Untersuchung tiber die Flechten. Justus Lieb Ann Chern 49:103–124

Kokubun T, Shiu W, Gibbons S (2007) Inhibitory activities of lichen-derived compounds against methicillin- and multidrug-resistant *Staphylococcus aureus*. Planta Med 73(2):176–179. https://doi.org/10.1055/s-2006-957070

Koparal AT, Ulus G, Zeytinoğlu M, Tay T, Türk AO (2010) Angiogenesis inhibition by a lichen compound olivetoric acid. Phytother Res 24:754–758

Krishna DR, Venkataramana D (1992) Pharmacokinetics of D(+)-usnic acid in rabbits after intravenous and oral administration. Drug Metab Dispos 20(6):909–911

Kumar KC, Müller K (1999) Lichen metabolites. 2. Antiproliferative and cytotoxic activity of gyrophic, usnic, and diffractaic acid in human keratinocyte growth. J Nat Prod 62(6):821–823. https://doi.org/10.1021/np980378z

Kumar K, Upreti DK (2001) *Parmelia* spp. (lichens) in ancient medicinal plant lore of India. Econ Bot 55(3):458–459

Lauterwein M, Oethinger M, Belsner K, Peters T, Marre R (1995) In vitro activities of the lichen secondary metabolites vulpinic acid, (+)-usnic acid, and (−)-usnic acid against aerobic and anaerobic microorganisms. Antimicrob Agents Chemother 39(11):2541–2543. https://doi.org/10.1128/aac.39.11.254

Lawrey JD (1995) The chemical ecology of lichen mycoparasites. Can J Bot 73(Suppl. 1):603–608

Lee SH (2015) Pharmaceutical composition for preventing or treating bone related diseases. KR 1020150100331 (Patent)

Liu H, Liu Y, Liu Y, Xu A, Young CYF, Yuan H, Lou H (2010) A novel anticancer agent, retigeric acid B, displays proliferation inhibition, S phase arrest and apoptosis activation in human prostate cancer cells. Chem Biol Interact 188(3):598–606. https://doi.org/10.1016/j.cbi.2010.07.024

Llano GA (1948) Economic uses of lichen. Econ Bot 2:15–45

Lücking R, Hodkinson BP, Leavitt SD (2016) The 2016 classification of lichenized fungi in the Ascomycota and Basidiomycota – approaching one thousand genera. Bryologist 119(4):361–416. https://doi.org/10.1639/0007-2745-119.4.361

Luzina OA, Salakhutdinov NF (2018) Usnic acid and its derivatives for pharmaceutical use: a patent review *(2000-2017)*. Expert Opin Ther Pat 28(6):477–491. https://doi.org/10.1080/13543776.2018.1472239

Mao AA, Dash SS (2019) Plant discoveries. Botanical Survey of India, Kolkata

Martins MCB, de Lima MJG, Silva FP, Azevedo-Ximenes E, da Silva NH, Pereira EC (2010) Cladia aggregata (lichen) from Brazilian northeast: chemical characterization and antimicrobial activity. Braz Arch Biol Technol 53(1):115–122. https://doi.org/10.1590/s1516-89132010000100015

Mayer M, Ma O'n, Ke M, Ns S-M, Ama C-L, Am T, Vcl A (2005) Usnic acid: a non-genotoxic compound with anticancer properties. Anti-Cancer Drugs 16(8):805–809

Mitrović T, Stamenković S, Cvetković V, Nikolić M, Tošić S, Stojičić D (2011) Lichens as source of versatile bioactive compounds. Biol Nyssana 2(1):1–6

Molnár K, Farkas E (2010) Current results on biological activities of lichen secondary metabolites: a review. Z Naturforsch C 65C:157–173

Najdenova V, Lisickov K, Zoltán D (2001) Antimicrobial activity and stability of usnic acid and its derivatives in some cosmetic products. Olaj, Szappan, Kozmetika 50:158–160. (in Hungarian)

Nash TH III (ed) (2008) Lichen biology, 2nd edn. Cambridge University Press, Cambridge. viii +486

Nayaka S, Upreti DK, Khare R (2010) Medicinal lichens of India. In: Trivedi PC (ed) Drugs from plants. Avishkar, Jaipur, pp 1–38

Nylander W (1866) Circa novum in studio lichenum criterium chemicum. Flora 49:198–201

Odabasoglu F, Cakir A, Suleyman H, Aslan A, Bayir Y, Halici M, Kazaz C (2006) Gastroprotective and antioxidant effects of usnic acid on indomethacin-induced gastric ulcer in rats. J Ethnopharmacol 103(1):59–65. https://doi.org/10.1016/j.jep.2005.06.043

Ögmundsdóttir HM, Zoëga GM, Gissurarson SR, Ingólfsdóttir K (1998) Natural products: anti-proliferative effects of lichen-derived inhibitors of 5-Lipoxygenase on malignant cell-lines and mitogen-stimulated lymphocytes. J Pharm Pharmacol 50(1):107–115. https://doi.org/10.1111/j.2042-7158.1998.tb03312

Okuyama E (1995) Usnic acid and diffractaic acid as analgesic and antipyretic components of *Usnea diffracta*. Planta Med 61:113–115

Qu CH, Tu P, Zhao YU (2015) Usnic acid nanometer suspension, and preparation method and use thereof. CN104398477 (Patent)

Ranković B, Mišić M (2008) The antimicrobial activity of the lichen substances of the lichens *Cladonia furcata, Ochrolechia androgyna, Parmelia caperata* and *Parmelia conspersa*. Biotechnol Biotechnol Equip 22:1013–1016

Ranković B, Mišić M, Sukdolak S (2008) The antimicrobial activity of substances derived from the lichens *Physcia aipolia, Umbilicaria polyphylla, Parmelia caperata* and *Hypogymnia physodes*. World J Microbiol Biotechnol 24:1239–1242

Rikkinen J (1995) What's behind the pretty colours?: A study on the photobiology of lichens. Helsinki, Finnish Bryological Society

Rundel PW (1978) Ecological relationships of desert fog zone lichens. Bryologist 81(2):277–293

Russo A (2006) Pannarin inhibits cell growth and induces cell death in human prostate carcinoma DU-145 cells. Anticancer Drugs 17(10):1163–1169. https://doi.org/10.1097/01.cad.0000236310.66080.ed

Russo A (2008) Lichen metabolites prevent UV light and nitric oxide-mediated plasmid DNA damage and induce apoptosis in human melanoma cells. Life Sci 83:468–474

Russo A (2012) Effect of vicanicin and protolichesterinic acid on human prostate cancer cells: role of Hsp70 protein. Chem Biol Interact 195(1):1–10. https://doi.org/10.1016/j.cbi.2011.10.005

Shibata S, Taguchi H (1967) Occurrence of isousnic acid in lichens with reference to "isodihydrousnic acid" derived from dihydrousnic acid. Tetrahedron Lett 8(48):4867–4871. https://doi.org/10.1016/s0040-4039(01)89620-9

Shrestha G, St. Clair LL (2013) Lichens: a promising source of antibiotic and anticancer drugs. Phytochem Rev 12:229–244

Shukla V, Joshi GP, Rawat MSM (2010) Lichens as a potential natural source of bioactive compounds: a review. Phytochem Rev 9:303–314. https://doi.org/10.1007/s11101-010-9189-6

Sinha GP, Nayaka S, Joseph S (2018) Additions to the checklist of Indian lichens after 2010. Cryptogam Biodivers Assess 197–206 (Special Issue)

Takai M, Uehara Y, Ja B (1979) Usnic acid derivatives as potential antineoplastic agents. J Med Chem 22(11):1380–1384

Tokiwano T, Satoh H, Obara T, Hirota H, Yoshizawa Y, Yamamota Y (2009) A lichen substance as an antiproliferative compound against HL-60 human leukemia cells: 16-O-acetyl-leucotylic acid isolated from *Myelochroa aurulenta*. Biosci Biotechnol Biochem 73:2525–2527

Türk H, Yilmaz M, Tay T, Türk AO, Kivanç M (2006) Antimicrobial activity of extracts of chemical races of the lichen pseudevernia furfuracea and their physodic acid, chloroatranorin, atranorin, and olivetoric acid constituents. Z Naturforsch C, J Biosci 61:499–507

Wachtmeister CA (1952) Studies on the chemistry of lichens. I. Separation of depside components by paper chromatography. Acta Chem Scand 6(6):818–825

Yılmaz M, Türk AÖ, Tay T, Kıvanç M (2004) The Antimicrobial activity of extracts of the lichen Cladonia foliacea and Its (−)-usnic acid, atranorin, and fumarprotocetraric acid constituents. Zeitschrift Für Naturforschung C 59(3–4):249–254. https://doi.org/10.1515/znc-2004-3-423

Yuan C, Zhang ZJ, Guo YH, Sun LY, Ren Q, Zhao ZT (2010) Antibacterial compounds and other constituents of evernia divaricata (L.) Ach. J Chem Soc Pak 32(2):189–193

Zambare VP, Christopher LP (2012) Biopharmaceutical potential of lichens. Pharm Biol 50(6):778–798. https://doi.org/10.3109/13880209.2011.633089

Zopf W. (1907) Die Flechtenstoffe in chemischer, botanischer, pharmakologischer und technischer Beziehung.. Jena Verlag von Gustav Fischer

Algal Metabolites and Phyco-Medicine

13

Lakshmi Mangattukara Vidhyanandan, Suresh Manalilkutty Kumar, and Swapna Thacheril Sukumaran

Abstract

Research on bioproducts is an exciting area for scientists over the decades due to global applications in daily life. A large number of modern drugs have been isolated from natural resources, especially both higher and lower plants and different microorganisms. The discovery of these drugs is mainly based on traditional knowledge among people. It is a fact that the majority of developing countries depend on traditional herbal medicines for various disorders. The pharmacological properties of medicinally important plants mainly depend on their metabolites. The detection of the bioactive compounds provides unlimited scope for new drug leads of our interest. Thus the isolation and purification of plant-derived novel drugs have a significant role in the form of food additives, nutraceuticals, medicines, cosmetics, and other value-added compounds. Algae are a great source of bioactive molecules of diverse therapeutic value and have broad commercial applications. Both microalgae and seaweeds are a rich source of primary and secondary metabolites such as carbohydrates, fatty acids, carotenoids, lectins, mycosporine-like amino acids, polyphenols, alginic acid, agar, carrageenan, etc. There are still many algae that have not been explored. Many algal species have gained significant attention in recent years because of their beneficial health impact. Phyco-medicine denotes the use of algal sources as herbal medicine with therapeutic potential. There is an increased curiosity to develop algae-based medicine and its commercial application. This chapter highlights the importance of bioactive molecules present in different groups of algae and its applications in the pharmaceutical and food industry.

L. M. Vidhyanandan · S. T. Sukumaran (✉)
Department of Botany, University of Kerala, Thiruvananthapuram, Kerala, India

S. M. Kumar
Laboratory of Genetics and Genomics, National Cancer Institute—NIH, Bethesda, MD, USA

© Springer Nature Singapore Pte Ltd. 2020
S. T. Sukumaran et al. (eds.), *Plant Metabolites: Methods, Applications and Prospects*, https://doi.org/10.1007/978-981-15-5136-9_13

Keywords

Bioproducts · Nutraceutical · Pharmaceutical · Traditional medicine · Phytochemicals · Phyco-medicine · Secondary metabolites

13.1 Introduction

Early in the history of life on the planet, algae have been consumed in the human diet. Algae are found all over the globe and in every ecological niche conceivable (Bhattacharjee 2016). They can also grow as epiphyte and endophyte. They are genetically diverse autotrophic plants ranging from unicellular to multicellular forms. Different forms of algae are identified mainly based on their pigments, storage products, nature of cell wall, cell division, and mode of reproduction. Generally, aquatic algae are commonly seen in freshwater, brackish water, and marine water. Microalgae are microorganisms usually unicellular photosynthetic forms, whereas macroalgae are macroscopic, multicellular, and not differentiated into true leaves, stem, and root. The most critical factors that modulate the growth of algae are nutrients, light, pH, salinity, turbulence, and temperature. Algae provide about half of the planet's oxygen and thus support the life of both human beings and animals (Chapman 2013). They are the primary producers which are rich sources of vitamins, minerals, proteins, pigments, omega-3 fatty acids, polyphenols, plant growth hormones, etc. (Pulz and Gross 2004). They are an essential constituent of Asian diets. Research on development of novel plant-based pharmaceuticals has a worldwide scope in the current market due to the presence of unique potent bioactive components. Approximately 1000 bioactive compounds were characterized worldwide from marine sources in 2013, and they are effective against organisms including bacteria, fungi, viruses, and ailments like cancer, hypertension, high cholesterol, and many other diseases (Shannon and Abu-Ghannam 2016). There is a massive interest in the commercial production of medicinally active compounds from algae. The pharmaceutical potential of different algae has been started to be explored (Bhattacharjee 2016). Organic compounds derived from living organisms are produced through primary or secondary metabolic pathways and metabolites are the chemical products from these pathways. Primary metabolites such as carbohydrates, organic and amino acids, fats, oil, alcohol, etc. are directly involved in the growth, developments, and reproduction, whereas, secondary metabolites like phenolics, flavonoids, alkaloids, etc. are involved in the plant defense and stress adaptation. These secondary metabolites also show much pharmacological and nutritional importance. Primary metabolites are found all over the plant kingdom. The plant secondary metabolites can be found in the leaves, stem, root, fruit, seed, or the bark of the plant depending on the type of secondary metabolite that is produced (Hill 1952). Algae exhibit wide range of biological activities like antioxidant (Saravana et al. 2016), hypoglycemic (Shan et al. 2016), antiviral (Santoyo et al. 2011; Huheihel et al. 2002), antitumor (Ye et al. 2008), anti-inflammatory (Garcia-Vaquero et al. 2017), antimicrobial, and antihypertensive activity (Sivagnanam

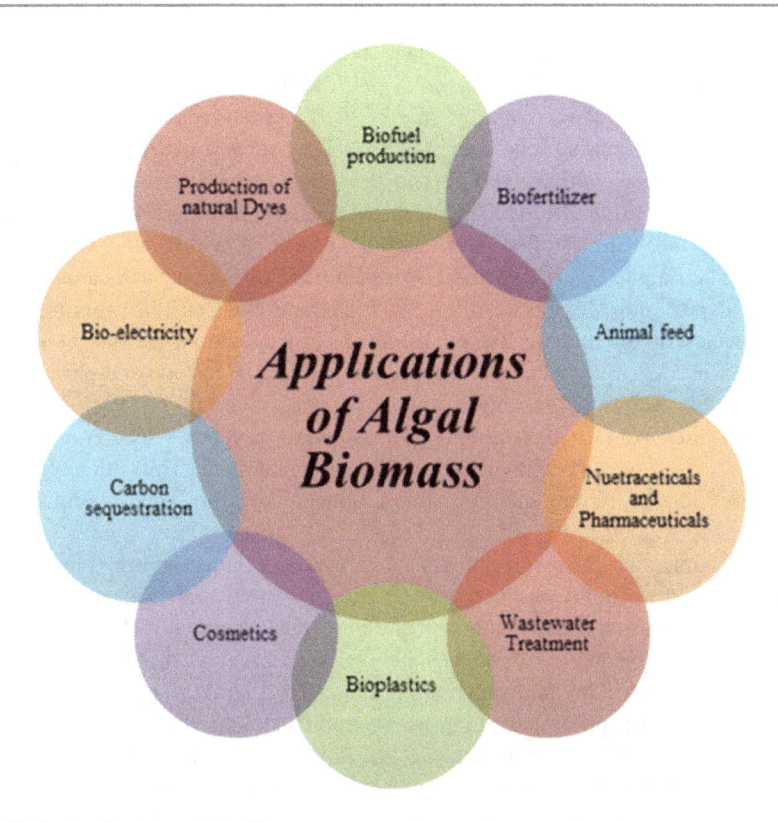

Fig. 13.1 Applications of algal biomass

et al. 2015) due to the presence of fucoidan, laminarin, carotenoids, etc. (Fig. 13.1). The metabolic diversity of algae with special reference to its application in day-to-day life is highlighted here.

13.2 Economic Impact in Pharmaceutical Industry

The distribution of metabolites in any plant tissue is not uniform and any plant group rich in metabolites is widely used in traditional medicine for the treatment of diseases or the commercial production of natural compounds. The secondary metabolites carry out several protective functions in the human body; it can boost the immune system, protect the body from free radicals, kill pathogenic germs, and keep the body fit (Anulika et al. 2016). Research on algal metabolites has shown that algae are promising groups of organisms for providing both novel biochemically active substances and essential compounds for human nutrition (Tringali 1997; Burja et al. 2001; Mayer and Hamann 2004).

When Earth's environment formed over 3 billion years ago, microalgae existed in Earth's oceans (Sathasivam et al. 2019). They are one of the earliest photosynthetic microforms typically found in freshwater, marine water, wet rocks, and extreme environmental conditions like heat, cold, pH, moisture, salinity, etc. Research on microalgae is an attractive topic of interest due to the large-scale production of biofuel from the biomass (Schenk et al. 2008). They show a worldwide distribution and have the capacity to convert carbon dioxide into bioactive compounds. The commercial production of microalgae is approximately 5000 tons/year of dry matter (Raja et al. 2008). Chinese first used microalgae as food, and later, the commercial forms of microalgae (*Spirulina* and *Chlorella*) were consumed as healthy food in Japan, Taiwan, and Mexico (Tamiya 1957; Durand-Chastel 1980; Soong 1980). Many species of microalgae like *Chlorella, Spirulina, Botryococcus, Dunaliella, Porphyridium, Nitzschia,* etc. have high commercial value in food and pharmaceutical industries. Spirulina has high protein content, vitamins (A, B1, B2, B12), pigments (carotenoids and xanthophylls), essential fatty acids (Richmond 1988). Chlorella is also used as healthy food for humans and animals. Thus microalgae play an essential role in human and animal diet, aquaculture, food, and pharmaceutical industry.

Marine algae or seaweeds are not only used for consumption but also the production of functional products of our interest (Holdt and Kraan 2011). Our ancestors used marine algae for therapeutic purposes (Pal et al. 2014). Macroalgae are also rich in bioactive substances like polysaccharides, proteins, lipids, and polyphenols, which show antibacterial, antifungal, and antiviral properties (Stengel et al. 2011; Bourgougnon and Stiger-Pouvreau 2011). Based on the structure and composition of pigments, marine algae are classified into three main categories: green algae (Chlorophyceae), brown algae (Phaeophyceae), and red algae (Rhodophyceae) (Khan et al. 2009). Among these classes, red algae are the largest producers of bioactive compounds (Abdel-Raouf et al. 2015). The most important bioactive compounds from marine macroalgae are polysaccharides, carotenoids, polyunsaturated fatty acids, and polyphenols (Holdt and Kraan 2011). The extraction of fatty acids and pigments is mainly derived from the biomass of microalgae (Messyasz et al. 2018). The brown algae *Laminaria japonica* and *Undaria pinnatifida*; red algae *Porphyra, Eucheuma, Kappapycus,* and *Gracilaria*; green algae *Monotroma* and *Enteromorpha* are mostly cultivated and used commercially for the production of agar, alginates, carrageenan, etc. (Lüning and Pang 2003). Some of the economically important marine macro algae can be seen in Fig. 13.2.

13.3 Extraction of Metabolites

Pharmacological activities of any plant are based on the presence of phytochemicals present in it. A few strategies have been implemented to identify some potential algae for therapeutic purposes. The phytochemical studies in any plant start with appropriate extraction methods and effective solvents to separate bioactive molecules through standard procedures. Different algal extraction methods such as enzyme-assisted extraction, microwave-assisted extraction, ultrasound-assisted extraction, supercritical fluid extraction, pressurized liquid extraction, etc. have

Caulerpa taxifolia

Caulerpa racemosa

Chaetomorpha antennina

Ulva lactuca

Padina tetrastromatica

Stoechospernum marginatum

Fig. 13.2 Selected economically important marine macro algae

Gracilaria corticata

Sargassum wightii

Hypnea musciformis

Gelidium micropterum

Fig. 13.2 (continued)

been discovered to increase the yield and to lower the time of extraction. Algae show high phytochemical diversity along with structural diversity. Many of them possess significant pharmacological properties. The production of biofuel from algal oil has been extensively investigated. The traditional extraction methods like soxhlet extraction, aqueous-alcoholic extraction, hydrodistillation, etc. are time-consuming, and it may affect the structural properties of the compounds present in it. The separation of the final compound from the extractant constitutes an environmental issue due to the release of large amounts of production-related chemical wastes (EPA 2014). To reduce the large-scale use of organic solvents for extraction, scientists have reported many novel approaches for processing under high pressure to enhance the extract quality and purity. The modern algal metabolite extraction techniques are given in Table 13.1.

Among these techniques, SFE and UAE are most widely adapted for large-scale production of algal derivatives. There are different extraction methods for deriving algal proteins: aqueous, acidic, and alkaline methods followed by several rounds of centrifugation and recovery using techniques such as ultrafiltration, precipitation, or

Table 13.1 Methods for the extraction of algal metabolites

Sl no:	Extraction methods	Metabolites
1.	Pressurized liquid extraction (PLE)	Phenolic compounds, fatty acids, and xanthophylls
2.	Supercritical fluid extraction (SFE)	Fatty acids and carotenoids
3.	Microwave-assisted extraction (MAE)	Phenolic compounds and carbohydrates
4.	Ultrasound-assisted extraction (UAE)	Phenolic compounds and minerals
5.	Enzyme-assisted extraction (EAE)	Carbohydrates and antioxidants
6.	SFE and MAE	Oil and proteins
7.	Pulsed electric field (PEF)	Proteins and lipids
8.	SWE and SFE	Lipids

chromatography (Bleakley and Hayes 2017). Some examples of novel protein extraction methods include ultrasound-assisted extraction, pulsed electric field, and microwave-assisted extraction (Kadam et al. 2013). Membrane technologies such as microfiltration, nanofiltration, ultrafiltration, and reverse osmosis are promising methods of enriching algal proteins and developing novel bioactive molecules (Bleakley and Hayes 2017). The biorefinery approach in the extraction of industrially valuable products is now advancing to reduce the cost of biodiesel.

13.4 Primary Metabolites in Algae

Primary metabolites generally have shared biological purposes across all species (Kabera et al. 2014). Carbohydrates, proteins, and lipids are the essential primary metabolites (Wen et al. 2015). The distribution of metabolites in different categories of algae and their medicinal importance are discussed here.

13.4.1 Carbohydrates

The first product of photosynthesis, carbohydrates are the universal component present in all living organisms. The majority of the phytochemical pathways start with the carbohydrates and they incorporate many phytochemicals through glycosidation linkages. Usually, carbohydrates can be seen in different forms such as cellulose, hemicellulose, starch, and other polysaccharides. The most abundant carbohydrates are glucose, rhamnose, xylose, and mannose (Markou et al. 2012). Carbohydrates have a defensive role in membrane thickness and they store food and water (Hussein and El-Anssary 2018). Generally, carbohydrates are rich in algae as it can produce high-energy biofuels through large-scale fermentation technology. Leonard et al. (2010) reported that sulfated polysaccharides inhibit the activity of many bacteria and viruses. So polysaccharides with sulfated groups possess significant economic value in food, pharmaceutical, and nutraceutical industries (Bayu and Handayani 2018). Wijffels and Barbosa (2010) studied the biofuel production in microalgae and reported that microalgal products were composed of 40% lipids, 50% proteins, and 10% carbohydrates.

The most important polysaccharides from algae are alginic acid, laminarin, galactans, and fucoidan, etc. Microalgal polysaccharides have a remarkable role in the cosmetic industry as hygroscopic agents and antioxidants (Raposo et al. 2013a). Certain microalgae are the ideal candidates for bioethanol production as carbohydrates from microalgae can be extracted to produce fermentable sugars (Nguyen 2012). Depending on the type of carbohydrates present in microalgae, they are classified into many groups; green microalgae having cellulose in the cell wall, amylose and amylopectin as storage product, red microalgae like *Porphyridium* have polysaccharide that is encapsulated in the gel state (composed of sulfated polysaccharides), brown algae have a polysaccharide storage product laminarin, (Villarruel-López et al. 2017). Diatoms have chrysolaminarin, which lacks mannitol residue at the reducing end of polysaccharide (Pulz and Gross 2004; Chiovitti et al. 2004). Glucans present in the class Prymnesiophyta can be classified as laminarin, paramylon, or chrysolaminarin based on the chain length and presence of mannitol end groups (Villarruel-López et al. 2017). The cryptophyte microalgae have a periplast instead of the cell wall, and some produce extracellular polysaccharides. Glaucophyta lacks a cell wall and accumulates starch in the cytosol (Villarruel-López et al. 2017). The reserve carbohydrate, paramylon starch, is present in the class Euglenophyta that is a β-1,3-polyglycan (Pulz and Gross 2004).

Carbohydrates in microalgae have demonstrated many biological activities. Beta-glucans are polysaccharides that have immune-stimulatory activity and they act as dietary fibers (Villarruel-López et al. 2017). Chou et al. (2012) reported the importance of polysaccharides in unicellular green algae *Chlorella sorokiniana* against viral infection and cancer. Sulfated polysaccharides are another class of polysaccharides present in the cell wall of algae (Ibrahim et al. 2017). Besides the storage properties, sulfated polysaccharides possess some important biological properties like antioxidant and antitumoral activity. Laminarin in brown algae modulates the response to systemic infection and hepatic inflammation (Neyrinck et al. 2007; Kraan 2012). Kidgell et al. (2019) have critically reviewed the biological activities of ulvan (a sulfated polysaccharide) such as antioxidant, anticancer, anticoagulant, antiviral, antihyperlipidemic, and immunomodulating activities in different species of *Ulva*. Jiao et al. (2011) made an elaborated review on structural chemistry and bioactivities of important sulfated polysaccharides such as galactans, ulvans, and fucans.

Stiger-Pouvreau et al. (2016) described seaweeds as the most abundant source of polysaccharides such as alginates, agar, carrageenans, etc. Polysaccharides present in seaweeds are effective and nontoxic antioxidants (Li and Kim 2011; Souza et al. 2012). The fibrillar skeleton in the algal cell wall is of different types: xylan in red algae and xylan along with mannan present in green algae (Stiger-Pouvreau et al. 2016). The cell wall in brown algae shows similarity with that of plants and animals in having cellulose and sulfated fucans, respectively (Michel et al. 2010a, b). The anti-inflammatory, antiviral, antitumor, and antioxidative activities are the most important activities reported for fucoidans present in brown algae (Song et al. 2012; Synytsya et al. 2010). Laminarin is one of the crucial polysaccharides present in brown algae having antiviral and antibacterial properties (O'Doherty et al. 2010). Alginates possess vigorous antibacterial and anti-inflammatory activities (Chojnacka et al. 2012). They also activate the growth of some symbiotic fungi in the rhizosphere (Kuda and Ikemori 2009; Heo et al. 2005; Kuda et al. 2007; Rioux et al. 2007). Several biomass conversion technologies were reported for the production of biofuel from carbohydrates such as anaerobic digestion, anaerobic fermentation, and biological biohydrogen production (Markou et al. 2012).

13.4.2 Proteins

Algae, especially microalgae, are rich sources of proteins (Michalak and Chojnacka 2015). *Spirulina platensis* is considered as a food supplement due to its high protein content (Templeton and Laurens 2015). *Aphanizomenon flos-aquae, Chlorella* sp., *Dunaliella salina* (*D. salina*), *Dunaliella tertiolecta* (*D. tertiolecta*), and *Spirulina plantensis* (*S. plantensis*) are some of the microalgae species widely used in human diet due to the presence of high protein content and nutritive value (Soletto et al. 2005; Rangel-Yagui et al. 2004). Many studies on algae reported that they are potential sources for feed supplement and substitute for conventional protein sources like soybean meal and fish meal (Becker 1994; Spolaore et al. 2006). The protein percentage in microalgae varies based on the cell wall features and mechanism of disruption (Servaites et al. 2012). However, the content of protein and amino acids is comparatively low in brown macroalgae, and higher protein content can be seen in green and red algae (Lourenço et al. 2002; Dawczynski et al. 2007). Lectins are essential proteins that take part in various biological activities such as antibacterial, antiviral, and anti-inflammatory activities (Cunningham and Joshi 2010) and can be used as medical diagnostic tools since it is helpful in the detection of many diseases related to glycosylation pattern changes and identification of microbial agents in medical microbiology. They are usually bound with carbohydrates and participate in intercellular communication (Chojnacka et al. 2012).

13.4.3 Lipids

The majority of the research on algal biofuel production based on lipids has been concentrated on microalgae. They play an essential role in improving the immune system, lipid metabolism, gut function, and stress resistance (Shields and Lupatsch 2012). Microalgae are an excellent source of lipids as long-chain polyunsaturated fatty acids (PUFA) and carotenoids (Banskota et al. 2019). Membrane lipids such as glycosylglycerides, phosphoglycerides, betaine ether lipids, and storage lipids like triacylglycerols are the vital forms of lipids in algae (Li-Beisson et al. 2019). Macroalgae are also an emerging source of lipids for biofuel production.

Solovchenko et al. (2008) reported that green algae *Parietochloris incisa* synthesize high amounts of total lipid contents and AA. Mainly two types of lipids: phospholipids and glycolipids are present in algae. Polyunsaturated fatty acids are usually abundant in algal species that live in cold climate than other algae (Holdt and Kraan 2011). Stabili et al. (2014) carried out the fatty acid profile of *Cladophora rupestris* through gas chromatography and revealed the presence of palmitic acid, myristic acid, oleic acid, α-linolenic acid, palmitoleic acid, and linoleic acid.

13.4.4 Pigments

Pigments in algae are of different types: chlorophyll a, b and c, carotenes, xanthophylls phycobilins fucoxanthin, violaxanthin, lutein, zeaxanthin, neoxanthin β-carotene, etc. Lutein is a major pigment belongs to xanthophyll and its content was reported in many green algae such as *Chlorella* sp., *Scenedesmus dimorphus, Scenedesmus obliquus, Desmodesmus* sp., and *Ankistrodesmus* sp. (Sallehudin et al. 2018). Different types of carotenoids are present in different species of algae and they display very strong antioxidant activity (Chojnacka et al. 2012). Phycobilibroteins produced mostly by cyanobacteria have anti-inflammatory, antiviral, and neuroprotective activities (Holdt and Kraan 2011). Moreau et al. (2006) have studied the antiproliferative activity of fucoxanthin in microalgae such as *Odontella aurita, Chaetoceros* sp, *Isochrisys aff. galbana* by using cell lines such as Bronchopulmonary A549, NSCLC-N6, and epithelial SRA 01/04. Astaxanthin is an important member of the xanthophyll family, derived mainly from the green unicellular microalgae, *Haematococcus pluvialis* (Panis and Carreon 2016). They have wide biological activities against cardiovascular diseases and cancers (Ambati et al. 2014). Pigments such as C-phycocyanin, allophycocyanin, phycoerythrin, chlorophyll a, and carotenoids in fresh and differently processed spirulina biomass were reported by Papalia et al. (2019). All these pigments show various health benefits and many therapeutic properties. Jerez-Martel et al. (2017) carried out phenolic profiling and determined the antioxidant activity of aqueous and methanolic extracts of several microalgae (*Ankistrodesmus* sp., *Spirogyra* sp., *Euglena cantabrica*, and *Caespitella pascheri*) and cyanobacteria (*Nostoc* sp., *Nostoc commune, Nodularia spumigena, Leptolyngbya protospira, Phormidiochaete* sp., and *Arthrospira platensis*). The strain *E. cantabrica* displayed the highest content of phenolic compounds and exhibited the highest antioxidant activity.

13.4.5 Vitamins

Vitamins are consumed not only for health but also for their strong antioxidant activity (Galasso et al. 2019). They are abundant in algal food. Microalgae are rich in vitamins and minerals such as vitamins A, B1, B2, B3, B12, C, D, and E, minerals like iodine, potassium, iron, magnesium, and calcium. The content of vitamin C and vitamin E was observed in some seaweeds such as *Laminaria* spp., *Porphyra umbilicalis, Himanthalia elongata,* and *Palmaria palmate* (Ferraces-Casais et al. 2012). A number of blue–green algae are used for food supplements since it contains pseudo vitamin B12, which is inactive in humans (Watanabe 2007).

13.5 Secondary Metabolites of Algae

A wide variety of bioactive compounds have been produced from algae and they contribute to the development of novel drugs in the pharmaceutical industry. Algae are producing these secondary metabolites to get adaptation to changing environmental conditions. The majority of the secondary metabolites synthesized by marine algae have cytotoxic properties (Manilal et al. 2009) and antimicrobial activities (Bultel-Poncé et al. 2002; Etahiri et al. 2007). The presence of phenolics in different algal species has been scientifically validated and reported in numerous research articles (Raposo et al. 2013b). Sudhakar et al. (2019) considered microalgae as a significant source of high-valued products and reported that microalgae contribute more bioactive compounds than macroalgae. The majority of microalgae accumulate bioactive compounds in the biomass, but some others excrete metabolites into the medium; these are known as exometabolites (Bhattacharjee 2016). The secondary metabolites of algae, like alkaloids, phenolics, flavonoids, tannins, sterols, terpenoids, etc. display significant pharmacological properties.

13.5.1 Phenolics

Most of the phenolics are pharmacologically important and shows strong antioxidant and anti-inflammatory properties. Phenolic compounds are rich in vegetables, fruits, legumes, spices, etc. Algae are also acting as a natural source of polyphenols. High phenolic content is present in brown seaweeds in comparison with green and red algae. Polyphenolic compounds discovered from macroalgae are pharmaceutically important for the prevention and management of neurogenerative diseases (Barbosa et al. 2014). Jerez-Martel et al. (2017) carried out profiling of simple phenolics like gallic acid, syringic acid, protocatechuic acid, and chlorogenic acid in several microalgae (*Ankistrodesmus* sp., *Spirogyra* sp., *Euglena cantabrica,* and *Caespitella pascheri*) and cyanobacteria (*Nostoc* sp., *Nostoc commune, Nodularia spumigena, Leptolyngbya protospira, Phormidiochaete* sp., and *Arthrospira platensis*). The results of their study show the highest concentration of phenolic compounds in *Euglena cantabrica,* with effective antioxidant activity. It has been reported that microalgae have many antioxidants such as tocopherols (Vitamin E) as well as ascorbic acid and phenolic compounds (Koller et al. 2014). Tocopherol is widely used to treat cancer, heart diseases, Alzheimer's, Parkinson's, etc. (Pham-Huy et al. 2008). Waghmode and Khilare (2018) conducted quantitative analysis of major phenolics present in some brown algae such as *Sargassum cinereum, S. Ilicifolium, S. tenerrimum,* and *S. wightii* by RP-HPLC method and revealed the presence of gallic acid and *p*-hydroxybenzoic acid. The separation and characterization of phenolic compounds in algae have not extensively studied due to the difficulty in the fractionation of polyphenols (Montero et al. 2018). Machu et al. (2015) identified some phenolic compounds like gallic acid, hydroxyl benzoic acid, catechin, epicatechin, catechin gallate, epigatechin gallate, epigallocatechin gallate in four brown algae, *Laminaria japónica, Eisenia bicycli, Hizikia fusiformis,* and *Undaria pinnatifida*), two red algae (*Porphyra tenera* and *Palmaria palmata*) and one green alga (*Chlorella pyrenoidosa*). *Dunaliella tertiolecta* shows the presence of gentisic acid, catechin, epicatechin, and display high antioxidant activity (López et al. 2015). Montero et al. (2018) made an extensive review of the different phenolics present in microalgae and their bioactivities such as antioxidant, antiproliferative, anti-obesity, and antidiabetic activities. Phenolic compounds such as gallic acid, catechin, caffeic acid, *p*-hydroxybenzoic acid, *p*-coumaric acid, ferulic acid, quercetin, genistein, and kaempferol were identified in differently stored spirulina biomass (Papalia et al. 2019). The impact of different storage methods applied in spirulina biomass on the nutritional and bioactive contents was studied. Chlorella produces many compounds like fatty acids and phenolic compounds with antimicrobial activities (Jorgensen 1962). Generalić Mekinić et al. (2019) made an extensive review of the phenolic content of brown algae with special reference to information on its extraction, identification, and quantification by considering all the previous reports. The chemical and biological properties of marine algae as natural antioxidants and their protective effects in biological systems were reviewed by Freile-Pelegrín and Robledo (2013). The presence of a strong quantity of phenolic compounds in the red algae, *Gracilaria burs-pastoris* can be correlated with high antioxidant activity (Ramdani et al. 2017). El-Baky et al. (2009) reported the presence of phenolic compounds in microalgae with the potential to fight with the free radicals.

13.5.2 Flavonoids

Flavonoids are another class of important secondary metabolites predominant in terrestrial plants. Microalgae synthesize polyphenols in a small amount when compared to brown and red algae (Zolotareva et al. 2019). The biological source, structure, and bioactivities of different flavonoid compounds such as 4B-chloro-2-hydroxyaurone, 4B-chloroaurone, morin, hesperidin, myricetin, scutellarein 40-methyl ether, quercetin, catechin, apigenin, etc. were documented by Ibrahim et al. (2017). Kazłowska et al. (2010) reported the presence of rutin and hesperidin in *Porphyra dentate* and the results of their study revealed the capacity to suppress NO production in LPS-stimulated macrophages via NF-kappaB-dependent iNOS gene transcription. Goiris et al. (2014) have constructed flavonoid pathways like that of higher plants in microalgae such as *Haematococcus pluvialis* and *Pavlova lutheri*. Flavonoids were isolated and identified from microalgae like Oscillatoria and Lyngbya having antioxidant and antimicrobial activity (Baviskar and Khandelwal 2015). A flavone compound, apigenin, was isolated from the marine red algae *Acanthophora spicifera* showed strong analgesic, anti-inflammatory, and antiproliferative activities (Shoubaky et al. 2016). Ethyl acetate extracts of *Nannochloropsis oculata* and *Gracilaria gracilis* algae show high levels of phenolic and flavonoid compounds (Ebrahimzadeh et al. 2018). The flavonoid compound morin has a positive effect on obesity, suppressing lipogenesis, gluconeogenesis, inflammation, and oxidative stress, tending to modify the concentration of triglycerides in the liver (Meneses and Flores 2019). Sabina and Aliya (2009) isolated a flavone compound, scutellarein 4′-methyl ether from the red alga *Osmundea pinnatifida*.

13.5.3 Tannins

Tannins are a group of polyphenols that can precipitate protein (Hussein and El-Anssary 2018). Phlorotannins are a subgroup of tannins present in algae, commonly known as algal polyphenols. They are classified into six different groups: fuhalols, eckols, carmalols, fucols, phlorethols, and fucophlorethols based on a number of hydroxyl groups and phloroglucinol units (Sudhakar et al. 2019). Phlorotannins show various biological activities such as antioxidant (Plaza et al. 2008), antibiotic (Chandini et al. 2008), antidiabetic (Kang et al. 2003), anticancer (Lim et al. 2002), hepatoprotective (Li et al. 2007), anti-inflammatory (Li et al. 2008), anti-HIV (Zubia et al. 2008), antibacterial (Zou et al. 2008), and antiallergic (Sampath-Wiley et al. 2008) activities. They are also used for protection against vascular diseases, glucose-induced oxidative stress, and treatment of arthritis (Sudhakar et al. 2019). Phlorotannins are abundant in brown algae, which possess strong antioxidant effects (Guedes et al. 2019). Fauzi and Satriani Lamma (2018) studied the total tannin content in *Padina* and *Sargassum* sp. and reported that *Padina* sp. shows a more effective antibleeding activity than *Sargassum* sp. Dieckol is a phlorotannin derived from the brown algae *Ecklonia cava*, which can prevent ovarian cancer (Ahn et al. 2015). Singh and Sidana (2013) summarized antimicrobial, antioxidant, anticancer, and other medicinal properties of phlorotannins and isolation techniques used in the purification of compounds. *Sargassum muticum* samples collected from North-Atlantic Coasts showed good cytotoxic potential as they are rich in phlorotannins (Montero et al. 2016).

13.5.4 Alkaloids

Alkaloids are a group of nitrogen-containing secondary metabolites that have a heterocyclic ring in its structure. They are found in plants, animals, microorganisms, and marine species (Alghazeer et al. 2013). They are relatively rare in algae as compared to higher plants (Güven et al. 2010). Morphine was the first isolated and extracted alkaloid from the tarry poppy (*Papaver somniferum*) seed juice by Serturner in 1805 (Krishnamurti and Rao 2016). Hordenine was the first alkaloid from the marine algae *Phyllophora nervosa* (Guven et al. 1969; Guven et al. 1970). The isolated alkaloids in macroalgae are of different types; 2-phenylethylamine, indole, halogenated indole, and 2,7-naphthyridine derivatives (Barbosa et al. 2014). Halogenated alkaloids are mostly seen in Chlorophyta, whereas indole alkaloids are accumulated in Rhodophyta. Phenylethylamine alkaloid which is derived from the amino acid tyrosine was first isolated from terrestrial plants (Güven et al. 2013). The highest amount of phenylethylamine was reported in *Gelidium crinale* and lowest in *Phyllophora crispa* (Percot et al. 2009a, b). They act as hormones, stimulants, hallucinogens, entactogenes, anorectics, bronchodilators, and antidepressants (Saavedra 1978). *N*-acetylphenylethylamine (*N*-ACPEA), an alkaloid first isolated from the red algae *G. crinale* (Percot et al. 2009a, b). The bisindole alkaloid, caulerpin has anti-inflammatory and antioxidant activity (de Souza et al. 2009). Racemosin A and B were known to have neuroprotective activity (Liu et al. 2013). Lophocladine A displayed an affinity for NMDA receptors and was found to be a δ-opioid receptor antagonist, whereas lophocladine B exhibited cytotoxicity to NCI-H460 human lung tumor and MDA-MB-435 breast cancer cell lines. Some alkaloids identified from algae are listed in Table 13.2.

Several nonprotein amino acids like pyrrolidine-2,5-dicarboxylic acid and N-methylmethionine sulfoxide have been observed in 18 macroscopic marine algae that belong to Rhodophyta and their structure was elucidated (Impellizzeri et al. 1975).

13.5.5 Terpenoids

Terpenoids are the largest and diverse class of secondary compounds, derived from five-carbon isoprene units. Reddy and Urban (2009) isolated three merodi250terpenoids, fallahydroquinone, fallaquinone, and fallachromenoic acid from the Australian marine brown algae, *Sargassum fallax* and they also characterized sargaquinoic acid and sargahydroquinoic acid using 2D NMR for the first time. Red algae also produce different types of terpenoids. Many sesquiterpenes such as Laurecomin, Glanduliferol, Okamurene E, Rogiolol, Obtusol, Elatol, Hurgadol, Dendroidiol, Dendroidone, etc. were isolated from different species of the genus Laurencia (Harizani et al. 2016). Shushizadeh (2019) analyzed mixtures of terpenoids and other compounds such as aromatic esters and phenol by GC-MS analysis. Chinese chemist Tu Youyou got Nobel Prize in Medicine for the discovery of artemisinin, an antimalarial drug that belongs to the category of sesquiterpenoid from *Artemisia annua* (Miller and Su 2011). Two diterpenes, 6(R)-6-hydroxydichotoma3,14-dieno-1,17-dial (denominated Da-1) and the natural acetate 6(R)-6-acetoxi-dichotoma-3,14-dieno-1,17-dial (denominated AcDa-1) were isolated from Brazilian marine brown alga *Dictyota menstrualis* and display anti-HI1 activity (Pereira et al. 2004). Sumayya and Murugan (2019) identified four different terpenoids like hexadecanoic acid, methyl ester, *n*-hexadecanoic acid, octadecanoic acid, and phytol from *Gracillaria dura* by GCMS analysis. Knott et al. (2005) isolated three compounds, Plocoralide A, B, and C from the red alga *Plocamium corallorhiza*. Two novel aromatic valerenane-type sesquiterpenes, Caulerpal A and B, were isolated from the

Table 13.2 Alkaloids from algae

Sl no:	Name of algae	Alkaloid identified	Reference
1.	*Nodularia harveyana*	Norharmane	Volk (2005)
2.	*Nostoc* sp.	Nostocarboline	Becher et al. (2005)
3.	*Calothrix*	Calothrixin	Rickards et al. (1999)
4.	*Hapalosiphon fontinalis*	Hapalindoles	Moore et al. (2002)
5.	*Nostoc muscorum*	Muscoride A	Nagatsu et al. (1995)
6.	*Desmerestia aculeate, Desmerestia viridis*	Phenethylamine	Steiner and Hartmann (1968)
7.	*Ceramium rubrum, Cystoclonium purpureum, Delesseria sanguine, Dumontia incrassata, Polysiphonia urceolata, Polyides rotundus*	Phenethylamine	Steiner and Hartmann (1968)
8.	*Scenedesmus acutus*	Phenethylamine	Güven et al. (2010)
9.	*Gelidium crinale, Gracilaria bursa-pastoris, Halymenia floresii, Phyllophora crispa, Polysiphonia morrowii, Polysiphonia tripinnata.*	Phenethylamine	Rolle et al. (1977)
10.	*Laminaria saccharina, Chondrus crispus* and *Polysiphonia urceolata*	Tyramine or TYR, 4-hydroxyphenylethylamine	Kneifel et al. (1977)
11.	*Scenedesmus acutus*	Tyramine or TYR, 4-hydroxyphenylethylamine	Percot et al. (2009a, b)
12.	*Caulerpa racemosa*	Caulerpin, Racemosin A and B	De Souza et al. (2009) Liu et al. (2013)
13.	*Acanthophora spicifera*	Acanthophoraine	Lin et al. (2019)
14.	*Lophocladia* sp.	Lophocladines A and B	Gross et al. (2006)

Chinese green alga *Caulerpa taxifolia* along with a known metabolite caulerpin (Mao et al. 2006).

13.5.6 Saponins

Saponins are naturally occurring steroid or triterpenoid glycosides with foaming properties (Desai et al. 2009). They are mainly present in plants, animals, lower forms like some algae and bacteria. The foaming and frothing properties of saponin are due to the presence of water-soluble sugar and soluble substance like aglycone. Saponin is a good frother for the harvest of microalgae by the flotation method (Kurniawati et al. 2014). Feroz (2018) made an elaborate review of the saponin contents extracted from different seaweed species and its applications in various fields. Leelavathi and Prasad (2015) made a quantitative analysis of saponin in the seaweeds such as *Gracilaria crassa, G. edulis, G. foliifera, Cymodoceae rotundata, C. serrulata, Ulva lactuca, U. reticulate, Gelidiella acerosa, Kappaphycus alvarezii*. The various biological properties of saponin have been documented, such as antifungal, anti-inflammatory, cytotoxic, antiangiogenic activities, etc. Champa et al. (2016) investigated the total saponin content in *Spirogyra* sp. using ultrasonic-assisted extraction. Triterpenoid saponin, 3-*O*-[α-L-arabinopyranosyl (1 → 2)-β-D-quinovopyranosyl]-quinovic acid-27-*O*-[β-D-glucopyranosyl] ester, along with two known compounds, 3-*O*-[β-D-glucopyranosyl]-β-sitosterol and erythrodiol-3-caffeate were isolated from *Zygophyllum propinquum by* Ahmad and Uddin (1992).

13.5.7 Steroids

Steroids are a type of lipids present in both plants and animals, but they differ from triglycerides and phospholipids in structure and function (Ibrahim et al. 2017). The major groups of steroids present in plants are phytosterols, brassinosteroids, withanolides, phytoecdysteroids, steroidal alkaloids, etc. Phytosterols are a unique class of cholesterol-like compounds that are mainly present in cellular membranes of plants and algae (Lopes et al. 2013). Algae produce different types of phytosterols such as brassicasterol, sitosterol, stigmasterol, campesterol, β-sitosterol, etc. They were reported to have wide applications in food and pharmaceutical industries (Luo et al. 2015). They are reported to have anticancer, antioxidant, and hypocholesterolemic activity. Ali et al. (2002) identified a new steroid namely iyengadione, two new steroidal glycosides (iyengaroside A and B), Clerosterol galactoside from a marine green algae *Codium iyengarii*. Fasya et al. (2019) isolated and identified steroid compounds such as cholesterol, campesterol, stigmasterol, and β-sitosterol from the red algae *Eucheuma cottonii* using column chromatography. These steroid isolates show strong antioxidant activity. Fucosterol was the major sterol reported in two brown algae, *Padina pavonica* and *Hormophysa triquetra* (Shoubaky and Salem 2014). Lopes et al. (2013) highlighted the most common sterols such as desmosterol, cholesterol, campesterol, fucosterol/isofucosterol, β-sitosterol, stigmasterol, ergosterol present in macroalgae and their effects in human health management. Kapetanović et al. (2005) conducted a study on the qualitative and quantitative sterol content in two green (*Ulva lactuca* and *Codium dichotomum*) as well as two brown algae (*Cystoseira adriatica* and *Fucus virsoides*). They reported the sterols like cholesterol and isofucosterol in *U. lactuca*, cholesterol and stigmast-5-en-3ß-ol in *Cystoseira adriatica*. The consumption of edible blue green algae *Nostoc commune* shows hypocholesterolemic effect in mice by reducing the intestinal cholesterol absorption (Rasmussen et al. 2009). Bakar et al. (2019) detected various sterol compounds such as stigmasterol, campesterol, β-sitosterol, fucosterol, epicoprostanol, coprostanol, 5β-cholestan-3-one from two macroalgae like *Dictyota dichotoma* and *Sargassum granuliferum*. Among these sterols, fucosterol and epicoprostanol showed strong antibacterial potential and the remaining sterols exhibited good antifouling activity. Palanisamy et al. (2019) identified all the metabolites from three seaweeds, *Ulva reticulata, Sargassum wightii, Gracilaria* sp. using GC-MS. Cholesterols (83.58%), cholest-5-en-3-ol,24-propylidene-, 3β (63.60%), and cholest-5-en-3-ol, 24-propylidene-, 3β (15.27%) are the major metabolites detected from these species. Lee et al. (2004) investigated the antidiabetic activity of fucosterol isolated from *Pelvetia siliquosa* and the results demonstrated a highly significant hyperglycemic response. Holken Lorensi et al. (2019) reported the presence of phytosterols and the association of these compounds with the entomotoxic potential of *Prasiola crispa* (Antarctic algae). Omer and Attar (2013) reported many steroid compounds like stigmasterol, campesterol, β-sitosterol, avenasterol, and clerosterol from *Chara* and *Spirigyra* sp. The steroids derived from algae can be marketed as a dietary supplement.

13.6 Conclusions

As stated in the data mentioned above, many valuable phytochemicals in algae are responsible for diverse biological activities. Algae provide a rich source of carbohydrates, vitamins, pigments, fatty acid, protein, and secondary metabolites like polyphenols, alkaloids, steroids, phlorotannins, etc. These metabolites enrich the nutritional efficacy of the human diet system. The practice of using medicine should be strictly based on the elaborative investigation of pharmacological studies. The phytochemicals present in both macro and microalgae were found to possess antimicrobial, antioxidative, anticancer, anti-inflammatory, and antiviral activities. This review conveys substantial data for broad applications in human health and the food system. The different extraction techniques in algae were well described in this review. SFE and UAE are the most widely accepted extraction technique for the large-scale production of algal metabolites. It can be concluded that algae provide a good source for the production of biofuels, bioactive medicinal constituents, food supplements, and cosmetics. With the advancements in different analytical and isolation strategies, still many bioactive molecules can be identified and characterized.

References

Abdel-Raouf N, Al-Enazi NM, Al-Homaidan AA, Ibraheem IBM, Al-Othman MR, Hatamleh AA (2015) Antibacterial -amyrin isolated from *Laurencia microcladia*. Arab J Chem 8:32–37

Ahmad VU, Uddin S (1992) A triterpenoid saponin from *Zygophyllum propinquum*. Phytochemistry 31(3):1051–1054

Ahn JH, Yang YI, Lee KT, Choi JH (2015) Dieckol, isolated from the edible brown algae *Ecklonia cava*, induces apoptosis of ovarian cancer cells and inhibits tumor xenograft growth. J Cancer Res Clin Oncol 141(2):255–268

Alghazeer R, Whida F, Abduelrhman E, Gammoudi F, Naili M (2013) In vitro antibacterial activity of alkaloid extracts from green, red and brown macroalgae from western coast of Libya. Afr J Biotechnol 12(51):7086–7091

Ali MS, Saleem M, Yamdagni R, Ali MA (2002) Steroid and antibacterial steroidal glycosides from marine green alga *Codium iyengarii* Borgesen. Nat Prod Lett 16(6):407–413

Ambati R, Phang SM, Ravi S, Aswathanarayana R (2014) Astaxanthin: sources, extraction, stability, biological activities and its commercial applications—a review. Mar Drugs 12(1):128–152

Anulika NP, Ignatius EO, Raymond ES, Osasere O-I, Abiola AH (2016) The chemistry of natural product: plant secondary metabolites. Int J Technol Enhance Emerg Eng Res 4(8). ISSN 2347-4289

Bakar K, Mohamad H, Tan HS, Latip J (2019) Sterols compositions, antibacterial, and antifouling properties from two Malaysian seaweeds: *Dictyota dichotoma* and *Sargassum granuliferum*. J Appl Pharm Sci 9(10):047–053

Banskota AH, Sperker S, Stefanova R, McGinn PJ, O'Leary SJ (2019) Antioxidant properties and lipid composition of selected microalgae. J Appl Phycol 31(1):309–318

Barbosa M, Valentão P, Andrade P (2014) Bioactive compounds from macroalgae in the new millennium: implications for neurodegenerative diseases. Mar Drugs 12(9):4934–4972

Baviskar JW, Khandelwal SR (2015) Extraction, detection and identification of flavonoids from microalgae: an emerging secondary metabolite. Int J Curr Microbiol App Sci 2:110–117

Bayu, A., & Handayani, T. (2018, December). High-value chemicals from marine macroalgae: opportunities and challenges for marine-based bioenergy development. IOP Conf Ser Earth Environ Sci 209, no. 1, 012046.

Becher PG, Beuchat J, Gademann K, Jüttner F (2005) Nostocarboline: isolation and synthesis of a new cholinesterase inhibitor from Nostoc 78-12A. J Nat Prod 68(12):1793–1795

Becker EW (1994) Microalgae biotechnology and microbiology. Cambridge University Press, Cambridge

Bhattacharjee M (2016) Pharmaceutically valuable bioactive compounds of algae. Asian J Pharm Clin Res 9:43–47

Bleakley S, Hayes M (2017) Algal proteins: extraction, application, and challenges concerning production. Foods 6(5):33

Bourgougnon N, Stiger-Pouvreau V (2011) Chemodiversity and bioactivity within red and brown macroalgae along the French coasts, metropole and overseas departments and territories. In: Kim S-K (ed) Handbook of marine macroalgae. Wiley, Chichester, pp 58–105

Bultel-Poncé V, Etahiri S, Guyot M (2002) New ketosteroids from the red alga *Hypnea musciformis*. Bioorg Med Chem Lett 12(13):1715–1718

Burja AM, Banaigs B, Abou-Mansour E, Burgess JG, Wright PC (2001) Marine cyanobacteria—a prolific source of natural products. Tetrahedron 57(46):9347–9377

Champa P, Whangchai N, Jaturonglumlert S, Nakao N, Whangchai K (2016) Determination of phytochemical compound from Spirogyra sp. using ultrasonic assisted extraction. Int J GEOMATE 11(24):2391–2396

Chandini SK, Ganesan P, Bhaskar N (2008) In vitro antioxidant activities of three selected brown seaweeds of India. Food Chem 107(2):707–713

Chapman RL (2013) Algae: the world's most important "plants"—an introduction. Mitig Adapt Strat Glob Chang 18(1):5–12

Chiovitti A, Molino P, Crawford SA, Teng R, Spurck T, Wetherbee R (2004) The glucans extracted with warm water from diatoms are mainly derived from intracellular chrysolaminaran and not extracellular polysaccharides. Eur J Phycol 39(2):117–128

Chojnacka K, Saeid A, Witkowska Z, Tuhy Ł (2012) Biologically active compounds in seaweed extracts - the prospects for the application. Open Conf Proc J 3(Suppl 1-M4):20–28

Chou NT, Cheng CF, Wu HC, Lai CP, Lin LT, Pan I, Ko CH (2012) Chlorella sorokiniana-induced activation and maturation of human monocyte-derived dendritic cells through NF-κB and PI3K/MAPK pathways. Evid Based Complement Alternat Med 2012:1–12

Cunningham S, Joshi L (2010) In: Kole C (ed) Transgenic crop plants. Springer, Berlin, pp 343–357
Dawczynski C, Schubert R, Jahreis G (2007) Amino acids, fatty acids, and dietary fibre in edible seaweed products. Food Chem 103(3):891–899
de Souza ET, de Lira DP, de Queiroz AC, da Silva DJ, de Aquino AB, Mella EA, Lorenzo VP, de Miranda GE, de Araújo-Júnior JX, Chaves MC, Barbosa-Filho JM, de Athayde-Filho PF, Santos BV, Alexandre-Moreira MS (2009) The antinociceptive and anti-inflammatory activities of caulerpin, a bisindole alkaloid isolated from seaweeds of the genus *Caulerpa*. Mar Drugs 7 (4):689–704
Desai SD, Desai DG, Kaur H (2009) Saponins and their biological activities. Pharm Times 41 (3):13–16
Durand-Chastel H (1980) Production and use of Spirulina in Mexico. In: Shelef G, Soeder CJ (eds) Algae Biomass. Elsevier North Holland Biomedical Press, Amsterdam, pp 51–64
Ebrahimzadeh MA, Khalili M, Dehpour AA (2018) Antioxidant activity of ethyl acetate and methanolic extracts of two marine algae, *Nannochloropsis oculata* and *Gracilaria gracilis*-an in vitro assay. Braz J Pharm Sci 54(1)
El-Baky HH, El Baz FK, El-Baroty GS (2009) Production of phenolic compounds from *Spirulina maxima* microalgae and its protective effects. Afr J Biotechnol 8(24):7059–7067
Etahiri S, El Kouri A, Butel-Poncé V, Guyot M, Assobhei O (2007) Antibacterial bromophenol from the marine red algae *Pterosiphonia complanata*. Nat Prod Commun 2(7):749–752
Fasya AG, Baderos A, Madjid ADR, Amalia S, Megawati DS (2019, July) Isolation, identification and bioactivity of steroids compounds from red algae *Eucheuma cottonii* petroleum ether fraction. AIP Conf Proc 2120(1):030025
Fauzi A, Satriani Lamma MR (2018) Total tannin levels analysis of brown algae (*Sargassum* sp. and *Padina* sp.) to prevent blood loss in surgery. J Dentomaxillofacial Sci 3(1):37–40
Feroz B (2018) Saponins from marine macroalgae: a review. J Mar Sci Res Dev 8(4). ISSN: 2155-9910
Ferraces-Casais P, Lage-Yusty MA, De Quirós ARB, López-Hernández J (2012) Evaluation of bioactive compounds in fresh edible seaweeds. Food Anal Methods 5(4):828–834
Freile-Pelegrín Y, Robledo D (2013) Bioactive phenolic compounds from algae. In: Hernández-Ledesma B, Herrero M (eds) Bioactive compounds from marine foods: plant and animal sources. Wiley, Chichester, pp 113–129
Galasso C, Gentile A, Orefice I, Ianora A, Bruno A, Noonan DM, Sansone C, Albini A, Brunet C (2019) Microalgal derivatives as potential nutraceutical and food supplements for human health: a focus on cancer prevention and interception. Nutrients 11(6):1226
Garcia-Vaquero M, Rajauria G, O'Doherty J, Torres S (2017) Polysaccharides from macroalgae: recent advances, innovative technologies and challenges in extraction and purification. Food Res Int 99:1011–1020
Generalić Mekinić I, Skroza D, Šimat V, Hamed I, Čagalj M, Popović Perković Z (2019) Phenolic content of Brown algae (Pheophyceae) species: extraction, identification, and quantification. Biomolecules 9(6):244
Goiris K, Muylaert K, Voorspoels S, Noten B, De Paepe D, Baart GJE, De Cooman L (2014) Detection of flavonoids in microalgae from different evolutionary lineages. J Phycol 50 (3):483–492
Gross H, Goeger DE, Hills P, Mooberry SL, Ballantine DL, Murray TF, Valeriote FA, Gerwick WH (2006) Lophocladines, bioactive alkaloids from the red alga Lophocladia sp. J Nat Prod 69 (4):640–644
Guedes AC, Amaro HM, Sousa-Pinto I, Malcata FX (2019) Algal spent biomass—a pool of applications. In: Pandey A (ed) Biofuels from Algae. Elsevier, London, pp 397–433
Guven KC, Bora A, Sunam G (1969) Alkaloid content of marine algae. I. *Hordenine from Phyllophora nervosa*. Eczacılık Bul 11:177–184
Guven KC, Bora A, Sunam G (1970) Hordenine from the alga *Phyllophora nervosa*. Phytochemistry 9:1893
Güven KC, Percot A, Sezik E (2010) Alkaloids in marine algae. Mar Drugs 8(2):269–284

Güven K, Coban B, Sezik E, Erdugan H, Kaleağasıoğlu F (2013) Alkaloids of marine macroalgae. In: Ramawat KG, Mérillon J-M (eds) Natural products: phytochemistry, botany and metabolism of alkaloids, phenolics and terpenes. Springer, Berlin, pp 25–37

Harizani M, Ioannou E, Roussis V (2016) The Laurencia paradox: an endless source of chemodiversity. Prog Chem Org Nat Prod 102:91–252

Heo SJ, Park EJ, Lee KW, Jeon YJ (2005) Antioxidant activities of enzymatic extracts from brown seaweeds. Bioresour Technol 96(14):1613–1623

Hill AF (1952) Economic Botany. A textbook of useful plant and plant Products, 2nd edn. MC-Graw-Hill Book, New York. 743p

Holdt SL, Kraan S (2011) Bioactive compounds in seaweed: functional food applications and legislation. J Appl Phycol 23:543–597

Holken Lorensi G, Soares Oliveira R, Leal AP, Zanatta AP, Moreira de Almeida CG, Barreto YC, Batista Pereira A (2019) Entomotoxic activity of *Prasiola crispa* (Antarctic algae) in *Nauphoeta cinerea* cockroaches: identification of Main steroidal compounds. Mar Drugs 17(10):573

Huheihel M, Ishanu V, Tal J, Arad S (2002) Activity of *Porphyridium* sp. polysaccharide against herpes simplex in vitro and in vivo. J Biochem Biophys Methods 20:189–200

Hussein RA, El-Anssary AA (2018) Plants secondary metabolites: the key drivers of the pharmacological actions of medicinal plants. In: Herbal medicine. IntechOpen

Ibrahim M, Salman M, Kamal S, Rehman S, Razzaq A, Akash SH (2017) Algae-based biologically active compounds. In: Algae based polymers, blends, and composites. Elsevier, pp 155–271

Impellizzeri G, Mangiafico S, Oriente G, Piattelli M, Sciuto S, Fattorusso E, Magno S, Santacroce C, Sica D (1975) Amino acids and low-molecular-weight carbohydrates of some marine red algae. Phytochemistry 14(7):1549–1557

Jerez-Martel I, García-Poza S, Rodríguez-Martel G, Rico M, Afonso-Olivares C, Gómez-Pinchetti JL (2017) Phenolic profile and antioxidant activity of crude extracts from microalgae and cyanobacteria strains. J Food Qual 2017:2924508

Jiao G, Yu G, Zhang J, Ewart H (2011) Chemical structures and bioactivities of sulfated polysaccharides from marine algae. Mar Drugs 9(2):196–223

Jorgensen EG (1962) Antibiotic substances from cells and culture solutions of unicellular algae with special reference to some chloropbyll derivatives. Plant Physiol 15:530–545

Kabera JN, Semana E, Mussa AR, He X (2014) Plant secondary metabolites: biosynthesis, classification, function and pharmacological properties. J Pharm Pharmacol 2:377–392

Kadam SU, Tiwari BK, O'Donnell CP (2013) Application of novel extraction technologies for bioactives from marine algae. J Agric Food Chem 61:4667–4675

Kang K, Park Y, Hwang HJ, Kim SH, Lee JG, Shin HC (2003) Antioxidative properties of brown algae polyphenolics and their perspectives as chemopreventive agents against vascular risk factors. Arch Pharm Res 26(4):286–293

Kapetanović R, Sladić DM, Popov S, Zlatović MV, Kljajić Z, Gašić MJ (2005) Sterol composition of the Adriatic Sea algae Ulva lactuca, Codium dichotomum, Cystoseira adriatica and Fucus virsoides. J Serb Chem Soc 70(12):1395–1400

Kazłowska K, Hsu T, Hou C-C, Yang W-C, Tsai G-J (2010) Anti-inflammatory properties of phenolic compounds and crude extract from Porphyra dentata. J Ethnopharmacol 128(1):123–130

Khan W, Rayirath UP, Subramanian S, Jithesh MN, Rayorath P, Hodges DM, Critchley AT, Craigie JS, Norrie J, Prithiviraj B (2009) Seaweed extracts as biostimulants of plant growth and development. J Plant Growth Regul 28:386–399

Kidgell JT, Magnusson M, de Nys R, Glasson CR (2019) Ulvan: a systematic review of extraction, composition and function. Algal Res 39:101422

Kneifel H, Meinicke M, Soeder ÇJ (1977) Analysis of amines in algae by high performance liquid chromatography. J Phycol 13:36

Knott MG, Mkwananzi H, Arendse CE, Hendricks DT, Bolton JJ, Beukes DR (2005) Plocoralides A–C, polyhalogenated monoterpenes from the marine alga *Plocamium corallorhiza*. Phytochemistry 66(10):1108–1112

Koller M, Muhr A, Braunegg G (2014) Microalgae as versatile cellular factories for valued products. Algal Res 6:52–63

Kraan S (2012) Algal polysaccharides, novel applications and outlook. In: Carbohydrates-comprehensive studies on glycobiology and glycotechnology, IntechOpen

Krishnamurti C, Rao SC (2016) The isolation of morphine by Serturner. Indian J Anaesth 60 (11):861

Kuda T, Ikemori T (2009) Minerals, polysaccharides and antioxidant properties of aqueous solutions obtained from macroalgal beach-casts in the Noto Peninsula, Ishikawa, Japan. Food Chem 112(3):575–581

Kuda T, Kunii T, Goto H, Suzuki T, Yano T (2007) Varieties of antioxidant and antibacterial properties of *Ecklonia stolonifera* and *Ecklonia kurome* products harvested and processed in the Noto peninsula, Japan. Food Chem 103(3):900–905

Kurniawati HA, Ismadji S, Liu JC (2014) Microalgae harvesting by flotation using natural saponin and chitosan. Bioresour Technol 166:429–434

Lee YS, Shin KH, Kim BK, Lee S (2004) Anti-diabetic activities of fucosterol from *Pelvetia siliquosa*. Arch Pharm Res 27(11):1120–1122

Leelavathi MS, Prasad MP (2015) Comparitive analysis of phytochemical compounds of marine algae isolated from Gulf of Mannar. World J Pharm Pharm Sci 4(5):640–654

Leonard SG, Sweeney T, Pierce KM, Bahar B, Lynch BP, O'Doherty JV (2010) The effects of supplementing the diet of the sow with seaweed extracts and fish oil on aspects of gastrointestinal health and performance of the weaned piglet. Livest Sci 134:135–138

Li K, Li XM, Ji NY, Wang BG (2007) Natural bromophenols from the marine red alga *Polysiphonia urceolata* (Rhodomelaceae): structural elucidation and DPPH radical-scavenging activity. Bioorg Med Chem 15(21):6627–6631

Li YX, Kim SK (2011) Utilization of seaweed derived ingredients as potential antioxidants and functional ingredients in the food industry: an overview. Food Sci Biotechnol 20(6):1461–1466

Li, Y., Qian, Z. J., Le, Q. T., Kim, M. M., & Kim, S. K. (2008). Bioactive phloroglucinoi derivatives isolated from an edible marine brown alga, *Eckionia cava*. In: 13th International Biotechnology Symposium and Exhibition (第 13 届 IUPAC 国际生物工程会议) (pp. 578-578). 大连理工大学.

Li-Beisson, Y., Thelen, J.J., Fedosejevs, E. and Harwood, J.L., 2019. The lipid biochemistry of eukaryotic algae. Prog Lipid Res74:31-68. doi: https://doi.org/10.1016/j.plipres.2019.01.003.

Lim SN, Cheung PCK, Ooi VEC, Ang PO (2002) Evaluation of antioxidative activity of extracts from a brown seaweed, *Sargassum siliquastrum*. J Agric Food Chem 50(13):3862–3866

Lin JL, Liang YQ, Liao XJ, Yang JT, Li DC, Huang YL, Jiang ZH, Xu SH, Zhao BX (2019) Acanthophoraine A, a new pyrrolidine alkaloid from the red alga *Acanthophora spicifera*. Nat Prod Res 1–6. doi:https://doi.org/10.1080/14786419.2019.1569008

Liu DQ, Mao SC, Zhang HY, Yu XQ, Feng MT, Wang B, Feng LH, Guo YW (2013) Racemosins A and B, two novel bisindole alkaloids from the green alga *Caulerpa racemosa*. Fitoterapia 91:15–20

Lopes G, Sousa C, Valentão P, Andrade PB (2013) Sterols in algae and health. In: Hernández-Ledesma B, Herrero M (eds) Bioactive compounds from marine foods: plant and animal sources. Wiley, Chichester, pp 173–191

López A, Rico M, Santana-Casiano JM, González AG, González-Dávila M (2015) Phenolic profile of *Dunaliella tertiolecta* growing under high levels of copper and iron. Environ Sci Pollut Res 22(19):14820–14828

Lourenço SO, Barbarino E, De-Paula JC, Pereira LODS, Marquez UML (2002) Amino acid composition, protein content and calculation of nitrogen-to-protein conversion factors for 19 tropical seaweeds. Phycol Res 50(3):233–241

Lüning K, Pang S (2003) Mass cultivation of seaweeds: current aspects and approaches. J Appl Phycol 15(2-3):115–119

Luo X, Su P, Zhang W (2015) Advances in microalgae-derived phytosterols for functional food and pharmaceutical applications. Mar Drugs 13(7):4231–4254

Machu L, Misurcova L, Vavra Ambrozova J, Orsavova J, Mlcek J, Sochor J, Jurikova T (2015) Phenolic content and antioxidant capacity in algal food products. Molecules 20(1):1118–1133

Manilal A, Sujith S, Kiran GS, Selvin J, Shakir C, Gandhimathi R, Panikkar MVN (2009) Bio-potentials of seaweeds collected from southwest coast of India. J Mar Sci Technol 17:67–73

Mao SC, Guo YW, Shen X (2006) Two novel aromatic valerenane-type sesquiterpenes from the Chinese green alga *Caulerpa taxifolia*. Bioorg Med Chem Lett 16(11):2947–2950

Markou G, Angelidaki I, Georgakakis D (2012) Microalgal carbohydrates: an overview of the factors influencing carbohydrates production, and of main bioconversion technologies for production of biofuels. Appl Microbiol Biotechnol 96(3):631–645

Mayer AMS, Hamann MT (2004) Marine pharmacology in 2000: marine compounds with antibacterial, anticoagulant, antifungal, anti-inflammatory, antimalarial, antiplatelet, antituberculosis, and antiviral activities; affecting the cardiovascular, immune, and nervous system and other miscellaneous mechanisms of action. Mar Biotechnol 6:37–52

Meneses MM, Flores MEJ (2019) Flavonoids: a promising therapy for obesity due to the high-fat diet. In: Flavonoids-a coloring model for cheering up life. IntechOpen

Messyasz B, Michalak I, Łęska B, Schroeder G, Górka B, Korzeniowska K, Lipok J, Wieczorek P, Rój E, Wilk R, Dobrzyńska-Inger A, Górecki H, Chojnacka K (2018) Valuable natural products from marine and freshwater macroalgae obtained from supercritical fluid extracts. J Appl Phycol 30(1):591–603

Michalak I, Chojnacka K (2015) Algae as production systems of bioactive compounds. Eng Life Sci 15(2):160–176

Michel G, Tonon T, Scornet D, Cock JM, Kloareg B (2010a) Central and storage carbon metabolism of the brown alga *Ectocarpus siliculosus*: insights into the origin and evolution of storage carbohydrates in eukaryotes. New Phytol 188(1):67–81

Michel G, Tonon T, Scornet D, Cock JM, Kloareg B (2010b) The cell wall polysaccharide metabolism of the brown alga *Ectocarpus siliculosus*. Insights into the evolution of extracellular matrix polysaccharides in eukaryotes. New Phytol 188(1):82–97

Miller LH, Su X (2011) Artemisinin: discovery from the Chinese herbal garden. Cell 146(6):855–858

Montero L, del Pilar Sánchez-Camargo A, Ibáñez E, Gilbert-López B (2018) Phenolic compounds from edible algae: bioactivity and health benefits. Curr Med Chem 25(37):4808–4826

Montero L, Sánchez-Camargo AP, García-Cañas V, Tanniou A, Stiger-Pouvreau V, Russo M, Rastrelli L, Cifuentes A, Herrero M, Ibáñez E (2016) Anti-proliferative activity and chemical characterization by comprehensive two-dimensional liquid chromatography coupled to mass spectrometry of phlorotannins from the brown macroalga *Sargassum muticum* collected on North-Atlantic coasts. J Chromatogr A 1428:115–125

Moore RE, Cheuk C, Yang X, Patterson GML, Bonjouklian R, Smitka TA, Mynderse JS, Foster RS, Jones ND, Swartzendruber JK, Deeter JB (2002) Hapalindoles, antibacterial and antimycotic alkaloids from the cyanophyte Hapalosiphon fontinalis. J Org Chem 52(6):1036–1043

Moreau D, Tomasoni C, Jacquot C, Kaas R, Le Guedes R, Cadoret JP, Muller-Feuga A, Kontiza I, Vagias C, Roussis V, Roussakis C (2006) Cultivated microalgae and the carotenoid fucoxanthin from *Odontella aurita* as potent anti-proliferative agents in bronchopulmonary and epithelial cell lines. Environ Toxicol Pharmacol 22(1):97–103

Nagatsu A, Kajitani H, Sakakibara J (1995) Muscoride a: a new oxazole peptide alkaloid from freshwater cyanobacterium *Nostoc muscorum*. Tetrahedron Lett 36(23):4097–4100

Neyrinck AM, Mouson A, Delzenne NM (2007) Dietary supplementation with laminarin, a fermentable marine β (1–3) glucan, protects against hepatotoxicity induced by LPS in rat by modulating immune response in the hepatic tissue. Int Immunopharmacol 7(12):1497–1506

Nguyen THM (2012) Bioethanol production from marine algae biomass: prospect and troubles. J Vietnam Environ 3(1):25–29

O'Doherty JV, Dillon S, Figat S, Callan JJ, Sweeney T (2010) The effects of lactose inclusion and seaweed extract derived from Laminaria spp. on performance, digestibility of diet components and microbial populations in newly weaned pigs. Anim Feed Sci Technol 157(3-4):173–180

Omer TA, Attar T (2013) Isolation and identification of some chemical constituents in two different types of fresh water macro-algae in Bestansur Village in Suleiman city Kurdistan region (North Iraq) by HPLC technique. J Appl Chem (IOSR-JAC) 4(3):45–55

Pal A, Kamthania MC, Kumar A (2014) Bioactive compounds and properties of seaweeds—a review. Open Access Libr J 1:1–17

Palanisamy SK, Arumugam V, Rajendran S, Ramadoss A, Nachimuthu S, Magesh Peter D, Sundaresan U (2019) Chemical diversity and anti-proliferative activity of marine algae. Nat Prod Res 33(14):2120–2124

Panis G, Carreon JR (2016) Commercial astaxanthin production derived by green alga *Haematococcus pluvialis*: a microalgae process model and a techno-economic assessment all through production line. Algal Res 18:175–190

Papalia T, Sidari R, Panuccio MR (2019) Impact of different storage methods on bioactive compounds in *Arthrospira platensis* biomass. Molecules 24(15):2810

Percot A, Güven KC, Aysel V, Erduğan H, Gezgin T (2009a) N-acetyltyramine from phyllophora crispa (Hudson) ps Dixon and n-acetylphenylethylamine from *Gelidium crinale* (hare ex turner) craillon. Acta Pharm Sci 51(1)

Percot A, Yalçın A, Aysel V, Erdugan H, Dural B, Güven KC (2009b) β-Phenylethylamine content in marine algae around Turkish coasts. Bot Mar 52(1):87–90

Pereira HS, Leão-Ferreira LR, Moussatché N, Teixeira VL, Cavalcanti DN, Costa LJ, Diaz R, Frugulhetti IC (2004) Antiviral activity of diterpenes isolated from the Brazilian marine alga *Dictyota menstrualis* against human immunodeficiency virus type 1 (HIV-1). Antiviral Res 64(1):69–76

Pham-Huy LA, He H, Pham-Huy C (2008) Free radicals, antioxidants in disease and health. Int J Biomed Sci 4(2):89

Plaza M, Cifuentes A, Ibáñez E (2008) In the search of new functional food ingredients from algae. Trends Food Sci Technol 19(1):31–39

Pulz O, Gross W (2004) Valuable products from biotechnology of microalgae. Appl Microbiol Biotechnol 65(6):635–648

Raja R, Hemaiswarya S, Ashok Kumar N, Sridhar S, Rengasamy R (2008) A perspective on the biotechnological potential of microalgae. Crit Rev Microbiol 34:77–88

Ramdani M, Image P, Elasri O, Image P, Saidi N, Image P, Elkhiati N, Image P, Taybi AF, Image P, Mostareh M, Image P, Zaraali O, Image P, Haloui B, Image P, Ramdani M (2017) Evaluation of antioxidant activity and total phenol content of *Gracilaria bursa-pastoris* harvested in Nador lagoon for an enhanced economic valorization. Chem Biol Technol Agric 4(1):28

Rangel-Yagui CO, Godoy-Danesi ED, Carvalho JCM, Sato S (2004) Chlorophyll production from *Spirulina platensis*: cultivation with urea addition by fed-batch process. Bioresour Technol 92:114–133

Raposo de Jesus MF, de Morais RMSC, de Morais AMMB (2013a) Health applications of bioactive compounds from marine microalgae. Life Sci 93(15):479–486

Raposo de Jesus MF, de Morais RMSC, de Morais AMMB (2013b) Bioactivity and applications of sulphated polysaccharides from marine microalgae. Mar Drugs 11:233–252

Rasmussen HE, Blobaum KR, Jesch ED, Ku CS, Park YK, Lu F, Carr TP, Lee JY (2009) Hypocholesterolemic effect of *Nostoc commune* var. sphaeroides Kützing, an edible blue-green alga. Eur J Nutr 48(7):387–394

Reddy P, Urban S (2009) Meroditerpenoids from the southern Australian marine brown alga *Sargassum fallax*. Phytochemistry 70(2):250–255

Richmond A (1988) Spirulina. In: Borowitzka A, Borowitzka L (eds) Microalgal biotechnology. Cambridge University Press, Cambridge, pp 83–121

Rickards RW, Rothschild JM, Willis AC, de Chazal NM, Kirk J, Kirk K, Smith GD (1999) Calothrixins a and B, novel pentacyclic metabolites from Calothrix cyanobacteria with potent activity against malaria parasites and human cancer cells. Tetrahedron 55(47):13513–13520

Rioux LE, Turgeon SL, Beaulieu M (2007) Characterization of polysaccharides extracted from brown seaweeds. Carbohydr Polym 69:530–537

Rolle I, Hobucher HE, Kneifel H, Paschold B, Riepe W, Soeder CJ (1977) Amines in unicellular green algae: 2. Amines in *Scenedesmus acutus*. Anal Biochem 77(1):103–109

Saavedra JM (1978) β-Phenylethylamine: is this biogenic amine related to neuropsychiatic disease. Mod Pharmacol Toxicology 12:139–157

Sabina H, Aliya R (2009) Seaweed as a new source of flavone, scutellarein 4′-methyl-ether. Pak J Bot 41(4):1927–1930

Sallehudin NJ, Raus RA, Mustapa M, Othman R, Mel M (2018) Screening of lutein content in several fresh-water microalgae. Int Food Res J 25(6)

Sampath-Wiley P, Neefus CD, Jahnke LS (2008) Seasonal effects of sun exposure and emersion on intertidal seaweed physiology: fluctuations in antioxidant contents, photosynthetic pigments and photosynthetic efficiency in the red alga *Porphyra umbilicalis* Kützing (Rhodophyta, Bangiales). J Exp Mar Biol Ecol 361(2):83–91

Santoyo S, Plaza M, Jaime L, Ibanez E, Reglero G, Senorans J (2011) Pressurized liquids as an alternative green process to extract antiviral agents from the edible seaweed *Himanthalia elongata*. J Appl Phycol 23:909–917

Saravana PS, Cho YJ, Park YB, Woo HC (2016) Structural, antioxidants and emulsifying activities of fucoidan from *Saccharina japonica* using pressurized liquid extraction. Carbohydr Polym 153:518–525

Sathasivam R, Radhakrishnan R, Hashem A, Abd_Allah EF (2019) Microalgae metabolites: a rich source for food and medicine. Saudi J Biol Sci 26(4):709–722

Schenk PM, Thomas-Hall SR, Stephens E, Marx UC, Mussgnug JH, Posten C, Hankamer B (2008) Second generation biofuels: high-efficiency microalgae for biodiesel production. Bioenergy Res 1(1):20–43

Servaites JC, Faeth JL, Sidhu SS (2012) A dye binding method for measurement of total protein in microalgae. Anal Biochem 421(1):75–80

Shan X, Liu X, Hao J, Cai C, Fan F, Dun Y, Zhao X, Liu X, Li C, Yu G (2016) In vitro and in vivo hypoglycemic effects of brown algal fucoidans. Int J Biol Macromol 82:249–255

Shannon E, Abu-Ghannam N (2016) Antibacterial derivatives of marine algae: an overview of pharmacological mechanisms and applications. Mar Drugs 14(4):81

Shields RJ, Lupatsch I (2012) Algae for aquaculture and animal feeds. Technikfol Theorie Praxis 21:2337

Shoubaky GA, Salem EA (2014) Terpenes and sterols composition of marine brown algae *Padina pavonica* (Dictyotales) and *Hormophysa triquetra* (Fucales). Int J Pharmacogn Phytochem Res 6(4):894–900

Shoubaky GAE, Abdel-Daim MM, Mansour MH, Salem EA (2016) Isolation and identification of a flavone apigenin from marine red alga *Acanthophora spicifera* with antinociceptive and anti-inflammatory activities. J Exp Neurosci 10:JEN-S25096

Shushizadeh MR (2019) Gas chromatography-mass evaluation of Terpenoids from Persian gulf *Padina tetrastromatica* sp. Asian J Pharm (AJP) 12(04)

Singh IP, Sidana J (2013) Phlorotannins. In: Functional ingredients from algae for foods and nutraceuticals. Woodhead, pp 181–204

Sivagnanam SP, Yin S, Choi JH, Park YB, Woo HC, Chun BS (2015) Biological properties of fucoxanthin in oil recovered from two brown seaweeds using supercritical CO_2 extraction. Mar Drugs 13:3422–3442

Soletto D, Binaghi L, Lodi A, Carvalho JCM, Converti A (2005) Batch and fedbatch cultivations of *Spirulina platensis* using ammonium sulphate and urea as nitrogen sources. Aquaculture 243:217–224

Solovchenko AE, Khozin-Goldberg I, Didi-Cohen S, Cohen Z, Merzlyak MN (2008) Effects of light intensity and nitrogen starvation on growth, total fatty acids and arachidonic acid in the green microalga *Parietochloris incise*. J Appl Phycol 20:225–245

Song MY, Ku SK, Han JS (2012) Genotoxicity testing of low molecular weight fucoidan from brown seaweeds. Food Chem Toxicol 50(3-4):790–796

Soong P (1980) Production and development of chlorella and Spirulina in Taiwan. In: Shelef G, Soeder CJ (eds) Algae biomass. Elsevier, Amsterdam, pp 97–113

Souza BWS, Cerqueira MA, Bourbon AI, Pinheiro AC, Martins JT, Teixeira JA, Coimbra MA, Vicente AA (2012) Chemical characterization and antioxidant activity of sulfated polysaccharide from the red seaweed *Gracilaria birdiae*. Food Hydrocoll 27(2):287–292

Spolaore P, Joannis-Cassan C, Duran E, Isambert A (2006) Commercial applications of microalgae. J Biosci Bioeng 101(2):87–96

Stabili L, Acquaviva MI, Biandolino F, Cavallo RA, De Pascali SA, Fanizzi FP, Narracci M, Cecere E, Petrocelli A (2014) Biotechnological potential of the seaweed *Cladophora rupestris* (Chlorophyta, Cladophorales) lipidic extract. N Biotechnol 31(5):436–444

Steiner M, Hartmann T (1968) Über Vorkommen und Verbreitung flüchtiger Amine bei Meeresalgen. Planta 79(2):113–121

Stengel DB, Connan S, Popper ZA (2011) Algal chemodiversity and bioactivity: sources of natural variability and implications for commercial application. Biotechnol Adv 29(5):483–501

Stiger-Pouvreau V, Bourgougnon N, Deslandes E (2016) Carbohydrates from seaweeds. In: Fleurence J, Levine I (eds) *Seaweed in health and disease prevention*. Academic, London, pp 223–274

Sudhakar MP, Kumar BR, Mathimani T, Arunkumar K (2019) A review on bioenergy and bioactive compounds from microalgae and macroalgae-sustainable energy perspective. J Clean Prod. https://doi.org/10.1016/j.jclepro.2019.04.287

Sumayya, S. S., & Murugan, K. (2019). Antioxidant potentialities of marine red algae *Gracillaria dura*: a search.

Synytsya A, Kim W-J, Kim S-M, Pohl R, Synytsya A, Kvasnicka F, Copíková J, Park YI (2010) Structure and antitumour activity of fucoidan isolated from sporophyll of Korean brown seaweed *Undaria pinnatifida*. Carbohydr Polym 81(1):41–48

Tamiya H (1957) Mass culture of algae. Annu Rev Plant Physiol 8:309–344

Templeton DW, Laurens LM (2015) Nitrogen-to-protein conversion factors revisited for applications of microalgal biomass conversion to food, feed and fuel. Algal Res 11:359–367

Tringali C (1997) Bioactive metabolites from marine algae: recent results. Curr Org Chem 1:375–394

United States Environmental Protection Agency (EPA) (2014) 2014 Toxics release inventory national analysis complete report. https://www.epa.gov/toxics-release-inventory-tri-program/2014-trinational-analysis-complete-report. Accessed 4 Sept 2017

Villarruel-López A, Ascencio F, Nuño K (2017) Microalgae, a potential natural functional food source—a review. Polish J Food Nutr Sci 67(4):251–264

Volk RB (2005) Screening of microalgal culture media for the presence of algicidal compounds and isolation and identification of two bioactive metabolites, excreted by the cyanobacteria *Nostoc insulare* and *Nodularia harveyana*. J Appl Phycol 17(4):339–347

Waghmode AV, Khilare CJ (2018) RP-HPLC profile of major phenolics from brown marine macro algae. J Appl Pharm 10(262):2

Watanabe F (2007) Vitamin B12 sources and bioavailability. Exp Biol Med 232(10):1266–1274

Wen W, Li K, Alseekh S, Omranian N, Zhao L, Zhou Y, Xiao Y, Jin M, Yang N, Liu H, Florian A, Li W, Pan Q, Nikoloski Z, Yan J, Fernie AR (2015) Genetic determinants of the network of primary metabolism and their relationships to plant performance in a maize recombinant inbred line population. Plant Cell 27(7):1839–1856

Wijffels RH, Barbosa MJ (2010) An outlook on microalgal biofuels. Science 329:796–799

Ye H, Wang K, Zhou C, Liu J, Zeng X (2008) Purification, antitumor and antioxidant activities in vitro of polysaccharides from the brown seaweed *Sargassum pallidum*. Food Chem 111:428–432

Zolotareva EK, Mokrosnop VM, Stepanov SS (2019) Polyphenol compounds of macroscopic and microscopic algae. Int J Algae 21(1)

Zou Y, Qian ZJ, Li Y, Kim MM, Lee SH, Kim SK (2008) Antioxidant effects of phlorotannins isolated from Ishige okamurae in free radical mediated oxidative systems. J Agric Food Chem 56(16):7001–7009

Zubia M, Payri C, Deslandes E (2008) Alginate, mannitol, phenolic compounds and biological activities of two range-extending brown algae, *Sargassum mangarevense* and *Turbinaria ornata* (Phaeophyta: Fucales), from Tahiti (French Polynesia). J Appl Phycol 20(6):1033–1043

Bioactive Metabolites in Gymnosperms

14

Athira V. Anand, Vivek Arinchedathu Surendran, and Swapna Thacheril Sukumaran

Abstract

Exploration of flora for health-stimulating agents has a long history, and plants and plant-derived compounds continue to augment modern medicine. The taxa Gymnosperms are a unique collection of plants with ethnomedicinal and economic implications. This chapter aims at bringing together the attempts made by researches across the world to investigate the biological effects of Gymnosperm metabolites. Plants were assessed under the groupings *viz* Cycadales, Coniferales and Gnetales. It was observed that Cycadales and Gnetales were least explored and their medicinal chemistry could be exploited further. Along with other therapeutic phytochemicals, specific compounds *viz* cycasin, were reported from Cycadales. Like Coniferales being the most diverse Gymnosperms, diverse metabolites were identified from its members through numerous therapeutic investigations. *Araucaria, Cupressus, Pinus, Podocarpus, Taxus, Cephalotaxus,* etc. are some of the well-explored genera. Cephalotaxine–alkaloids, Homoharringtonine, Taxol and Taxine alkaloids are some alleviative compounds in modern medicine, originated from Coniferales. Numerous other molecules are reported, but their medicinal possibilities are not investigated to the fullest. Studies on compounds like isorhapontigenin, piceatannol, gnetol etc., identified from *Gnetum* and *Ephedra* reported multifaceted bioactivities. Gymnosperms have the potential to contribute promising ligands to drug discovery but successful preliminary studies are not followed up in many plants, which results in anonymity regarding the active compound. Research in this area can lead to the discovery of numerous prospective drug molecules in the future.

A. V. Anand · V. Arinchedathu Surendran · S. T. Sukumaran (✉)
Department of Botany, University of Kerala, Thiruvananthapuram, Kerala, India

> **Keywords**
>
> Bioactivity · Metabolites · Gymnosperm · Homoharringtonine · Cycadales · Coniferales · Gnetales

14.1 Introduction

The 'Gymnospermae' is the appellation of a group of plants with ancient (395–359 million years ago) origin (Biswas and Johri 1997). Like the term gymnospermae suggests, this group of plants possess naked ovules. Going with the evolution of the term Gymnospermae to Gymnosperms, there is a series of gymnosperm classification starting from Engler (1886) to Bierhorst (1971).

The members of Gymnosperms are having economic, therapeutic, aesthetic and ecological significance. This taxa contain the tallest, largest, thickest and oldest plants on earth. The therapeutic potential of Gymnosperms or any plant is attributed to the diverse phytochemicals they possess, biologically active metabolites in particular. Here bioactive metabolites identified and isolated from Gymnosperms by researchers across the world are analysed.

14.2 Bioactive Metabolites from Natural Sources

A bioactive compound is simply a substance that has a biological activity (Chin et al. 2009). Such compounds can have a positive or negative impact on some life processes that can influence the functioning of the biological system. Many bioactive compounds have been isolated from microorganisms and are being administered widely against many diseases. The discovery of antibiotic penicillin from fungus in 1928 inspired the discovery of 'cephalosporin' cefprozil, antidiabetic agent 'acarbose' and anticancer agent 'epirubicin' (Chin and Balunas 2006). Over the years, antibacterial drugs have been developed from natural compounds like β-lactams (cephalosporins), tetracyclines (Glycylcyclines), macrolides (Erythromycin and Rifamycin analogues), Spectinomycins and Glycopeptides-like 'Vancomycin' and 'Teicoplanin' analogues (Shu 1998).

There are reports regarding the use of *Huperzia serrate* leaves in China for boosting the memory power. An acetylcholinesterase inhibitor, huperzine-A, isolated from this plant is effective in the treatment of neurodegenerative disorders. Bioactive principles from many animals are also being used in the modern medicine. An alkaloid 'Epibatidine' isolated from an Ecuadorian poison frog *Epipedobates tricolor* was found to be more potent than morphine (Spande et al. 1992). The anticoagulant 'Heparin' and many vaccines like Rotarix, RotaTeq, Zostavax etc. are derived from porcine. Zootoxins from poisonous animals have also proved to be important pharmacophores. Venom from a Brazilian viper, Teprotide, has led to the development of two anti-hypertensive agents Cilazapril and Captopril. The most depending source of bioactive metabolites having beneficial health effects in man and animals are of plant origin (Zhao et al. 2015).

Plant-based compounds form a major share of the natural products used in medicine. Most of the presently administered drugs are either directly formulated from natural products or derived from them through the application of chemical synthesis method. There are numerous bioactive molecules isolated from plants that are effective anticancer, antibacterial, antifungal antimalarial, hepatoprotective, larvicidal, anti-inflammatory and antidiuretic agents.

Plants contain immense phytochemicals with consequential biological activities. A large proportion of mankind still relies on plants or plant-based products for curing various ailments. There are different traditional systems of medicine in different parts of the world that are widely accepted and are mainly banking on natural products. WHO stated that around 65–80% of the world's population depends mainly on plant-based traditional remedies for their primary health care (Farnsworth et al. 1985).

The bioactive metabolites in plants are the secondary metabolites derived from the primary metabolites through various pathways. The evolution of enzymatic machinery of living entities by means of evolutionary pressure leads to the synthesis of new secondary metabolites from adjusted biosynthetic pathways, to cope with varying living conditions and stress. Other than reasons like competition and coevolution, need of exquisite communication, defence, synergism and predation became a cause to the production of variety classes of secondary metabolites in the living system (Bohlin et al. 2012). In 1981, Kossel introduced the term 'Secondary' to the field of bio-science. He discriminated the primary metabolites from the 'metabolome' of the organism. Like the term secondary, the importance of this category of compounds is secondary to the plant in its lifecycle. Hence this compound directly does not curtail the existence of the plant even though it dilatorily affects the existence of the organism (Tiwari and Rana 2015; Thirumurugan et al. 2018). The secondary metabolites exist as a transitional zone since the C and N stored in plant could be brought back to primary metabolites, when on demand (Collin 2001; Thirumurugan et al. 2018).

Secondary metabolites serve the functions other than the growth of the plant, like defence. Each plant contains unique secondary metabolites that provide distinct characters to the plant. These compounds if enter into the human circulatory system modify the metabolic processes. The activity of such metabolic regulators includes antioxidant, inhibition or stimulation of enzymes, inhibition of receptor accomplishments and induction and inhibition of gene expression (Correia et al. 2012).

Plants around the globe are prominent components of traditional and modern systems of medicine, nutraceuticals, food supplements, pharmaceutical intermediates and chemical entities for the development of synthetic drugs (Ncube et al. 2008). Scientists across the world investigated plants for their ability to cure various ailments and management of chronic wounds due to the presence of life-supporting phytoconstituents in them (Nayak 2006). Currently, the quest for novel bioactive molecules is on a different path where ethnobotany and ethnopharmacognosy are being used as a guide, which will lead towards the discovery of diverse sources and classes of potent compounds (Gurib-Fakim 2006).

The traditional knowledge plays an important role here. Majority of research works justify the use of specific plants by traditional healers to cure various diseases, based on the activity of the respective bioactive molecule present. Indigenous knowledge-based screening of plants for active molecules leads to the discovery of novel lead molecules. So the identification of bioactive molecules from plants can augment the drug discovery programs. Even after being the main source of crude drugs used to alleviate human sickness for many centuries, plants play an equally important role in drug development process in the present era of medicine engineering (Saklani and Kutty 2008).

Many compounds isolated from plants from different parts of the world are being clinically used as drugs in modern medicine. Isolation of Quinine from *Cinchona* bark, which is used against malaria, also led to the development of analogues like chloroquine, primaquine, mepacrine and mefloquine (Coatney et al. 1953). Ancient Chinese medical literature described the plant *Artemisia annua* as a cure for malaria. Later in 1960s antimalarial Artemisinin (van Agtmael et al. van Agtmael et al. 1999) was isolated from *A. annua*. Its analogues are artemether (Haynes and Vonwiller 1994) and artesunate (Lin et al. 1987). Podophyllotoxin, isolated from *Podophyllum peltatum,* and its analogues Etoposide and teniposide are important anticancer agents. Indole alkaloids from *Catharanthus roseus* include the anticancer agents: vincristine and vinblastine. This plant is also a source of an antihypertensive agent ajmalicine. Camptothecin isolated from *Camptotheca acuminata*, and its derivatives topotecan and irinotecan are effective anticancer agents. Paclitaxel isolated from *Taxus brevifolia* (Wani et al. 1971) and epothilones isolated from *Sorangium cellulosum* have also proved to be potential anticancer drugs. The compounds like Homoharringtonine are derived from *Cephalotaxus harringtonia,* Topotecan or Irinotecan from *Camptotheca acuminate,* Combretastatins from *Combretum caffrum* and Flavopiridol synthesised through modification of rohutikine from *Dysoxylum binectariferum* are some other potent anticancer agents identified and isolated from plant sources. Cabergoline, an indole alkaloid isolated from *Claviceps purpurea.* Cabergoline, is a potent, dopamine D2 receptor agonist and it is used in the treatment of Parkinson's disease. Other examples include the cardiotonic digitalin isolated from *Digitalis purpurea,* Aspirin from *Salix* species, Antihypertensive agent reserpine from *Rauwolfia serpentina,* Memory boosting agent Physostigmine was isolated from *Physostigma venenosum* and muscle-relaxing agent Tubocurarine from *Chondrodendron* spp. etc.

The effect of bioactive compounds may or may not affect the whole body, in some case, it will be tissue specific and even cytospecific. Like a micro or macro nutrient intake preference, there is no need for any compelled intake of the bioactive molecule (Gibney et al. 2009). Some bioactive compounds will influence metabolic pathways through food chain (Gibney et al. 2009). These compounds include non-pro-vitamin A, carotenoids and polyphenols, phytosterols, fatty acids and peptides (Astley and Finglas 2016). Scientists are curious about the mode of action of most of the bioactive compounds in the plant. Some of them act as a scavenger, some other act as a barrier against the attack of the foreign body. Some relevant bioactive compounds are now available in market as nutraceutical, dietary supplement or food supplement under different trade name with prescribed dosages. The improper diet and successive disorders force people to consume nutraceuticals and other dietary supplements. Fruits are rich in bioactive compounds and through the consumption of fruits and seeds, one could reduce or avoid the consumption of dietary supplements (Correia et al. 2012).

The bioavailability of each bioactive compound differs greatly, and the most abundant compounds in fruits are not necessarily leading to the increased concentrations of active metabolites in target tissues (Manach et al. 2005). While studying the role of bioactive compounds in human health, bioavailability is not always well known (Carbonell-Capella et al. 2014). The knowledge about plant-derived bio active compounds is nearly 20% of the available plants (Nayak 2006). Free radical attack and successive damage are the reason behind the production or synthesis of a bioactive compound in plant and such a compound is responsible for the scavenging or barrier activity against free radicals. The accumulation of bioactive compounds varies upon plant and even the distribution varies within a species; it may be in leaf for one and seed for another. Bioactive compounds have a role in shelf life also. The antioxidant compounds inhibit the reactive oxygen species (ROS) in fruits, vegetables and increase the shelf life (Ames et al. 1993).

Similar to the individual life forms having unique features, their biochemistry and chemistry will also be unique. To an extent, the morphological discrimination of life forms to specific taxa (Bryophytes, Pteridophytes, Gymnosperms and Angiosperms) is indirectly based on the genetic encryption via biochemical and chemical constituents. Even in Phanerozoic period (since 54 million years, till date) gymnosperms sustain with evolving status and such sustainability of a taxa over time would be due to its physiology and biochemistry and that character makes them a prominent candidate for drug discovery.

14.3 Traditional Knowledge on Gymnosperms

14.3.1 Cycadales

Order Cycadales constitutes the cone-bearing dioecious plants in gymnosperms (Nagalingum et al. 2011). The era of Cycads during the periods of Jurassic period extend over a millennia and these plants still exist under the closed canopy of tropical rainforests with woody stout stem, pinnately compound leaves and their giant cones filled with brilliantly coloured seeds. The predominant mestizos in 90% of Hondurans and indigenous ethnic groups utilise *Dioon mejiae* (Tiusintes) as a source of food in maze–bean diet. The people of different parts in Kerala, India still use the flour of *Cycas circinalis* seeds to prepare different traditional dishes. To avoid the toxic effect of cycad seeds, the traditional practice of 'seeds leeching' is performed. The use of unleached seed might lead to poisoning and continuous vomiting (Singh and Singh 2008; Saneesh 2009). The Mexican diet like tamales and tortillas are made from the processed seeds of *D. mejiae* female cones. Christians of Honduras still use the leaves and cones of the plant for their ritual celebration. Currently, the ethno medical practice with this plant is getting reduced day by day due to improper timber extraction, pastoralism and Swidden agriculture (Euraque 2003; Bonta et al. 2006).

The accumulation of toxic secondary metabolites in cycads is believed to have happened in the period of their evolution. Carboniferous fossil records corroborate the communication of beetle-like insects with cycads. This interaction of beetles with cycads was like mutualism (Jolivet 1998; Praz et al. 2008). The fossil records also favour the interaction of cycads with insects and it was also a mode of defence against non-specific host.

The continent of Africa is a dale of cycads. The local peoples of different countries use it for various purposes. People in Africa mainly use it as a source of starch during droughts and famines. In the current scenario, South African cycads are on the verge of extinction due to the exploratory mode of horticultural cycad collection (Cousins et al. 2011). Mozambique is the only nation currently using cycads as a famine crop (Donaldson 2003). The people near Mphaphuli in Africa use the parts of *Encephalartos transvenosus* for medicinal and magical purposes, ornamentation etc. (Ravele and Makhado 2010).

India is one of the nations with cycad food culture. Assam and Meghalaya are two states in India that exploit cycads for food, medicine and ornamentation purposes. In Kerala, India, people use cycads for gardening purposes and processed seeds are utilised in food preparation (Singh and Singh 2008). Alzheimer's and Parkinson's diseases observed in the Pacific island are due to the dietary nature of the peoples where they use untreated cycad flour for food preparation, which is one of the reasons reported for the occurrence of disease in the local inhabitants (Borenstein et al. 2007).

14.3.2 Coniferales

Coniferales being the predominant class of plants in gymnospermae are the highly demanded class among gymnospermae in therapeutic potential too. Many of the members form inevitable components of various traditional systems of medicine across the world. There are numerous literatures on the application of *Araucaria* members in traditional healing process. *Araucaria bidwillii* is reported to be a part of herbal medicine for its antiulcer and antipyretic properties (Anderson 1986). *Araucaria angustifolia* commonly known as Brazilian Pine is a critically endangered tree with edible seeds and various parts of the plant are used in Brazilian folk medicine (Souza et al. 2014). The plant is traditionally used for the treatment of dried skin, wounds, shingles and sexually transmitted diseases (Freitas et al. 2009). Medicinal preparations of the bark are employed in the treatment of muscle strains and the resin syrup can be used to treat respiratory tract infections. The extract of the needles is applied to cure fatigue, scrofula and anaemia (Marquesini 1995; Franco and Fontana 1997). *Araucaria araucana* has gastroprotective and wound-healing activities

(Schmeda-Hirschmanna et al. 2005), *A. angustifolia* and *A. cunninghamii* are important sources of antimicrobial agents.

The Mediterranean cypress, *Cupressus sempervirens,* has medicinally important leaves and cones. The plant is part of the traditional system of medicine and its dried leaves are used in the treatment of stomach pain, diabetes, inflammation, toothache, laryngitis and as contraceptive (Selim et al. 2014). The cones and leaves were used internally as an astringent and a decoction of them was applied to cure haemorrhoids. Also, the plant extract was used to cure varicose veins and venous circulation disorders. The essential oil extracted from the plant was used as antiseptic and an antispasmodic for persistent cough (Rawat et al. 2010). The plant was reported to have diuretic properties and can improve venous circulation to the kidneys and bladder area, bladder tone and can function as a co-adjuvant in urinary incontinence and enuresis therapy (Mahmood et al. 2013). Essential oil from the needles of *Cupressus torulosa* is an effective astringent and it is used for relief from whooping cough and rheumatism (Sellappan et al. 2007).

Turpentine is an important product from Pinus species. Oil of turpentine has the therapeutic potential like anti-pathogenic analgesic, dis-infectant, revulsive properties (Valnet 2002). *Pinus sylvestris* L. commonly called as 'Scot's Pine' or 'Scotch Pine' is used for its antiseptic activity and to cure respiratory disorders.

Communities in different parts of the world consume the fruits of many *Podocarpus* species like *P. dacrydioides* A. Rich., *P. neriifolius, P. nivalis* Hook., *P. salignus, P. totara, P. nagi* (Abdillahi et al. 2010). The younger leaves of *P. nagi* are also edible (Weiner 1980; Facciola 1990). The fleshy stem and fruits of *P. elatus* R. Br. ex Endl. are eaten by the Aborigine peoples of Australia. The genus *Podocarpus sensu latissimo* is a part of the traditional healing system of many communities and is used for the treatment of fevers, asthma, coughs, cholera, distemper, chest complaints and venereal diseases (Abdillahi et al. 2010). The reported ethnomedicinal uses of podocarpus species in different parts of the world were summarised by Abdillahi et al. (2010). *P. henkelii* Stapf ex Dallim. and B. D. Jacks., *P. falcatus* (Thunb.) R. Br. Ex Mirb., *P. ferrugineus* Don. (Miro), *P. latifolius* (Thunb.) R. Br. Ex Mirb., *P. macrophyllus* (Thunb.) Sweet, *P. nagi* (Thunb.) Zoll. and Moritz., *P. nakaii* Hayata, *P. neriifolius* D. Don., *P. totara* G. Bennett ex D. Don etc. are some other important *Podocarpus* species with significant ethnomedicinal importance.

The poisonous nature of *Taxus* species was known since ages and people in Europe and India used decoctions of yew leaf as an abortifacient (Bryan-Brown 1932). There are records in European cultural group, Celts who used yew plant leaf extract as a poison for committing suicide and was applied on the tips of the weapons used in the Gaelic wars to make it more fatal (Foster and Duke 1990). *Taxus baccata* has toxicity and medicinal applications like anti-spasmodic, cardiotonic, expectorant, narcotic and purgative properties. The arriles from the seeds have diuretic and laxative property. It was used medicinally to treat viper bites, hydrophobia (rabbies) and as an abortifacient.

14.3.3 Gnetales

Gnetales is the most advanced taxa under gymnosperms. *Gnetum ula, G. parvifolium* and *G. gnemon* are few ethnomedicinally important species in Gnetales. The genus *Ephedra* was previously included in Gnetales and later constituted into a new order Ephedrales. The plant *Ephedra alata* is an elixir for folk medicine. Folk practice includes its role as stimulant (as a decoction) and uses it as a deobstruent for kidney, bronchi, circulatory system, gastro-intestinal system disorders, to relieve asthma attack and for treatment of cancer. Also, the plant stems are chewed for the treatment of bacterial and mycological infections (Jaradat et al. 2015).

The *Ephedra sinica* Stapf. is also a traditional Chinese medicine. The plant has diaphoretic, anti-asthmatic and diuretic activities. It was also used to cure bronchial asthma, anaphylatic reaction and as anti-hypotensive agent (Miyazawa et al. 1997; NCI Thesaurus 2019). The compound 'Ephedrine' is the cause of the eponym Ephedra and was discovered about a century ago. Ephedrine belongs to the group of alkaloid compounds. It has bronchodilation activity, which is one of the reasons for the anti-asthmatic property of the plant *E. sinica* (Miyazawa et al. 1997).

14.3.4 Metabolites from Cycadales

Cycasin is one of the naturally occurring Azoxyglycoside in Cycads (Spencer et al. 2015), and it is observed to be throughout in all cycad species. Cycasin and some of the analogues were characterised from the seeds of cycad species, like *Cycas revolute,* and *C. circinalis.* Cycasin is a unique compound specific to genus cycadales and not found in any other plants (Moretti et al. 1983; Harborne et al. 1994; Osbome 2002).

The order cycadales is a rich source of metabolites including toxic and poisonous azoxyglycosides (cycasin and macrozamin), which is major class of compounds present in the various plant parts of different species. The α-amino-β-methylaminopropionic acid (MAM) is a highlight of the genus *Cycas* and it is an amino acid with non-protein origin. There is uncertainty about the role of this toxic substance in complex diseases like Parkinsonism and dementia (Kurland 1988; Osbome 2002). All the metabolites in major and minor quantity have its own role in the species and this variation in chemical composition could act as a chemotaxonomic key.

Most of the Macrozamia and Cycas species consist of either or both 'primeveroside macrozamin' and 'glucoside cycasin, which are the precursor of Cycasin. Cycas contains Uridine 5′-diphospho-glucuronosyltransferase, a cytosolic glycosyltransferase enzyme that synthesises cycasin from its precursor called methylazoxymethanol (MAM) (Spencer et al. 2015). The linked form of MAM to primeverose sugar residue was also reported in a compound called macrozamian by Cooper in 1941, and it was the first report of the compound macrozamin. Macrozamin was identified from the seed of an Australian cycad called *Macrozamia spiralis* (Osbome 2002). An average concentration of 2–4% (W/V) is to be observed in plants like cycads, and the cycasin is easily hydrolysed under acidic condition, hence it gets easily hydrolysed in methanol, formaldehyde, nitrogen and glucose in stomach. 1-BMAA (1-beta-methylamino-L-alanine) or 2-amino-3-(methylamino)-propanoic acid is a neurotoxic compound found in cycads. This non-protein amino acid is expected to be produced due to the presence of cyanobacteria in the leaves of cycads (Snyder and Marler 2011). The study of Lee and McGeer (2012) proved the weak cytotoxicity of BMAA through in vitro study and reframed the existing concept about neurotoxicity of BMAA. The studies on the toxicology of MAM elucidating the enzymatic conversion of MAM to methyl azoxy-formaldehyde (MAMAL), and is a spasm to cellular macromolecules (Zedeck et al. 1979). One-hundred and twenty-two species of cycads confirmed the Phenolic compounds like caffeic, protocatechuic, p-coumaric, p-hydroxybenzoic, ferulic and vanillic acids etc. The reports pointed out that the dihydroxybenzoic acid is restricted to the species of *Encephalartos* and *Bowenia serrulata* (Osbome 2002).

Coniferyl alcohol, *p*-coumaryl alcohol and sinapyl alcohol units were found from the lignin isolated from *Stangeria* spp. (Towers and Gibbs 1953). The high amount of vaniline and least amount of syringic aldehyde (from sinapyl alcohol) were reported by Logan and Thomas (1985) from the Cycad spp. The similarity of lignin composition in *Encephalartos* and *Zamia* with *Ginkgo* and conifers was also reported (Nishida et al. 1955; Riggs 1956; Osbome 2002).

The presence of leucoanthocyanidins is another key to the *Dioon* and *Macrozamia,* but it is absent in *Stangeria.* Proanthocyanidins belonging to tannins also have chemotaxonomic relevance (Bate-Smith and Lerner 1954; Osbome 2002). Cytotoxicity of this compound to the tumour cell lines was identified (Kolodziej et al. 1994), but the activity of the compound is not studied in detail (Makabe 2013). *Cycas* is the major source of Proanthocyanidins, which is also reported from the genus *Stangeria* but was absent in Zamiceae members (Osbome 2002). The identification of apigenin and luteolin (Proanthocyanidins) from *Dioon spinulosum* was the first report from cycads (Carson and Wallace 1972).

Biflavanoids are the chief attraction of the order cycadales. Methyl ether of amentoflavone, sotesuflavone is a compound isolated from *C. revolute.* Mentoflavone and hinokiflavone (17) (a compound with a substitution of apigenin to the phenoxy group at sixth position of 4-(5, 7-dihydroxy-4-oxo-4H-chromen-2-yl) are two other reports of biflavanoids from *C. revolute* (Geiger and de Groot Pfleiderer 1971; Lin et al. 1989; Osbome 2002).

The presence of biflavones varies in different genera and family; cycadaceae is the richest source of biflavones, Zamiaceae also possess biflavones. The Stangeriaceae is one of the families devoid of the biflavone compounds (Dossaji et al. 1975; Osbome 2002). Takagi and Itabashi (1982) isolated amino acids from the lipids of 20 species of gymnosperms and the study revealed that all the 20 species analysed contain trace amounts of nonmethylene-interrupted polyenoic (NMIP), palmitic, oleic, linoleic and α-linolenic acids.

Sciadopitysin is a biflavonoid compound reported in *Encephalotous* spp. and *Ginko biloba.* It was also reported in some members of zamiaceae and Taxales (Harborne et al. 1994). The studies of Suh et al. (2013) described that pretreatment with sciadopitysin could reduce antimycin A-induced cell damage. The role of sciadopitysin in prevention of osteoblasts degeneration is proved. Sciadopitysin isolated from the *T. baccata* and *G. biloba* has antifungal activity as well (Krauze-Baranowska and Wiwart 2003).

Amentoflavone is also a biflavonoid compound that displays various beneficial effects including anti-inflammatory, anti-oxidative and anticancer properties. The compound also has an inhibitory activity against human UDP-glucuronosyltransferases (Lv et al. 2018). The bright colouration of cycad seed is due to the presence of carotenoid compounds and this helps for attracting insects (Bouchez et al. 1970).

Foster and Gifford referred to coniferales as the most dominant and conspicuous gymnosperms in the floras of the modern world. It is the most diverse class in gymnosperms with evergreen trees and is commonly seen in the temperate regions. The earlier studies on the chemistry of secondary metabolites extracted from wood and bark of medicinally important gymnosperm species indicated the presence of many compounds like terpenes and lignans (Silva and Bittner 1986; Céspedes et al. 2000; Flores et al. 2001; Calderón et al. 2001; Torres et al. 2003). Generally, Conifers contain secondary metabolites like phenolics, alkaloids, terpenoids including hemiterpenes (C_5), monoterpenes (C_{10}), sesquiterpenes (C_{15}) and diterpenes (C_{20}) are reported by Bohlman et al. (2000).

14.3.5 Metabolites from Coniferales

Araucariaceae under coniferales consist of evergreen trees classified into three genera *viz Agatis, Araucaria* and *Wollemia*. Species in Araucariaceae are endomycorrhizal (McKenzie et al. 2002). Reports suggested the presence of biflavanoid, isoflavanoids, phenyl propanoids, furans, lignans, protein, terpenes and polysaccharides in *Araucaria species* (Aslam et al. 2013). Proanthocyanidins or oligomeric flavonoids were present in many plants in Araucariaceae. These compounds are prominent antioxidant agents and have other bioactivities also. Proanthocyanidins are reported to be the major bioactive compound in *A. angustifolia* needles. Freitas et al. (2009) demonstrated the antiherpes activity of *A. angustifolia* leaves that justified the application of this plant in folk medicine system. Further investigation revealed the presence of proanthocyanidins and the biflavonoids like bilobetin, II-7-*O*-methyl-robustaflavone and cupressuflavone in the plant. Studies by Yamaguchi et al. 2005 also proved the presence of proanthocyanidins and biflavonoids in needles of *A. angustifolia* and other compounds like amentoflavone, mono-*O*-methyl amentoflavone, di-*O*-methyl amentoflavone, Ginkgetin, Tri-*O*-methylamento-flavone and tetra-*O*-methylamentoflavone.

Even the lifeless bark of *A. angustifolia* has proved to have high concentration of anthocyanins and proanthocynidins, and significant antioxidant effects on liposomes and rat microsomes (Seccon et al. 2010). The biflavonoid fraction from needles of *A. angustifolia* was proved to be capable of protecting calf thymus DNA from damage caused by UV radiation (Yamaguchi et al. 2009). It was quantified by high-performance liquid chromatography coupled to tandem mass spectrometry (HPLC-MS/MS) in a multiple reaction monitoring mode (MRM) and through HPLC-coulometric detection. The seed of *A. angustifolia* contains lectins with anti-inflammatory and antibacterial activities (Santi-Gadelha et al. 2006).

Araucariaceae members are important sources of diverse phenolic compounds. Souza et al. (2014) demonstrated the antioxidant and antigenotoxic activities of *A. angustifolia* bract extract and polyphenolic profiling of *A. angustifolia* indicated the presence of Catechin, epicatechin, quercetin and apigenin as the major compounds. The extract proved to have the potential to protect the MRC5 cell line (Medical Research Council cell strain 5) from damage (Souza et al. 2014). The phenolic profile of the bract extract of *A. angustifolia* proved to have in vitro and in vivo antioxidant activity and antimutagenic properties (Michelon et al. 2012). Catechin, epicatechin and rutin were the main phenolic compounds found in the extract. Presence of phenolic compounds is always related to the antioxidant and anticancer activities possessed by the plant (Michelon et al. 2012).

Presence of alkaloids, flavonoids, sterols, cardiac glycosides, saponins, tannins, phenols and terpenoids in *A. heterophylla* leaf extract was proved by Reddy et al. (2017). *Araucaria columnaris* bark peel extracts were subjected to phytochemical profiling and free radical scavenging activity was studied by Jadav and Gowda in 2017. Antioxidant activity of the methanol extract was due to the presence of flavonoids, tannins and phenolic compounds. Aslam et al. (2014) investigated the phytochemistry of aerial parts of *A. columnaris* and reported the presence of Tannins and cardiac glycosides. The biologically active diterpene–abietane is a compound isolated from *A. columnaris* (Cox et al. 2007).

Benzoic acid, 1H-N-Hydroxynaphth (2,3) imidazole-6,7-dicarboximide, 2-Propenoic acid, 3-(4-methoxyphenyl), 1H-N-Hydroxynaphth (2,3-d) imidazole-6,7-dicarboximi were identified from *A. columnaris* bark extract through GC-MS analysis by Devi et al. (2015). Toxic effect of the extract on kidney was analysed through human embryonic kidney cell line that revealed minimal toxicity and recommended the application of this plant-derived molecules for pharmaceutical purpose (Devi et al. 2015).

Resin from conifers is important source of diverse metabolites. Lignans were reported from the resin of *A. angustifolia* including secoisolariciresinol acetates, six lariciresinol acetates, two 7′-hydroxylariciresinol acetates and an isolariciresinol acetate. Other lignans reported were Shonanin and 7′-hydroxy lariciresinol, 5-methoxy lariciresinol-9-acetate, 5′-methoxy lariciresinol-9-acetate, methoxy pinoresinol dimethyl ether and 5-methoxy pinoresinol (Yamamoto et al. 2004). Lignans containing syringyl moieties, characteristic for angiosperms, occur in the resin of *A. angustifolia* (Yamamoto et al. 2004).

Antibacterial and antifungal potential of the Methanol extract and lignans namely secoisolariciresinol, pinoresinol, eudesmin, lariciresinol and 4-methoxy-pinoresinol (2.96%) from the methanol extract of the heartwood of *A. araucana* tree were studied by Céspedes et al. (2006). The methanol extract showed the highest inhibitory activity against the tested Gram-positive bacteria and the compounds secoisolariciresinol and pinoresinol were proved to be responsible for the antifungal potential of the heartwood extract of *A. araucana*.

Gas chromatography–mass spectrometry (GC-MS) analysis of *Agathis borneensis* revealed the presence of cyclohexane, farnesol, germacrene D, β-caryophyllene and δ-cadinene as the major compounds. Caryophyllene oxide and β-caryophyllene show antiplasmodial activity (Adam et al. 2017). Venditti et al. (2017) identified agathisflavone, 7″-*O*-methyl-agathisflavone, cupressuflavone, rutin, shikimic acid and (2S)-1, 2-Di-O-[(9Z, 12Z,15Z)-octadeca-9,12,15-trienoyl]-3-*O*-β-d-galactopyranosyl glycerol from the leaves of *Agathis robusta* (C. Moore ex F. Muell.) F.M. Bailey. *Agathis* resins are rich sources of diterpenoids (Cox et al. 2007). Volatile materials from the resin of *Agathis atropurpurea* and *Agathis philippinensis* oils were studied and limonene content dominated both the resins (Lassak and Brophy 2008; Garrison et al. 2016).

The male cones of *Wollemia nobilis* were investigated for the phytochemical composition leading to the identification of several diterpenoids of chemosystematic significance, namely isocupressic acid, acetyl-isocupressic acid, wollemol, methyl (E)-communate and sandaracopimaric acid, 7-4‴-dimethoxyagathisflavone, shikimic acid (Venditti et al. 2017).

The Australian plant *W. nobilis* contains 2-propylphenol, 3, 4-dimethoxyphenol, 2-methoxybenzoic acid, vanillyl alcohol and isovanillic acid in the leaf extracts (Seal et al. 2010).

Brophy et al. 2000 compared the volatile oil composition of *W. nobilis* with other Araucariaceae members. The leaf essential oil of *W. nobilis* or Wollemi Pine consists of (+)-16-kaurene (60%), alpha-pinene (9%) and germacrene-D (8%) as the major compounds whereas, *A. atropurpurea* oil contains phyllocladene, 16-kaurene, alpha-pinene and delta-cadinene, *A. robusta* oil contained spathulenol and rimuene and *A. australis* oil contained 16-kaurene, sclarene and germacrene-D. *A. heterophylla* oil contains alpha-pinene and phyllocladene.

Cupressaceae is another important and large family of coniferales that consist of nearly 30 genera. The genus *Cupressus* consists of nearly 20 species distributed mostly in the northern hemisphere, including Western North America, Central America, Northwest Africa, Asia and Mexico (Allemand 1979; Pierre-Leandri et al. 2003; Wang and Ran 2014). The phytochemical screening of *C. sempervirens* indicated the presence of alkaloids (0.7%), flavonoids (0.22%), tannin (0.31%), saponins (1.9%), phenols (0.067%) and essential oils along with other phytoconstituents (Hassanzadeh Khayyat et al. 2005, Selim et al. 2014). Selim et al. (2014) used GC/MS system for profiling the chemical constituents of the hydrodistilled essential oil of *C. sempervirens*. Out of the 20 constituents recognised, α-pinene (48.6%), δ-3-carene (22.1%), limonene (4.6%) and α-terpinolene (4.5%) were the main components that constituted 79.8% of the oil. *C. sempervirens* showed strong antibacterial activity and the essential oil and methanol extract showed considerable antibiofilm activity against *Klebsiella pneumoniae*. The antimicrobial activity of *C. sempervirens* may be due to the presence of phenolics, alkaloids, flavonoids, terpenoids and polyacetylenes.

C. torulosa commonly known as Himalayan cypress was much studied for the medicinal properties. Regarding the phytochemistry, the needles are rich sources of biflavones like amentoflavone, cupressuflavone, hinokiflavone and apigenin, α-pinene, δ-3-carene, limonene and sabinene (Natarajan et al. 1970; Lohani et al. 2012; Padalia et al. 2013). Studies have indicated the presence of Mono, sesqui and di-terpenes in the needles of this plant (Cool et al. 1998). Antioxidant activity of volatile oil and the antimicrobial activity of extracts of *C. torulosa* were thoroughly studied by Dhanabal et al. (2000) and Joshi et al. (2014). The polar extracts of *C. torulosa* contained higher quantity of phenols, flavonoids and hence showed higher antioxidant activity (Khulbe et al. 2016).

According to Ismail et al. (GC/MS) analysis of the essential oils obtained by hydrodistillation from leaves, branches and female cones of Tunisian *C. sempervirens* revealed the presence of 52 phytoconstituents. The oils contained abundant monoterpene hydrocarbons, and the major components identified were α-pinene, α-cedrol, δ-3-carene and germacrene D. On the bioactivity part, the essential oils of *C. sempervirens* exhibited significant phytotoxicity on the germination of four weeds: *Sinapis arvensis* L., *Trifolium campestre* Schreb (dicots), *Lolium rigidum* Gaud and *Phalaris canariensis* L. (monocots). Also, the essential oil of *C. sempervirens* showed significant antifungal activity against 10 fungal species affecting cultivated crops and proved its potential to be used as a bio-herbicide and fungicide.

Miloš et al. (1998) isolated glycosides from fresh cones of *C. sempervirens* and identified 3-hydroxybenzoic acid methyl ester and Thymoquinone (2-isopropyl-5-methyl-1,4-benzoquinone). Significant aglycones were perillyl alcohol, p-cymen-8-ol, 2-phenylethanol and carvacrol. Diterpenes like 6-deoxytaxodione (11-hydroxy-7, 9(11), 13-abietatrien-12-one), taxodione, ferruginol, sugiol, trans-communic acid, 1,5-acetoxy imbricatolic acid and imbricatolic acid were isolated by Tumen et al. (2012) from *C. sempervirens*. Boukhris et al. (2012) demonstrated *C. sempervirens* extracts to have potent antibacterial activity against the strains like *Staphylococcus aureus, Bacillus subtilis, Pseudomonas aeruginosa, Escherichia coli, K. pneumoniae* and *Salmonella typhimurium*.

Pinus is a very prominent member of the taxa Coniferales and the genus includes around 105 species distributed mostly in the Northern Hemisphere. Pine resin is a key forest product obtained from the exudate after tapping pine tree bark. This is a mixture of acidic and neutral diterpenes and volatile compounds (Rezzi et al. 2005). Pine resin has industrial application and in countries like Java, it is subjected to distillation to collect turpentine oil (volatile compound) and rosin (diterpenes).

Gerard (1978) isolated some flavonoids from the needles of *Pinus jeffreyi* namely Kaempferol, the 3-glucosides of Kaempferol namely Quercetin, Isorhamnetin, Laricitrin, Syringetin (Kaempferol-3-(p-coumarylglucoside) and Kaempferol-3-(Ferulylglucoside). Myricetin-3-rhamnoside was also identified from *P. jeffreyi*. Shumailova (1971) isolated a number of catechins namely catechin, epicatechin, gallocatechin, epigallocatechin, epigallocatechin gallate and epicatechin gallate from the needles of *P. sibirica* (Niemann 1977).

Cedrus deodara is a component of traditional Chinese medicine. The needles of the plant were subjected to study the antibrowning and antimicrobial activities by Zeng et al. (2011). Phenols and flavonoids are present in a considerable amount in the extract and it showed potent free radical scavenging activity and antibacterial activity. The study recommended the application of pine needles of *C. deodara* as antibrowning and antimicrobial agents in food preservation.

Pinus densiflora needle extracts showed antioxidant activity in DPPH (1,1-diphenyl-2-picrylhydrazyl) radical method and ROS inhibition activity in MC3T3 E-1 cells due to the presence of polymeric proanthocyanidins and monomeric catechins (Park et al. 2011). Shibuya et al. (1978) investigated the solvent-extractable lipids in *P. densiflora* pollen and the *cis*- and *trans*-isomers of 1, 16-dioxo-, 1-hydroxy-16-oxo- and 1, 16-dihydroxyhexadecan-7-yl *p*-coumarates were identified. Leaf essential oil of *Pinus kesiya* was investigated for the mosquito larvicidal activity against malaria, dengue and lymphatic filariasis vectors namely *Anopheles stephensi, Aedes aegypti* and *Culex quinquefasciatus* respectively. The essential oil of *P. kesiya* evaluated through gas chromatography–mass spectroscopy indicated the presence of 18 compounds, α-pinene, β-pinene, myrcene and germacrene D were important among them. In acute toxicity assays for larvicidal activity, the essential oil exhibited remarkable toxicity against early third-stage larvae of *A. stephensi, A. aegypti* and *C. quinquefasciatus* and indicated the presence of larvicidal compounds in Pinaceae plants against malaria, dengue and filariasis (Govindarajan et al. 2016).

The trees like *Pinus massoniana, P. strobus* and *P. palustris* are resistant to the pine wood nematode, *Bursaphelenchus xylophilus*. The heartwood of *P. massoniana* contains the phytoconstituents like α-humulene, pinosylvin monomethylether and (−)-nortrachelogenin and the bark indicated the presence of nematicides, methyl ferulate and (+)-pinoresinol. Pinosylvin monomethylether had the highest nematicidal activity and it was also present in the heartwood of *P. strobus* and heartwood and bark of *P. palustris* (Suga et al. 1993).

P. massoniana Lamb is an ethnomedicinally important Chinese red pine species and its bark extract has anticancer property. A study by Wu et al. (2011) proved that *P. massoniana* bark extract can inhibit the migration of human cervical cancer cell line, HeLa cells. B-type procyanidin is a major constituent of the plant extract.

Turpentine oils, gum oleoresins and rosins collected from *Pinus merkusii* Jungh et de Vries were analysed by gas chromatograph mass spectrometry (Batu 2006), and neutral fractions from the gum oleoresins and turpentine oils indicated the presence of α-pinene, Δ-3-carene and β-pinene. The acidic fractions and rosins contained sandaracopimaric acid, isopimaric acid, palustric acid, dehydroabietic acid, abietic acid, neoabietic acid and merkusic acid. Acidic fractions contained palustric acid as the major component, whereas abietic acid was the major component of rosin.

Analysis of the essential oils from the resins of *Pinus brutia* and *Pinus pinea* through GC-MS and GC/FID analysis revealed the presence of α-pinene, β-pinene and caryophyllene as the major compounds (Ulukanli et al. 2014). The bioactivities of the oils were also analysed and they proved to have remarkable antimicrobial effect on *Micrococcus luteus* and *B. subtilis*, insecticidal effect on *Ephestia kuehniella* eggs, phytotoxicity on *Lactuca sativa, Lepidium sativum* and *Portulaca oleracea*, and antioxidant activity.

Pine bark extracts are much studied by the researchers across the world for their therapeutic potential and phytochemistry. *Pinus pinaster, P. roxburghii, P. radiata, P. massoniana, P. brutia, P. cembra, P. caribaea, P. siberian* etc. are some of the well-explored species. These extracts have potent antioxidant, anti-inflammatory, anticancer, cardioprotective and neuroprotective activities. Studies have proved that proanthocyanidins are among the most abundant constituents in the pine bark extract and it includes Pycnogenol and Enzogenol (Li et al. 2015). Ince et al. reported that Taxifolin from the *P. brutia* bark extract had anti-inflammatory property (Ince et al. 2009). Tannins from *P. caribaea* Morelet bark extract are reported to possess antioxidant, antilipid peroxidation and antigenotoxic activity (Fuentes et al. 2006). Procyanidine present in bark extract of *Pinus koraiensis* exhibited antioxidant and anticancer activity (Li et al. 2007). *P. cembra* L. extract showed antioxidant and antimicrobial properties, which is attributed to the presence of phenolics, flavonoids, proanthocyanidins in it (Apetrei et al. 2011). Arabinogalactan sulfate from the bark extract of *P. siberian* cedar is an anticoagulant (Drozd et al. 2008). The diuretic, laxative, antispasmodic, antihypertensive, analgesic and anti-inflammatory activities possessed by *P. roxburghii* sarg bark extract is due to the flavonoid content (Kaushik et al. 2012a: Kaushik et al. 2012b).

Yue et al. (2013) evaluated three 6-C-methyl flavonoids identified from *Pinus densata* for their inhibitory effect on the human leukaemia cell line, HL-60. The compounds were 5,4'-dihydroxy-3,7,8-trimethoxy-6-C-methylflavone, 5,7,4'-trihydroxy -3,8-dimethoxy-6-C-methylflavone and 5,7,4'-trihydroxy-3-methoxy-6-C-methylflavone. The analysis suggested that among these compounds 5, 4'-dihydroxy-3, 7, 8-trimethoxy-6-C-methylflavone has remarkable anticancer potential as it prevented proliferation of HL-60 cells in a concentration dependent manner and showed an IC_{50} value of 7.91 µM. The molecule can induce apoptosis of HL-60 cells through mitochondrial caspase-3-dependent apoptosis pathway.

The greatest diversity of Podocarpaceae family members is observed in Malesia and Australasia where 17 of the 19 living *podocarpus* are found. Anticancer agents have been isolated from many *Podocarpus* species. Taxol is such a compound reported from *Podocarpus gracilior* Pilger that can inhibit proliferation of HeLa cell line (Stahlhut et al. 1999). Urdilactone A, B and C isolated from *Podocarpus purdieanus* Hook. exhibited in vitro cytotoxicity in mouse lymphocytic leukaemia and human lung carcinoma, breast adenocarcinoma and colon adenocarcinoma cell lines (Wang et al. 1997). Norditerpenes and totarols from *Podocarpus* are known to have cytotoxic activities against murine leukaemia cell line (Park et al. 2003). Totarol is also a potent antibacterial agent reported from *P. sensu latissimo* (Abdillahi et al. 2010).

In the study of Shrestha et al. (2001), *P. neriifolius* D. Don ethanolic extract showed antiproliferative activity against two major tumour cell lines, viz. human HT-1080 fibrosarcoma and murine colour 26-L5 carcinoma. Nagilactone C identified from *P. totara* and *P. neriifolius* has potent antiproliferative activity against human fibrosarcoma and murine colon carcinoma tumour cell lines. Nagilactone F and nagilactone G reported from *P. milanjianus* Rendle and *P. sellowii* Klotzsch ex Endl. also exhibited cytotoxicity against various cell lines (Hembree et al. 1979).

Compounds like Imuene and Nubigenol were reported from *Podocarpus saligna*, and the structures of three new norditerpene dilactones isolated from the plant *viz* salignones K, L and M were elucidated by Matlin et al. (1984). Sanchez et al. (1970) reported the isolation of Ponasterone A, Sequoyitol and four norditerpene dilactones from the leaf ethanol extract of one of a Brazilian *Podocarpus* species, *P. sellowii* Klotzsch (Sanchez et al. 1970).

Phytochemical investigation of polyphenolic contents of *P. gracilior* Pilger leaves (Kamal et al. 2012) helped to identify three known polyphenolic compounds namely Apigenin 8-C-β-D-glucopyranosyl-($1'2'$)-O-β-D-glucopyranoside (Vitexin $2'$-O-β-D-glucopyranoside), Quercetin 3-O-β-D glucopyranoside (Isoquercetin) and II-4', I-7-dimethoxy amentoflavone (Podocarpusflavone B). Also, unsaturated fatty acids proved to be higher than saturated fatty acids in *P. gracilior* and the leaf methanol extracts of *P. gracilior* exhibited weak cytotoxicity to the breast adenocarcinoma cell line, MCF-7. Antioxidant and antimicrobial properties of the extract were also analysed against *E. coli, S. aureus, Aspergillus flavus, Candida albicans* and exhibited antioxidant, antibacterial properties and no antifungal activity. The isolation and identification of polyphenolic compounds were done through acid hydrolysis, comparative PC, UV, ESI-MS, 1H-, 13C-NMR, 2D-NMR spectroscopy etc. Previously Kuo et al. (2008) identified Quercetin 3-O-β-D glucopyranoside from the bark of *P. fasciculus*. II-4', I-7-dimethoxy amentoflavone (Podocarpusflavone B) was reported from the species like *P. neriifolius, P. fasciculus, P. fleuryi* and *P. elongates* (Rizvi et al. 1974; Xu et al. 1990, 1993; Xu and Fang 1991; Kuo et al. 2008; Faiella et al. 2012).

Flavonoids have the diversity to be used as a chemotaxonomic tool in *Podocarpus* species as biflavonoids like amentoflavone and hinokiflavone have proved to be reliable taxonomic markers in *P. sensu latissimo* (Roy et al. 1987). Presence or absence of monomer flavonoid glycosides can be a prominent factor in the identification and approval of new plants. *Dacrycarpus*, endemic to New Zealand showed the presence of 3-methoxyflavones, while the presence of flavonol 3-oglycosides and flavone C-glycosides is the feature of *Prumnopitys and Podocarpus*, respectively (Markham et al. 1985).

Cephalotaxaceae family of conifers consists of three genera like Cephalotaxus, Amentotaxus and Torreya and around 20 species. Cephalotaxus alkaloids are important class of plant secondary metabolites. Activity of Cephalotaxus extracts against leukaemia in mice was reported earlier. Cephalotaxine is the major alkaloid of this series was isolated from *Cephalotaxus drupacea* species by Paudler et al. (1963). This has led to the identification of many compounds with prominent bioactivities from the genus. The botanical distribution of the Cephalotaxane alkaloids is limited to the Cephalotaxus. Antileukemic alkaloid Homoharringtonine or Omacetaxine mepesuccinate isolated from *Cephalotaxus hainanensis* was effective in curing orphan myeloid leukaemia and was approved by European Medicine Agency and by US Food and Drug Administration (Pérard-Viret et al. 2017). The compound is a translation inhibitor that binds to the 80S ribosome in eukaryotic cells and inhibits protein synthesis by interfering with chain elongation. Drug screening studies have demonstrated omacetaxine mepesuccinate or homoharringtonine as a compound having synergism with FMS-like tyrosine kinase-3 (FLT3) inhibitors in acute myeloid leukaemia subtype (Leung et al. 2017). Omacetaxine is a semisynthetic form of homoharringtonine, having considerable subcutaneous bioavailability and is now approved by FDA of the United States for the treatment of chronic myeloid leukaemia refractory to tyrosine kinase inhibitors (Lü and Wang 2014). Cephalotaxine-type alkaloids are the anticancer components in twigs, leaves, roots and seeds of *Cephalotaxus fortunine* and it includes Cephalotaxine, *Epi*-wilsonine and Acetylcephalotaxine, Drupacine, Wilsonine and Fortunine (Liu et al. 2009).

Twenty-eight Cephalotaxus tropone analogues and their structures were identified from *Cephalotaxus fortunei* Hook. var. alpina H. L. Li and *C. lanceolata* K. M. Feng (Ni et al. 2018). They showed significant anticancer activity and the structure–activity relationship analysis revealed that the tropone moiety and the lactone ring were responsible for this (Ni et al. 2018). Biologically active *Cephalotaxus tropones* have been identified from *Cephalotaxus* Sieb. and Zucc. Ex Endl (Abdelkafi and Nay 2012). Buta et al. (1978) reported the non-alkaloid Harringtonolide (Hainanolide), which is a Cephalotaxus tropone from *C. harringtonia* and Sun et al. (1979) identified it from *C. hainanensis* H. L. Li. Hainanolidol (Xue et al. 1982), fortunolide A/B (Du et al. 1999) and 10 hydroxyhainanolidol (11-hydroxyhainanolidol) (Yoon et al. 2007) were isolated from *C. hainanensis, C. fortunei* Hook. var. alpnia H. L. Li and *C. koreana* Nakai, respectively.

Taxus species are the prominent members in the family Taxaceae and are commonly called as Yews. The toxicity of this plant species is due to the presence of taxine alkaloids except the scarlet aril (Bryan-Brown 1932). Investigation of the therapeutic potential of taxine alkaloids revealed their cardiovascular activity, adverse effects in involuntary muscle, contraction of the uterus in situ, relaxation of the intestines, contraction of the duodenum and ileum etc. (Bryan-Brown 1932; Vohora 1972). Volatile oil, tannic acid, gallic acid and resinous substances are reported from the leaves of *T. baccata* and are reported to be used in the treatment of asthma, bronchitis, hiccups, indigestion, rheumatism and epilepsy (Sharma et al. 2014).

In 1856, Lucas, a pharmacist extracted an alkaloid powder from the foliage of *T. baccata* L. and named it Taxine. Taxine alkaloids commonly called as Taxine is the source molecule for the development of efficient chemotherapy drugs like Paclitaxel (Fig. 14.1) used in the treatment of ovarian cancer, breast cancer, lung cancer, Kaposi sarcoma, cervical cancer and pancreatic cancer and Docetaxel for treatment of breast, prostate, stomach, head and neck and non-small cell lung cancers.

The amount of Taxine alkaloids varies in taxus species and *T. baccata* and *Taxus cuspidata* have recorded high amounts of Taxine (Wilson et al. 2001). The toxicity of taxus species is due to the presence of toxic alkaloids (Taxine B, Paclitaxel,

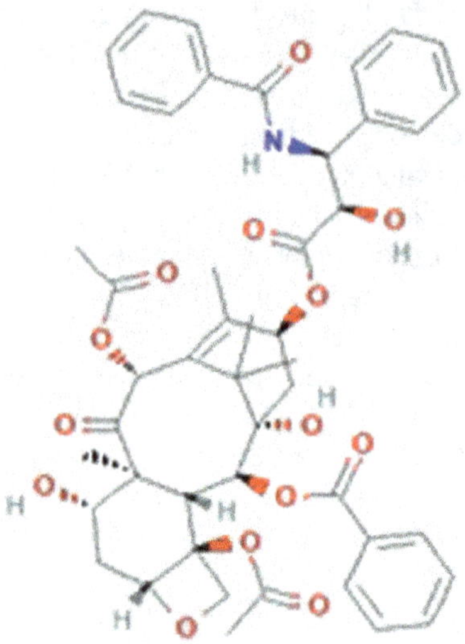

Fig. 14.1 Chemical structure of paclitaxel

Isotaxine B, taxine A), glycosides (Taxicatine) and taxane derivates (taxol A, taxol B) (Ramachandran 2014).

Taxol is one of the taxane-derived chemotherapeutic agents used to treat breast, ovary, lung and Kaposi's sarcoma originated from *T. brevifolia*. *Taxodium distichum* cone extract has been used in traditional medicine for curing the parasitic disease, malaria. *T distichum* cones are reported to have in vitro antileishmanial activities. The essential oil of the cone contained limonene and α-pinene as the major compounds and the leaf and branch oils were also rich in α-pinene (Flamini et al. 2000).

14.3.6 Metabolites from Gnetales

Different species of *Ephedra* are widening in both the hemisphere, nevertheless the species having the high degree of Ephedrine alkaloid content is restricted in Eurasia. The quantification of ephedrine alkaloid (EA) is comparatively easy and there are techniques to quantify the EA from other than plant extract, and an erudite protocol for quantification of EA extract is far away. Choi et al. (1999) standardised the protocol for the isolation of ephedrine from *E. sinica*, using mixtures of CO_2, diethylamine and methanol. The chemical fingerprinting could also be archived through reverse phase high-performance liquid chromatography (RP-HPLC) incorporated into photodiode array detection (Schaneberg et al. 2003). The ephedrine (Fig. 14.2) is not only a medicine but also a dietary supplement in many countries and is responsible for body weight. Ephedrine has a negative impact too. The improper dosage and misuse of ephedrine may lead to reduced blood pressure via weakening heart muscle. If the dose is high, the heart beat will increase abruptly resulting in increased culatory stimulation. Ephedrine also has a mydriatic effect

Fig. 14.2 Chemical structure of ephedrine

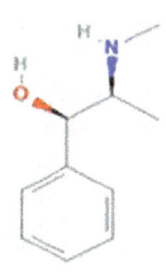

higher than that of cocaine and its activity makes it applicable in retina examination. Ephedrine is a circulatory stimulant but the excess concentration of the ephedrine causes constriction of blood vessels that supply blood to vasoconstrictor nerves. This can severely affect the splanchnic area than the limb (Chen and Schmidt 1924).

E. alata Decne is a perennial seedy herb that belongs to Gnetales, and is closely related to angiosperms. The study by Jaradat et al. (2015) described that the plant *E. alata* has a wide range of distribution throughout Iran, Algeria, Iraq, Chad, Egypt, Palestine, Lebanon, Jordan, Saudi Arabia, Morocco, Syrian Arab Republic, Libya, Mauritania, Mali, Somalia and Tunisia. Boiled 'dried tender stem' of *E. alata* was used as a beverage as well as medicine in traditional medicine. *E. alata* is a rich source of phytochemicals like cardiac glycosides, reducing sugars, flavonoids, phenolics and alkaloids. Flavonoid containing taxa, the phenolics is the most relevant phytochemical group having inevitable role in the therapeutic potential of a species. *E. alata* is a repository of phytoconstituents like phenolics. The DPPH assay substantiated the anti-oxidant activity of *E. alata*. Efficient isolation protocol for bioactive compounds from the plant could be an avenue for the bio prospecting of medicinally less evaluated other species of the genus. Chlorogenic acid, Catechin, Quercetin and Coumaric acid are some of the compounds reported from *E. alata* (Hegazi and El-Lamey 2011).

Coumaric acid (4-hydroxycinnamic acid), a phenolic acid is produced through shikimate pathway. The compound is the precursor for other phenolic compounds. In plants, coumaric acid exists in a conjugate or free form. Antioxidant, anticancer, antimicrobial, antivirus, anti-inflammatory, antiplatelet aggregation, anxiolytic, antipyretic, analgesic and anti-arthritis activities are some of the bioactivities reported to be possessed by the compound. Even though it has extensive bioactivity, the low absorption of the compound makes it as inappropriate candidate drug molecule in the field of drug development (Pei et al. 2016).

E. sinica is another plant that belongs to Gnetales. Traditional Chinese medicinal system followed *E. sinica* for more than 5000 years as stimulant and antiasthmatic tea. Western countries utilise the extracts of *Ephedra intermedia and E. sinica*, as a dietary supplement, stimulant and to reduce weight gain (Schaneberg et al. 2003). Chemical constituents of volatile oil from various *Ephedra sp.* are given in Table 14.1.

G. gnemon L. is another species with a phytochemical background. It is pool of flavonoid compounds such as isovitexin, its 7-O-glucoside, vicenin II, the 7-U-methyl-C-glucosylflavones, swertisin its X″-O-glucoside, isowertisin,

Table 14.1 Volatile compounds reported from *Ephedra* sp. (Kobayashi et al. 2005)

Oil of *Ephedra distachya*	
Ethyl benzoate	46.90%
Benzaldehyde	8.00%
Cis-calamenene	3.60%
Oil of *E. fragilis*	
Pentacosane	5.20%
E-phytol	10.10%
6,10,14-Trimethyl-2-pentadecanone	5.30%
Cis-thujopsene	3.50%
α-Terpineol	3.00%
Oil of *E. major*	
Eugenol	4.30%
α-Terpineol	3.70%
Methyl linoleate	3.50%

swertiajaponin and isoswertiajaponin (Wallace and Morris 1978). In addition to this, compounds like C-glycosylapigenins are also found in trace amount. Flavonol O-glycosides are derivatised form of vitexin, also found in *Larix laricirrc*. Biflavone aglycones are the foremost flavonoid compound in *D. spinulosum*, and Vitexin and orientin are few compounds present in the least. Biflavones are the compounds reported from cycadales and conifers. The isolatable form of biflavons, flavons or flavanols is absent in the genera Gnetum, hence the chemotaxonomic approach for the classification of the genera is insignificant (Wallace and Morris 1978). C-Glycosylflavones a compound reported from the leaves of *G. gnemon* (Wallace and Morris 1978). The in vitro studies with the *Pennisetum millet* (L.) leeke proved that the C-glycosylflavones have a goitrogenic and antithyroid activity (Gaitan et al. 1989).

3, 4-Dimethoxychlorogenic acid, resveratrol and 3-methoxyresveratrol (5-[2-(4-Hydroxyphenyl) ethenyl]-3-methoxycyclohexa-1, 5-diene-1,3-diol) are the three bioactive phenolic compounds identified from *G. gnemon*. 3, 4-Dimethoxychlorogenic acid is a derivative of chlorogenic acid (CGA); CGAs are also abundant in coffee (Sinisi et al. 2015). They are the group of compounds associated to better health, including hepatoprotective and bile induction activity (choleretic activities), antioxidant, antiviral, antibacterial, anticancer and anti-inflammatory properties, modulation of gene expression of antioxidant enzymes and it can suppress the chances of cardiovascular diseases by reducing the expression of p-selectin on platelets. These CGAs have a role in the non-insulin-dependent diabetes mellitus (NIDDM) and senile dementia (Ludwig et al. 2014; Wianowska and Gil 2019). CGAs are responsible for the reduction in food craving leading to reduced calorie and increase fat loss by increased burning of energy (Garg 2016). The free radical scavenging is another bioactivity of CGAs, it can inhibit the generation of reactive oxygen group even in vitro (Kweon et al. 2001). Natural source, like plants, consists of only fewer amounts of CGAs and its isolation is very difficult due to its high susceptibility to environment. The CGA is sensitive to light, temperature and pH. The complex molecular synthesis is very difficult and the isolation procedure of the same from natural sources is far off.

All the studies pointed to phenol–Stilbenoid as the compound responsible for the pharmacological attributes of *G. gnemon*. Also, the nutritional factors and fat-burning metabolites in it makes the plant and its by-products valuable in international market. Along with the traditional and ethno medicinal uses, phytochemistry and biological activities of *G. gnemon* with clinical and toxicity data could possibly make recommendations for further research (Haloi and Barua 2015).

The study conducted with *G. parvifolium* also substantiated the presence of flavonoids and stilbenes (Lan et al. 2014) and it revealed that the highest flavonoid content is observed in leaves of the plant among seedlings fruit flesh and seed. The accumulation of stilbenes is observed in root of seedlings and trees or mature plant. Resveratrol, piceatannol, isorhapontigenin and gnetol are few of the stilbenes identified from particular tissues. Resveratrol, Isorhapontigenin and Piceatannol are found from the root of seedlings of *G. parvifolium* (Deng et al. 2016).

14.4 Conclusions

Gymnospermae is a medicinally important class of plants which is not completely exploited for its therapeutic potential. Cycadales, coniferales and gnetales are the important orders in the class, most of them are used in traditional medicine. Bioactivities of the plants were studied by the researchers across the world and bioactive metabolites have been isolated from many plants.

Cycadale members are mostly used by ethnic communities in different part of world for their diet. Cycasin is a major compound find in order cycadales. Phenolics, and acids like caffeic, protocatechuic, p-coumaric, p-hydroxybenzoic, ferulic and vanillic are reported from the order, some are specific compounds like dihydroxybenzoical restricted to *Encephalartos* and *B. serrulata, and* Leucoanthocyanidins reported from *Dioon* and *Macrozamia*. Biflavonoids are one of the dominant compounds in the order. Sciadopitysin and Amentoflavone are the major biflavanoids reported.

Coniferales form the predominant part of Gymnospermae and the members of the order are components of many traditional systems of medicine. Araucaria, *Cupressus, Pinus, Podocarpus, Taxus, Cephalotaxusre* are some of the therapeutically exploited Genera of Coniferales. The properties like anticancer, antibacterial, antifungal antimalarial, hepatoprotective, larvicidal, anti-inflammatory properties of various plant species in the order were analysed through studies by different researchers. Many of the plants proved to have significant medicinal potential. With many bioactive molecules identified from these plants, advanced studies have exploited some of these molecules to be used in the pharmaceutical industry. Cephalotaxine, homoharringtonine, taxol and other taxine alkaloids are some examples for the molecules from Coniferales with potential to be used in modern medicine. In Gnetales, some genera like *Gnetum* and *Ephedra* are medicinally very much exploited. Studies have demonstrated the medicinal importance of the different plant species that justified their use in traditional medicine. Bioactive molecules like Isorhapontigenin, Piceatannol, and Gnetol are some of the important and Piceatanuol is therapeutic due to its activity against cancer, ageing, cellular senescence, cardiac aging, stress and prion-mediated neurodegeneration. Gymnosperms provide a wide pool of bioactive molecules as many of such molecules have proved their potential to be used in drug development.

References

Abdelkafi H, Nay B (2012) Natural products from Cephalotaxus sp.: chemical diversity and synthetic aspects. Nat Prod Rep 29(8):845–869

Abdillahi HS, Stafford GI, Finnie JF, Van Staden J (2010) Ethnobotany, phytochemistry and pharmacology of *Podocarpus sensu latissimo* (sl). S Afr J Bot 76(1):1–24

Adam AZ, Juiling S, Lee SY, Jumaat SR, Mohamed R (2017) Phytochemical composition of *Agathis borneensis* (Araucariaceae) and their biological activities. Malays For 80(2):169–177

van Agtmael MA, Eggelte TA, van Boxtel CJ (1999) Artemisinin drugs in the treatment of malaria: from medicinal herb to registered medication. Trends Pharmacol Sci 20(5):199–205

Allemand P (1979) Relations phylogeniques dans le genre Cupressus (Cupressaceae). In: Grasso V, Raddi P (eds) Il cipresso: Malattie e difesa, 23-24 November 1979. AGRIMED, Commission of the European communities, Firenze, pp 51–67

Ames BN, Shigenaga MK, Hagen TM (1993) Oxidants, antioxidants, and the degenerative diseases of aging. Proc Natl Acad Sci 90(17):7915–7922

Anderson EF (1986) Ethnobotany of hill tribes of northern Thailand. II. Lahu medicinal plants. Econ Bot 40:442–450

Apetrei CL, Tuchilus C, Aprotosoaie AC, Oprea A, Malterud KE, Miron A (2011) Chemical, antioxidant and antimicrobial investigations of *Pinus cembra* L. bark and needles. Molecules 16 (9):7773–7788

Aslam MS, Choudhary BA, Uzair M, Ijaz AS (2013) Phytochemical and ethno-pharmacological review of the genus Araucaria–review. Trop J Pharm Res 12(4):651–659

Aslam MS, Choudhary BA, Uzair M, Ijaz AS (2014) Phytochemistry of aerial parts of *Araucaria columnaris*. J Appl Pharm 6(1):114–120

Astley S, Finglas P (2016) Nutrition and health. Ref Mod Food Sci:1–6. https://doi.org/10.1016/B978-0-08-100596-5.03425-9

Bate-Smith EC, Lerner NH (1954) Leuco-anthocyanins. 2. Systematic distribution of leuco-anthocyanins in leaves. Biochem J 58(1):126

Batu JG (2006) Chemical Composition of Indonesian *Pinus merkusii* Turpentine Oils, Gum Oleoresine and Rosins from Sumatra and Java. Pak J Biol Sci 9(1):7–14

Bierhorst DW (1971) Morphology of vascular plants (No. Sirsi) a266918. MacMillan, New York

Biswas C, Johri BM (1997) The Gymnosperms. Springer, Berlin, pp 1–3

Bohlin L, Alsmark C, Göransson U, Klum M, Wedén C, Backlund A (2012) Strategies and methods for a sustainable search for bioactive compounds. Planta Med 78(11):IL13

Bonta M, Pinot OF, Graham D, Haynes J, Sandoval G (2006) Ethnobotany and conservation of tiusinte (*Dioon mejiae* Standl. & LO Williams, Zamiaceae) in northeastern Honduras. J Ethnobiol 26(2):228–258

Borenstein AR, Mortimer JA, Schofield E, Wu Y, Salmon DP, Gamst A, Olichney J, Thal LJ, Silbert L, Kaye J, Craig UL, Schellenberg GD, Galasko DR (2007) Cycad exposure and risk of dementia, MCI, and PDC in the Chamorro population of Guam. Neurology 68(21):1764–1771

Bouchez MP, Arpin N, Deruaz D, Guilluy R (1970) Chemotaxonomy of vascular plants. 20. Chemical study of *Cycas revoluta*; seed pigments. Plantes Med Phytother 4:117–125

Boukhris M, Regane G, Yangui T, Sayadi S, Bouaziz M (2012) Chemical composition and biological potential of essential oil from Tunisian *Cupressus sempervirens* L. J Arid Land Stud 22(1):329–332

Brophy JJ, Goldsack RJ, Wu MZ, Fookes CJ, Forster PI (2000) The steam volatile oil of *Wollemia nobilis* and its comparison with other members of the Araucariaceae (Agathis and Araucaria). Biochem Syst Ecol 28(6):563–578

Bryan-Brown T (1932) The pharmacological actions of taxine. Q J Pharm Pharmacol 5:205–219

Buta JG, Flippen JL, Lusby WR (1978) Harringtonolide, a plant growth inhibitory tropone from *Cephalotaxus harringtonia* (Forges) K. Koch. J Org Chem 43(5):1002–1003

Calderón JS, Céspedes CL, Rosas R, Gómez-Garibay F, Salazar JR, Lina L, Aranda E, Kubo I (2001) Acetylcholinesterase and Insect Growth Inhibitory Activities of *Gutierrezia microcephala* on Fall Army worm *Spodoptera frugiperda* JE Smith. Z Naturforsch C 56 (5-6):382–394

Carbonell-Capella JM, Buniowska M, Barba FJ, Esteve MJ, Frígola A (2014) Analytical methods for determining bioavailability and bioaccessibility of bioactive compounds from fruits and vegetables: A review. Compr Rev Food Sci Food Saf 13(2):155–171

Carson JL, Wallace JW (1972) Detection of C-glycosylflavones in *Dioon spinulosum*. Phytochemistry 11(2):842–843

Céspedes CL, Calderón JS, Lina L, Aranda E (2000) Growth inhibitory effects on fall armyworm *Spodoptera frugiperda* of some limonoids isolated from *Cedrela* spp. (Meliaceae). J Agric Food Chem 48(5):1903–1908

Céspedes CL, Avila JG, García AM, Becerra J, Flores C, Aqueveque P, Bittner M, Hoeneisen M, Martinez M, Silva M (2006) Antifungal and antibacterial activities of *Araucaria araucana* (Mol.) K. Koch heartwood lignans. Z Naturforsch C 61(1-2):35–43

Chen KK, Schmidt CF (1924) The action of ephedrine, the active principle of the Chinese drug Ma Huang. J Pharmacol Exp Ther 24(5):339–357

Chin YW, Balunas MJ (2006) Chai hB and Kinghorn AD: Drug discovery from natural sources. AAPS J 8:e239–e253

Chin Y, Balunas MJ, Chai HB, Kinghorn AD (2009) Dictionary of food science and technology, 2nd edn. International Food Information Service (IFIS Editor), Reading, pp 47–48

Choi YH, Kim J, Kim YC, Yoo KP (1999) Selective extraction of ephedrine from *Ephedra sinica* using mixtures of CO_2, diethylamine, and methanol. Chromatographia 50(11-12):673–679

Coatney GR, Cooper WC, Eddy NB, Greenbeeg J (1953) Survey of antimalarial agents. Chemotherapy of *Plasmodium gallinaceum* infections; toxicity; correlation of structure and action. Bull Med Libr Assoc 42(2):281–282

Collin HA (2001) Secondary product formation in plant tissue cultures. Plant Growth Regul 34 (1):119–134

Cool LG, Hu ZL, Zavarin E (1998) Foliage terpenoids of Chinese Cupressus species. Biochem Syst Ecol 26(8):899–913

Cooper JM (1941) Isolation of a toxic principle from the seeds of *Maccrozamia spiralis*. J Proc Roy Soc New South Wales 74:450–454

Correia RT, Borges KC, Medeiros MF, Genovese MI (2012) Bioactive compounds and phenolic-linked functionality of powdered tropical fruit residues. Food Sci Technol Int 18(6):539–547

Cousins SR, Williams VL, Witkowski ET (2011) Quantifying the trade in cycads (*Encephalartos species*) in the traditional medicine markets of Johannesburg and Durban, South Africa. Econ Bot 65(4):356–370

Cox RE, Yamamoto S, Otto A, Simoneit BR (2007) Oxygenated di-and tricyclic diterpenoids of southern hemisphere conifers. Biochem Syst Ecol 35(6):342–362

Deng N, Chang E, Li M, Ji J, Yao X, Bartish IV, Liu J, Ma J, Chen L, Jiang Z, Shi S (2016) Transcriptome characterization of *Gnetum parvifolium* reveals candidate genes involved in important secondary metabolic pathways of flavonoids and stilbenoids. Front Plant Sci 7:174

Devi KS, Sruthy PB, Anjana JC, Rathinamala J (2015) Identification of bioactive compounds and toxicity study of Araucaria columnaris bark extract on human embryonic kidney cell line. Asian J Biotechnol 7(3):129–136

Dhanabal SP, Manimaran S, Subburaj T, Elango K, Kumar EP, Dhanaraj SA (2000) Evaluation of antimicrobial and anti-inflammatory activity of volatile oil from *Cupressus*. Drug Lines 3 (1):9–12

Donaldson JS (ed) (2003) Cycads: status survey and conservation action plan. IUCN–the World Conservation Union, p 16

Dossaji SF, Mabry TJ, Bell EA (1975) Biflavanoids of the Cycadales. Biochem Syst Ecol 2 (3-4):171–175

Drozd NN, Kuznetsova SA, Lapikova ES, Davydova AI, Makarov VA, Kuznetsov BN, Butylkina AI, Vasil'eva NI, Skvortsova GP (2008) Anticoagulant activity of arabinogalactane sulfate and cedar bark extract studied in vitro. Eksp Klin Farmakol 71(4):30–34

Du J, Chiu MH, Nie RL (1999) Two new lactones from *Cephalotaxus fortunei* var. alpnia. J Nat Prod 62(12):1664–1665

Engler, A. (1886). Führer durch den Königlich botanischen garten der universität zu Breslau, hrsg. von dr. Adolf Engler

Euraque D (2003) The threat of blackness to the Mestizo nation: race and ethnicity in the Honduran banana economy, 1920s and 1930s. In: Striffler S, Moberg M (eds) Banana Wars: Power, Production, and History in the Americas. Duke University Press, Durham, pp 229–249

Facciola S (1990) Cornucopia—a source book of edible plants. Kampong Publications. isbn:0-9628087-0-9

Faiella L, Temraz A, De Tommasi N, Braca A (2012) Diterpenes, ionol-derived, and flavone glycosides from *Podocarpus elongates*. Phytochemistry 76:172–177

Farnsworth NR, Akerele O, Bingel AS, Soejarto DD, Guo Z (1985) Medicinal plants in therapy. Bull World Health Organ 63(6):965

Flamini G, Cioni PL, Morelli I (2000) Investigation of the essential oil of feminine cones, leaves and branches of *Taxodium distichum* from Italy. J Essent Oil Res 12(3):310–312

Flores C, Alarcón J, Becerra J, Bittner M, Hoeneisen M, Silva M (2001) Extractable compounds of native tree, chemical and biological study. Bol Soc Chil Quím 46(1). https://doi.org/10.4067/S0366-16442001000100010

Foster S, Duke JA (1990) American yew. In: Eastern/ Central Medicinal Plants. Houghton Miffin, Boston, p 226

Franco IJ, Fontana VL (1997) Ervas e plantas: a medicina dos simples. Livraria Vida Limited

Freitas AM, Almeida MTR, Andrighetti-Fröhner CR, Cardozo FTGS, Barardi CRM, Farias MR, Simões CMO (2009) Antiviral activity-guided fractionation from *Araucaria angustifolia* leaves extract. J Ethnopharmacol 126(3):512–517

Fuentes JL, Vernhe M, Cuetara EB, Sánchez-Lamar A, Santana JL, Llagostera M (2006) Tannins from barks of *Pinus caribaea* protect Escherichia coli cells against DNA damage induced by γ-rays. Fitoterapia 77(2):116–120

Gaitan E, Lindsay RH, Reichert RD, Ingbar SH, Cooksey RC, Legan J, Meydrech EF, Hill J, Kubota KJ (1989) Antithyroid and goitrogenic effects of millet: role of C-glycosylflavones. Clin Endocrinol Metab 68(4):707–714

Garg SK (2016) Green coffee bean. In: Nutraceuticals. Academic Press, pp 653–667

Garrison MS, Irvine AK, Setzer WN (2016) Chemical composition of the resin essential oil from *Agathis atropurpurea* from North Queensland, Australia. Am J Essent Oils Nat Prod 4(4):04–05

Geiger H, de Groot Pfleiderer W (1971) Über 2, 3-dihydrobiflavone in *Cycas revolta*. Phytochemistry 10(8):1936–1938

Gibney MJ, Lanham-New SA, Cassidy A, Vorster HH (eds) (2009) Introduction to human nutrition (The Nutrition Society Textbook), 2nd edn. Wiley-Blackwell

Govindarajan M, Rajeswary M, Benelli G (2016) Chemical composition, toxicity and non-target effects of *Pinus kesiya* essential oil: an eco-friendly and novel larvicide against malaria, dengue and lymphatic filariasis mosquito vectors. Ecotoxicol Environ Saf 129:85–90

Gurib-Fakim A (2006) Medicinal plants: traditions of yesterday and drugs of tomorrow. Mol Aspects Med 27(1):1–93

Haloi P, Barua IC (2015) *Gnetum gnemon* Linn.: a comprehensive review on its biological, pharmacological and pharmacognostical potentials. Int J Pharmacogn Phytochem Res 7(3):531–539

Harborne JB, Baxter H, Webster FX (1994) Phytochemical dictionary: a handbook of bioactive compounds from plants. J Chem Ecol 20(3):815–818

Hassanzadeh Khayyat M, Emami SA, Rahimizadeh M, Fazly-Bazzaz BS, Assili J (2005) Chemical constituents of *Cupressus sempervirens* L. cv. Cereiformis Rehd. essential oils. Iran J Pharm Sci 1(1):39–42

Haynes RK, Vonwiller SC (1994) Extraction of artemisinin and artemisinic acid: preparation of artemether and new analogues. Trans R Soc Trop Med Hyg 88:23–26

Hegazi GAE, El-Lamey TM (2011) In vitro production of some phenolic compounds from *Ephedra alata* Decne. J Appl Environ Biol Sci 1(8):158–163

Hembree JA, Ching-Jer Chang, McLaughlin JL, Cassady JM, Watts DJ, Wenkert E, Fonseca SF. Jayr De Paiva Campelloc (1979). The cytotoxic norditerpene dilactones of Podocarpus milanjianus and Podocarpus sellowii. Phytochemistry, 18(10), 1691-1694.

Ince I, Yesil-Celiktas O, Karabay-Yavasoglu NU, Elgin G (2009) Effects of *Pinus brutia* bark extract and Pycnogenol in a rat model of carrageenan induced inflammation. Phytomedicine 16:1101–1104

Jadav KM, Gowda KNN (2017) Preliminary phytochemical analysis and in vitro antioxidant activity of *Araucaria columnaris* bark peel and cosmos Sulphureus flowers. Int J Curr Pharm Res 9(4):96–99

Jaradat N, Hussen F, Al Ali A (2015) Preliminary phytochemical screening, quantitative estimation of total flavonoids, total phenols and antioxidant activity of *Ephedra alata* Decne. J Mater Environ Sci 6(6):1771–1778

Jolivet P (1998) Interrelationship between insects and plants. CRC Press, Boca Raton, pp 183–184

Joshi S, Kumar P, Sati SC (2014) Evaluation and screening of antibacterial activity of Kumaun Himalayan *Cupressus torulosa* D. Don leaf extracts. Oaks 10:74–78

Kamal AM, Abdelhady M, Elmorsy EM, Mady MS, Abdelkhalik SM (2012) Phytochemical and biological investigation of leaf extracts of *Podocarpus gracilior* and *Ruprechtia polystachya* resulted in isolation of novel polyphenolic compound. Life Sci J 9(4):1126–1135

Kaushik D, Kumar A, Kaushik P, Rana AC (2012a) Analgesic and anti-inflammatory activity of *Pinus roxburghii* Sarg. Adv Pharm Sci 2012:245431. https://doi.org/10.1155/2012/245431

Kaushik D, Kumar A, Kaushik P, Rana AC (2012b) Anticonvulsant activity of alcoholic extract of bark of *Pinus roxburghii* Sarg. J Chinese Integr Med 10(9):1056–1060

Khulbe K, Verma U, Pant P (2016) Determination of phytochemicals and *in vitro* antioxidant of different extracts of Himalayan Cypress (*Cupressus torulosa* D. DON) needles. Int J Adv Biol Res 6(2):259–266

Kolodziej H, Haberland C, Woerdenbag HJ, Konings AWT (1994) Moderate cytotoxicity of proanthocyanidins to human tumour cell lines. Phytother Res 9:410–415

Krauze-Baranowska M, Wiwart M (2003) Antifungal activity of biflavones from *Taxus baccata* and *Ginkgo biloba*. Z Naturforsch C 58(1-2):65–69

Kuo YJ, Hwang SY, Wu MD, Liao CC, Liang YH, Kuo YH, Ho HO (2008) Cytotoxic constituents from *Podocarpus fasciculus*. Chem Pharm Bull 56(4):585–588

Kurland LT (1988) Amyotrophic lateral sclerosis and Parkinson's disease complex on Guam linked to an environmental neurotoxin. Trends Neurosci 11(2):51–54

Kweon MH, Hwang HJ, Sung HC (2001) Identification and antioxidant activity of novel chlorogenic acid derivatives from bamboo (*Phyllostachys edulis*). J Agric Food Chem 49(10):4646–4655

Lan Q, Liu JF, Shi SQ, Chang EM, Deng N, Jiang ZP (2014) Nutrient and medicinal components in *Gnetum parvifolium* seeds. For Res 27:441

Lassak EV, Brophy JJ (2008) The steam-volatile oil of commercial "almaciga" resin (*Agathis Philippinensis* warb.) from the Philippines. J Essent Oil Bear Plants 11(6):634–637

Lee M, McGeer PL (2012) Weak BMAA toxicity compares with that of the dietary supplement beta-alanine. Neurobiol Aging 33(7):1440–1447

Leung GM, Zhang C, To AW, Lam SS, Kwong YL, Leung A (2017) Combination of Omacetaxine Mepesuccinate (homoharringtonine) and sorafenib as an effective regimen for acute myeloid leukaemia (AML) carrying FLT3-ITD. Blood 130:3849

Li K, Li Q, Li J, Zhang T, Han Z, Gao D, Zheng F (2007) Antitumor activity of the procyanidins from *Pinus koraiensis* bark on mice bearing U14 cervical cancer. Yakugaku Zasshi 127(7):1145–1151

Li YY, Feng J, Zhang XL, Cui YY (2015) Pine bark extracts: nutraceutical, pharmacological, and toxicological evaluation. J Pharmacol Exp Ther 353(1):9–16

Lin AJ, Klayman DL, Milhous WK (1987) Antimalarial activity of new water-soluble dihydroartemisinin derivatives. J Med Chem 30(11):2147–2150

Lin YM, Chen FC, Lee KH (1989) Hinokiflavone, a cytotoxic principle from *rhus succedanea* and the cytotoxicity of the related biflavonoids. Planta Med 55(02):166–168

Liu Z, Du Q, Wang K, Xiu L, Song G (2009) Completed preparative separation of alkaloids from *Cephaltaxus fortunine* by step-pH-gradient high-speed counter-current chromatography. J Chromatogr A 1216(22):4663–4667

Logan KJ, Thomas BA (1985) Distribution of lignin derivatives in plants. New Phytol 99(4):571–585

Lohani H, Gwari G, Andola C, Bhandari U, Chauhan N (2012) α-Pinene rich volatile constituents of *Cupressus torulosa* D. Don from Uttarakhand Himalaya. Indian J Pharm Sci 74(3):278–280

Lü S, Wang J (2014) Homoharringtonine and omacetaxine for myeloid hematological malignancies. J Hematol Oncol 7(1):2

Lucas H (1856) Ueber ein in den Blättern von *Taxus baccata* L. enthaltenes Alkaloid (das Taxin). Arch Pharm 135(2):145–149

Ludwig IA, Clifford MN, Lean ME, Ashihara H, Crozier A (2014) Coffee: biochemistry and potential impact on health. Food Funct 5(8):1695–1717

Lv X, Zhang JB, Wang XX, Hu WZ, Shi YS, Liu SW, Hao DC, Zhang DW, Ge GB, Yang L (2018) Amentoflavone is a potent broad-spectrum inhibitor of human UDP-glucuronosyltransferases. Chem Biol Interact 284:48–55

Mahmood Z, Ahmed I, Saeed MUQ, Sheikh MA (2013) Investigation of physico-chemical composition and antimicrobial activity of essential oil extracted from lignin-containing *Cupressus sempervirens*. Bio Resources 8(2):1625–1633

Makabe H (2013) Recent syntheses of proanthocyanidins. Heterocycles 87:2225–2248

Manach C, Williamson G, Morand C, Scalbert A, Rémésy C (2005) Bioavailability and bioefficacy of polyphenols in humans. I. Review of 97 bioavailability studies. Am J Clin Nutr 81(1):230S–242S

Markham KR, Webby RF, Whitehouse LA, Molloy BP, Vilain C, Mues R (1985) Support from flavonoid glycoside distribution for the division of Podocarpus in New Zealand. N Z J Bot 23(1):1–13

Marquesini NR (1995) Plantas usadas como medicinais pelos índios do Paraná e Santa Catarina, sul do Brasil: guarani, kaingang, xokleng, ava-guarani, kraô e cayuá. Universidade Federal do Paraná, Curitiba

Matlin SA, Prazeres MA, Bittner M, Silva M (1984) Norditerpene dilactones from *Podocarpus saligna*. Phytochemistry 23(12):2863–2866

McKenzie EHC, Buchanan PK, Johnston PR (2002) Checklist of fungi on kauri (*Agathis australis*) in New Zealand. N Z J Bot 40(2):269–296

Michelon F, Branco CS, Calloni C, Giazzon I, Agostini F, Spada PKW, Salvador M (2012) *Araucaria angustifolia*: A potential nutraceutical with antioxidant and antimutagenic activities. Curr Nutr Food Sci 8(3):155–159

Miloš M, Mastelić J, Radonić A (1998) Free and glycosidically bound volatile compounds from cypress cones (*Cupressus sempervirens* L.). Croat Chem Acta 71(1):139–145

Miyazawa M, Minamino Y, Kameoka H (1997) Volatile components of Ephedra sinica Stapf. Flavour Fragr J 12(1):15–17

Moretti A, Sabato S, Gigliano GS (1983) Taxonomic significance of methyl azoxymethanol glycosides in the cycads. Phytochemistry 22(1):115–117

Nagalingum NS, Marshall CR, Quental TB, Rai HS, Little DP, Mathews S (2011) Recent synchronous radiation of a living fossil. Science 334(6057):796–799

Natarajan S, Murti VVS, Seshadri TR (1970) Biflavones of some Cupressaceae plants. Phytochemistry 9(3):575–579

Nayak S (2006) Influence of ethanol extract of Vinca rosea on wound healing in diabetic rats. J Biol Sci 6(2):51–55

NCI Thesaurus. (2019), Ephedrine, compound summary, U.S. National Library of Medicine, National Center for Biotechnology Information. https://pubchem.ncbi.nlm.nih.gov/compound/9294

Ncube NS, Afolayan AJ, Okoh AI (2008) Assessment techniques of antimicrobial properties of natural compounds of plant origin: current methods and future trends. Afr J Biotechnol 7(12)

Ni L, Zhong XH, Chen XJ, Zhang BJ, Bao MF, Cai XH (2018) Bioactive norditerpenoids from *Cephalotaxus fortunei* var. alpina and *C. lanceolata*. Phytochemistry 151:50–60

Niemann GJ (1977) Flavonoids and Related Compounds in Leaves of Pinaceae. II*. *Cedrus atlantica cv Glauca*. Z Naturforsch C 32(11-12):1015–1017

Nishida K, Kobayashi A, Nagahama T (1955) Cycasin, a new toxic glycoside of *Cycas revoluta* Thunb. I. Isolation and structure of cycasin. Bull Agr Chem Soc Japan 19:77–84

Osborne R (2002) Chemistry of Cycadales. In: Pant DD (ed) An Introduction to Gymnosperms, Cycas and Cycadales, BSIP Monograph No. 4. Birbal Sahni Institute of Palaeobotany, Lucknow, p 345

Padalia RC, Verma RS, Chauhan A, Chanotiya CS (2013) Essential oil compositions of branchlets and cones of *Cupressus torulosa* D. Don. J Essent Oil Res 25(4):251–256

Park HS, Takahashi Y, Fukaya H, Aoyagi Y, Takeya K (2003) S_R-Podolactone D, a new sulfoxide-containing norditerpene dilactone from *Podocarpus macrophyllus* var. *maki*. J Nat Prod 66 (2):282–284

Park YS, Jeon MH, Hwang HJ, Park MR, Lee SH, Kim SG, Kim M (2011) Antioxidant activity and analysis of proanthocyanidins from pine (*Pinus densiflora*) needles. Nutr Res Pract 5 (4):281–287

Paudler WW, Kerley GI, McKay J (1963) The Alkaloids of *Cephalotaxus drupacea* and *Cephalotaxus fortunei*. J Org Chem 28(9):2194–2197

Pei K, Ou J, Huang J, Ou S (2016) p-Coumaric acid and its conjugates: dietary sources, pharmacokinetic properties and biological activities. J Sci Food Agric 96(9):2952–2962

Pérard-Viret J, Quteishat L, Alsalim R, Royer J, Dumas F (2017) Cephalotaxus alkaloids. Alkaloids Chem Biol 78:205–352

Pierre-Leandri C, Fernandez X, Lizzani-Cuvelier L, Loiseau AM, Fellous R, Garnero J, Oli CA (2003) Chemical composition of cypress essential oils: volatile constituents of leaf oils from seven cultivated Cupressus species. J Essent Oil Research 15(4):242–247

Praz CJ, Müller A, Dorn S (2008) Specialized bees fail to develop on non-host pollen: do plants chemically protect their pollen. Ecology 89(3):795–804

Ramachandran MS (2014) Heart and toxins. Elsevier, Academic Press

Ravele AM, Makhado RA (2010) Exploitation of *Encephalartos transvenosus* outside and inside Mphaphuli cycads nature reserve, Limpopo Province, South Africa. Afr J Ecol 48(1):105–110

Rawat P, Khan MF, Kumar M, Tamarkar AK, Srivastava AK, Arya KR, Maurya R (2010) Constituents from fruits of *Cupressus sempervirens*. Fitoterapia 81(3):162–166

Rezzi S, Bighelli A, Castola V, Casanova J (2005) Composition and chemical variability of the oleoresin of *Pinus nigra* ssp. laricio from Corsica. Ind Crop Prod 21(1):71–79

Riggs NV (1956) Glucosyloxyazoxymethane, a constituent of the seeds of *Cycas circinalis* L. Chem Ind 35:926

Rizvi SHM, Rahman W, Okigawa M, Kawano N (1974) Biflavones from *Podocarpus neriifolius*. Phytochemistry 13:1990

Roy SK, Qasim MA, Kamil M, IIyas M (1987) Biflavones from the genus Podocarpus. Phytochemistry 26:1985–1987

Saklani A, Kutty SK (2008) Plant-derived compounds in clinical trials. Drug Discov Today 13 (3-4):161–171

Sanchez WE, Brown KS, Nishida T, Durham LJ, Duffield AM (1970) Hydrophilic chemical constituents of *Podocarpus sellowii* Klotzsch. Anais Acad. Bras Cienc An 42:77–85

Saneesh CS (2009) Bread from the Wild'-*Cycas circinalis* L. Endemic, Endangered, and Edible. Cycad Newslet 32(1):4

Santi-Gadelha T, de Almeida Gadelha CA, Aragão KS, de Oliveira CC, Mota MRL, Gomes RC, de Freitas Pires A, Toyama MH, de Oliveira Toyama D, de Alencar NMN, Criddle DN, Assreuy AMS, Cavada BS (2006) Purification and biological effects of *Araucaria angustifolia* (Araucariaceae) seed lectin. Biochem Biophys Res Commun 350(4):1050–1055

Schaneberg BT, Crockett S, Bedir E, Khan IA (2003) The role of chemical fingerprinting: application to *Ephedra*. Phytochemistry 62(6):911–918

Schmeda-Hirschmanna G, Astudillo L, Sepulveda B, Rodriguez JA, Theoduloz C, Yanez T, Palenzuela JA (2005) Gastroprotective effect and cytotoxicity of natural and semisynthetic labdane diterpenes from *Araucaria araucana* resin. Z Naturforsch C 60:511–522

Seal AN, Pratley JE, Haig TJ, An M, Wu H (2010) Plants with phytotoxic potential: Wollemi pine (*Wollemia nobilis*). Agr Ecosyst Environ 135(1-2):52–57

Seccon A, Rosa DW, Freitas RA, Biavatti MW, Creczynski-Pasa TB (2010) Antioxidant activity and low cytotoxicity of extracts and isolated compounds from *Araucaria angustifolia* dead bark. Redox Rep 15(6):234–242

Selim SA, Adam ME, Hassan SM, Albalawi AR (2014) Chemical composition, antimicrobial and antibiofilm activity of the essential oil and methanol extract of the Mediterranean cypress (*Cupressus sempervirens* L.). BMC Comple Altern Med 14(1):179

Sellappan M, Palanisamy D, Joghee N, Bhojraj S (2007) Chemical composition and antimicrobial activity of the volatile oil of the cones of *Cupressus torulosa* D. DON from Nilgiris, India. Asian J Tradit Med 2(6):206–211

Sharma AK, Dhyani S, Kour GD (2014) *Taxus baccata* Linn.: a mystical herb. Unique J Pharm Biol Sci 2(2):68–70

Shibuya T, Funamizu M, Kitahara Y (1978) Novel p-coumaric acid esters from *Pinus densiflora* pollen. Phytochemistry 17(5):979–981

Shrestha K, Banskota AH, Kodata S, Shrivastava SP, Strobel G, Gewali MB (2001) An antiproliferative norditerpene dilactone, Nagilactone C, from *Podocarpus neriifolius*. Phytomedicine 8(6):489–491

Shu YZ (1998) Recent natural products based drug development: a pharmaceutical industry perspective. J Nat Prod 61(8):1053–1071

Shumailova MP (1971) Chemical study of the shell and needles of the cedar pine. Nachnye Trudy Irkutskii Meditsinskii Institut 113:20–22

Silva M, Bittner M (1986) Terpenes of Podocarpus species from Chile. Bol. Soc. Chil. Quím. 31 (1):19–35

Singh KJ, Singh R (2008) The ethnobotany of Cycas in the states of Assam and Meghalaya, India. In: Proceedings of Cycad 2008. The 8th International Conference on Cycad Biology, Panama City, Panama, 13–15 January 2008

Sinisi V, Forzato C, Cefarin N, Navarini L, Berti F (2015) Interaction of chlorogenic acids and quinides from coffee with human serum albumin. Food Chem 168:332–340

Snyder LR, Marler TE (2011) Rethinking cycad metabolite research. Commun Integr Biol 4 (1):86–88

Souza M, Branco C, Sene J, DallAgnol R, Agostini F, Moura S, Salvador M (2014) Antioxidant and antigenotoxic activities of the Brazilian pine *Araucaria angustifolia* (Bert.) O. Kuntze. Antioxidants 3(1):24–37

Spande TF, Garraffo HM, Edwards MW, Yeh HJ, Pannell L, Daly JW (1992) Epibatidine: a novel (chloropyridyl) azabicycloheptane with potent analgesic activity from an Ecuadoran poison frog. J Am Chem Soc 114(9):3475–3478

Spencer PS, Garner CE, Palmer VS, Kisby GE (2015) Environmental neurotoxins linked to a prototypical neurodegenerative disease. In: Environmental factors in neurodevelopmental and neurodegenerative disorders. Academic Press, pp 211–252

Stahlhut R, Park G, Petersen R, Ma W, Hylands P (1999) The occurrence of the anti-cancer diterpene Taxol in *Podocarpus gracilior* Pilger (Podocarpaceae). Biochem Syst Ecol 27 (6):613–622

Suga T, Ohta S, Munesada K, Ide N, Kurokawa M, Shimizu M, Ohta E (1993) Endogenous pine wood nematicidal substances in pines, *Pinus massoniana*, *P. strobus* and *P. palustris*. Phytochemistry 33(6):1395–1401

Suh KS, Lee YS, Kim YS, Choi EM (2013) Sciadopitysin protects osteoblast function via its antioxidant activity in MC3T3-E1 cells. Food Chem Toxicol 58:220–227

Sun NJ, Xue Z, Liang XT, Huang L (1979) Studies on the structure of a new tropone from *Cephalotaxus harringtonia* (Forbes) K. Koch. J Org Chem 43

Takagi T, Itabashi Y (1982) cis-5-Olefinic unusual fatty acids in seed lipids of Gymnospermae and their distribution in triacylglycerols. Lipids 17(10):716–723

Thirumurugan D, Cholarajan A, Raja SS, Vijayakumar R (2018) An introductory chapter: secondary metabolites. Sources and applications, secondary metabolites. Intech, p 1

Tiwari R, Rana CS (2015) Plant secondary metabolites: a review. Int J Eng Res Gen Sci 3 (5):661–670

Torres P, Avila JG, De Vivar AR, Garcia AM, Marin JC, Aranda E, Céspedes CL (2003) Antioxidant and insect growth regulatory activities of stilbenes and extracts from *Yucca periculosa*. Phytochemistry 64(2):463–473

Towers GHN, Gibbs RD (1953) Lignin chemistry and the taxonomy of higher plants. Nature 172 (4366):25

Tumen I, Senol FS, Orhan IE (2012) Evaluation of possible in vitro neurobiological effects of two varieties of *Cupressus sempervirens* (Mediterranean cypress) through their antioxidant and enzyme inhibition actions. Turk J Biochem 37(1)

Ulukanli Z, Karabörklü S, Bozok F, Burhan ATES, Erdogan S, Cenet M, Karaaslan MG (2014) Chemical composition, antimicrobial, insecticidal, phytotoxic and antioxidant activities of Mediterranean *Pinus brutia* and *Pinus pinea* resin essential oils. Chin J Nat Med 12 (12):901–910

Valnet J (2002) Phytotherapy: treatment of diseases by plants. Rahe-kamal Pub, Tehran, pp 358–361. Translated to Persian by: Emami A, Shams-Ardekani MR, Nekoei-naeini N

Venditti A, Frezza C, Sciubba F, Foddai S, Serafini M, Bianco A (2017) Terpenoids and more polar compounds from the male cones of *Wollemia nobilis*. Chem Biodivers 14(3)

Vohora SB (1972) Studies on *Taxus baccata*–II. Pharmacological investigation of the total extract of leaves. Planta Med 22(05):59–65

Wallace JW, Morris G (1978) C-Glycosylflavones in *Gnetum gnemon*. Phytochemistry 17 (10):1809–1810

Wang X-Q, Ran J-H (2014) Evolution of biogeography of gymnosperms. Mol Phylogenet Evol 75:24–40

Wang X, Cai P, Chang CJ, Ho DK, Cassady JM (1997) Three new cytotoxic norditerpenoids dilactones from *Podocarpus purdieanus* Hook. Nat Prod Lett 10(1):59–67

Wani MC, Taylor HL, Wall ME, Coggon P, McPhail AT (1971) Plant antitumor agents. VI. Isolation and structure of taxol, a novel antileukemic and antitumor agent from *Taxus brevifolia*. J Am Chem Soc 93(9):2325–2327

Weiner MA (1980) Earth medicine-earth food: plant remedies, drugs, and natural foods of the North American Indians. MacMillan

Wianowska D, Gil M (2019) Recent advances in extraction and analysis procedures of natural chlorogenic acids. Phytochem Rev 18(1):273–302

Wilson CR, Sauer JM, Hooser SB (2001) Taxines: a review of the mechanism and toxicity of yew (*Taxus* spp.) alkaloids. Toxicon 39(2–3):175–185

Wu DC, Li S, Yang DQ, Cui YY (2011) Effects of *Pinus massoniana* bark extract on the adhesion and migration capabilities of HeLa cells. Fitoterapia 82(8):1202–1205

Xu YM, Fang SD (1991) The structure of a new biflavone from *Podocarpus fleuryi*. Acta Bot Sin 33 (2):162–163

Xu YM, Fang SD, He QM (1990) Chemical constituents of *Podocarpus fleuryi*. Acta Botanica Sin 32(4):302–306

Xu LZ, Chen ZHEN, Sun NJ (1993) Studies on chemical compositions of *Podocarpus neriifolius* D. Don. Acta Bot Sin (Chinese Edition) 35:138–138

Xue Z, Sun NJ, Liang XT (1982) Studies on the structure of hainanolidol. Acta Pharm Sin 17:236–237

Yamaguchi LF, Vassão DG, Kato MJ, Di Mascio P (2005) Biflavonoids from Brazilian pine *Araucaria angustifolia* as potentials protective agents against DNA damage and lipoperoxidation. Phytochemistry 66(18):2238–2247

Yamaguchi LF, Kato MJ, Di Mascio P (2009) Biflavonoids from *Araucaria angustifolia* protect against DNA UV-induced damage. Phytochemistry 70(5):615–620

Yamamoto S, Otto A, Simoneit BR (2004) Lignans in resin of Araucaria angustifolia by gas chromatography/mass spectrometry. J Mass Spectrom 39(11):1337–1347

Yoon KD, Jeong DG, Hwang YH, Ryu JM, Kim J (2007) Inhibitors of osteoclast differentiation from *Cephalotaxus koreana*. J Nat Prod 70(12):2029–2032

Yue R, Li B, Shen Y, Zeng H, Li B, Yuan H, He Y, Shan L, Zhang W (2013) 6-C-Methyl flavonoids isolated from *Pinus densata* inhibit the proliferation and promote the apoptosis of the HL-60 human promyelocytic leukaemia cell line. Planta Med 79(12):1024–1030

Zedeck MS, Frank N, Wiessler M (1979) Metabolism of the colon carcinogen methylazoxymethanol acetate. Front Gastrointest Res 4:32

Zeng WC, Jia LR, Zhang Y, Cen JQ, Chen X, Gao H, Feng S, Huang YN (2011) Antibrowning and antimicrobial activities of the water-soluble extract from pine needles of *Cedrus deodara*. J Food Sci 76(2):318–323

Zhao Y, Wu Y, Wang M (2015) Bioactive substances of plant origin. In: Cheung P, Mehta B (eds) Handbook of food chemistry. Springer, Heidelberg, p 967

Flavonoids for Therapeutic Applications

15

Thirukannamangai Krishnan Swetha, Arumugam Priya, and Shunmugiah Karutha Pandian

Abstract

Flavonoids constitute a large group of plant phenolic metabolites with diverse structural compounds exhibiting multiple biological activities. Flavonoids have been used over centuries in folk medicine for tackling various human ailments and promoting the human health. With centuries old historical background, flavonoids still hold the valour to be captivated by researchers and clinicians for reframing the current medications to recuperate the equilibrium in human health. In this context, a vast range of biological activities of flavonoids has been documented by various research groups. These findings unwind the multi-targeting potential of flavonoids in various clinical conditions, which hints the ability of flavonoids to gratify the need of current treatment strategy mandating the handling of other complications accompanying a diseased condition. Moreover, the ubiquitous dietary sources of flavonoids underscore their innocuous nature as well as the likelihood to be used in clinical settings. This also enlightens that daily consumption of flavonoids from various dietary sources could act as better nutraceuticals for nourishing the health and assist the risk management of many complications. With this, a comprehensive overview on therapeutic applications of multipotent flavonoids has been provided in this chapter.

Thirukannamangai Krishnan Swetha and Arumugam Priya contributed equally.

T. K. Swetha · A. Priya · S. K. Pandian (✉)
Department of Biotechnology, Science Campus, Alagappa University, Karaikudi, Tamil Nadu, India

Keywords

Flavonoids · Phenolic metabolites · Nutraceuticals · Therapeutic applications · Human health

15.1 Introduction

In the modern era, diverse life style behaviours, inclined environmental pollutions and constant work-associated pressure have made human being vulnerable to numerous ailments. Currently available therapeutic regimens could not effectively combat the sufferings. Upsurge in ethnopharmacological research paved way for numerous bioactive components with curative potential against several ailments. Among divergent bioactive elements, plant phenolic compounds constitute one of the major classes of secondary metabolites. Flavonoids are characterized as the most common and ubiquitously distributed group of phenolics that are abundant in dietary supplements. The name 'Flavonoid' was derived from the Latin word 'flavus', which signifies yellow. More than 10,000 varieties of flavonoids have been identified so far with many beneficial effects that surpass the chemical therapy. Almost every part of the plant, such as fruits, vegetables, nuts, seeds, stem, and flowers, is highly engrossed with flavonoids. The general structure of flavonoids includes a 15 carbon skeleton containing two phenyl rings (A and B rings) and a heterocyclic ring (C ring). Six major subclasses of flavonoids include flavones, flavonols, flavanones, flavan-3-ols, anthocyanidins and isoflavones. This classification is based on the differences in the general structure of C ring, functional groups and the position at which B ring is attached to C ring. Individual flavonoid within each subclass is differentiated by the pattern of hydroxylation, methoxylation, glycosylation, glucronidation and conjugation (Gil and Couto 2013; Panche et al. 2016). Several epidemiological studies have postulated an opposing correlation between consumption of flavonoid-rich foods and progression of various age-associated disorders including diabetes, cardiovascular disease, neurodegenerative disorders, cancers and osteoporosis (Arts and Hollman 2005; Murphy et al. 2019). The dietary intake of flavonoids may range anywhere between 10 and 1000 mg/day which vastly relies on the dietary habit of a particular realm. Consumption of flavonoid-rich foods contributes to numerous health benefits. Flavonoids are widely used in nutraceutical, pharmaceutical, medical and cosmetic applications owing to their ability to modulate key regulatory enzymes and cellular metabolic processes (Havsteen 1983). Flavonoids, by virtue of their low-molecular weight, could act on multiple cellular targets simultaneously and mediate their beneficial and protective effects on vital organs of the human body including heart, liver, brain, kidney, colon etc. The primary activities that underscore the biological actions of almost all flavonoids are antioxidant and radical scavenging activities (Gil and Couto 2013). Due to their numerous health-promoting potential, flavonoids are being considered as a foundation for the synthesis of drug molecules that can play a pivotal role in the discovery of novel therapeutics. Hence, major therapeutic applications of flavonoids such as anticancer, antimicrobial, antidiabetic, anti-inflammatory and antiallergic activities, neuroprotective, cardioprotective and hepatoprotective effects are reviewed in this chapter.

15.2 Therapeutic Applications

15.2.1 Flavonoids in Cancer Chemotherapy and Prevention

Cancer, a pervasive disease in the contemporary life style, has aggrandized as the second root cause of death universally (Saranath and Khanna 2014). According to WHO reports, approximately one in six deaths occur owing to cancer and enormous of which occur in indigent countries (www.who.int/news-room/fact-sheets/detail/cancer). Low consumption of fruits and vegetables stands as one of the leading dietetic risk factors for the progression of cancer. Numerous epidemiological studies have evidenced an inverse correlation between intake of fruits and vegetables and risk of most forms of cancer, especially in the epithelial cancers of respiratory and alimentary tracts (Steinmetz and Potter 1991). In past decades, plant folklore has been explored extensively for excavating novel chemoprotective antitumour agents, which prevailed in the discovery of most effective anticancer drugs such as vinblastine, vincristine, irinotecan, topotecan, etoposide, paclitaxel, docetaxel etc. Among the numerous anticancer medications available to date, around 69% of approved drugs are either natural products or synthetic drugs that are designed with the knowledge acquired from natural products (Sak 2014). Amidst diverse bioactive plant derivatives, non-nutritive dietary flavonoids have been demonstrated to have potential role in preventing or delaying the progression of carcinogenesis (Havsteen 2002). The primary mechanisms through which flavonoids exert anticancer potential includes cell cycle blockage, induction of apoptosis, constrainment of cell proliferation and differentiation, disruption of mitotic spindle formation or inhibition of angiogenesis and metastasis and conspicuous reversal of multidrug resistance (Kuntz et al. 1999; Beutler et al. 1998; Mojzis et al. 2008; Ravishankar et al. 2013). Besides, flavonoids also act as antioxidants, free-radical scavengers, inhibitors of enzymes/hormones etc. Flavonoids act through various mechanisms for confronting different stages of cancer, which are detailed below.

15.2.1.1 Mechanistic Basis of Anticancerous Activity of Flavonoids

Aversion of Carcinogen Metabolic Activation
Metabolism of carcinogenic agents inside the human body has critical role in influencing the incidence of different forms of cancer. Among various drug metabolism enzymes, cytochrome P450 plays an imperative part in carcinogen activation, which metabolically activates various procarcinogens to form reactive intermediates that sensitize the immune cells and eventually elicit carcinogenesis. Flavonoids with the proficiency to interact with phase I metabolizing enzymes namely P450 can

defend induction of cellular damage caused due to the carcinogen activation. Besides this primary mechanism, flavonoids are known to exert their anticancerous activity through induction of phase II metabolizing enzymes such as glutathione-S-transferase, UDP–glucuronyl transferase and quinone reductase through which the carcinogens are detoxified and readily eliminated from the body. In vitro and in vivo studies have demonstrated that certain flavonoids such as apigenin, quercetin, ellagic acid, curcumin etc., can alter the metabolism and disposition of carcinogens and preclude cancer (Kato et al. 1983; Chang et al. 1985; Huang et al. 1987; Verma et al. 1988; Wei et al. 1990).

Natural flavanones such as naringenin, apigenin, luteolin, eriodictyol, 7-hydroxyflavanone, 7-hydroxyflavone etc., and several synthetic flavanones have been reported to act as potent anti-breast cancer agents by inhibiting the aromatase activity in the steroidogenesis pathway of oestrogens which is responsible for the conversion of androgens to oestrogens (Pouget et al. 2002).

Antiproliferation

Abnormal proliferation of normal cells is the critical feature of neoplasia that transforms benign to malignant tumours. The antiproliferative propensity of flavonoids arose from the inhibition of prooxidant process, which is associated with tumour progression. Tumour promoters namely arachidonate-metabolizing enzymes, cyclooxygenases (COX) and lipoxygenases (LOX) can induce or activate the prooxidation enzymes. Flavonoids are effectual in inhibiting xanthine oxidase, COX and/or LOX and thereby inhibit tumour cell proliferation (Chang et al. 1993; Mutoh et al. 2000). Another contributing mechanism of antiproliferative activities of flavonoids includes inhibition of polyamine biosynthesis which is associated with the rate of DNA synthesis and cell proliferation. Experimental evidences reveal that flavonoids such as hesperidin, diosmin, chalcone, 2-hydroxychalcone, quercetin etc., inhibit ornithine decarboxylase, a rate-limiting enzyme in the polyamine biosynthesis, which subsequently decreases polyamine and inhibits DNA/protein biosynthesis (Tanaka et al. 1997; Makita et al. 1996). Signal transducing enzymes such as protein tyrosine kinase (PTK), protein kinase C (PKC), phosphoinositide 3-kinases (PIP_3) involved in cell cycle proliferation are effectively inhibited by the action of flavonoids (Ferry et al. 1996; Lin et al. 1997a, b; Sato et al. 2002).

Cell Cycle Arrest

Cyclin-dependent kinases (CDKs) are the master regulators of cell cycle progression. Modulation in the activity of CDKs is a key hallmark for the commencement of neoplasm. Several types of cancers are known to be associated with dysregulation of CDKs which occurs as a result of mutation in CDK genes or CDK inhibitor genes. Flavonoids such as quercetin, kaempferol, apigenin, silymarin, epigallocatechin 3-gallate, luteolin, genistein, daidzein etc., have been found to regulate cell cycle checkpoints at both G1/S and G2/M phase. In addition, experimental studies have revealed the ability of flavopiridol to induce cell cycle arrest either at G1 or G2/M phase by inhibiting all CDKs (Zi et al. 1998; Casagrande and Darbon 2001).

Induction of Apoptosis

Though several flavonoids have been shown to induce apoptosis, the exact molecular mechanism is still indefinite. Multitude of mechanisms including inhibition of DNA topoisomerase activity, reduction in the reactive oxygen species (ROS) level, modulation of signalling pathways, regulation of the expression of heat shock proteins and nuclear transcription factors, activation of caspase systems and downregulation of Bcl expression, activation of endonucleases etc., are recounted to be accomplished by flavonoids for exhibiting apoptotic activity (Wang et al. 1999; Lee et al. 2002).

Modulation of Multi-Drug Resistance

Foremost hindrance to the successful cancer chemotherapy is the development of multi-drug resistance due to P-glycoprotein (Pgp) or multi-drug resistance-associated protein (MRP). Various flavonoids including apigenin, baicalein, kaempferol, naringenin, luteolin, morin, quercetin, myricetin, silybin, genistein, biochanin A, silymarin, diosmin etc., have been demonstrated to reduce Pgp and MRP-mediated drug efflux action by diminishing the overexpression of multi-drug resistance gene-1 (MDR-1), binding with the nucleotide-binding domain of Pgp, inhibiting ATPase activity and hydrolysis of nucleotide etc. (Zhang and Morris 2003; Yoo et al. 2007; Wesołowska 2011). With unique property of reversing multi-drug resistance, flavonoids help in preventing the multi-drug resistance tumours.

Impediment of Angiogenesis

Angiogenesis, the process of capillary vessels growth from existing blood vessels, is mandatory for the growth and progression of solid tumours. Angiogenesis requires extracellular matrix degradation which is mediated through extracellular proteolytic enzymes including matrix metalloproteinases (MMP) and serine proteases. Flavonoids such as genistein, apigenin and 3-hydroxyflavone are demonstrated to inhibit the in vitro angiogenesis by preventing the expression of MMP and urokinase-type plasminogen activator (uPA), activating pro-MMP-2 and modulating their inhibitors. The secreted vascular endothelial growth factor (VEGF) was found to be decreased upon treatment with silymarin. Thus, the anti-angiogenic potential of flavonoids contributes essentially to the cancer chemopreventive efficacy (Jiang et al. 2000; Kim 2003) (Fig. 15.1).

The anticarcinogenic effect of flavonoids is enormous as they can interfere with all phases of cancer viz. initiation, development and progression by modulating cellular proliferation, differentiation, apoptosis, angiogenesis and multi-drug resistance without sparing the normal cells. With the potential to quench almost every possible drawback of present therapeutic regimen, flavonoids could be explored more for cancer chemoprevention and chemotherapy.

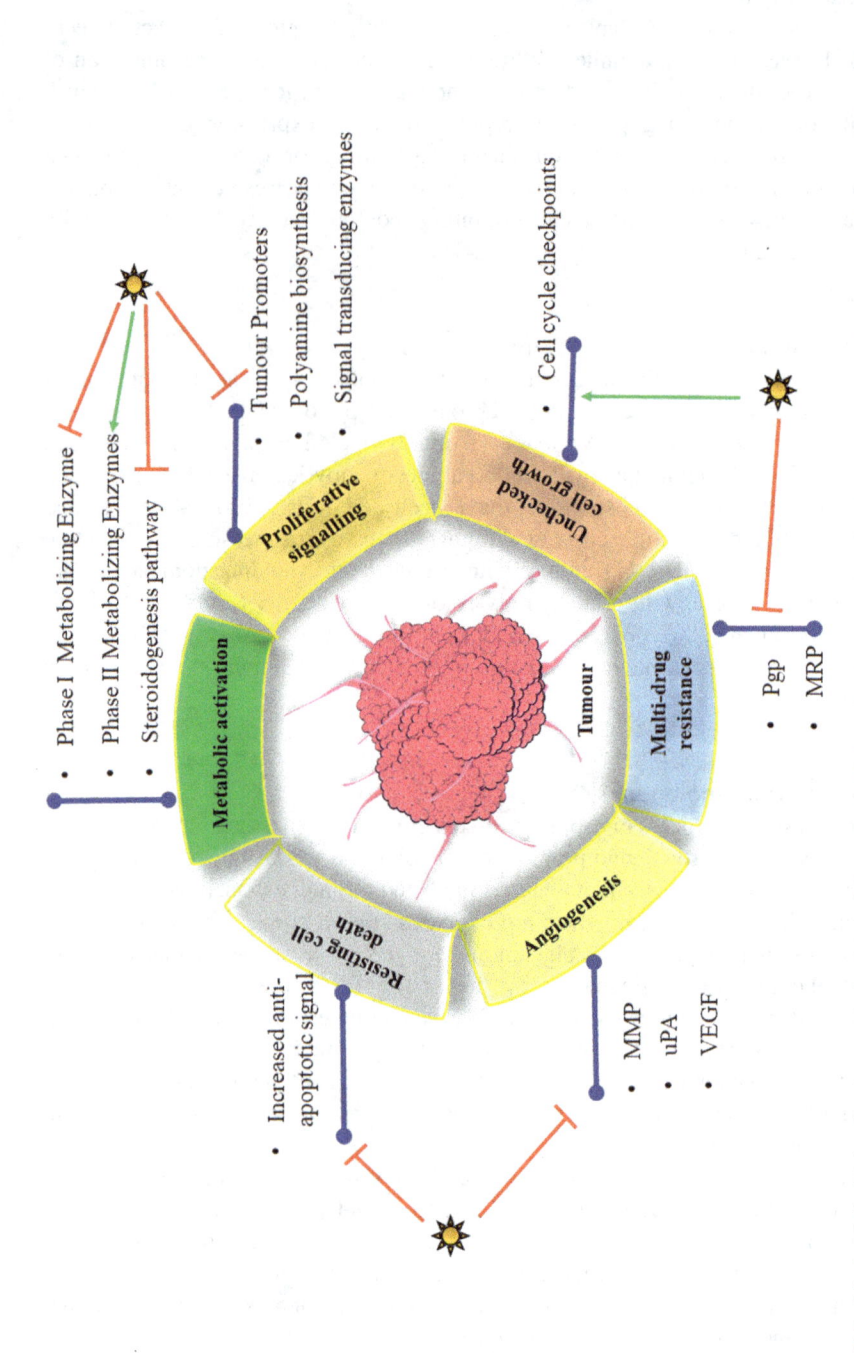

Fig. 15.1 Mechanistic basis of anticancerous activity of flavonoids. Hallmarks of tumorous cell includes activation of drug-metabolizing enzymes, increased proliferative signalling and anti-apoptotic signalling, uncontrolled cell growth, resistance to chemotherapy, angiogenesis and metastasis. Flavonoids can impede

the phase I metabolizing enzymes and induce the phase II metabolizing enzymes through which the carcinogens are detoxified and readily eliminated from body. In addition, flavonoids can inhibit the proliferative signals, increase the apoptosis and regulates the cell cycle checkpoints through which uncontrolled cell mass is prevented. P-glycoprotein (Pgp) and multi-drug resistance-associated protein (MRP)-mediated drug efflux action are reduced by the action of flavonoids. Expression of matrix metalloproteinases (MMP), urokinase-type plasminogen activator (uPA) and vascular endothelial growth factor (VEGF) which are essential for the angiogenesis was demonstrated to be inhibited by flavonoids. (☀) Flavonoid

15.2.2 Flavonoids as Antimicrobials

Antimicrobial resistance (AMR) has materialized as a major global issue, which imparts adverse effects on human health and world economy via augmented rates of mortality and morbidity and inflated treatment expenses (Padiyara et al. 2018). The inappropriate and or overuse of antibiotics along with increased frequency and rapidity of microbial evolution to hostile environment through multiple modes are anticipated to underlay the advent of AMR (McGettigan et al. 2018; Santos-Lopez et al. 2019). By the year 2050, AMR is posited to outstrip the death rate of cancer by 10 million (Garrett 2019), which mandates the necessity of immediate probing of new strategies to combat AMR crisis.

In quest of efficient approaches, flavonoid and its various classes with multiple pharmacological and health benefits have lend a hand to confront the encumbrance of AMR. Commencing several decades ago, the contributory role of antimicrobial flavonoids in traditional as well as modern medicine for human health has indeed a long stretch of history (Osonga et al. 2019). Being the part of natural defence system in plants that affords protection from pathogenic attack (Piasecka et al. 2015), the antimicrobial activity of plant flavonoids holds importance in human health also.

Strikingly, the antimicrobial activity of certain flavonoids recorded at nano-molar level illuminates their potency to surpass the activity of several conventional antibiotics (in terms of minimum inhibitory concentration (MIC)) by several folds. For instance, MIC of isobavachalcone against Gram-positive bacteria was found to be fourfold lesser than that of gentamycin and MIC of piliostigmol against Gram-negative bacteria was observed to be threefolds lesser than that of amoxicillin (Mbaveng et al. 2008; Babajide et al. 2008). In addition to their remarkable activity, flavonoids were manifested to act synergistically with antibiotics and in certain cases; the synergistic action of flavonoid blended antibiotic combination was testified to incredibly reverse AMR. The synergistic combination of quercetin and amoxicillin will serve as a fair example wherein, the resistance of amoxicillin-resistant *Staphylococcus epidermidis* was stated to be reversed (Siriwong et al. 2016). These versatile roles of flavonoids underscore the feasibility of relieving antibiotic load to certain extent in clinical settings, which might assist in tackling the catastrophes of AMR (Xie et al. 2015).

15.2.2.1 Mechanism of Action of Antimicrobial Flavonoids

Covering a vast range of structurally diverse compounds, flavonoid and its classes have been recorded to exhibit antimicrobial activity against a broad spectrum of pathogens including Gram-positive bacteria, Gram-negative bacteria, fungi and viruses (Orhan et al. 2010). Nevertheless, flavonoids have been proposed to shadow various modes to execute their antimicrobial activity against different pathogens. Broadly, their antibacterial action has been documented to comply with some well-studied mechanisms of conventional antibiotics such as inhibition of nucleic acid and cell wall synthesis, impairment of cytoplasmic membrane function and membrane potential and abatement of other vital metabolic functions. Besides, some flavonoids are known to challenge the confronting pathogens by other mechanisms such as inhibition of quorum sensing, biofilm formation, cellular attachment, microbial virulence, efflux pump etc., that are unique in its ploy than conventional drugs. These distinctive mechanisms are anticipated to aid the reversal of AMR and also highlight the propensity of flavonoids to act as novel class of antimicrobial drugs (Górniak et al. 2019).

Cell Membrane Disruption

Flavonoids that partake in cell membrane disruption are identified to accomplish their effect by (1) sandwiching between the lipid bilayers, which sequentially causes membrane disarrangement and disorientation, (2) pulling together the cell membrane, thereby causing cellular leakage and aggregation and (3) producing ROS which in turn, affects membrane integrity and permeability. Catechins, one of the flavonol class compounds, are more often categorized under flavonoids that implement their antimicrobial action through interactions with bacterial cell membrane (Cushnie et al. 2008; Sirk et al. 2009; Górniak et al. 2019). Other flavonoids that execute cell membrane-mediated antimicrobial effect include quercetin, naringenein, galangin, sophoraflavanone G etc. (Mirzoeva et al. 1997; Tsuchiya and Iinuma 2000; Cushnie and Lamb 2005).

Moreover, flavonoid and cell membrane interactions are known to occur through different ways based on the hydrophobic/hydrophilic nature of the flavonoids. Hydrophobic flavonoids were stated to interact by partitioning in the hydrophobic region of lipid membrane, while hydrophilic flavonoids interconnect with the polar heads of lipid membrane via hydrogen bond formation (Górniak et al. 2019). Flavonoids with amphipathic nature (i.e. expressing both hydrophobic and hydrophilic characteristics together) are also stated to perform a critical role in antibacterial properties. For instance, galangin displays hydrophilic nature at ring A due to the presence of substituents and lipophilic nature at ring B that is devoid of substituents. This amphipathic molecule is stated to affect cytoplasmic membrane by creating potassium loss and cell aggregation in *Staphylococcus aureus* (Echeverría et al. 2017).

Inhibition of Nucleic Acid Synthesis

Nucleic acid synthesis directed antimicrobial activity of flavonoids is also reported wherein, crucial enzymes such as DNA gyrase, topoisomerases, HIV-I reverse transcriptase, helicases, DNA and RNA polymerases etc., involved in the replication process are frequently targeted (Ono et al. 1989; Ono et al. 1990; Ohemeng et al. 1993, Bernard et al. 1997; Lin et al. 1997a; Xu et al. 2001). Along with these, quercetin (a pentahydroxyflavone) is also known to block DNA synthesis by interfering with GyrB subunit of DNA gyrases in *Escherichia coli* (Plaper et al. 2003). HIV-I integrase, an important enzyme of HIV life cycle, is reported to be impeded by flavonoids such as robinetin, baicalein for protracting their antiviral activity (Fesen et al. 1994). Another important enzyme, dihydrofolate reductase (DHFR) of folic acid synthesis pathway, which supplies precursors (purines and pyrimidines) for nucleic acid synthesis, is also stated to be a target of flavonoids such as epigalocatechin (Navarro-Martínez et al. 2005; Spina et al. 2008; Raju et al.

2015). Apart from the enzyme inhibitory activity, flavonoids such as myricetin, robinetin and (−)-epigallocatechin are related to intercalate with DNA or form hydrogen bond with nucleic acid bases to stall nucleic acid synthesis (Mori et al. 1987). Taken together, the multi-targeted effect of flavonoids elucidates their potency as well as comprehensive role in the inhibition of nucleic acid synthesis.

Attenuation of Cell Wall Synthesis

Antimicrobial flavonoids that affect other vital metabolic functions (e.g. cell wall synthesis) to mitigate the microbial growth are reported to employ the components of fatty acid pathway as their potential targets. Flavonoids such as sakuranetin, apigenin, quercetin, taxifolin, naringenin, 5-hydroxy-4′,7-dimethoxyflavone, epigalocatechin are chronicled to perturb the components of fatty acid synthase II (FAS-II) including 3-hydroxyacyl-ACP dehydrase, 3-ketoacyl-ACP synthase, malonyl-CoA-acyl carrier protein transacylase, FabG and Fab-I reductases (Górniak et al. 2019). In silico docking studies, Jeong et al. 2009 proposed that the hydrogen bonding between hydroxyl substituents present at C-4′ and C-5′ positions of ring B in flavonoids and Arg38 and Phe308 residues of enzymes might be the plausible base for the antimicrobial activity of flavonoids. In addition to this, flavonoids such as quercetin and apigenin target precursors involved in peptidoglycan synthesis such as D-alanine: D-alanine ligase to annihilate the cell envelope synthesis (Wu et al. 2008; Singh et al. 2013).

Other Mechanisms

Apart from some conventionally described mode of actions, antimicrobial flavonoids are also engrossed in impairing the pathogenicity of microorganisms. Flavonoids (e.g. baicalein, catechin) are stated to ruffle the quorum sensing (QS) ability of microbial communication system, which is one of the imperative sources of AMR crisis. Besides, flavonoids (e.g. epigallocatechin gallate, proanthocyanidins) are also identified to hamper the microbial biofilm formation, which is another notable arsenal of microbial pathogenicity. On the other hand, flavonoids reduce pathogenic traits such as adhesion, bacterial toxins and virulence enzymes such as sortases, ureases, etc. (Cushnie et al. 2008). With this, the ability of certain flavonoids (e.g. biochanin, sarothrin, quercetin, epigallocatechin) to act as efflux pump inhibitors (Górniak et al. 2019) has also been reported, which is envisioned to provide an advantageous credit in managing the AMR crisis.

With growing burden of AMR in clinical settings and imperative need to probe novel antimicrobials, the multi-potent flavonoids grip the suitable features to be captivated in the antimicrobial research. The structure–activity relationship (SAR) studies in flavonoids have disclosed that the presence of suitable substituents at favourable positions of aromatic rings is essential for the antimicrobial activity of flavonoids (Xie et al. 2015). Though natural flavonoids have been reported to exhibit good antimicrobial activity, the evolution of semi-synthetic and synthetic flavonoids is still welcoming, as it is envisaged to drop down the effectual concentration of flavonoids to nano-level as well as boost antimicrobial strength for combating AMR in a more effective manner (Fig. 15.2).

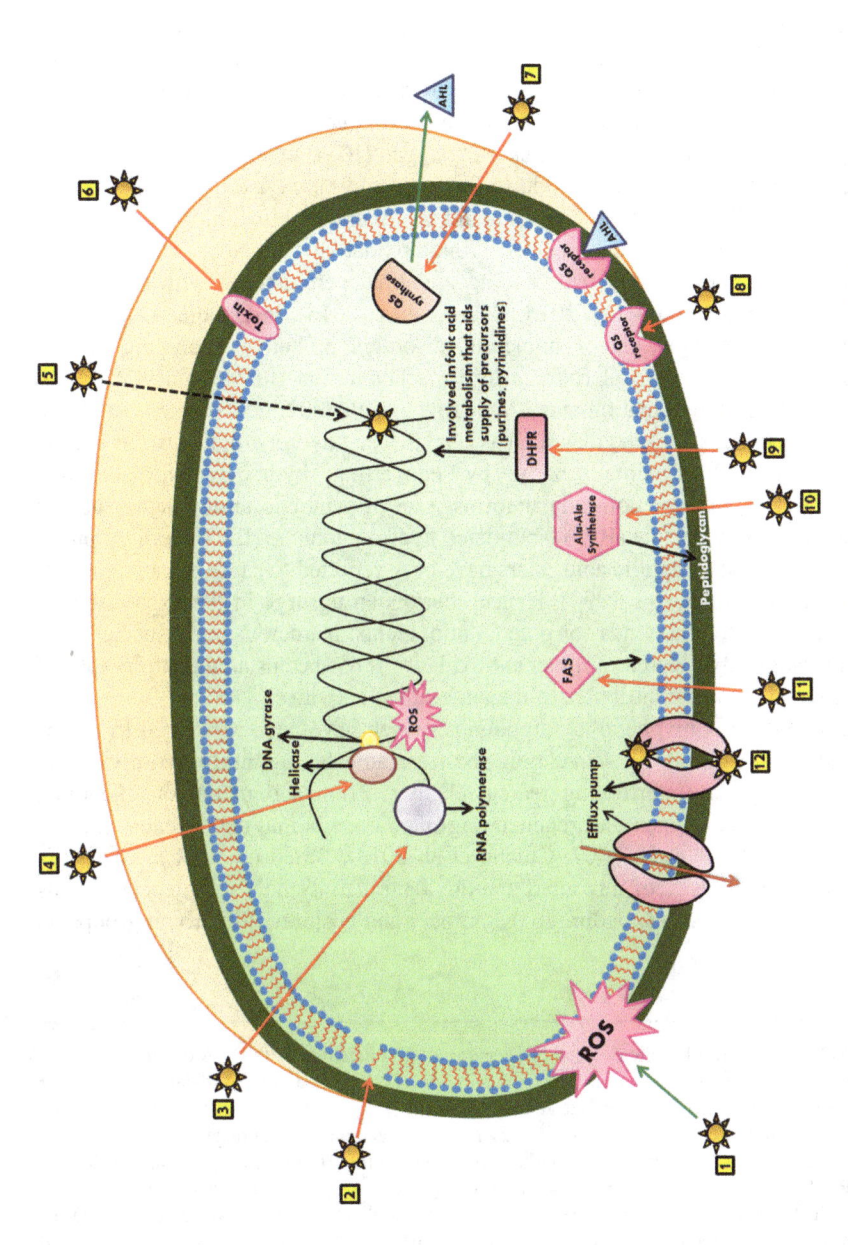

Fig. 15.2 Schematic representation of plausible mode of action of antimicrobial flavonoids. Flavonoids (1) induce ROS that could disrupt membrane or cause DNA damage; (2) disorient and disarrange the lipid bilayers of membrane; impede nucleic acid replication components such as (3) RNA polymerase,

15.2.3 Flavonoids as Antioxidants

During cellular metabolism, expenditure of oxygen for generation of energy results in excessive free radical liberation due to imbalance in oxidants and antioxidants ratio. In addition to metabolic by-products, exogenous factors such as environmental pollutants, radiation, chemicals, toxins, deep-fried foods, spicy foods and physical stress furthermore contribute to the formation of free radicals (Pourmorad et al. 2006). Hydrogen peroxide (H_2O_2), singlet oxygen ($^1O_2$), superoxide radical ($O2\bullet^-$), hydroxyl radical ($OH^\bullet$), nitric oxide ($NO\bullet$) and lipid peroxyl ($LOO\bullet$) are the constantly formed ROS during metabolism and are ascertained to contribute to cellular ageing, mutagenesis or carcinogenesis through oxidative damage to DNA, proteins, lipids and enzymes by covalent binding and lipid peroxidation with subsequent tissue injury (Baba and Malik 2015; Cherrak et al. 2016). Free radicals have been implicated in pathogenesis of numerous disorders in human including cancer, neurodegeneration, atherosclerosis, arthritis, ischemia, reperfusion injury of several tissues, gastritis, inflammation and AIDS (Braca et al. 2002; Pourmorad et al. 2006; Pham-Huy et al. 2008). Enzymes such as catalases and hydroperoxides act as natural antioxidants within human body by converting hydrogen peroxide and hydroperoxides to their non-radical forms. Use of synthetic antioxidants such as butylated hydroxy anisole (BHA), butylated hydroxy toluene (BHT), tertiary butylated hydroquinon and gallic acid esters have been reported to cause negative impact on human health (Barlow 1990). Hence, there is an upsurge in the exploration of natural resources/compounds for potent antioxidants. Numerous plant species have been investigated in search of novel antioxidants. It has been stated that presence of phenolic compounds contributes to the antioxidant activities.

Flavonoids are polyphenolic compounds, which have been acclaimed to evince antioxidant proficiency by scavenging the free radicals, inhibiting hydrolytic and oxidative enzymes, interrupting radical chain reaction and metal ion chelation, quenching singlet oxygen and regenerating membrane-bound antioxidants such as α-tocopherol (Amic et al. 2007; Chiang et al. 2013). Based on SAR studies, it is predicted that intensity of flavonoid's antioxidant activity intensely depends on its chemical structure corresponding to the number and position of hydroxyl groups on

---

Fig. 15.2 (continued) (4) helicases and DNA gyrases; (5) intercalate with DNA or form hydrogen bond with nucleic acid bases to perturb replication process (Dashed arrow indicates the entry of flavonoids into the cell); (6) attenuate toxins; (7) suppress quorum sensing (QS) synthases that produce acyl homoserine lactone (AHL) essential for bacterial communication; (8) block the binding site of AHL by competitive binding to QS receptor; (9) inhibit dihydrofolate reductase (DHFR) of folic acid synthesis pathway that supplies precursors essential for nucleic acid synthesis; (10) mitigate precursors involved in peptidoglycan synthesis such as Ala–Ala synthetases; (11) impede the components of fatty acid synthase (FAS) II to perturb cell wall synthesis and (12) acts as efflux pump inhibitors to exhibit antimicrobial activity and aid reversal of antimicrobial resistance; ※ represents flavonoids, red arrows indicate inhibition and green arrows indicate production

A and B rings as well as the stretching stretch between B and C rings (Amic et al. 2007).

Direct scavenging of ROS by donation of hydrogen atom is considered to be the primary mechanism of antioxidant action of flavonoids. Flavonoids react with oxidizing free radicals, which ensues in generation of more stable and less reactive radicals (Amic et al. 2007). Naringin was reported to exhibit strong radical scavenging activity, whereas quercetin was well recognized as OH scavenger (Fig. 15.3).

Alternate antioxidant mechanism includes inhibition of enzymes involved in ROS generation, that is, glutathione S-transferase, mitochondrial succinoxidase, microsomal monooxygenase and nicotinamide adenine dinucleotide hydrate oxidase. Flavonoids primarily affects the function of enzyme systems involved in the inflammatory processes, particularly tyrosine and serine threonine protein kinases (Aslani and Ghobadi 2016).

Several flavonoids efficiently chelate metal ions such as Fe^{2+}, Cu^+ which play an important role in the metabolism of oxygen and free radical formation. The predicted binding site of metal ions in flavonoids is 3′, 4′-diOH moiety of B ring. Besides, C3 and C5–OH groups and carbonyl group influence the metal chelation (Amic et al. 2007). Inhibition of prooxidant enzymes such as COX, LOX and inducible nitric oxide synthase which are accountable for generation of nitric oxide, prostanoids, leukotrienes and other inflammatory mediators such as cytokines and chemokines similarly contribute to the antioxidant activity of flavonoids (Tunon et al. 2009).

Naringin displays antioxidant activity by intensifying the activity of enzymes such as catalase, superoxide dismutase, glutathione peroxidase, paraoxonase and other antioxidant enzymes (Mamdouh and Monira 2004). Hesperidin is reported for its DPPH scavenging activity and Cu^{2+} metal ion chelation (Toumi et al. 2009). Naringenin, another citrus flavonoid, is recounted to inhibit β oxidation of fatty acids in liver by regulating the enzymes involved in fatty acid oxidation process such as carnitine palmitoyltransferase, 3-hydroxy-3-methyl-glutaryl-CoA reductase and paraoxonase (Jung et al. 2006). Rutin has been recorded to exhibit strong DPPH radical scavenging activity and lipid peroxidation inhibitory activity (Yang et al. 2008). Taxifolin, a flavanol abundant in citrus fruits and onion was demonstrated as an influential antioxidant and antiradical compound with metal chelating activities (Topal et al. 2016). Flavonoids in ginger such as pyrogallol, ferulic acid and coumaric acid are stated to be potent antioxidants with DPPH and DMPD scavenging activity and metal chelation activity (Tohma et al. 2017). Catechin, epicatechin, quercetin, rutin and resveratrol from red grapes were reported to have significant antiradical mediated antioxidant activity (Lacopini et al. 2008). Flavonoids such as ferulic acid and gallic acid have been proved to have xanthine oxidase and Cyclo oxygenase activity which alter the metabolic processes involved in generation of ROS including hydrogen peroxide, superoxide ions (Nile et al. 2016).

Several dietary flavonoids such as galangin, pinobanksin, pinocembrin, D-glucopyranoside and glabranin were also observed to exhibit antioxidant activity through inhibition of xanthine oxidase enzyme (Nile et al. 2018).

In current circumstances, the incidence of exogenous factors influencing free radical generation is excessive, which subsequently contributes to the deterioration

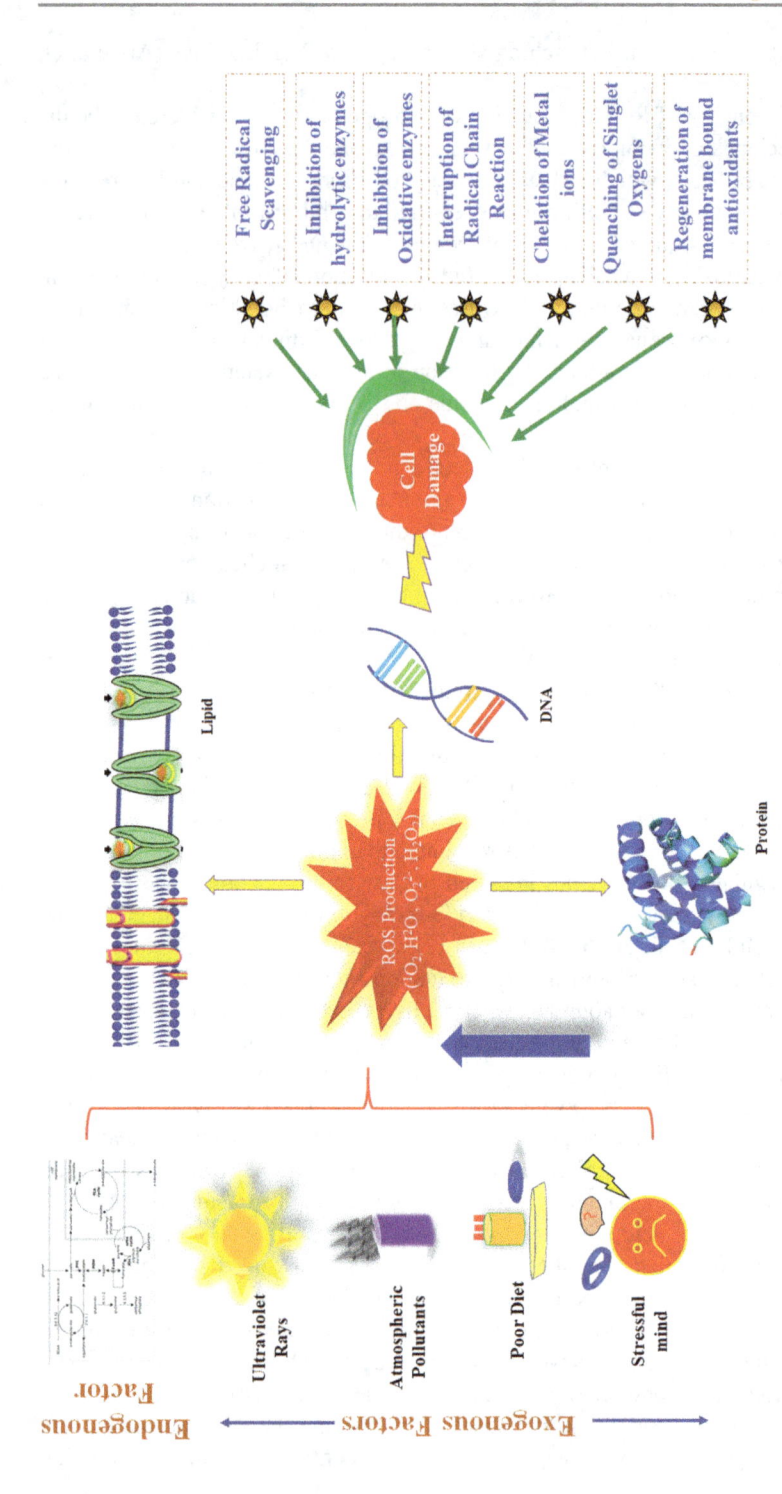

Fig. 15.3 Antioxidant potential of flavonoids. Various exogenous (Ultraviolet radiation, atmospheric pollutants, poor diet, stress) and endogenous (metabolic by-products) factors can contribute to the increased reactive oxygen species (ROS) within the cell which can lead to oxidative damage of DNA, protein and lipids. Free radicals have been implicated in pathogenesis of numerous disorders. Flavonoids exhibit antioxidant proficiency through numerous mechanisms; indicates flavonoid. Mechanisms through which flavonoids exert antioxidant potential are mentioned within the box

of immune defense as well as culminate in different maladies. In addition, several synthetic antioxidants have negative impact on human health. Thus, consuming antioxidants as free radical scavengers from dietary sources is essential for a sustainable life.

15.2.4 Flavonoids as Neuroprotectant

Neurodegenerative disorder (ND) is a generic term covering a broad range of debilitating pathological conditions in which, neurons of central nervous system (CNS) are progressively affected or deteriorated (Maher 2019). Free radical generation and oxidative stresses are considered to be the primary causes of NDs. However, ageing, neuro-inflammation and abnormal aggregation of aberrant proteins are also stated to be the other depictions of ND's pathogenesis (de Andrade Teles et al. 2018). Though hundreds of NDs are reported, the hiking incidences and prevalence of NDs such as Alzheimer's disease (AD), Huntington's disease (HD), Parkinson's disease (PD) and amyotrophic lateral sclerosis (ALS) mark their clinical significance. Since ageing brain is associated with copious fluctuating changes, pointing a single change remains unreliable for designing an effective therapeutic regimen for NDs. Thus, multi-targeting compounds which could influence diverse age-related changes are of current need (Maher 2019). With growing pile of therapeutic potentials, flavonoids are unsurprisingly being engrossed by researchers as a credible source for handling NDs. Besides, the jeopardy of neurodegeneration as well as cognitive decline has been documented to be allayed by high consumption of foods and beverages rich in flavonoids (Letenneur et al. 2007). Hence, intensive research on temporal nature of dietary flavonoids that protract neuroprotective effects and aid in reversing the complications and delaying the onset of age-related NDs are expected to come up with potential therapy.

Hitherto, flavonoids have been documented to present multifarious pharmacological effects such as anticholinesterase (Adedayo et al. 2015), anti-inflammatory (Ashafaq et al. 2012), free radical scavengers, antioxidant (Li et al. 2012), metal chelators (Mandel et al. 2004), antiamyloidogenic (Jiménez-Aliaga et al. 2011), antiapoptosis (Kong et al. 2017) and neurotrophic (Matsuzaki et al. 2008) effects that help subsiding the progression of NDs. In furtherance, flavonoids have also been reported to impede microglia activation, arbitrate inflammatory process in CNS (de Andrade Teles et al. 2018) and improve learning and memory (Kim et al. 2009). Being capable of traversing blood–brain barrier upon short- or long-term administration, flavonoids represent the feasibilities to be used as prospective neuroprotectants (Elbaz et al. 2016) (Fig. 15.4).

There are numerous flavonoids reported for single ND, which implement their pharmacological role through various modes. AD, being the most prevalent form of ND, is stated to be counteracted by flavonoids such as quercetin, rutin, hesperidin, naringinin, silibinin, anthocyanins etc. via different mechanisms. Quercetin is reported to extend its antagonistic effect by ameliorating AMP-activated protein kinase activity through down regulation of tau phosphorylation and suppression of

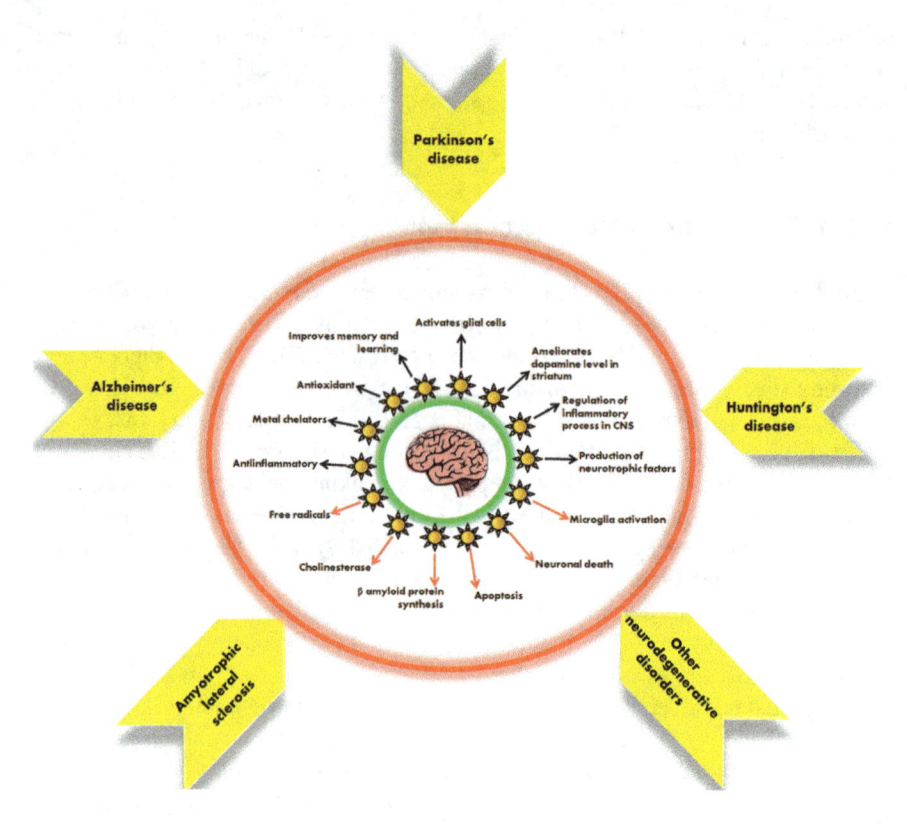

Fig. 15.4 Representative image of activities exhibited by flavonoids as neuroprotectants; ☀ represents flavonoids, black arrows depict the activities exhibited by flavonoids and red arrows indicate the inhibitory effects of flavonoids. Green circle is drawn as a depiction of neuroprotective effects exhibited by flavonoids. Red circle is drawn as a depiction of inhibitory potential of flavonoids against major neurogenerative disorders (NDs) such as Alzheimer's disease, Huntington's disease, Parkinson's disease, Amyotrophic lateral sclerosis and other NDs

apoptosis and ROS. While various researches propose that naringinin reduces neuronal death, hesperidin enhances neuronal differentiation, rutin decreases oxidative stress, anthocyanins mitigate β-amyloid protein production and silibinin impedes inflammatory responses and ROS to deliver their neuroprotective effects against AD. Similarly, in case of PD (the second most prevalent ND), previous studies suggest that naringinin decreases ROS, silibinin attenuates neuronal death, rutin suppresses the expression of nitric oxide synthase, baicalein ameliorates the dopamine level in striatum and activates glial cells and hesperidin exhibits antioxidant effect in striatum to repress the progression of PD. In parallel, flavonoids such as chrysin, quercetin, hesperidin, rutin, genistein and kaempferol were reported to adjourn the progression of HD via attenuation of oxidative stress and inflammation markers, reduction of striatal neuron loss and mitigation of motor deficits (de Andrade Teles et al. 2018).

With this, the beneficial role of flavonoids has been documented in various other NDs also. Though many dietary flavonoids are stated to participate in tackling various NDs, the bioavailability details of flavonoids in human system remain scarce. Hence, studies featuring the trajectory of dietary flavonoids from ingestion till it reaches brain and their metabolism in digestive tract would be helpful in exploring their potential therapeutic applications.

15.2.5 Flavonoids as Antidiabetics

Diabetes mellitus (DM), a complex metabolic disorder that occurs due to deficits in insulin secretion or utilization, has become a serious illness globally with 8.5% prevalence among world's adult population (www.who.int/news-room/fact-sheets/detail/diabetes). Amid two major classifications of DM, type I DM is linked with the insulin deficiency due to the devastation of pancreatic β-cells, whereas type II DM is allied with insulin resistance, anomalous insulin secretion, β-cell apoptosis and elevated production of hepatic glucose (Akkati et al. 2011). Hyperglycaemia dramatically increases the micro and macrovascular abnormalities in vital organs such as heart, kidney, brain, nerves etc. (Lotfy et al. 2017). Lifestyle refinement remains as the first-line cure for early-stage diabetes. Despite this, for type II DM, medical intervention remains to be the efficacious strategy for confronting the ailment. However, the oral antidiabetic drugs have posed a massive list of side effects. The after effects of sulphonylureas include hypoglycaemia, weight gain, skin inflammations, acute porphyria and hyponatremia. Whereas, biguanides drug that acts as insulin sensitizers cause lactic acidosis. Beyond this, thiazolidinediones can exhibit adverse effects such as oedema, cardiac failure, liver toxicity and anaemia (Bösenberg and van Zyl 2014). Consequently, unearthing new antidiabetic drug that does not enforce any diabetic complications is imperative. With numerous evidences, flavonoids have been recognized to have promising beneficial effects on diabetes by improving and harmonizing the glycaemic level, lipid profile and antioxidant activity (Vinayagam and Xu 2015).

Natural plant-derived flavonoids such as quercetin, kaempferol, hesperidin, myricetin, epicatechin, hesperitin, apigenin, naringenin, epigallocatechin, luteolin, genistein, baicalein, diosmetin, diadzein, chrysin, shamimin etc., are stated as eminent leads for DM remedy (Mukherjee et al. 2006). Various mechanisms through which flavonoids present their antidiabetic activities are as follows (Fig. 15.5).

15.2.5.1 Mechanism of Action of Antidiabetic Flavonoids

Flavonoids as Insulin Secretagogues
Various flavonoids such as quercetin, catechin, rutin, genistein, epicatechins, daidzein and epigallocatechin–gallate have been evidenced in both in vitro and in vivo systems to increase insulin secretion. Modulation of β-cell proliferation could be one of the possible mechanisms through which flavonoids such as genistein, puerarine, epigallocatechin–gallate induce insulin secretion (Pinent et al. 2008). Besides insulin

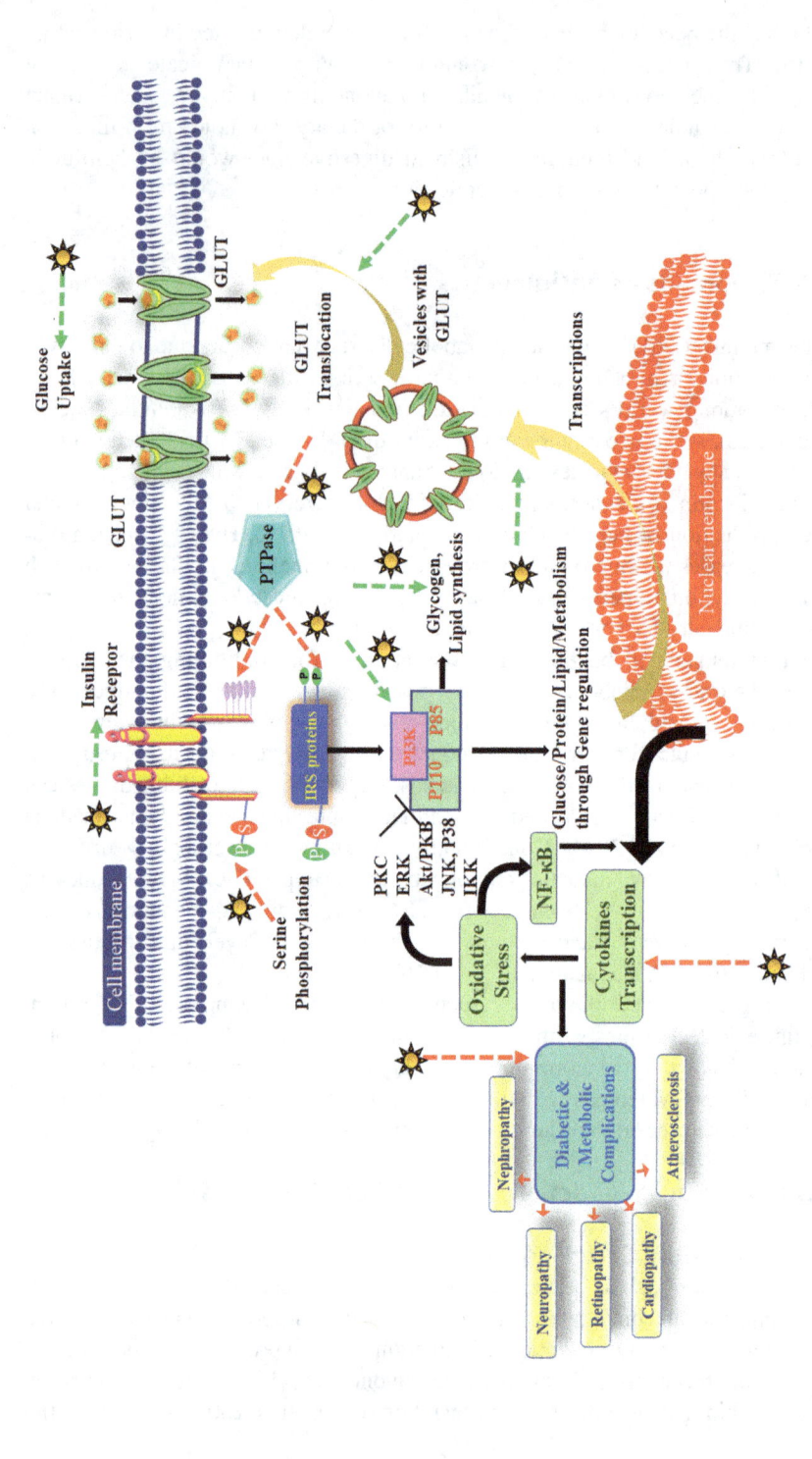

Fig. 15.5 Modulatory effect of flavonoids on glucose metabolism pathway. Insulin signal transduction cascade has been reported to be modulated by flavonoids. *GLUT* glucose transporter, *PTPase* phosphotyrosine phosphatase, *IRS* insulin receptor substrate, *PI3K* phosphoinositide 3-kinase, *PKC* protein

secretion, quercetin and epigallocatechin–gallate can defend insulin-producing INS-1 cells from oxidative stress by inducing antiapoptotic signals (Kim et al. 2010). Through stimulation of several signalling pathways, genistein has been evidenced to induce proliferation of pancreatic β-cell in mice. Furthermore, genistein induces expression of a major cell cycle regulator cyclin-D1, which is essential for β-cell growth (Fu et al. 2010).

Flavonoids as Insulin Sensitizers

Insulin sensitivity can be increased by either of two major pathways (1) activation of adenosine 5′-monophosphate-activated protein kinase (AMPK) or (2) activation of peroxisome proliferator-activated receptor gamma (PPARγ). The flavonoid naringenin increases the uptake of glucose from skeletal muscle cells by AMPK-dependent manner, whereas epigallocatechin–gallate and rutin subdue the glucotoxicity in pancreatic β-cells by activating the insulin receptor substrate and AMPK signalling (Cai and Lin 2009; Zygmunt et al. 2010). Also, naringenin was found to activate PPARγ (Goldwasser et al. 2010). Hesperidin glycosides have been found to increase insulin sensitivity by altering the expression of genes coding PPARγ, 3-hydroxy-3-methyl-glutaryl coenzyme A (HMGCoA) reductase and LDL-receptor in experimental rats (Akiyama et al. 2009). In insulin signal transduction pathway, downstream of insulin receptor kinase such as phosphotyrosine phosphatase type-1B and inhibitory kappaB kinase-beta (IKK-β) have been modulated by the action of flavonoid quercetin (Peet and Li 1999).

Flavonoids as Incretin Potentiators

Incretins are peptide hormones secreted rapidly in response to food intake, which in turn, stimulates pancreatic β-cells to secrete insulin. The two major incretins are glucose-dependent insulinotropic polypeptide and glucagon-like peptide-1 (GLP-1). Dipeptidylpeptidase-IV (DPP-IV) is an enzyme that inactivates incretins and halts its insulinotropic activity. Several flavonoid-containing extracts have been found to inhibit DPP-IV (Bansal et al. 2012). Flavonoids have also been testified to escalate the action of incretin and act as agonist to GLP-1 receptor. Quercetin is reported to act as an allosteric ligand for GLP-1R (Koole et al. 2010).

Flavonoids as Modulators of Carbohydrate Absorption from Gastrointestinal Tract

Certain antidiabetic drugs, such as acarbose, exhibit their action through inhibition of α- glucosidase enzyme which consequently decreases the breakdown of oligo and disaccharides into monosaccharides prior to adsorption (Krentz and Bailey 2005). Quercetin, myricetin and luteolin have been evidenced to be effective α-glucosidase

---

Fig. 15.5 (continued) kinase C, *ERK* extracellular signal-regulated kinases, *Akt/PKB* protein kinase B, *JNK* Janus kinase, *P38* mitogen-activated protein kinase p38, *IKK* inhibitory κB kinase, *NF-κB* nuclear factor κB; ☼ indicates flavonoid. Red dotted arrows indicate inhibition by flavonoids. Green dotted arrows indicate activation by flavonoids

inhibitor (Tadera et al. 2006). In addition to natural flavonoids, synthetic flavonoid derivates were also found to be potent inhibitors of α-glucosidase enzyme (Wang et al. 2010).

Flavonoids in Carbohydrate Metabolism/Transport
Antidiabetic activity of flavonoids is also materialized through the regulation of glucose metabolism. For instance, naringin and hesperidin have been shown to significantly increase the glucokinase activity. Naringin has also been stated to lower the transcriptional expression of phosphoenol-pyruvate carboxykinase in liver, which has been demonstrated in diabetic mice (Jung et al. 2006). By reducing the elevated levels of glucose-6-phosphate and fructose-1,6-bisphosphate in liver, rutin was found to induce significant hypoglycemic effect (Prince and Kamalakkannan 2006). Quercetin was found to increase the activity of glucokinase and hexokinase enzymes in diabetic rats without modulating the normal metabolism in control rats. Flavones such as pectolinarigenin and pectolinarin were known to improve glucose and lipid homeostasis by modulating the expression of adiponectin and leptin (Liao et al. 2010).

The antidiabetic potential of flavonoids is huge because of their modulatory effects on blood sugar transporter such as augmentation of insulin secretion, stimulation of pancreatic β-cells proliferation and mitigation of insulin resistance, oxidative stress, inflammation and apoptosis by various pathways. Broad range of antidiabetic potential of flavonoids provides new insights into the development of drugs against hyperglycaemia and their associated complications. With rapid increase in the incidence of diabetes globally, there is an instant need for the development of antidiabetic medication. As antidiabetic potential of numerous flavonoids has been demonstrated in animal models, clinical trials are required to further potentiate their antidiabetic drugability in the future.

15.2.6 Other Therapeutic Applications

15.2.6.1 Flavonoids as Anti-inflammatory and Antiallergics
Inflammation is the intricate form of biological response that is elicited upon encountering a tissue damage and or microbial infection (Nathan 2002). It is closely concatenated with immune system and becomes quintessential for rectifying the infection as well as reversing the equilibrium in homeostasis. The characteristic effects of inflammatory response are thus transitory. Nevertheless, in case of inflammatory disorders, this condition surpasses the threshold level and transpires as a chronic effect that produces more mutilation in host than the microbial infections (Barton 2008). This hostile chronic inflammation exacerbates the clinical conditions such as NDs, obesity, cardiovascular diseases, arteriosclerosis, arthritis, diabetes and also cancer (García-Lafuente et al. 2009). Despite the current usage of steroid-based anti-inflammatory drugs in acute inflammation, these drugs are less effective against chronic inflammation (Kim et al. 2004). In light of this situation, several flavonoids such as quercetin, genistein, rutin, hesperidin, morin, glycitin etc., have been

recorded to act as a supportive tool in mitigating the adverse effects of chronic inflammation. Several mechanisms manifested to underlie the anti-inflammatory property of flavonoids include (1) alteration of the activity of phospholipase A2, LOX and COX (involved in arachidonic acid metabolism) and nitric acid synthase, (2) antioxidant and free radical scavenging, (3) regulation of cellular functions of inflammatory cells and (4) alteration of pro-inflammatory gene expression and generation of pro-inflammatory molecules (García-Lafuente et al. 2009). For example, genistein has been observed to protract its anti-inflammatory effect by perturbing the enzymatic activity of tyrosine protein kinases that are engaged in the elicitation of inflammatory response (Akiyama et al. 1987). While, morin, luteolin and galangin were testified to act as LOX or COX inhibitors (Baumann et al. 1980) and quercetin was identified to obstruct the release of interleukin-6 and tumour necrosis factor-α, which are the major contributors of chronic inflammatory responses (Cho et al. 2003).

On the other hand, the term 'allergy' defines the abnormal proclivity of certain individuals to develop hypersensitive biological response upon confronting certain substances called allergens. Allergic or atopic disorders include allergic asthma, allergic rhinitis, eczema and several food allergies. Moreover, allergic disorders are manifested to be the prolonged consequences of chronic allergic inflammation occurred due to the repeated acquaintance with allergens. This finding has prompted the need to design new strategies with enhanced immune tolerance to confronting allergens and sketch the immune response for precluding the onset of allergic disorders (Galli et al. 2008). In this regard, flavonoids such as quercetin, naringenin, naringenin chalcone (Escribano-Ferrer et al. 2019) and luteolin-7-O-rutinoside (Inoue et al. 2002) were chronicled to act as antiallergens.

Thus, the role of flavonoids in acute and chronic inflammation and allergic inflammation suggests their prospective applications as prophylactic and therapeutic regimens. However, in-depth studies focusing bioavailability and other crucial factors are still required for developing flavonoids as anti-inflammatory and antiallergic drugs.

15.2.6.2 Flavonoids as Cardioprotectants

Cardiovascular diseases (CVDs) represent clinical conditions with damaged heart and blood vessels such as coronary heart disease (CHD), arterial restenosis, atherosclerosis, peripheral arterial disease, acute coronary syndrome, myocardial ischemia-reperfusion injury etc. With mounting incidence and prevalence, CVD has been recognized as one of the precarious diseases that potentiates the risk of high mortality rates (Tijburg et al. 1997; García-Lafuente et al. 2009). In this context, the constructive role of flavonoids and flavonoid-rich foods in CVD prevention has been evidenced in several epidemiological studies (Hertog et al. 1993; Knekt et al. 1996; Knekt et al. 2002; Sesso et al. 2003; Mink et al. 2007). The cardioprotective property of flavonoids has been documented to be attributed to their antiplatelet, anti-inflammatory, antioxidant effects along with their potential to ameliorate endothelial function and enhance high-density lipoprotein (García-Lafuente et al. 2009). The cardioprotective role of soy isoflavone in CVD with chronic inflammation has been demonstrated to be driven through their anti-inflammatory potential, which was accomplished by the suppression of inflammatory mediators (Droke et al. 2007). In addition, isoflavones have been testified to perturb the adhesion of monocyte and endothelial cells for wielding its cardioprotective effect in inflammatory vascular diseases (Chacko et al. 2007). In parallel, the cardioprotective potential of dietary flavonoids such as chrysin, quercetin, kaempferol and apigenin has been evinced to be associated with the reduction of adhesion molecules expression in human aortic endothelial cells (Lotito and Frei 2006). Quercetin has been identified to interrupt atherosclerotic plaques and suppress metalloproteinase-1 expression for tumbling the risk of CHD (Osiecki 2004). Though several flavonoids have been recorded to possess cardioprotective effect, the correlation between consumption of flavonoid-rich foods and risk of CVD is still inconsistent. Hence, further investigations are required to unravel the cardioprotective potential of flavonoids to be used as potent therapeutic regimen.

15.2.6.3 Flavonoids as Hepatoprotectants

Hepatic diseases are concerned with the diseases that greatly affect the structural and functional homeostasis of liver. They are not only limited to hepatitis and liver cirrhosis but also include various pathological conditions that occur as a sequel of congestive heart failure and inflammatory, infectious, metabolic, toxic and neoplastic diseases (Williams 1983). In this regard, many flavonoids have been recounted to display hepatoprotective effects against many hepatic diseases. For instance, hesperidin is reported to exert hepatoprotective effect by repressing the expression and activity of acetaldehyde-induced matrix metalloproteinase-9 in HepG2 cells, which is envisaged to control the progression of hepatocellular carcinoma in alcoholic patients (Yeh et al. 2009). Quercetin is recounted to protect liver function of experimental rats with carbon tetrachloride prompted liver cirrhosis via partial reduction of oxidative stress and collagen accumulation (Pavanato et al. 2003). Naringenin is stated to exhibit hepatoprotective effect against cholestatic liver disease by mitigating the liver damage in bile duct ligated rodent model (Sánchez-Salgado et al. 2019). Silymarin was testified to attenuate hepatitis C virus instigated liver disease in vitro by expressing anti-inflammatory and immunomodulatory effects (Morishima et al. 2010). Along with this, several other flavonoids such as chrysin (Pingili et al. 2019), gossypin, kolaviron, hispidulin (Di Carlo et al. 1999) and apigenin (Di Carlo et al. 1993) were also documented to possess hepatoprotective effect.

15.2.6.4 Flavonoids in Chronic Obstructive Pulmonary Disease (COPD)

COPD, encompassing more complicated and progressive lung diseases such as refractory asthma, chronic bronchitis and emphysema, is recounted to create higher risk of morbidity and mortality as well as projected to hit third place in the list of primary causes of mortality by 2020 (Vestbo et al. 2013). Chronic inflammatory response incited by tissue damage that has occurred due to the prolonged exposure of lungs to pollutants is stated to prompt COPD. The inflammatory mediators instigated as the result of tissue damage are identified to exacerbate the severity and iterate the series of tissue damage and inflammatory response even in the absence of exposure to the pollutants (Tuder and Petrache 2012; Rovina et al. 2013; Simpson et al. 2013; Lago et al. 2014; Zuo et al. 2014). These events represent a strong correlation between inflammation and COPD (Man et al. 2012). Apart from inflammation, COPD is also closely linked with oxidative stress (Rezaeetalab and Dalili 2014). The correlation between flavonoid-rich diet intake and reduced risk of COPD has been documented by multiple researches involving animal models, which enunciates the positive role of flavonoids in COPD (Ganesan et al. 2010; Guan et al. 2012; Yang et al. 2012; Bao et al. 2013; Huang et al. 2015). The anti-inflammatory and antioxidant effects of flavonoids have been envisaged to partially reflect in the abridged COPD symptoms and risks. Flavonoids such as hyperoside, quercitrin and afzelin have been identified to suppress the inflammatory response in mouse model with lipopolysaccharides-driven acute lung injury (Lee et al. 2015). Also, in animal models with cigarette prompted lung inflammation, flavonoids such as liqueritin apioside and apple polyphenol have been observed to wield antioxidant and anti-inflammatory potential for pulmonary protection (Guan et al. 2012; Bao et al. 2013). Although strong evidence for harmony between increased flavonoid intake and reduced COPD risk is not available so far, further in-depth investigation is envisioned to elucidate the potential therapeutic application of flavonoids in allaying COPD.

15.3 Conclusions

The advancements made in the medical field still remain deficit to endure the upsurge in number of morbidity and mortality cases at clinical settings. The influence of multi-drug-resistant microbes in diseased conditions still worsens the severity of existing harrowing scenario and slackens the efficacy of prevailing medications. This standpoint impels the inevitable requirement to identify new valuable sources for confronting the clinical complications and or to probe molecules/techniques that upgrade the efficacy of current medications. The reprise of traditional medical practice in modern medicine with suffice refinements is perceived to relax the existing complexities. In this conjunction, the nature's bountiful benefits in the form of flavonoids in myriad plant and dietary sources with multi-targeting potential have opened up a new platform for the researchers and clinicians to fine-tune the therapeutic regimen. Discussion in this chapter with information gathered from distinct research groups evidence the spectrum of biological activities exhibited by structurally diverse compounds of flavonoid group. The plausibility of innocuous nature of flavonoids owing to their presence in various dietary sources encourages the possibility of using them in clinical settings.

Number of epidemiological studies with the intent to unwind the reliability of flavonoid-rich diet intake and risk management of various clinical conditions has been made. However, inconsistency between the observations of several research groups confounds the understanding of therapeutic role of flavonoids. Also, upon consumption, the complex nature of flavonoids drives them to undergo various metabolic changes in the host system. Thus, delineation of course of events taking place within the host system upon flavonoids intake and identification of bioavailability of flavonoids for imparting their beneficial role are essential to be decrypted. Connecting these missing links with further in-depth researches is envisioned to culminate in unpinning the clinical applications of flavonoids. Further, appropriate formulation of flavonoids trailed by adequate clinical trials for advancing the previously researched flavonoids to clinical use may act as future endeavours.

Acknowledgement The authors thankfully acknowledge the support extended by DST-FIST [Grant No. SR/FST/LSI-639/2015(C)], UGC-SAP [Grant No. F.5-1/2018/DRS-II (SAP-II)] and DST-PURSE [Grant No. SR/PURSE Phase 2/38 (G)]. The authors sincerely acknowledge the computational and bioinformatics facility provided by Bioinformatics Infrastructure Facility (funded by DBT, GOI; File No. BT/BI/25/012/2012, BIF). SKP is thankful to UGC for Mid-Career Award [F.19-225/2018(BSR)] and RUSA 2.0 [F.24-51/2014-U, Policy (TN Multi-Gen), Dept. of Edn, GoI].

References

Adedayo BC, Oboh G, Oyeleye SI, Ejakpovi II, Boligon AA, Athayde ML (2015) Blanching alters the phenolic constituents and in vitro antioxidant and anticholinesterases properties of fireweed (*Crassocephalum crepidioides*). J Taibah Univ Med Sci 10(4):419–426

Akiyama T, Ishida J, Nakagawa S, Ogawara H, Watanabe SI, Itoh N, Shibuya M, Fukami Y (1987) Genistein, a specific inhibitor of tyrosine-specific protein kinases. J Biol Chem 262 (12):5592–5595

Akiyama S, Katsumata SI, Suzuki K, Nakaya Y, Ishimi Y, Uehara M (2009) Hypoglycemic and hypolipidemic effects of hesperidin and cyclodextrin-clathrated hesperetin in Goto-Kakizaki rats with type 2 diabetes. Biosci Biotechnol Biochem 73(12):2779–2782

Akkati S, Sam KG, Tungha G (2011) Emergence of promising therapies in diabetes mellitus. J Clin Pharmacol 51(6):796–804

Amic D, Davidovic-Amic D, Beslo D, Rastija V, Lucic B, Trinajstic N (2007) SAR and QSAR of the antioxidant activity of flavonoids. Curr Med Chem 14(7):827–845

de Andrade Teles, R.B., Diniz, T.C., Pinto, C., Coimbra, T., de Oliveira Júnior, R.G., Gama e Silva, M., de Lavor, É.M., Fernandes, A.W.C., de Oliveira, A.P., de Almeida Ribeiro, F.P.R. and da Silva, A.A.M., 2018. Flavonoids as therapeutic agents in Alzheimer's and Parkinson's diseases: a systematic review of preclinical evidences. Oxid Med Cell Longev, 2018

Arts IC, Hollman PC (2005) Polyphenols and disease risk in epidemiologic studies. Am J Clin Nutr 81(1):317S–325S

Ashafaq M, Raza SS, Khan MM, Ahmad A, Javed H, Ahmad ME, Tabassum R, Islam F, Siddiqui MS, Safhi MM, Islam F (2012) Catechin hydrate ameliorates redox imbalance and limits inflammatory response in focal cerebral ischemia. Neurochem Res 37(8):1747–1760

Aslani BA, Ghobadi S (2016) Studies on oxidants and antioxidants with a brief glance at their relevance to the immune system. Life Sci 146:163–173

Baba SA, Malik SA (2015) Determination of total phenolic and flavonoid content, antimicrobial and antioxidant activity of a root extract of *Arisaema jacquemontii* Blume. J Taibah Univ Med Sci 9(4):449–454

Babajide OJ, Babajide OO, Daramola AO, Mabusela WT (2008) Flavonols and an oxychromonol from *Piliostigma reticulatum*. Phytochemistry 69(11):2245–2250

Bansal P, Paul P, Mudgal J, Nayak PG, Pannakal ST, Priyadarsini KI, Unnikrishnan MK (2012) Antidiabetic, antihyperlipidemic and antioxidant effects of the flavonoid rich fraction of *Pilea microphylla* (L.) in high fat diet/streptozotocin-induced diabetes in mice. Exp Toxicol Pathol 64 (6):651–658

Bao MJ, Shen J, Jia YL, Li FF, Ma WJ, Shen HJ, Shen LL, Lin XX, Zhang LH, Dong XW, Xie YC (2013) Apple polyphenol protects against cigarette smoke-induced acute lung injury. Nutrition 29(1):235–243

Barlow, S.M., 1990. Toxicological aspects of antioxidants used as food additives. In: Food antioxidants (pp. 253-307). Springer, Dordrecht

Barton GM (2008) A calculated response: control of inflammation by the innate immune system. J Clin Invest 118(2):413–420

Baumann J, Bruchhausen FV, Wurm G (1980) Flavonoids and related compounds as inhibitors of arachidonic acid peroxidation. Prostaglandins 20(4):627–639

Bernard FX, Sable S, Cameron B, Provost J, Desnottes JF, Crouzet J, Blanche F (1997) Glycosylated flavones as selective inhibitors of topoisomerase IV. Antimicrob Agents Chemother 41(5):992–998

Beutler JA, Hamel E, Vlietinck AJ, Haemers A, Rajan P, Roitman JN, Cardellina JH, Boyd MR (1998) Structure–activity requirements for flavone cytotoxicity and binding to tubulin. J Med Chem 41(13):2333–2338

Bösenberg LH, van Zyl DG (2014) The mechanism of action of oral antidiabetic drugs: a review of recent literature. J Endocrinol Metab Diabetes S Afr 13(3):80–88

Braca A, Sortino C, Politi M, Morelli I, Mendez J (2002) Antioxidant activity of flavonoids from *Licania licaniaeflora*. J Ethnopharmacol 79(3):379–381

Cai EP, Lin JK (2009) Epigallocatechin gallate (EGCG) and rutin suppress the glucotoxicity through activating IRS2 and AMPK signaling in rat pancreatic β cells. J Agric Food Chem 57 (20):9817–9827

Casagrande F, Darbon JM (2001) Effects of structurally related flavonoids on cell cycle progression of human melanoma cells: regulation of cyclin-dependent kinases CDK2 and CDK1. Biochem Pharmacol 61(10):1205–1215

Chacko BK, Chandler RT, D'Alessandro TL, Mundhekar A, Khoo NK, Botting N, Barnes S, Patel RP (2007) Anti-inflammatory effects of isoflavones are dependent on flow and human endothelial cell PPAR γ. J Nutr 137(2):351–356

Chang RL, Huang MT, Wood AW, Wong CQ, Newmark HL, Yagi H, Sayer JM, Jerina DM, Conney AH (1985) Effect of ellagic acid and hydroxylated flavonoids on the tumorigenicity of benzo [a] pyrene and ($\pm$)-7β, 8α-dihydroxy-9α, 10α-epoxy-7, 8, 9, 10-tetrahydrobenzo [a] pyrene on mouse skin and in the new born mouse. Carcinogenesis 6(8):1127–1133

Chang WS, Lee YJ, Lu FJ, Chiang HC (1993) Inhibitory effects of flavonoids on xanthine oxidase. Anticancer Res 13(6A):2165–2170

Cherrak, S.A., Mokhtari-Soulimane, N., Berroukeche, F., Bensenane, B., Cherbonnel, A., Merzouk, H. and Elhabiri, M., 2016. In vitro antioxidant versus metal ion chelating properties of flavonoids: a structure-activity investigation. PLoS one, 11(10), p.e0165575

Chiang CJ, Kadouh H, Zhou K (2013) Phenolic compounds and antioxidant properties of gooseberry as affected by in vitro digestion. LWT - Food Science and Technology 51(2):417–422

Cho SY, Park SJ, Kwon MJ, Jeong TS, Bok SH, Choi WY, Jeong WI, Ryu SY, Do SH, Lee CS, Song JC (2003) Quercetin suppresses proinflammatory cytokines production through MAP kinases and NF-κB pathway in lipopolysaccharide-stimulated macrophage. Mol Cell Biochem 243(1–2):153–160

Cushnie TT, Lamb AJ (2005) Detection of galangin-induced cytoplasmic membrane damage in Staphylococcus aureus by measuring potassium loss. J Ethnopharmacol 101(1–3):243–248

Cushnie TPT, Taylor PW, Nagaoka Y, Uesato S, Hara Y, Lamb AJ (2008) Investigation of the antibacterial activity of 3-O-octanoyl-(−)-epicatechin. J Appl Microbiol 105(5):1461–1469

Di Carlo G, Autore G, Izzo AA, Maiolino P, Mascolo N, Viola P, Diurno MV, Capasso F (1993) Inhibition of intestinal motility and secretion by flavonoids in mice and rats: structure-activity relationships. J Pharm Pharmacol 45(12):1054–1059

Di Carlo G, Mascolo N, Izzo AA, Capasso F (1999) Flavonoids: old and new aspects of a class of natural therapeutic drugs. Life Sci 65(4):337–353

Droke EA, Hager KA, Lerner MR, Lightfoot SA, Stoecker BJ, Brackett DJ, Smith BJ (2007) Soy isoflavones avert chronic inflammation-induced bone loss and vascular disease. J Inflamm 4(1):17

Echeverría J, Opazo J, Mendoza L, Urzúa A, Wilkens M (2017) Structure-activity and lipophilicity relationships of selected antibacterial natural flavones and flavanones of Chilean flora. Molecules 22(4):608

Elbaz A, Carcaillon L, Kab S, Moisan F (2016) Epidemiology of Parkinson's disease. Rev Neurol 172(1):14–26

Escribano-Ferrer E, Queralt Regué J, Garcia-Sala X, Boix Montañés A, Lamuela-Raventos RM (2019) In vivo anti-inflammatory and Antiallergic activity of pure Naringenin, Naringenin Chalcone, and Quercetin in mice. J Nat Prod 82(2):177–182

Ferry DR, Smith A, Malkhandi J, Fyfe DW, deTakats PG, Anderson D, Baker J, Kerr DJ (1996) Phase I clinical trial of the flavonoid quercetin: pharmacokinetics and evidence for in vivo tyrosine kinase inhibition. Clin Cancer Res 2(4):659–668

Fesen MR, Pommier Y, Leteurtre F, Hiroguchi S, Yung J, Kohn KW (1994) Inhibition of HIV-1 integrase by flavones, caffeic acid phenethyl ester (CAPE) and related compounds. Biochem Pharmacol 48(3):595–608

Fu Z, Zhang W, Zhen W, Lum H, Nadler J, Bassaganya-Riera J, Jia Z, Wang Y, Misra H, Liu D (2010) Genistein induces pancreatic β-cell proliferation through activation of multiple signaling pathways and prevents insulin-deficient diabetes in mice. Endocrinology 151(7):3026–3037

Galli SJ, Tsai M, Piliponsky AM (2008) The development of allergic inflammation. Nature 454(7203):445

Ganesan S, Faris AN, Comstock AT, Chattoraj SS, Chattoraj A, Burgess JR, Curtis JL, Martinez FJ, Zick S, Hershenson MB, Sajjan US (2010) Quercetin prevents progression of disease in elastase/LPS-exposed mice by negatively regulating MMP expression. Respir Res 11(1):131

García-Lafuente A, Guillamón E, Villares A, Rostagno MA, Martínez JA (2009) Flavonoids as anti-inflammatory agents: implications in cancer and cardiovascular disease. Inflamm Res 58(9):537–552

Garrett L (2019) Seven circles of antimicrobial hell. Lancet 393(10174):865–867

Gil ES, Couto RO (2013) Flavonoid electrochemistry: a review on the electroanalytical applications. Rev Bras 23(3):542–558

Goldwasser, J., Cohen, P.Y., Yang, E., Balaguer, P., Yarmush, M.L. and Nahmias, Y., 2010. Transcriptional regulation of human and rat hepatic lipid metabolism by the grapefruit flavonoid naringenin: role of PPARα, PPARγ and LXRα. PloS one, 5(8), p.e12399

Górniak I, Bartoszewski R, Króliczewski J (2019) Comprehensive review of antimicrobial activities of plant flavonoids. Phytochem Rev 18(1):241–272

Guan Y, Li FF, Hong L, Yan XF, Tan GL, He JS, Dong XW, Bao MJ, Xie QM (2012) Protective effects of liquiritin apioside on cigarette smoke-induced lung epithelial cell injury. Fundam Clin Pharmacol 26(4):473–483

Havsteen B (1983) Flavonoids, a class of natural products of high pharmacological potency. Biochem Pharmacol 32(7):1141–1148

Havsteen BH (2002) The biochemistry and medical significance of the flavonoids. Pharmacol Ther 96(2–3):67–202

Hertog MG, Feskens EJ, Kromhout D, Hollman PCH, Katan MB (1993) Dietary antioxidant flavonoids and risk of coronary heart disease: the Zutphen Elderly Study. Lancet 342(8878):1007–1011

Huang MT, Smart RC, Conney AH (1987) Inhibition of tumour promotion by curcumin, a major constituent of the food additive, turmeric. Proc Am Assoc Cancer Res 28:173

Huang R, Zhong T, Wu H (2015) Quercetin protects against lipopolysaccharide-induced acute lung injury in rats through suppression of inflammation and oxidative stress. Arch Med Sci 11(2):427

Inoue T, Sugimoto Y, Masuda H, Kamei C (2002) Antiallergic effect of flavonoid glycosides obtained from Mentha piperita L. Biol Pharm Bull 25(2):256–259

Jeong KW, Lee JY, Kang DI, Lee JU, Shin SY, Kim Y (2009) Screening of flavonoids as candidate antibiotics against Enterococcus faecalis. J Nat Prod 72(4):719–724

Jiang C, Agarwal R, Lü J (2000) Anti-angiogenic potential of a cancer chemopreventive flavonoid antioxidant, silymarin: inhibition of key attributes of vascular endothelial cells and angiogenic cytokine secretion by cancer epithelial cells. Biochem Biophys Res Commun 276(1):371–378

Jiménez-Aliaga K, Bermejo-Bescós P, Benedí J, Martín-Aragón S (2011) Quercetin and rutin exhibit antiamyloidogenic and fibril-disaggregating effects in vitro and potent antioxidant activity in APPswe cells. Life Sci 89(25–26):939–945

Jung UJ, Lee MK, Park YB, Kang MA, Choi MS (2006) Effect of citrus flavonoids on lipid metabolism and glucose-regulating enzyme mRNA levels in type-2 diabetic mice. Int J Biochem Cell Biol 38(7):1134–1145

Kato R, Nakadate T, Yamamoto S, Sugimura T (1983) Inhibition of 12-O-tetradecanoylphorbol-13-acetate-induced tumour promotion and ornithine decarboxylase activity by quercetin: possible involvement of lipoxygenase inhibition. Carcinogenesis 4(10):1301–1305

Kim MH (2003) Flavonoids inhibit VEGF/bFGF-induced angiogenesis in vitro by inhibiting the matrix-degrading proteases. J Cell Biochem 89(3):529–538

Kim HP, Son KH, Chang HW, Kang SS (2004) Anti-inflammatory plant flavonoids and cellular action mechanisms. J Pharmacol Sci:0411110005–0411110005

Kim DH, Kim S, Jeon SJ, Son KH, Lee S, Yoon BH, Cheong JH, Ko KH, Ryu JH (2009) Tanshinone I enhances learning and memory, and ameliorates memory impairment in mice via the extracellular signal-regulated kinase signalling pathway. Br J Pharmacol 158(4):1131–1142

Kim MK, Jung HS, Yoon CS, Ko JH, Chun HJ, Kim TK, Kwon MJ, Lee SH, Koh KS, Rhee BD, Park JH (2010) EGCG and quercetin protected INS-1 cells in oxidative stress via different mechanisms. Front Biosci (Elite Ed) 2:810–817

Knekt P, Jarvinen R, Reunanen A, Maatela J (1996) Flavonoid intake and coronary mortality in Finland: a cohort study. BMJ 312(7029):478–481

Knekt P, Kumpulainen J, Järvinen R, Rissanen H, Heliövaara M, Reunanen A, Hakulinen T, Aromaa A (2002) Flavonoid intake and risk of chronic diseases. Am J Clin Nutr 76(3):560–568

Kong D, Liu Q, Xu G, Huang Z, Luo N, Huang Y, Cai K (2017) Synergistic effect of tanshinone IIA and mesenchymal stem cells on preventing learning and memory deficits via anti-apoptosis, attenuating tau phosphorylation and enhancing the activity of central cholinergic system in vascular dementia. Neurosci Lett 637:175–181

Koole C, Wootten D, Simms J, Valant C, Sridhar R, Woodman OL, Miller LJ, Summers RJ, Christopoulos A, Sexton PM (2010) Allosteric ligands of the glucagon-like peptide 1 receptor (GLP-1R) differentially modulate endogenous and exogenous peptide responses in a pathway-selective manner: implications for drug screening. Mol Pharmacol 78(3):456–465

Krentz AJ, Bailey CJ (2005) Oral antidiabetic agents. Drugs 65(3):385–411

Kuntz S, Wenzel U, Daniel H (1999) Comparative analysis of the effects of flavonoids on proliferation, cytotoxicity, and apoptosis in human colon cancer cell lines. Eur J Nutr 38(3):133–142

Lacopini P, Baldi M, Storchi P, Sebastiani L (2008) Catechin, epicatechin, quercetin, rutin and resveratrol in red grape: content, in vitro antioxidant activity and interactions. J Food Compos Anal 21(8):589–598

Lago J, Toledo-Arruda A, Mernak M, Barrosa K, Martins M, Tibério I, Prado C (2014) Structure-activity association of flavonoids in lung diseases. Molecules 19(3):3570–3595

Lee WR, Shen SC, Lin HY, Hou WC, Yang LL, Chen YC (2002) Wogonin and fisetin induce apoptosis in human promyeloleukemic cells, accompanied by a decrease of reactive oxygen species, and activation of caspase 3 and Ca^{2+}-dependent endonuclease. Biochem Pharmacol 63(2):225–236

Lee JH, Ahn J, Kim JW, Lee SG, Kim HP (2015) Flavonoids from the aerial parts of *Houttuynia cordata* attenuate lung inflammation in mice. Arch Pharm Res 38(7):1304–1311

Letenneur L, Proust-Lima C, Le Gouge A, Dartigues JF, Barberger-Gateau P (2007) Flavonoid intake and cognitive decline over a 10-year period. Am J Epidemiol 165(12):1364–1371

Li JK, Jiang ZT, Li R, Tan J (2012) Investigation of antioxidant activities and free radical scavenging of flavonoids in leaves of Polygonum multiflorum thumb. China Food Additives 2:69–74

Liao Z, Chen X, Wu M (2010) Antidiabetic effect of flavones from Cirsium japonicum DC in diabetic rats. Arch Pharm Res 33(3):353–362

Lin YM, Anderson H, Flavin MT, Pai YHS, Mata-Greenwood E, Pengsuparp T, Pezzuto JM, Schinazi RF, Hughes SH, Chen FC (1997a) In vitro anti-HIV activity of biflavonoids isolated from *Rhus succedanea* and *Garcinia multiflora*. J Nat Prod 60(9):884–888

Lin JK, Chen YC, Huang YT, Lin-Shiau SY (1997b) Suppression of protein kinase C and nuclear oncogene expression as possible molecular mechanisms of cancer chemoprevention by apigenin and curcumin. J Cell Biochem 67(S28–29):39–48

Lotfy M, Adeghate J, Kalasz H, Singh J, Adeghate E (2017) Chronic complications of diabetes mellitus: a mini review. Curr Diabetes Rev 13(1):3–10

Lotito SB, Frei B (2006) Dietary flavonoids attenuate tumour necrosis factor α-induced adhesion molecule expression in human aortic endothelial cells structure-function relationships and activity after first pass metabolism. J Biol Chem 281(48):37102–37110

Maher P (2019) The potential of flavonoids for the treatment of neurodegenerative diseases. Int J Mol Sci 20(12):3056

Makita H, Tanaka T, Fujitsuka H, Tatematsu N, Satoh K, Hara A, Mori H (1996) Chemoprevention of 4-nitroquinoline 1-oxide-induced rat oral carcinogenesis by the dietary flavonoids chalcone, 2-hydroxychalcone, and quercetin. Cancer Res 56(21):4904–4909

Mamdouh MA, Monira AAEK (2004) The influence of naringin on the oxidative state of rats with streptozotocin-induced acute hyperglycaemia. Z Naturforsch C 59(9–10):726–733

Man SP, Van Eeden S, Sin DD (2012) Vascular risk in chronic obstructive pulmonary disease: role of inflammation and other mediators. Can J Cardiol 28(6):653–661

Mandel S, Weinreb O, Amit T, Youdim MB (2004) Cell signaling pathways in the neuroprotective actions of the green tea polyphenol (−)-epigallocatechin-3-gallate: implications for neurodegenerative diseases. J Neurochem 88(6):1555–1569

Matsuzaki K, Miyazaki K, Sakai S, Yawo H, Nakata N, Moriguchi S, Fukunaga K, Yokosuka A, Sashida Y, Mimaki Y, Yamakuni T (2008) Nobiletin, a citrus flavonoid with neurotrophic action, augments protein kinase A-mediated phosphorylation of the AMPA receptor subunit, GluR1, and the postsynaptic receptor response to glutamate in murine hippocampus. Eur J Pharmacol 578(2–3):194–200

Mbaveng AT, Ngameni B, Kuete V, Simo IK, Ambassa P, Roy R, Bezabih M, Etoa FX, Ngadjui BT, Abegaz BM, Meyer JM (2008) Antimicrobial activity of the crude extracts and five flavonoids from the twigs of *Dorstenia barteri* (Moraceae). J Ethnopharmacol 116(3):483–489

McGettigan P, Roderick P, Kadam A, Pollock A (2018) Threats to global antimicrobial resistance control: centrally approved and unapproved antibiotic formulations sold in India. Br J Clin Pharmacol 85(1):59–70

Mink PJ, Scrafford CG, Barraj LM, Harnack L, Hong CP, Nettleton JA, Jacobs DR Jr (2007) Flavonoid intake and cardiovascular disease mortality: a prospective study in postmenopausal women. Am J Clin Nutr 85(3):895–909

Mirzoeva OK, Grishanin RN, Calder PC (1997) Antimicrobial action of propolis and some of its components: the effects on growth, membrane potential and motility of bacteria. Microbiol Res 152(3):239–246

Mojzis J, Varinska L, Mojzisova G, Kostova I, Mirossay L (2008) Antiangiogenic effects of flavonoids and chalcones. Pharmacol Res 57(4):259–265

Mori A, Nishino C, Enoki N, Tawata S (1987) Antibacterial activity and mode of action of plant flavonoids against *Proteus vulgaris* and *Staphylococcus aureus*. Phytochemistry 26 (8):2231–2234

Morishima C, Shuhart MC, Wang CC, Paschal DM, Apodaca MC, Liu Y, Sloan DD, Graf TN, Oberlies NH, Lee DYW, Jerome KR (2010) Silymarin inhibits in vitro T-cell proliferation and cytokine production in hepatitis C virus infection. Gastroenterology 138(2):671–681

Mukherjee PK, Maiti K, Mukherjee K, Houghton PJ (2006) Leads from Indian medicinal plants with hypoglycemic potentials. J Ethnopharmacol 106(1):1–28

Murphy KJ, Walker KM, Dyer KA, Bryan J (2019) Estimation of daily intake of flavonoids and major food sources in middle-aged Australian men and women. Nutr Res 61:64–81

Mutoh M, Takahashi M, Fukuda K, Komatsu H, Enya T, Matsushima-Hibiya Y, Mutoh H, Sugimura T, Wakabayashi K (2000) Suppression by flavonoids of cyclooxygenase-2-promoter-dependent transcriptional activity in colon cancer cells: structure-activity relationship. Jpn J Cancer Res 91(7):686–691

Nathan C (2002) Points of control in inflammation. Nature 420(6917):846

Navarro-Martínez MD, Navarro-Perán E, Cabezas-Herrera J, Ruiz-Gómez J, García-Cánovas F, Rodríguez-López JN (2005) Antifolate activity of epigallocatechin gallate against *Stenotrophomonas maltophilia*. Antimicrob Agents Chemother 49(7):2914–2920

Nile SH, Ko EY, Kim DH, Keum YS (2016) Screening of ferulic acid related compounds as inhibitors of xanthine oxidase and cyclooxygenase-2 with anti-inflammatory activity. Rev Bras 26(1):50–55

Nile, S.H., Keum, Y.S., Nile, A.S., Jalde, S.S. and Patel, R.V., 2018. Antioxidant, anti-inflammatory, and enzyme inhibitory activity of natural plant flavonoids and their synthesized derivatives. Journal of biochemical and molecular toxicology, 32(1), p.e22002

Ohemeng KA, Schwender CF, Fu KP, Barrett JF (1993) DNA gyrase inhibitory and antibacterial activity of some flavones (1). Bioorg Med Chem Lett 3(2):225–230

Ono K, Nakane H, Fukushima M, Chermann JC, Barré-Sinoussi F (1989) Inhibition of reverse transcriptase activity by a flavonoid compound, 5, 6, 7-trihydroxyflavone. Biochem Biophys Res Commun 160(3):982–987

Ono K, Nakane H, Fukushima M, Chermann JC, Barré-Sinoussi F (1990) Differential inhibitory effects of various flavonoids on the activities of reverse transcriptase and cellular DNA and RNA polymerases. Eur J Biochem 190(3):469–476

Orhan DD, Özçelik B, Özgen S, Ergun F (2010) Antibacterial, antifungal, and antiviral activities of some flavonoids. Microbiol Res 165(6):496–504

Osiecki H (2004) The role of chronic inflammation in cardiovascular disease and its regulation by nutrients. Altern Med Rev 9(1)

Osonga FJ, Akgul A, Miller RM, Eshun GB, Yazgan I, Akgul A, Sadik OA (2019) Antimicrobial activity of a new class of phosphorylated and modified flavonoids. ACS Omega 4(7):12865–12871

Padiyara P, Inoue H, Sprenger M (2018) Global governance mechanisms to address antimicrobial resistance. Infect Dis (Auckl) 11:1178633718767887

Panche AN, Diwan AD, Chandra SR (2016) Flavonoids: an overview. J Nutr Sci 5

Pavanato A, Tuñón MJ, Sánchez-Campos S, Marroni CA, Llesuy S, González-Gallego J, Marroni N (2003) Effects of quercetin on liver damage in rats with carbon tetrachloride-induced cirrhosis. Dig Dis Sci 48(4):824–829

Peet GW, Li J (1999) IκB kinases α and β show a random sequential kinetic mechanism and are inhibited by staurosporine and quercetin. J Biol Chem 274(46):32655–32661

Pham-Huy LA, He H, Pham-Huy C (2008) Free radicals, antioxidants in disease and health. Int J Biomed Sci 4(2):89

Piasecka A, Jedrzejczak-Rey N, Bednarek P (2015) Secondary metabolites in plant innate immunity: conserved function of divergent chemicals. New Phytol 206(3):948–964

Pinent M, Castell A, Baiges I, Montagut G, Arola L, Ardévol A (2008) Bioactivity of flavonoids on insulin-secreting cells. Compr Rev Food Sci Food Saf 7(4):299–308

Pingili RB, Pawar AK, Challa SR, Kodali T, Koppula S, Toleti V (2019) A comprehensive review on hepatoprotective and nephroprotective activities of chrysin against various drugs and toxic agents. Chem Biol Interact

Plaper A, Golob M, Hafner I, Oblak M, Šolmajer T, Jerala R (2003) Characterization of quercetin binding site on DNA gyrase. Biochem Biophys Res Commun 306(2):530–536

Pouget C, Fagnere C, Basly JP, Besson AE, Champavier Y, Habrioux G, Chulia AJ (2002) Synthesis and aromatase inhibitory activity of flavanones. Pharm Res 19(3):286–291

Pourmorad F, Hosseinimehr SJ, Shahabimajd N (2006) Antioxidant activity, phenol and flavonoid contents of some selected Iranian medicinal plants. Afr J Biotechnol 5(11)

Prince PSM, Kamalakkannan N (2006) Rutin improves glucose homeostasis in streptozotocin diabetic tissues by altering glycolytic and gluconeogenic enzymes. J Biochem Mol Toxicol 20 (2):96–102

Raju A, Degani MS, Khambete MP, Ray MK, Rajan MGR (2015) Antifolate activity of plant polyphenols against *Mycobacterium tuberculosis*. Phytother Res 29(10):1646–1651

Ravishankar D, Rajora AK, Greco F, Osborn HM (2013) Flavonoids as prospective compounds for anti-cancer therapy. Int J Biochem Cell Biol 45(12):2821–2831

Rezaeetalab F, Dalili A (2014) Oxidative stress in COPD, pathogenesis and therapeutic views. Rev Clin Med 1(3):115–124

Rovina N, Koutsoukou A, Koulouris NG (2013) Inflammation and immune response in COPD: where do we stand? Mediators Inflamm 2013

Sak K (2014) Cytotoxicity of dietary flavonoids on different human cancer types. Pharmacogn Rev 8(16):122

Sánchez-Salgado JC, Estrada-Soto S, García-Jiménez S, Montes S, Gómez-Zamudio J, Villalobos-Molina R (2019) Analysis of flavonoids bioactivity for cholestatic liver disease: systematic literature search and experimental approaches. Biomol Ther 9(3):102

Santos-Lopez A, Marshall CW, Scribner MR, Snyder D, Cooper VS (2019) Evolutionary pathways to antibiotic resistance are dependent upon environmental structure and bacterial lifestyle. bioRxiv:581611

Saranath D, Khanna A (2014) Current status of cancer burden: global and Indian scenario. Biomed Res J 1(1):1–5

Sato T, Koike L, Miyata Y, Hirata M, Mimaki Y, Sashida Y, Yano M, Ito A (2002) Inhibition of activator protein-1 binding activity and phosphatidylinositol 3-kinase pathway by nobiletin, a polymethoxy flavonoid, results in augmentation of tissue inhibitor of metalloproteinases-1 production and suppression of production of matrix metalloproteinases-1 and-9 in human fibrosarcoma HT-1080 cells. Cancer Res 62(4):1025–1029

Sesso HD, Gaziano JM, Liu S, Buring JE (2003) Flavonoid intake and the risk of cardiovascular disease in women. Am J Clin Nutr 77(6):1400–1408

Simpson JL, McDonald VM, Baines KJ, Oreo KM, Wang F, Hansbro PM, Gibson PG (2013) Influence of age, past smoking, and disease severity on TLR2, neutrophilic inflammation, and MMP-9 levels in COPD. Mediators Inflamm 2013

Singh SP, Konwarh R, Konwar BK, Karak N (2013) Molecular docking studies on analogues of quercetin with D-alanine: D-alanine ligase of helicobacter pylori. Med Chem Res 22 (5):2139–2150

Siriwong S, Teethaisong Y, Thumanu K, Dunkhunthod B, Eumkeb G (2016) The synergy and mode of action of quercetin plus amoxicillin against amoxicillin-resistant *Staphylococcus epidermidis*. BMC Pharmacol Toxicol 17(1):39

Sirk TW, Brown EF, Friedman M, Sum AK (2009) Molecular binding of catechins to biomembranes: relationship to biological activity. J Agric Food Chem 57(15):6720–6728

Spina M, Cuccioloni M, Mozzicafreddo M, Montecchia F, Pucciarelli S, Eleuteri AM, Fioretti E, Angeletti M (2008) Mechanism of inhibition of wt-dihydrofolate reductase from E. coli by tea epigallocatechin-gallate. Proteins 72(1):240–251

Steinmetz KA, Potter JD (1991) Vegetables, fruit, and cancer. I. Epidemiology. Cancer Causes Control 2(5):325–357

Tadera K, Minami Y, Takamatsu K, Matsuoka T (2006) Inhibition of α-glucosidase and α-amylase by flavonoids. J Nutr Sci Vitaminol 52(2):149–153

Tanaka T, Makita H, Kawabata K, Mori H, Kakumoto M, Satoh K, Hara A, Sumida T, Tanaka T, Ogawa H (1997) Chemoprevention of azoxymethane-induced rat colon carcinogenesis by the naturally occurring flavonoids, diosmin and hesperidin. Carcinogenesis 18(5):957–965

Tijburg LBM, Mattern T, Folts JD, Weisgerber UM, Katan MB (1997) Tea flavonoids and cardiovascular diseases: a review. Crit Rev Food Sci Nutr 37(8):771–785

Tohma H, Gülçin İ, Bursal E, Gören AC, Alwasel SH, Köksal E (2017) Antioxidant activity and phenolic compounds of ginger (*Zingiber officinale* Rosc.) determined by HPLC-MS/MS. J Food Measur Characteriz 11(2):556–566

Topal F, Nar M, Gocer H, Kalin P, Kocyigit UM, Gülçin İ, Alwasel SH (2016) Antioxidant activity of taxifolin: an activity–structure relationship. J Enzyme Inhib Med Chem 31(4):674–683

Toumi ML, Merzoug S, Boutefnouchet A, Tahraoui A, Ouali K, Guellati MA (2009) Hesperidin, a natural citrus flavanone, alleviates hyperglycaemic state and attenuates embryopathies in pregnant diabetic mice. J Med Plant Res 3(11):862–869

Tsuchiya H, Iinuma M (2000) Reduction of membrane fluidity by antibacterial sophoraflavanone G isolated from *Sophora exigua*. Phytomedicine 7(2):161–165

Tuder RM, Petrache I (2012) Pathogenesis of chronic obstructive pulmonary disease. J Clin Invest 122(8):2749–2755

Tunon MJ, Garcia-Mediavilla MV, Sanchez-Campos S, Gonzalez-Gallego J (2009) Potential of flavonoids as anti-inflammatory agents: modulation of pro-inflammatory gene expression and signal transduction pathways. Curr Drug Metab 10(3):256–271

Verma AK, Johnson JA, Gould MN, Tanner MA (1988) Inhibition of 7, 12-dimethylbenz (a) anthracene-and N-nitrosomethylurea-induced rat mammary cancer by dietary flavonol quercetin. Cancer Res 48(20):5754–5758

Vestbo J, Hurd SS, Agustí AG, Jones PW, Vogelmeier C, Anzueto A, Barnes PJ, Fabbri LM, Martinez FJ, Nishimura M, Stockley RA (2013) Global strategy for the diagnosis, management, and prevention of chronic obstructive pulmonary disease: GOLD executive summary. Am J Respir Crit Care Med 187(4):347–365

Vinayagam R, Xu B (2015) Antidiabetic properties of dietary flavonoids: a cellular mechanism review. Nutr Metab 12(1):60

Wang IK, Lin-Shiau SY, Lin JK (1999) Induction of apoptosis by apigenin and related flavonoids through cytochrome c release and activation of caspase-9 and caspase-3 in leukaemia HL-60 cells. Eur J Cancer 35(10):1517–1525

Wang S, Yan J, Wang X, Yang Z, Lin F, Zhang T (2010) Synthesis and evaluation of the α-glucosidase inhibitory activity of 3-[4-(phenylsulfonamido) benzoyl]-2H-1-benzopyran-2-one derivatives. Eur J Med Chem 45(3):1250–1255

Wei H, Tye L, Bresnick E, Birt DF (1990) Inhibitory effect of apigenin, a plant flavonoid, on epidermal ornithine decarboxylase and skin tumour promotion in mice. Cancer Res 50(3):499–502

Wesołowska O (2011) Interaction of phenothiazines, stilbenes and flavonoids with multidrug resistance-associated transporters, P-glycoprotein and MRP1. Acta Biochim Pol 58(4)

Williams RL (1983) Drug administration in hepatic disease. N Engl J Med 309(26):1616–1622

Wu D, Kong Y, Han C, Chen J, Hu L, Jiang H, Shen X (2008) D-alanine: D-alanine ligase as a new target for the flavonoids quercetin and apigenin. Int J Antimicrob Agents 32(5):421–426

Xie Y, Yang W, Tang F, Chen X, Ren L (2015) Antibacterial activities of flavonoids: structure-activity relationship and mechanism. Curr Med Chem 22(1):132–149

Xu H, Ziegelin G, Schröder W, Frank J, Ayora S, Alonso JC, Lanka E, Saenger W (2001) Flavones inhibit the hexameric replicative helicase RepA. Nucleic Acids Res 29(24):5058–5066

Yang J, Guo J, Yuan J (2008) In vitro antioxidant properties of rutin. LWT—Food Science and Technology 41(6):1060–1066

Yang IA, Clarke MS, Sim EH, Fong KM (2012) Inhaled corticosteroids for stable chronic obstructive pulmonary disease. Cochrane Database Syst Rev 7:

Yeh MH, Kao ST, Hung CM, Liu CJ, Lee KH, Yeh CC (2009) Hesperidin inhibited acetaldehyde-induced matrix metalloproteinase-9 gene expression in human hepatocellular carcinoma cells. Toxicol Lett 184(3):204–210

Yoo HH, Lee M, Chung HJ, Lee SK, Kim DH (2007) Effects of diosmin, a flavonoid glycoside in citrus fruits, on P-glycoprotein-mediated drug efflux in human intestinal Caco-2 cells. J Agric Food Chem 55(18):7620–7625

Zhang S, Morris ME (2003) Effect of the flavonoids biochanin A and silymarin on the P-glycoprotein-mediated transport of digoxin and vinblastine in human intestinal Caco-2 cells. Pharm Res 20(8):1184–1191

Zi X, Feyes DK, Agarwal R (1998) Anticarcinogenic effect of a flavonoid antioxidant, silymarin, in human breast cancer cells MDA-MB 468: induction of G1 arrest through an increase in Cip1/p21 concomitant with a decrease in kinase activity of cyclin-dependent kinases and associated cyclins. Clin Cancer Res 4(4):1055–1064

Zuo L, He F, Sergakis GG, Koozehchian MS, Stimpfl JN, Rong Y, Diaz PT, Best TM (2014) Interrelated role of cigarette smoking, oxidative stress, and immune response in COPD and corresponding treatments. Am J Physiol Lung Cell Mol Physiol 307(3):L205–L218

Zygmunt K, Faubert B, MacNeil J, Tsiani E (2010) Naringenin, a citrus flavonoid, increases muscle cell glucose uptake via AMPK. Biochem Biophys Res Commun 398(2):178–183

Plant-Based Pigments: Novel Extraction Technologies and Applications

16

Juan Roberto Benavente-Valdés, Lourdes Morales-Oyervides, and Julio Montañez

Abstract

In nature we can find a wide variety of colors, these colors are due to the presence of pigments. This chapter presents the basic information about plant pigments: chlorophylls, carotenoids, anthocyanins, and betalains; focusing attention on economic importance, chemical structure, sources, and main properties for health benefits. Particular emphasis is placed on the use of novel extraction technologies such as ultrasound, pulsed electric fields, ohmic heating, enzymatic, supercritical fluid, high pressure, and microwave on plant-based pigments recovery. Also, the most recent advances in their applications as coloring, therapeutic, and energy production agents are described.

Keywords

Natural pigments · Novel extraction technologies · Applications

16.1 Introduction

Pigments are responsible for imparting color to living materials and beings, giving unique characteristics to each organism. Plants are the primary producers, and their pigments are found in leaves, fruits, vegetables, and flowers. In addition, they are present in the skin, eyes, fur, and other animal structures, as well as in bacteria, fungi, and algae. Natural and synthetic pigments are widely used in food, drugs, textiles, and cosmetics. The natural color of plant-based pigments is mainly due to four groups of pigments: green chlorophylls, yellow–orange–red carotenoids, red–blue–purple anthocyanins, and yellow–red betalains (Rodriguez-Amaya 2019).

J. R. Benavente-Valdés · L. Morales-Oyervides · J. Montañez (✉)
Department of Chemical Engineering, Universidad Autónoma de Coahuila, Coahuila, Mexico
e-mail: julio.montanez@uadec.edu.mx

© Springer Nature Singapore Pte Ltd. 2020
S. T. Sukumaran et al. (eds.), *Plant Metabolites: Methods, Applications and Prospects*, https://doi.org/10.1007/978-981-15-5136-9_16

Natural colorants obtaining from plants are an eco-friendly alternative that has been gradually replacing the synthetic colorants used in diverse industries (Yusuf et al. 2017). This growth is partly explained by the concern about the adverse effects that some synthetic dyes would have on human health and the negative impact on the environment. Likewise, advances of new sources for their obtention and the development of environmentally friendly extraction technologies enable obtaining natural dyes at a lower cost and in a more efficient way (Barba et al. 2015). However, the use of such pigments has been restricted by their poor stability under various conditions, such as temperature, pH, light, water, and enzymatic activity, as well as the presence of oxygen and/or metals. So, the implementation of technologies that aid in increasing their stability and preserving their activities is a crucial aspect for their commercial use (Horuz and Belibağlı 2018, 2019; Zardini et al. 2018; Medeiros et al. 2019).

Accordingly, this chapter describes the economic importance and the main characteristics of plant-based pigments. Likewise, it describes the most recent advances in the use of nonconventional technologies for the extraction of these compounds and their applications as coloring, therapeutic, and energy production agents.

16.2 Economic Importance

The demand for natural dyes is growing in the food, cosmetic, textile, and pharmaceutical industry displacing the use of dyes of synthetic origin. This global trend is led by the European Community, where the demand for natural dyes has meant a significant increase in the market, with an estimated growth of around 6.4% until 2020 (AINIA 2019; www.marketsandmarkets.com).

The global food colors market is projected to $5.12 billion by 2023. In 2018, the global market of natural dyes was estimated at $3.88 billion in 2018, in which 81% was occupied by plant-based pigments, while those of animal and mineral origin occupied 13% and 6%, respectively (https://www.marketsandmarkets.com/Market-Reports/food-colors-market-36725323.html). The food sector is the fastest-growing sector with $1.6 billion dollars, followed by the textile sector with an increase of $0.28 billion dollars, pharmaceutical with $0.28 billion dollars, and cosmetics with $0.22 billion dollars. Figure 16.1 illustrates the natural pigment market segmented by end users. The food and beverage industries occupy more than half of the pigment market, followed by textile, pharmaceutical, and cosmetic industries. Therefore, plant-based pigments are of great importance in the world market. The primary pigments found in plants are chlorophyll, carotenoids, anthocyanins, and betalains. Of these, chlorophyll is the most abundant in plant species. In 2018, the global chlorophyll market was established in $279.5 million, with a projection for 2025 of $463.7 million (www.valuemarketresearch.com, 2019). On the other hand, the global carotenoids market was valued at $1577 million in 2017 and is projected to reach $2098 million by 2025 (www.alliedmarketresearch.com, 2019). So, these pigments are the ones with the highest demand and profits worldwide. Likewise, anthocyanins and betalains are pigments with a growing market. The global anthocyanin market was $291.7 million in 2014, which is expected in $387.4 million for

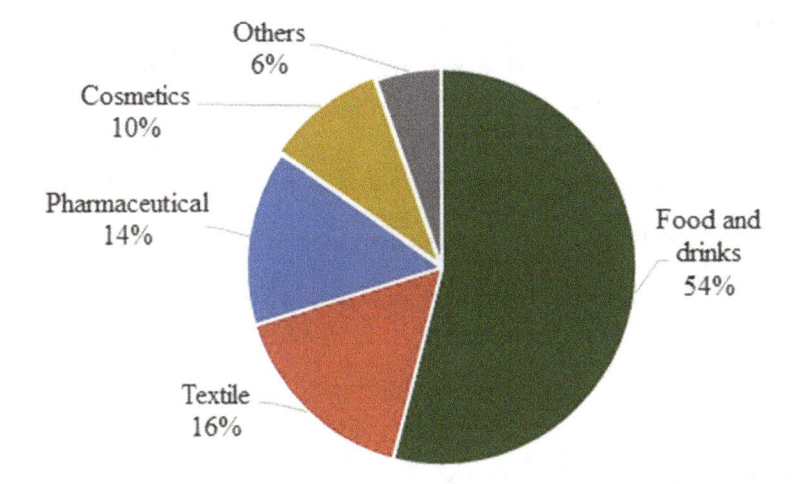

Fig. 16.1 Natural pigments world market segmented by end users in 2018

2021 (Appelhagen et al. 2018). Beetroot is the principal source of commercial betalains. This powder market was in $15 billion in 2016 and is expected to reach $23 billion by 2027 (Ciriminna et al. 2015). The previous data highlight the potential of the natural dye industry, which demands the development of new extraction techniques that are more efficient, sustainable, and economical, which allow a competitive market for this type of pigment.

16.3 Principal Pigments Found in Plants

In plants, vegetables, and fruits, we can find a wide variety of colors; these colors are due to the presence of pigments in their cells. Plant pigments are chromophores found in the chloroplast thylakoids, and their fundamental function is to absorb sunlight from photosynthesis. The principal pigment found in plants is chlorophyll, which is responsible for photosynthetic activity. In addition, accessory pigments such as carotenoids can be found. The most abundant pigments present in plants include many molecules such as chlorophyll, carotenoids, anthocyanins, and betalains (Table 16.1). In this section, their chemical structure, the occurrence, and their most common applications of these pigments will be described.

16.4 Chlorophyll

Chlorophyll is the unique pigment with a green color that allows the process of photosynthesis to be carried out, and it is found in plants, some algae, microalgae, cyanobacteria, and bacteria (Vernon and Seely 2014; Da Silva Ferreira and Sant'Anna 2017). This pigment was discovered in 1817 by Caventou and Pelletier,

Table 16.1 Occurrence of pigments in nature

Pigment	Types	Principal source	Reference
Chlorophyll	Chlorophyll a	All photosynthetic plants (excluding bacteria)	Hörtensteiner and Kräutler (2011), Pareek et al. (2017)
	Chlorophyll b	Higher plants and various green algae	Kräutler (2016)
	Chlorophyll c	Brown algae and diatoms	Kuczynska et al. (2015)
	Chlorophyll d	Various red algae, reported in *Rhodophyta*	Strain (1958)
	Chlorophyll e	Golden-yellow algae (*Vaucheria hamata* and *Tribonema bombycinum*)	Eugene and Govindjee (1969), Pareek et al. (2017)
	Chlorophyll f	Cyanobacteria	Chen et al. (2010)
Carotenoids	β-Carotene	Apricots, asparagus, carrots, broccoli, Chinese cabbage, grapefruits, chili powder, paprika Buriti, tucuma, acerola, mango, some varieties of pumpkin and nuts	Gul et al. (2015), Mezzomo and Ferreira (2016)
	Lycopene	Tomato, cherry, guava, watermelon, and papaya	Bruno et al. (2016), Mezzomo and Ferreira (2016)
	Lutein and Zeaxanthin	Green and dark green leafy vegetables (broccoli, Brussels sprouts, spinach, and parsley)	Mezzomo and Ferreira (2016)
	Astaxanthin	Single-cell alga *Haematococcus pluvialis*	Capelli et al. (2019)
Anthocyanins	Pelargonidin	Radish, cherry, sour cherry, pomegranate, and barberry	Martínez and Ventura (2016), Yari and Rashnoo (2017)
	Cyanidin	*Arbutus unedo* fruits	López et al. (2019)
	Delphinidin	*Solanum melongena* and maqui berry	Casati et al. (2019); Alvarado et al. (2016)
	Peonidin	Purple sweet potato	Sun et al. (2018)
	Petunidin	Black goji and purple potato	Tang (2018)
	Malvidin	Red grape	Setford et al. (2018)
Betalains	Betanin	Red beet	Esatbeyoglu et al. (2015)
	Indicaxanthin	Cactus pear	Tesoriere et al. (2015)

who isolated it from the leaves of the plants and named it from the Greek *chloros*, which means "green" and *phyllon*, which means "leaf." Chlorophyll is found within chloroplasts in almost every green part of plants like leaves and stems.

The molecule of chlorophyll contains a porphyrin ring with a central magnesium ion (Mg^{2+}) and usually a long hydrophobic chain which gives the fat-soluble properties. This pigment absorbs blue light (430 nm) and red light (660 nm) of solar radiation, and it reflects the green spectrum (Inanc 2011). Additionally, there have been elucidated six different types of chlorophyll (*a, b, c, d, e,* and *f*), and their molecular structure varies by one or several side-chain substitutions. Chlorophylls *a* and *b* are the most abundant in nature because they are the main components of photosystems in photosynthetic organisms. These two forms exist in an approximate ratio of 3:1, being chlorophyll *a* the most abundant (Christaki et al. 2015).

In recent years, the primary sources of plant-based chlorophyll are spinach (Leite et al. 2017; Derrien et al. 2018; Li et al. 2019), alfalfa leaves (Saberian et al. 2018; Ningrum et al. 2019), and other green leafy vegetables as Chinese kale, Chinese flowering cabbage, water spinach, cassava leaf, green leaf lettuce, Chinese cabbage pak-choi, basil, sword-leaf lettuce, and head cabbage (Limantara et al. 2015). Chlorophyll pigments have different properties like anticancer, antibacterial, antioxidant, and energizing properties; they help to oxygenate the blood and detoxify our body and also to reduce the high levels of cholesterol and triglycerides (Pareek et al. 2017). Therefore, the use of chlorophylls is varied; they can be used as an ingredient in the food industry or even for medicinal purposes.

16.5 Carotenoids

Carotenoids are natural lipophilic pigments that are synthesized by plants, algae, and photosynthetic bacteria. These pigments are responsible for the vast majority of the yellow, orange, or red colors present in plants, vegetables, fruits, algae, and microorganisms, and also the orange colors of various animal foods. In green plants and vegetables, they are found in chloroplasts, forming part of the photosynthesis system, being accessory pigments of the process. In some cases, these can be more abundant and visible, coloring some roots, fruits, and flowers.

Most carotenoids are hydrocarbon chains containing 40 carbon atoms and two terminal rings. There are more than 700 different carotenoids known, which have been identified as derivatives of the same basic C40 isoprenoid skeleton (Mezzomo and Ferreira 2016). These are divided into two basic types: carotenes, which are hydrocarbons (α-carotene, β-carotene, and lycopene), and xanthophylls, their oxygenated derivatives (lutein and zeaxanthin). In addition, they can be classified by their activity, recognizing two groups, with and without provitamin A (Olson 1989).

The main carotenoids with applications in different industries are β-carotene, lycopene, lutein, zeaxanthin, and astaxanthin (Gul et al. 2015; Mezzomo and Ferreira 2016; Bruno et al. 2016; Capelli et al. 2019). β-carotene was the first purified carotenoid, isolated in 1831 by Wackenroder in a crystalline form from the carrot, receiving the name from the Latin name of this vegetable (*Daucus carota*).

β-carotene is one of the main carotenoids of commercial interest. The main sources are apricots, asparagus, carrots, broccoli, Chinese cabbage, grapefruits, chili powder, and paprika (Gul et al. 2015). In recent years, new sources of β-carotene have been studied. Crude palm oil, palm pressed fiber, and empty fruit bunches have been studied recently with yields of 3790, 1414, and 702 ppm, respectively (Kupan et al. 2016). Banana peel has also been studied as an alternative source of this pigment, with yields of 0.84 μg/100 g (Yan et al. 2016). In the same way, the investigation of beta-carotene extraction processes from carrots continues, being one of the sources with the highest content of this pigment (Sharmin et al. 2016; Hasan et al. 2019; Dai and Row 2019).

Tomatoes and their derived products are the major sources of lycopene (Ladole et al. 2018; Catalkaya and Kahveci 2019; Silva et al. 2019). Other sources of lycopene are watermelon, pink guava, papaya, and grapefruit (Bruno et al. 2016). On the other hand, the main sources for obtaining lutein and zeaxanthin are marigold petals, green and dark green leafy vegetables like broccoli, brussels sprouts, spinach, and parsley. In recent years, research has focused on the extraction of these pigments from nonconventional sources as corn (Jiao et al. 2018), corn gluten meal (Wang et al. 2019; Cobb et al. 2018), and distillers dried grain (Li and Engelberth 2018). Lastly, astaxanthin in recent years has been obtained from microalgae, specifically from *Haematococus pluvialis,* with a maximum reported productivity of 3.25×10^2 mg/L/d with a genetically modified strain (Galarza et al. 2018).

Carotenoid pigments have unique coloring properties, so they are widely used in the food, pharmaceutical, cosmetic, and animal feed industry. They are also used for the benefit of human health, such as food fortification due to provitamin A, antioxidant properties, reduced risk of degenerative diseases, obesity, and hypolipidemic diseases (Milani et al. 2017).

16.6 Anthocyanins

Anthocyanins are vacuolar pigments present in nature that give the red, purple, or blue color to the leaves, flowers, and fruits. The name of these pigments originates from the Greek *anthos*, a flower, and *kyanos*, dark blue. These pigments are flavonoids based on a C15 skeleton with a chromane ring bearing a second aromatic ring B in position 2 (C6–C3–C6) and with one or more sugar molecules bonded at different hydroxylated positions of the basic structure (Delgado-Vargas et al. 2000). This chemical structure is responsible for producing color variations by combination with glycosides and acyl groups. Anthocyanins can occur in different molecules, the main forms are pelargonidin, cyanidin, delphinidin, peonidin, petunidin, and malvidin (Casati et al. 2019).

Anthocyanins production is based on the extraction from plants, being the most common sources grape skin, black carrots, red cabbage, sweet potato, apple, cherry, fig, peach, and berries (Sun et al. 2018). Also, this pigment is present in some ornamental plants as *Dianthus, Petunia, Rosa, Tulipa,* and *Verbena* (Delgado-Vargas et al. 2000).

In recent years, new extraction techniques have been studied that have allowed obtaining anthocyanins from other sources such as mulberry (Guo et al. 2019), blackberries (Du et al. 2018), corn kernels (Li et al. 2018), *Arbutus unedo* L. fruits (López et al. 2018), *Hibiscus sabdariffa* L. (Miranda-Medina et al. 2018), *Ficus carica* L peel (Backes et al. 2018) and blue and purple corn (Somavat et al. 2018). In general, most fruits and vegetables contain anthocyanins from 0.1 to 1% dry weight.

Anthocyanins are of broad commercial interest for two main reasons. The first, due to their great coloring power, giving sensory characteristics to foods with vibrant red, purple, and blue colors (Aguilera-Otíz et al. 2011; Narayana et al. 2019). The second, due to their properties with favorable impacts for human health, such as antioxidant, antiallergic, anti-inflammatory, antiviral, antiproliferative, antimicrobial, antimutagenic, antitumor, and as adjuvants for the improvement of disorders such as obesity (Wrolstad 2004; Diaconeasa et al. 2015; Abdel-Aal et al. 2018). Similarly, it has been observed that they prevent cardiovascular diseases and cancer (De Rosas et al. 2019; Rodríguez-Amaya 2019). However, the use of anthocyanins as food dyes and functional ingredients has been limited by their low stability to changes in pH and temperature, as well as to the interaction with other compounds (Zapata et al. 2019).

16.7 Betalains

Betalains are secondary nitrogen metabolites of plants, water-soluble that impart colors, from yellow and orange to red and violet. These pigments can be classified into two broad groups: betacyanins and betaxanthines. Betacyanins have red and violet tones resulting from different substitution patterns. As for betaxanthines, the color yellow and orange are defined by the different chains of amino acids or amines in the molecule. The chemical structure of these pigments is based on a betalamic acid, which is the chromophore group responsible for the coloring of the batalains. The type of residue that is added to the betalamic acid molecule determines the classification of the pigment as betacyanin or betaxanthine. Betacyanins absorb around 540 nm and betaxanthins at 480 nm (Strack et al. 2003; Azeredo 2009).

Betalains can be found in roots, fruits, and flowers (Strack et al. 2003). These are the main pigments of the beetroot and other species, such as Malabar spinach, amaranth, pitaya, and cactus pear (*Opuntia* sp.) (Esatbeyoglu et al. 2015; Tesoriere et al. 2015). Beetroot represents the primary commercial source of concentrated betalains (Mancha et al. 2019). A peculiar characteristic of these pigments is that betalains and anthocyanins are mutually exclusive, so it has never been reported a plant source with both (Stafford 1994). The new technologies of extraction have allowed extracting betalains of colored quinoa (Laqui-Vilca et al. 2018), red beet stalks (Dos Santos et al. 2018), and fermented red dragon fruit (*Hylocereus polyrhizus*) (Choo et al. 2018).

In the same way as the other pigments in plants, they perform functions of attraction of pollinators and dispersers, also absorption of ultraviolet light and protection against herbivores (Azeredo 2009). The principal use of betalains is as a natural dye in the food, pharmaceutical, and cosmetic industry, with stability at a wide pH range, and they are not toxic or allergen, so they are an alternative to synthetic colorants (Esquivel, 2016). In addition, these pigments provide properties with a beneficial effect on health due to their antioxidant, antidiabetic, anti-inflammatory, and anticancer activities (Mancha et al. 2019).

16.8 Novel Extraction Technologies

Conventionally, the recovery of plant-based molecules such as pigments is carried out through a series of unit operations (drying, grinding, solid–liquid extraction, filtration, concentration, and purification) (Chemat et al. 2019). Among these operations, the solid–liquid extraction step is critical because is a time-demanding process with high energy consumption, and a significant volume of harmful solvents is required. In recent years, the use of novel technologies to solve these disadvantages has been addressed by the scientific community. In this scenario, much work has been done in order to intensify this process. Based on that concept, new emerging technologies are now being explored in order to develop intense and greener processes for the recovery of added-value molecules such as natural pigments, which are widely applied in different sectors.

16.9 Ultrasound-Assisted Extraction (UAE)

Ultrasound-assisted extraction is an alternative extraction technology that has been used in many operations, highlighting the solid–liquid extraction process. This technology presents some advantages over conventional processes like short time process, mild temperatures, low solvent volume consumption, and less expensive (Roselló-Soto et al. 2016). The ultrasound irradiation creates a cavitation phenomenon, leading to high shear forces in the media, which disrupts the cell wall facilitating the leaching of bioactive compounds from the food matrix to the solvent extraction. However, UAE effect occurs not only due to cavitation but also other independent or combined effects of fragmentation, erosion, sonoporation, detexturation, and capillarity (Chemat et al. 2017). Several authors have addressed the use of UAE for solid–liquid extraction, food pasteurization, fermentation, among others. However, a considerable amount of reports using UAE for recovery of bioactive compounds from plant-based products, agro-industrial wastes, yeast, and microalgae, among others can be found in literature. Koubaa et al. (2016) studied the use of UAE pretreatment on prickly-pear peels and pulp for betalains aqueous extraction, their study demonstrated that this pretreatment allowed higher pigments yields compared to conventional extraction process. In another study, Laqui-Vilca et al. (2018) optimized the extraction process of betalains by UAE from quinoa hulls, finding that when UAE was applied the extraction time was 9.2, meanwhile in conventional extraction (without UAE) the processing time was 30 min. Righi Pessoa da Silva et al. (2018) demonstrated that UAE process allowed obtaining higher betalains yields from red beet in comparison with a conventional method. Goula et al. (2017) developed a UAE process for the recovery of carotenoids from pomegranate peel using vegetable oil as solvent, obtaining a high extraction efficiency of carotenoids. Regarding anthocyanins extraction, Velmurugan et al. (2017) applied the eco-friendly UAE for the recovery of anthocyanins from purple sweet potato obtaining higher extraction yields. Also, Ferarsa et al. (2018) reported the application of UAE for the recovery of anthocyanins from purple eggplant reduced the extraction time process and higher yields compared to conventional extraction process. In the same way, Agcam et al. (2017) optimized the anthocyanins recovery from black carrot pomace by using UAE combined with heat (thermosonication) finding a synergistic effect increasing the yield by 20%. Ravi et al. (2018) reported high yields of anthocyanins by ultrasound-assisted extraction using ethanol as solvent. Leichtweis et al. (2019) reported a rapid, efficient, and low-cost process based on UAE for the recovery of anthocyanins from *Prunis spinose* L fruit epicarp, observing an optimal extraction time of 5 min. Backes et al. (2018) tested different recovery processes of anthocyanins from *F. carica* L. peel, concluding that UAE was the most effective technique. Recently, Pinela et al. (2019) reported the use of UAE on the recovery of anthocyanins from *H. sabdariffa* calyces, observing that UAE showed improved yields in comparison with heat-assisted extraction. Similarly, Sharmila et al. (2019) reported the use of UAE for the extraction of yellow pigment from *Tecoma castanifolia* floral petals, showing a short extraction time of 15 min when UAE was employed. In conclusion, the application of UAE is a promising technology with high efficiency for the recovery of natural pigments from plant-based matrices. Boukroufa et al. (2017) reported an interesting strategy for the recovery of carotenoids from citrus fruit wastes by using UAE. First, essential oil was removed by solvent-free microwave extraction, then carotenoids were recovered by UAE by using the D-limonene obtained from the essential oil recovered. Finally, they obtained 40% higher yields following this strategy in comparison with conventional extraction process. The combination of this strategy allowed them to develop a green approach for the total valorization of citrus wastes.

16.10 Pulsed Electric Fields (PEF)

Pulsed electric fields technology is a novel technology based on the application of high-strength electric fields, by millisecond or microsecond, that induce nonreversible electroporation of the cell membranes; thus, facilitating the migration of intracellular compounds (Luengo et al. 2016). Recently, the application of PEF technology for the recovery of plant-based pigments has raised, demonstrating the potential for further industrial applications based on short processing times with low energy costs. Several authors have reported the application of PEF technology for the recovery of betanins and betalains. Luengo et al. (2016) reported the use of a PEF pretreatment for the recovery of betanins from red beet, demonstrating that PEF resulted in better yields (six to seven times) in comparison with control samples process yield (independently of the pulse duration, milliseconds or microseconds). Most recently, Nowacka et al. (2019) reported a positive effect of PEF technology on efficient extraction of betalains from beetroot. Also, Koubaa et al. (2016) studied the use of PEF pretreatment on prickly-pear peels and pulp for betalains aqueous extraction, where their findings demonstrated the potential of this technology for increasing colorants extraction yield and purity. Also, PEF processes have been reported for the extraction of anthocyanins. In this regard, Yammine et al. (2018) reported the potential of PEF technology for anthocyanins recovery from winemaking by-products. Pataro et al. (2017) applied PEF pretreatment technology in fresh blueberries, highlighting that pretreated samples increase the juice yields and also anthocyanins content and antioxidant activity in comparison with non-pretreated blueberries.

In the same way, He et al. (2018) reported that PEF technology allowed increasing recovery yields of anthocyanins by decreasing the processing time from *Vitis amurensis* Rupr. Also, carotenoids, molecules with high commercial value, have been extracted through the implementation of PEF technology. Zhang et al. (2017) demonstrated that the application of PEF improved the extraction yield of pigments from spinach (chlorophyll and carotenoids); moreover, the antioxidant capacity of such molecules was increased. Based on the above, PEF technology shows a promising future for the recovery of natural pigments. Nevertheless, more studies addressing the use of this technology on natural pigments recovery from plant-based products are needed.

16.11 Ohmic Heating (OH)

Among the emerging technologies, ohmic heating has potential applications in different fields like microbial inactivation, sterilization, pasteurization, distillation, fermentation, blanching, solid–liquid extraction, and thawing among others. It is based on the passage of alternating electrical current through a semi-conductive material allowing the generation of internal heat due to the inherent electrical resistance of the product (Gavahian et al. 2016; Pereira et al. 2016). Among the advantages of OH is that rapid, uniform, and precise heating of products can be achieved with high energy efficiency. Then ohmic heating is of great interest for the application in solid–liquid processes for the recovery of bioactive compounds. Several reports have shown the potential of OH in the recovery of natural pigments. Loypimai et al. (2015) reported the application of OH for the recovery of anthocyanins from black rice. Their results showed that high yields were obtained when OH was used. Later, Pereira et al. (2016) applied OH technology for the recovery of anthocyanins from colored potato, reporting that with the application of OH, high yields of these compounds were obtained at short processing time; thus, energy savings were higher. OH technology is a promising technology for recovery of natural pigments, showing a high potential due to the low-cost investment.

16.12 Supercritical Fluid Extraction (SCFE)

Supercritical fluid extraction is considered a green extraction process and successful applications in plant metabolites have been widely demonstrated. It consists of the use of CO_2 in its critical conditions, allowing to extract nonpolar compounds. Several authors have reported the use of this technology for natural pigment recovery from plants, fruits, and vegetable by-products. Carotenoids have been one of the most reported molecules recovered by SCFE. Kehili et al. (2017) studied the extraction process by using SCFE in the recovery of lycopene and B-carotene from tomato peels by-product, showing that these molecules presented a high antioxidative effect. Xie et al. (2019) reported good yields of ethanol-modified SCFE for recovery of astaxanthin from *Camelina sativa* seeds, providing a green extraction process in comparison with traditional extraction process using hexane. Sánchez-Camargo et al. (2019) reported the use of SCFE for the recovery of carotenoids from mango peels, demonstrating the high efficiency of the process. Paula et al. (2018) extracted anthocyanins from *Arrabidaea chica* Verlot using SCFE. Derrien et al. (2018) studied the extraction process of SCFE of lutein and chlorophylls from spinach by-products, obtaining high yields of these pigments. Regarding the extraction of anthocyanins, Garcia-Mendoza et al. (2017) reported the use of SCFE for the recovery of anthocyanins from juçara, showing promising yields by the use of this green technology. SCFE is a worth mentioning technology, it is an environmental safe technique, also the use of organic solvents is low. A lot of efforts have been done regarding this technology for the recovey of bioactive compounds; however, some work needs to be done regarding process intensification before the application of SCFE at industrial scale.

Nowadays, the combination of SCFE with ultrasound is a promising alternative to decrease extraction time and increasing extractions yields. Santos-Zea et al. (2019) reported the use of SCFE + US for the recovery of phenolic compounds from *Agave salmiana* bagasse. However, the application of this hybrid emerging technology for pigments recovery from plant-based matrices has not been reported yet.

16.13 Advances in Plant-Based Pigments Applications

16.13.1 Application as Colorants

Dyes have been of interest in various industries because they confer special features and make food, textiles, or cosmetics more attractive, as well as aids with distinguishing of drugs and medications. Plant-based pigments such as chlorophyll, carotenoids, anthocyanins, and betalains are the natural dyes most used due to their natural origin with benefits in human health and low environmental impact. In this section, we will present the most recent advances in the use of these pigments as dyes in food and textiles.

The most widely used natural dyes in the food industry are carotenoids and recently they have been used in industries where they were not commonly used.

The World Health Organization has related the utilization of nitrous compounds to improve red color and prevent oxidation of meat with carcinogenesis in humans. Consequently, natural dyes, such as carotenoids, are being used to replace toxic compounds. Annato seeds contain high levels of carotenoids and have been evaluated as replacers of nitrous compounds in the meat industry with excellent results (Bolognesi and Garcia 2018). Additionally, encapsulation techniques have been developed in order to preserve the color and bioactive properties of pigments. Medeiros et al. (2019) evaluated the nano-encapsulation of carotenoids extracted from Cantaloupe melon in a porcine gelatin matrix and reported an increment of water solubility by three times, which facilitated their incorporation into food matrices. In addition, the coloration was stable for 60 days when the nano-encapsulated carotenoids were evaluated as an additive in yogurt. Lycopene extracted from tomatoes has been encapsulated in nanostructured lipid carriers for the enrichment of orange drinks, improving coloration, and being well accepted by the consumer (Zardini et al. 2018). Similarly, the nanoencapsulation of carotenoids by electrospinning has shown that it improves water solubility and stability (Horuz and Belibağlı 2018, 2019).

Like carotenoids, anthocyanins and betalains have been encapsulated for application in different edible matrices, maintaining the stability of these compounds and their antioxidant properties (Mancha et al. 2019). De Moura et al. (2018) encapsulated anthocyanins by ionic gelation showed higher stability when stored at refrigeration temperatures. In 2018, Amjadi and collaborators (2018) encapsulated betanin in liposomes for incorporation into a gummy candy showing double antioxidant activity in comparison with gummy candy with free pigment.

The textile industry has also opted for the use of natural dyes for dyeing clothes. Carotenoids have been used for staining of clothes such as wool, silk, and polyamide (Baaka et al. 2017). An extract rich in carotenoids from *Tagetes erecta* (Marigold) flower was used for the coloring of wool fibers, obtaining a wide range of bright tones with satisfactory colorimetric properties when applied with metallic mordants (Shabbir et al. 2018).

Anthocyanins and betalains have also been used for fabric coloring. Tidder et al. (2018) used purified blackcurrant extract to dying silk and wool fabrics giving pink, purple, green, and brown colorations, depending on the type of mordant used. Likewise, anthocyanins from purple sweet potato and betalains from wild *Phytolaccaceae* berries were used to dye silk (Yin et al. 2017, 2018). On the other hand, chlorophyll has been used as a biomordant for the coloring of wool fibers with betanin, enabling the fixing of betanin by using high concentrations of the green pigment (Guesmi et al. 2013).

The commercial application of plant pigments for food and textile processing is limited by their instability; however, different encapsulation and mordant techniques have shown the potential use of these natural dyes.

16.14 Therapeutic Applications

Apart from imparting color to food and textiles, pigments extracted from plants have shown beneficial effects for the improvement of human health. The most recent advances in the use of these pigments for therapeutic effects are described here. Currently, chlorophyll is marketed as a natural supplement either as a liquid or powder extract with promising health benefits. Li et al. (2019) found that a spinach extract with higher content of chlorophyll reduces body weight gain, inflammation, and increases glucose tolerance. Also, chlorophyll exhibits antiproliferative effects in pancreatic cancer cell lines (PaTu-8902, MiaPaCa-2, and BxPC-3), with a dose of 10–125 μmol/L, a significant reduction of pancreatic tumor size in mice was shown (Vaňková et al. 2018).

Of all natural pigments, carotenoids are the most studied due to their high antioxidant activity and their function as provitamin A. These pigments act as immunoenhancement agents, reducing the risk of developing chronic degenerative disease (Rodriguez-Amaya 2019). Lycopene has shown capacity as a preventive protector of cancer due to the antioxidant, anti-inflammatory, and antiproliferative activities. Lian and Wang (2019) demonstrated that a diet rich in lycopene reduces the incidence of lung cancer. This same effect has been reported for breast (Arathi et al. 2018) and prostate cancer (Moran et al. 2019). Lycopene has not only shown anticancer properties, in a recent study, it showed prebiotic effects by increasing the population of *Bifidobacterium adolescentis* and *Bifidobacterium longum* (Wiese et al. 2019). On the other hand, lutein and zeaxanthin have shown positive effects on the prevention of irreversible blindness in cravings, reducing macular degeneration (Dinu et al. 2019). In contrast, a recent study has shown that the intake of β-carotene by smokers increases the incidence of lung cancer regardless of the tar or nicotine content of smoked cigarettes (Middha et al. 2019).

Like carotenoids and chlorophylls, recent studies have demonstrated the anticancer effect of anthocyanins and betalains for colon, breast, bladder, prostate, and cervical cancer (Fernández et al. 2018; Mazzoni et al. 2019; Lechner and Stoner 2019).

Cardiovascular disease can also be prevented by supplementing plant-based pigments. Rahimi et al. (2019) researched the impact of supplementation with 50 mg betalain/betacyanin in patients with coronary artery disease. They found a significant decrease in homocysteine, glucose, total cholesterol, triglyceride, and LDL concentrations. Supplementation with β-carotene and lycopene mitigates oxidative stress, which causes cardiovascular diseases (Ghose et al. 2019; Cheng et al. 2019).

Carotenoids, chlorophylls, anthocyanins, and betalains are essential compounds for human health with recognized potential protective effects against chronic degenerative diseases. Therefore, the intake of sources containing these compounds is crucial for health promotion and disease treatment.

16.14.1 Dye-Sensitized Solar Cells

The search for renewable energy sources with a low impact on the environment has resulted in the implementation of new technologies that take advantage of them efficiently. One of these new technologies is dye-sensitized solar cells (DSSC) that are constituted by a mesoporous working electrode. In addition, compared to conventional cells, they are easy to manufacture at a low cost. In these devices, the dye or also called photosensitizer is the only active component that absorbs photons from the solar spectrum, generating electrical energy (Grätzel 2003). Schematically a DSCC consists of a photoanode, which is a broadband semiconductor oxide like titanium dioxide (TiO_2). The photoanode is photosensitized by the dye. The dye is commonly a metal complex based on ruthenium (Ru) (Espinoza Pizarro 2019), however, recent studies have shown that the implementation of plant-based pigments can replace Ru-based pigments.

Ridwan et al. (2018) developed a DSCC using chlorophyll extracted from sargassum, obtaining an efficiency of 1.50% and a filling factor of 43.2%. However, comparing the efficiency of DSCCs made with synthetic pigments, these are superior to those obtained with natural pigments such as chlorophyll. Siddick et al. (2018) evaluated a DSCC based on graphene–titania (TrGO) photoanode using conventional N719 and natural green chlorophyll dye. They found that N719 dye showed 3.95% efficiency, whereas natural chlorophyll green dye exhibited 0.67% of efficiency.

β-carotene also has recently been used in organic solar cells, with a fill factor around 35% and average power conversion efficiencies of 0.58% (Vohra et al. 2019). Similarly, Halidun et al. (2018) used β-carotene as colorant, reporting a fill factor of 68.89% and overall efficiency of 0.074%.

In the same way, anthocyanins extracted from fruits have been used as dyes in DSCC. Anthocyanin extract from fruits (*Melastoma malabthricum* L) using the technique of co-pigmentation showed a maximum efficiency of 1.32% (Indra 2019), whereas anthocyanin extracted from blueberry resulted in electrical energy conversion of 0.8% (Mizuno et al. 2018).

The low efficiency of the DSCC is due to the weak interaction between the natural dye molecule and the TiO_2 surface. Nevertheless, the use of natural dyes in DSCC is a promising option for the cost reduction of this type of equipment.

16.15 Conclusions

Today the society is more health conscious looking for "fresh-like" products where the use of natural products is preferred. Indeed plant-based dyes are of great value due to their properties. However, their applications will depend on developing a cost-effective and sustainable process for their obtention, and also, the improvement of their stability under the processing conditions that such pigments can be submitted is crucial. The efficient, fast, and greener extraction processes for the recovery of natural pigments from plant-based matrices are of paramount importance in food and pharmaceutical industries. Emerging technologies are being used for a long time; however, there are still locally raw materials that have not been analyzed. Also, new hybrid technologies combining emerging technologies need to be tested in order to improve existing extraction capabilities. There is no ideal emerging technology for pigments extraction; however, its selection will depend on the final application, product market price, and available resources. Finally, a life cycle analysis (LCA) is highly recommended in order to find the most viable and environmentally friendly process.

References

Abdel-Aal ESM, Hucl P, Rabalski I (2018) Compositional and antioxidant properties of anthocyanin-rich products prepared from purple wheat. Food Chem 254:13–19

Agcam E, Akyıldız A, Balasubramaniam VM (2017) Optimization of anthocyanins extraction from black carrot pomace with thermosonication. Food Chem 237:461–470

Aguilera-Otíz M, del Carmen Reza-Vargas M, Chew-Madinaveita RG, Meza-Velázquez JA (2011) Propiedades funcionales de las antocianinas. Biotecnia 13:16–22

Alvarado JL, Leschot A, Olivera-Nappa Á, Salgado AM, Rioseco H, Lyon C, Vigil P (2016) Delphinidin-rich maqui berry extract (Delphinol®) lowers fasting and postprandial glycemia and insulinemia in prediabetic individuals during oral glucose tolerance tests. Biomed Res Int 2016:9070537. https://doi.org/10.1155/2016/9070537

Amjadi S, Ghorbani M, Hamishehkar H, Roufegarinejad L (2018) Improvement in the stability of betanin by liposomal nanocarriers: Its application in gummy candy as a food model. Food Chem 256:156–162

Appelhagen I, Wulff-Vester AK, Wendell M, Hvoslef-Eide AK, Russell J, Oertel A, Martens S, Mock HP, Martin C, Matros A (2018) Colour bio-factories: Towards scale-up production of anthocyanins in plant cell cultures. Metab Eng 48:218–232

Arathi BP, Raghavendra-Rao Sowmya P, Kuriakose GC, Shilpa S, Shwetha HJ, Kumar S, Lakshminarayana R (2018) Fractionation and characterization of lycopene-oxidation products by LC-MS/MS (ESI)+: elucidation of the chemopreventative potency of oxidized lycopene in breast-cancer cell lines. J Agric Food Chem 66:11362–11371

Azeredo HM (2009) Betalains: properties, sources, applications, and stability–a review. Int J Food Sci Technol 44:2365–2376

Baaka N, El Ksibi I, Mhenni MF (2017) Optimisation of the recovery of carotenoids from tomato processing wastes: application on textile dyeing and assessment of its antioxidant activity. Nat Prod Res 31:196–203

Backes E, Pereira C, Barros L, Prieto MA, Genena AK, Barreiro MF, Ferreira IC (2018) Recovery of bioactive anthocyanin pigments from *Ficus carica* L. peel by heat, microwave, and ultrasound based extraction techniques. Food Res Int 113:197–209

Barba FJ, Grimi N, Vorobiev E (2015) New approaches for the use of non-conventional cell disruption technologies to extract potential food additives and nutraceuticals from microalgae. Food Eng Rev 7:45–62

Bolognesi VJ, Garcia CE (2018) Annatto carotenoids as additives replacers in meat products. In: Alternative and replacement foods. Academic Press, pp 355–384

Boukroufa M, Boutekedjiret C, Chemat F (2017) Development of a green procedure of citrus fruits waste processing to recover carotenoids. Res Eff Technol 3:252–262

Bruno RS, Wildman RE, Schwartz SJ (2016) Lycopene: food sources, properties, and health. In: Handbook of nutraceuticals and functional foods. CRC Press, pp 66–83

Capelli B, Talbott S, Ding L (2019) Astaxanthin sources: Suitability for human health and nutrition. Funct Foods Health Dis 9:430–445

Casati L, Pagani F, Fibiani M, Scalzo RL, Sibilia V (2019) Potential of delphinidin-3-rutinoside extracted from *Solanum melongena* L. as promoter of osteoblastic MC3T3-E1 function and antagonist of oxidative damage. Eur J Nutr 58:1019–1032

Catalkaya G, Kahveci D (2019) Optimization of enzyme assisted extraction of lycopene from industrial tomato waste. Sep Purif Technol 219:55–63

Chemat F, Abert-Vian M, Fabiano-Tixier AS, Strube J, Uhlenbrock L, Gunjevic V, Cravotto G (2019) Green extraction of natural products. Origins, current status, and future challenges. TrAC Trends Anal Chem 118:248–263

Chemat F, Rombaut N, Sicaire A-G, Meullemiestre A, Fabiano-Tixier A-S, Abert-Vian M (2017) Ultrasound assisted extraction of food and natural products. Mechanisms, techniques, combinations, protocols and applications. A review. Ultrason Sonochem 34:540–560

Chen M, Schliep M, Willows RD, Cai ZL, Neilan BA, Scheer H (2010) A red-shifted chlorophyll. Science 329:1318–1319

Cheng HM, Koutsidis G, Lodge JK, Ashor AW, Siervo M, Lara J (2019) Lycopene and tomato and risk of cardiovascular diseases: A systematic review and meta-analysis of epidemiological evidence. Crit Rev Food Sci Nutr 59:141–158

Choo KY, Kho C, Ong YY, Thoo YY, Lim RLH, Tan CP, Ho CW (2018) Studies on the storage stability of fermented red dragon fruit (Hylocereus polyrhizus) drink. Food Sci Biotech 27:1411–1417

Christaki E, Bonos E, Florou-Paneri P (2015) Innovative microalgae pigments as functional ingredients in nutrition. In: Handbook of marine microalgae. Academic Press, pp 233–243

Ciriminna R, Fidalgo A, Avellone G (2015) Economic and technical feasibility of betanin and pectin extraction from *Opuntia ficus-indica* peel via microwave-assisted hydro-diffusion. ACS Omega 4:12121–12124

Cobb BF, Kallenbach J, Hall CA, Pryor SW (2018) Optimizing the supercritical fluid extraction of lutein from corn gluten meal. Food Bioproc Tech 11:757–764

Dai Y, Row KH (2019) Isolation and determination of beta-carotene in carrots by magnetic chitosan beta-cyclodextrin extraction and high-performance liquid chromatography (HPLC). Analyt Lett 52:1828–1843

de Moura SC, Berling CL, Germer SP, Alvim ID, Hubinger MD (2018) Encapsulating anthocyanins from *Hibiscus sabdariffa* L. calyces by ionic gelation: Pigment stability during storage of microparticles. Food Chem 241:317–327

De Rosas MI, Deis L, Martínez L, Durán M, Malovini E, Cavagnaro JB (2019) Anthocyanins in nutrition: biochemistry and health benefits, Psychiatry and neuroscience update. Springer, Cham, pp 143–152

Delgado-Vargas F, Jiménez AR, Paredes-López O (2000) Natural pigments: carotenoids, anthocyanins, and betalains—characteristics, biosynthesis, processing, and stability. Crit Rev Food Sci Nutr 40:173–289

Derrien M, Aghabararnejad M, Gosselin A, Desjardins Y, Angers P, Boumghar Y (2018) Optimization of supercritical carbon dioxide extraction of lutein and chlorophyll from spinach by-products using response surface methodology. LWT Food Sci 93:79–87

Diaconeasa Z, Leopold L, Rugină D, Ayvaz H, Socaciu C (2015) Antiproliferative and antioxidant properties of anthocyanin rich extracts from blueberry and blackcurrant juice. Int J Mol Sci 16:2352–2365

Dinu M, Pagliai G, Casini A, Sofi F (2019) Food groups and risk of age-related macular degeneration: a systematic review with meta-analysis. Eur J Nutr 58:2123–2143

dos Santos CD, Ismail M, Cassini AS, Marczak LDF, Tessaro IC, Farid M (2018) Effect of thermal and high-pressure processing on stability of betalain extracted from red beet stalks. J Food Sci Technol 55:568–577

Du W, Zhu Z, Bai YL, Yang Z, Zhu S, Xu J, Fang J (2018) An anionic sod-type terbium-MOF with extra-large cavities for effective anthocyanin extraction and methyl viologen detection. Chem Commun 54:5972–5975

Esatbeyoglu T, Wagner AE, Schini-Kerth VB, Rimbach G (2015) Betanin—A food colorant with biological activity. Mol Nutr Food Res 59:36–47

Espinoza Pizarro DJ (2019) Estudio de celdas solares sensibilizadas por colorantes acopladas a fósforos inorgánicos. PhD Thesis. Universidad de Chile

Esquivel P (2016) Betalains. In: Handbook on Natural Pigments in Food and Beverages (81–99). Woodhead Publishing, Cambridge

Eugene R, Govindjee (1969) Photosynthesis. Willey, New York

Ferarsa S, Zhang W, Moulai-Mostefa N, Ding L, Jaffrin MY, Grimi N (2018) Recovery of anthocyanins and other phenolic compounds from purple eggplant peels and pulps using ultrasonic-assisted extraction. Food Bioprod Process 109:19–28

Fernández J, García L, Monte J, Villar C, Lombó F (2018) Functional anthocyanin-rich sausages diminish colorectal cancer in an animal model and reduce pro-inflammatory bacteria in the intestinal microbiota. Genes 9:133

Galarza JI, Gimpel JA, Rojas V, Arredondo-Vega BO, Henríquez V (2018) Over-accumulation of astaxanthin in *Haematococcus pluvialis* through chloroplast genetic engineering. Algal Res 31:291–297

Garcia-Mendoza M, Espinosa-Pardo FA, Baseggio AM, Barbero GF, Junior MRM, Rostagno MA, Martínez J (2017) Extraction of phenolic compounds and anthocyanins from juçara (*Euterpe edulis* Mart.) residues using pressurized liquids and supercritical fluids. J Supercrit Fluids 119:9–16

Gavahian M, Farahnaky A, Sastry S (2016) Ohmic-assisted hydrodistillation: A novel method for ethanol distillation. Food Bioprod Process 98:44–49

Ghose S, Varshney S, Chakraborty R, Sengupta S (2019) Dietary antioxidants in mitigating oxidative stress in cardiovascular diseases, Oxidative stress in heart diseases. Springer, Singapore, pp 83–139

Goula AM, Ververi M, Adamopoulou A, Kaderides K (2017) Green ultrasound-assisted extraction of carotenoids from pomegranate wastes using vegetable oils. Ultrason Sonochem 34:821–830

Grätzel M (2003) Dye-sensitized solar cells. J Photochem Photobiol C Photchem Rev 4(2):145–153

Guesmi A, Ladhari N, Hamadi NB, Msaddek M, Sakli F (2013) First application of chlorophyll-a as biomordant: sonicator dyeing of wool with betanin dye. J Clean Prod 39:97–104

Gul K, Tak A, Singh AK, Singh P, Yousuf B, Wani AA (2015) Chemistry, encapsulation, and health benefits of β-carotene-A review. Cogent Food Agric 1:1018696

Guo N, Jiang YW, Wang LT, Niu LJ, Liu ZM, Fu YJ (2019) Natural deep eutectic solvents couple with integrative extraction technique as an effective approach for mulberry anthocyanin extraction. Food Chem 296:78–85. https://doi.org/10.1016/j.foodchem.2019.05.196

Halidun WONS, Prima EC, Yuliarto B (2018) Fabrication dye sensitized solar cells (DSSCs) Using β-carotene pigment based natural dye. In: MATEC Web of Conferences, vol 159. EDP Sciences, p 02052

Hasan HM, Mohamad AS, Aldaaiek GA (2019) Extraction and determination of Beta carotene content in carrots and tomato samples collected from some markets at El-Beida City, Libya. EPH-Int J Appl Sci:1, 105–110. *(ISSN: 2208-2182)*

He Y, Wen L, Liu J, Li Y, Zheng F, Min W, Pan P (2018) Optimisation of pulsed electric fields extraction of anthocyanin from *Beibinghong* Vitis Amurensis Rupr. Nat Prod Res 32:23–29

Horuz Tİ, Belibağlı KB (2018) Nanoencapsulation by electrospinning to improve stability and water solubility of carotenoids extracted from tomato peels. Food Chem 268:86–93

Horuz Tİ, Belibağlı KB (2019) Nanoencapsulation of carotenoids extracted from tomato peels into zein fibers by electrospinning. J Sci Food Agri 99:759–766

Hörtensteiner S, Kräutler B (2011) Chlorophyll breakdown in higher plants. Biochim Biophys Acta 1807:977–988

İnanç AL (2011) Chlorophyll: structural properties, health benefits and its occurrence in virgin olive oils. Academic Food Journal/Akademik GIDA

Indra A (2019) Preparation of dye sensitized solar cell (DSSC) using isolated anthocyanin from fruit sat (*Melastoma malabathricum* L) dicopimented with salicylic acid as dye. J Phys Conf Ser 1317(1):012028

Jiao Y, Li D, Chang Y, Xiao Y (2018) Effect of freeze-thaw pretreatment on extraction yield and antioxidant bioactivity of corn carotenoids (lutein and zeaxanthin). J Food Qual:1–9

Kehili M, Kammlott M, Choura S, Zammel A, Zetzl C, Smirnova I, Allouche N, Sayadi S (2017) Supercritical CO_2 extraction and antioxidant activity of lycopene and β-carotene-enriched oleoresin from tomato (*Lycopersicum esculentum* L.) peels by-product of a Tunisian industry. Food Bioprod Process 102:340–349

Koubaa M, Barba FJ, Grimi N, Mhemdi H, Koubaa W, Boussetta N, Vorobiev E (2016) Recovery of colorants from red prickly pear peels and pulps enhanced by pulsed electric field and ultrasound. Innov Food Sci Emerg Technol 37:336–344

Kräutler B (2016) Breakdown of chlorophyll in higher plants—phyllobilins as abundant, yet hardly visible signs of ripening, senescence, and cell death. Angew Chem Int Ed 55:4882–4907

Kuczynska P, Jemiola-Rzeminska M, Strzalka K (2015) Photosynthetic pigments in diatoms. Mar Drugs 13:5847–5881

Kupan S, Hamid H, Kulkarni A, Yusoff M (2016) extraction and analysis of Beta-carotene recovery in cpo and oil palm waste by using HPLC. ARPN J Eng Appl Sci 11:2184–2188

Ladole MR, Nair RR, Bhutada YD, Amritkar VD, Pandit AB (2018) Synergistic effect of ultrasonication and co-immobilized enzymes on tomato peels for lycopene extraction. Ultrason Sonochem 48:453–462

Laqui-Vilca C, Aguilar-Tuesta S, Mamani-Navarro W, Montano-Bustamante J, Condezo-Hoyos L (2018) Ultrasound-assisted optimal extraction and thermal stability of betalains from colored quinoa (*Chenopodium quinoa* Willd) hulls. Ind Crop Prod 111:606–614

Lechner JF, Stoner GD (2019) Red beetroot and betalains as cancer chemo-preventative agents. Molecules 24:1602

Leichtweis MG, Pereira C, Prieto MA, Barreiro MF, Baraldi IJ, Barros L, Ferreira IC (2019) Ultrasound as a rapid and low-cost extraction procedure to obtain anthocyanin-based colorants from *Prunus spinosa* L. fruit epicarp: comparative study with conventional heat-based extraction. Molecules 24:573

Leite AC, Ferreira AM, Morais ES, Khan I, Freire MG, Coutinho JA (2017) Cloud point extraction of chlorophylls from spinach leaves using aqueous solutions of nonionic surfactants. ACS Sustain Chem Eng 6:590–599

Li J, Engelberth AS (2018) Quantification and purification of lutein and zeaxanthin recovered from distillers dried grains with solubles (DDGS). Biores Bioprocess 5:32

Li Q, de Mejia EG, Singh V, Somavat P, West M, West L, Donahue P (2018) U.S. Patent Application No. 15/739,528

Li Y, Cui Y, Lu F, Wang X, Liao X, Hu X, Zhang Y (2019) Beneficial effects of a chlorophyll-rich spinach extract supplementation on prevention of obesity and modulation of gut microbiota in high-fat diet-fed mice. J Funct Foods 60:103436

Lian F, Wang XD (2019) Lycopene and lung cancer. In: Preedy VR, Watson RR (eds) Lycopene: nutritional, medicinal and therapeutic properties. CRC Press, Boca Raton, p 365

Limantara L, Dettling M, Indrawati R, Brotosudarmo THP (2015) Analysis on the chlorophyll content of commercial green leafy vegetables. Proc Chem 14:225–231

López CJ, Caleja C, Prieto MA, Barreiro MF, Barros L, Ferreira IC (2018) Optimization and comparison of heat and ultrasound assisted extraction techniques to obtain anthocyanin compounds from Arbutus unedo L. fruits. Food Chem 264:81–91

López CJ, Caleja C, Prieto MA, Sokovic M, Calhelha RC, Barros L, Ferreira IC (2019) Stability of a cyanidin-3-O-glucoside extract obtained from *Arbutus unedo* L. and incorporation into wafers for colouring purposes. Food Chem 275:426–438

Loypimai P, Moongngarm A, Chottanom P, Moontree T (2015) Ohmic heating-assisted extraction of anthocyanins from black rice bran to prepare a natural food colourant. Innov Food Sci Emerg Technol 27:102–110

Luengo E, Martínez JM, Álvarez I, Raso J (2016) Effects of millisecond and microsecond pulsed electric fields on red beet cell disintegration and extraction of betanines. Ind Crop Prod 84:28–33

Mancha MAF, Rentería AL, Chávez A (2019) Estructura y estabilidad de las betalaínas. Interciencia 44:318–325

Martínez ACB, Ventura ID (2016) Pelargonidina extraída del rábano como sustituto de indicadores de pH ácido-base de origen sintético. Portal de la Ciencia 10:93–104

Mazzoni L, Giampieri F, Suarez JMA, Gasparrini M, Mezzetti B, Hernandez TYF, Battino MA (2019) Isolation of strawberry anthocyanin-rich fractions and their mechanisms of action against murine breast cancer cell lines. Food Funct 10:7103–7120

Medeiros AKDOC, de Carvalho GC, de Araújo Amaral MLQ, de Medeiros LDG, Medeiros I, Porto DL, Passos TS (2019) Nanoencapsulation improved water solubility and color stability of carotenoids extracted from Cantaloupe melon (Cucumis melo L.). Food Chem 270:562–572

Mezzomo N, Ferreira SR (2016) Carotenoids functionality, sources, and processing by supercritical technology: a review. J Chem 2016:3164312. https://doi.org/10.1155/2016/3164312

Miranda-Medina A, Hayward-Jones PM, Carvajal-Zarrabal O, de Guevara LDAL, Ramírez-Villagómez YD, Barradas-Dermitz DM, Aguilar-Uscanga MG (2018) Optimization of Hibiscus sabdariffa L. (Roselle) anthocyanin aqueous-ethanol extraction parameters using response surface methodology. Sci Study, Res Chem, Chem Eng Biotechnol, Food Ind 19:53–62

Middha P, Weinstein SJ, Männistö S, Albanes D, Mondul AM (2019) β-carotene supplementation and lung cancer incidence in the alpha-tocopherol, Beta-carotene cancer prevention study: the role of tar and nicotine. Nicotine Tob Res 21(8):1045–1050. https://doi.org/10.1093/ntr/nty115

Milani A, Basirnejad M, Shahbazi S, Bolhassani A (2017) Carotenoids: biochemistry, pharmacology and treatment. Br J Pharmacol 174:1290–1324

Mizuno A, Yamada G, Ohtani N (2018) Natural dye-sensitized solar cells containing anthocyanin dyes extracted from frozen blueberry using column chromatography method. In 2018 IEEE 7th World Conference on Photovoltaic Energy Conversion (WCPEC) (A Joint Conference of 45th IEEE PVSC, 28th PVSEC & 34th EU PVSEC) (pp. 1129-1131). IEEE

Moran NE, Thomas-Ahner JM, Fleming JL, JP ME, Mehl R, Grainger EM, Riedl KM, Toland AE, Schwartz SJ, Clinton SK (2019) Single nucleotide polymorphisms in β-Carotene oxygenase 1 are associated with plasma lycopene responses to a tomato-soy juice intervention in men with prostate cancer. J Nutr 149:381–397

Narayana SDTU, Wedamulla NE, Wijesinghe WAJP, Rajakaruna RAMAT, Wijerama HJKSS (2019) Extraction of anthocyanin from Hinembilla (*Antidesma alexiteria*) fruit as a natural food colorant. Uva Wellassa University of Sri Lanka

Ningrum EZ, Fajri LF, Oktaviana D (2019) Effect of liquid chlorophyll from alfalfa leaves (*Medicago sativa* L.) as a support if supplement to the performance of broiler chickens. Bantara J Anim Sci 1(1)

Nowacka M, Tappi S, Wiktor A, Rybak K, Miszczykowska A, Czyzewski J, Drozdzal K, Witrowa-Rajchert D, Tylewicz U (2019) The impact of pulsed electric field on the extraction of bioactive compounds from beetroot. Foods 8:244

Olson JA (1989) Provitamin A function of carotenoids: the conversion of β-carotene into vitamin A. J Nutr 119:105–108

Pareek S, Sagar NA, Sharma S, Kumar V, Agarwal T, González-Aguilar GA, Yahia EM (2017) Chlorophylls: chemistry and biological functions. Fruit and Vegetable Phytochemicals:269–284

Pataro G, Bobinaitė R, Bobinas Č, Šatkauskas S, Raudonis R, Visockis M, Ferrari G, Viškelis P (2017) Improving the extraction of juice and anthocyanins from blueberry fruits and their by-products by application of pulsed electric fields. Food Bioproc Tech 10:1595–1605

Paula JT, Sousa IM, Foglio MA, Cabral FA (2018) Selective fractionation of extracts of *Arrabidaea chica* Verlot using supercritical carbon dioxide as antisolvent. J Supercrit Fluids 133:9–16

Pereira RN, Rodrigues RM, Genisheva Z, Oliveira H, de Freitas V, Teixeira JA, Vicente AA (2016) Effects of ohmic heating on extraction of food-grade phytochemicals from colored potato. LWT 74:493–503

Pinela J, Prieto MA, Pereira E, Jabeur I, Barreiro MF, Barros L, Ferreira IC (2019) Optimization of heat-and ultrasound-assisted extraction of anthocyanins from *Hibiscus sabdariffa* calyces for natural food colorants. Food Chem 275:309–321

Rahimi P, Mesbah-Namin SA, Ostadrahimi A, Abedimanesh S, Separham A (2019) Effects of betalains and betacyanins on atherogenic risk factors in patients with atherosclerotic cardiovascular disease. Food Funct. https://doi.org/10.1039/C9FO02020A

Ravi HK, Breil C, Vian MA, Chemat F, Venskutonis PR (2018) Biorefining of bilberry (*Vaccinium myrtillus* L.) pomace using microwave hydro-diffusion and gravity, ultrasound-assisted, and bead-milling extraction. ACS Sust Chem Eng 6:4185–4193

Ridwan MA, Noor E, Rusli MS (2018) Fabrication of dye-sensitized solar cell using chlorophylls pigment from sargassum. IOP Conf Ser Earth Environ Sci 144(1):012039

Righi Pessoa da Silva H, da Silva C, Bolanho BC (2018) Ultrasonic-assisted extraction of betalains from red beet (*Beta vulgaris* L.). J Food Proc Eng 41:e12833

Rodriguez-Amaya DB (2019) Update on natural food pigments-A mini-review on carotenoids, anthocyanins, and betalains. Food Res Int 124:200–205

Roselló-Soto E, Parniakov O, Deng Q, Patras A, Koubaa M, Grimi N, Boussetta N, Tiwari BK, Vorobiev E, Lebovka N, Barba FJ (2016) Application of non-conventional extraction methods: Toward a sustainable and green production of valuable compounds from mushrooms. Food Eng Rev 8:214–234

Saberian H, Hosseini F, Bolourian S (2018) Optimizing the extraction condition of chlorophyll from alfalfa and investigating its qualitative and quantitative properties in comparison to different plant resources. Food Sci Technol 14:47–57

Sánchez-Camargo A, Gutiérrez LF, Vargas SM, Martinez-Correa HA, Parada-Alfonso F, Narváez-Cuenca CE (2019) Valorization of mango peel: Proximate composition, supercritical fluid extraction of carotenoids, and application as an antioxidant additive for an edible oil. J Supercrit Fluids 152:104574

Santos-Zea L, Gutiérrez-Uribe JA, Benedito J (2019) Effect of ultrasound intensification on the supercritical fluid extraction of phytochemicals from *Agave salmiana* bagasse. J Supercrit Fluids 144:98–107

Setford P, Jeffery D, Grbin P, Muhlack R (2018) Modelling the mass transfer process of malvidin-3-glucoside during simulated extraction from fresh grape solids under wine-like conditions. Molecules 23:2159

Shabbir M, Bukhari MN, Khan MA, Mohammad F (2018) Extraction of carotenoid colorants from *Tagetes erecta* (Marigold) flowers and application on textile substrate for coloration. Curr Smart Mater 3:124–135

Sharmila G, Muthukumaran C, Suriya E, Keerthana RM, Kamatchi M, Kumar NM, Jeyanthi J (2019) Ultrasound aided extraction of yellow pigment from *Tecoma castanifolia* floral petals: Optimization by response surface method and evaluation of the antioxidant activity. Ind Crop Prod 130:467–477

Sharmin T, Ahmed N, Hossain A, Hosain MM, Mondal SC, Haque MR et al (2016) Extraction of bioactive compound from some fruits and vegetables (pomegranate peel, carrot and tomato). Am J Food Nutr 4:8–19

Siddick SZ, Lai CW, Juan JC (2018) An investigation of the dye-sensitized solar cell performance using graphene-titania (TrGO) photoanode with conventional dye and natural green chlorophyll dye. Mater Sci Semicond Process 74:267–276

Silva Ferreira V, Sant'Anna C (2017) Impact of culture conditions on the chlorophyll content of microalgae for biotechnological applications. World J Microb Biot 33:20

Silva YP, Ferreira TA, Jiao G, Brooks MS (2019) Sustainable approach for lycopene extraction from tomato processing by-product using hydrophobic eutectic solvents. J Food Sci Technol 56:1649–1654

Somavat P, Kumar D, Singh V (2018) Techno-economic feasibility analysis of blue and purple corn processing for anthocyanin extraction and ethanol production using modified dry grind process. Ind Crop Prod 115:78–87

Stafford HA (1994) Anthocyanins and betalains: evolution of the mutually exclusive pathways. Plant Sci 101:91–98

Strack D, Vogt T, Schliemann W (2003) Recent advances in betalain research. Phytochemistry 62:247–269
Strain HH (1958) Chloroplast pigments and chromatographic analysis, vol 32. Pennsylvania State University, State College
Sun H, Zhang P, Zhu Y, Lou Q, He S (2018) Antioxidant and prebiotic activity of five peonidin-based anthocyanins extracted from purple sweet potato (*Ipomoea batatas* (L.) Lam.). Sci Rep 8:5018
Tang P (2018) Petunidin derivatives from black goji and purple potato as promising natural colorants, and their co-pigmentation with metals and isoflavones. Doctoral dissertation, The Ohio State University
Tesoriere L, Attanzio A, Allegra M, Livrea MA (2015) Dietary indicaxanthin from cactus pear (*Opuntia ficus-indica* L. Mill) fruit prevents eryptosis induced by oxysterols in a hypercholesterolaemia-relevant proportion and adhesion of human erythrocytes to endothelial cell layers. Br J Nutr 114:368–375
Tidder A, Benohoud M, Rayner CM, Blackburn RS (2018) Extraction of anthocyanins from blackcurrant (*Ribes nigrum* L.) fruit waste and application as renewable textile dyes. In: 91st Textile Institute World Conference Book of Abstracts. Textile Institute, pp 18–19
Vaňková K, Marková I, Jašprová J, Dvořák A, Subhanová I, Zelenka J, Novosádová I, Rasl J, Vomastek T, Sobotka R, Muchová L, Vítek L (2018) Chlorophyll-mediated changes in the redox status of Pancreatic cancer cells are associated with its anticancer effects. Oxid Med Cell Longev 6:1–11. https://doi.org/10.1155/2018/4069167
Velmurugan P, Kim JI, Kim K, Park JH, Lee KJ, Chang WS, Oh BT (2017) Extraction of natural colorant from purple sweet potato and dyeing of fabrics with silver nanoparticles for augmented antibacterial activity against skin pathogens. J Photochem Photobiol B Biol 173:571–579
Vernon LP, Seely GR (2014) The chlorophylls. Academic Press, Cambridge, Massachusetts
Vohra V, Uchiyama T, Inaba S, Okada-Shudo Y (2019) Efficient ultrathin organic solar cells with sustainable β-carotene as electron donor. ACS Sust Chem Eng:4376–4381. https://doi.org/10.1021/acssuschemeng.8b06255
Wang L, Lu W, Li J, Hu J, Ding R, Lv M, Wang Q (2019) Optimization of ultrasonic-assisted extraction and purification of zeaxanthin and lutein in corn gluten meal. Molecules 24:2994
Wiese M, Bashmakov Y, Chalyk N, Nielsen DS, Krych Ł, Kot W, Klochkov V, Pristensky D, Bandaletova T, Chernyshova M, Kyle N, Petyaev I (2019) Prebiotic effect of lycopene and dark chocolate on gut microbiome with systemic changes in liver metabolism, skeletal muscles and skin in moderately obese persons. Biomed Res Int 2019:4625279. https://doi.org/10.1155/2019/4625279
Wrolstad RE (2004) Anthocyanin pigments—Bioactivity and coloring properties. J Food Sci 69: C419–C425
Xie L, Cahoon E, Zhang Y, Ciftci ON (2019) Extraction of astaxanthin from engineered *Camelina sativa* seed using ethanol-modified supercritical carbon dioxide. J Supercrit Fluids 143:171–178
Yammine S, Brianceau S, Manteau S, Turk M, Ghidossi R, Vorobiev E, Mietton-Peuchot M (2018) Extraction and purification of high added value compounds from by-products of the winemaking chain using alternative/nonconventional processes/technologies. Crit Rev Food Sci Nutr 58:1375–1390
Yan L, Fernando WM, Brennan M, Brennan CS, Jayasena V, Coorey R (2016) Effect of extraction method and ripening stage on banana peel pigments. Int J Food Sci Technol 51:1449–1456
Yari A, Rashnoo S (2017) Optimization of a new method for extraction of cyanidin chloride and pelargonidin chloride anthocyanins with magnetic solid phase extraction and determination in fruit samples by HPLC with central composite design. J Chromatogr B 1067:38–44
Yin Y, Fei L, Wang C (2018) Optimization of natural dye extracted from phytolaccaceae berries and its mordant dyeing properties on natural silk fabric. J Nat Fibers 15:69–79
Yin Y, Jia J, Wang T, Wang C (2017) Optimization of natural anthocyanin efficient extracting from purple sweet potato for silk fabric dyeing. J Clean Prod 149:673–679

Zapata IC, Felipe Álzate A, Zapata K, Arias JP, Puertas MA, Rojano B (2019) Effect of pH, temperature and time of extraction on the antioxidant properties of *Vaccinium meridionale* Swartz. J Berry Res 9(1):39–49

Zardini AA, Mohebbi M, Farhoosh R, Bolurian S (2018) Production and characterization of nanostructured lipid carriers and solid lipid nanoparticles containing lycopene for food fortification. J Food Sci Technol 55:287–298

Yusuf M, Shabbir M, Mohammad F (2017) Natural colorants: Historical, processing and sustainable prospects. Nat Prod Bioprospect 7:123–145

Zhang ZH, Wang LH, Zeng XA, Han Z, Wang MS (2017) Effect of pulsed electric fields (PEFs) on the pigments extracted from spinach (*Spinacia oleracea* L.). Innov Food Sci Emerg Technol 43:26–34

Plant Lectins: Sugar-Binding Properties and Biotechnological Applications

17

P. H. Surya, M. Deepti, and K. K. Elyas

Abstract

Lectins are of prime importance due to the reversibility and specificity in their sugar-binding properties, which differ from enzymes or antibodies in being non-catalytic and mostly weak in nature with broader specificity. There are several approaches which help to characterize these interactions in terms of thermodynamic parameters like affinity constant, binding stoichiometry, free energy change, enthalpy change, and the change in entropy of the interaction. These techniques range from the simple hapten inhibition assay to determine the sugar specificity to the more advanced methods like fluorescence titration and surface plasmon resonance that helps to determine the binding parameters.

Sugars are mysterious molecules of biological systems due to the challenges in predicting their sequence and the complexities in terms of chain length, type of sugar, branching, and anomerism. There are a variety of glycans on cell surfaces of both prokaryotes and eukaryotes, which can be targeted to identify the cell types. Lectin binding to cell surface may induce downstream pathways, leading to cell death via complement-mediated lysis, leakage of nutrients, poor nutrient absorption, apoptosis, or cell cycle arrest. This property makes them very good antimicrobial, anti-insect, and anticancer agents. Some lectins can induce cell division, as observed in case of certain mitogenic lectins. This chapter addresses the techniques used widely to understand the thermodynamic parameters of lectin–sugar interaction and discuss the role of this binding in the cytotoxic and mitogenic potential of these proteins. Analytical uses of lectins as affinity matrixes, lectin blotting, lectin microarray, enzyme-linked lectin assay (ELLA), and lectin-based biosensor are also briefly outlined.

P. H. Surya · M. Deepti · K. K. Elyas (✉)
Department of Biotechnology, University of Calicut, Malappuram, Kerala, India

Keywords

Lectin · Hemagglutination · Blood grouping · Hapten inhibition · Lectin blotting

17.1 Introduction

Lectins are simply proteins that can bind with sugars. Sugar residues are also represented by the term glycan, which is defined as any polysaccharide or oligosaccharide, and in many cases both terms are used interchangeably. The interesting feature of this binding lies upon its reversibility and specificity. Type of glycan can range from simple sugars to polysaccharides in solution or on cell surfaces. Plants are reported to accumulate these proteins in their storage tissues like seeds or in the vegetative parts like leaves. The first lectin to be reported was "ricin" from the castor bean *Ricinus communis* by Stillmark in 1888. The protein is still of therapeutic importance because of its chimeric nature in having a catalytic domain in addition to the carbohydrate-binding site. The glycan-binding domain gives this protein specificity and the catalytic region functions as the ribosomal inactivating site.

Overall structure-based classification of lectins divides them into four major classes namely merolectins, hololectins, superlectins, and chimerolectins (Table 17.1). This classification is based on the number of sugar-binding sites in the lectin and the differences in their sugar determinant(s). Evolutionary relationship among lectins is made use in another system to classify lectins as legume lectins, monocot mannose-binding lectins, chitin-binding proteins containing hevein domain, type 2 RIPs (ribosomal-inactivating proteins), Cucurbitaceae phloem lectins, jacalin family, and Amaranthaceae lectins (Table 17.2). Specificity of lectins to simple sugars also became a criterion to classify them as glucose/mannose binding, galactose and *N*-acetyl-D-galactosamine (GalNAc) binding, *N*-acetyl-D-glucosamine (GlcNAc) binding, fucose (Fuc) binding, and sialic acid (Sia) binding lectins. When a lectin is specific to more complex sugars than mono or disaccharides, it is then classified as the lectin with a "complex" specificity (Peumans and Van Damme 1998).

For each lectin, the carbohydrate recognition domain (CRD) will be made up of a unique amino acid sequence, a characteristic fold, and the binding region. Based on similarity in these amino acid sequences and fold structures, plant lectins are again classified into 11 families (Table 17.3). Amaranthin domain, homolog of class V chitinases, cyanovirin domain, *Euonymus europaeus* lectin domain, *Galanthus nivalis* agglutinin domain, hevein domain, jacalin domain, legume domain, Lys M domain, nictaba-like domain, and ricin-B domain. But lectin specificity cannot be determined merely from the CRD structure as the same sugar can be detected by different CRDs (Van Damme 2014; Lannoo and Van Damme 2014; Hendrickson and Zherdev 2018).

Usually, the terminal sugar residue of complex biomolecules is targeted by lectins, while others bind glycan residues within a single carbohydrate chain. As reviewed by Hendrickson and Zherdev, the typical affinity constant, K_a of lectin–

Table 17.1 Classification of lectins based on the number and similarity of sugar-binding sites (Peumans and Van Damme 1998)

Sl. No.	Class of lectin	Key feature	Remarks	Example	Sugar specificity
1	Merolectins	Single glycan-binding domain. Monovalent.	Incapable of agglutinating cells and precipitating glycoconjugates	Hevein from the latex of *Hevea brasiliensis*	Chitin
2	Hololectins	At least two very similar or identical glycan-binding domain. Di- or multivalent.	Perform agglutination and precipitation reactions	Concanavalin A from *Canavalia ensiformis*	Glucose/mannose
3	Superlectins	At least two nonidentical glycan-binding domain that recognize structurally different sugars. Di- or multivalent.	Perform agglutination and precipitation reactions	Tulip bulb lectin (TxLC) from *Tulipa gesneriana*	Mannose and N-acetyl galactosamine
4	Chimerolectins	At least one glycan-binding domain and an unrelated domain with a characteristic biological activity. Can be merolectins or hololectins based on number of sugar-binding sites.	Depends on the number of glycan-binding sites	Ricin from *Ricinus communis*	Galactose

sugar interaction is 10^3–10^7 M^{-1} (Ashraf and Khan 2003; Lebed et al. 2006; Gabius et al. 2011). Lectin–carbohydrate binding is basically non-covalent in nature involving hydrogen bonding, hydrophobic (Sharon and Lis 2004), and Van der Waals interactions (Coelho et al. 2017) with rare involvement of electrostatic forces (Kumar et al. 2012). Glycan structures cannot be predicted directly from the genome of the organism making its exploration a difficult task. They can vary in the type of sugars, chain length, branching, or in the spatial conformation, which all are specifically targeted by lectins. Several lectins can bind specific anomeric form of the sugar (α or β) and some are even specific to a definite sequence of glycan residues. Although the affinity of lectin to monosaccharides is low, most lectins

Table 17.2 Classification of lectin based on evolutionary relatedness (Peumans and Van Damme 1998)

Sl. No.	Lectin group	Occurrence	Specificity
1	Legume lectins	Legumes	Diverse
2	Monocot mannose-binding lectins	Liliales	Mannose
3	Chitin-binding proteins containing hevein domain	Monocots, dicots	GlcNAc or (GlcNAc)n
4	Type 2 RIPs	Monocots, dicots	Gal, GalNAc, or Neu5Acα(2,6)Gal/GalNAc
5	Cucurbitaceae phloem lectins	Cucurbitaceae	(GlcNAc)n
6	Jacalin family	Moraceae, Convolvulaceae	Gal, mannose/maltose
7	Amaranthaceae lectins	Amaranthaceae	GalNAc

have a much higher selectivity for oligosaccharides and branched carbohydrates than simple sugars and the reasons behind this are the length of binding site and the polyvalency of lectin-binding sites. Here, usually the terminal glycan residue interacts with the major binding site of lectin and other sugar residues interact with secondary sites along the carbohydrate chain (Hendrickson and Zherdev 2018). In some cases, the same sugar will be recognized by structurally unrelated lectins. Specificity of this binding also accounts for their preferential binding to some blood groups. Erythrocytes have a wide range of glycoconjugates on their cell surface and their glycan heads specify the blood groups. Lectins can bind specifically to these sugar structures causing the cells to clump together resulting in hemagglutination (Khan et al. 2002) which is the most common criterion to screen tissue extracts for the presence of lectins. Terminal sugar residues added to a common disaccharide unit form the antigenic determinants in ABH and related blood groups. The disaccharide, L-Fuc-(1–2) D-Gal-β1 is the H antigen and the addition of D-GalNAc to the galactose of H antigen forms A antigen. *Ulex europeus* lectin is specific to the H antigenic determinant and the lectin from *Phaseolus limenis* bind specifically to the A antigen. If a α-D-galactose unit is added to the chain instead D-GalNAc, it forms the B blood group determinant which can be targeted by a lectin from *Griffonia simplicifolia*. The sugar determinants of ABH blood grouping are briefly outlined in Fig. 17.1. The aforementioned lectins are given as examples only and many others are also reported to possess blood group specificity (Khan et al. 2002).

Purification of lectin generally starts with aqueous or saline extraction of tissue homogenate using conventional protein purification methods. A pre-extraction with organic solvents like methanol or petroleum ether is followed to eliminate lipid contamination that can cause aggregation or agglutination of cells. Total protein can be fractionated through salting out using ammonium sulfate or by precipitating with organic solvents like alcohol and acetone. Affinity chromatography is generally opted for the purification of lectin by using immobilized sugars as adsorbent matrix (Hamid et al. 2013). The bound lectin can then be eluted by using any of the methods

Table 17.3 Classification of plant lectins based on the structural and sequence similarity in the lectin carbohydrate recognition domain. (Van Damme 2014; Lannoo and Van Damme 2014; Hendrickson and Zherdev 2018) and the Lectins Glyco-3D database (Pérez et al. 2015)

Sl. No	Lectin family	Fold structure	Example	Sugar specificity	PDB code
1	Amaranthin domain	β- trefoil	*Amaranthus caudatus* agglutinin	D-Gal, GalNAc	lJLX
2	Homolog of class V chitinases	TIM barrel	*Robinia pseudoacacia* agglutinin	GalNAc	1FNY 1FNZ
3	Cyanovirin domain	Triple-stranded β-sheet and a β- hairpin	*Ceratopteris richardii* lectin	Man	2JZJ
4	*Euonymus europaeus* lectin domain	Unknown structure	*Euonymus europaeus* agglutinin	GalNAc, Gal, Fuc, Man, NeuAc, GlcNAc, Xyl, Glc	Not available
5	*Galanthus nivalis* agglutinin domain	β- barrel	*Galanthus nivalis* agglutinin	D-Man	lJPC
6	Hevein domain	Hevein fold	*Phytolacca americana* agglutinin	GlcNAc	1UHA 1ULK 1ULM 1ULN
7	Jacalin domain	β-prism	*Artocarpus integrifolia* lectin	Glc, Man, Gal, GalNAc	1J4T 1J4S 1J4U 1VBO 1VBP 1JAC 1KU8 1KUJ 1M26 1PXD 1UGW 1UGX 1UGY 1UH0 1UH1
8	Legume domain	β-sandwich	*Canavalia ensiformis* lectin	Ara, Man, Glc, GlcNAc,	1APN 1BXH 1C57 1CES 1CJP 1CN1 1CON 1CVN 1DQ0 1DQ1 1DQ2 1DQ3 1DQ4 1DQ5 1DQ6 1ENQ 1ENR 1ENS 1GIC 1GKB 1HQW 1I3H 1JBC 1JN2 1JOJ 1JUI 1JW6 1JYC 1JYI 1NLS 1NXD 1ONA 1QDC 1QDO 1QGL 1QNY 1SCR 1SCS 1TEI 1VAL 1VAM 1VLN 1XQN 2CNA 2CTV 2ENR 2UU8 3CNA 3D4K 3ENR 5CNA ND1

(continued)

Table 17.3 (continued)

Sl. No	Lectin family	Fold structure	Example	Sugar specificity	PDB code
9	Lysine motif (Lys M) domain	β–α–α–β-structure	*Medicago truncatula* lectin	Glc/Man	Not available
10	Nictaba-like domain	Unknown structure	*Nicotiana tabacum* lectin	GlcNAc	Not available
11	Ricin-B domain	β-trefoil	*Ricinus communis* agglutinin	Gal, Glc	1RZO 2AAI

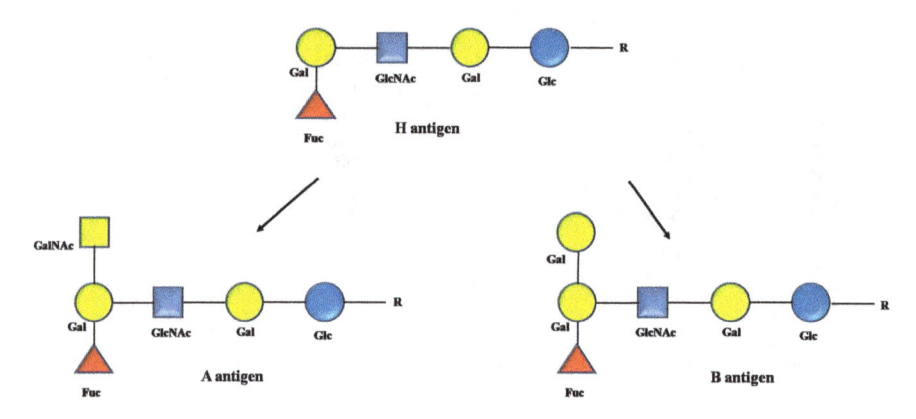

Fig. 17.1 Schematic representation of the sugar determinants of ABH blood grouping system. H antigen is characterized by the terminal disaccharide Fuc–Gal and the addition of GalNAc to the Gal residue of H antigen forms A determinant. The B antigen is characterized by the presence of Gal in place of GalNAc in A antigen. Fuc–fucose, Gal–galactose, GlcNAc-*N*-acetylglucosamine, Glc-glucose, GalNAc-*N*-acetylgalactosamine (Khan et al. 2002)

like addition of competing sugar, disturbing the interaction of lectin with the stationary phase by changing of pH or chaotropic agents like urea and guanidine hydrochloride supplemented in mobile phase. Sugar specificity of the lectin can be determined by a simple hapten inhibition assay in which various sugars will be tested for their ability to inhibit agglutination. For each lectin, the method of purification has to be standardized with minimal number of steps to retain maximum activity and yield. The lectin from *Arisaema utile* was purified using asialofetuin-linked amino activated silica beads. The bound lectin was eluted with 100 mM glycine–HCl buffer, pH 2.5 with immediate neutralization of the fractions with 2 M Tris–HCl buffer, pH 8.3 (Dhuna et al. 2010). Whereas a lectin from the latex of *Euphorbia tirucalli* was purified by a two-step chromatographic method involving DEAE–sephacel column followed by immobilized D-galactose–agarose affinity chromatography (Palharini et al. 2017). Some lectins strongly bind the chromatographic matrix making their purification a difficult task. For example, wheat germ agglutinin could not be eluted from GlcNAc immobilized affinity matrix even at higher concentrations of GlcNAc or di-N-acetylchitobiose. It was easily eluted when 0.05 M HCl was used as the eluent at which the lectin existed in its highly soluble monomeric form (Bouchard et al. 1976). The strong binding is attributed to the self-association of wheat germ agglutinin mediated by the immobilized sugar and the strong avidity of the interaction. The binding of oligovalent lectins with oligosaccharides or glycoproteins can also form cross-linked lattices (Monsigny et al. 2000).

In plants, lectins play the role of antibodies involving in the defensive mechanisms to combat infections mediated by bacteria, fungi, virus, and insects. Apart from agglutination, lectin-mediated complement activation and subsequent clearing of bacteria is a well-known defense mechanism in animals and humans which is mostly mediated by mannose-binding lectins. In plants also, there are numerous reports on the discovery of mannose-binding lectins which could be used as potential molecules to activate complement-mediated lysis of bacteria. A lectin from *Bauhinia monandra* leaf when incorporated in artificial diet resulted mortality in the insects *Zabrotes subfaciatus* and *Callosobruchus maculates* (Macedo et al. 2007). It is reported that growth of fungi *Coprinus comatus, Rhizoctonia solani,* and *Valsa mali were* inhibited by *Phaseolus vulgaris* seed lectin (Ye et al. 2001; Ang et al. 2014). Principle behind this protective role will be detailed in the section entitled "antimicrobial properties of lectin." Legume lectins are reported to bind with species-specific glycans on rhizobium cell surface aiding in their colonization and nodulation. This is mediated by species-specific interaction of lipo-chito oligosaccharide signal from bacteria known as nodulation factors (Nods) and plant root lectin (Kalsi and Etzler 2000). In a study, soybean agglutinin (SBA) agglutinated most strains of *Bradyrhizobium japonicum* that nodulate soybeans but not non-nodulating Bradyrhizobial strains (Sharon and Lis 2004). They also perform the functions as carrier molecules for sugar transport and as storage proteins. Globulin, the primary storage protein of oat (*Arena satina* L.) seeds, was reported to agglutinate rabbit erythrocytes and the hemagglutination was inhibited by laminarin and by the carbohydrate cleaved from the native globulin (Langston-Unkefer and Gade 1984).

Lectins are also useful in diagnostics and research applications. Specificity of lectins to sugars is used to tag cell surface morphology in histochemical sections and aberrant expression of glycosylated structures on cancer cell surface. Many of the lectins known so far are reported to be antimicrobial, anti-insect, or antiproliferative with some of them capable of inducing mitosis in lymphocytes. Apart from these applications, lectins are used as affinity matrix for the purification of glyco-conjugates, for visualizing glycoproteins on western blot and glycan analysis by lectin microarray. Such applications and uses of lectins will be detailed in the subsequent sections.

17.2 Analysis of Lectin–Sugar Interactions

Lectin interacts with sugar or glycan residues with a unique binding domain known as the carbohydrate recognition domain (CRD) comprised of about 120–160 amino acids. They can hold more than one simple sugar in their binding sites because of their structural similarity. For example, GlcNAc and N-acetylneuraminic acid (Neu5Ac) can both interact with the CRD of wheat germ agglutinin (Monsigny et al. 1980). The physicochemical and biological properties of lectin–sugar interaction are dependent on the density of recognition domain, structure, and spatial conformation of sugar residues (Monsigny et al. 2000). Each subunit of an oligomeric lectin usually has a single CRD. Most plant lectins are multivalent in nature because of their dimeric or tetrameric nature. Such multivalency allows ligand crosslinking with increased avidity (Cummings et al. 2017). When a monovalent lectin binds to a sugar, at equilibrium, the affinity constant is defined as association constant or K_a (Eq. 17.1). When multivalent interactions are concerned, a set of equilibrium constants has to be considered to represent the overall avidity of the reaction (Cummings et al. 2017).

$$K_a = \frac{[\text{Lectin} - \text{sugarcomplex}]}{[\text{Lectin}][\text{Sugar}]} \qquad (17.1)$$

$$K_d = \frac{1}{K_a} \qquad (17.2)$$

$$\Delta G° = -RT \ln K_a = RT \ln K_d = \Delta H° - T\Delta S° \qquad (17.3)$$

K_a: Affinity constant of lectin–sugar interaction at equilibrium
K_d: Dissociation constant of lectin–sugar interaction at equilibrium
$\Delta G°$: Standard free energy change associated with the binding of lectin to sugar molecule
R: Universal gas constant (0.00198 kcal/mol-K)
T: Absolute temperature in Kelvin
$\Delta H°$: Enthalpy change involved in lectin–sugar interaction
$\Delta S°$: Entropy change involved in lectin–sugar interaction

Lectin interacts with sugars mainly through hydrogen bonding and hydrophobic interactions (Del Carmen et al. 2012). Apart from amines and carboxyl groups of substituted carbohydrates, the occurrence of numerous hydroxyl groups plays an important role to establish hydrogen bonding with the lectin receptor. They can act as both donors and acceptors of hydrogen bonding. Together with that, the same hydroxyl group can sometimes act as both hydrogen bond donor and acceptor, a feature known as cooperative hydrogen bonding which also occurs in lectin–sugar interaction. On the lectin side, the side chains of polar residues, mostly, aspartic acid, glutamic acid, asparagine, glutamine, arginine, and serine will be involved in this binding. The backbone amine and carbonyl groups of this protein also take part in hydrogen bonding (Piotukh et al. 1999).

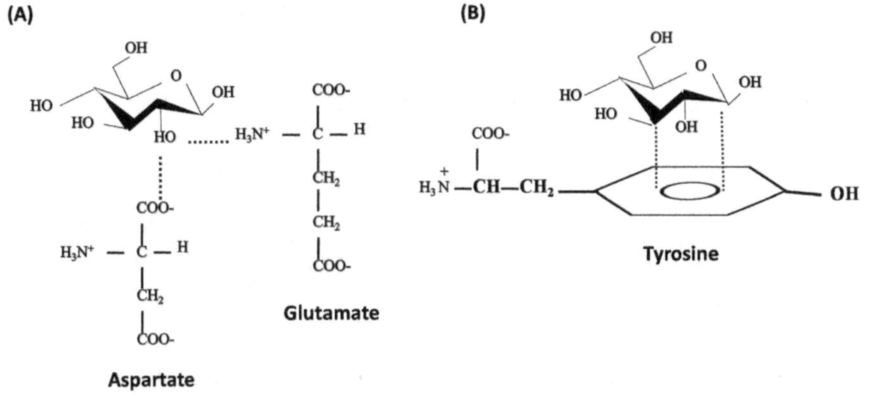

Fig. 17.2 Illustration of possible interactions between sugar and lectin. (**a**) Cooperative hydrogen bonding between D-glucose and amino acid residues of lectin such as aspartate and glutamate in which the same hydroxyl group of sugar act as both hydrogen bond donor and acceptor. (**b**) Weak electro-attractive forces between the π-system of amino acid residues such as tyrosine and the partially positively charged C–H groups in D-glucose. This type of interaction is expected to stabilize hydrophobic interactions between lectin and sugar

In sugars, hydrophobic regions are formed by the positioning of C–H groups together by the steric repulsion of hydroxyl groups. This can then interact with the hydrophobic interactive domains in lectin, effecting lectin–sugar interaction. This usually includes the aromatic rings from tyrosine, tryptophan, and phenylalanine (Jiménez-Barbero et al. 2006). The interaction is stabilized by weak electro-attractive forces between the π- system of aromatic ring and the partially positively charged C–H groups in sugar residue (Fig. 17.2) (Fernández-Alonso et al. 2005).

There are numerous methods to explore the mechanisms of lectin–sugar interaction. This ranges from the basic hapten inhibition assay to detect sugar specificity of lectin to the advanced methods like surface plasmon resonance, some of which are briefly outlined below.

17.2.1 Hapten Inhibition Assay

Sugar specificity of lectin gives them the potential to agglutinate erythrocytes which is generally used as the screening assay for lectins. Agglutination of erythrocyte surface sugar chains can be blocked by supplementing simple sugars in solution which shows more affinity to the lectin and the method is known as hapten inhibition assay or hemagglutination inhibition assay. In this method, serial twofold dilutions of the sugar are made in a microtiter plate (U-bottom) to which an equal volume of lectin with four hemagglutination units (HU) is added and incubated at room temperature for 1 h. One HU is defined as the inverse of highest dilution of lectin showing visible agglutination. The incubation period gives enough time for the lectin to interact efficiently with sugar dilutions after which an equal volume of

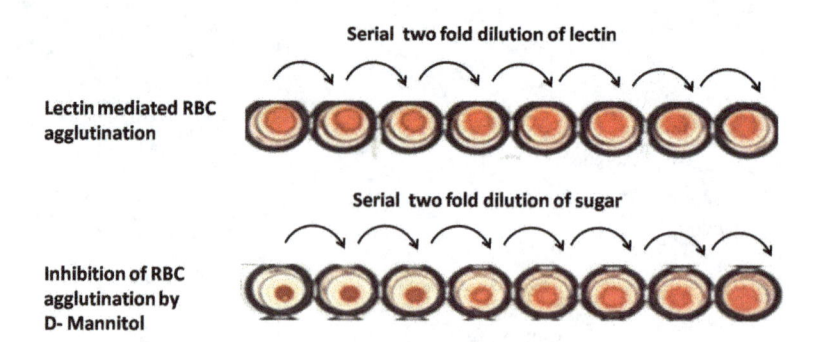

Fig. 17.3 Hemagglutination inhibition of *Calycopteris floribunda* lectin with D-mannitol. The top figure represents agglutination of red blood cells (RBC) mediated by *C. floribunda* lectin and the bottom one shows inhibition of agglutination in the initial four wells. For the inhibition assay, the sugar was serially twofold diluted in phosphate-buffered saline (PBS) and treated with the lectin having four HU for 1 h at room temperature. The inhibition is visualized by incubating the mixture with 2% RBC. D-mannitol showed a minimum inhibitory concentration of about 16 mM. (Unpublished data)

2% erythrocyte suspension in normal saline or phosphate-buffered saline (PBS pH 7.4) is added to all the wells and kept at room temperature for 30 min to develop agglutination. If the initially supplemented sugar is specific to lectin, it saturates the binding sites of this protein, inhibiting its ability to bind with surface sugars of erythrocytes. This inhibited agglutination develops as red button on the bottom of microtiter plate, whereas settling of erythrocyte as a mat represents agglutination by lectin (Fig. 17.3). The hemagglutination inhibition activity is expressed as the lowest concentration of the sugar solution which can completely inhibit hemagglutination and is defined as minimum inhibitory concentration (Valadez-Vega et al. 2011).

17.2.2 Equilibrium Dialysis Method

In this method, lectin and sugar will be placed in individual chambers separated by a membrane that allows only the diffusion of simple sugar units into the lectin chamber. Once equilibrium is reached, the concentration of sugar in both chambers is detected to analyze lectin–sugar affinity. Radioactive labeled or chromophore-labeled sugar molecules are generally used for the trouble-free determination of sugar concentration. Affinity of the reaction is given by Eq. (17.4).

$$\frac{r}{[S]_f} = K_a n - K_a r \tag{17.4}$$

$$r = \frac{[S]_b}{[L]} \tag{17.5}$$

r: Molar ratio of sugar bound to lectin
$[S]_f$: Concentration of free sugar

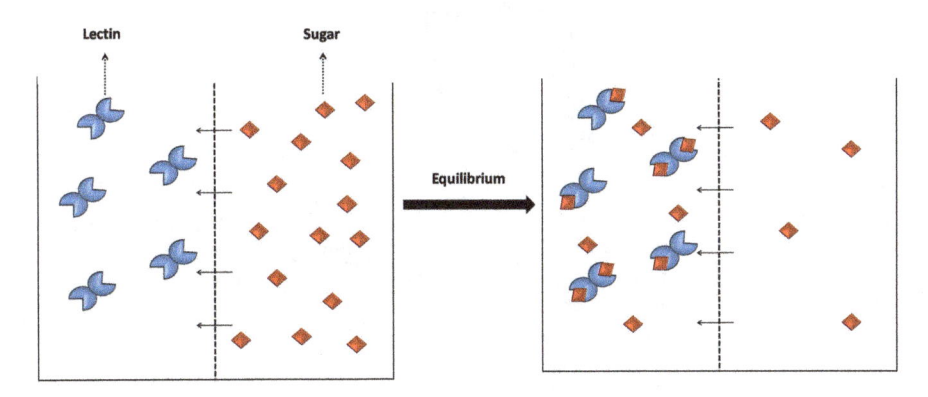

Fig. 17.4 Schematic representation of the principle of equilibrium dialysis. Lectin and sugar solutions are placed in two individual chambers partitioned by a dialysis membrane of appropriate molecular weight cutoff which allows only the passage of sugar molecules. Once equilibrium is reached, some sugar molecules will cross the membrane and bind with the lectin. The amount of free sugar in both the chambers will be same at this point but the total sugar content of lectin chamber will be higher than the chamber containing free sugar solution. By measuring the concentration of sugar in both the chambers, concentration of bound sugar and valency, i.e. number of sugar-binding sites of lectin, can be calculated (Eqs. 17.4 and 17.5) (Hatakeyama 2014)

$[S]_b$: Concentration of sugar bound with lectin
K_a: Binding constant of lectin–sugar interaction
n: Number of sugar-binding sites per lectin molecule
$[L]$: Concentration of lectin used for the study

In the experimental approach, when $r/[S]_f$ is plotted against r at different sugar concentrations, the slope of the line represents $-K_a$, and n is defined as the r intercept at an infinite sugar concentration. The concentration of bound sugar, $[S]_b$ is determined by subtracting the concentration of free sugar, $[S]_f$ from the total concentration of sugar in the lectin chamber (Hatakeyama 2014) (Fig. 17.4).

17.2.3 Isothermal Titration Calorimetric (ITC) Method

This method is highly useful in analyzing the binding parameters of lectin–sugar interaction by measuring the enthalpy change associated with the reaction. A fixed concentration of lectin is titrated with increasing concentrations of specific sugar. The heat absorbed or evolved during the binding reaction is recorded at regular intervals and compared with a reference cell. The heat absorbed or evolved during binding per mole of the sugar molecule in kcal/mol is plotted against molar ratio $[S]/[L]$ of total sugar concentration to draw the binding isotherm. $[S]$ refers to the total sugar concentration and $[L]$ represents total lectin concentration (Cummings et al. 2017). Lectin–sugar interaction is mostly exothermic because of the involvement of several hydrogen bonds (de Bentzmann et al. 2014). Nonlinear least squares analysis

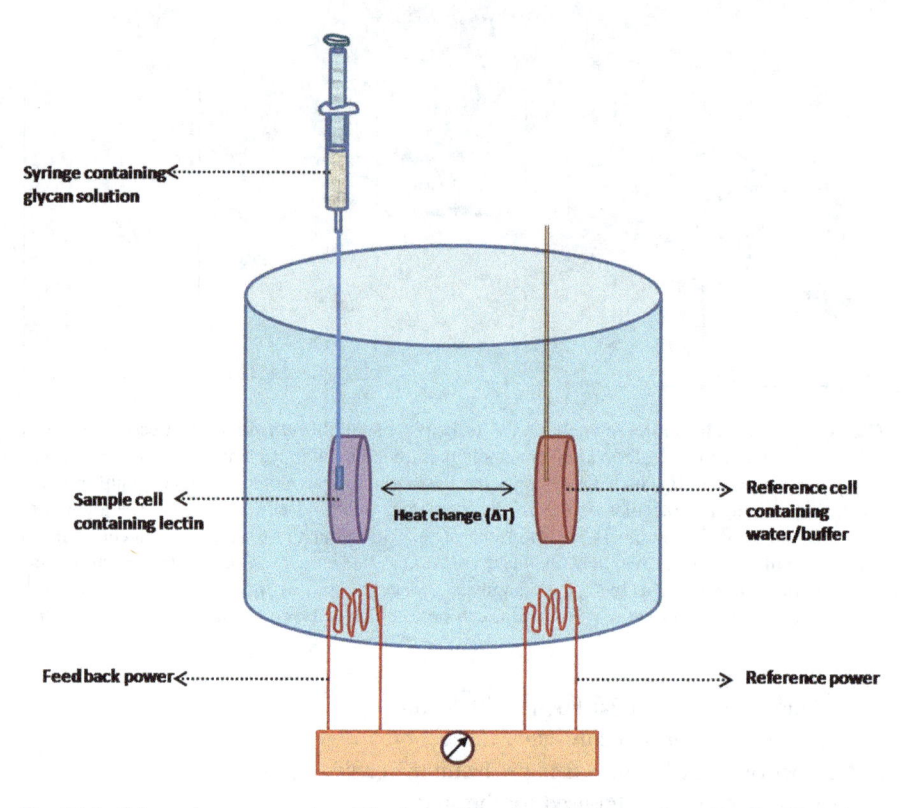

Fig. 17.5 Schematic representation of isothermal titration calorimetry. Sample cell will be filled with lectin solution which will be mixed evenly with the sugar solution released from the syringe at regular intervals. Heat change associated with the reaction is measured with reference to water/buffer used for the experiment and is converted to thermograms (Takeda and Matsuo 2014)

of the binding isotherm could be used to get the thermodynamic binding properties of lectin–sugar interaction. From Eq. (17.3), $\Delta G°$ and $\Delta S°$ can be calculated directly. The number of sugar-binding sites per lectin molecule is described by the Eq. 17.6 (Fisher and Singh 1995; Dam and Brewer 2002) (Fig. 17.5).

$$q = nV\Delta H°[LS_n] \qquad (17.6)$$

Q: Heat absorbed or evolved during binding
N: Number of binding sites per lectin molecule
V: Cell volume
$[LS]$: Concentration of bound sugar

17.2.4 Surface Plasmon Resonance Analysis

Lectin–carbohydrate interactions are a vital feature of all multicellular organisms and the functional basis of these interactions can be investigated by using the technique of surface plasmon resonance (SPR). SPR biosensor system is a new assay system for studying the affinity measurements between a lectin and sugar on a solid surface. SPR is a label-free technique for not only the real-time study of these interactions but also helps us to analyze the kinetic parameters simultaneously. SPR measures the changes of refractive index in the vicinity of the sensor–metal (gold) surface on which the ligand–sugar is immobilized when the analyte–lectin is introduced in the flow cell (Fig. 17.6). As the lectin binds to the ligand, a change in the refractive index on the working surface is recorded which results in a corresponding shift in the SPR frequency in real time (Duverger et al. 2003).

This change in the refractive index at the binding site where the interaction occurs is measured as resonance units (RU). Binding curves which are generated from the resulting RU values obtained are verified by certain algorithms which compare the data obtained to standard binding models and thus help us to determine the various thermodynamics parameters such as specificity, affinity, kinetics, and binding stoichiometry of the interaction under study.

Since the interpretation of lectin and sugar exhibits rapid association and dissociation rates in the form of square-pulse shaped sensorgrams (Fig. 17.7), determination

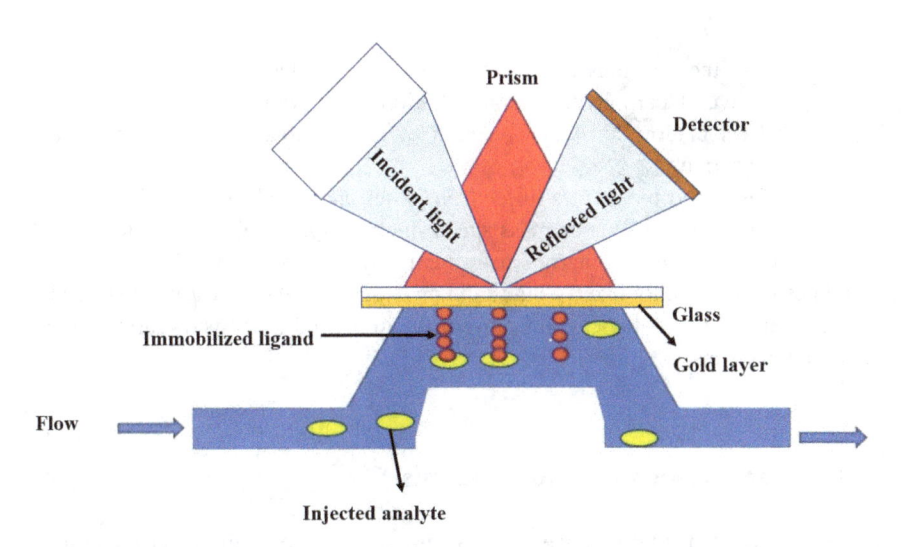

Fig. 17.6 Surface plasmon resonance (SPR) Instrumentation. Prism covered with a sensor chip coated underneath with a gold film on which the sugar is immobilized. Light is irradiated on the metal surface at different angles. Energy from the incident light is transferred present on the metal surface which leads to a reduction in the intensity of reflected light. The angle at which the reflected light is transmitted corresponds to the refractive index at the site of interaction between the ligand and the analyte (Shinohara and Furukawa 2014)

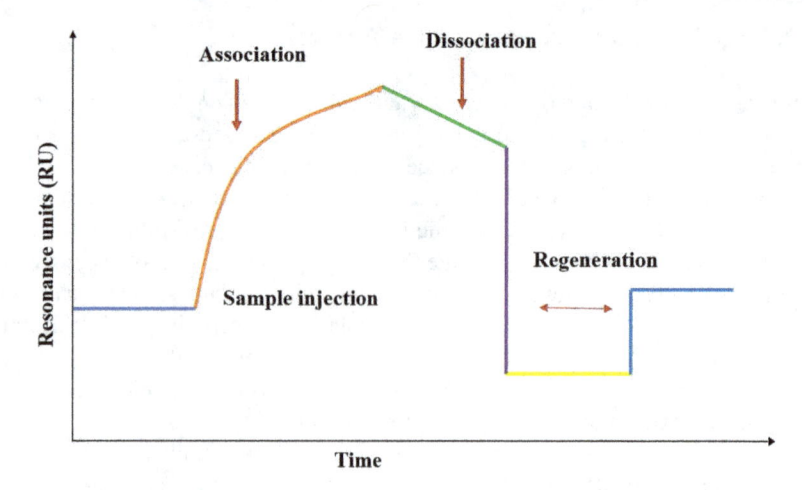

Fig. 17.7 Sensorgram depicting the various phases of a surface plasmon resonance (SPR) experiment. As the lectin is injected into the flow cell, it binds with the immobilzed sugar resulting in an increase in the resonance unit and the change in refractive index is measured (Shinohara et al. 1994)

of rate constant may be difficult due to the speed at which the interaction takes place. Hence, K_a and R_{max} can be determined by the Scatchard Eq. (17.7).

$$R_{eq}/C_o = K_a R_{max} - K_a R_{eq} \qquad (17.7)$$

Here, C_o is the constant concentration of the injected lectin. R_{eq} values are collected at several carbohydrate concentrations. R_{eq} values are plotted against R_{eq}/C_o. The association constant K_a and R_{max} are calculated from the slope and intercept, respectively.

Hence SPR-based biosensors differ from other analytical methods for studying lectin–sugar interaction such as mass spectrometry, high-resolution NMR, fluorescence spectroscopy, and isothermal calorimetry by way that either the lectin or the ligand needs to be immobilized on the solid phase. Hence, it can be asserted that SPR provides a novel technological platform for the improved analysis of various bimolecular interactions.

17.2.5 Fluorescence Titration Analysis

Intrinsic tryptophan fluorescence plays an important role in the determination of lectin–ligand interaction. Since most of the sugar-binding proteins such as lectins and immunoglobulins contain tryptophan, its innate fluorescence changes upon interaction with the sugar (Lee 1997). These conformational or chemical changes in the microenvironment of the tryptophan residues result in the deviations of fluorescence properties which can be monitored by Fluorescence titration analysis.

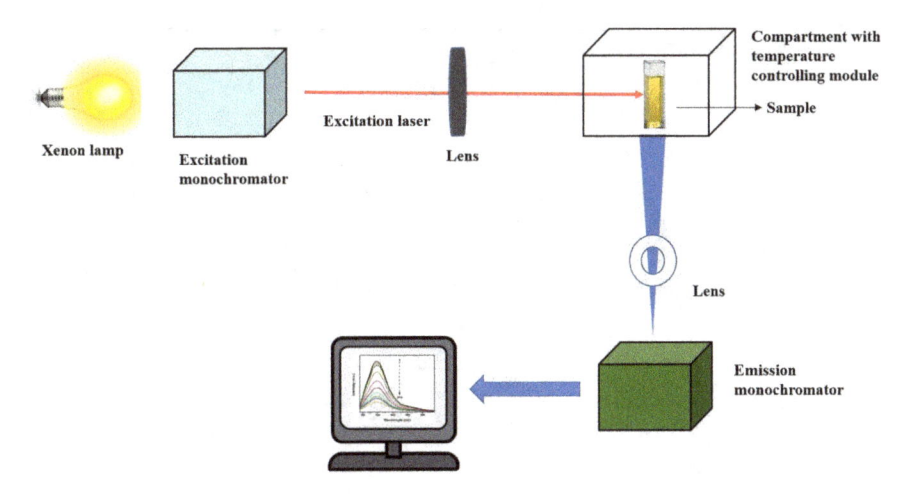

Fig. 17.8 Fluorescence spectrophotometer instrumentation. A typical spectrophotometer consists of two monochromators—one of which is used for excitation which passes only a selected wavelength and an emission monochromator which helps in the fluorescence emission analysis. The data represented are collected at varying concentration of the quencher

In the fluorescence titration analysis, the lectin under study is scanned in the excitation wavelength range varying between 200 and 700 nm. The wavelength at which maximum intensity is observed would be selected for further sugar titration experiments using a fluorescence spectrophotometer. The lectin at a particular concentration is titrated with varying sugar concentrations and the change in the fluorescence emission spectra of lectin would then be noted (Fig. 17.8).

The spectra would be constructed by plotting wavelength (nm) versus fluorescence intensity (Fig. 17.9). The fluorescence spectra would be further elucidated by means of a Scatchard plot using Eq. (17.8) (Fig. 17.10a), which helps us to determine the binding constant (K_a). Quenching constant is determined from the Stern Volmer (SV) plot (Fig. 17.10b) by considering the average integral fluorescence life time of tryptophan to be 4.31×10^{-9} S (Bardhan et al. 2011) using the Eq. (17.9).

$$\log \left[\frac{FO - F}{F}\right] = \log K_a + n \log [Q] \tag{17.8}$$

F: Steady-state fluorescence emission intensity in the absence of the quencher
F: Steady-state fluorescence emission intensity in the presence of the quencher
K_a: Binding constant
N: Slope of the Scatchard plot = Number of Binding stoichiometry
$[Q]$: Molar concentration of the quencher
   (Scatchard 1949)

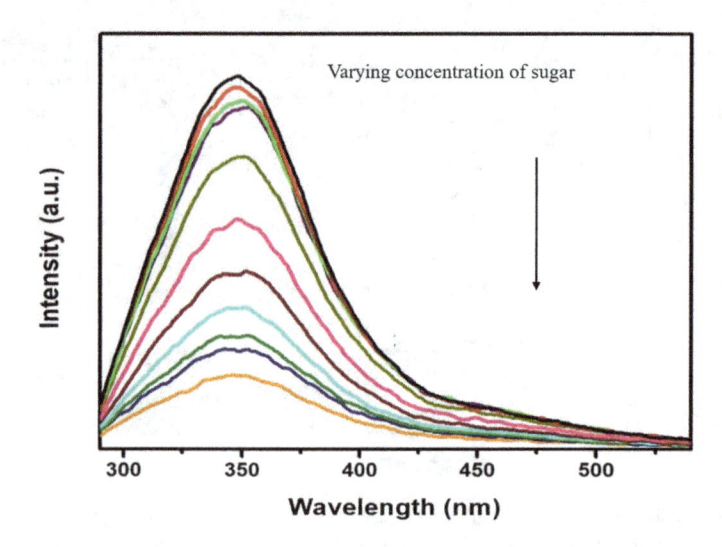

Fig. 17.9 Fluorescence emission spectra of *Leucaena leucocephala* lectin (0.2 mg/mL) in the presence of Glucose at different concentrations ranging from 0.4 mM to 4.8 mM. Each peak represents the fluorescence intensity at different concentration of sugar. The fluorescence emission peak was observed to be quenched regularly with an increase in glucose concentration up to 4.8 mM as depicted here

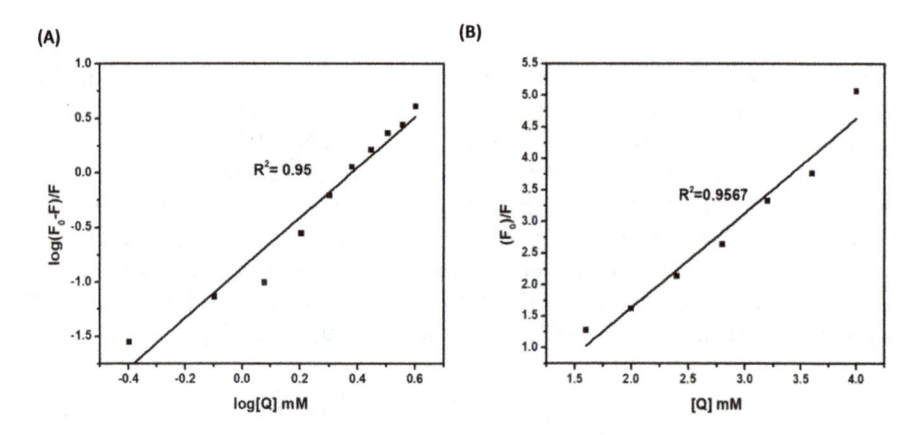

Fig. 17.10 (**a**) Scatchard plot for the binding of D-glucose to *Leucaena leucocephala* lectin and the binding was monitored at 298 K with excitation wavelength set at 280 nm. The binding was monitored at 280 nm for D-glucose. From the slope, K_a value was calculated. (**b**) Stern Volmer plot for the determination of quenching constant and binding stoichiometry which were found to be 3.55×10 M^{-1} S^{-1} and 2.3 respectively (Carvalho et al. 2015). The Stern Volmer constant was found to be 1.53×10^3 M^{-1}

$$\frac{F_O}{F} = 1 + K_{SV}[Q] = 1 + Kq\tau_0[Q] \qquad (17.9)$$

K_{SV}: Quenching constant determined from the slope of the Stern–Volmer plot at lower concentration of the quencher.
K_q: Bimolecular rate constant of the quenching constant
τ_O: Average integral fluorescence life time of tryptophan (4.31×10^{-9} S)
(Chen et al. 2016; Bardhan et al. 2011)

Fluorescence titration is a label-free technique to study lectin–sugar interaction and helps in the determination of binding constant (K_a) and number of binding sites (n). The changes in the fluorescence emission intensity upon sugar binding can be rapidly and efficiently monitored using a small amount of sample. Thus, it is a valuable tool in elucidating the lectin sugar interaction.

Other common approaches to study lectin–sugar interaction includes atomic force microscopy (AFM) which measures the force required to separate a bound sugar molecule from immobilized lectin, NMR spectroscopy that can measure the interaction in real time in solution without separating the lectin–sugar complex from free molecules and finally, mass spectrometry analysis. For a detailed description of these techniques, see Gabius et al. (2011).

17.3 Lectin Bioassays and Other Applications

17.3.1 Antibacterial Activity of Lectins

At present, a large number of antimicrobial agents are available commercially with broad spectrum effect. But the emergence of increased resistance among the microbes against these drugs necessitates the development of new powerful alternatives from natural resources like lectins. These proteins, due to their specificity in sugar binding, are becoming an excellent tool to target microbial surface sugars.

Antibacterial activity of lectins toward Gram-positive bacteria occurs through the interaction with *N*-acetylglucosamine, teichoic acid, teichuronic acid, *N*-acetylmuramic acid, and tetrapeptides linked to *N*-acetylmuramic acid present in the cell wall. Binding of lectins to cell wall lipopolysaccharides is reported in case of Gram-negative bacteria (Ayouba et al. 1994). Isolectin I from *Lathyrus ochrus* seeds interact with the muramic acid and muramyl dipeptide of bacterial cell wall through hydrogen bonds between sugar hydroxyl oxygen and CRD of lectin and hydrophobic interactions with the side chains of residues Tyr100 and Trp128 in the lectin (Bourne et al. 1994). Lectin–bacteria interaction being extracellular, it is thought that they neither alter the membrane structure or permeability nor interfere the normal intracellular processes but exert an indirect effect through these interactions (Peumans and Van Damme 1995). However, there is evidence for the *Araucaria angustifolia* lectin-promoting morphologic alterations, including the formation of bubbling on the cell wall of Gram-negative bacteria and pores in the membrane of Gram-positive bacteria (Santi-Gadelha et al. 2012). The treatment of *Serratia marcescens* with *Moringa oleifera* lectin resulted in the loss of cell integrity and strong leakage of intracellular proteins in a dose-dependent way. *Vicieae* tribe lectins interact with the bacterial cell wall components, muramic acid, *N*-acetylmuramic acid, and muramyl dipeptides (Ayouba et al. 1994). There are reports of lectins from *Dolichos lablab* L., *Triticum vulgare,* and *Bauhinia variegata* L. acting against *Mycobacterium rhodochrous, Bacillus cereus, B. megaterium, B. sphaericus, Escherichia coli, S. marcescens, Corynebacterium xerosis,* and *Staphylococcus aureus* (Sammour and El-Shanshoury 1992). These interactions can also crosslink bacterial cells effecting in their agglutination (Velayutham et al. 2017). A study on *Apuleia leiocarpa* seed lectin (ApulSL) showed the lectin to be bacteriostatic in nature against the Gram-positive bacteria *S. aureus, Streptococcus pyogenes, Enterococcus faecalis, Micrococcus luteus, Bacillus subtilis,* and *B. cereus* and on the Gram-negative bacteria *Xanthomonas campestris, Klebsiella pneumoniae, E. coli, Pseudomonas aeruginosa* and *S. enteritidis*. ApulSL also showed bactericidal effect on three varieties of *X. campestri Zs* (Carvalho et al. 2015). The intercellular communication mechanisms of bacteria known as Quorum sensing (QS) are also targeted by some lectins (Klein et al. 2015; Jayanthi et al. 2017). Bacterial biofilms, which are highly resistant to antibiotics, are also dependent on these QS systems. It is evidenced that lectins from *Canavalia ensiformis, P. vulgaris,* and *Pisum sativum* inhibit biofilm formation by *Streptococcus mutans* (Islam and Khan 2012). A lectin from *Solanum tuberosum* L. was demonstrated to inhibit biofilm formation by *P. aeruginosa* (Hasan et al. 2014). The *Canavalia maritima* (ConM) lectin was shown to reduce expression of genes related with biofilm formation in *S. mutans* cells (Cavalcante et al. 2013).

17.3.2 Antifungal Activity of Lectins

The presence of a rigid and thick cell wall in fungus impairs lectin from penetrating cytoplasm or binding with membrane glycoconjugates and their antifungal action is therefore mediated by indirect mechanisms resulting from the attachment of lectins to chitin or other cell wall glycans (Peumans and van Damme 1995; Adams 2004; Wong et al. 2010). This can include inhibition of fungal growth as a result of poor nutrient absorption and by impairment of spore germination process (Lis and Sharon 1981). They can also induce different morphological changes including swollen hyphae, vacuolization of the cell content, and improved lysis of hyphal cell wall which all can make the fungi more susceptible to different stress conditions (Ciopraga et al. 1999). There are reports of small antifungal lectins too capable of penetrating the fungal cell wall to reach the cell membrane where it can block active sites of enzymes modifying cell wall morphogenesis (Van Parijs et al. 1991; Ciopraga et al. 1999).

A chitinase-free lectin from *Urtica dioica* impeded fungal growth. Cell wall synthesis was interrupted by modification of chitin synthesis and/or deposition (Van Parijs et al. 1991). Antifungal lectins may be grouped as two—the first one includes merolectins with one chitin-binding domain and the latter covers chimerolectins belonging to class I chitinases. Hevein, a 43-amino acid polypeptide from the latex of *Hevea brasiliensis* (Van Parijs et al. 1991) and the 30-amino acid chitin-binding polypeptide from *Amaranthus caudatus* seeds, belongs to the first group and is unable to kill the fungi (Broekaert et al. 1992). Wheat germ agglutinin (WGA) on the other hand could inhibit spore germination and hyphal growth of *Trichoderma viride* (Schlumbaum et al. 1986). Lectins belonging to class I chitinases are only considered as fungicidal because of their enzymatic property. Hence the antifungal action of these proteins relies on their catalytic domain rather than the glycan recognition domain (Hamid et al. 2013). *S. tuberosum* lectin also shows activity against fungi *Rhizopus* sp., *Penicillium* sp. and *A. niger* (Hasan et al. 2014). A galactose-specific lectin from *B. monandra* roots (BmoRoL) showed antifungal activity on plant pathogenic species of *Fusarium*, with more inhibition against *Fusarium solani* (Souza et al. 2011). A jacalin-related, mannose-specific lectin from sunflower seedlings (Helja) could agglutinate *S. cerevisiae* cells and inhibit growth of the pathogenic *Candida tropicalis* and *Pichia genera* (Regente et al. 2014). The treatment of *C. tropicalis* cells with Helja also resulted in the production of reactive oxygen species (ROS). The *Calliandra surinamensis* lectin (CasuL) could inhibit the growth of *Candida krusei* through a variety of mechanisms including retraction of cytoplasmic content and the presence of ruptured cells or cellular debris (Procopio et al. 2017). The authors also observed incomplete budding/division of cells on lectin treatment.

17.3.3 Antiviral Activity of Lectins

The surface of retroviruses like human immunodeficiency virus (HIV) has glycoproteins like gp120 and gp41. The lectins wheat germ agglutinin, *Vicia faba* agglutinin, concanavalin A, *Lens culinaris* agglutinin, and *P. sativum* agglutinin could bind to gp120 and inhibited fusion of HIV-infected cells with CD4 cells (Hansen et al. 1989). Cross linking of these surface glycoconjugates can be mediated by lectins preventing virus interaction with co-receptors (Charungchitrak et al. 2011). Mechanisms underlying the antiviral nature of lectins differ based on their sugar specificity. Mannose-specific lectins can interfere with virus attachment in the early stage of coronavirus replication and also inhibit viral development by binding during the final stage of viral replication cycle (Keyaerts et al. 2007). Also, the extra-long autumn purple bean lectin (Fang et al. 2010a) and *Glycine max* agglutinins (Fang et al. 2010b) were able to inhibit HIV-1 reverse transcriptase activity.

17.3.4 Anti-Insect Activity of Lectins

The conventional insect control agents need to be replaced because of their potency to cause pollution and disturb food chain. Plant lectins offer an excellent tool against insect pests and have been engineered successfully into a variety of crops including potatoes and cereals like wheat and rice. This strategy could be included in the integrated pest management approaches to defeat pest attack (Hamid et al. 2013). An ideal insecticidal agent must resist the proteolytic cleavage within the insect gut. Fortunately, many lectins are resistant to proteolytic inactivation and retain their properties in a wide pH range (Kumar et al. 2012; Lagarda-Diaz et al. 2017). Lectins may delay development and/or adult emergence, increase mortality, and reduce fecundity. Mostly, these effects are analyzed by feeding larvae with lectin-containing food supplements or with transgenic plants over expressing the lectin (Michiels et al. 2010). According to Peumans and van Damme (1995), lectin interacts with insect in three possible ways namely (a) binding with chitin in the peritrophic membrane (PM), (b) binding to the glycoconjugates exposed on the epithelial cells along the digestive tract, and (c) binding of lectins to glycosylated digestive enzymes. Binding to enzymes may result in orchestration of enzymatic activity which can be lethal to the insect. An increase in activity of esterases and a decrease in phosphatase and alkaline phosphatase activity were seen in insects after treatment with different lectins (Hamid et al. 2013). They can also exercise their lethal effects through binding to the peritrophic gel (PG) or the brush-border microvilli of epithelial cells (Dandagi et al. 2006). PM or PG is a film of chitin and other proteins that surrounds the food bolus in many insects. Lectins from *R. solani* and *Sambucus nigra* can cross the PM of the red flour beetle, *Tribolium castaneum* and this property of reaching the endoperitrophic space is highly dependent on the size of lectin and the charge distribution and dimensions of PM pores (Walski et al. 2014). When the insect lacks a PM, then the lectin can directly interact with the epithelial surface glycoconjugates (Roy et al. 2014). Sometimes the lectins are internalized by interaction with glycosylated metabolic enzymes or proteins.

A lectin from garlic leaf was one among them which was internalized by binding to a glycosylated alkaline phosphatase anchored to *Helicoverpa armigera* midgut membrane affecting the survival and development of the moth. Also, the *G. nivalis* lectin was able to cross the midgut epithelium of *Nilaparvata lugens* and it was co-localized in the fat body and hemolymph. This co-localization was presumed via a carrier–receptor, ferritin. The *Colocasia esculenta* tuber agglutinin was also reported to be internalized into *Lipaphis erysimi* hemolymph. The receptors involved in this reaction include vacuolar ATP synthase, heat shock protein 70, and clathrin heavy chain (Roy et al. 2014). The *Arum maculatum* tuber lectin acted against *L. erysimi* and *Aphis craccivora* by binding to the gut brush border membrane vesicle proteins (Majumder et al. 2005). An artificial diet containing *B. monandra* leaf lectin caused mortality in *Z. subfaciatus* and *Callosobruchus maculatus*. The lectin was found to bind with the midgut proteins of *C. maculates*. A 40% reduction in the weight of *A. kuehniella* larvae was also observed with the same lectin (Macedo et al. 2007). The leaves of transgenic tobacco plants expressing *Allium sativum* lectins resulted in a reduction in the weight gain, development, and metamorphosis of *Spodoptera littoralis* larvae (Sadeghi et al. 2008). The receptors involved in the antimicrobial and anti-insect activities of lectins and the associated responses are summarized in Fig. 17.11.

17.3.5 Mitogenic Activity of Lectins

Lectins have garnered the attention due to its immense potential with respect to various biological activities, which includes their ability to proliferate several cell types. The ability of the lectins to interact with the various cell types has been attributed to the presence of carbohydrates on the surface of cells. Due to their ability to recognize carbohydrate structures present on the cell surface, they can be used as tools to study the alterations in the number and distribution of receptors associated with cell growth, proliferation, and cell–cell interactions.

The unique property of lectins to be mitogenic agents has opened a new area of research for scientists to investigate the role of lectins in cell growth and development. Certain plant lectins are deemed to be mitogenic owing to the fact that they can stimulate the transformation of cells from the resting phase to blast like cells which may undergo subsequent mitotic division (Ashraf and Khan 2003). Lectins can either positively or negatively stimulate the lymphocytes, which are the usual targets to evaluate the mitogenic capacity of lectins. A number of lectins have been characterized from plants which possess the unique property to stimulate cell growth. The best examples of such lectins include Phytohemagglutinin (PHA), Concanavalin A (Con A), *Morus alba* lectin (MAL), *G. max,* and *Arachis hypogaea* lectin. On the other hand, certain lectins have been observed to inhibit cell growth such as the lectin from *Vigna sesquipedalis*, wheat germ agglutinin, and *Viscum album* agglutinin. Majority of the lectins studied are mitogenic against T lymphocytes, whereas the Pokeweed mitogen (Pa-1) is found to stimulate both the T cells and B cells (Bekeredjian et al. 2012). Hence the study of lectin–lymphocyte interaction has facilitated in the better understanding of the mechanism of lymphocyte activation and control.

It has been postulated that the stimulation of lymphocytes by lectins occurs in a non-preferential manner, i.e., the lectin binds to all lymphocytes possessing the same sugar moieties under appropriate conditions (Carvalho et al. 2018). A complex series of events occurs as a result of lectin–lymphocyte binding. The earliest noticeable changes occur with respect to increased plasma membrane permeability for a number of metabolites such as amino acids, glucose, and ions such as K^+ and Ca^{2+}. The activation and proliferation of lymphocytes can be observed by the in vitro culturing of these cells in the presence of lectins. It has been postulated that lectins activate lymphocytes by cross linking to T-cell receptor (TCR), irrespective of the antigen. Lectin-induced lymphocyte activation can result in the stimulation of certain molecules such as p38 mitogen-activated protein kinase (MAPK) which in turn leads to the activation of a family of transcription factors such as signal transducers and activators of transcription (STAT) pathway (Pujari et al. 2010). In case the lectins

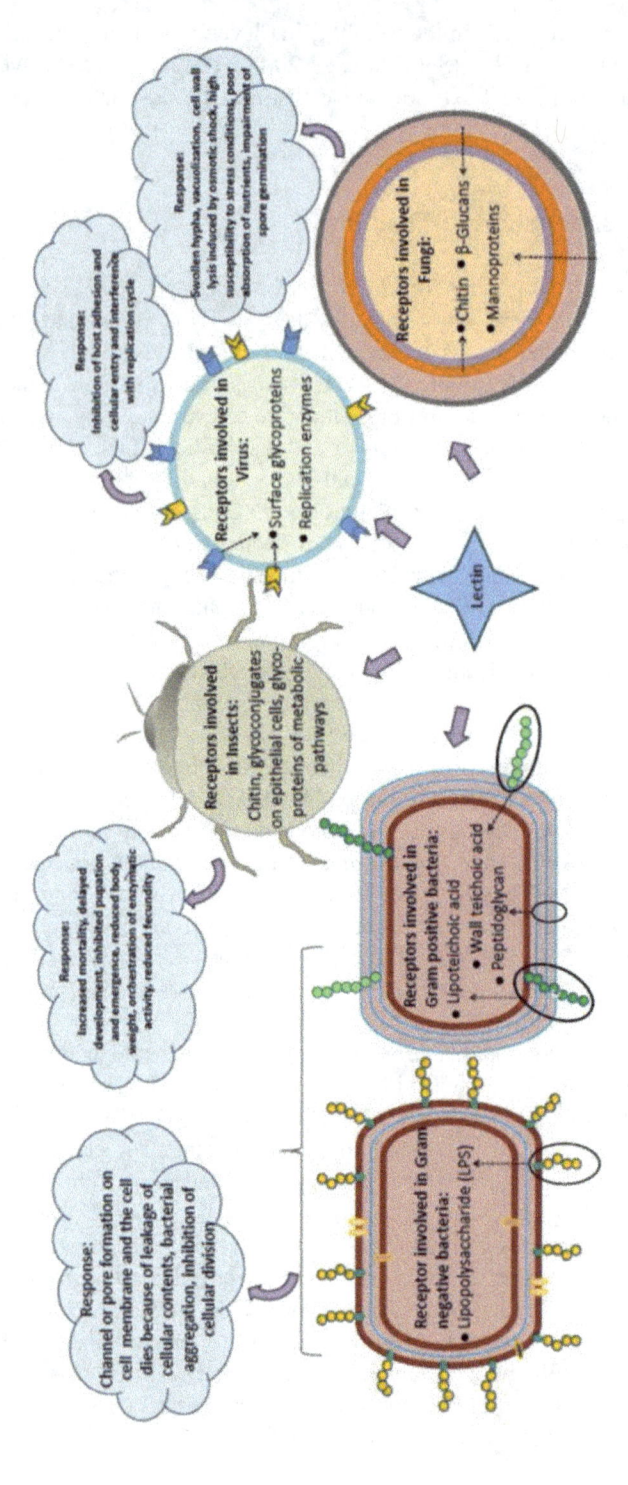

Fig. 17.11 Schematic representation of the antimicrobial and anti-insect properties of lectin. Lectin interacts with the glycoconjugates on the cell surface of bacteria, fungi and viruses and impairs with their host attachment or invasion mechanisms. In insects, they additionally can alter the enzymatic activity by binding with glycosylated enzymes or proteins in their metabolic pathways ultimately alarming the insect life

are pretreated with their specific glycoconjugates, they lose the property of inducing cell proliferation thereby confirming the role of sugar specificity in mitogenic ability. This property was well illustrated in the case of a seed lectin from *Artocarpus lingnanensis* (ALL) that promoted human T lymphocyte proliferation by a receptor-like tyrosine phosphatase CD45 signaling pathway. CD45 is a glycosylated receptor and a vital factor of the TCR signaling pathway. Binding of ALL to T lymphocytes was abolished after the treatment of the specific carbohydrate, N-Acetyl-D-galactosamine (GalNAc) (Cui et al. 2017). In the case of the best-studied mitogenic lectins such as Con A and PHA, their binding on the T-cell receptor (TCR) occurs in a similar manner as the receptor-mediated recognition of their APC surface ligands. As a result of the binding of the mitogenic lectin to the TCR, the tyrosine kinase domains would autophosphorylate and become activated which as a result, phosphorylates several tyrosine phosphoproteins which include p36/p38 which upon activation associates to the GRB2. GRB2 binds to SOS and forms the p36/38-GRB2–SOS complex which in turn activates RAS. Activated RAS stimulates a cascade of phosphorylation reactions performed by MAPK (Fig. 17.12).

As mentioned previously, a change in the chemical environment of the cell–surface receptors occurs as a consequence of lectin binding. The initial changes involve a difference with respect to the membrane phosphoinositide turnover. Due to the hydrolysis of phosphatidyl inositol 4,5-diphosphate (PIP2), secondary messengers such as diacylglycerol (DAG) and inositol 1,4,5-triphosphate are produced, thereby leading to an increased cytosolic concentration of calcium ions. Lectin-induced mitogenic activity also leads to the increased synthesis of specific biologically active peptides such as lymphokines. One of the well-studied examples of this mechanism is observed in the case of wheat germ agglutinin, which leads to the production of interleukin 2 (IL-2) which is one of the thoroughly studied lymphokine (Kawakami et al. 1988). In certain instances, an alteration of lymphocyte surface by glycosidases is a prerequisite for stimulation by certain agglutinins. For example, peanut agglutinin lectin (PNA) is capable of mitogenic stimulation of rat and human lymphocytes along with HT-29 and SW-1222 cells. PNA acts by binding to Thomsen–Friedenreich (TF) oncofetal carbohydrate antigen which is observed abundantly in the case of colon cancers and inflammatory bowel disease (Zhao et al. 2014). Therefore, an in-depth understanding of the molecular signaling pathway and biochemical cascade is a prerequisite for elucidating the general mechanism of action of lectins.

Hence, mitogenic stimulation by lectins can be used as a diagnostic tool for the detection of congenital and acquired immunological deficiencies by the detection of sensitization caused due to infectious agents and autoimmune diseases.

17.3.6 Lectin Histochemistry

Plant lectins have indicated to play a significant role as markers in histochemical studies involving various cell types. Due to their inherent glycan-binding properties, lectins have replaced the conventional antibodies used as immunochemical markers

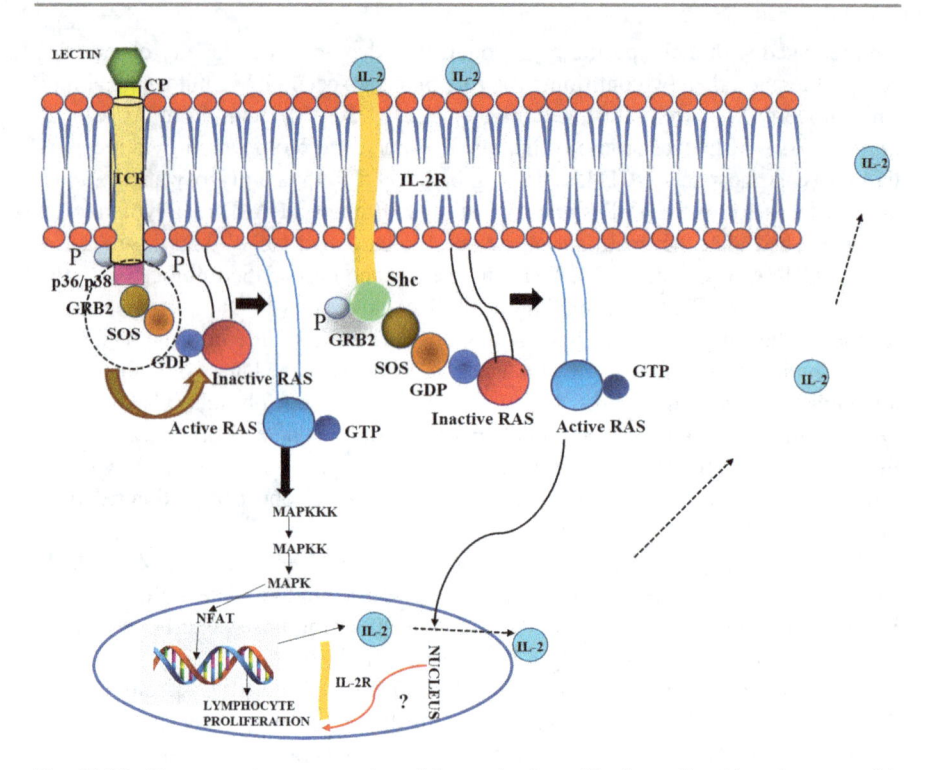

Fig. 17.12 Diagrammatic representation of the mechanism of lectin mediated lymphocyte proliferation: CP: carbohydrate portion; TCR: T-cell receptor; P: phosphate group; p36/p38: p36/38 proteins; GRB2-receptor-mediated protein; SOS–SOS protein; GDP: guanosine diphosphate; RAS–RAS protein; GTP: guanosine triphosphate; MAPK: mitogen-activated protein kinase; NFAT: nucleated factor of activated T-cells; IL-2: interleukin 2; Shc protein: Shc protein. Lectin upon binding to the carbohydrate portion of the T-cell receptor results in the activation of various proteins of the STAT pathway. P-36/38–GRB2–SOS complex activates inactive RAS to active RAS which further activates the MAPK pathway. This stimulates NFAT which results in the transcription of IL-2 genes. IL-2 binds to IL-2R which in turn phosphorylates Shc protein which binds to GRB2 which in turn binds to SOS. This complex further activates RAS which further stimulates a series of events in the cell, the mechanism of which needs further understanding (Carvalho et al. 2018)

to characterize normal and diseased cell types. A number of lectins especially those purified from seeds are used as markers (Table 17.4). The best-studied examples of plant lectins used in the histochemical studies include Con A (Concanavalin A from *C. ensiformis*), *L. culinaris* lectin, Peanut agglutinin (PNA), *Dolichos biflorus* (DBA), and *Helix pomatia* lectin (HPA).

Changes in glycosylation patterns are the hallmarks of metastasis and cancer progression. Hence in this context, lectins have been used as candidates in the prognosis of the disease. Lectin histochemistry is a microscopy-based method for the visualization of cell surface changes using lectins instead of antibodies. Due to their apparent glycan-binding properties, lectins are used in the detection of glycan

Table 17.4 A short list of plant lectins used as histochemical markers

Plant lectin	Saccharide specificity	Cancer type	Reference
Parkia pendula lectin (PpeL)	Glucose/mannose	Meningothelial tumor	Beltrão et al. (2003)
Ricinus communis (RCA-1)	Lactose	Microglial cells in human brain	Mannoji et al. (1986)
Lycopersicon esculentum lectin (LEA)	*N*-acetyl glucosamine	Normal and tumoral blood vessels of central nervous system	Mazzetti et al. (2004)
Concanavalin A (con A)	Glucose/mannose	Mucoepidermoid carcinoma, detection of glycosylation levels of CA15-3 antigen in breast cancer	Sobral et al. (2010), Choi et al. (2018)
Ulex europeus lectin (UEA)	Fucose, chitobiose	Mucoepidermoid carcinoma	Sobral et al. 2010
Lens culinaris (α-fetoprotein-AFP)	Glucose/mannose	Diagnostic and prognostic marker for hepatocellular carcinoma (HCC)	Leerapun et al. (2007)
Peanut agglutinin (PNA)	Galactose/lactose	Histochemical marker for pheochromocytomas	Moorghen and Carpenter (1991)

moieties on the cell surface as well as in the cellular compartments. Thus lectins play a pivotal role in the detection of diseases. A number of glycosylation changes are associated with cancer progression such as sialylation, fucosylation, increased branching of *N*-glycans, and overexpression of mucin type-O glycans (Pinho and Reis 2015).

Labeling using lectins can be of two main types—(1) Direct labeling method and the (2) indirect labeling method. In the direct labeling method, the lectin is directly bound to the fluorophore, enzyme, or colloidal metal. In the case of indirect labeling method, lectin is conjugated to biotin or digoxigenin and thereafter detected by using enzyme-linked streptavidin or anti-digoxigenin (Fig. 17.13). The tissue sections have to be processed in an efficient manner for fixation and tissue embedding. Chemical fixatives such as paraffin and formaldehyde should be avoided for tissue fixation because it can denature and weaken lectin binding.

Cancer being one of the most prevalent diseases, the search for new biomarkers is still prevalent. There are only a handful of biomarkers available for the detection of cancer and hence needs approval from US Food and Drug Administration (FDA) to be used in the clinical background. CA-125 is used as a biomarker for ovarian cancer and is found to be elevated in other cancers of the breast, lung, colon, and rectum. Serum/plasma is the most widely used sample used for clinical analysis. However, due to the large abundance of proteins present in the serum, specialized techniques in genomics, proteomics, and glycomics need to be combined for the identification of more sensitive and specific cancer biomarkers.

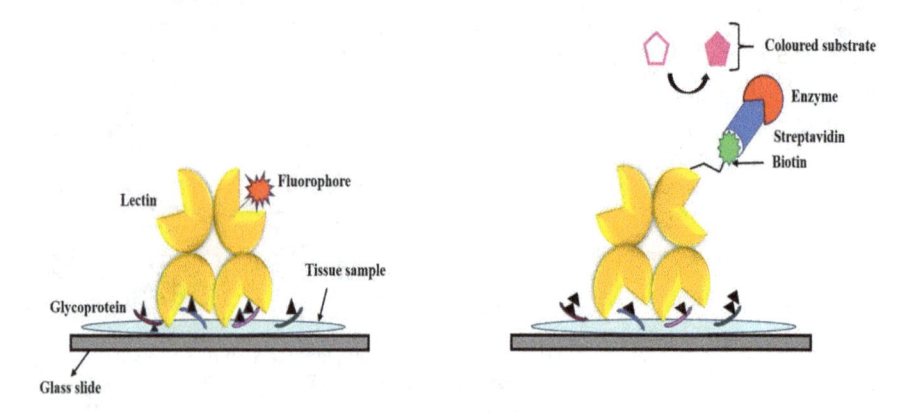

Fig. 17.13 Schematic representation of lectin histochemistry. (**a**) Direct labeling method. Lectin is covalently linked to fluorophore, colloidal gold or enzyme, and the glycoproteins present on the tissue surface are detected by the respective microscopic method. (**b**) Indirect labeling method. Lectin is conjugated to biotin or digoxigenin which is then ultimately recognized by the corresponding enzyme-linked streptavidin or anti-digoxigenin (Hashim et al. 2017)

17.3.7 Immobilized Lectin-Based Affinity Chromatography

Interaction between lectin and glycan is generally weak involving mainly non covalent interactions with a dissociation constant in the range 10^{-3} to approximately 10^{-6} M (Liang et al. 2007). Hence, it can be efficiently used for the separation of glycoconjugates or glycans and the bound substance can be easily eluted using a competitive binding molecule. Also, lectin being non catalytic in nature, modification of the bound substance is restricted. The purified glycan could be then analyzed by mass spectrometry which in cancer research aids to identify the potential glycan marker associated with cancer (Hashim et al. 2017). In 1970, Donnelly and Goldstein developed the first lectin-based affinity column by immobilizing Con A on Bio-Gel P-10. The lectin was cross-linked using glutaraldehyde treatment before mixing them with Bio-Gel P-10 (Donnelly and Goldstein 1970). Con A is having specificity toward *N*-linked α-mannose and α-glucose and hence can be used for the separation of glycoconjugates containing these residues. A general workflow of the lectin-based affinity chromatography is given in Fig. 17.14.

Lectins are not as specific as antibodies and can bind with two or more sugars in an affinity column (Durham and Regnier 2006). So, a serial lectin affinity chromatography (SLAC) could be used to further fractionate them. A combination of Con A (specific to mannose) and Jacalin (specific for GalNAc in O-glycosylation and mannose *N*-type glycans) columns was selected for the study of O-glycosylation sites on human serum proteins (Durham and Regnier 2006). Affinity columns made of Con A, *P. sativum* (PS) lectin, WGA, phytohemagglutinin E4 (PHA-E4), and phytohemagglutinin L4 (PHA-L4) were used in series to compare the sugar-chain structural changes of serum PSA between prostate carcinoma (PCA) and benign prostatic hyperplasia (BPH). The lectin column, separated PSA into seven fractions

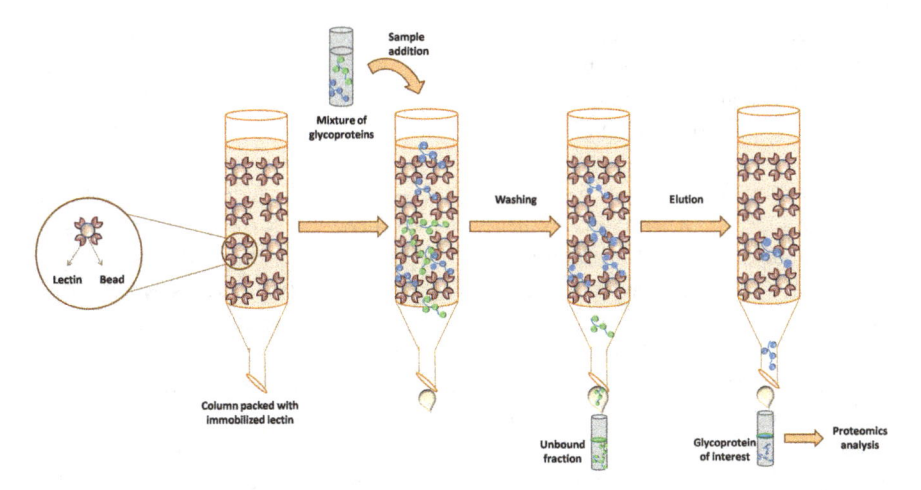

Fig. 17.14 Schematic work flow of immobilized-lectin-affinity chromatography. Lectin of known sugar specificity is immobilized on a gel matrix. After equilibration of the lectin matrix with the working buffer, sample glycoconjugates are added. Unbound fractions were collected by washing the column with working buffer and the bound glycoprotein is eluted with specific competing glycan solution which can be then subjected to proteomics analysis like mass spectrometry (Hashim et al. 2017)

and the relative amount of a multiantennary complex type PSA with branched GlcNAc β (1 → 4) mannose, was observed to be high in PCA (Sumi et al. 1999). A high level of the serum marker α-fetoprotein (AFP) is used for the diagnosis of hepatocellular carcinoma (HCC), but the method is found to be less sensitive (Debruyne and Delanghe 2008). Lectins are used for the isolation and characterization of different glycoforms of this marker which is presumed to increase the specificity. Con A affinity chromatography separated the AFP into AFP-C1 and AFP-C2, *L. culinaris* agglutinin-(LCA) based matrix separated AFP into AFP-L1, AFP-L2, and AFP-L3, whereas the *P. vulgaris* erythro-agglutinating phytohemagglutinin (PHA-E) chromatography fractionated the protein into AFP-P1, AFP-P2, AFP-P3, AFP-P4, and AFP-P5 (Taketa 1990). Among these, AFP-L3 which can be recognized by LCA is found to be specifically produced by cancer cells (Li et al. 2001). The method is found to have a sensitivity and specificity of 80–90% and more than 95% respectively over the method of total serum AFP detection method with a sensitivity of 40–65% and specificity of 76–96% (Debruyne and Delanghe 2008).

17.3.8 Lectin Blotting

Lectin blotting is similar to immunoblotting with the difference that specific antibodies are used in the latter approach instead of lectins. Lectin blotting serves as a reliable method to identify glycoproteins based on the specificity of lectin to a particular glycan. Glycoproteins after electrophoretic separation are blotted on to nitrocellulose membrane or polyvinylidene fluoride (PVDF) membrane, treated initially with biotinylated specific lectin and then with streptavidin-conjugated enzyme solution. The bands are visualized by supplementing suitable substrate (Morgan et al. 2013). Visualization of lectin complex is also possible by using conjugates of fluorescent dyes, colloidal gold, or radioactive isotopes. The technique could be used for the analysis of specific glycosylation patterns associated with cancer progression. Lectin blotting in combination with mass spectrometry helps for a better understanding of the identity of glycoprotein being analyzed.

In a study by Rambaruth et al. (2012), the lectin stained bands were further analyzed by mass spectrometry and demonstrated an increased expression of heat shock protein 27 (Hsp27), D-like HnRNP, HnRNP A2/B1, and enolase 1 (ENO1) in the metastatic cells of MCF7 and T47D. Cells in G1 phase were reported to be more sensitive to Influenza virus infection than S/G2/M phase cells. The underlying mechanism behind this was explored making use of lectin blotting to analyze sialic acid expression in cell membranes as this is the glycan determinant of Influenza surface hemagglutinin. Sialic acid specific, digoxigenin-labeled *Maackia amurensis* lectin (MAL) and *S. nigra* agglutinin (SNA) were used to detect α2–3 and α2–6 linked sialic acid molecules by lectin blotting. The study could demonstrate increased expression of both α2–3 and α2–6 linked sialic acid in G1-phase cells (Ueda et al. 2013). A study by Qiu et al. (2008) used biotinylated SNA lectin for the analysis of differential *N*-linked sialylation of complement C3 in colorectal cancer patients, compared to normal subjects. Aberrant fucosylation of haptoglobin β chain associated with colon cancer progression was studied by another research group using *Lotus tetragonolobus* lectin (Park et al. 2012).

17.3.9 Lectin Microarray

Lectin microarrays were first described in 2005 and became an important analytical tool in glycan analysis (Hirabayashi et al. 2014). In this method, several lectins of known specificity are immobilized as microdots on a solid support like glass slide by physical adsorption or covalent bonding. Pretreatment of glass slides is usually done with *N*-hydroxy succinimidyl esters (Hsu and Mahal 2006) or epoxides (Kuno et al. 2005). The prepared slide is kept in place by a multi-well gasket, allowing the addition of samples into each well (Hashim et al. 2017). Analytes can include glycoproteins, cells, or bacteria which are usually labeled with a fluorophore or antibody before treating with the microarray slide. The samples will bind to the respective lectin on microarray chips depending on the type of glycan residue present and are visualized by a confocal-type fluorescence scanner (Fig. 17.15). Unbound molecules are removed by washing which needs to be done with care as it can decrease the sensitivity due to the most weak interactions between lectin and glycoprotein (Liang et al. 2007). Another detection approach is evanescent-field-activated fluorescence detection. Here, excitation light is applied at a critical angle from both sides of the glass slide to attain total internal reflection at the glass–liquid interface. An evanescent wave is thus generated from the surface in a limited space

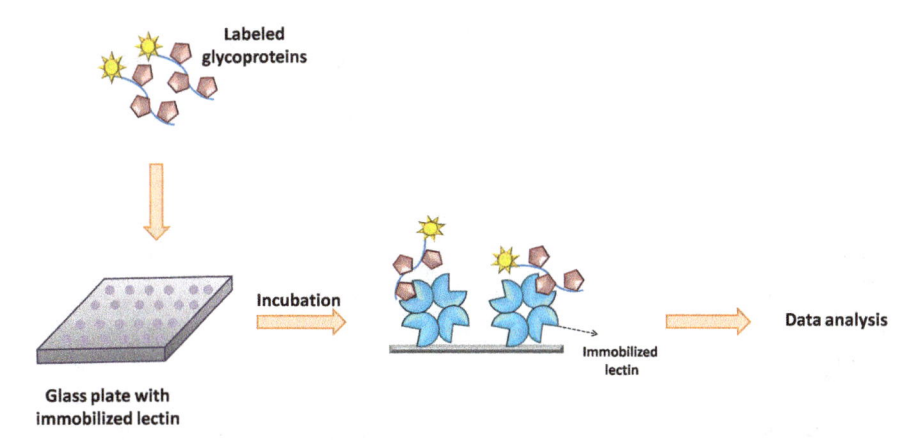

Fig. 17.15 Diagrammatic representation of lectin microarray. Specific lectins are immobilized on a solid support like glass surface by covalent bonding or physical adsorption. Fluorescent labeled analytes are then incubated with the microarray chip and analyzed by a suitable detection system like confocal type fluorescence scanner and proteins having glycan residues specific for the lectin will give a signal

(near-field optic) enabling the observation of liquid phase. The approach could be used for the analysis of glycosylated samples with increased sensitivity and specificity (Hirabayashi 2014).

Lectin microarray was used in a study to analyze glycosylation differences in prostate-specific antigen (PSA) molecules of healthy persons and oncological patients. They demonstrated that the lectins, SNA, and Jacalin could efficiently discriminate aggressive cancer forms (Li et al. 2011). Another study by Tateno et al. used lectin microarray systems for the complete analysis of the glycan content of human-induced pluripotent stem cells (iPSCs) from four types of somatic cells (SCs). Result was compared with the glycan profiles of nine types of human embryonic stem cells (ESCs). The study illustrated the difference between the glycosylation of somatic and embryonic stem cells and also concluded that on induction of pluripotency, SCs attain glycans specific for embryonic stem cells (Tateno et al. 2011). Lectin microarrays could also help to discover new biomarkers. For example, for predicting the progression of diabetic nephropathy, fetuin-A was discovered as a possible urine marker by applying samples from patients to lectin microarrays (Inoue et al. 2013).

17.3.10 Enzyme-Linked Lectin Assay (ELLA)

Enzyme-linked lectin assay was introduced by McCoy et al. (McCoy Jr et al. 1983) that works on the same principle of enzyme-linked immunosorbent assay (ELISA) with the difference of using enzyme-conjugated lectin in the assay procedure. In the direct assay method (Fig. 17.16a), samples containing glycoconjugates are coated directly on the wells of a microtiter plate, followed by addition of an

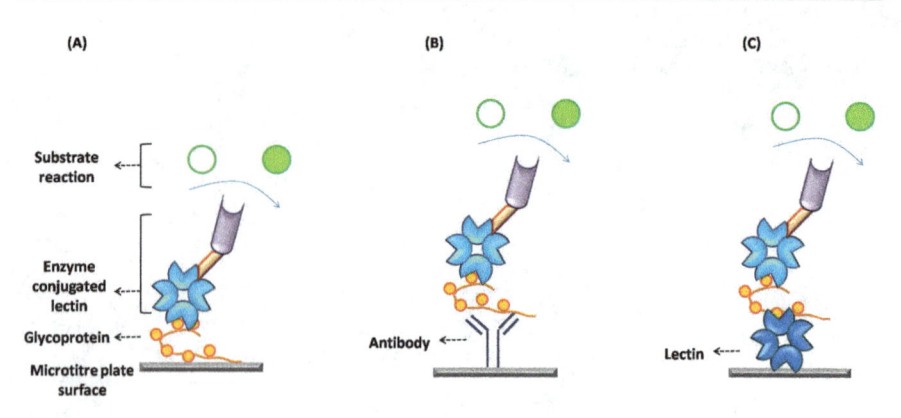

Fig. 17.16 Diagrammatic representation of the three different approaches for enzyme-linked lectin assay. (**a**) Direct assay method in which analyte is immobilized on microtitre plate well and incubated with enzyme-conjugated specific lectin. (**b**) In this hybrid method, antibody specific for a protein of interest is initially coated on the well surface which will be treated with the sample mixture. Only the protein of interest will be retained by the antibody and unbound fractions are washed off. The complex is then treated with specific lectin to analyze the glycosylation pattern of the bound protein. (**c**) Sandwich enzyme-linked lectin assay involves the use of two different lectins. The first lectin will be immobilized on the well surface which will act as the capturing agent while the second lectin plays the role of detection. For all these methods, the bound glycoprotein is visualized by enzyme-conjugated lectin which when treated with a suitable substrate gives a characteristic color reaction (Hashim et al. 2017)

enzyme-conjugated lectin specific to the glycan immobilized. On addition of a suitable substrate, the enzyme converts the colorless substrate to a colored product, which can be quantified by a spectrophotometer and is directly proportional to the amount of coated glycoconjugates that reacted with lectin (Kuzmanov et al. 2013). In 2014, Couzens et al. used ELLA for the determination inhibitory antibodies to the Influenza virus surface neuraminidase in human sera. These antibodies are thought to inhibit release and spread of new virions from infected cells by selectively impairing the activity of viral neuraminidase. Galactose-specific peanut agglutinin is used for this purpose which can selectively bind to the terminal galactose exposed by the removal of sialic acid residue through the action of neuraminidase (Couzens et al. 2014).

A study by Reddi et al. (2000) reported the use of enzyme-linked peanut agglutinin (PNA) assay for the estimation of the levels of Thomsen–Friendenreich antigen (T-Ag) in sera of patients with squamous cell carcinoma of the uterine cervix. The study concluded the presence of higher levels of T-Ag in the sera of uterine cervical cancer patients compared to normal ones and the expression was directly proportional to the aggressiveness of the cancer. Antibody and lectin can be used in combination in a variant of ELLA known as the hybrid approach (Fig. 17.16b) to analyze glycosylation of a specific protein (Kim et al. 2008). Here, an antibody specific to a protein is coated directly on the wells of a microtiter plate which helps to pre-capture the protein of interest from a complex mixture. Enzyme-conjugated lectin is then added to detect glycosylation of the bound protein.

One of the drawbacks of this technique is binding of lectin to the glycosylated residues of antibody which can result in high background staining reducing sensitivity. Another approach which is described by Lee et al. (2013) follows a sandwich enzyme-linked assay which uses two different lectins both binding to O-glycan structures of glycoproteins (Fig. 17.16c). The assay used Champedak (*Artocarpus integer*) galactose-binding lectin (CGB) lectin as capturing reagent and enzyme-conjugated jacalin as detection probe. When the assay was used to measure the levels of mucin-type O-glycosylated proteins, an increased expression of O-glycosylated proteins was found in the sera of both stage 0 and stage I breast cancer patients compared to normal individuals. (Lee et al. 2016).

The method of ELLA could also be used for the analysis of glycoproteins in tissue lysates. When breast cancer tissue lysates were analyzed, an increased interaction was seen with the lectins ConA, *R. communis* Agglutinin I, AAL, and *M. amurensis* lectin II (MAL II) compared to normal subjects demonstrating enhanced mannosylation, galactosylation, sialylation, and fucosylation of glycoproteins in the breast cancer tissues (Wi et al. 2016).

17.3.11 Lectins as Biosensors

Detection of pathological conditions and onset of diseases can be monitored by glycoproteins and glycan-based biomarkers from biological fluids and cells. Lectin-based biosensor method is based on the measurement of signals generated as a result of lectin–carbohydrate interaction. They differ from the conventional enzyme-based and fluorescent-tagged systems in the non-requirement of any labels for the interaction studies. Despite the availability of different biosensing methods, electrochemical biosensors are more superior and user-friendly and find numerous applications in detecting pathogens.

Electrochemical-based biosensors use electrodes coated with polymers and nanoparticles as sensing surfaces for immobilizing lectins and detect the lectin–ligand interaction by measuring the changes in charge transfer resistance after interactions by electrical impedance spectroscopy (EIS) (Fig. 17.17).

Con-A lectin has been widely used for various biosensor experiments for the detection of glucose even at a low concentration in a linear range from 1.0×10^{-6} to 1.0×10^{-4} M. Hence this biosensor can be used to detect the blood glucose levels in diabetic individuals. The main benefit in using this Con A-based biosensors lies in the fact that certain enzymes such as glucose oxidase (GOx) and Horse-radish peroxidase (HRP) can be immobilized without the requirement of any labels due to the presence of intrinsic hydrocarbon chains. Lectins not only serve as molecular glue for protein immobilization but also as recognition elements in biosensors. In recent times, food and water safety and detection of pathogenic microorganisms are of great importance. Lectin-based biosensors have replaced the conventional culturing and colony-counting methods due to their efficiency and being less laborious. An *E. coli* sensor containing an Au electrode coated with a quinone-fused poly (thiophene) film followed by substituting with mannose residues has been recently

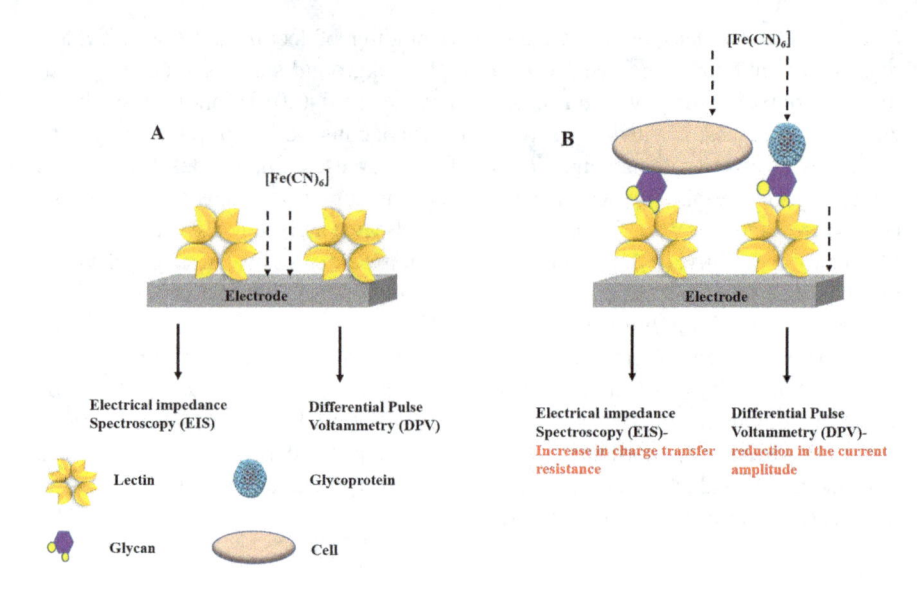

Fig. 17.17 Diagrammatic representation of lectin-based electrode surface (**a**) before and (**b**) after binding. In this system, lectin–glycan interactions are measured using an electrode coated with a redox probe $[Fe(CN)_6]^{3-/4-}$. The reduction or oxidation conditions produce signals in the form of charge transfer resistance for electrochemical impedance spectroscopy (EIS) and current for differential pulse voltammetry (DPV) thereby promotes monitoring electrode surface interactions– EIS (Coelho et al. 2017)

reported (Ma et al. 2015). In this biosensor, *E. coli* can be captured on the gold (Au) electrode by either direct binding or Con A mediated binding. The bacteria can be detected by either electrochemical or gravimetric measurement. In the electrochemical response mode of detection, the increase in the *E. coli* concentration is measured by the corresponding decrease in the peak current of the sensor, whereas in the gravimetric method, the resonance frequency of the sensor altered based on the *E. coli* concentration (Fig. 17.18).

Bacterial toxins can also be detected by lipopolysaccharides (LPS) or teichoic acid mediated lectin binding due to the presence of these sugar moieties on the surface of Gram-negative and Gram-positive bacteria respectively. Con-A and poly (aniline) film-coated electrodes have been developed which is sensitive to the lipopolysaccharide from *E. coli* and lipoteichoic acid (LTA) from *S. aureus* (da Silva et al. 2014). Dengue virus from the serum samples of patients has also been detected by electrochemical impedance sensors by coating the Au electrode surface with a lipid membrane coated with Con A. The sensors have the capability of displaying different responses to the four serotypes of dengue virus (Luna et al. 2014). Lectin-modified electrodes have also been constructed for detecting cells by utilizing the selective binding of the lectins to hydrocarbon chains expressed on the surface of cancer cells. Con A modified Au electrode coupled with Au nanoparticles coated

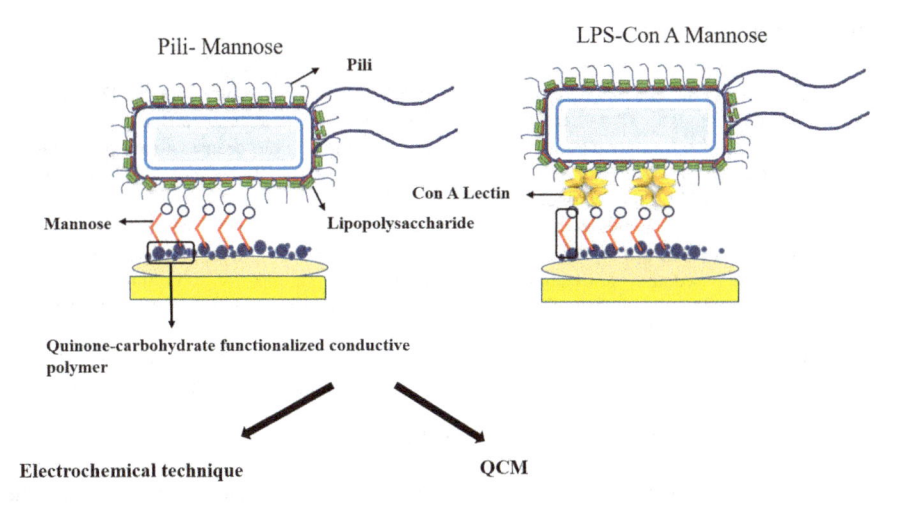

Fig. 17.18 Con A-based lectin sensor for the detection of *E. coli* based on a Quinone fused poly (thiophene) film-coated Au electrode. The surface of the Au electrode coated with polythiophene film was substituted with mannose residues. *E.coli* was trapped by either direct method or Con A facilitated binding (Ma et al. 2015)

with Con A and ferrocenylhexanethiol (Fc-SH) are constructed for their ability to bind K562 leukemic cells (Ding et al. 2011).

Due to the recent advancement of nanomaterials, lectin-based nano-sensors, provide an ultrasensitive detection of pathogenic bacteria and cancer cells with efficient reproducibility and reliability.

17.4 Conclusions

Plant lectins owing to their unique and varied properties have drawn immense interest for biochemists, cancer biologists, and glycobiologists. This chapter concludes the major properties of plant lectins with a focus on their interaction with sugar, applications in biotechnology and medicine. Lectins, apart from their use as signaling molecules or cytotoxic agents, can also serve as analytical tools for the detection, purification, and analysis of glycoconjugates from biological preparations including cell extracts and serum from certain disease conditions like cancer. Lectin microarray and enzyme-linked lectin assay (ELLA) use immobilized lectins on glass plate or microtiter plate respectively, which could be used for analysis of labeled glycoconjugates either free in solution or on cell surfaces. Aberrant glycosylation, branching, or increased fucosylation associated with cancer progression are well targeted by this method. In lectin blotting, electrophoresed glycoproteins are transferred to nitrocellulose membrane and visualized by specific lectins conjugated to suitable enzyme or fluorophore. Lectin-based biosensors offer a sensitive tool for the study of this interaction. The analytical applications and cytotoxic potential of lectin are broad, keeping in pace with their expanding list.

References

Adams DJ (2004) Fungal cell wall chitinases and glucanases. Microbiology 150(7):2029–2035
Ang ASW, Cheung RCF, Dan X et al (2014) Purification and characterization of a glucosamine binding antifungal lectin from *Phaseolus vulgaris* cv. Chinese pinto beans with antiproliferative activity towards nasopharyngeal carcinoma cells. Appl Biochem Biotechnol 172(2):672–686
Ashraf MT, Khan RH (2003) Mitogenic lectins. Med Sci Monit 9(11):265–269
Ayouba A, Causse H, Van Damme EJM et al (1994) Interactions of plant-lectins with the components of the bacterial-cell wall peptidoglycan. Biochem Syst Ecol 22(2):153–159
Bardhan M, Chowdhury J, Ganguly T (2011) Investigations on the interactions of aurintricarboxylic acid with bovine serum albumin: steady state/time resolved spectroscopic and docking studies. J Photochem Photobiol 102(1):11–19
Bekeredjian DI, Foermer S, Kirschning CJ et al (2012) Poke weed mitogen requires toll like receptor ligands for proliferative activity in human and murine B lymphocytes. PLoS One 7(1):e29806
Beltrão EIC, Medeiros PL, Rodrigues OG (2003) *Parkia pendula* lectin as histochemistry marker for meningothelial tumour. Eur J Histochem 47(2):139–142
de Bentzmann S, Varrot A, Imberty A (2014) Monitoring lectin interactions with carbohydrates. In: Filloux A, Ramos JL (eds) Pseudomonas methods and protocols, Methods Mol bio (methods and protocols), vol 1149. Humana Press, New York, pp 403–414
Bouchard P, Moroux Y, Tixier R et al (1976) An improved method for the purification of wheat germ agglutinin (lectin) by affinity chromatography. Biochimie 58(10):1247–1253
Bourne Y, Ayouba A, Rouge P et al (1994) Interaction of a legume lectin with two components of the bacterial cell wall. J Biol Chem 269(13):9429–9435
Broekaert WF, Mariën W, Terras FR et al (1992) Antimicrobial peptides from *Amaranthus caudatus* seeds with sequence homology to the cysteine/glycine-rich domain of chitin-binding proteins. Biochemistry 31(17):4308–4314
del Carmen FAM, Díaz D, Álvaro BM et al (2012) Protein-carbohydrate interactions studied by NMR: from molecular recognition to drug design. Curr Protein Pept Sci 13(8):816–830
Carvalho AS, da Silva MV, Gomes FS et al (2015) Purification, characterization and antibacterial potential of a lectin isolated from *Apuleia leiocarpa* seeds. Int J Biol Macromol 75:402–408
Carvalho EV, Oliveira WF, Coelho LC et al (2018) Lectins as mitosis stimulating factors: briefly reviewed. Life Sci 207:152–157
Cavalcante TTA, Carneiro VA, Neves CC et al (2013) A ConA-like lectin isolated from *Canavalia maritima* seeds alters the expression of genes related to virulence and biofilm formation in *Streptococcus mutans*. Adv Biosci Biotechnol 4(12):1073–1078
Charungchitrak S, Petsom A, Sangvanich P et al (2011) Antifungal and antibacterial activities of lectin from the seeds of *Archidendron jiringa* Nielsen. Food Chem 126(3):1025–1032
Chen YC, Wang HM, Niu QX (2016) Binding between saikosaponin C and human serum albumin by fluorescence spectroscopy and molecular docking. Molecules 21(2):153
Choi JW, Moon BI, Lee JW (2018) Use of CA15-3 for screening breast cancer: an antibody-lectin sandwich assay for detecting glycosylation of CA15-3 in sera. Oncol Rep 40(1):145–154
Ciopraga J, Gozia O, Tudor R et al (1999) *Fusarium* sp growth inhibition by wheat germ agglutinin. BBA-Gen Subj 1428(2–3):424–432
Coelho LC, Silva PM, Lima VL et al (2017) Lectins, interconnecting proteins with biotechnological/pharmacological and therapeutic applications. Evid Based Complement Alternat Med 2017:1594074. https://doi.org/10.1155/2017/1594074
Couzens L, Gao J, Westgeest K et al (2014) An optimized enzyme-linked lectin assay to measure influenza a virus neuraminidase antibody titers in human sera. J Virol Methods 210:7–14
Cui B, Li L, Zeng Q et al (2017) A novel lectin from *Artocarpus lingnanensis* induces proliferation and Th1/Th2 cytokine secretion through CD45 signaling pathway in human T lymphocytes. J Nat Med 71(2):409–421

Cummings RD, Schnaar RL, Esko JD et al (2017) Principles of glycan recognition. In: Varki A, Cummings RD, Esko JD et al (eds) Essentials of Glycobiology, 3rd edn. Cold Spring Harbor, New York, pp 373–385

Dam TK, Brewer CF (2002) Thermodynamic studies of lectin-carbohydrate interactions by isothermal titration calorimetry. Chem Rev 102(2):387–429

Dandagi P, Mastiholimath V, Patil M et al (2006) Biodegradable microparticulate system of captopril. Int J Pharm 307(1):83–88

Debruyne EN, Delanghe JR (2008) Diagnosing and monitoring hepatocellular carcinoma with alpha fetoprotein: new aspects and applications. Clin Chim Acta 395(1–2):19–26

Dhuna V, Dhuna K, Singh J et al (2010) Isolation, purification and characterization of an N-acetyl-D-lactosamine binding mitogenic and anti-proliferative lectin from tubers of a cobra lily *Arisaema utile* Schott. Adv Biosci and Biotechnol 1:79–90

Ding C, Qian S, Wang Z (2011) Electrochemical cytosensor based on gold nanoparticles for the determination of carbohydrate on cell surface. Anal Biochem 414(1):84–87

Donnelly EH, Goldstein IJ (1970) Glutaraldehyde-insolubilized concanavalin a: an adsorbent for the specific isolation of polysaccharides and glycoproteins. Biochem J 118(4):679–680

Durham M, Regnier FE (2006) Targeted glycoproteomics: serial lectin affinity chromatography in the selection of O-glycosylation sites on proteins from the human blood proteome. J Chromatogr A 1132(1–2):165–173

Duverger E, Frison N, Roche AC (2003) Carbohydrate-lectin interactions assessed by surface plasmon resonance. Biochimie 85(1–2):167–179

Fang EF, Lin P, Wong JH et al (2010a) A lectin with anti- HIV-1 reverse transcriptase, antitumor, and nitric oxide inducing activities from seeds of *Phaseolus vulgaris* cv. Extra long autumn purple bean. J Agric Food Chem 58(4):2221–2229

Fang EF, Wong JH, Lin P et al (2010b) Biochemical and functional properties of a lectin purified from Korean large black soybeans a cultivar of glycine max. Protein Pept Lett 17(6):690–698

Fernández-Alonso MC, Cañada FJ, Jiménez-Barbero J et al (2005) Molecular recognition of saccharides by proteins. Insights on the origin of the carbohydrate-aromatic interactions. J Am Chem Soc 127(20):7379–7386

Fisher HF, Singh N (1995) Calorimetric methods for interpreting protein-ligand interactions. Methods Enzymol 259:194–221

Gabius HJ, Andre S, Jimenez-Barbero J et al (2011) From lectin structure to functional glycomics: principles of the sugar code. Trends Biochem Sci 36(6):298–313

Hamid R, Masood A, Wani IH et al (2013) Lectins: proteins with diverse applications. J App Pharm Sci 3:93–103

Hansen JE, Nielsen CM, Nielsen C et al (1989) Correlation between carbohydrate structures on the envelope glycoprotein gp120 of HIV-1 and HIV-2 and syncytium inhibition with lectins. AIDS 3(10):635–641

Hasan I, Ozeki Y, Kabir SR (2014) Purification of a novel chitin-binding lectin with antimicrobial and antibiofilm activities from a Bangladeshi cultivar of potato (*Solanum tuberosum*). Indian J Biochem Biophys 51(2):142–148

Hashim OH, Jayapalan JJ, Lee CS (2017) Lectins: an effective tool for screening of potential cancer biomarkers. Peer J 5:e3784

Hatakeyama T (2014) Equilibrium dialysis using chromophoric sugar derivatives. In: Hirabayashi J (ed) Lectins, Methods Mol Biol (Methods and Protocols), vol 1200. Humana Press, New York, pp 165–171

Hendrickson OD, Zherdev AV (2018) Analytical application of lectins. Crit Rev Anal Chem 48 (4):279–292

Hirabayashi J (2014) Lectin-based glycomics: how and when was the technology born? In: Hirabayashi J (ed) Lectins, Methods mol biol (methods and protocols), vol 1200. Humana Press, New York, pp 225–242

Hirabayashi J, Kuno A, Tateno H (2014) Development and applications of the lectin microarray. In: Gerardy-Schahn R, Delannoy P, von Itzstein M (eds) SialoGlyco chemistry and biology II top Curr Chem, vol 367. Springer, Cham, pp 105–124

Hsu KL, Mahal LK (2006) A lectin microarray approach for the rapid analysis of bacterial glycans. Nat Protoc 1(2):543–549

Inoue K, Wada J, Eguchi J et al (2013) Urinary fetuin-a is a novel marker for diabetic nephropathy in type 2 diabetes identified by lectin microarray. PLoS One 8(10):e77118

Islam B, Khan AU (2012) Lectins: to combat infections. In: Ahmad R (ed) Protein purification. IntechOpen, Rijeka, pp 167–188

Jayanthi S, Ishwarya R, Anjugam M et al (2017) Purification, characterization and functional analysis of the immune molecule lectin from the haemolymph of blue swimmer crab *Portunus pelagicus* and their antibiofilm properties. Fish Shellfish Immunol 62:227–237

Jiménez-Barbero J, Cañada FJ, Asensio JL et al (2006) Hevein domains: an attractive model to study carbohydrate-protein interactions at atomic resolution. Adv Carbohydr Chem Biochem 60:303–354

Kalsi G, Etzler ME (2000) Localization of a nod factor-binding protein in legume roots and factors influencing its distribution and expression. Plant Physiol 124(3):1039–1048

Kawakami K, Yamamoto Y, Onoue K (1988) Effect of wheat germ agglutinin on T lymphocyte activation. Microbiol Immunol 32(4):413–422

Keyaerts E, Vijgen L, Pannecouque C et al (2007) Plant lectins are potent inhibitors of coronaviruses by interfering with two targets in the viral replication cycle. Antivir Res 75 (3):179–187

Khan F, Khan RH, Sherwani A et al (2002) Lectins as markers for blood grouping. Med Sci Monit 8 (12):293–300

Kim HJ, Lee SJ, Kim HJ (2008) Antibody-based enzyme-linked lectin assay (ABELLA) for the sialylated recombinant human erythropoietin present in culture supernatant. J Pharm Biomed Anal 48(3):716–721

Klein RC, Fabres-Klein MH, Oliveira LL et al (2015) A C-type lectin from *Bothrops jararacussu* venom disrupts staphylococcal biofilms. PLoS One 10(3):e0120514

Kumar KK, Chandra KL, Sumanthi J et al (2012) Biological role of lectins: a review. J Orofac Sci 4 (1):20–25

Kuno A, Uchiyama N, Koseki-Kuno S et al (2005) Evanescent-field fluorescence assisted lectin microarray: a new strategy for glycan profiling. Nat Methods 2(11):851–856

Kuzmanov U, Kosanam H, Diamandis EP (2013) The sweet and sour of serological glycoprotein tumor biomarker quantification. BMC Med 11:31

Lagarda-Diaz I, Guzman-Partida AM, Vazquez-Moreno L (2017) Legume lectins: proteins with diverse applications. Int J Mol Sci 18(6):1242

Langston-Unkefer PJ, Gade W (1984) A seed storage protein with possible self-affinity through lectin-like binding. Plant Physiol 74(3):675–680

Lannoo N, Van Damme EJ (2014) Lectin domains at the frontiers of plant defense. Front Plant Sci 5:397

Lebed K, Kulik AJ, Forro L et al (2006) Lectin-carbohydrate affinity measured using a quartz crystal microbalance. J Colloid Interface Sci 299(1):41–48

Lee YC (1997) Flourescence spectrometry in studies of carbohydrate-protein interactions. J Biochem 121(5):818–825

Lee CS, Muthusamy A, Abdul-Rahman PS et al (2013) An improved lectin-based method for the detection of mucin-type O-glycans in biological samples. Analyst 138(12):352–3529

Lee CS, Taib NA, Ashrafzadeh A et al (2016) Unmasking heavily O-glycosylated serum proteins using perchloric acid: identification of serum proteoglycan 4 and protease C1 inhibitor as molecular indicators for screening of breast cancer. PLoS One 11(2):e0149551

Leerapun A, Suravarapu SV, Bida JP (2007) The utility of *Lens culinaris* agglutinin reactive α-fetoprotein in the diagnosis of hepatocellular carcinoma: evaluation in a United States referral population. Clin Gastroenterol Hepatol 5(3):394–402

Li D, Mallory T, Satomura S (2001) AFP-L3: a new generation of tumor marker for hepatocellular carcinoma. Clin Chim Acta 313(1–2):15–19

Li Y, Tao SC, Bova GS et al (2011) Detection and verification of glycosylation patterns of glycoproteins from clinical specimens using lectin microarrays and lectin-based immunosorbent assays. Anal Chem 83(22):8509–8516

Liang PH, Wang SK, Wong CH (2007) Quantitative analysis of carbohydrate-protein interactions using glycan microarrays: determination of surface and solution dissociation constants. J Am Chem Soc 129(36):11177–11184

Lis H, Sharon N (1981) Lectins in higher plants. In: Marcus A (ed) The biochemistry of plants, vol 6. Academic Press Inc, New York, pp 371–447

Luna DM, Oliveira MD, Nogueira ML (2014) Biosensor based on lectin and lipid membranes for detection of serum glycoproteins in infected patients with dengue. Chem Phys Lipids 180:7–14

Ma F, Rehman A, Liu H (2015) Glycosylation of quinone-fused polythiophene for reagentless and label-free detection of *E. coli*. Anal Chem 87(3):1560–1568

Macedo MLR, Freire MGM, da Silva MBR et al (2007) Insecticidal action of *Bauhinia monandra* leaf lectin (BmoLL) against *Anagasta kuehniella* (Lepidoptera: Pyralidae), *Zabrotes subfasciatus* and *Callosobruchus maculatus* (Coleoptera: Bruchidae). Comp Biochem Physiol A Mol Integr Physiol 146(4):486–498

Majumder P, Mondal HA, Das S (2005) Insecticidal activity of *Arum maculatum* tuber lectin and its binding to the glycosylated insect gut receptors. J Agric Food Chem 53(17):6725–6729

Mannoji H, Yeger H, Becker LE (1986) A specific histochemical marker (lectin *Ricinus communis* agglutinin-1) for normal human microglia, and application to routine histopathology. Acta Neuropathol 71(3–4):341–343

Mazzetti S, Frigerio S, Gelati M (2004) *Lycopersicon esculentum* lectin: an effective and versatile endothelial marker of normal and tumoral blood vessels in the central nervous system. Eur J Histochem 48(4):423–428

McCoy JP Jr, Varani J, Goldstein IJ (1983) Enzyme-linked lectin assay (ELLA): use of alkaline phosphatase-conjugated *Griffonia simplicifolia* B4 isolectin for the detection of alpha-D-galactopyranosyl end groups. Anal Biochem 130(2):437–444

Michiels K, Van Damme EJ, Smagghe G (2010) Plant–insect interactions: what can we learn from plant lectins? Arch Insect Biochem Physiol 73(4):193–212

Monsigny M, Roche AC, Sené C et al (1980) Sugar-lectin interactions: how does wheat germ agglutinin bind sialoglycoconjugates? Eur J Biochem 104(1):147–153

Monsigny M, Mayer R, Roche AC (2000) Sugar-lectin interactions: sugar clusters, lectin multivalency and avidity. Carbohydr Lett 4(1):35–52

Moorghen M, Carpenter F (1991) Peanut lectin: a histochemical marker for phaeochromocytomas. Virchows Arch A Pathol Anat Histopathol 419(3):203–207

Morgan GW, Kail M, Hollinshead M et al (2013) Combined biochemical and cytological analysis of membrane trafficking using lectins. Anal Biochem 441(1):21–31

Palharini JG, Richter AC, Silva MF et al (2017) Eutirucallin: a lectin with antitumor and antimicrobial properties. Front Cell Infect Microbiol 7:136

Park SY, Lee SH, Kawasaki N et al (2012) Alpha1-3/4 fucosylation at Asn 241 of beta haptoglobin is a novel marker for colon cancer: a combinatorial approach for development of glycan biomarkers. Int J Cancer 130(10):2366–2376

Pérez S, Sarkar A, Rivet A et al (2015) Glyco3D: a portal for structural glycosciences. Methods Mol Biol 1273:241–258. http://glyco3d.cermav.cnrs.fr/search.php?type=lectin

Peumans WJ, Van Damme E (1995) Lectins as plant defense proteins. Plant Physiol 109 (2):347–352

Peumans WJ, Van Damme EJM (1998) Plant lectins: versatile proteins with important perspectives in biotechnology. Biotechnol Genet eng 15(1):199–228

Pinho SS, Reis CA (2015) Glycosylation in cancer: mechanisms and clinical implications. Nat Rev Cancer 15(9):540–555

Piotukh K, Serra V, Borriss R et al (1999) A. Protein-carbohydrate interactions defining substrate specificity in bacillus 1,3-1,4-beta-D-glucan-4-glucanohydrolases as dissected by mutational analysis. Biochemistry 38(49):16092–16104

Procopio TF, Patriota LLS, Moura MC et al (2017) CasuL: a new lectin isolated from *Calliandra surinamensis* leaf pinnulae with cytotoxicity to cancer cells, antimicrobial activity and antibiofilm effect. Int J Biol Macromol 98:419–429

Pujari R, Nagre NN, Chachadi VB et al (2010) *Rhizoctonia bataticola* lectin (RBL) induces mitogenesis and cytokine production in human PBMC via p38 MAPK and STAT-5 signaling pathways. Biochim Biophys Acta 1800(12):1268–1275

Qiu Y, Patwa TH, Xu L et al (2008) Plasma glycoprotein profiling for colorectal cancer biomarker identification by lectin glycoarray and lectin blot. J Proteome Res 7(4):1693–1703

Rambaruth ND, Greenwell P, Dwek MV (2012) The lectin *Helix Pomatia* agglutinin recognizes O-GlcNAc containing glycoproteins in human breast cancer. Glycobiology 22(6):839–848

Reddi AL, Sankaranarayanan K, Arulraj HS et al (2000) Enzyme- linked PNA lectin binding assay of serum T-antigen in patients with SCC of the uterine cervix. Cancer Lett 149(1–2):207–211

Regente M, Taveira GB, Pinedo M et al (2014) A sunflower lectin with antifungal properties and putative medical mycology applications. Curr Microbiol 69(1):88–95

Roy A, Gupta S, Hess D et al (2014) Binding of insecticidal lectin *Colocasia esculenta* tuber agglutinin (Cea) to midgut receptors of *Bemisia tabaci* and *Lipaphis erysimi* provides clues to its insecticidal potential. Proteomics 14(13–14):1646–1659

Sadeghi A, Smagghe G, Broeders S et al (2008) Ectopically expressed leaf and bulb lectins from garlic (*Allium sativum* L.) protect transgenic tobacco plants against cotton leafworm (*Spodoptera littoralis*). Transgenic Res 17(1):9–18

Sammour RH, El-Shanshoury A (1992) Antimicrobial activity of legume seed proteins. Bot Bull Acad Sin 31:185–190

Santi-Gadelha T, Rocha BAM, Gadelha CAA et al (2012) Effects of a lectin-like protein isolated from *Acacia Farnesiana* seeds on phytopathogenic bacterial strains and root-knot nematode. Pestic Biochem Phys 103(1):15–22

Scatchard G (1949) The attractions of proteins for small molecules and ions. Ann N Y Acad Sci 51(4):660–672

Schlumbaum A, Mauch F, Vogeli U et al (1986) Plant chitinases are potent inhibitors of fungal growth. Nature 324:365–367

Shinohara Y, Kim F, Shimizu M et al (1994) Kinetic measurement of the interaction between an oligosaccharide and lectins by a biosensor based on surface plasmon resonance. Eur J Biochem 223(1):189–194

Shinohara Y, Furukawa J (2014) Surface plasmon resonance as a tool to characterize lectin-carbohydrate interactions. Methods Mol Biol 1200:185–205

Sharon N, Lis H (2004) History of lectins: from hemagglutinins to biological recognition molecules. Glycobiology 14(11):53–62

da Silva JS, Oliveira MD, de Melo CP (2014) Impedimetric sensor of bacterial toxins based on mixed (Concanavalin a)/polyaniline films. Colloids Surf B: Biointerfaces 117:549–554

Sobral APV, Rego MJ, Cavalacanti CL (2010) ConA and UEA-I lectin histochemistry of parotid gland mucoepidermoid carcinoma. J Oral Sci 52(1):49–54

Souza JD, Silva MBS, Argolo ACC et al (2011) A new *Bauhinia monandra* galactose specific lectin purified in milligram quantities from secondary roots with antifungal and termiticidal activities. Int Biodeterior Biodegradation 65(5):696–702

Sumi S, Arai K, Kitahara S et al (1999) Serial lectin affinity chromatography demonstrates altered asparagine-linked sugar-chain structures of prostate-specific antigen in human prostate carcinoma. J Chromatogr B Biomed Sci Appl 727(1–2):9–14

Takeda Y, Matsuo I (2014) Isothermal calorimetric analysis of lectin-sugar interaction. In: Hirabayashi J (ed) Lectins, Methods Mol bio (methods and protocols), vol, vol 1200. Humana Press, New York, pp p207–p214

Taketa K (1990) Alpha-fetoprotein: Reevaluation in hepatology. Hepatology 12(6):1420–1432

Tateno H, Toyota M, Saito S et al (2011) Glycome diagnosis of human induced pluripotent stem cells using lectin microarray. J Biol Chem 286(23):20345–20353

Ueda R, Sugiura T, Kume S et al (2013) A novel single virus infection system reveals that influenza virus preferentially infects cells in g1 phase. PLoS One 8(7):e67011

Valadez-Vega C, Guzmán-Partida AM, Soto-Cordova FJ et al (2011) Purification, biochemical characterization, and bioactive properties of a lectin purified from the seeds of white Tepary bean (*Phaseolus Acutifolius* variety Latifolius). Molecules 16(3):2561–2582

Van Damme EJ (2014) History of plant lectin research. Methods Mol Biol 1200:3–13

Van Parijs J, Broekaert WF, Goldstein IJ et al (1991) Hevein: an antifungal protein from rubber-tree (*Hevea brasiliensis*) latex. Planta 183(2):258–264

Velayutham V, Shanmugavel S, Somu C et al (2017) Purification, characterization, and analysis of antibacterial activity of a serum lectin from the grub of rhinoceros beetle. Oryctes rhinoceros Process Biochem 53:232–244

Walski T, Van Damme EJ, Smagghe G (2014) Penetration through the peritrophic matrix is a key to lectin toxicity against *Tribolium castaneum*. J Insect Physiol 70:94–101

Wi GR, Moon BI, Kim HJ et al (2016) A lectin-based approach to detecting carcinogenesis in breast tissue. Oncol Lett 11(6):3889–3895

Wong JH, Ng TB, Cheung RC et al (2010) Proteins with antifungal properties and other medicinal applications from plants and mushrooms. Appl Microbiol Biotechnol 87(4):1221–1235

Ye XY, Ng TB, Tsang PWK et al (2001) Isolation of a homodimeric lectin with antifungal and antiviral activities from red kidney bean (*Phaseolus vulgaris*) seeds. J Protein Chem 20(5):367–375

Zhao Q, Duckworth CA, Wang W et al (2014) Peanut agglutinin appearance in the blood circulation after peanut ingestion mimics the action of endogenous galectin-3 to promote metastasis by interaction with cancer-associated MUC1. Carcinogenesis 35(12):2815–2821

Plant Metabolites as Immunomodulators

Sony Jayaraman and Jayadevi Variyar

Abstract

The immune system defends the body from various pathogens. Occasionally immune mechanisms fail in responding to antigens or they respond to self-antigens leading to severe complications like hypersensitivity, autoimmune disorders, cancer and AIDS. These conditions necessitate the use of compounds capable of modifying the immune system called immunomodulators. A large number of synthetic compounds are currently used for augmenting the immune system causing variety of side effects. Using natural biomolecules which are benign and economical can lead to safe immunomodulation. Application of plant metabolites for activating or suppressing the immune system is currently a fascinating area of research. This rational approach of immunomodulation has led to the discovery of many phytocompounds from known medicinal plants. Berberine, leonurine, piperine, gelselegine, chelerythrine and pseudo coptisine are some immunomodulatory agents derived from plants. Exploring plants with the use of innovative techniques will increase the choice of immunomodulators.

Keywords

Immunomodulation · Phytocompounds · Interleukins · Macrophages · Leukocytes

S. Jayaraman · J. Variyar (✉)
Department of Biotechnology & Microbiology, Kannur University, Dr. E K Janaki Ammal Campus, Kannur, Kerala, India

© Springer Nature Singapore Pte Ltd. 2020
S. T. Sukumaran et al. (eds.), *Plant Metabolites: Methods, Applications and Prospects*, https://doi.org/10.1007/978-981-15-5136-9_18

18.1 Introduction

The immune system constantly monitors invading pathogens and mount an attack when it encounters them. Proteins, cells and organs together constitute the immune system of the body. The immune system eliminates antigen through a cascade of pathways. Any incompetence in these pathways leads to serious complications resulting in a diseased state. Immunomodulators are employed to help the immune system. They act by either suppressing or activating the immune system and hence are of two types: immunosuppressors and immunostimulators. Ayurveda, the Indian traditional system of medicine, emphasizes on the use of medicinal plants for longevity and improvement of health. Plant-derived immunomodulators are relatively safe and cheaper when compared to the synthetic immunomodulators.

18.1.1 Immune System

Daily encounters of pathogens like bacteria, virus, allergens, and fungus elicit an immune response involving immune cells acting through various pathways for the destruction of the pathogen. The immune cells originate in the bone marrow, mature within the bone marrow and thymus and remain in the secondary lymphoid organs or circulate in the lymphatic system. The white blood cells or leukocytes circulate in the blood guarding against pathogens. Macrophages, monocytes, mast cells and neutrophils are the major leukocytes which destroy the cell via phagocytosis. Lymphocytes are the other class of leukocytes which consist of B lymphocytes and T lymphocytes. TH2 lymphocytes activate B lymphocytes which produce antibodies specific to the antigen. Antibodies destroy foreign body through different mechanisms. Antibodies or immunoglobulins are large Y-shaped proteins and are of five types: IgG, IgA, IgM, IgE and IgD. The immune system has two different defence mechanisms: Innate immunity, which is inborn like skin and mucosal membrane of gut and acquired immunity, which is gained after exposure to antigens leading to immunological memory. Spleen, lymph node and mucosal-associated lymphoid tissues are three main peripheral lymphoid tissues which collect pathogens from blood, site of infection and epithelial surface of the body. After an encounter with antigen, the T helper cells are induced to proliferate and activate B cells and T cells which differentiate into antibody-secreting cells and effector T cells respectively finally leading to the destruction of antigen.

18.1.2 Immune-Related Disorders

The network of the immune mechanism is very intricate and even a minute error in this multiplex system results in the breakdown of the defence mechanism leading to a variety of ailments. Primary immunodeficiencies are caused by the mutation in specific genes like the X-linked disorders. Almost 200 primary immunodeficiency disorders are known. Autoimmune lymphoproliferative syndrome, BENTA disease, chronic granulomatous disease, common variable immunodeficiency disease, hyperimmunoglobulin M syndrome, leukocyte adhesion deficiency are some examples of primary immunodeficiency disorders. Severe combined immunodeficiency disorder is a fatal primary immunodeficiency disorder characterized by the absence of both T and B lymphocytes. Secondary immunodeficiency diseases are caused by prolonged infections like diabetes and cancer, intake of drugs or alcohol and by radiations. Diabetic patients have a malfunctioned WBCs because of high sugar concentration in their blood. Human immunodeficiency virus infection leads to acquired immunodeficiency syndrome which compromises the immune system of the host by depleting the $CD4^+$ T cells. Immunodeficiency disorders can harm either the lymphoid or myeloid cells resulting in the failure of one or more immune mechanisms. Cancers like leukaemia and lymphoma destruct the function of bone marrow causing insufficient number of B and T lymphocytes. Under-nutrition and ageing also impair the function of immune system. Patients subjected to chemo and radiation therapies have a suppressed immune system. Immunostimulators are used to activate the immune system of such patients.

Hypersensitivity reaction is a condition in which the immune system overreacts. Autoimmune reaction and allergies are two different aspects of hypersensitivity. Type I hypersensitivity reactions are immediate allergic reactions. Type II is cytotoxic and Type III involves immune complex-mediated reactions. Type IV is delayed and cell-mediated hypersensitivity response. In autoimmunity, the immune cells recognize its own component as foreign and induce an immune response. It may either be localized within a specific organ like Hashimoto's thyroiditis, Grave's disease affecting thyroid, multiple sclerosis affecting brain or it will affect the whole body like systemic lupus erythematosus. Organ transplantation is a life-saving approach for various fatal diseases. But graft rejection is a common problem associated with the same. Immunosuppressants can be used in such cases to prevent and treat rejection.

18.1.3 Immunomodulators

Immunomodulators suppress or activate the immune system by acting on immune cells or interfering in its pathway. In the fifteenth century, Chinese and Turks used dried crusts of smallpox for inducing variolation. The father of immunology, Edward Jenner inoculated cowpox pustules for preventing smallpox. A new era of vaccination was marked by Louis Pasteur. Research in the area of immunomodulatory drugs was mainly aimed to treat cancer (Agarwal and Singh 1999). According to World Health Organization, by 2020 the efficacy of the currently known antibiotics will be very low and microbial infections will be hard to control. In this scenario, immunomodulation can play a very important role. Improving one's immunity can help in increasing the resistance to various infections (Dhama et al. 2015). Modulation can be achieved by either stimulating the immune system to treat infections, cancers and AIDS or suppressing the immune system to avoid graft rejection and treating hypersensitivity syndromes. They are derived synthetically or from nature.

The concept of Rasayana in Ayurveda describes the importance of immunomodulators. *Glycyrrhiza glabra*, *Tinospora cordifolia*, *Centella asiatica*, *Convolvulus pluricaulis* are some plants mentioned in the group of Rasayanas. Compounds isolated from microorganisms like *Bacille calmette* Guerin, Lipopolysaccharide, β-glucans, zymosan are proven immunomodulators (Novak and Vetvicka 2008). Neuroendocrine hormones like somatostatin, oxytocin, vasopressin and glucocorticoids modulate the immune system by increasing or decreasing the T and B cell growth, macrophage activity and antibody production, leukocytes. Thymosin, thymulin, thymic humoral factor, splenopentin and thymopentin regulate the immune system by increasing the production of lymphocytes and interferons, differentiating T cells, increases cGMP and increasing the cytotoxicity of lymphoid cells (Singh et al. 1998). Majority of the cytokines also exhibit immunomodulatory effects. Among the synthetic drugs, thalidomide came into the market for the first time against erythema nodosum leprosum and later was used for treating multiple myeloma. Its derivatives, Lenalidomide and pomalidomide, were shown to have higher activity than the parent molecule (Chanan-Khan et al. 2013). Levamisole, avridine, thiabendazole, ciclosporin and dapsone are the synthetic drugs currently available in the market. But all these drugs have various side effects like renal dysfunction, hyperuricemia, nephrotoxicity, bone necrosis, leucopenia, vomiting and thrombocytopenia (Jantan et al. 2015). An ideal immunomodulator must be capable of stimulating both specific and non-specific immune response, should be economical, non-toxic and orally effective. So a rapid research in the development of immunomodulatory drugs from plants is now on the way.

18.1.4 Mechanism of Immunomodulators

Clinically, a compound capable of reconstituting the immune dysfunctions or suppressing the overactive immune response acts as potential immunomodulators. The immunostimulants activate the suppressed immune system of the host suffering from cancer, AIDS and microbial infections. They target mainly T or B lymphocytes because the infections compromise their lymphocytes (Singh et al. 2016). They can stimulate the immune response in the following ways:

- Increasing the lymphocyte production.
- Increasing the antibody production.
- Increasing cGMP.
- Co-stimulation of T cells.
- Neutrophil chemotaxis and activation.
- T cell proliferation and differentiation.
- Proliferation of Natural Killer cells.
- Complement activation.
- Monocyte proliferation, differentiation and activation.
- Secretion of interleukins.

- Increasing macrophage activity.
- Increasing delayed type hypersensitivity response.
- Increases blastogenic response to T and B cells.

In autoimmune diseases, graft rejections, hypersensitivity reactions, the immune system responds in an exaggerated manner. The immunosuppressors control these pathological conditions by:

- Inhibition of calcineurin and folic acid.
- Decreasing cytokine production.
- Inhibition of leukocytes.
- Decreasing blastogenic response of lymphocytes.
- Inhibiting antibody synthesis.
- Decreasing the activity of macrophages.
- Inhibiting the production of Nitric oxide synthase.
- Inhibiting human peripheral blood mononuclear cells.

18.2 Plants in Therapeutics

Drug discovery from natural sources specifically plants are increasing in a rapid manner. Seventy-eight percent of drugs from natural sources are phytocompounds. Traditional medicine like Ayurveda in India, Chinese traditional medicine, Kampoo in Japan, African traditional medicine and Unani system of medicine in East Asia use plant resources. Plants enlisted in these have been used since ancient times by common people for treating various ailments. The use of herbs for treatment has been approved in many regions of the world. Depending on this, the value of herbs in International Trade Market is increasing. A total of 28,187 plant species have been recorded as being used for medicine (Allkin 2017). Roots, stems, flowers, leaves, tubers, fruits or the whole plant are used for the preparation of medicine. Many of the plant extracts used in traditional medicinal system are being evaluated for their medicinal properties and the phytoconstituents responsible for these properties are being isolated for use in pharma industry. Various phytocompounds possessing anti-cancerous, anti-inflammatory, anti-diabetic, wound healing, analgesic, anti-bacterial, anti-fungal and immunomodulatory properties have been isolated. Research in this area of phytomedicine is now ongoing all over the world. India is a country with its indigenous plants and traditional medicinal system of Ayurveda. Exploring our own heritage of plants can lead to the development of bioactive compounds. This can help in relying less on the developed countries for exporting pharmaceutical drugs and this approach is safe and economical. Assessing the safety of these plant species is also necessary because of the blind use of the plants largely for different disorders. Plant metabolites like flavonoids, terpenoids, tannins, polysaccharides and alkaloids have been known to possess immunomodulatory properties.

18.2.1 Plants as Immunomodulators

In Ayurveda, a large number of plants have been identified to possess immunomodulatory properties and they have been grouped into Rasayanas. *Withania somnifera*, *Allium sativum*, *Azadirachta indica*, *Piper longum*, *Asparagus racemosus*, *G. glabra*, *Aloe vera*, *Gmelina arborea* and *T. cordifolia* are some ayurvedic plants showing immunomodulatory properties. The ethyl acetate extract of the aerial parts of *Leucas aspera* showed a dose-dependent increase in neutrophil adhesion to the nylon fibres, antibody titre and phagocytic index. The extract renders protection from cyclophosphamide-induced myelosuppression and induces both humoral- and antibody-mediated immune response (Augustine et al. 2014). *T. cordifolia* protects animals against infections in immunosuppressed states induced by hemi splenectomy or surgery. The polymorphonuclear and monocyte-macrophage system is activated and induces granulocyte macrophage colony stimulating factor and IL-1 in a concentration-dependent manner. It also acts as an adjuvant to increase the efficacy of conventional chemotherapy in treating jaundice, tuberculosis etc. Syringin and cordial isolated from this plant inhibited the C3-convertase of the complement pathway and also increased IgG antibodies in the serum. Macrophage activation was induced by bioactive components like cordioside, cordiofolioside A and cordial (Dahanukar et al. 2000). The hydro-alcoholic extract of *Hibiscus rosa* stimulated the immune response in Wistar albino rats. The phagocytic index, humoral antibody titre and delayed type hypersensitivity response were induced by the extract indicating that they possess potent immunostimulatory compound (Gaur et al. 2009).

The methanolic stem bark extract from *Pouteria cambodiana* induced the macrophage phagocytic index and increased activity of lysosomal enzymes. The proliferation of splenocytes was also increased (Manosroi et al. 2005). *Cleome gynandra* Linn extract decreased the phagocytic index, humoral antibody titre and delayed type hypersensitivity in rat models exerting an immunosuppressive effect (Gaur et al. 2009). A myrecitin derivative isolated from *Mimosa pudica* was found to induce proliferation in lymphocytes (Jose et al. 2016). Crude polysaccharide isolated from *Callicarpa macrophylla* induced the proliferation of lymphocytes isolated from humans (Jayaraman and Jayadevi Variyar 2015). *Lepidagathis cuspidate* and *Phaseolus trilobus* are plants used in Ayurveda for curing boils, blisters and treating liver diseases. At 100 µg/mL, *Lepidagathis cuspidata* induced IL-2, IL-4, IL-6 and IL-12 at varying levels. At the same concentration, *P. trilobus* triggered the production of IL-2, IL-6 and IL-12 (Jayaraman et al. 2015). The purified triterpenoid from *L. cuspidate* and the flavonoid isolated from *P. trilobus* could increase the production of Il-2 and IL-12, but no significant difference was seen in IL-4 and IL-6 in the spleen cells.

The crude extracts of plants like *Capparis zeylanica*, *Nelumbo nucifera*, *T. cordifolia* and *Caesalpinia bonducella* have been reported to have immunomodulatory activity, but the active compounds from these plants are yet to be identified (Kumar et al. 2011). *C. zeylanica* leaf ethyl acetate extract and *n*-butanol extract increased the hemagglutination antibody titre contributing to a rise in the humoral immunity. Cell-mediated immunity was also induced by these extracts as they increased delayed type hypersensitivity response (Agarwal et al. 2010). The fruits of *Randia dumetorum* stimulated the immune system by increasing the antibody titre, delayed type hypersensitivity and WBC count of cyclophosphamide-induced myelosuppression model (Satpute et al. 2009). Flavonoids in *Calendula arvensis* induced the proliferation of lymphocytes proving them to be immunoboosters (Attard and Cuschieri 2009). *Matricaria chamomilla* and *Cichorium intybus* showed a stimulatory index of 2.18 and 1.70, respectively, in mixed lymphocyte reaction revealing their immunostimulatory capacity (Amirghofran et al. 2000). IL-12 stimulates proliferation of PHA-activated lymphoblasts and is also a growth factor for activated human T cells of both CD4+ and CD8+ subsets and natural killer cells. It has the capacity to enhance the proliferation of resting PBMS (Gately et al. 1991). High molecular weight polysaccharides from *Juniperus scopulorum* cones helps in the stimulation of immune system by stimulating NO production, enhancing respiratory burst and inducing the production of interleukins like IL-1, IL-12, TNF-α and IL-6 in J774.A1 macrophages (Schepetkin et al. 2005). Nitric oxide synthase, IFN-γ, TNF-α, IL-1, IL-10, IL-12 and IL-18 were induced by *Pelargonium sidoides* extract validating their immunomodulating effect (Kołodziej and Kiderlen 2007).

18.2.2 Bioactive Plant Secondary Metabolites

Plants synthesize secondary metabolites to protect them from predators and also from different types of stress. They contribute to plant characteristics like colour and help in maintaining an innate balance with the environment. They are classified into the following:

18.2.2.1 Terpenes
Terpenes are hydrocarbons which help in escaping from predators. They are classified based on the isoprene units into monoterpenes, diterpenes, triterpenes, sesquiterpenes, polyterpenes and norisoprenoids. D-Limonene, 1,8-cineole, boswellic acid, betulinic acid, β-sitosterol and ursolic acid are terpenes with biological functions. Boswellic acid induces the influx of Ca^{2+} and also the phosphorylation of p38 MAPK eliciting the central signalling pathway in human platelets. Betulinic acid suppresses the catalytic activity of topoisomerase I and decreases the expression of p21 protein in glioblastoma cells. The proinflammatory mediators like NF-κB, TNF-α and IL-6 expression were downregulated by ursolic acid (Singh and Sharma 2015). Carvone, limonene and perillic acid increased WBC count, bone marrow cellularity and total antibody production (Raphael and Kuttan 2003). Structures of some of the immunomodulatory terpenes are given in Fig. 18.1.

18.2.2.2 Alkaloids
Alkaloids are nitrogenous compounds present in plants. Narcotine was the first alkaloid to be isolated. Alkaloid-rich extracts are used for treatment for fever, insanity and snake bite. They are divided into non-heterocyclic and heterocyclic

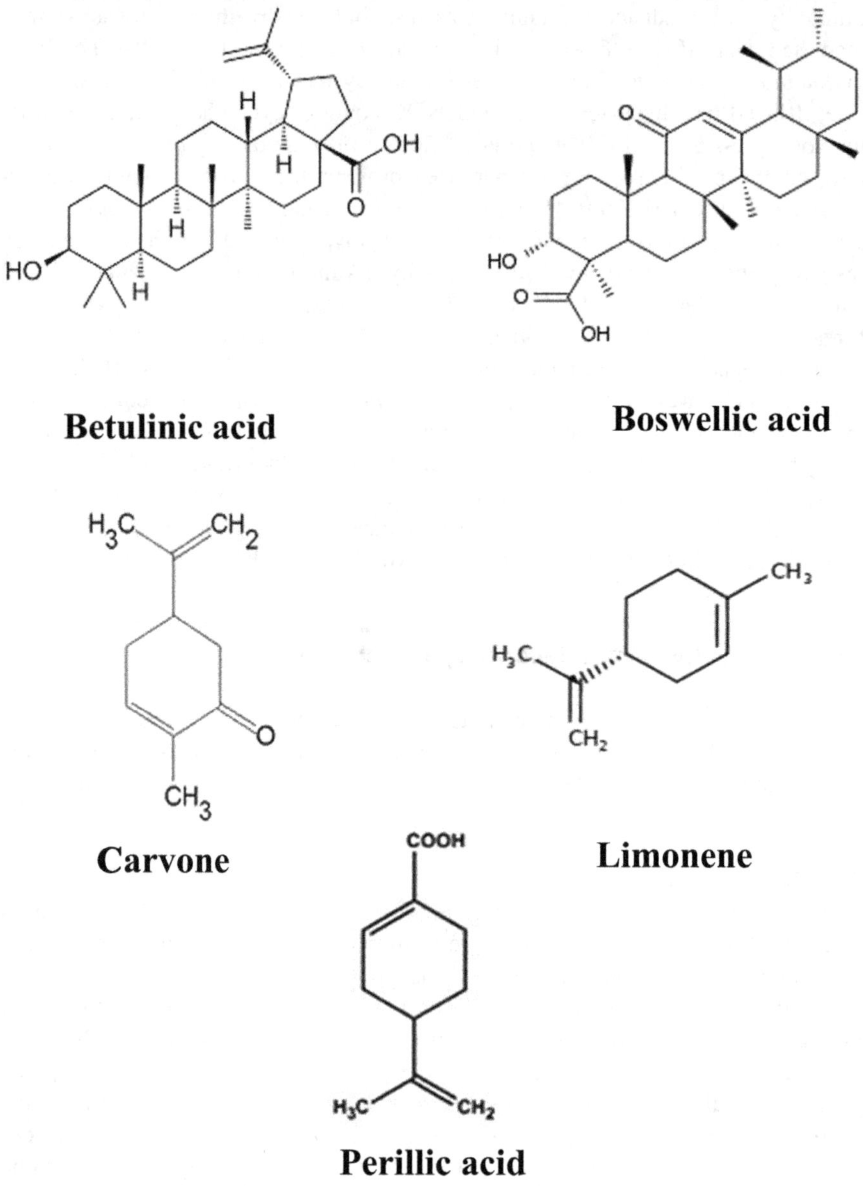
H
OH
H
O
H
HO
H
O
HO
O
OH
Betulinic acid
Boswellic acid
H3C
CH2
CH3
O
CH3
CH3
H3C
CH2
Carvone
COOH
Limonene
H3C
CH2
Perillic acid

Fig. 18.1 Chemical structures of terpenes with immunomodulatory activity

alkaloids. Cocaine, an immunosuppressive agent, inhibits leukocyte chemotaxis and mononuclear cell maturation (Delafuente and Devane 1991). Alkaloid cocaine isolated from *Erythroxylon coca* suppressed haemolytic plaque assay and also delayed type hypersensitivity response. Five other alkaloids isolated from the same plant also exhibited this suppression, but the level of suppression was low compared to cocaine (Watson et al. 1983). Sinomenine, isolated from *Sinomenium acutum*, exhibited anti-inflammatory and antirheumatic effects. Experiments in cardiac transplant models showed infiltration of mononuclear cells with edema, microvascular platelet and fibrin deposition prolonging the allograft survival (Jun et al. 1992). Tetrandrine, a bis-benzylisoquinoline, isolated from *Stephania tetrandra*, is being used for the treatment of autoimmune diseases like rheumatoid arthritis (Lai 2002). Berberine decreases TNF-α and IL-2 and also IL-4 production. Piperine downregulates proinflammatory cytokines like IL-1β, IL-6 and TNF-α. Structures of some of the immunomodulatory terpenes are given in Fig. 18.2.

18.2.2.3 Phenolic Compounds

Flavonoids, tannins, lignins, anthraquinones, stilbenes, xanthones and naphthoquinones come under phenolic compounds. These secondary metabolites are synthesized largely by medicinal plants. Phenolic acid consists of a p-hydroxybenzoic and hydroxycinnamic acid groups. They are abundantly present in blueberry, cherry, potato, tea, coffee, beans, spinach and lettuce. Butein, a flavonoid isolated from *Semecarpus anacardium*, hinders the translocation of NF-κB and decreases nitric oxide production. Shikonin, a quinone isolated from *Lithospermum erythrorhizon*, suppresses Th1 cytokine expression and NF-κB activity. Piceatannol, a stilbene isolated from grapes and nuts, reduces iNOS expression and inhibits transcription factors like NF-κB, ERK and STAT3. Wogonin is an O-methylated flavone isolated from *Scutellaria baicalensis*. Wogonin suppressed the production of IL-17 and reduced neutrophil infiltration in neutrophilic airway inflammation mouse models. It also inhibited dendritic cell-mediated Th17 differentiation. Chrysin isolated from *Citrus nobilis* blocks NF-κB and MAPK signalling pathways thereby inhibits pro-inflammatory mediators, COX-2 and iNOS expression. Emodin-8-*O*-β-D glucoside is a quinone isolated from *Polygonum amplexicaule* stimulates the proliferation and differentiation of osteoblasts. It also inhibits PGE2 production increasing alkaline phosphatase expression in MC3T3-1 (Jantan et al. 2015). The structure of phenolic compounds with immunomodulatory properties are given in Fig. 18.3.

18.2.2.4 Polysaccharides

They are primary metabolites controlling cell division, regulating cell growth and maintaining normal metabolism. Many polysaccharides isolated from plants possess antibacterial, immunomodulatory, antitumour and antioxidant properties. Polysaccharides isolated from *Ganoderma lucidum* modulate macrophages, T helper, natural killer cells and other effector cell activities. In peritoneal macrophages, it induces the production of IL-1α and TNF-α. The expression of TNF-α was induced by this polysaccharide in vivo and in vitro and it was also found to increase the influx of calcium ions and scavenge free radicals. It also increased protein kinase C, p38 mitogen-activated protein kinase and hematopoietic cell line activities in human neutrophils. Astragalus polysaccharides from *Astragalus membranaceus* increases the serum antibody titre and induced the release of cytokines like IL-1, IL-6 and TNF-α. It induced a significant increase in IL-12 and

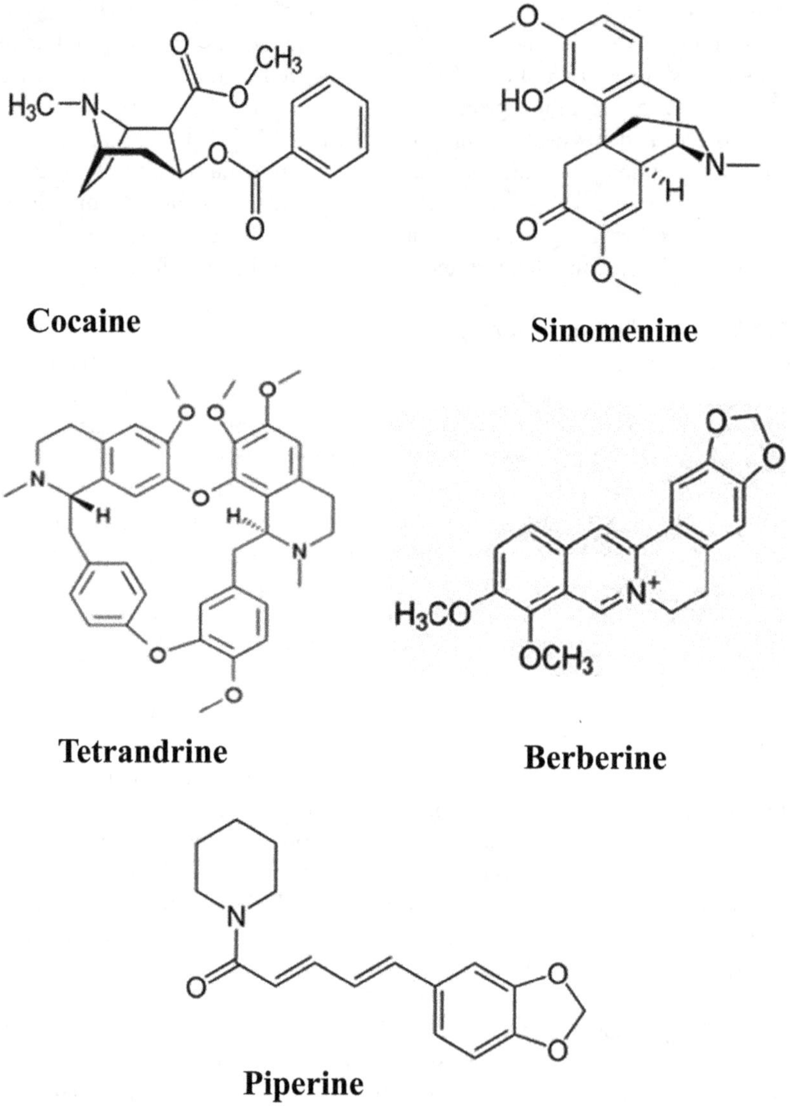

Cocaine

Sinomenine

Tetrandrine

Berberine

Piperine

Fig. 18.2 Chemical structures of some alkaloids with immunomodulatory activity

in mouse macrophages, it also induced the production of NO and also iNOS transcription. Astragalus polysaccharide helps in treating arthritis by reducing cell accumulation, swelling and arthritic index of the joints (He et al. 2012). Polysaccharide PST001 isolated from seed kernel of *Tamarindus indica* increased total WBC, $CD4^+$ T cell population and bone marrow cellularity in mice models (Aravind et al. 2011). Administration of β-1,3;1–6 glucans in patients with allergic rhinitis led to a decrease in IL-4, IL-5 and eosinophil count, whereas in patients with canker sores, it

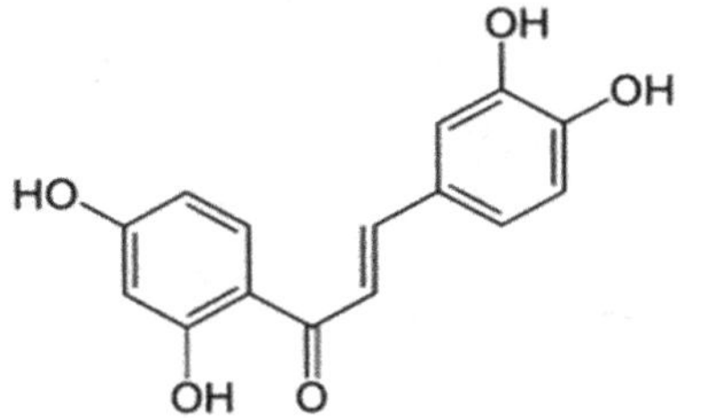

Butein

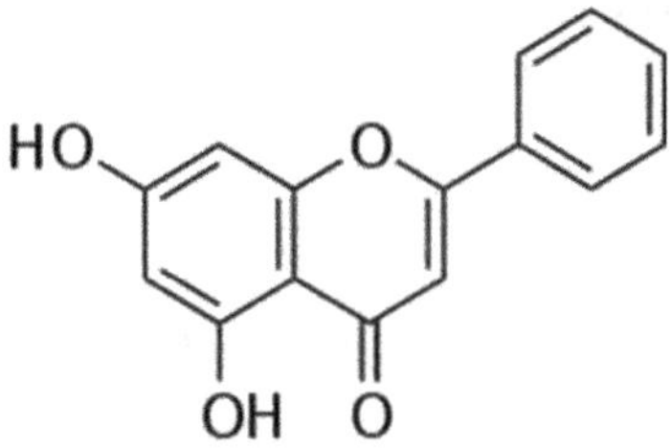

Chrysin

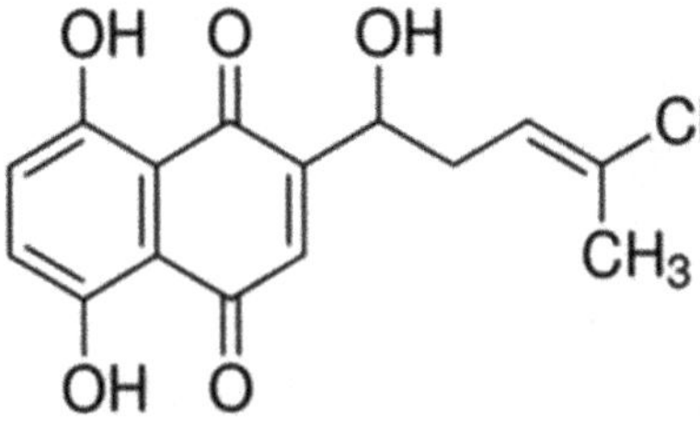

Shikonin

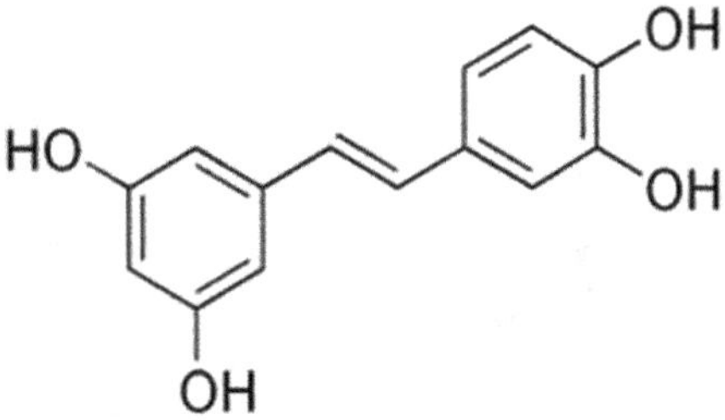

Piceatannol

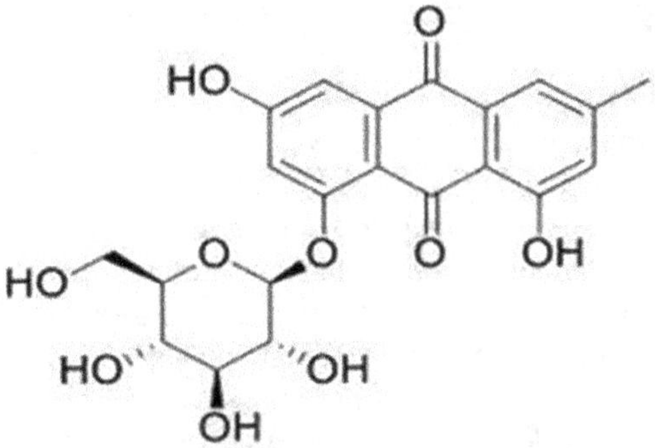

Emodin-8-O-β-D-glucoside

Fig. 18.3 Chemical structures of some phenolic acids possessing immunomodulatory activity

induced lymphocyte proliferation and decreased ulcer sores. Laminarin (Fig. 18.4) from *Laminaria digitata* increased macrophage expression of dectin-1 in gut-associated lymphoid tissue cells and TLR2 expression in Peyer's patch dendritic cells (Ramberg et al. 2010). Immunomodulatory active polysaccharides have been isolated from *Radix bupleuri, Cinnamomum cortex, Echinacea purpurea* and

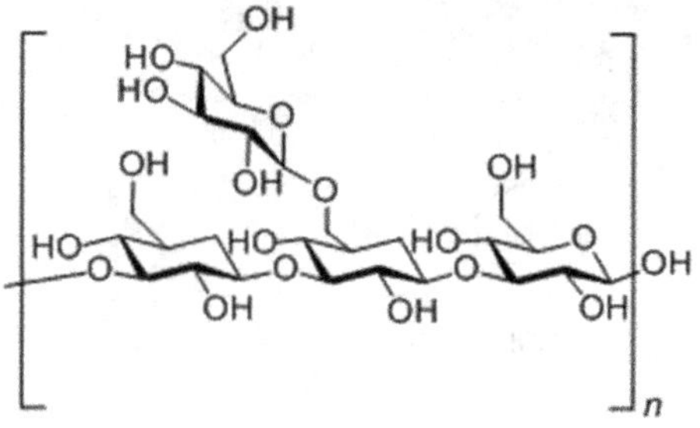

Fig. 18.4 Chemical structure of laminarin polysaccharide

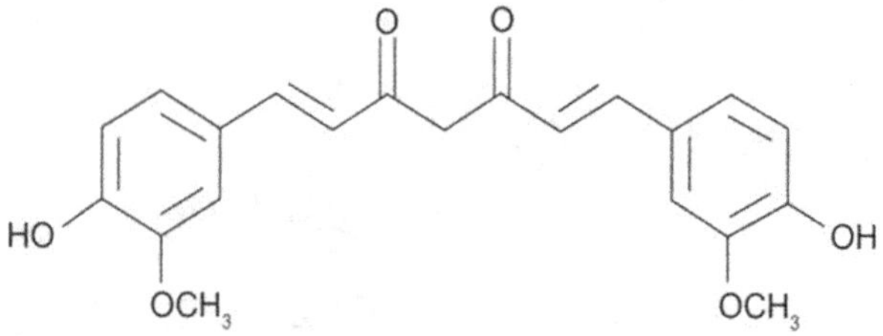

Fig. 18.5 Chemical structure of curcumin

Lentinus edodes possessing the capacity to activate macrophages, cytotoxic T lymphocytes, natural killer cells and lymphokine-activated killer cells (Wong et al. 2011).

18.2.3 Phytocompounds as Immunomodulators

18.2.3.1 Curcumin

Curcumin isolated from the plant *Curcuma longa* L. is used as a flavouring agent in foods. It is a diarylheptanoid existing in enolic form in solvents and as keto form in water (Fig. 18.5). It has been reported to have antioxidant, hepatoprotective and anticancer activity and inhibits macrophage activation, cyclooxygenase-2 and lipoxygenase, inducible nitric oxide synthase and TNF-α-activated macrophages (Surh et al. 2001). Curcumin has been proven to modulate both cellular and humoral immune response of mice infected with acute *Schistosomiasis mansoni*. Liver pathology and parasitic burden have also been shown to decrease (Singh et al. 2016). Curcumin also prevents degradation and phosphorylation of 1 kappa B thereby inhibiting NF-κB activation in PMA-activated myelo monoblastic cell lines. In LPS-activated alveolar mononuclear cells, curcumin inhibited IL-1β, IL-8, TNF-α and monocyte chemotactic protein-1. The production of IL-1 and TNF-α was decreased in LPS-activated human macrophage cell lines by 5 μM curcumin. They have shown good activity against adjuvant-induced inflammation in rats by reducing the level C-reactive protein and IL-1β. They also down regulate Toll-like receptors 2. Their wound healing approach has been proved in mice models suppressing chronic inflammation and boosting acute inflammation. The angiogenesis growth factors, enzymes and transcription factors including AP-1, angiopoetin-1 and

Table 18.1 Effect of curcumin on the immune cells (Jagetia and Aggarwal 2007)

Immune cells	Effect of curcumin
T cells	1. Inhibits proliferation of activated human spleen cells, T cells and peripheral blood morphonuclear cells. 2. Inhibits IL-2 expression 3. Inhibits proliferation of con-A stimulated thymocytes in rat. 4. Inhibits proliferation of HTLV-1 virus-infected T cells and decreases the expression of cyclin D1, Cdc25c and Cdk1.
B cells	1. Inhibits Epstein Barr virus induced B cell proliferation and induce cell death. 2. Increases B-cell proliferation in intestinal mucosa of mice.
Macrophages	1. Increases macrophage phagocytosis. 2. The reactive oxygen species and lysosomal secretion are down-regulated. 3. Reduces Th1 and nitric oxide production. 4. Inhibition of TNF-α, IL-1, IL-6 and IL-12 in activated macrophages
Polymorphonuclear neutrophils	1. Increase in WBCs, circulating antibody titre and plaque-forming cells. 2. Inhibit 5- hydroxyeicosatetraenoic acid in neutrophils.
Dendritic cells	1. Inhibition of IL-12, IL-6, TNF-α and IL-1β in activated dendritic cells. 2. Suppress expression of CD80, CD86 and MHC class II antigens in stimulated dendritic cells.

2, matrix metalloproteinase-9 and vascular endothelial growth factor have been shown to decrease by curcumin treatment (Banerjee et al. 2003). In anterior uveitis patients, curcumin was given daily three times for 12 weeks and observed 100% marked improvement after 2 weeks. Curcumin also impacts patients with gastric ulcer, kidney rejection and acute pancreatitis (Cundelland and Wilkinson 2014).

Curcumin exhibits chemoprotective activity by inhibiting angiogenesis and has shown to inhibit NF-κB pathway by inhibiting p65 subunit and also NF-κB inducing kinase and 1κB kinase enzymes. Mitogen-activated protein kinase induced IL-8 gene expression and luciferase expression was also inhibited by curcumin (Jobin et al. 2011). Curcumin downregulated the allergic response in murine models by blocking histamine release from mast cells (Kurup and Barrios 2008). Curcumin inhibits PMA or PHA-activated human spleen cells, PBMCs and T cells. It also suppresses the IL-2 expression in T cells. In Balb/c mice, it showed potent immunostimulatory activity, increased the antibody titre, WBC count, spleen cells, bone marrow cellularity and macrophage phagocytosis (Antony et al. 1999). The effect of curcumin on various immune cells is enlisted in Table 18.1.

18.2.3.2 Resveratrol

It is a stilbenoid belonging to polyphenols in plants like grapes, berries and peanuts. 3,5,4'-trihydroxystilbene exist in *trans* and cis form (Fig. 18.6). Resveratrol is a phytoalexin produced by plants in response to bacterial and fungal infections. It is an anti-inflammatory and antioxidizing agent with the capability to reduce blood sugar. Resveratrol is used as a fat burner and it has gained importance as a good

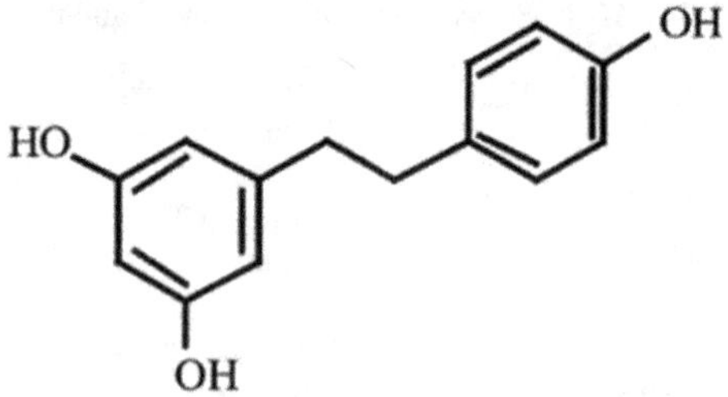

Fig. 18.6 Chemical structure of resveratrol

nutraceutical. Resveratrol reduces inflammation and cancer growth by downregulating AKT, MAPK and NF-κB signalling pathways (Berman et al. 2017). They exhibit cytotoxicity in multiple myeloma cell lines by inhibiting STAT3 and NF-κB. At low concentrations, they stimulate IFN-γ, IL-2, CD8+ and CD4+ T cells and suppress these components of the immune system at high concentrations. Similarly, at high concentration, it shows inhibition of cytotoxic T lymphocytes and natural killer cells and vice versa at low concentration (Falchetti et al. 2001).

Resveratrol induced the release of IFN-γ and suppressed TNF-α, thereby regulating humoral immune response. In piglets, it has proved to be a good adjuvant for enhancing the immune response to vaccines (Fu et al. 2018). It can stimulate macrophages and NK cells. Resveratrol inhibits the inflammatory factors by activating sirtuin-1 (Saiko et al. 2008). It has been shown to render protection against autoimmune encephalomyelitis and PHA-stimulated poly blood mononuclear cells. TNF-α and IFN-γ were inhibited by resveratrol. In PMA-stimulated mouse skin, resveratrol causes a pronounced reduction in the c-fos and TGF-β1expression (Boscolo et al. 2003). In autoimmune diseases, they exert a suppressive effect on T cells, but in the case of cancer models, it reduces the suppressive function of T regulatory cells inhibiting the tumour growth. Table 18.2 narrates the effect of resveratrol on immune cells.

18.2.3.3 Quercetin

Quercetin, a flavonoid found in green tea, apples, berries, broccoli, red wine and onions, has a profound influence in the pharmacological industry. It is a pentahydroxyflavone having the five hydroxyl groups placed at the 3-, 3'-, 4'-, 5- and 7-positions (Fig. 18.7). It is usually found in conjugated form with sugars like glucose, rhamnose and galactose. They act as a natural antihistamine and were found to fight allergies. They are known for its antioxidant activity by scavenging free radicals. They suppress pro-inflammatory cytokines, production of leukotrienes and IL-4. The Th1/Th2 balance can be revamped by quercetin as it inhibited the increase in Th2 cytokine in asthma mice model and increased Th1 in quercetin-administered mice (Park et al. 2009). It has the capability to suppress lipoxygenase, peroxidase enzymes. Eosinophils and inflammatory mediators which occur prominently during allergy are suppressed by quercetin. Quercetin can be effectively applied in the treatment of bronchial asthma, allergic rhinitis and anaphylactic reactions. In mast

18 Plant Metabolites as Immunomodulators

Table 18.2 Effect of resveratrol on immune cells (Malaguarnera 2019)

Immune cells	Effect of resveratrol
T lymphocytes	1. Inhibit T cell activation. 2. Reduce cytokine production 3. Inhibit Th17 cells, inducers of chronic inflammation. 4. Suppress CD4 + and `CD25+ cells. 5. Increase IL-17+/IL-10+ T cells and CD4-IFN-γ+ 6. Downregulates TGF-β and IFN-γ expression in CD8+ T cells 7. Suppress CD28 and CD80 8. Increase expression of Sirtuin-1 in turn inhibiting T cell.
Macrophages	Decrease NF-κB activation and COX-2 expression in LPS-induced RAW264.7 Induces TLR4-TRAF6, MAPK and AKT pathways in LPS-induced macrophages. Blocks the increase of acetylated α-tubulin caused by mitochondrial damage in macrophages. Modulates transcription of NF-κB elements, STAT1 and cAMP- responsive element binding protein 1. It affects TNF-α induced activation of NF-κB. Decreased CD14 and interleukin-1 receptor-associated kinase expression. Suppress granulocyte–macrophage colony-stimulating factor. Decreases IL-1β and IL-6 expression in macrophages.
Natural killer cells	Ameliorates the proliferation of NK cells. Inhibits STAT3 signalling pathway. Increases apoptosis by caspase signalling pathway. Upregulates NKG2D and IFN-γ In NK-92 treatment, phosphorylation of ERK-1/2 and JNK was induced along with perforin expression. Increases DR4 and DR5 in carcinoma cell lines facilitating NK cell-mediated killing. Induce CD95L expression in leukaemic and breast cancer cell lines mediating NK-cell dependent apoptosis.
Platelets	Inhibit platelet aggregation by inhibiting integrin gp IIb/IIIa on the platelet membrane. Decrease TxA2, a platelet-activating factor.
B lymphocytes	Low dose inhibits B16 melanoma. Inactivate B regulatory cell's ability to express TGFβ expression. Inactivate STAT3 thereby inhibiting B regulatory cells. Reduces B cells in spleen and bone marrow in lupus nephritis. Induces Sirt1 inhibiting B cell proliferation and autoantibody production. In SLE patients, it upregulates FcγRIIb in B cells

Fig. 18.7 Chemical structure of quercetin

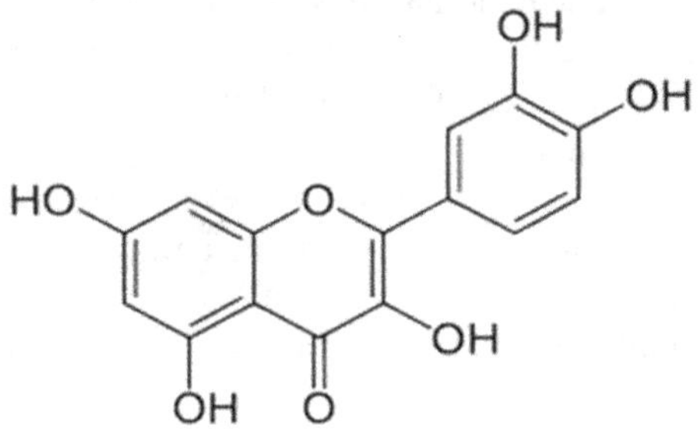

Table 18.3 Effect of quercetin on immune cells (Li et al. 2016; Nickel et al. 2011)

Immune cells	Effect of quercetin
Macrophages	Inhibits TNF-α production. Retards P13K tyrosine phosphorylation and TLR4/MyD88 and P13K complex formation restricting LPS induced inflammation in RAW cells. Increase nitric oxide synthase and nitric oxide. Upregulates IL-10, IL-1, IL-8 and lactate dehydrogenase.
Mast cells	Inhibit FcεRI-mediated release of proinflammatory cytokines. Suppress calcium influx and phosphor–protein kinase C. Hinders IL-6 secretion stimulated by IL-1. Possess mast cell stabilizing activity. Inhibit histamine and serotonin release. Inhibits tryptase release, MCP-1 and histidine decarboxylase
T cells	Regulate extracellular regulated kinase 2 mitogen-activated protein kinase signal pathway. Blocks IL-12 induced tyrosine phosphorylation of JAK2, TYK2, STAT3 and STAT4 Modulate the differentiation of naive glycoprotein CD4 T cells by inhibiting aryl hydrocarbon receptor.
Human peripheral blood mononuclear cells	Induce Th-1 derived cytokine Induce IFN-γ Inhibit Th-2 derived cytokine.
Dendritic cells	Decreased dendritic cell adhesion NF-κB activity reduced Lowers oxLDL-induced upregulation of NK-κB activity. BDCA-2 protein expression reduced. Suppression of DC migration

cells, quercetin inhibited prostaglandin D2 and granulocyte macrophage-colony stimulating factor. It decreased Ca^{2+} influx and suppresses IL-8 and IL-6 (Mlcek et al. 2016).

In human blood, quercetin is found conjugated as glycosides. The poor oral bioavailability of quercetin may be due to its low absorption, ample metabolism and quick elimination. Derivatives of quercetin are better absorbed than quercetin. They may be differently absorbed based on the sugar molecule attached to them. It has the capacity to stabilize mast cells and protect gastrointestinal activity (Li et al. 2016). In glial cells, quercetin can inhibit TNF-α and IL-1α resulting in a low level of apoptotic neuronal cell death (Bureau et al. 2008). In human umbilical vein endothelial cells, quercetin downregulates the expression of vascular cell adhesion molecule 1 and CD80. Oral administration of quercetin induced splenocyte proliferation in mice models. Pro-inflammatory cytokines like TNF-α, IL-1β and IL-6 showed a decrease in blood samples of the treated mice. The decreased protein and albumin level in radiation-induced mice was significantly increased by quercetin administration proving that they can improve immunity and nutritional status (Jung et al. 2012). Table 18.3 enlists effect of quercetin on immune cells.

18.2.3.4 Capsaicin

Capsaicin is a component of chilli pepper, mainly used as a food additive. Capsaicin is 8-methyl-N-vanillyl-6-nonenamide is hydrophobic, colourless and highly pungent (Fig. 18.8). It is known to improve cardiovascular functions, relieve migraines and joint pains, improves metabolism and reduces the risk of cancer, flu, colds and other infections. It exerted an inhibitory effect on intestinal mucosal mast cells (Gottwald et al. 2003).

Capsaicin-induced T cell mitogen and lymphocyte proliferation and increased antigen-producing B cells and serum IgG and IgM values. The TNF-α level was shown to increase in treated mice macrophages (Yu et al. 1998). Neuropathic pain can be eased with the help of capsaicin treatment. Sciatic nerve inflammation was reduced by the oral administration of capsaicin in rats. T-cell infiltration was significantly reduced by 50 μg/day administration of capsaicin where TNF-α and IFN-γ were also reduced. Il-4 was induced by the same concentration (Motte et al. 2018). Autoreactive T cells are proliferated by the oral administration of capsaicin in pancreatic lymph nodes giving protection from the autoimmune disease and type 1 diabetes (Nevius et al. 2012). Perineural application of capsaicin induced the level of IL-1β, IL-6 and nerve growth factors. Increase in the TNF-α level was not as prominent as other cytokines (Saadé et al. 2002).

18.2.3.5 Epigallocatechin-3-Gallate

Epigallocatechin gallate, a catechin, is the ester of epigallocatechin and gallic acid (Fig. 18.9). It is abundantly found in green, black and white tea. Apple skin, plums,

Fig. 18.8 Chemical structure of capsaicin

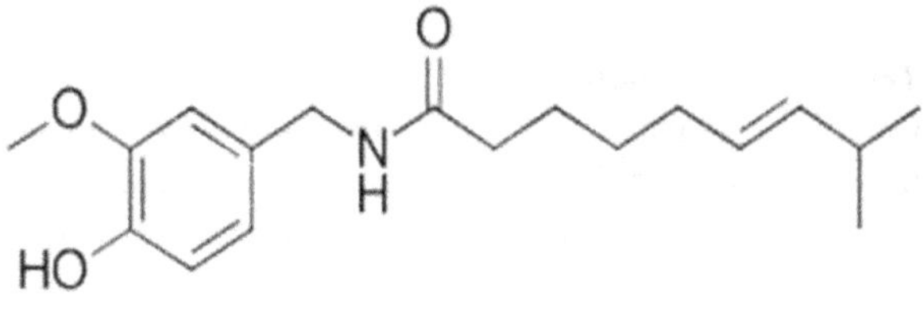

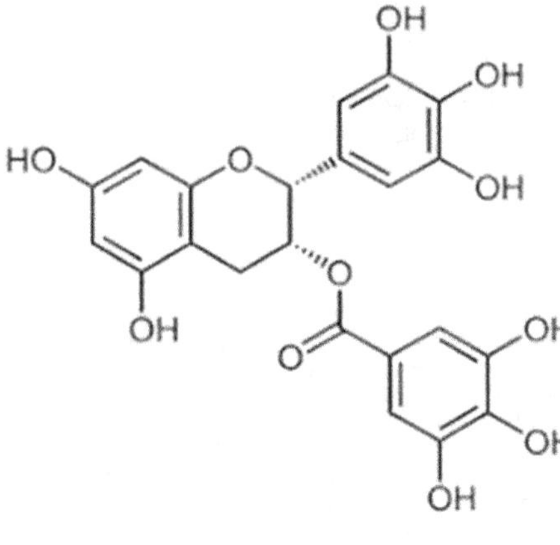

Fig. 18.9 Chemical structure of epigallocatechin-3-gallate

onions, hazelnuts and pecans contain trace amounts of this catechin. It has gained a lot of attention due to its beneficiary effect on human health. They act as potent antioxidant agents protecting from the adverse effects of reactive oxygen species. The antioxidant potential of epigallocatechin gallate has shown to have an impact in the treatment of cancer (Huynh 2017).

Epigallocatechin-3-gallate acts on T cell activation, proliferation and differentiation. It induces the production of cytokines. The differentiation of $CD4^+$ T cells into different effector cells has been attained by epigallocatechin-3-gallate proving them to be useful in treating T-cell mediated autoimmune disorders (Pae and Wu 2013). Epigallocatechin gallate restored the movement behaviour of Parkinson's disease mouse model. The ratio of $CD3,^+$ $CD4,^+$ to $CD3,^+$ $CD8^+$ T lymphocytes increased in epigallocatechin gallate-treated mice and it reduced TNF-α and IL-6 in the serum (Zhou et al. 2018). Skin inflammation caused by psoriasis could be treated with the help of epigallocatechin-3-gallate as it has the ability to infiltrate $CD11C^+$ dendritic cells, reduce IL-17A, IL-17F, IL-22 and IL-23 cytokines in the plasma. The $CD4^+$ T cells were seen to increase along with the bioactivities of superoxide dismutase and catalase. Elevated IgE levels in asthma patients were reduced by treating with epigallocatechin-3-gallate (Huynh 2017). Epigallocatechin gallate can induce T_{reg} cells in vitro and in vivo thereby triggering immune system. Foxp3, a promoter gene involved in regulating T cell function, is upregulated by treating with epigallocatechin-3-gallate. It also increased the Treg cells in spleen, pancreatic lymph nodes and mesenteric lymph nodes. This treatment can help in preventing and prolonging graft rejection (Wong et al. 2011). Epigallocatechin-3-gallate induce indoleamine 2,3-dioxygenase producing dendritic cells, elevating regulatory T cells and in turn activating Nrf-2 antioxidant pathway (Min et al. 2015).

18.2.3.6 Colchicine

Colchicine isolated from *Colchicum autumnale* is an alkaloid. It is used to treat gout, rheumatism, familial Mediterranean fever and Behcet's disease. It has been approved as an anti-inflammatory agent in the United States since 1961. Heavy dose is highly poisonous. Structurally they possess exocyclic N-atoms and one stereogenic centre at C_7 and chirality axis as they exhibit atropisomerism (Kurek 2017). The structure of colchicine is given in Fig. 18.10.

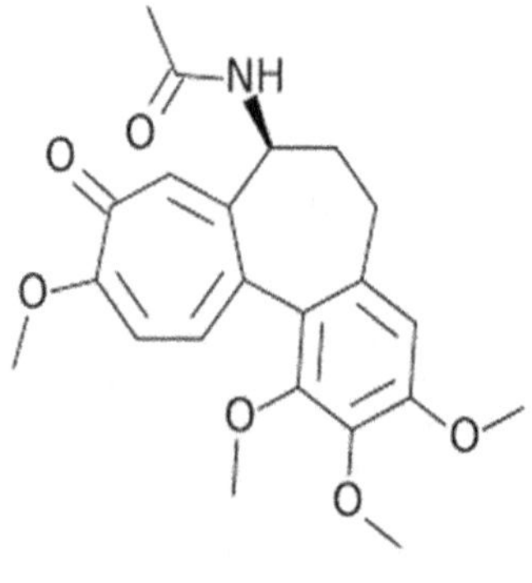

Fig. 18.10 Chemical structure of colchicine

Colchicine binds to tubulin, blocking the assembly and polymerization of microtubules. They block the cells in metaphase stage of cell cycle. It inhibits angiogenesis and retards ATP influx into mitochondria. Colchicine inhibits neutrophil activation and release of IL-1, Il-8 and superoxide. It can induce dendritic cell maturation and suppress vascular endothelial growth factor and proliferation. In LPS induced liver macrophages, it has been shown to induce TNF-α. NO and IL-1β release was lowered in peritoneal macrophages by treatment with colchicine (Leung et al. 2015). Colchicine improved liver function test in patients with the autoimmune disorder, primary biliary cirrhosis and also increased the life expectancy (Kaplan 1997). Administration of colchicine in atopic asthma patients showed a significant increase in concanavalin-A induced suppressor cell function and decreased serum IgE thus reducing the daily number of inhalations of albuterol (Schwarz et al. 1990). The percentage of OKT4+ helper T cells, OKT3+ total T cells and OKT8+ cytotoxic T cells was significantly reduced by the oral administration of colchicines in healthy volunteers (Ilfeld et al. 1984).

18.2.3.7 Genistein

Genistein, an isoflavone isolated from *Genista tinctoria,* is an angiogenesis inhibitor. It inhibits topoisomerase, tyrosine kinases, GLUT 1, glycine receptor, DNA methyltransferase and nicotine acetylcholine receptor. Genistein is a phytoestrogen with oestrogen receptor modulator protein. Genistein is 7- dihydroxyisoflavone with additional hydroxyl groups at the 5′ and 4′ position. The structure of genistein is shown in Fig. 18.11.

Genistein suppresses humoral- and cell-mediated immunity. When administered in ovariectomized mice, they induce thymic atrophy and suppressed delayed type hypersensitivity response (Yellayi et al. 2003). Genistein could induce T cell cytotoxic activity along with IL-2 induced natural killer cell activity. The basal splenocyte proliferation was also induced (Guo et al. 2001). Treatment of genistein on mice models exhibited an increase in lymphocyte proliferation, IFN-γ and LDH release (Ghaemi et al. 2012). It increases IL-4 production and suppresses IFN-γ thus modulating Th1 immune response in rheumatoid arthritis rat. Proinflammatory mediators, IL-1β, TNF-α, COX-2 and iNOS, were inhibited by treatment with genistein. The clinical trials for cancer, osteoporosis and Alzheimer's disease are going on for the development of genistein as a potent drug (Jantan et al. 2015).

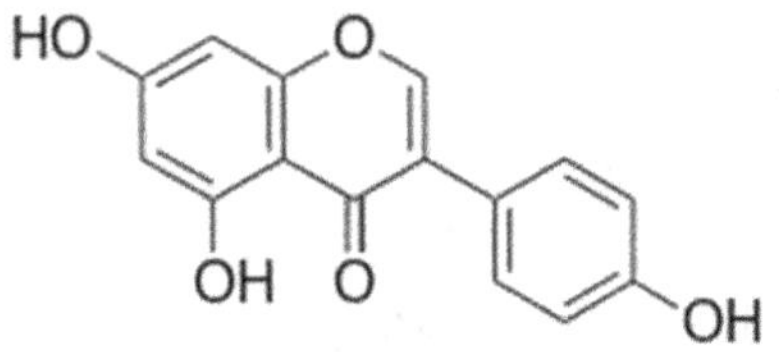

Fig. 18.11 Chemical structure of genistein

18.2.3.8 Andrographolide

The diterpenoid, isolated from *Andrographis paniculata*, is chemically (3-[2-[decahydro-6-hydroxy-5-(hydroxymethyl)-5,8-dimethyl-2-methylene-1-naphthalenyl] ethylidene] dihydro-4-hydroxy-2(3H)-furanone (Fig. 18.12).

It has a profound influence on cell signalling and immunomodulatory pathways. Andrographolide suppresses cell cycle progression by inducing cell-cycle inhibitory proteins p16, p21 and p27 and also by decreasing the expression of cyclin A, cyclin D, CDK4 and CDK2. It activates caspase-3 and caspase-8 inducing apoptosis of cancer cells and induced the production of cytotoxic T lymphocytes and human peripheral blood lymphocytes thereby enhancing IL-2 production. It has been found to exhibit anti-inflammatory property by inhibiting NF-κB binding to DNA, thereby inhibiting COX-2 and nitric oxide synthase. The chemotactic migration of macrophages is inhibited by treatment with andrographolide by suppressing the Erk1/2 and Akt signalling (Varma et al. 2011). Andrographolide inhibited CD40 and CD86 maturation markers and hinders the uptake and presentation of antigen leading to reduced antibody production. Treatment with andrographide reduces IL-12/IL-10 ratio and inhibited IL-23. It decreased endothelial cell proliferation and adhesion molecule ICAM-1 exhibiting angiogenesis. Modulation of immune system by andrographolide is achieved by both alternative and classical activation of macrophages and by synchronizing antibody and splenocyte production (Jantan et al. 2015).

18.2.3.9 ArtinM

ArtinM is a mannose-binding lectin isolated from the seeds of *Artocarpus heterophyllus*. It activates antigen-presenting cells, induce cytokine production and Th1 immunity. Artocarpin, a Gal/GalNAc-binding lectin, is able to selectively bind to human IgA1. It is a tetramer of four carbohydrate binding domain and 13-kDa subunits. In neutrophils, they induce signal transduction via G protein, phosphorylates tyrosine, increases TLR2 expression and induce phagocytic activity causing cell activation and haptotaxis. Mast cell recruitment and degranulation are induced by artinM by TNF-α release and mast cell recruitment from bone marrow.

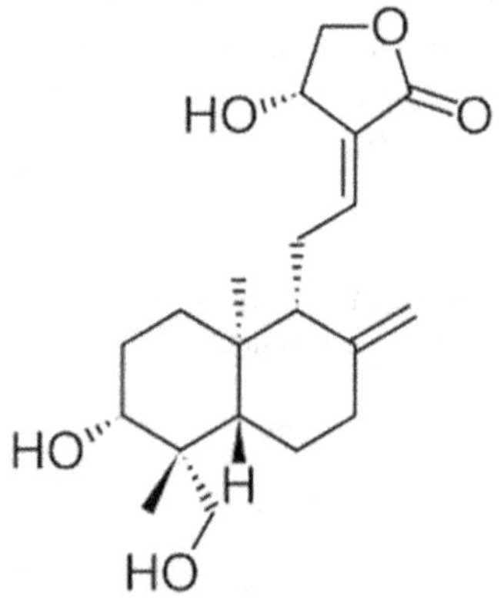

Fig. 18.12 Chemical structure of andrographolide

This in turn leads to neutrophil chemotaxis. ArtinM induces signal transduction via MYD88, NF-κB and IL-12 production in macrophages contributing to Th1-mediated immunity. In dendritic cells, it increases CD80 and CD86 expression and IL-12 production modulating cell maturation event and Th1 immunity (Souza et al. 2013).

18.3 Conclusions

Immunomodulators are used currently in clinical practices to either suppress or induce the immune function of an individual. Plants are natural sources which are safe and economical for immunomodulatory drugs. Many secondary metabolites like terpenes, alkaloids, flavonoids, lignin, lectin and polysaccharides with immunomodulatory properties have been identified and isolated. Only a few phytocompounds like curcumin, quercetin, capsaicin, colchicines, artinM, genistein, andrographolide have been studied vastly for their effect on immune cells. Some of these are also under clinical trials for immune-related disorders. A large variety of plants are known to possess immunomodulatory activity, but the scientific validation of the bioactive compound responsible for the activity is yet to be done. Also, the mechanism and mode of action of immunomodulatory phytocompounds need to be validated for the development of drugs in clinical practice.

References

Agarwal SS, Singh VK (1999) Medicinal plants; immunomodulators review of studies on Indian medicinal plants and synthetic peptides. Proc Indian Natl Sci Acad B65:179–204

Agarwal SS, Khadase SC, Talele GS (2010) Studies on immunomodulatory activity of *Capparis zeylanica* leaf extracts. Int J Pharm Sci Nanotechnol 3(1):887–892

Amirghofran Z, Azadbakht M, Karimi MH (2000) Evaluation of the immunomodulatory effects of five herbal plants. J Ethnopharmacol 72(1–2):167–172

Antony S, Kuttan R, Kuttan G (1999) Immunomodulatory activity of Curcumin. Immunol Invest 28 (5–6):291–303

Attard E, Cuschieri A (2009) *In vitro* immunomodulatory activity of various extracts of Maltese plants from the Asteraceae family. J Med Plant Res 3(6):457–461

Augustine BB, Dash S, Lahkar M, Amara VR, Samudrala PK, Thomas JM (2014) Evaluation of immunomodulatory activity of ethyl acetate extract of *Leucas aspera* in Swiss albino mice. Int J Green Pharm:84–89

Banerjee M, Tripathi LM, Srivastava VML, Puri A, Shukla R (2003) Modulation of inflammatory mediators by ibuprofen and curcumin treatment during chronic inflammation in rat. Immunopharmacol Immunotoxicol 25:213–224

Berman AY, Motechin RA, Wiesenfeld MY, Holz MK (2017) The therapeutic potential of resveratrol: a review of clinical trials (Review Article). Precis Oncol 35:1–9

Bob Allkin (2017) Useful plants—medicines. State of the World's Plants 2017. Royal Botanic Gardens

Boscolo P, del Signore A, Sabbioni E, Di Gioacchino M, Di Giampaolo L, Reale M, Conti P, Paganelli R, Giaccio M (2003) Effects of resveratrol on lymphocyte proliferation and cytokine release. Ann Clin Lab Sci 33(2):226–231

Bureau G, Longpre F, Martinoli MG (2008) Resveratrol and quercetin, two natural polyphenols, reduce apoptotic neuronal cell death induced by neuroinflammation. J Neurosci Res 86:403–410

Chanan-Khan AA, Swaika A, Paulus A et al (2013) Pomalidomide: the new immunomodulatory agent for the treatment of multiple myeloma. Blood Cancer J 3(9):e143. https://doi.org/10.1038/bcj.2013.38

Cundelland DR, Wilkinson F (2014) Curcumin: powerful immunomodulator from turmeric. Curr Immunol Rev 10:122–132. https://doi.org/10.2174/1573395510666141029233003

Dahanukar SA, Kulkarni RA, Rege NN (2000) Pharmacology of medicinal plants and natural products. Indian J Pharm 32:S81–S118

Delafuente JC, Devane CL (1991) Immunologic effects of cocaine and related alkaloids. Immunopharmacol Immunotoxicol 13:1–2

Dhama K, Saminathan M, Jacob SS, Mithilesh S, Karthik K, Amarpal RT, Sunkara LT, Malik YS, Singh RK (2015) Effect of immunomodulation and immunomodulatory agents on health with some bioactive principles, modes of action and potent biomedical applications. Int J Pharm 11:253–290

Falchetti R, Fuggetta MP, Lanzilli G, Tricarico M, Ravagnan G (2001) Effects of resveratrol on human immune cell function. Life Sci 70(1):81–96

Fu Q, Cui Q, Yang Y, Zhao X, Song X, GuangxiWang LB, Chen S, Tian Y, Zou Y, Li L, GuizhouYue RJ, Yin Z (2018) Effect of resveratrol dry suspension on immune function of piglets. Evid Based Complement Alternat Med:5952707. https://doi.org/10.1155/2018/5952707

Gately MK, Desai BB, Wolitzky AG, Quinn PM, Dwyer CM, Podlaski FJ, Familletti PC, Sinigaglia F, Chizzonite R, Gubler U (1991) Regulation of human lymphocyte proliferation by a heterodimeric cytokine, IL-12 (cytotoxic lymphocyte maturation factor). J Immunol 147 (3):874–882

Gaur K, Kori ML, Nema RK (2009) Comparative screening of immunomodulatory activity of hydro-alcoholic extract of *Hibiscus rosa sinensis* Linn. And Ethanolic extract of *Cleome gynandra* Linn. Global J Pharmacol 3(2):85–89

Ghaemi A, Soleimanjahi H, Razeghi S, Gorji A, Tabaraei A, Moradi A, Alizadeh A, Vakili MA (2012) Genistein induces a protective immunomodulatory effect in a mouse model of cervical Cancer. Iran J Immunol 9(2):119–127

Gottwald T, Lhotak S, Stead RH (2003) Effect of truncal vagotomy and capsaicin on mast cells and IgA-positive plasma cells in rat jejunal mucosa. Neurogastroenterol Motil 9(1)

Guo TL, McCay JA, Zhang LX, Brown RD, You L, Karrow NA, Germolec DR, White KL Jr (2001) Genistein modulates immune responses and increases host resistance to B16F10 tumor in adult female B6C3F1 mice. J Nutr 131(12)

He X, Niu X, Li J, Xu S, Lu A (2012) Immunomodulatory activities of five clinically used Chinese herbal polysaccharides. J Exp Integr Med 2(1):15–27

Huynh NB (2017) The immunological benefits of green tea (*Camellia sinensis*). Int J Biol 9 (1):10–17

Ilfeld D, Feierman E, Kuperman O, Kivity S, Topilsky M, Netzer L, Pecht M, Trainin N (1984) Effect of colchicine on T cell subsets of healthy volunteers. Immunology 53(3):595–598

Jagetia GC, Aggarwal BB (2007) "Spicing up" of the immune system by Curcumin. J Clin Immunol 27(1):1–19

Jantan I, Ahmad W, Bukhari SNA, Jantan I, Ahmad W, Bukhari SNA (2015) Plant-derived immunomodulators: an insight on their preclinical evaluation and clinical trials. Front Plant Sci 6:655

Jayaraman S, Jayadevi Variyar E (2015) Evaluation of immunomodulatory and antioxidant activities of polysaccharides isolated from *Callicarpa macrophylla Vahl*. Int J Pharm Pharm Sci 7(9):1–4

Jayaraman S, Sumesh Kumar TM, Jayadevi Variyar E (2015) Evaluation of immunomodulation by *Lepidagathis cuspidate* and *Phaseolus trilobus*. Intl J Innov Pharm Sci Res 3(7):1–10

Jobin C, Bradhan CA, Russo MP, Juma B, Narula AS, Brenner DA, Sartor RB (2011) Curcumin blocks cytokine mediated NF-kappa B activation and pro-inflammatory gene expression. J Immunol 50(3):163–172

Jose J, Dhanya AT, Haridas KR, Sumesh Kumar TM, Jayaraman S, Variyar EJ, Sudhakaran S (2016) Structural characterization of a novel derivative of myricetin from *Mimosa pudica* as an antiproliferative agent for the treatment of cancer. Biomed Pharmacother 84:1067–1077

Jun W, Pei-Gen X, Shi-Ying L, Gao P (1992) The inhibitory effect of Sinomenine on immunological function in mice. Phytother Res 6(3):117–120

Jung J-H, Kang J-I, Kim H-S (2012) Effect of quercetin on impaired immune function in mice exposed to irradiation. Nutr Res Pract 6(4):301–307

Kaplan MM (1997) The use of methotrexate, colchicine, and other immunomodulatory drugs in the treatment of primary biliary cirrhosis. Semin Liver Dis 17(2):129–136

Kołodziej H, Kiderlen AF (2007) In vitro evaluation of antibacterial and immunomodulatory activities of *Pelargonium reniforme, Pelargonium sidoides* and the related herbal drug preparation EPs® 7630. Phytomedicine 14(Suppl 6):18–26

Kumar SV, Kumar SP, Rupesh D, Nitin K (2011) Immunomodulatory effects of some traditional medicinal plants. J Chem Pharm Res 3(1):675–684

Joanna Kurek (2017) Cytotoxic colchicine alkaloids: from plants to drugs, Cytotoxicity, Tülay Aşkın Çelik, IntechOpen: https://doi.org/10.5772/intechopen.72622. https://www.intechopen.com/books/cytotoxicity/cytotoxic-colchicine-alkaloids-from-plants-to-drugs

Kurup VP, Barrios CS (2008) Immunomodulatory effects of curcumin in allergy. Mol Nutr Food Res 52(9):1031–1039

Lai JH (2002) Immunomodulatory effects and mechanisms of plant alkaloid tetrandrine in autoimmune diseases. Acta Pharmacol Sin 23(12):1093–1101

Leung YY, Hui LLY, Kraus VB (2015) Colchicine—update on mechanisms of action and therapeutic uses. Semin Arthritis Rheum 45(3):341–350

Li Y, Yao J, Han C, Yang J, Chaudhry MT, Wang S, Liu H, Yin Y (2016) Quercetin, inflammation and immunity. Nutrients 8(3):167

Malaguarnera L (2019) Review-influence of resveratrol on the immune response. Nutrients 11(5):94–96

Manosroi A, Saraphanchot Witthaya A, Manosroi J (2005) In vitro immunomodulatory effect of *Pouteria cambodiana* (Pierre ex Dubard) Baehni extract. J Ethnopharmacol 101:90–94

Min S-Y, Yan M, Kim SB, Ravikumar S, Kwon S-R, Vanarsa K, Kim H-Y, Davis LS, Mohan C (2015) Green tea Epigallocatechin-3-Gallate suppresses autoimmune arthritis through Indoleamine-2,3-Dioxygenase expressing dendritic cells and the nuclear factor, Erythroid 2-like 2 antioxidant pathway. J Inflamm 12(53)

Mlcek J, Jurikova T, Skrivankova S, Sochor J (2016) Quercetin and its anti-allergic immune response. Molecules 21(5)

Motte J, Ambrosius B, Grüter T, Bachir H, Sgodzai M, Pedreiturria X, Pitarokoili K, Gold R (2018) Capsaicin-enriched diet ameliorates autoimmune neuritis in rats. J Neuroinflammation 15:122

Nevius E, Srivastava PK, Basu S (2012) Oral ingestion of capsaicin, the pungent component of chili pepper, enhances a discrete population of macrophages and confers protection from autoimmune diabetes. Mucosal Immunol 5(1):76–86

Novak M, Vetvicka V (2008) β-Glucans, history, and the present: Immunomodulatory aspects and mechanisms of action. J Immunotoxicol 5:47–57

Pae M, Wu D (2013) Immunomodulating effects of epigallocatechin-3-gallate from green tea: mechanisms and applications. Food Funct 4(9):1287–1303

Park HJ, Lee CM, Jung ID, Lee JS, Jeong YI, Chang JH, Chun SH, Kim MJ, Choi IW, Ahn SC, Shin YK, Yeom SR, Park YM (2009) Quercetin regulates Th1/Th2 balance in a murine model of asthma. Int Immunopharmacol 3:261–267

Ramberg JE, Nelson ED, Sinnott RA (2010) Immunomodulatory dietary polysaccharides: a systematic review of the literature. Nutr J 9(54):1–22

Raphael TJ, Kuttan G (2003) Immunomodulatory activity of naturally occurring monoterpenes carvone, limonene, and perillic acid. Immunopharmacol Immunotoxicol 25(2):285–294

Saadé NE, Massaad CA, Ochoa-Chaar CI, Jabbur SJ, Safieh-Garabedian B, Atweh SF (2002) Upregulation of proinflammatory cytokines and nerve growth factor by intraplantar injection of capsaicin in rats. J Physiol 545(1):241–253

Saiko P, Szakmary A, Jaeger W, Szekeres T (2008) Resveratrol and its analogs: Defense against cancer, coronary disease and neurodegenerative maladies or just a fad? Mutat Res Rev 658:68–94

Satpute KL, Jadhav MM, Karodi RS, Katare YS, Patil MJ, Rub R, Bafna AR (2009) Immunomodulatory activity of fruits of *Randia dumetorum* Lamk. J Pharmacogn Phytother 1:1–5

Schepetkin IA, Faulkner CL, Nelson-Overton LK, Wiley JA, Quinn MT (2005) Macrophage immunomodulatory activity of polysaccharides isolated from *Juniperus scopulorum*. Int Immunopharmacol 5(13):1783–1799

Schwarz YA, Kivity S, Ilfeld DN, Schlesinger M, Greif J, Topilsky M, Garty MS (1990) A clinical and immunologic study of colchicines in asthma. J Allergy Clin Immunol 85(3):578–582

Singh B, Sharma RA (2015) Plant terpenes: defense responses, phylogenetic analysis, regulation and clinical applications. 3 Biotech Biotech 5(2):129–151

Singh VK, Biswas S, Mathur KB, Haq W, Garg SK, Agarwal SS (1998) Thymopentin and splenopentin as immunomodulators. Current status (review). Immunol Res 17(3):345–368

Singh N, Tailang M, Mehta SC (2016) A review on herbal plants as immunomodulators. Int J Pharm Sci Res 7(9):3602–3610

Souza MA, Carvalho FC, Ruas LP, Ricci-Azevedo R, Roque-Barreira MC (2013) The immunomodulatory effect of plant lectins: a review with emphasis on Artin M properties. Glycoconj J 30:641–657

Surh YJ, Chun KS, Cha HH, Han SS, Keum Y, Park KK, Lee SS (2001) Molecular mechanisms underlying chemopreventive activities of anti-inflammatory phytochemicals: down-regulation of COX-2 and iNOS through suppression of NF-kappa B activation. Mutat Res:43–68

Varma A, Padh H, Shrivastava N (2011) Andrographolide: a new plant-derived antineoplastic entity on horizon. Evid Based Complement Alternat Med 2011:815390. https://doi.org/10.1093/ecam/nep135

Watson ES, Murphy JC, ElSohly HN, ElSohly MA, Turner CE (1983) Effects of the administration of coca alkaloids on the primary immune responses of mice: interaction with delta 9-tetrahydrocannabinol and ethanol. Toxicol Appl Pharmacol 71(1):1–13

Wong CP, Nguyen LP, Noh SK, Bray TM, Bruno RS, Ho E (2011) Induction of regulatory T cells by green tea polyphenol EGCG. Immunol Lett 139(1):7–13

Yellayi S, Zakroczymski MA, Selvaraj V, Valli VE, Ghanta V, Helferich WG, Cooke PS (2003) The phytoestrogen genistein suppresses cell-mediated immunity in mice. J Endocrinol 176 (2):267–274

Yu R, Park JW, Kurata T, Erickson KL (1998) Modulation of select immune responses by dietary capsaicin. Int J Vitam Nutr Res 68(2):114–119

Zhou T, Zhu M, Liang Z (2018) (−)-Epigallocatechin-3-gallate modulates peripheral immunity in the MPTP-induced mouse model of Parkinson's disease. Mol Med Rep 17(4):4883–4888

Polyphenols: An Overview of Food Sources and Associated Bioactivities

19

Alejandro Zugasti-Cruz, Raúl Rodríguez-Herrera, and Crystel Aleyvick Sierra-Rivera

Abstract

Polyphenolic compounds constitute one of the largest and most important groups of plant secondary metabolites because of their widespread occurrence in nature, with more than 8000 identified phenolic structures. Polyphenols are compounds with one or more hydroxyl groups attached to the benzene ring. One of the main properties of polyphenols is the antioxidant capacity, which has been related to the presence of hydroxyl groups in their chemical structure. This activity has been demonstrated in both in vitro and in vivo models aimed at preventing or treating diseases related to oxidative stress, such as cancer, diabetes, cardiovascular and neurological disorders and inflammatory diseases, among others. Due to the promising results in experimental models in recent years, some clinical trials have been conducted in patients with various conditions treated with polyphenolic compounds. However, to date, there are few protocols completed. Likewise, the biological effect of polyphenols depends on their bioavailability. In the present manuscript, data were collected about biological activities, health beneficial properties and possible therapeutic applications of the polyphenolic compounds.

A. Zugasti-Cruz
Laboratory of Immunology and Toxicology, Autonomous University of Coahuila, Saltillo, Coahuila, Mexico

Food Research Department, Faculty of Chemistry, Autonomous University of Coahuila, Saltillo, Coahuila, Mexico

R. Rodríguez-Herrera (✉)
Food Research Department, Faculty of Chemistry, Autonomous University of Coahuila, Saltillo, Coahuila, Mexico
e-mail: raul.rodriguez@uadec.edu.mx

C. A. Sierra-Rivera
Laboratory of Immunology and Toxicology, Autonomous University of Coahuila, Saltillo, Coahuila, Mexico

Keywords

Polyphenols · Biological activity · Cancer · Anti-inflammatory · Phytoestrogens

19.1 Introduction

The World Health Organization (WHO) declared in 2016 that worldwide cardiovascular diseases, chronic obstructive disease (COPD), respiratory infections cancer and diabetes caused 25 million deaths (WHO 2019). In this context, scientific research is focused on the search for natural compounds that promote an improvement in the quality of life of patients suffering from various diseases or conditions that cause high rates of morbidity and mortality.

Plant metabolism is performed through chemical reactions to synthesize complex substances from simpler ones or to degrade complex ones and obtain simple substances (Matsuura et al. 2018). Primary metabolites (PM) are called carbon and nitrogen compounds that contribute to cell function, including amino acids, nucleotides, sugars and lipids. These are present in all plants and are involved in the photosynthetic, reproduction and respiratory processes, nutrient assimilation, transport, protein carbohydrates and lipid synthesis. On the other hand, there are numerous chemical compounds derived from primary metabolic pathways that do not have a direct function in the essential processes of the plant cell which are called secondary metabolites (SM). The SM is involved in the ecological interactions between the plant and its environment, defence functions, attracting pollinators and are used as allelopathic agents, drugs and insecticides (Pagare et al. 2015). Currently, biomedicine has a strong interest in the discovery of the biological properties of SM. For this reason, the pharmaceutical industry is promoting research on isolation, identification and purification of biologically active compounds to develop novel drugs with analgesic action, antipyretic, antimicrobial and antitumour potential. SM is classified based on chemical structures including alkaloids, saponins, terpenes, lipids, carbohydrates and polyphenols (Hussein and El-Anssary 2018). The main objective of this article is to integrate the scientific evidence developed in recent years about the importance of polyphenols, their sources and possible therapeutic applications.

19.2 Polyphenolic Compounds: Sources and Classification

Polyphenols are secondary metabolites most frequently found in plants and these metabolites contain a hydroxyl group in the aromatic ring called Phenol (Pagare et al. 2015). There are more than 8000 identified phenolic structures (Tanase et al. 2019). These compounds are distributed in different parts of the plant such as stem, roots, seeds, bark, flowers and leaves and contribute to the pigmentation, flavour and biological activity of various foods and beverages prepared with herbs, vegetables or fruits. Some sources rich in polyphenols are apple, aloe, cranberry, banana, broccoli, carrot, cucumber, ginger, lemon, onion, pear, pomegranate, red grapes and spinach containing about 200–300 mg polyphenols per 100 g fresh weight. A glass of red wine or a cup of coffee has about 100 mg of polyphenols (Srivastava and Kumar-Mishra 2015).

Several researchers have highlighted the biological activities of polyphenols including the potential antibacterial, anti-inflammatory, antioxidant, antitumoural activities. Based on the chemical structure polyphenols can be classified into sub-groups according to the number of phenol rings. These are phenolic acids, flavonoids stilbenes, and lignans. Tables 19.1–19.4 show the precise classification as well as the most studied compounds (Tanase et al. 2019).

19.2.1 Phenolic Acids

The number and position of hydroxyl groups in the aromatic ring cause the structural variation of phenolic acids. The chemical classification of phenolic acids consists of the hydroxycinnamic and hydroxybenzoic acid (Tanase et al. 2019). Hydroxycinnamic acids are present in many products including cauliflower, coffee, olive oil, tea, cocoa, mushrooms, pear and wine (El-Seedi et al. 2018). For example, a cup of coffee contains up to 350 mg and blueberries contain 2 g of hydroxycinnamic acids/kg fresh weight (Archivio et al. 2007). On the other hand, black radishes, onions, tea leaves are rich sources of free hydroxybenzoic acid (Table 19.1) (Papuc et al. 2017). Also, red fruits such as blackberries and raspberry contain around 100–270 mg/kg fresh weight of gallic acid (Archivio et al. 2007).

19.2.2 Flavonoids

Flavonoids are low-molecular-weight compounds composed of two aromatic rings (A and B) linked by an oxygenated heterocyclic ring (C) constituted of three carbons atoms (C6–C3–C6). Depending on the type of hydroxylation and differences in the ring C, flavonoids can be divided into subfamilies, including flavonols, flavanones, flavanonols, flavones, flavan-3-ols or catechins (Zamora-Ros et al. 2014), isoflavones (Szeja et al. 2017), anthocyanin (Khoo et al. 2017), chalcones (Panche et al. 2016) and tannins (Khanbabaee and Van Ree 2001) – Table 19.2. Although dietary habits are very different in the world, it is estimated that the average value of flavonoid intake is about 23 mg/day and quercetin is the predominant flavonoid. Among the main food sources of flavonoids are black tea, onions, apples, black pepper, which contain around 4 g/kg of quercetin (Salminen and Karonen 2011).

19.2.3 Stilbenes

Stilbenes' chemical structure contains two aromatic rings (ring A and B) linked with a methylene bridge (Tanase et al. 2019, Lim and Koffas 2010). These compounds

Table 19.1 Phenolic acids: classification, sources and biological activity

Sub group: Hydoxycinnamic acid		
Compound and source	Biological activity	References
Caffeic acid—apple, aubergine, blueberries, ciders, coffee beverage.	Antioxidant, antitumour, anti-ischaemia, anti-thrombotic, antihypertension, antifibrosis, antivirus.	Magnani et al. (2014), Kanimozhi and Prasad (2015), Papuc et al. (2017)
Ferulic acid—Banana, bean, beetroot, citrus juices, coffee, nuts, peanut, pineapple, rice, seeds of coffee, spinach, wheat.	Antioxidant, anticancer, anti-inflammatory, anti-thrombotic, antidiabetic, antimicrobial, cardio- and neuroprotection.	Kumar and Pruthi (2014), Mancuso and Santangelo (2014), Song et al. (2014), Kanimozhi and Prasad (2015), Wang et al. (2016)
p-Coumaric acid—beet fibre, berries, cereals brans, grapes, onions, orange, potatoes, spinach, sugar, tomato.	Antioxidant, anticancer, anti-inflammatory, antiplatelet aggregation, anxiolytic, antipyretic, analgesic, anti-arthritis antidiabetic, antimicrobial, antivirus.	Kanimozhi and Prasad (2015), Pei et al. (2016)
Synaptic acid—broccoli, citrus juices.	Antioxidant, anticancer, antimutagenic, anti-inflammatory	Martinović and Abramovi (2014)
Subgroup: Hydroxybenzoic acid		
Gallic acid—apple, blueberries, black and green tea, raspberry, red wine walnuts, watercress.	Antioxidant, anticancer, antihyperlipidemic, antidiabetic cardio and neuroprotection, antimicrobial.	Ow and Stupans (2003), Archivio et al. (2007), Locatelli et al. (2013)
Protocatechuic acid—grain brown rice, onion, plums, gooseberries, grapes, nuts, almonds, cinnamon.	Antioxidant, anticancer, antiulcer, antidiabetic, antiaging, antifibrotic, antiviral, anti-inflammatory, analgesic, antiatherosclerotic, cardiac-. Hepato-, neuro- and nephroprotective, antibacterial.	Kakkar and Bais (2014), Khan et al. (2015)
Syringic acid—olive, date, pumpkin, grapes, honey, red wine.	Anticancer, anti-inflammatory, antidiabetic, antiangiogenesis, antimicrobial.	Srinivasulu et al. (2018)
Vanillic acid—Vanilla, whole wheat, berry, mango pulp.	Anti-inflammatory, anti-diabetic, anti-hypertensive, anti-asthmatic and liver protective.	Itoh et al. (2010), Kumar et al. (2011), Palafox et al. (2012), Chang et al. (2015)

can be divided into two subgroups: monomeric and oligomeric stilbenes. Monomeric stilbenes are classified into four subclasses: stilbenes, bibenzyls, bisbibenzyls and phenanthrenoids. On the other hand, oligomeric stilbenes are divided into three groups, the first group containing at least one to five-membered oxygen heterocycle, the second group containing no oxygen heterocycle and finally, the third group

Table 19.2 Flavonoids: classification, sources and biological activity

Subgroup flavonols		
Compound and source	Biological activity	References
Fisetin—apple, cucumber, grape, kiwi, onion, peach, strawberry, tomato.	Antioxidant, anti-inflammatory, antitumourigenic, antiangiogenic, antidiabetic, neuroprotective, cardioprotective.	Khan et al. (2013), Pal et al. (2016)
Kaempferol—apple, beans, broccoli, cassava, strawberries, tea.	Antioxidant, anti-inflammatory, anti-metastasis/anti-angiogenesis.	Chen and Chen (2013), Rajendran et al. (2014), Baião et al. (2017)
Morin—almond hulls, apple, cereal grains, coffee, guava, onion, red wine, seaweeds, tea.	Antioxidant, anti-inflammatory, antitumoural, antidiabetic, antihypertensive, hypouricemic, neuroprotective effects, antibacterial.	Caselli et al. (2016), Zhou et al. (2017a, b)
Myricetin—berry, coffee, red wine, tea, nut.	Antioxidant, anticancer, anti-inflammatory, antidiabetic, neuroprotection.	Somerset and Johannot (2008), Semwal et al. (2016), Seydi et al. (2016)
Quercetin—apple, beans, broccoli, cocoa, curly kale, cherry tomato, grape, leek, onion, red leaf lettuce, romaine lettuce, asparagus, green pepper.	Anti-inflammatory, antihypertensive, vasodilator effects, antiobesity, antihypercholesterolemic and antiatherosclerotic activities.	Manach et al. (2004), Somerset and Johannot (2008), Nishimuro et al. (2015), Anand David et al. (2016), Baião et al. (2017)
Rutin—apple, black tea, cassava, citrus, flower buds of *Saccharina japonica*, *Sophora* sp.	Antioxidant, cytoprotective, vasoprotective, anticarcinogenic, neuro and cardioprotective.	Atanassova and Bagdassarian (2009), Panthati et al. (2011), Ganeshpurkar and Saluja (2017)
Subgroup Flavanones		
Hesperetin—lemon, mandarin, sweet orange.	Anti-inflammatory, cardio and neuro protective, anti-allergic dermatitis-inhibiting effects, hypolipidemic, antimicrobial.	Somerset and Johannot (2008), Liu et al. (2011), Wang et al. (2017)
Hesperidin—grapefruit juice sweet orange.	Antioxidant, neuroprotective, anti-allergic, antihyperlipidemic, cardioprotective, antihypertensive, antidiabetic, antimicrobial.	Liu et al. (2011), Zanwar et al. (2014a, b)
Naringin—grapefruit, orange, pummelo.	Antioxidant, anti-inflammatory, anti-apoptotic, anti-ulcer, anti-osteoporotic, anti-carcinogenic properties.	Alam et al. (2014), Chen et al. (2016)
Naringenin—beans, bergamot, cocoa, grapefruit, lemon juice tomato.	Antioxidant, antitumour, antiviral, antibacterial, anti-inflammatory, cardioprotective effects.	Manach et al. (2004), Somerset and Johannot (2008), Salehi et al. (2019)

(continued)

Table 19.2 (continued)

Subgroup flavonols		
Compound and source	Biological activity	References
Subgroup Flavanonols		
Taxifolin—Citrus fruits, cloudberry, cocoa, grapes, green tea, macadamia nut, olive oil, onions, wine.	Anti-inflammatory, analgesic, antioxidant, antipyretic, platelet inhibitory, anticancer actions.	Yang et al. (2016), Asmi et al. (2017)
Subgroup flavones		
Apigenin—parsley, celery, chamomile, cherry, grapefruit onion, orange.	Atherogenesis, hypertension, cardiac hypertrophy, ischemia/reperfusion-induced heart injury, and autoimmune myocarditis, inhibits asthma, improves pancreatitis, type 2 diabetes and its complication, osteoporosis, collagen-induced arthritis.	Somerset and Johannot (2008), Medina et al. (2014), Pápay et al. (2015), Zhou et al. (2017a, b)
Baicalein—roots of *Scutellaria baicalensis*, *Scutellaria* radax, *Scutellaria lateriflora*, *Oroxylum indicum*.	Antioxidant, anticancer, anti-inflammatory, cardio and neuro-protective properties.	de Oliveira et al. (2015), Bie et al. (2017)
Diosmetin—bergamot and lemon juice, olive leaves.	Antioxidant, anti-inflammatory, anti-allergic, oestrogenic, antitumoural, hepatoprotective, antithrombotic, antibacterial, antiviral.	Patel et al. (2013a), Poór et al. (2014), Qiao et al. (2016), Bie et al. (2017)
Luteolin—broccoli, *Capsicum pepper*, celery, chamomile tea, English spinach, green peppers, oregano, thyme.	Antioxidant, anticarcinogenic, anti-inflammatory, antidepressant, antinociceptive, anxiolytic-like effects.	Somerset and Johannot (2008), Medina et al. (2014), Chen et al. (2016)
Subgroup Flavan-3-ols or catechins		
Catechin—apple, bean, beer, blueberry, cacao, cassava chocolate, gooseberry, grape seed, green tea, kiwi, red wine, strawberry.	Antioxidative, antihypertensive, anti-inflammatory, antiproliferative, antithrombogenic, anti-hyperlipidemic antiangiogenesis.	Zanwar et al. (2014a, b), Baião et al. (2017)
Epigallocatechin—cocoa, cherry green tea, red wine, walnut.	Antioxidant, anticancer, anti-inflammatory, antidiabetic, cardio and neuroprotective.	Medina et al. (2014), Naumovski et al. (2015), Granja et al. (2017), Cai et al. (2018)
Epicatechin—apple, apricot, blackberry, black tea, cider, cocoa, cherry, grape, green tea, peach, red wine.	Antioxidant, anticancer, antiangiogenesis, antidiabetic	Manach et al. (2004), Medina et al. (2014), Abdulkhaleq et al. (2017)

(continued)

Table 19.2 (continued)

Compound and source	Biological activity	References
Subgroup flavonols		
Theaflavin—black tea	Anticancer, anti-inflammatory, anti-obesity, anti-atherosclerotic, anti-osteoporotic, anti-dental caries properties, antiviral, antibacterial.	Kosińska and Andlauer (2014), Takemoto and Takemoto (2018)
Subgroup Isoflavones		
Afromosin—heartwood of *Peterocarpus erinaceus*, *Wistaria brachybotrys*.	Antitumoural	Konoshima et al. (1992), Noufou et al. (2012)
Betavulgarin—beetroot	Anti-tumour, antioxidant	Medina et al. (2014)
Daidzein—soybean, root of *Sophora flavescens*.	Estrogenic activity, antioxidant, anticancer, cardiovascular disease, diabetes, osteoporosis, skin disease, neurodegenerative disease.	Manach et al. (2004), Panthati et al. (2011), Rivera et al. (2013), Sun et al. (2016)
Daidzin—soybean.	Estrogenic activity and anticancer.	Medina et al. (2014), Poschner et al. (2017)
Genistein—miso, soybeans.	Estrogenic activity, antioxidant, anticancer, antiangiogenic and anthelmintic.	Manach et al. (2004), Sail and Hadden (2013), Jeevan et al. (2018)
Luteone—seed of *Lupinus luteus*, root of *Lupinus albus*.	Antioxidant and antibacterial.	Jeevan et al. (2018)
Pratensin—root of *Sophora secundiflora*.	Antiproliferative effects inhibit prostatic smooth muscle contractions.	Brandli et al. (2010), Panthati et al. (2011), Akter et al. (2016)
Subgroup anthocyanin		
Cyanidin—blackberry, cassava, cherry, purple grape, soybean, taro.	Antioxidant, anticancer, antidiabetic, reduce atherosclerosis, neuro, cardio-, and hepato-protective.	Manach et al. (2004) Medina et al. (2014), Chen et al. (2018a, b, c)
Delphinidin—black grape, blueberry, soybean, taro.	Antioxidant, anticancer, anti-inflammatory neuro and hepato-protective.	Manach et al. (2004), Patel et al. (2013b), Medina et al. (2014)
Pelargonidin—black currant.	Antioxidant, anti-carcinogenic, anti-inflammatory, antidiabetic, cardio and hepato-protective.	Manach et al. (2004), Duarte et al. (2018), Lee et al. (2018)
Peonidin—blueberry, cherry.	Antioxidant, anticancer	Manach et al. (2004), Chen et al. (2005)
Malvidin—blueberry, cherry, plum, red wine, red cabbage, rhubarb, strawberry.	Antioxidant, anticancerogenic and anti-inflammatory. Antihypertensive, cardioprotective	Manach et al. (2004), Somerset and Johannot (2008), Huang et al. (2016)

(continued)

Table 19.2 (continued)

Subgroup flavonols		
Compound and source	Biological activity	References
Petunidin—blueberry	Osteoblastogenesis	Somerset and Johannot (2008), Nagaoka et al. (2019)
Subgroup Chalcones		
Arbutin—pear skin, strawberry, wheat, *Bergenia crassifolia*, leaves of *Arbutus unedo*, *Arctostaphylos uvaursi*.	Skin-whitening agent	Inoue et al. (2013), Reis (2013), Jurica et al. (2015)
Chalconaringenin—cherry tomato.	Antioxidant	Slimestad and Verheul (2005)
Phloretin—apple tree bark, apple pulp and peel.	Antioxidant anti-inflammatory	Reis (2013), Rana and Bhushan (2016), Aliomrani et al. (2016)
Phloridzin—apple.	Produce renal glycosuria and block intestinal glucose absorption—diabetes mellitus	Ehrenkranz et al. (2005), Reis (2013)
Subgroup tannins		
Gallotannins—apple, berry, grape juice, pomegranate, strawberry	Antioxidant antimicrobial, anti-cancer, and neuro and cardio-protective properties	Medina et al. (2014), Kujawski et al. (2016)
Ellagitannins—apple, berry, grape juice, pomegranate, strawberry	Antioxidant, anti-inflammatory, chemopreventive, anticarcinogenic, and antiproliferative activities	Ismail et al. (2016)
Complex tannins—grape, green tea, persimmon leaves, jelly fig (*Ficus awkeotsang*), *Quercus acutissima*	Antioxidant, anticancer, hepatoprotective	Mamat et al. (2013), Quideau et al. (2013), Wang et al. (2019)
Condensed tannins (*Proanthocyanidins*)—apple, apricot, beer, berry, cider, cranberry, chocolate grape, kaki, peach, pear, tea, wine	Antioxidant, anticancer, anti-inflammatory, immunomodulatory, cardio-protective and antithrombotic properties. Antibacterial and antiviral activities	Manach et al. (2004), Lamy et al. (2016), Smeriglio et al. (2017)

consists of resveratrol oligomers including isorhapontigenin, piceatannol and resveratrol. This last group has the main biological activities reported – Table 19.3, (Valavanidis and Vlachogianni 2013).

Stilbenes are synthesized naturally by several plants such as *Vitis vinifera*, *Arachis hypogaea*, *Sorghum bicolor*, *Rhodomyrtus tormentosa*, *Rheum undulatum*, *Euphorbia lagascae*, *Vaccinium myrtillus*, *Pinus spp*, *Picea spp*, among others. Stilbenes act as a defense agent against phytopathogens and stress mediated by UV light (Reinisalo et al. 2015). Also, various studies have shown that stilbenes

Table 19.3 Stilbenes: classification, sources and biological activity

Compound and source	Biological activity	References
Resveratrol—bilberry, cocoa, cranberry, grape, highbush blueberry, red and white wine, *Polygonum cuspidatum*, peanut, rhubarbs, sugar cane, tomato.	Antioxidant, anticancer, cardio-protection, inflammation, ageing process, diabetes, neurological dysfunction.	Xia et al. (2010), Medina et al. (2014), Park and Pezzuto (2015), Reinisalo et al. (2015), Tsai et al. (2017), El Khawand et al. (2018)
Piceatannol—grape, highbush blueberry, red wine, rhubarbs, sugar cane.	Antioxidant, anticancer, anti-inflammatory, anti-atherogenic, estrogenic activity, cardioprotective, antimicrobial.	Piotrowska et al. (2012), Tang and Chan (2014), Reinisalo et al. (2015)
Isorhapontigenin—*Picea mariana*, wine grapes.	Antioxidant, anti-inflammatory anticancer, and cardio-protection.	Fernández et al. (2012), Reinisalo et al. (2015), Dai et al. (2018), Tanase et al. (2019)
Pterostilbene—Highbush blueberry.	Anti-inflammatory, including hypolipidemic, antidiabetic, and mechanisms cardio-protection.	Reinisalo et al. (2015), Tsai et al. (2017)
Pinosylvin—heartwood *Pinus densiflora*, *Pinus thunbergii*.	Anti-inflammatory	Kodan et al. (2002), Laavola et al. (2015), Reinisalo et al. (2015)

have anti-obesity, anti-diabetic, anticancer, anti-inflammatory effects and have a cardio and neuroprotective activity (El Khawand et al. 2018).

Resveratrol is one of the main stilbene molecules that has antioxidant and antiproliferative activity and it also inhibits amyloid-beta peptide aggregation which suggests that it could have beneficial effects in patients with Alzheimer's disease (Németh et al. 2017). It is found only in low quantities in human diet (Lim and Koffas 2010) due to poor bioavailability after oral intake. Currently, there are dietary supplements available that contain resveratrol. In human trails, it has been observed that resveratrol is well tolerated in doses of 25–150 mg. However, repeated doses of 2.5 g and 5 g of resveratrol caused moderate adverse effects such as diarrhoea (Almeida et al. 2009). However, the intake values are variable with the factors like sex, age and lifestyle. For example, a study realized by Zamora-Ros et al. (2008) found a greater amount of stilbenes in the diet of men and older people.

19.2.4 Lignans

Lignans are phenylpropanoid dimers and the phenylpropane units are linked by the central carbon. Lignans are compounds found in many foods such as grains, legumes, vegetables, seeds, tea, coffee and wine (Table 19.4). Lignans are classified into eight subgroups: furofuran, furan, dibenzylbutyrolactone, aryletralin,

Table 19.4 Lignans: classification, sources and biological activity

Compound Source	Biological activity	References
Sesamin—sesame seeds	Anticancer, anti-inflammatory, anti- and pro-angiogenic.	Seth et al. (2015), Majdalawieh et al. (2017), Rodríguez et al. (2019)
Matairesinol—flaxseed, kale	Antioxidant, hormone replacement therapy, estrogenic, or anti-estrogenic activities and reduce the risk of hormone-dependent cancer. Reduce hypercholesterolemia, atherosclerosis and antidiabetic.	Tomás-Barberán et al. (2009), Choi et al. (2014), Medina et al. (2014), Seth et al. (2015), Rodríguez et al. (2019)
Pinoresinol—Beer, brassica cabbage, bread, green bean, tea, wheat potato.	Hormone replacement therapy, anticancer, reduce hypercholesterolemia, atherosclerosis and diabetes. Antimicrobial	Milder et al. (2005), Tomás-Barberán et al. (2009), Hwang et al. (2010), Seth et al. (2015), Rodríguez et al. 2019
Lariciresinol—apricot, broccoli, brussel sprout, cashew nut, cauliflower, carrot, cucumber, flaxseed, grapefruit, green bean, green sweet pepper, kale, peanut, potato, red and white cabbage, red sweet pepper, strawberry, tomato, zucchini.	Hormone replacement therapy, anticancer, reduce hypercholesterolemia, atherosclerosis and diabetes.	Tomás-Barberán et al. (2009), Medina et al. (2014), Seth et al. (2015), Rodríguez et al. (2019)
Secoisolariciresinol diglucoside—broccoli, carrot, cashew nut, flaxseed, kale, walnut, wheat bran, sauerkraut.	Cardiovascular diseases, diabetes, cancer and mental stress.	Medina et al. (2014), Seth et al. (2015), Kezimana et al. (2018), Rodríguez et al. (2019)

arylnaphthalene, dibenzocyclooctadiene, dibenzylbutyrolactol. Its classification is based on the chemical structure in levels of oxidation of aromatic rings and propyl side chains. In humans, secoisolariciresinol diglucoside, matairesinol, pinoresinol and lariciresinol lignan compounds are metabolized by gut microbiota to enterolactone and enterodiol (Seth et al. 2015). Its biological activities include anti-inflammatory, apoptotic, antiviral, antioxidant effects and induce estrogen receptor activation so they are called phytoestrogens (Milder et al. 2005). Flaxseeds are the foodstuff with the highest content of lignan, it contains around 294.21 mg/100 g lignin. Cashew nuts are also rich in lignans containing 56.33 mg/100 g. Sesaminol is the most abundant constituent of lignan in sesame seeds (containing 538.08 mg/100 g) (Rodríguez et al. 2019).

19.3 Plant Polyphenols in Therapy

19.3.1 Cancer

Cancer is caused by multifactorial events such as mutations in protooncogenes or tumour suppressor genes, caused by exposure to physical, chemical and biological agents (IARC 2018). This leads to the loss of regulation of the cell cycle, causing uncontrolled and autonomous proliferation, as well as the inability to induce a terminal differentiation and consequently the loss of cell division capacity (WHO 2018).

Cancer is one of the leading causes of death in the world. According to the National Institute of Cancer (NIC) 14.1 million new cases and 8.8 million deaths were related to cancer in 2015. It is expected that by 2030 morbidity will increase to about 23.6 million cases. The main types of cancer that caused morbidity and mortality are breast cancer, lung and bronchial cancer, prostate cancer, colon and rectum cancer, skin melanoma, bladder cancer, non-Hodgkin lymphoma, kidney and renal pelvis cancer, endometrial cancer, leukaemia, pancreatic cancer, thyroid cancer and liver cancer (NCI 2018). Currently, there are several treatments to combat this disease, including surgery, radiotherapy, chemotherapy, hormone therapy and immunotherapy. The choice of treatment will depend on the type of cancer, localization, extent, age and secondary diseases of the patient. However, some of these treatments induce adverse effects such as vomiting, diarrhoea, fatigue, immunosuppression and tissue toxicity (WHO 2018).

For this reason, nowadays natural compounds obtained from plants, vegetables or fruits that prevent the development of cancer or reduce tumour growth are sought.

Several investigations have determined the antitumour effects in vitro and in vivo of polyphenols, the results found are summarized in Tables 19.5 and 19.6. In a study by Xia et al. (2010) found that flavones reduced the proliferation of human adenocarcinoma cells HT-29 at 54.8 μmol/L and catechin decreased the cell viability of breast cancer cells MCF-7 at 30 μg/mL. On the other hand, proanthocyanidins significantly inhibited cell proliferation and viability in a concentration-dependent manner on metastatic breast cancer cells 4T1. In the same way, breast cancer MDA-MB-231 cell line was treated with epigallocatechin gallate or gallic acid, which has been associated with proteasome inhibition and leads to the death of tumour cells (Li et al. 2014). It was also associated with the presence of mutations in the p53 tumour suppressor gene (TP53) and are frequently found in human cancers (Kumar and Pandey 2013) representing an early stage in tumourigenesis (Perri et al. 2016). It has been found that flavonoids decreased the expression of mutant TP53 in human breast cancer cell lines (Kumar and Pandey 2013).

On the other hand, Song et al. (2016) determined an arrest in the G1 phase of the cell cycle and induction of apoptosis on hepatoma cells HepG2 treated with pomegranate. Similarly, Bai et al. (2010) found that resveratrol induced cell death by apoptosis on human bladder tumour cell line T24 (Cutrim and Sloboda, 2018) and human leukaemia cells HL60 (Xia et al. 2010). Besides, it has been described that quercetin, kaempferol, myricetin and rutin act as anticancer activity in vitro and

Table 19.5 Antitumoural and anti-inflammatory potential of polyphenols in in vitro models

Polyphenol	Concentration	Cellular model	Therapeutic target/ effects	References
Caffeic acid	10 mM	HTB-34	↑AMP-K	Tyszka et al. (2017)
Ferulic acid	150 μM	143B MG63	↑Arrest cell cycle Caspase-3	Wang et al. (2016)
Protocatechuic acid	2–8 μmol/L	MCF-7 A549 HepG2 HeLa LNCaP	DNA fragmentation ↑Caspase-3 ↓IL-6 IL-8	Yin et al. (2009)
Protocatechuic acid	0.5 μmol/L	PBMs	↓CCR2	Wang and Zou (2011)
Ellagic and gallic acid	50–200 μg/mL	RAW264.7	↓NO, PGE-2 and IL-6	BenSaad et al. (2017)
Kaempferol	100 μg/mL	MCF-7 SGC-7901 HeLa A549	↓Cell proliferation ↑Apoptosis	Liao et al. (2016)
Quercetin	50 μM	HepG2	↓NFκB ↑Caspase-3	Granado et al. (2010)
Hesperidin	200 μM	GBC	↑Caspase-3 Cell cycle arrest G2/M	Pandey et al. (2019)
Naringenin	24 μM	HL-60, HeLa, MCF-7, MDA-MB-231	↑Apoptosis ↑By caspase 3 Sub-G1 phase	Bodduluru et al. (2016)
Apigenin	100 μM	HNSCC	↓Cell proliferation and EGFR	Masuelli et al. (2011)
Epigallocatechin-3-gallate (EGCG)	100 μM	SW780	↑Apoptosis by caspase 3	Luo et al. (2017)
Phloridzin	100 μM	HepG2 MDA-MB-231	↓DNA topoisomerase Cellular cycle arrest ↑Caspase 3	Nair et al. (2014)
Malvidin-3-glucoside	100 μM	HUVECs	↓TNF-α ↑MCP-1	Huang et al. (2014)
Anthocyanins	20 μg	HT-29	↑ROS TNF-a, IL-1β, ↓IL-6, and NFκB	Venancio et al. (2017)
Cyanidin	20 μmol/L	SK-N-SH	↓NFκB, iNOS and NO	Thummayot et al. (2018)
8-Hydroxydaidzein	50 μM	RAW264.7	↓iNOS, COX-2, TNF-α	Kim et al. (2018)
Pterostilbene	100 μM	Huh-7, HepG-2	↓Proliferative activity ↑Autophagy	Yu et al. (2019)
Resveratrol	50 μM	B16	↓Cell viability, COX-2 and VEGF	Lee et al. (2015)

(continued)

Table 19.5 (continued)

Polyphenol	Concentration	Cellular model	Therapeutic target/ effects	References
Isorhapontigenin	100 µM	A549	↓IL-6, CXCL8, NFκB and AP-1	Yeo et al. (2017)
Piceatannol	100 µM	AH109A	↑Apoptosis ↓ROS	Kita et al. (2012)
Pinosylvin	0.1 µmol/L	THP-1	↑LOX and ALOX15 expression	Kwon et al. (2018)
Resveratrol	50 µM	SW-480	↓iNOS and NFkB	Panaro et al. (2012)
Sesamin	10 µM	Chondrocytic cells	↓NO, PGE2, MMP1, MMP3, MMP13 and NFκB	Kong et al. (2016)
Sesamin	100 µM	PC3	↓Proliferation cell, cyclin D1, COX-2, Bcl-2, TNF-α, and NFκB	Xu et al. (2015)
Mataresinol	4.27 µM	CCRF-CEM	Arrest cell cycle in S phase, ↑apoptosis, by Bax, Bad, caspase 9	Su et al. (2015)
Mataresinol	25 µM	BV2	↓Microglia activation, NO, COX-2, NFκB.	Xu et al. (2017)
Pinoresinol	100 µM	Caco-2	↓IL-6, MCP-1, NFkB and COX-2	During et al. (2012)
Pinoresinol	40 µM	SKOV-3	↓Viability cell, invasion Capacity ↑Autophagy	Ning et al. (2019)
Secoisolariciresinol diglucoside	10 µmol/L	GL261, F98	↓Cell viability	Sehm et al. (2014)

AMP-K adenosine monophosphate -activated protein kinase, *HTB-34* human epidermoid cervical carcinoma, *143B and MG-63* human osteosarcoma, *MCF-7* human breast cancer, *A549* human lung cancer, *HeLa* human cervix cancer, *LNCaP* human prostate cancer, *PBMs* isolated peripheral blood monocytes, *RAW264.7* macrophage murine, *CCR-2* chemokine receptor, *SGC-7901* Human stomach carcinoma, *HepG2* human hepatoma, *GBC* gall bladder carcinoma, *HL-60* human acute promyelocytic leukaemia, *MDA-MB-231* human breast cancer, *HNSCC* Head and neck squamous cell carcinomas, *SW780* bladder cancer, *HUVECs* Human umbilical vein endothelial cells, *HT-29* human colon tumour, *SK-N-SH* human neuroblastoma, *Huh-7* human hepatocarcinoma, *B16 Mus musculus* skin melanoma, *VEGF* vascular endothelial growth factor, *CXCL8* chemokine CXC-8, *AP1* activator protein 1, *AH109A* hepatoma, *THP-1* monocytic leukaemia, *LOX* lipoxygenase, *ALOX15* arachidonate 15-lipoxygenase, *SW-480* colorectal adenocarcinoma, *MMP* metalloproteinase, *PC3* prostate adenocarcinoma, *CCRF-CEM* acute lymphoblastic leukemia, *BV2* murine microglia cells, *Caco-2* human colorectal adenocarcinoma, *MCP-1* monocyte chemotactic protein, *SKOV-3* human ovarian cancer

in vivo (Saxena et al. 2012). Mahmoud et al. (2000) found that caffeic acid, quercetin, and rutin decreased tumour growth in melanoma murine by 63% and reduced the oncoprotein β-catenin (Cutrim and Sloboda, 2018). β-catenin signalling

Table 19.6 Antitumoural and anti-inflammatory potential of polyphenols in in vivo models

Polyphenol	Dose/duration	In vivo model	Observations	References
Gallic acid	50 mg/kg 5 weeks	Hepatocellular carcinoma in rats	↓α-Fetoprotein and HSP-gp96	Aglan et al. (2017)
Protocatechuic acid	0.05% 35 weeks	Esophageal cancer in rats	↓COX-2, iNOS	Peiffer et al. (2014)
p-Coumaric acid	20 mM	LPS in rats	Modulation of TNF-α and IL-6	Maryam et al. (2019)
Taxifolin	25 mg/kg 24 days	Tumour xenograft U2OS	↓55% in volume of tumours	Chen et al. (2017)
Apigenin	30 mg/kg 21 days	Tumour xenograft ACHN	↓Tumour growth	Shuai et al. (2017)
Epigallocatechin-3-gallate	100 mg/kg 3 weeks	Mice-bearing SW780 tumour	↓68.4% tumour weight, NFκB and MMP-9	Luo et al. (2017)
Fisetin	80 mg/kg 7 days	LPS in mice	↓IL-1β, IL-6, TNF-α, iNOS and NFκB	Yu et al. (2016)
Genistein	10 mg/kg 7 days	Gastric injury in rats	↓TNF-α, MPO, MMP-9	Hegab et al. (2018)
Pterostilbene	112 mg/kg 6 weeks	SK-Hep-1 tumour xenograft	↓Tumour weight	Yu et al. (2019)
Resveratrol	10 mg/kg 10 days	B16 tumour mice	↓Angiogenesis	Lee et al. (2015)
Piceatannol	0.005% 20 days	Hepatoma bearing rats	↓Tumour size and weight	Kita et al. (2012)
Pinosylvin	100 mg/kg 1 day	Mice	↓80% of inflammation of paw ↓iNOS	Laavola et al. (2015)
Sesamin	10 mg/kg 20 days	Nude mice inoculated with PC3	Suppress tumour growth	Xu et al. (2015)
Secoisolariciresinol	100 mg/kg 8 weeks	E0771	↓Tumour volume and NFκB	Bowers et al. (2019)
Pinoresinol	40 mg/kg 6 weeks	Xenografted mice	↓Tumour volume and weight	Ning et al. (2019)
Pinoresinol	200 mg/kg 2 h	CCl_4 induced liver injury in mice	↓TNF-α, iNOS, COX-2, NFκB, AP-1	Kim et al. (2010)

(continued)

Table 19.6 (continued)

Polyphenol	Dose/duration	In vivo model	Observations	References
Secoisolarici-resinol	100 mg/kg twice daily	Aseptic meningitis/encephalitis in mice	↓Leukocyte adhesion, prevented BBB permeability, TNF-α, IL-1β, HMGB1	Rom et al. (2018)

HSP-gp96 Heat shock protein gp96, *U2OS* osteosarcoma, *ACHN* renal cell adenocarcinoma, *SW780* bladder cancer, *MPO* myeloperoxidase, *SK-Hep-1* liver adenocarcinoma, *B16 Mus musculus* skin melanoma, *PC3* prostate adenocarcinoma, *E0771* mammary adenocarcinoma cells, *CCl4* carbon tetrachloride, *AP1* activator protein 1, *BBB* Blood-brain barrier

is involved in embryonic development, but the aberrant high expression promotes the transcription of oncogenes (Jun, c-Myc, cyclin D). This leads to an alteration in the cell cycle resulting in the development of breast, colorectal, hepatocellular, pancreatic, melanoma, leukaemia, ovarian and lung cancer (Shuang et al. 2017). Also, hesperidin inhibited colon and mammary cancer in rat models (Kumar and Pandey 2013). The epidermal growth factor receptor (EGFR) when it binds to its ligand activates the signal transduction that culminates with the transcription of genes involved in cell proliferation, angiogenesis and invasion. The production and activation of metalloproteinases (MMP-2 and MMP-9) are vital for tumour cell dissemination, a process known as metastasis. Therefore, EGFR and MMP are therapeutic targets in the treatment against cancer (Fantini et al. 2015). It has been shown that quercetin and luteolin are potential antitumour agents inhibiting EGFR tyrosine kinase pathway and MMP (Hassan et al. 2017).

19.3.2 Inflammatory Conditions

Inflammation is a normal response initiated by innate immunity in response to biological, physical or chemical agents. During the inflammation process, mediators release substances such as prostaglandins, histamine and serotonin, which induce an increase in vascular permeability. This leads to the transmigration of leukocyte cells that support the capture and elimination of microorganisms or foreign agents and initiates tissue repair mechanism—a process known as acute inflammation. Normally, inflammation should decrease within a period of 72 h, but if the inflammation does not stop, this becomes a chronic inflammation (Chen et al. 2018a, b, c).

Nitric oxide (NO) is a key molecule involved in several biological functions in the human body. Inducible nitric oxide synthase (iNOS), endothelial nitric oxide synthase (eNOS) and neuronal nitric oxide synthase (nNOS) are the enzymes responsible for generating nitric oxide from the amino acid L-arginine. iNOS-derived NO regulates the process of inflammation and infection (Luiking et al. 2010). For example, NO-induced inflammatory response when interacting with reactive oxygen species (ROS) forming peroxynitrite (Li et al. 2014).

Chronic inflammation is characteristic of various pathologies such as cardiovascular, bowel, autoimmune diseases. The increase in the production of interleukin-1 (IL-1) allows the transcription of genes that code for cyclooxygenase type 2 (COX-2), inducible nitric oxide synthetase (iNOS) and together with overexpression of nuclear factor-κB (NF-κB leads to an increase in the production of pro-inflammatory cytokines such as interleukin-2 (IL-2), interleukin-6 (IL-6), tumour necrosis factor-alpha (TNF-α), and contribute to the appearance of manifestations such as flushing, pain, swelling, oedema and loss of function (Luiking et al. 2010).

Polyphenols have shown significant anti-inflammatory effects in in vitro and in vivo models (Tables 19.5 and 19.6) (Li et al. 2014). For example, Hou et al. (2005) showed that low concentration of anthocyanins such as delphinidin and cyanidin on macrophage murine RAW cells treated with lipopolysaccharide (LPS) inhibited COX-2 and iNOS expression and suppressing NF-kB activation. These results reflect the anti-inflammatory potential of polyphenols (Miguel 2011).

Quercetin and luteolin have been used for the treatment of periodontal inflammation, interstitial cystitis and contact dermatitis in humans (Srivastava and Kumar-Mishra 2015). On the other hand, EGCG reduced the expression of both iNOS and COX-2 in the liver cells. Kaempferol caused a significant inhibition of pro-inflammatory cytokines (Hassan et al. 2017). Blood cells were stimulated with LPS in the presence of caffeic acid or oleuropein at a concentration of 10^{-4} M. The results show a decrease in the production of IL-1β by 40% and 80%, respectively. Also, treatment with kaempferol decreased the production of prostaglandin E (PGD-E) by 95%. Similar results were reported by Kao et al. (2009) when they used polyphenols obtained from *Hibiscus sabdariffa* and treated RAW264.7 cells. The results obtained showed a decrease in ON and PGD-E (Li et al. 2014). Similarly, the oral administration (50 mg/kg) of hydrocaffeic acid reduced the expression of IL-1β, IL-8, TNF-α (Li et al. 2014). Some of the research on anti-inflammatory effects in vivo models are shown in Table 19.6.

19.3.3 Menopause

The menstrual cycle starts the reproductive stage in women and is triggered by the presence of estrogen. The increase in estrogen levels induces the appearance of sexual characteristics and intervenes in the growth of the endometrium, mammary glands and maturation of the ovary (Tungmunnithum et al. 2018). Studies have demonstrated that menstrual cycle change during the reproductive stage, decline after 45 years, the average age of onset of menopause (Harlow and Paramsothy 2011). Menopause is defined as the permanent absence of menstruation, after 12 months consecutive of amenorrhoea, without pathological causes. This absence causes alterations in sexual hormones, such as follicle-stimulating hormone (FSH) and estrogen. Thus, when estrogen levels decrease, different signs and symptoms occur such as irregular bleeding, hot flashes, breast pain, tiredness, vaginitis, interrupted sleep, dry skin, anxiety, mood swings, bone mineral loss and loss of libido (Sandoval et al. 2017). Besides this, it has been shown that menopausal women who have metabolic syndrome increase the risk of developing cardiovascular diseases and osteoporosis (Tungmunnithum et al. 2018). The therapeutic strategy consists of hormone replacement therapy (HRT), selective modulators of estrogen receptors (SERM) such as raloxifene and tamoxifen. However, these treatments have adverse effects including ovarian and breast cancer, thrombosis and other cardiovascular diseases (Sandoval et al. 2017). Hence, clinical investigations have been conducted to evaluate the effect of the administration of polyphenols on symptomatology in menopausal patients (Table 19.7). For example, isoflavones contain phytoestrogens, which are compounds that are chemically and biologically similar to estrogens (Patisaul and Jefferson 2010). Isoflavones are found in large quantities in berries, grains, nuts, grains, among others (Table 19.2). In an investigation the effects of the administration of genistein, daidzein or HRT during the first year in postmenopausal women were determined, finding that genistein and daidzein did not cause adverse effects and significantly improved bone health in comparison with HRT. Results from another where, in an 8-week, randomized, a double-blind trial showed that consumption of 8-prenylnaringenin (8-PN), which is found in considerable quantities in beer, resulted in a significant decrease in menopause symptoms (Sandoval et al. 2017).

As mentioned before, though polyphenols have benefits in various research models, few clinical trials show their effectiveness in the treatment of various diseases. An essential factor for the biological functioning of a compound is bioavailability. Bioavailability is defined as the proportion of the substance, nutrient or compound that can be digested, absorbed and metabolized (Fraga et al. 2019). It has been found that polyphenols have an affinity with various cellular and tissue proteins, pH changes can be affecting the capacity of digestion and biological activity. For example, in in vitro digestion model, catechin and epicatechin were eliminated by 43.9% and 85.3%, respectively after 2 h of incubation. Procyanidins after 2 h of incubation in an in vitro model were stable at pH 7 in the intestine medium, but at pH 7.4 was degraded by 20%. While epicatechin and catechin incubated to the same conditions above, at pH 7.5, 34% of epicatechin was found degraded but catechin was stable (Li et al. 2014).

19.4 Conclusions

The great benefits that polyphenols can generate in different health conditions are unquestionable. However, concentrations and solubility of polyphenols modulate their bioavailability. It is important to highlight that in oral consumption, the uptake of polyphenols into the body is not complete, and a certain percentage is not absorbed. Therefore, despite eating foods high in polyphenols, it may not be reflected by the biological benefits obtained under in vitro conditions, due to its low bioavailability in the body. For this reason, a prerequisite for polyphenols to have any in vivo systemic effects is that they must be absorbed from the gastrointestinal tract after food consumption and, subsequently, reach sufficiently high plasma

Table 19.7 Potential of polyphenols in human clinical trials

Study description	Objectives	Results/status	References
A phase III, randomized, parallel assignment. 300 ESCC patients (stage IV) Doses: 300 mg, tid of caffeic acid/3 months.	To investigate the efficiency and safety of caffeic acid in Chinese advanced esophageal squamous cell cancer.	Enrolling by invitation	Clinicaltrials.gov NCT03070262
Randomized clinical trial. Dietary supplement with curcumin, rutin or quercetin. Doses: 3 doses twice a day/10 week.	Determine the response of the colonic epithelium in normal volunteers at average or above average risk of colon cancer.	Completed	Clinicaltrials.gov NCT00003365
Randomized, parallel assignment. 382 participants with colorectal cancer. Doses: 20 mg apigenin and 20 mg EGCG as tablets/day.	Determine the effect of dietary supplementation with bioflavonoids (Flavo-Natin®) will diminish the recurrence rate of colonic neoplasia	Suspended	Clinicaltrials.gov NCT00609310
A phase II, randomized, parallel assignment. 31 participants treated with 800 mg EGCG or 1200 mg EGCG once daily for 14–28 days.	To compare the levels of EGCG in nonmalignant bladder tissue from patients with bladder cancer.	Absence of unacceptable toxicity. After finishing the treatment, the patients undergo trans-urethral resection of bladder tumour or cystectomy.	Clinicaltrials.gov NCT00666562
A phase II, randomized, parallel assignment. 1075 participants. Two green tea extract capsules (containing 800 mg EGCG/ day) twice daily for 1 year.	To investigate whether green tea prevents breast cancer in postmenopausal women with high breast density.	The results suggest that GTE supplementation significantly increases circulating estradiol concentrations in healthy postmenopausal women.	Clinicaltrials.gov NCT00917735 Samavat et al. (2019)
Randomized, parallel assignment. 89 participants. FO 3/day and oral GTE (EGCG) extract 2/day.	Determine the prostate cancer-preventing effects of fish oil supplementation and green tea extract use.	Results indicate FO or EGCG supplementation for a short duration may not be enough to produce biologically meaningful changes in cell proliferation by Ki67 levels in prostate tissue.	Clinicaltrials.gov NCT00253643 Zhang et al. (2016)

(continued)

Table 19.7 (continued)

Study description	Objectives	Results/status	References
Randomized, parallel assignment. 98 participants. Polyphenon E (containing 800 mg EGCG) once daily for 4 months.	Assess the effect of green tea extract (Polyphenon E®) in patients with HPV expression and low-grade CIN1.	The results indicate that Polyphenon E intervention did not promote the clearance of persistent high-risk HPV and related CIN1.	Clinicaltrials.gov NCT00303823 Garcia et al. (2014)
11 participants Ingestion of 80 g of GP per day/ 2 weeks	Determine if treatment with resveratrol modulates Wnt signalling in vivo in colon cancer and normal colonic mucosa.	GP, which contains low dosages of resveratrol, can inhibit the Wnt pathway in vivo and that this effect is confined to the normal colonic mucosa.	Clinicaltrials.gov NCT00256334 Nguyen et al. (2009)
Randomized, parallel assignment. 210 participants Isoflavone (daidzein) supplement, 60 mg once a day.	The study examines the effect of a new soy supplement, as compared to a placebo, in menopausal women on hot flash symptoms.	Completed	Clinicaltrials.gov NCT00179556
Randomized, parallel assignment. 203 participants. Soy protein with 96 mg isoflavones/day/ 9 months.	The aim of the present study was to investigate the effects of soy and soy plus isoflavones on BMD and physical performance in post-menopausal women.	The study did not find a significant positive effect of soy protein isolate supplemented with isoflavones on BMD and the serum lipid profile in early postmenopausal women.	Clinicaltrials.gov NCT00661856 Gallagher et al. (2004)
Randomized, double-blind. 138 participants 54 mg of genistein, daily/24 months.	To assess the safety profile of genistein on mammary and thyroid glands and endometrium and cardiovascular apparatus and its effects on bone metabolism after a 3-year therapy with pure, standardized genistein (54 mg/day) on women menopausal.	BMD had increased in genistein recipients. Genistein decreased urinary excretion of pyridinoline increased levels of bone-specific alkaline phosphatase and did not change endometrial thickness. Gastrointestinal adverse effects occur, and patients discontinued the study.	Clinicaltrials.gov NCT00626769 Marini et al. (2007)
Randomized, crossover assignment. 23 participants Healthy post-menopausal women. 100 mg gensitein from soy protein isolate for 50 days.	This study will evaluate the effect of soy isoflavones on calcium absorption and bone loss in post-menopausal women.	Soy isoflavones did not significantly affect calcium metabolism.	Clinicaltrials.gov NCT00244907 Spence et al. (2005)

(continued)

Table 19.7 (continued)

Study description	Objectives	Results/status	References
Randomized, parallel assignment. 96 participants. A total daily dose of 110 mg (NovaSoy®).	The purpose of this study is to examine the effects of soy (NovaSoy®) and estrogen on menopausal symptoms such as hot flashes, sleep disturbances, and mood alteration in perimenopausal women.	Completed	Clinicaltrials.gov NCT00997893
Randomized, parallel assignment. 58 participants. 22 g freeze-dried blueberry powder (including anthocyanins, phenolic acids, and pterostilbene) for 8 weeks.	To examine the effects of daily blueberry consumption on blood pressure and arterial stiffness in postmenopausal women with pre- and stage 1-hypertension.	Daily blueberry consumption may reduce blood pressure and arterial stiffness and increased NO production.	Clinicaltrials.gov NCT01686282 Johnson et al. (2015)
32 participants. Randomized, parallel assignment. SDG supplementation as 1.6 g/day of BeneFlax containing 600 mg SDG/24 weeks.	To examine whether a dietary intervention (flax seed containing lignan SDG) might decrease inflammation.	Preliminary results indicate that no safety concerns are associated with administering of SDG to adults between the ages of 60 and 80 years.	Clinicaltrials.gov NCT01846117 Alcorn et al. (2017)
Randomized, parallel assignment. 18 participants. 125 mL of 100% pomegranate juice twice daily for 12 weeks.	The aim of this research is to study the effects of a pomegranate juice on calprotectin levels in patients suffering of IBD in clinical remission	Polyphenols may thus be considered able to prevent or delay the progression of IBD. However, knowledge of the use of polyphenols in managing human IBD is still scanty.	Clinicaltrials.gov NCT03000101 Biasi et al. (2011)
Randomized, parallel assignment. 120 participants. EGCG at 200 mg twice daily, after 3 months 400 mg twice daily.	To determinate if green tea extract has anti-inflammatory and neuroprotective properties in patients with relapsing–remitting MS.	Completed	Clinicaltrials.gov NCT00525668
Randomized, parallel assignment. 40 participants. 20 mg/kg/day of fisetin, orally for 2 consecutive days.	To test the efficacy of the anti-inflammatory drug (Fisetin) in reducing markers of inflammation in blood in elderly adults.	Recruiting	Clinicaltrials.gov NCT03675724

(continued)

Table 19.7 (continued)

Study description	Objectives	Results/status	References
Randomized, parallel assignment. 34 participants. Nutrasorb from blueberry and green tea extracts (2136 mg/day gallic acid equivalents).	This study will use the Nutrasorb soy protein product that is matrixed with polyphenols from blueberries and green tea extract, and test for efficacy as a nutritional countermeasure to exercise-induced physiologic stress.	Supplementation did not alter established biomarkers for inflammation.	Clinicaltrials. gov NCT01775384 Nieman et al. (2013)

ESCC Esophageal Cancer, *Tid* three times a day, *EGCG* epigallocatechin gallate, *GTE* green tea extract, *FO* Oral fish oil, *GP* Resveratrol-containing freeze-dried grape powder, *BMD* bone mineral density, *NO* nitric oxide, *HPV* human papillomavirus, *CIN* cervical intraepithelial neoplasia, *SDG* Secoisolariciresinol diglucoside, *IBD* inflammatory bowel disease, *MS* multiple sclerosis

concentrations in the blood circulation to induce biological activity. It has been established that although many bioactive compounds obtained from plants had been resourceful against a range of diseases, few have been developed into drugs based on phytotherapeutics. A large number of them have not yet been scaled through clinical trials to determine their safety and efficacy. Therefore, it is expected that phytomedicine based on polyphenols can be adopted or integrated into the national health care system in many countries.

References

Abdulkhaleq LA, Assi MA, Noor MHM, Abdullah R, Saad MZ, Taufiq-Yap YH (2017) Therapeutic uses of epicatechin in diabetes and cancer. Vet World 10(8):869–872

Aglan HA, Ahmed HH, El-Toumy SA, Mahmoud NS (2017) Gallic acid against hepatocellular carcinoma: an integrated scheme of the potential mechanisms of action from in vivo study. Tumour Biol 39(6):1010428317699127. https://doi.org/10.1177/1010428317699127

Akter K, Barnes EC, Loa-Kum-Cheung WL, Yin P, Kichu M, Brophy JJ, Barrow RA, Imchen I, Vemulpad SR, Jamie JF (2016) Antimicrobial and antioxidant activity and chemical characterisation of *Erythrina stricta* Roxb. (Fabaceae). J Ethnopharmacol 185:171–181

Alam MA, Subhan N, Rahman MM, Uddin SJ, Reza HM, Sarker SD (2014) Effect of citrus flavonoids, naringin and naringenin, on metabolic syndrome and their mechanisms of action. Adv Nutr 5(4):404–417

Alcorn J, Whiting S, Viveky N, Di Y, Mansell K, Fowler S, Thorpe L, Almousa A, Cheng P, Jones J, Billinsky J, Hadjistavropoulos T (2017) Protocol for a 24-week randomized controlled study of once-daily oral dose of flax lignan to healthy older adults. JMIR Res Protoc 6(2):e14. https://doi.org/10.2196/resprot.6817

Aliomrani M, Sepand MR, Mirzaei HR, Kazemi AR, Nekonam S, Sabzevari O (2016) Effects of phloretin on oxidative and inflammatory reaction in rat model of cecal ligation and puncture induced sepsis. Daru 24(1):15

Almeida L, Vaz-da-Silva M, Falcão A, Soares E, Costa R, Loureiro AI, Fernandes-Lopes C, Rocha JF, Nunes T, Wright L, Soares-da-Silva P (2009) Pharmacokinetic and safety profile of trans-resveratrol in a rising multiple dose study in healthy volunteers. Mol Nutr Food Res 53(1):7–15

Anand David AV, Arulmoli R, Parasuraman S (2016) Overviews of biological importance of quercetin: a bioactive flavonoid. Pharmacogn Rev 10(20):84–89

Archivio M, Filesi C, Di Benedetto R, Gargiulo R, Giovannini C, Masella R (2007) Polyphenols, dietary sources and bioavailability. Ann Ist Super Sanita 43(4):348–361

Asmi KS, Lakshmi T, Balusamy SR, Parameswari R (2017) Therapeutic aspects of taxifolin—an update. J Adv Pharm Educ Res 7(3):187–189

Atanassova M, Bagdassarian V (2009) Rutin content in plant products. J Univ Chem Technol Metal 44(2):201–203

Bai Y, Mao Q-Q, Qin J, Zheng X-Y, Wang Y-B, Yang K, Shen H-F, Xie L-P (2010) Resveratrol induces apoptosis and cell cycle arrest of human T24 bladder cancer cells in vitro and inhibits tumor growth in vivo. Cancer Sci:101488–101493

Baião DD, de Freitas CS, Gomes LP, da Silva D, Correa A, Pereira PR, Aguila EMD, Paschoalin VMF (2017) Polyphenols from root, tubercles and grains cropped in Brazil: chemical and nutritional characterization and their effects on human health and diseases. Nutrients 9 (9):1044. https://doi.org/10.3390/nu9091044

BenSaad LA, Kim KH, Quah CC, Kim WR, Shahimi M (2017) Anti-inflammatory potential of ellagic acid, gallic acid and punicalagin a&B isolated from *Punica granatum*. BMC Complement Altern Med 17(1):47. https://doi.org/10.1186/s12906-017-1555-0

Biasi F, Astegiano M, Maina M, Leonarduzzi G, Poli G (2011) Polyphenol supplementation as a complementary medicinal approach to treating inflammatory bowel disease. Curr Med Chem 18 (31):4851–4865

Bie B, Sun J, Guo Y, Li J, Jiang W, Yang J, Huang C, Li Z (2017) Baicalein: a review of its anti-cancer effects and mechanisms in hepatocellular carcinoma. Biomed Pharmacother 93:1285–1291

Bodduluru LN, Kasala ER, Madhana RM, Barua CC, Hussain MI, Haloi P, Borah P (2016) Naringenin ameliorates inflammation and cell proliferation in benzo(a) pyrene induced pulmonary carcinogenesis by modulating CYP1A1, NFκB and PCNA expression. Int Immunopharmacol 30:102–110

Bowers LW, Lineberger CG, Ford NA, Rossi EL, Punjala A, Camp KK, Kimler BK, Fabian CJ, Hursting SD (2019) The flaxseed lignan secoisolariciresinol diglucoside decreases local inflammation, suppresses NFκB signaling, and inhibits mammary tumor growth. Breast Cancer Res Treat 173(3):545–557. https://doi.org/10.1007/s10549-018-5021-6

Brandli A, Simpson JS, Ventura S (2010) Isoflavones isolated from red clover (*Trifolium pratense*) inhibit smooth muscle contraction of the isolated rat prostate gland. Phytomedicine 17 (11):895–901

Cai ZY, Li XM, Liang JP, Xiang LP, Wang KR, Shi YL, Yang R, Shi M, Ye JH, Lu JL, Zheng XQ, Liang YR (2018) Bioavailability of tea Catechins and its improvement. Molecules 23(9):2346. https://doi.org/10.3390/molecules23092346

Caselli A, Cirri P, Santi A, Paoli P (2016) Morin: a promising natural drug. Curr Med Chem 23:1–18

Chang WC, Wu JS, Chen CW, Kuo PL, Chien HM, Wang YT, Shen SC (2015) Protective effect of Vanillic acid against Hyperinsulinemia, Hyperglycemia and Hyperlipidemia via alleviating hepatic insulin resistance and inflammation in high-fat diet (HFD)-fed rats. Nutrients 7 (12):9946–9959

Chen AY, Chen YC (2013) A review of the dietary flavonoid, kaempferol on human health and cancer chemoprevention. Food Chem 138(4):2099–2107

Chen PN, Chu SC, Chiou HL, Chiang CL, Yang SF, Hsieh YS (2005) Cyanidin 3-Glucoside and Peonidin 3-Glucoside inhibit tumor cell growth and induce apoptosis in vitro and suppress tumor growth in vivo. Nutr Cancer 53(2):232–243

Chen R, Qi QL, Wang MT, Li QY (2016) Therapeutic potential of naringin: an overview. Pharm Biol 54(12):3203–3210

Chen X, Gu N, Xue C, Li BR (2017) Plant flavonoid taxifolin inhibits the growth, migration and invasion of human osteosarcoma cells. Mol Med Rep 17:3239–3245. https://doi.org/10.3892/mmr.2017.8271

Chen J, Sun J, Jiang J, Zhou J (2018a) Cyanidin protects SH-SY5Y human Neuroblastoma cells from 1-Methyl-4-Phenylpyridinium-induced neurotoxicity. Pharmacology 102(3-4):126–132

Chen L, Deng H, Cui H, Fang J, Zuo Z, Deng J, Li Y, Wang X, Zhao L (2018b) Inflammatory responses and inflammation-associated diseases in organs. Oncotarget 9(6):7204–7218

Chen X, Gu N, Xue C, Li BR (2018c) Plant flavonoid taxifolin inhibits the growth, migration and invasion of human osteosarcoma cells. Mol Med Rep 17(2):3239–3245. https://doi.org/10.3892/mmr.2017.8271

Choi SW, Park KI, Yeon JT, Ryu BJ, Kim KJ, Kim SH (2014) Anti-osteoclastogenic activity of matairesinol via suppression of p38/ERK-NFATc1 signaling axis. BMC Complement Altern Med 14:35. https://doi.org/10.1186/1472-6882-14-35

Cutrim SC, Sloboda MA (2018) A review on polyphenols: classification, beneficial effects and their application in dairy products. Int J Dairy Technol 71(3):564–578. https://doi.org/10.1111/1471-0307.12515

Dai Y, Yeo SCM, Barnes PJ, Donnelly LE, Loo LC, Lin HS (2018) Pre-clinical pharmacokinetic and Metabolomic analyses of Isorhapontigenin, a dietary resveratrol derivative. Front Pharmacol 9:753. https://doi.org/10.3389/fphar.2018.00753

Duarte LJ, Chaves VC, Nascimento MVPDS, Calvete E, Li M, Ciraolo E, Ghigo A, Hirsch E, Simões CMO, Reginatto FH, Dalmarco EM (2018) Molecular mechanism of action of Pelargonidin-3-O-glucoside, the main anthocyanin responsible for the anti-inflammatory effect of strawberry fruits. Food Chem 247:56–65

During A, Debouche C, Raas T, Larondelle Y (2012) Among plant lignans, pinoresinol has the strongest antiinflammatory properties in human intestinal Caco-2 cells. J Nutr 142(10):1798–1805. Epub 2012 Sep 5

Ehrenkranz JR, Lewis NG, Kahn CR, Roth J (2005) Phlorizin: a review. Diabetes Metab Res Rev 21(1):31–38

El Khawand T, Courtois A, Valls J, Richard T, Krisa S (2018) A review of dietary stilbenes: sources and bioavailability. Phytochem Rev 17(5):1007–1029

El-Seedi HR, Taher EA, Sheikh BY, Anjum S, Saeed A, Alajmi MF, Moustafa MS, Al-Mousawi SM, Farag MA, Hegazy ME, Khalifa SA (2018) Hydroxycinnamic acids: natural sources, biosynthesis, possible biological activities, and roles in Islamic medicine. In: Atta-ur-Rahman (ed) Studies in natural products chemistry, vol 55. Elsevier, Amsterdam, pp 269–292

Fantini M, Benvenuto M, Masuelli L, Frajese GV, Tresoldi I, Modesti A, Bei R (2015) In vitro and in vivo antitumoral effects of combinations of polyphenols, or polyphenols and anticancer drugs: perspectives on cancer treatment. Int J Mol Sci 16(5):9236–9282. https://doi.org/10.3390/ijms16059236

Fernández MM, Guerrero RF, García Parrilla MC, Puertas B, Richard T, Rodriguez-Werner MA, Winterhalter P, Monti JP, Cantos VE (2012) Isorhapontigenin: a novel bioactive stilbene from wine grapes. Food Chem 35(3):1353–1359

Fraga CG, Croft KD, Kennedy DO, Tomás BF (2019) The effects of polyphenols and other bioactives on human health. Food Funct 10(2):514–528

Gallagher JC, Satpathy R, Rafferty K, Haynatzka V (2004) The effect of soy protein isolate on bone metabolism. Menopause 11(3):290–298

Ganeshpurkar A, Saluja AK (2017) The pharmacological potential of Rutin. Saudi Pharm J 25(2):149–164

Garcia FA, Cornelison T, Nuño T, Greenspan DL, Byron JW, Hsu CH, Alberts DS, Chow HH (2014) Results of a phase II randomized, double-blind, placebo-controlled trial of Polyphenon E in women with persistent high-risk HPV infection and low-grade cervical intraepithelial neoplasia. Gynecol Oncol 132(2):377–382. https://doi.org/10.1016/j.ygyno.2013.12.034

Granado SA, Martín MA, Bravo L, Goya L, Ramos S (2010) Quercetin modulates NF-kappa B and AP-1/JNK pathways to induce cell death in human hepatoma cells. Nutr Cancer 62(3):390–401

Granja A, Frias I, Neves AR, Pinheiro M, Reis S (2017) Therapeutic potential of epigallocatechin gallate nanodelivery systems. Biomed Res Int 2017:5813793. https://doi.org/10.1155/2017/5813793

Harlow S, Paramsothy P (2011) Menstruation and the menopause transition. Obstet Gynecol Clin North Am 38(3):595–607

Hassan R, Mohammad HF, Reza K (2017) Polyphenols and their benefits: a review. Int J Food Prop 20(2):1700–1741

Hegab II, Abd-Ellatif RN, Sadek MT (2018) The gastroprotective effect of N-acetyl cysteine and genistein in indomethacin-induced gastric injury in rats. Can J Physiol Pharmacol 96 (11):1161–1170. https://doi.org/10.1139/cjpp-2017-0730

Hou DX, Yanagita T, Uto T, Masuzaki S, Fujii M (2005) Anthocyanidins inhibit cyclooxygenase-2 expression in LPS-evoked macrophages: structure-activity relationship and molecular mechanism involved. Biochem Pharmacol 70:417–425

Huang WY, Liu YM, Wang J, Wang XN, Li CY (2014) Anti-inflammatory effect of the blueberry anthocyanins malvidin-3-glucoside and malvidin-3-galactoside in endothelial cells. Molecules 19(8):12827–12841

Huang W, Zhu Y, Li C, Sui Z, Min W (2016) Effect of blueberry Anthocyanins Malvidin and glycosides on the antioxidant properties in endothelial cells. Oxid Med Cell Longev 2016:1591803. https://doi.org/10.1155/2016/1591803

Hussein RA, El-Anssary AA (2018) Pharmacological actions of medicinal plants. IntechOpen Herbal Med. https://doi.org/10.5772/intechopen.76139

Hwang B, Lee J, Liu QH, Woo ER, Lee DG (2010) Antifungal effect of (+)-pinoresinol isolated from *Sambucus williamsii*. Molecules 15(5):3507–3516

Inoue Y, Hasegawa S, Yamada T, Date Y, Mizutani H, Nakata S, Matsunaga K, Akamatsu H (2013) Analysis of the effects of hydroquinone and arbutin on the differentiation of melanocytes. Biol Pharm Bull 36(11):1722–1730

International Agency for Research on Cancer (IARC). Monographs on the evaluation of carcinogenic risks to humans. Available in: https://monographs.iarc.fr/wp-content/uploads/2018/09/Classifications AlphaOrder.pdf [April 2018]

Ismail T, Calcabrini C, Diaz AR, Fimognari C, Turrini E, Catanzaro E, Akhtar S, Sestili P (2016) Ellagitannins in Cancer chemoprevention and therapy. Toxins 8(5):151. https://doi.org/10.3390/toxins8050151

Itoh A, Isoda KM, Kawase M, Watari A, Kobayashi M, Tamesada M, Yagi K (2010) Hepatoprotective effect of syringic acid and vanillic acid on CCl4-induced liver injury. Biol Pharm Bull 33(6):983–987

Jeevan KP, Stephen B, Michael W (2018) Analyzing ingredients in dietary supplements and their metabolites (chapter 24). In: Watson R, Preedy V, ZibadI S (eds) Polyphenols: mechanisms of action in human health and disease. Academic, San Diego, pp 337–346

Johnson SA, Figueroa A, Navaei N, Wong A, Kalfon R, Ormsbee LT, Feresin RG, Elam ML, Hooshmand S, Payton ME, Arjmandi BH (2015) Daily blueberry consumption improves blood pressure and arterial stiffness in postmenopausal women with pre- and stage 1-hypertension: a randomized, double-blind, placebo-controlled clinical trial. J Acad Nutr Diet 115(3):369–377. https://doi.org/10.1016/j.jand.2014.11.001

Jurica K, Karačonji IB, Šegan S, Opsenica DM, Kremer D (2015) Quantitative analysis of arbutin and hydroquinone in strawberry tree (*Arbutus unedo* L., Ericaceae) leaves by gas chromatography-mass spectrometry. Arch Ind Hyg Toxicol 66(3):197–202

Kakkar S, Bais S (2014) A review on protocatechuic acid and its pharmacological potential. ISRN Pharmacol 952943

Kanimozhi G, Prasad NR (2015) Anticancer effect of caffeic acid on human cervical cancer cells. Coffee Health Dis Prev:665–661

Kao ES, Hsu JD, Wang CJ, Yang SH, Cheng SY, Lee HJ (2009) Polyphenols extracted from *Hibiscus sabdariffa* L. inhibited lipopolysaccharide-induced inflammation by improving antioxidative conditions and regulating cyclooxygenase-2 expression. Biosci Biotechnol Biochem 73:385–390

Kezimana P, Dmitriev AA, Kudryavtseva AV, Romanova EV, Melnikova NV (2018) Secoisolariciresinol Diglucoside of flaxseed and its metabolites: biosynthesis and potential for Nutraceuticals. Front Genet 9:641. https://doi.org/10.3389/fgene.2018.00641

Khan N, Syed DN, Ahmad N, Mukhtar H (2013) Fisetin: a dietary antioxidant for health promotion. Antioxid Redox Signal 19(2):151–162

Khan AK, Rashid R, Fatima N, Mahmood S, Mir S, Khan S, Jabeen N, Murtaza G (2015) Pharmacological activities of protocatechuic acid. Acta Pol Pharm 72(4):643–650

Khanbabaee K, Van Ree T (2001) Tannins: classification and definition. Nat Prod Rep 18 (6):641–649

Khoo HE, Azlan A, Tang ST, Lim SM (2017) Anthocyanidins and anthocyanins: colored pigments as food, pharmaceutical ingredients, and the potential health benefits. Food Nutr Res 61 (1):1361779

Kim HY, Kim JK, Choi JH, Jung JY, Oh WY, Kim DC, Lee HS, Kim YS, Kang SS, Lee SH, Lee SM (2010) Hepatoprotective effect of pinoresinol on carbon tetrachloride-induced hepatic damage in mice. J Pharmacol Sci 112(1):105–112

Kim E, Kang YG, Kim JH, Kim YJ, Lee TR, Lee J, Kim D, Cho JY (2018) The antioxidant and anti-inflammatory activities of 8-Hydroxydaidzein (8-HD) in activated macrophage-like RAW264.7 cells. Int J Mol Sci 19(7):1828. https://doi.org/10.3390/ijms19071828

Kita Y, Miura Y, Yagasaki K (2012) Antiproliferative and anti-invasive effect of piceatannol, a polyphenol presents in grapes and wine, against hepatoma AH109A cells. J Biomed Biotechnol 672416. https://doi.org/10.1155/2012/672416

Kodan A, Kuroda H, Sakai F (2002) A stilbene synthase from Japanese red pine (*Pinus densiflora*): implications for phytoalexin accumulation and down-regulation of flavonoid biosynthesis. Proc Natl Acad Sci U S A 99(5):3335–3339

Kong P, Chen G, Jiang A, Wang Y, Song C, Zhuang J, Xi C, Wang G, Ji Y, Yan J (2016) Sesamin inhibits IL-1β-stimulated inflammatory response in human osteoarthritis chondrocytes by activating Nrf2 signaling pathway. Oncotarget 7(50):83720–83726. https://doi.org/10.18632/oncotarget.13360

Konoshima T, Kokumai M, Kozuka M, Tokuda H, Nishino H, Iwahima A (1992) Anti-tumor-promoting activities of afromosin and soyasaponin I isolated from *Wistaria brachybotrys*. J Nat Prod 55(12):1776–1778

Kosińska A, Andlauer W (2014) Antioxidant capacity of tea: effect of processing and storage (chapter 12). In: Preedy V (ed) Processing and impact on antioxidants in beverages. Elsevier, Amsterdam, pp 109–120

Kujawski R, Ma K, Ożarowski M, Baraniak J, Laskowska H, Nowocień T, Borowska M, Szulc M, Sobczak A, Mikołajczak P (2016) Perspectives for gallotannins neuroprotective potential – current experimental evidences. J Med Sci 85(4):313–318

Kumar S, Pandey AK (2013) Chemistry and biological activities of flavonoids: an overview. Sci World J 162750. https://doi.org/10.1155/2013/162750

Kumar N, Pruthi V (2014) Potential applications of ferulic acid from natural sources. Biotechnol Rep 4:86–93

Kumar S, Prahalathan P, Raja B (2011) Antihypertensive and antioxidant potential of vanillic acid, a phenolic compound in L-NAME-induced hypertensive rats: a dose-dependence study. Redox Rep 16:208–215

Kwon O, Seo Y, Park H (2018) Pinosylvin exacerbates LPS-induced apoptosis via ALOX 15 upregulation in leukocytes. BMB Rep 51(6):302–307. https://doi.org/10.5483/bmbrep.2018.51.6.024

Laavola M, Nieminen R, Leppänen T, Eckerman C, Holmbom B, Moilanen E (2015) Pinosylvin and monomethylpinosylvin, constituents of an extract from the knot of *Pinus sylvestris*, reduce inflammatory gene expression and inflammatory responses in vivo. J Agric Food Chem 63 (13):3445–3453. https://doi.org/10.1021/jf504606m

Lamy E, Pinheiro C, Rodrigues L, Capela-Silva F, Lopes OS, Moreira P, Tavares S, Gaspar R (2016) Determinants of tannin-rich food and beverage consumption: oral perception vs. psychosocial aspects. In: Combs CA (ed) Tannins: biochemistry, food sources and nutritional properties. Nova, New York, pp 29–58

Lee SH, Koo BS, Park SY, Kim YM (2015) Anti-angiogenic effects of resveratrol in combination with 5-fluorouracil on B16 murine melanoma cells. Mol Med Rep 12(2):2777–2783. https://doi.org/10.3892/mmr.2015.3675

Lee W, Lee Y, Kim J, Bae JS (2018) Protective effects of pelargonidin on lipopolysaccharide-induced hepatic failure. Nat Prod Commun 13(1):45–48

Li AN, Li S, Zhang YJ, Xu XR, Chen YM, Li HB (2014) Resources and biological activities of natural polyphenols. Nutrients 6(12):6020–6047

Liao W, Chen L, Ma X, Jiao R, Li X, Wang Y (2016) Protective effects of kaempferol against reactive oxygen species-induced hemolysis and its antiproliferative activity on human cancer cells. Eur J Med Chem 114:24–32

Lim CG, Koffas AG (2010) Bioavailability and recent advances in the bioactivity of flavonoid and Stilbene compounds. Curr Org Chem 14(16):1727–1751

Liu XR, Zhang Y, Lin ZQ (2011) Advances in studies on the biological activities of hesperidin and hesperetin. Chinese J New Drugs 20:329–333

Locatelli C, Filippin-Monteiro FB, Creczynski-Pasa TB (2013) Alkyl esters of gallic acid as anticancer agents: a review. Eur J Med Chem 60:233–239

Luiking Y, Engelen M, Deutz N (2010) Regulation of nitric oxide production in health and disease. Curr Opin Clin Nutr Metab Care 13(1):97–104

Luo KW, Chen W, Lung WY, Wei XY, Cheng BH, Cai ZM, Huang WR (2017) EGCG inhibited bladder cancer SW780 cell proliferation and migration both in vitro and *in vivo* via down-regulation of NF-κB and MMP-9. J Nutr Biochem 41:56–64

Magnani C, Isaac V, Correa MA, Salgado HR (2014) Caffeic acid: a review of its potential use in medications and cosmetics. Anal Methods 6:3203. https://doi.org/10.1039/c3ay41807c

Mahmoud NN, Carothers AM, Grunberger D, Bilinski RT, Churchill MR, Martucci C, Newmark HL, Bertagnolli MM (2000) Plant phenolics decrease intestinal tumors in an animal model of familial adenomatous polyposis. Carcinogenesis:21921–21927

Majdalawieh AF, Massri M, Nasrallah GK (2017) A comprehensive review on the anti-cancer properties and mechanisms of action of sesamin, a lignan in sesame seeds (*Sesamum indicum*). Eur J Pharmacol 815:512–521

Mamat SS, Kamarolzaman MF, Yahya F, Mahmood ND, Shahril MS, Jakius KF, Mohtarrudin N, Ching SM, Susanti D, Taher M, Zakaria ZA (2013) Methanol extract of Melastoma malabathricum leaves exerted antioxidant and liver protective activity in rats. BMC Complement Altern Med 13:326. https://doi.org/10.1186/1472-6882-13-326

Manach C, Scalbert A, Morand C, Rémésy C, Jiménez L (2004) Polyphenols: food sources and bioavailability. Am J Clin Nutr 79:727–747

Mancuso C, Santangelo R (2014) Ferulic acid: pharmacological and toxicological aspects. Food Chem Toxicol 65:185–195

Marini H, Minutoli L, Polito F, Bitto A, Altavilla D, Atteritano M, Gaudio A, Mazzaferro S, Frisina A, Frisina N, Lubrano C, Bonaiuto M, D'Anna R, Cannata ML, Corrado F, Adamo EB, Wilson S, Squadrito F (2007) Effects of the phytoestrogen genistein on bone metabolism in osteopenic postmenopausal women: a randomized trial. Ann Intern Med 146(12):839–847

Martinović N, Abramovi H (2014) Sinapic acid and its derivatives: natural sources and bioactivity. Compr Rev Food Sci Food Saf 13(1). https://doi.org/10.1111/1541-4337.12041

Maryam K, Mahin D, Mohammad B, Seyyed A, Vahid B (2019) P-Coumaric acid attenuates lipopolysaccharide-induced Lung inflammation in rats by scavenging ROS production: an in vivo and in vitro study. Inflammation. https://doi.org/10.1007/s10753-019-01054-6

Masuelli L, Marzocchella L, Quaranta A, Palumbo C, Pompa G, Izzi V, Canini A, Modesti A, Galvano F, Bei R (2011) Apigenin induces apoptosis and impairs head and neck carcinomas EGFR/ErbB2 signaling. Front Biosci 16:1060–1068

Matsuura HN, Malik S, De Costa F, Yousefzadi M, Hossein MM, Arroo R, Bhambra AS, Strnad M, Bonfill M, Fett-Neto AG (2018) Specialized plant metabolism characteristics and impact on target molecule biotechnological production. Mol Biotechnol 60:169

Medina RA, Rimbau AT, Valderas MP, Estruch R, Lamuela RR (2014) Polyphenol consumption and blood pressure. In: Polyphenols in human health and disease, vol 2. Academic, London, pp 971–987

Miguel MG (2011) Anthocyanins: antioxidant and/or anti-inflammatory activities. J Appl Pharm Sci 1(6):07–15

Milder IE, Feskens EJ, Arts IC, Bueno de Mesquita HB, Hollman PC, Kromhout D (2005) Intake of the plant lignans secoisolariciresinol, matairesinol, lariciresinol, and pinoresinol in Dutch men and women. J Nutr 135(5):1202–1207

Nagaoka M, Maeda T, Moriwaki S, Nomura A, Kato Y, Niida S, Kruger MC, Suzuki K (2019) Petunidin, a B-ring 5'-O-methylated derivative of Delphinidin, stimulates Osteoblastogenesis and reduces sRANKL-induced bone loss. Int J Mol Sci 20(11):2795. https://doi.org/10.3390/ijms20112795

Nair SV, Ziaullah, Rupasinghe HP (2014) Fatty acid esters of phloridzin induce apoptosis of human liver cancer cells through altered gene expression. PLoS One 9(9):e107149. https://doi.org/10.1371/journal.pone.0107149

National Cancer Institute (NCI) at the National Institutes of Health. 2018. Cancer statistics. https://www.cancer.gov/about-cancer/understanding/statistics

Naumovski N, Blades BL, Roach PD (2015) Food inhibits the Oral bioavailability of the major green tea antioxidant Epigallocatechin Gallate in humans. Antioxidants (Basel) 4(2):373–393

Németh G, Hegyi O, Dunai A, Kocsis L (2017) Stilbenes in the different organs of *Vitis vinifera* cv. Merlot grafted on TK5BB rootstock. Oeno One 51(3):323–328

Nguyen AV, Martinez M, Stamos MJ, Moyer MP, Planutis K, Hope C, Holcombe RF (2009) Results of a phase I pilot clinical trial examining the effect of plant-derived resveratrol and grape powder on Wnt pathway target gene expression in colonic mucosa and colon cancer. Cancer Manag Res 1:25–37

Nieman DC, Gillitt ND, Knab AM, Shanely RA, Pappan KL, Jin F, Lila MA (2013) Influence of a polyphenol-enriched protein powder on exercise-induced inflammation and oxidative stress in athletes: a randomized trial using a metabolomics approach. PLoS One 8(8):e72215. https://doi.org/10.1371/journal.pone.0072215. eCollection 2013

Ning Y, Fu YL, Zhang QH, Zhang C, Chen Y (2019) Inhibition of *in vitro* and *in vivo* ovarian cancer cell growth by pinoresinol occurs by way of inducing autophagy, inhibition of cell invasion, loss of mitochondrial membrane potential and inhibition Ras/MEK/ERK signalling pathway. J BUON 24(2):709–714

Nishimuro H, Ohnishi H, Sato M, Ohnishi-Kameyama M, Matsunaga I, Naito S, Ippoushi K, Oike H, Nagata T, Akasaka H, Saitoh S, Shimamoto K, Kobori M (2015) Estimated daily intake and seasonal food sources of quercetin in Japan. Nutrients 7(4):2345–2358

Noufou O, Wamtinga SR, André T, Christine B, Marius L, Emmanuelle HA, Jean K, Marie-Geneviève D, Pierre GI (2012) Pharmacological properties and related constituents of stem bark of *Pterocarpus erinaceus* Poir. Asian Pac J Trop Med 5(1):46–51

de Oliveira MR, Nabavi SF, Habtemariam S, Erdogan Orhan I, Daglia M, Nabavi SM (2015) The effects of baicalein and baicalin on mitochondrial function and dynamics: a review. Pharmacol Res 100:296–308

Ow YY, Stupans I (2003) Gallic acid and gallic acid derivates: effects on drug metabolizing enzymes. Curr Drug Metab 4(3):241–248

Pagare S, Bhatia M, Tripathi N, Pagare S, Bansal YK (2015) Secondary metabolites of plants and their role: overview. Curr Trends Biotechnol Pharm 9:293–304

Pal HC, Pearlman RL, Afaq F (2016) Fisetin and its role in chronic diseases. Adv Exp Med Biol 928:213–244

Palafox CH, Gil CJ, Sotelo MR, Namiesnik J, Gorinstein S, González AG (2012) Antioxidant interactions between major phenolic compounds found in "Ataulgo" mango pulp: chlorogenic, gallic, protocatechuic and vanillic acids. Molecules 17(11):12657–12664

Panaro MA, Carofiglio V, Acquafredda A (2012) Antiinflammatory effects of resveratrol occur via inhibition of lipopolysaccharide-induced NF-κB activation in Caco-2 and SW480 human colon cancer cells. Br J Nutr 108:1623–1632

Panche AN, Diwan AD, Chandra SR (2016) Flavonoids. An overview. J Nutr Sci 5:e47. https://doi.org/10.1017/jns.2016.41

Pandey P, Sayyed U, Tiwari RK, Siddiqui MH, Pathak N, Bajpai P (2019) Hesperidin induces ROS-mediated apoptosis along with cell cycle arrest at G2/M phase in human gall bladder carcinoma. Nutr Cancer 71:676–687

Panthati MK, Rao K, Sandhya S (2011) David Banji. A review on phytochemical, ethnomedical and pharmacological studies on genus *Sophora*, Fabaceae. Rev Bras Farmacogn 22(5):1145–1154

Pápay ZE, Balogh E, Zariwala MG, Somavarapu S, Antal I (2015) Apigenin and Naringenin natural sources, pharmacology and role in cancer prevention. In: Stacks NM (ed) Drug delivery approaches for apigenin: A review. Nova, New York, pp 1–21

Papuc C, Goran VG, Predescu NC, Nicorescu V, Stefan G (2017) Antibacterial agents for shelf-life extension of meat and meat products: classification, structures, sources, and action mechanism. Compr Rev Food Sci Food Saf 16(6):1243–1268

Park EJ, Pezzuto JM (2015) The pharmacology of resveratrol in animals and humans. Biochim Biophys Acta 1852(6):1071–1113

Patel K, Gadewar M, Tahilyani V, Patel DK (2013a) A review on pharmacological and analytical aspects of diosmetin: a concise report. Chin J Integr Med 19(10):792–800

Patel K, Jain A, Patel DK (2013b) Medicinal significance, pharmacological activities, and analytical aspects of anthocyanidins "delphinidin": a concise report. J Acute Dis 2(3):169–178

Patisaul H, Jefferson W (2010) The pros and cons of phytoestrogens. Front Neuroendocrinol 31(4):400–419

Pei K, Ou J, Huang J, Ou S (2016) P-coumaric acid and its conjugates: dietary sources, pharmacokinetic properties and biological activities. J Sci Food Agric 96(9):2952–2962

Peiffer DS, Zimmerman NP, Wang LS, Ransom BW, Carmella SG, Kuo CT, Siddiqui J, Chen J, Oshima K, Huang Y, Hecht S, Stoner GD (2014) Chemoprevention of esophageal cancer with black raspberries, their component anthocyanins, and a major anthocyanin metabolite, protocatechuic acid. Cancer Prev Res 7(6):574–584. https://doi.org/10.1158/1940-6207.CAPR-14-0003

Perri F, Pisconti S, Scarpati G (2016) P53 mutations and cancer: a tight linkage. Ann Transl Med 4(24):522. https://doi.org/10.21037/atm.2016.12.40

Piotrowska H, Kucinska M, Murias M (2012) Biological activity of piceatannol: leaving the shadow of resveratrol. Mutat Res 750(1):60–82

Poór M, Veres B, Jakus PB, Antus C, Montskó G, Zrínyi Z, Vladimir-Knežević S, Petrik J, Kőszegi T (2014) Flavonoid diosmetin increases ATP levels in kidney cells and relieves ATP depleting effect of ochratoxin a. J Photochem Photobiol B 132:1–9

Poschner S, Maier-Salamon A, Zehl M, Wackerlig J, Dobusch D, Pachmann B, Sterlini KL, Jäger W (2017) The impacts of Genistein and Daidzein on Estrogen conjugations in human breast Cancer cells: a targeted metabolomics approach. Front Pharmacol 8:699. https://doi.org/10.21010/ajtcam.v13i3.15

Qiao J, Liu J, Jia K, Li N, Liu B, Zhang Q, Zhu R (2016) Diosmetin triggers cell apoptosis by activation of the p53/Bcl-2 pathway and inactivation of the Notch3/NF-κB pathway in HepG2 cells. Oncol Lett 12(6):5122–5128

Quideau S, Jourdes M, Saucier C, Glories Y, Pardon P, Baudry C (2013) DNA topoisomerase inhibitor Acutissimin a and other Flavano-Ellagitannins in red wine. Angew Chem Int Ed 42:6012–6014

Rajendran P, Rengarajan T, Nandakumar N, Palaniswami R, Nishigaki Y, Nishigaki I (2014) Kaempferol, a potential cytostatic and cure for inflammatory disorders. Eur J Med Chem 86:103–112

Rana S, Bhushan S (2016) Apple phenolic as nutraceuticals: assessment, analysis and application. J Food Sci Technol 53(4):1727–1738

Reinisalo M, Kårlund A, Koskela A, Kaarniranta K, Karjalainen RO (2015) Polyphenol Stilbenes: molecular mechanisms of Defense against oxidative stress and aging-related diseases. Oxid Med Cell Longev 340520. https://doi.org/10.1155/2015/340520

Reis GM (2013) Food phenolic compounds: main classes, sources and their antioxidant power. In: Oxidative stress and chronic degenerative diseases -a role for antioxidants. IntechOpen. https://doi.org/10.5772/51687

Rivera P, Pérez-Martín M, Pavón FJ, Serrano A, Crespillo A, Cifuentes M, López-Ávalos MD, Grondona JM, Vida M, Fernández-Llebrez P, de Fonseca FR, Suárez J (2013) Pharmacological administration of the isoflavone daidzein enhances cell proliferation and reduces high fat diet-induced apoptosis and gliosis in the rat hipocampos. PLoS One 8(5):e64750. https://doi.org/10.1371/journal.pone.0064750

Rodríguez GC, Sánchez QC, Toledo E, Delgado RM, Gaforio JJ (2019) Naturally lignan-rich foods: a dietary tool for health promotion? Molecules 24(5):917

Rom S, Zuluaga RV, Reichenbach NL, Erickson MA, Winfield M, Gajghate S, Solomidou MC, Scuitto KJ, Persidsky Y (2018) Secoisolariciresinol diglucoside is a blood-brain barrier protective and anti-inflammatory agent: implications for neuroinflammation. J Neuroinflammation 15 (1):25. https://doi.org/10.1186/s12974-018-1065-0

Sail V, Hadden MK (2013) Identification of small molecule Hes1 modulators as potential anticancer chemotherapeutics. Chem Biol Drug Des 81(3):334–342

Salehi B, Fokou PVT, Sharifi-Rad M, Zucca P, Pezzani R, Martins N, Sharifi-Rad J (2019) The therapeutic potential of Naringenin: a review of clinical trials. Pharmaceuticals 12(1):11. https://doi.org/10.3390/ph12010011

Salminen JP, Karonen M (2011) Chemical ecology of tannins and other phenolics: we need a change in approach. Funct Ecol 25:325–338

Samavat H, Wu AH, Ursin G, Torkelson CJ, Wang R, Yu MC, Yee D, Kurzer MS, Yuan JM (2019) Green tea Catechin extract supplementation does not influence circulating sex hormones and insulin-like growth factor Axis proteins in a randomized controlled trial of postmenopausal women at high risk of breast Cancer. J Nutr 149(4):619–627. https://doi.org/10.1093/jn/nxy316

Sandoval RB, Lamuela RR, Estruch RS, Doménech M, Tresserra RA (2017) Beer polyphenols and menopause: effects and mechanisms-a review of current knowledge. Oxid Med Cell Longev 4749131. https://doi.org/10.1155/2017/4749131

Saxena M, Saxema J, Pradhan A (2012) Flavonoids and phenolic acids as antioxidants in plants and human health. Int J Pharm Sci Rev Res 16(2):130–134

Sehm T, Fan Z, Weiss R, Schwarz M, Engelhorn T, Hore N, Doerfler A, Buchfelder M, Eyüpoglu IY, Savaskan NE (2014) The impact of dietary isoflavonoids on malignant brain tumors. Cancer Med 3(4):865–877

Semwal DK, Semwal RB, Combrinck S, Viljoen A (2016) Myrcetin: a dietary molecule with diverse biological activities. Nutrients 8(2):90

Seth CY, Lancaster SM, Hullar MA, Lampe JW (2015) Gut microbial metabolism of plant lignans: influence on human health (chapter 7). In: Tuohy KM, Del Rio D (eds) Diet-microbe interactions in the gut. Elsevier, Amsterdam, pp 103–117

Seydi E, Rasekh HR, Salimi A, Mohsenifar Z, Pourahmad J (2016) Myricetin selectively induces apoptosis on cancerous hepatocytes by directly targeting their mitochondria. Basic Clin Pharmacol Toxicol 119(3):249–258

Shuai M, Yi Z, Jiang FL, Xiao W, Zhen L, Shi-QL XX, Hong C, Ben L, Xiang YZ, Li PX (2017) Apigenin inhibits renal cell carcinoma cell proliferation. Oncotarget 8(12):19834–19842

Shuang S, Fang H, Zhuo WH (2017) The regulation of β-catenin activity and function in cancer: therapeutic opportunities. Oncotarget 8(20):33972–33989

Slimestad R, Verheul M (2005) Content of chalconaringenin and chlorogenic acid in cherry tomatoes is strongly reduced during postharvest ripening. J Agric Food Chem 53(18):7251–7256

Smeriglio A, Barreca D, Bellocco E, Trombetta D (2017) Proanthocyanidins and hydrolysable tannins: occurrence, dietary intake and pharmacological effects. Br J Pharmacol 174(11):1244–1262. https://doi.org/10.1111/bph.13630

Somerset SM, Johannot L (2008) Dietary flavonoid sources in Australian adults. Nutr Cancer 60(4):442–449

Song Y, Wu TG, Yang XQ, Chen XY, Wang MF, Wang Y, Peng XC, Ou SY (2014) Ferulic acid alleviates the symptoms of diabetes in obese rats. J Funct Foods 9:141–147

Song B, Li J, Li J (2016) Pomegranate peel extract polyphenols induced apoptosis in human hepatoma cells by mitochondrial pathway. Food Chem Toxicol 93:158–166

Spence LA, Lipscomb ER, Cadogan J, Martin B, Wastney ME, Peacock M, Weaver CM (2005) The effect of soy protein and soy isoflavones on calcium metabolism in postmenopausal women: a randomized crossover study. Am J Clin Nutr 81(4):916–922

Srinivasulu C, Ramgopal M, Ramanjaneyulu G, Anuradha CM, Suresh KC (2018) Syring acid (SA)—a review of its occurrence, biosynthesis, pharmacological and industrial importance. Biomed Pharmacother 108:547–557

Srivastava T, Kumar-Mishra S (2015) Novel function of polyphenols in human health. Res J Phytochem 9(3):116–126

Su S, Cheng X, Wink M (2015) Cytotoxicity of arctigenin and matairesinol against the T-cell lymphoma cell line CCRF-CEM. J Pharm Pharmacol 67(9):1316–1323. https://doi.org/10.1111/jphp.12426

Sun MY, Ye Y, Xiao L, Rahman K, Xia W, Zhang H (2016) Daidzein: a review of pharmacological effects. Afr J Tradit Complement Altern Med 13(3):117–132

Szeja W, Grynkiewicz G, Rusin A (2017) Isoflavones, their glycosides and glycoconjugates. Synthesis and biological activity. Curr Org Chem 21(3):218–235

Takemoto M, Takemoto H (2018) Synthesis of Theaflavins and their functions. Molecules 23(4):918. https://doi.org/10.3390/molecules23040918

Tanase C, Coșarcă S, Muntean DL (2019) A critical review of phenolic compounds extracted from the bark of Woody vascular plants and their potential biological activity. Molecules 24(6):1182

Tang YL, Chan SW (2014) A review of the pharmacological effects of piceatannol on cardiovascular diseases. Phytother Res 28(11):1581–1588

Thummayot S, Tocharus C, Jumnongprakhon P, Suksamrarn A, Tocharus J (2018) Cyanidin attenuates Aβ25-35-induced neuroinflammation by suppressing NF-κB activity downstream of TLR4/NOX4 in human neuroblastoma cells. Acta Pharmacol Sin 39(9):1439–1452

Tomás-Barberán FA, Gil-Izquierdo A, Moreno DA (2009) Bioavailability and metabolism of phenolic compounds and glucosinolates. In: McClements DA, Decker EA (eds) Designing functional foods: measuring and controlling food structure breakdown and nutrient absorption. Woodhead, Cambridge, pp 194–229

Tsai HY, Ho CT, Chen YK (2017) Biological actions and molecular effects of resveratrol, pterostilbene, and 3'hydroxypterostilbene. J Food Drug Anal 25(1):134–147

Tungmunnithum D, Thongboonyou A, Pholboon A, Yangsabai A (2018) Flavonoids and other phenolic compounds from medicinal plants for pharmaceutical and medical aspects: an overview. Medicines 5(3):93. https://doi.org/10.3390/medicines5030093

Tyszka CM, Konieczny P, Majka M (2017) Caffeic acid expands anti-tumor effect of metformin in human metastatic cervical carcinoma HTB-34 cells: implications of AMPK activation and impairment of fatty acids De novo biosynthesis. Int J Mol Sci 18(2):462. https://doi.org/10.3390/ijms18020462

Valavanidis A, Vlachogianni T (2013) Plant polyphenols: recent advances in epidemiological research and other studies on Cancer prevention. Stud Nat Prod Chem 39:269–295

Venancio VP, Cipriano PA, Kim H, Antunes LM, Talcott ST, Mertens-Talcott SU (2017) Cocoplum (*Chrysobalanus icaco* L.) anthocyanins exert anti-inflammatory activity in human colon cancer and non-malignant colon cells. Food Funct 8(1):307–314. https://doi.org/10.1039/c6fo01498d

Wang D, Zou T (2011) Yang, Yan X, Ling W. Cyanidin-3-O-β-glucoside with the aid of its metabolite protocatechuic acid, reduces monocyte infiltration in apolipoprotein E-deficient mice. Biochem Pharmacol 82(7):713–719. https://doi.org/10.1016/j.bcp.2011.04.007

Wang T, Gong X, Jiang R, Li H, Du W, Kuang G (2016) Ferulic acid inhibits proliferation and promotes apoptosis via blockage of PI3K/Akt pathway in osteosarcoma cell. Am J Transl Res 8 (2):968–980

Wang X, Dong T, Zhang A, Sun H (2017) Pharmacokinetic–Pharmacodynamic study of Zhi Zhu wan. Serum Pharmacochemistry of traditional Chinese medicine. Technologies:171–183. https://doi.org/10.1016/B978-0-12-811147-5.00012-1

Wang ST, Feng YJ, Lai YJ, Su NW (2019) Complex tannins isolated from jelly fig achenes affect pectin gelation through non-specific inhibitory effect on pectin methylesterase. Molecules 24 (8):1601. https://doi.org/10.3390/molecules24081601

World Health Organization (WHO). Cancer. 2018. Available on http://www.who.int/mediacentre/factsheets/fs297/es. Access 10 Feb 2018)

World Health Organization (WHO). NCD mortality and morbidity. 2019. Available in https://www.who.int/gho/ncd/mortality_morbidity/en/

Xia EQ, Deng GF, Guo YJ, Li H (2010) Biological activities of polyphenols from grapes. Int J Mol Sci 11:622–646

Xu P, Cai F, Liu X, Guo L (2015) Sesamin inhibits lipopolysaccharide-induced proliferation and invasion through the p38-MAPK and NF-κB signaling pathways in prostate cancer cells. Oncol Rep 33(6):3117–3123. https://doi.org/10.3892/or.2015.3888

Xu P, Huang MW, Xiao CX, Long F, Wang Y, Liu SY, Jia WW, Wu WJ, Yang D, Hu JF, Liu XH, Zhu YZ (2017) Matairesinol suppresses Neuroinflammation and migration associated with Src and ERK1/2-NF-κB pathway in activating BV2 microglia. Neurochem Res 42(10):2850–2860. https://doi.org/10.1007/s11064-017-2301-1

Yang P, Xu F, Li HF, Wang Y, Li FC, Shang MY, Liu GX, Wang X, Cai SQ (2016) Detection of 191 taxifolin metabolites and their distribution in rats using HPLC-ESI-IT-TOF-MS(n). Molecules 21(9):1209. https://doi.org/10.3390/molecules21091209

Yeo S, Fenwick PS, Barnes PJ, Lin HS, Donnelly LE (2017) Isorhapontigenin, a bioavailable dietary polyphenol, suppresses airway epithelial cell inflammation through a corticosteroid-independent mechanism. Br J Pharmacol 174(13):2043–2059

Yin MC, Lin CC, Wu HC, Tsao SM, Hsu CK (2009) Apoptotic effects of protocatechuic acid in human breast, lung, liver, cervix, and prostate cancer cells: potential mechanisms of action. J Agric Food Chem 57(14):6468–6473

Yu X, Jiang X, Zhang X, Chen Z, Xu L, Chen L, Wang G, Pan J (2016) The effects of fisetin on lipopolysaccharide-induced depressive-like behavior in mice. Metab Brain Dis 31 (5):1011–1021

Yu CL, Yang SF, Hung TW, Lin CL, Hsieh YH, Chiou HL (2019) Inhibition of eIF2α dephosphorylation accelerates pterostilbene-induced cell death in human hepatocellular carcinoma cells in an ER stress and autophagy-dependent manner. Cell Death Dis 10(6):418. https://doi.org/10.1038/s41419-019-1639-5

Zamora-Ros R, Andres-Lacueva C, Lamuela-Raventós RM, Berenguer T, Jakszyn P, Martínez C, Sánchez MJ, Navarro C, Chirlaque MD, Tormo MJ, Quirós JR, Amiano P, Dorronsoro M, Larrañaga N, Barricarte A, Ardanaz E, González CA (2008) Concentrations of resveratrol and derivatives in foods and estimation of dietary intake in a Spanish population: European Prospective Investigation into Cancer and Nutrition (EPIC)-Spain cohort. Br J Nutr 100 (1):188–196

Zamora-Ros R, Forouhi NG, Sharp SJ, González CA, Buijsse B, Guevara M, van der Schouw YT, Amiano P, Boeing H, Bredsdorff L, Fagherazzi G, Feskens EJ, Franks PW, Grioni S, Katzke V, Key TJ, Khaw KT, Kühn T, Masala G, Mattiello A, Molina-Montes E, Nilsson PM, Overvad K, Perquier F, Redondo ML, Ricceri F, Rolandsson O, Romieu I, Roswall N, Scalbert A,

Schulze M, Slimani N, Spijkerman AM, Tjonneland A, Tormo MJ, Touillaud M, Tumino R, van der ADL, van Woudenbergh GJ, Langenberg C, Riboli E, Wareham NJ (2014) Dietary intakes of individual flavanols and flavonols are inversely associated with incident type 2 diabetes in European populations. J Nutr 144(3):335–343

Zanwar AA, Badole SL, Shende PS, Hegde MV, Bodhankar SL (2014a) Cardiovascular effects of hesperidin: a flavone glycoside. In: Polyphenols in human health and disease, vol 2. Academic, London, pp 989–992

Zanwar AA, Badole SL, Shende PS, Hegde MV, Bodhankar SL (2014b) Antioxidant role of Catechin in health and disease. In: Watson RR, Preedy VR, Zibadi S (eds) Polyphenols in human health and disease, vol 1. Academic, London, pp 267–271

Zhang Z, Garzotto M, Beer TM, Thuillier P, Lieberman S, Mori M, Stoller WA, Farris PE, Shannon J (2016) Effects of ω-3 fatty acids and Catechins on fatty acid synthase in the prostate: a randomized controlled trial. Nutr Cancer 68(8):1309–1319

Zhou X, Wang F, Zhou R, Song X, Xie M (2017a) Apigenin: a current review on its beneficial biological activities. J Food Biochem 41(4):e12376. https://doi.org/10.1111/jfbc.12376

Zhou Y, Cao ZQ, Wang HY, Cheng YN, Yu LG, Zhang XK, Sun Y, Guo XL (2017b) The anti-inflammatory effects of Morin hydrate in atherosclerosis is associated with autophagy induction through cAMP signaling. Mol Nutr Food Res 61(9). https://doi.org/10.1002/mnfr.201600966

Plant Metabolites Against Enteropathogens   20

Praseetha Sarath, Swapna Thacheril Sukumaran, Resmi Ravindran, and Shiburaj Sugathan

Abstract

Plants have been a source of medicinal products for many years and it provides an enormous reservoir of biologically active compounds with many useful properties, which can cure as well as prevent diseases. Plant-based drugs continue to play an important role in healthcare. So far, many secondary metabolites from plants have proven to be an excellent source of new drugs. Many anti-diarrheal agents have been isolated from different plants. Further attempts are progressing in various laboratories around the world to develop new anti-diarrheal and other drugs from plants. This chapter describes the mechanism of diarrhoea, etiological agents, ethnomedicinal plants with anti-diarrheal action and their molecular targets, ayurvedic concepts of diarrhoea and databases of naturally occurring anti-diarrheal agents. Curcumin, carvacrol, glaucarubin, capsaicin, thymol, gallic acid, eugenol, zingerone, berberine, piperine, geraniol, oleanolic acid and ayurvedic formulations are among the ones discussed.

P. Sarath
Division of Microbiology, Jawaharlal Nehru Tropical Botanic Garden and Research Institute, Thiruvananthapuram, Kerala, India

Department of Botany, University of Kerala, Thiruvananthapuram, Kerala, India

S. T. Sukumaran
Department of Botany, University of Kerala, Thiruvananthapuram, Kerala, India

R. Ravindran
Department of Pathology & Laboratory Medicine, University of California Davis Medical Center, Research Building III, Sacramento, CA, USA

S. Sugathan (✉)
Division of Microbiology, KSCSTE—Jawaharlal Nehru Tropical Botanic Garden and Research Institute, Thiruvananthapuram, Kerala, India

© Springer Nature Singapore Pte Ltd. 2020
S. T. Sukumaran et al. (eds.), *Plant Metabolites: Methods, Applications and Prospects*, https://doi.org/10.1007/978-981-15-5136-9_20

Keywords

Anti-diarrheal · Enteropathogens · ESKAPE · Cholera toxin · Galacto-glycerolipid

20.1 Introduction

Humans have been using natural products from plants since prehistoric times to treat various kinds of diseases. Plants are seen everywhere in every habitable environment. Being sedentary, plants are faced with many challenges and stresses from environmental as well as from animals, insects and microorganisms; therefore, to escape and flourish they have developed many biomolecules for their self-defence. These molecules give plants their colour, fragrances and many unique potentials. These biomolecules also make plants an attractive source of antimicrobial agents. And so plant-derived medicines have been used to treat many pathological conditions since time immemorial.

The increasing emergence of antibiotic-resistant microorganisms, the emergence of new pathogen and the resurrection of old ones are becoming a concern to the scientific community. Above all the lack of effective new therapeutics are worsening the problems. Thus, to improve human health, the need of the hour is to discover and develop new antimicrobial agents. The real value of medicinal flora can be understood only from the study of medicinal plants as antimicrobial agents. As a guideline for the search for new drugs, ethnobotanical plants are of great importance. Ethnobotanical plants are presently in great demand and accepted worldwide. Plants have given us many important medicines, from aspirin to Paclitaxel, the chemotherapy drug. Drugs derived from plants have been a part of human evolution and medicare for thousands of years. To date, many plant-derived bioactive molecules have already been proven to exert antimicrobial, antifungal and anticancer activities; still, the scientific community is continuing to identify, purify and characterize new natural compounds, thereby replacing synthetic chemical drugs. The requirement of a high quantity of drugs due to population growth has demanded the synthesis and use of chemical products. However, there are many noxious effects of synthetic substances like organic balance alteration, resistance induction, pollution of environmental systems and changes in living beings at the genetic level (Lelario et al. 2018). The recent universal trend is moving from synthetic to phytomedicine, i.e., "return to nature" with emphasis on the use of bioactive molecules produced from plants as they possess high levels of activity profile drugs.

The overuse, misuse and underuse of antibiotics by men have led to the development of antibiotic resistance, which is becoming a vital concern for public health. Many human pathogens have ineluctably evolved to become multidrug resistant (MDR) to various presently available drugs causing considerable mortality worldwide. Many life-threatening multidrug-resistant pathogens are increasing including methicillin-resistant *Staphylococcus aureus* (MRSA), MDR-*Mycobacterium tuberculosis* and malarial parasites. Globally, unfolding MDR pathogens also called

"ESKAPE", which include *Enterococcus* spp., *S. aureus*, *Klebsiella* spp., *Acinetobacter baumannii*, *Pseudomonas aeruginosa* and *Enterobacter* spp. Strains of *M. tuberculosis* are extremely drug resistant (XDR) and are resistant to all classes to antibiotics (Subramani et al. 2017). Ethanopharmacology is now the area of interest as many medicinal plants are used against infectious diseases. The secondary metabolites from plants act by a mechanism different from that of conventional antibiotics so that they can be used in the treatment of resistant bacteria. Secondary metabolites and their derivatives from natural products in the future can be used as pharmaceutical drugs. Innumerable bioactive compounds have been derived from plants like alkaloids, tannins, flavonoids, glycosides, quinones, saponins, terpenoids and steroids, glycoalkaloids and glucosinolates.

20.2 Enteropathogens

Members of enterobacteriaceae are a group of non-sporing, non-acid fast, gram-negative bacilli that are found in the gut of animals and humans. Whereas, enteropathogens are microorganisms with pathogenicity for the intestine. They include *Enterobacter* spp., *Enterococcus* spp., *Corynebacterium* spp., *Brachyspira hyodysenteriae* (swine dysentery), *Aeromonas* spp., *Lawsonia intracellularis*, *Campylobacter coli*, *C. jejuni*, *Clostridium difficile*, strains of pathogenic *Escherichia coli*, *Vibrio* spp., *Proteus* spp., *Salmonella enterica*, *S. enteritidis*, *S. heidelberg*, *S. typhymurium*, *Bacillus cereus*, *Cryptosporidium* spp., *Shigella dysenteriae*, *Klebsiella* spp., *Yersinia pestis*, *Helicobacter pylori*, *Giardi intestinalis*, Coronavirus, Rotavirus, Torovirus, Calicivirus, Astrovirus, Canine parvovirus, Rinderpest virus, Malignant catarrhal fever virus, Coccidia, Cryptosporidia etc. Enteropathogens are mostly acquired through oral faecal route or contaminated food and water or from person to person contact. People with some genetic defects are more prone to infections from enteropathogens.

Global estimates indicate that diarrhoea is a common disease amongst children and results in the highest rate of infant mortality worldwide, causing major health problems (Finkelstein 1996). According to WHO, the second leading cause of death (760,000 per year) in children below 5 years is due to diarrhoea (Shrivastava et al. 2017). From the 1885 outbreak of diarrhoea, *E. coli* was suspected to be the causative agent. Later in 1945, it was established that the causative agent of enteritis is a specific type of *E. coli*. Soon, many other enteropathogenic serotypes of *E. coli* came into light, which was responsible for diarrhoea. Currently six different types of diarrheagenic *E. coli* are recognized: Enteropathogenic *E. coli* (EPEC), Enterotoxigenic *E. coli* (ETEC), Enteroinvasive *E. coli* (EIEC), Enterohemorrhagic *E. coli* (EHEC), Enteroaggregative *E. coli* (EAEC) and Diffusely adherent *E. coli* (DAEC). Enteropathogens associated with traveller's diarrhoea include bacteria, viruses and parasites. Intensive research is going on in this area; however, still more remains to be known.

Enteropathogenic bacteria have developed multiple strategies to overcome the natural and acquired host defences. They cause various diseases, either due to the accidental introduction of environmental bacteria into the host or from a long-term adaptation of environmental bacteria to the new survival conditions of the digestive tract. Of the large proportion, only a few produce virulence factors that account for gastrointestinal or food-borne diseases.

The GI (gastrointestinal) tract is the best ecosystem for enteric pathogens mainly because of its mucous nature and presence of nutrients on the epithelial cell lining. Though it provides shelter, yet GI tracts are easily accessible by enteric pathogens. Some can be highly pathogenic when they invade the GI tract, thereby causing many gastrointestinal discomforts like gastroenteritis, diarrhoea, salmonellosis, shigellosis to highly fatal consequences (Omojate Godstime et al. 2014).

All strains of *E. coli* are not harmful. The commensal strains in the human gut are essential as they synthesize vitamin K_2. However, the elevation of antimicrobial resistance in strains of *E. coli* is having a considerable impact on the medical industry. With the advent of extended-spectrum, β-lactamase (ESBL) and carbapenem-resistant Enterobacteriaceae (CRE) strains have caused panic in doctors that these strains might become as dangerous as methicillin-resistant *S. aureus* (MRSA) strains. It is, therefore, essential that physicians keep themselves updated on these strains, their impact on the society, and the treatment methods to be adopted. All the pathogenic *E. coli* strains are equipped with potential virulence factors. They include capsule, flagella, fimbriae, cell wall lipopolysaccharide, outer membrane proteins, cytolysins, haemolysins, siderophores and toxins (in some strains) (Reygaert 2017).

Cholera is endemic and epidemic in places with poor sanitation. It is transmitted by the faecal–oral route. Cholera is a life-threatening diarrheal disease, causing an outflow of voluminous watery stools with vomiting and thus resulting in hypovolemic shock and acidosis. Until 1992, two serotypes, Inaba (AC) and Ogawa (AB), and two biotypes of toxigenic O group, *V. cholerae* O1 (classical) and *V. cholerae* El Tor were thought to be causing cholera. Later on, more serogroups were identified as *V. cholerae* 0139 (non-O1 strain) emerged as an epidemic in India and Bangladesh. Other *Vibrio* spp., include Halophilic *V. parahaemolyticus*, also causes enteritis due to the ingestion of improperly cooked seafood, *V. fetus*, now called *C. jejuni* causes dysentery-like gastroenteritis. *H. pylori,* formerly called *Campylobacter pylori* is a vibrio like organism causing gastritis (Waldor and Mekalanos 1994; Finkelstein 1996). Collectively they are known as enterotoxic enteropathogens. Cholera is still a major epidemic disease. The potential virulence factors are flagellar protein, outer membrane protein (Finkelstein 1996) and cholera toxin (CT).

Aeromonas spp. are becoming renowned as an enteric pathogen of serious public health. The main virulence factors associated with *Aeromonas* are surface polysaccharides, capsule, iron-binding systems, exotoxins, flagellar protein, extracellular enzymes, fimbriae and other nonfilamentous adhesins (Tomás 2012). Severe and chronic gastrointestinal sickness are caused by either bacterial, parasitic and viral enteropathogens. Table 20.1 shows major enteropathogens.

Table 20.1 Enteropathogens causing gastroenteritis

Enteropathogens	Mode of transmission	Common symptoms	Mode of action	References
EPEC	Contaminated water and food	Prolonged diarrhoea	Adhere to microvilli and inject effector proteins(T3SS)	Reygaert (2017)
ETEC	Contaminated water and food	Watery diarrhoea	Enterotoxins	Reygaert (2017)
EIEC	Contaminated water and food	Watery diarrhoea	Enter M cells and invade enterocytes	Reygaert (2017)
EAEC	Contaminated water and food	Non-bloody diarrhoea	Enterotoxins	Reygaert (2017)
EHEC serotype O157:H7	Uncooked meat, raw milk, raw vegetables	Bloody diarrhoea	Toxins	Reygaert (2017)
DAEC	Contaminated water and food	Diarrhoea	Bind to enterocytes	Reygaert (2017)
V. cholerae O1; El Tor and V cholera 0139	Contaminated water and food	Diarrhoea, cholera	Cholera enterotoxin	Reygaert (2017)
Giardia lamblia	Ingestion of cysts in contaminated water and food, faecal–oral route, human contact	Abdominal pain, nausea, persistent watery diarrhoea	Trophozoites interferes with nutrient absorption in small intestine, hyper permeability, and damage the brush border of the enterocyte.	Ross et al. (2013)
Entamoeba histolytica	Drinking water contaminated with faecal cysts, human contact	Abdominal pain, fever, persistent watery diarrhoea	Trophozoites bind to host mucin oligosaccharides.	Ross et al. (2013)
Strongyloides	Contaminated soil	Abdominal pain, persistent diarrhoea	Penetrate through the skin	Ross et al. (2013)
Schistosoma	Fresh-water contact where Schistosoma is endemic	Abdominal pain, persistent diarrhoea	Penetrate through the skin	Ross et al. (2013)
Shigella spp.	Human contact, food, drinking water	Diarrhoea, dysentery	Shiga toxins	Ross et al. (2013)
Salmonella enterica	Contaminated food	Watery diarrhoea	Inject effector proteins	Ross et al. (2013)
Campylobacter jejuni	Poultry, milk, drinking water	Bloody and watery diarrhoea	Cytolethal toxin, a nuclease, results in DNA damage and arrest of the cell cycle.	Ross et al. (2013)

(continued)

Table 20.1 (continued)

Enteropathogens	Mode of transmission	Common symptoms	Mode of action	References
Norovirus GI	Contaminated food & water, person to person	Diarrhoea	Binds intestinal lining	Robilotti et al. (2015)
Rotavirus	Person to person	Diarrhoea	Infect intestinal enterocytes,	Bernstein (2009)
Cryptosoridia	Contaminated food and water	Watery diarrhoea	Sporozoite membrane-associated protein, CP47, binds to p57 glycoprotein located on the ileal brush border	Leitch and He (2011)

20.3 Global Prevalence of Diarrhoea

Diarrhoea is a global health problem and the leading cause of morbidity and mortality in rural communities and developing, third world countries. It is either infectious or non-infectious. Infectious type is acquired by causative agents like bacteria, viruses or parasites and spreads through contaminated food and water. While the non-infectious type is contributed by toxins, chronic diseases or by antibiotics. The eighth leading cause of death (for more than 1.6 million) among all ages and the fifth leading cause of death in children below 5 years of age in 2016 was found to be diarrhoea. The leading etiological agent for mortality among children was found to be rotavirus. Mortality among adults older than 70 years was highest in Kenya (1877 deaths [1184–3029] per 100,000), Central African Republic (1282 deaths [680–2112] per 100,000), and India (1013 deaths [667–1578] per 100,000). About 90% of diarrheal deaths occurred in South Asia and sub-Saharan Africa are due to unsafe water, poor sanitation and childhood wasting. These findings suggest that though challenges exist, urgent need of the hour is to reduce the mortality due to diarrhoea, which can be achieved largely with new efforts to reduce diarrheal burden. Diarrhoea is globally present, but higher rates of morbidity and mortality occur in countries with low income as they have fewer robust infrastructure and resources to tackle the burden. In 2013, WHO and UNICEF came together and formed the Global action plan to reduce deaths due to diarrhoea in children by 2025 (Troeger et al. 2018).

Incidence of diarrhoea in younger children below 5 years decreased in most countries during this period, though at a slower rate from 2013 onwards. The driving force behind the reduction in mortality in diarrhoea is to reduce the risk of dying from the disease than reducing the risk of infection.

20.4 Ethnomedicinal Plants with Anti-Diarrheal Activity

From the beginning of the human race, plants have been used for the treatment of many health disorders and to cure diseases, including epidemics. Among the various diseases treated with traditional medicines, diarrhoea is the most common. Antidiarrheal plant extracts show a delay in gastrointestinal transit, suppress gut motility, inhibiting intestinal motility, antispasmodic properties, stimulates water adsorption, increasing colonic water and electrolyte reabsorption (de Wet et al. 2010).

Promising resources of new antidiarrheal drugs are Ethnobotanical plants. About 80% of people in developing countries are dependent on herbal medicines. Due to this reason, WHO is encouraging research in this field for the treatment and prevention of diarrheal diseases. Traditional medicines are good alternatives for the management of diarrhoea as there is an increasing occurrence of diarrhoea in developing countries and low availability of antidiarrheal drugs. From around the world, many extensive herbal medicines are reported with action against enteropathogens (Mekonnen et al. 2018).

From centuries ethnobotanical plants have been used as remedies for the treatment of human diseases as they have a surplus source of therapeutic components with antibacterial, antifungal and antiprotozoal activities. Plants contain many bioactive components that are used as folk medicines and some of which have been isolated and purified to form finished products called as phytomedicines. Using herbal medicines as an alternative form for healthcare was accepted only during the second half of the twentieth century. Researchers are trying to investigate the therapeutic potentials of these plants while at the same time extenuating the side effects caused by synthetic drugs. Among the wide varieties of therapeutic compounds produced by these plants, one important activity is antimicrobial which is evident from the dearth of diseases in plants itself. This shows how successful is the defence mechanism of plants. Plants with antimicrobial properties are being reported widely from different parts of the world. Researches are trying to investigate if the target sites of these plant extracts are other than the one used by antibiotics and if so, such bioactive components are then a potential source active against drug-resistant pathogens (Valle et al. 2015).

WHO has recommended the use of herbal formulations as complementary as well as alternative medicine. Many ethnobotanical plants have been proven to show activity against enteropathogens like *Holarrhena antidysenterica* (L.) Wall. ex A. DC., *Cassia fistula* Linn., *Desmodium puchellum* (L.) Desv., ***Terminalia arjuna*** Roxb.) Wight & Arn., *Euphorbia paralias* L., *Pentaclethra macrophylla* Benth., *Maranta arundinacea* Linn., *Terminalia alata* Heyne ex Roth, Paederia foetida (Forssk.) Boiss, *Justicia hypocrateriformis* Vahl., *Bidens bipinnata* L., *Schouwia thebaica* Webb., *Cynachum acutum* L., *Convolvulus fatmensis* Kunze., *Plantago major* L. etc.

The plant *H. antidysenterica* in Ayurveda is used in classical formulation, Kutajarishta, prescribed to control diarrhoea. The largest buyer of Kutaja is United States (USD 1644), followed by Canada and Singapore imported Kutaja worth USD

961 and USD 360, respectively. *H. antidysenterica* root bark decoction was used to study the anti-diarrheal activity on three strains of *E. coli* and it was found that it acts by inhibiting the stable toxin production and preventing its intestinal secretions, thereby decreasing the virulence of ETEC strains. Thus it gives protection at multiple stages of diarrhoea (Jamadagni et al. 2017). It causes activation of histamine receptors and blocks Ca^{2+} channel, thereby relaxing the gastrointestinal tract, which indicates its basic usefulness in gut motility related problems such as diarrhoea, colic and constipation. The in vitro activity of aqueous extracts of seeds of *H. antidysenterica* was found to be highly effective against *Shigella* spp., *S. aureus, E. coli* and *S. typhi,* pathogens responsible for diarrhoea (Niraj and Varsha 2016). The alcoholic and aqueous bark extracts of *H. antidysenterica* were found to be active against ten enteric pathogens at 200 mg/mL. The enteric pathogens studied were *V. cholerae* 01, *V. cholerae* 0139, EIEC, EPEC, *S. aureus, S. flexneri, S. boydii, S. typhimurium, S. enteritidis and P. aeruginosa* (Jamadagni et al. 2017). The bark contains alkaloids namely conkurchine, conessine, kurchine, holarrhenine, conkurchinine, lipid-binding, kurchicine etc. *C. fistula*, the *Indian laburnum,* is another plant with abundant secondary metabolites like its phenolic compound, flavonoids, tannins, alkaloids, glycosides, saponins and triterpenoids with many pharmacological properties. Apart from its medicinal use in diarrhoea, it is also used in the treatment of liver, abdominal tumours, convulsions, syphilis, epilepsy, skin diseases and leprosy (Rath and Padhy 2015).

Traditionally, a decoction of the bark of *Desmodium pulchellum* is used to cure diarrheal diseases. The methanolic and petroleum ether extract of this plant was proved to cure castor oil-induced diarrhoea. Ricinoleic acid present in castor oil induces diarrhoea. Ricinoleic acid after ingestion is released in the intestinal lumen by lipases. It induces diarrhoea in two ways, one by increasing peristaltic activity in the small intestine as a result of a change in the permeability of Na^+ and Cl^- ions and the other by stimulating prostaglandins of E series, which in turn is a diarrheagenic agent. Castor oil is metabolized in the gut to ricinoleic acid, which causes inflammation of the intestinal mucosa, thereby releasing inflammatory mediators like prostaglandins and histamine (Rahman et al. 2013). This is responsible for the contraction of intestinal smooth muscle by activation of EP3 receptors on the smooth muscles of the intestine, vasodilation, mucous secretion in the small intestine. The bioactive compounds present in this plant are bufotenin, β-carboline, *N,N*-dimethyltryptamine, gramine, 15 indole-3-alylamine derivatives, galactomannan, gramine, indole-3-alkylamine, betulin, α-amyrin. Significant reduction in intestinal volume and diarrheal severity was observed with the intake of oil (Velmurugan et al. 2014).

Rahman et al. (2013) proved the anti-diarrheal effect of methanolic extract of *M. arundinacea* L. leaves in castor oil-induced diarrhoea in rats. It reduced the secretion of water and electrolytes into the small intestine thereby inhibiting hypersecretion and inhibited intestinal motility. Flavonoids and polyphenols are mainly responsible for the antidiarrheal activity of the leaves of *M. arundinacea*. Flavanoids have shown to inhibit prostaglandin E2-mediated intestinal secretion. Polyphenols exhibit anti-diarrheal activity by inhibiting cytochrome P450 systems. Whole plant extract of *Euphorbia hirta* showed promising inhibition on acetylcholine and

KCl-induced contraction on guinea pig ileum which reduces diarrhoea (Atta and Mouneir 2005). Table 20.2 narrates the antibacterial activity of some ethnobotanical plants against aetiological agents of diarrhoea.

20.5 Pharmacological Evaluation of a Potential Anti-Diarrheal Agents

The basis of the pharmacological evaluation of a potential anti-diarrheal agent is that it should show a reduction in experimental diarrhoea and faecal output. The folkloric basis for the use of plant extract in controlling diarrhoea is because the extracts can give protection against castor oil–induced diarrhoea and reduce faecal output. The major mechanism by which many anti-diarrheal substances act is by reducing the gastrointestinal motility. As there was a noticeable reduction in the peristaltic movement of the gut in animals treated with plant extracts, which suggested the mode of action of plant extracts. Plant extracts showed both neurotropic and musculotropic inhibition of the smooth muscles. The neurotrophic effect of the extracts is suggested by its effect on transmitter release, as proved by inhibition of contractions induced by acetylcholine. The musculotropic effect of the extract is shown by its ability to reduce Ca^{2+} availability during the coupling reaction of excitation and contraction. Plant extracts contain many biologically active compounds with known gastrointestinal functions. Flavanoids inhibit the contraction of guinea pig ileum and inhibit small intestinal transit. Much more research needs to be done to understand the groups of phytochemicals with anti-diarrheal effects (Akah et al. 1999).

20.6 Bioactive Compounds in Plants

In spite of the progress made by the pharmaceutical industries in various fields, an effective restorative remedy for the treatment of gastrointestinal illness has not yet been found. There has been a remarkable increase in gastroenteritis in both developing and developed countries. Treatment failures due to rising levels of multidrug-resistant enteric pathogens are a big challenge. The extended emergence to the third and fourth generation beta-lactam antibiotics has worsened the situation. Vast reports are now published on the emergence of resistance among Enterobacteriaceae to extended-spectrum cephalosporin and other beta-lactam drugs in the community and the hospitals (Bisi-Johnson et al. 2017).

Plant bioactive compounds are derived from natural products and are compounds derived from primary metabolites such as amino acids, carbohydrates and fatty acids and are categorized as secondary metabolites. Secondary metabolites are generally not involved in metabolic activity. Plant species are a convenient source of bioactive molecules with promising microbicidal and microstatic activities against a range of enteropathogens and can be used as therapeutic agents. These phytochemicals include alkaloids, phenols, quinols, azadirachtin, flavonoids, tannins, terpenoids,

Table 20.2 Antibacterial activity of some ethnobotanical plants against etiological agents of diarrhoea

Plant Name and Family	Common name	Plant part	Traditional usage	References
Acacia mearnsii De Wild. Family: Leguminosae	Blackwood black wattle	Bark	Diarrhoea, dysentery, sore throat, coughs, children fever, tooth ache	Bisi-Johnson et al. (2017)
Acanthospermum glabratum (DC) Wild. Family: Asteraceae	Bristly starbur	Whole plant	Diarrhoea	de Wet et al. (2010)
Aloe arborescens Family: Xanthorrhoeaceae	Aloe	Leaves	Vomiting, skin ailments, diarrhoea, urinary complaints, rheumatism, tuberculosis	Bisi-Johnson et al. (2017)
Catharanthus roseus (L.) G. Don. Family: Apocynaceae	Graveyard plant	Root, stem or leaves	Diarrhoea	de Wet et al. (2010)
Chenopodium ambrosioides L. Family: Chenopodiaceae	Kattu ayamodakam	Leaves/seeds	Diarrhoea, intestinal worms, indigestion	de Wet et al. (2010)
Cassia fistula L. Family: Caesalpiniaceous	Sunari	Leaf/bark	Diarrhoea, constipation, convulsion, throat burns, epilepsy, leprosy	Rath and Padhy (2015)
Cissampelos hirta Klotzch. Family: Menispermaceae	Velvet leaf	Leaves	Diarrhoea	de Wet et al. (2010)
Cyathula uncinulata (Schrad.) Schinz Family: Amaranthaceae	NA	Leaves	Diarrhoea	Bisi-Johnson et al. (2017)
Garcinia livingstonei T. Anderson. Family: *Clusiaceae*	Mangosteen	Root or bark	Diarrhoea	de Wet et al. (2010)
Holarrhena antidysenterica L Wall. Family: Apocynaceae	Kutakappaala	Bark, leaves	Diarrhoea and dysentery	Rath and Padhy (2015)
Psidium guajava L. Family: Myrtaceae	Common guava	Leaves	Diarrhoea	Bisi-Johnson et al. (2017)

(continued)

Table 20.2 (continued)

Plant Name and Family	Common name	Plant part	Traditional usage	References
Eucomis autumnalis (Mill.) Chitt. Family: Asparagaceae	Common pineapple flower	Bulb	Colic and flatulence	Bisi-Johnson et al. (2017)
Terminalia alata Heyne ex Roth. Family: Combretaceae	Karimaridu	Leaf/bark	Diarrhoea and dysentery	Rath and Padhy (2015)
Hermbstaedtia odorata Wild cockscomb Family: Amaranthaceae	Rooi-aarbossie	Leaves	Cleansing stomach	Bisi-Johnson et al. (2017)
Terminalia arjuna (Roxb.) Wight & Arn. Family: Combretaceae	Arjuna	Leaf/bark	Acute diarrhoea, skin diseases, urinary infection, skin aliments including acne	Rath and Padhy (2015)
Hydnora africana Thunb. Family: Hydnoraceae	Warty jackal food, Jakkalskos Kanip	Tuber	Diarrhoea	Bisi-Johnson et al. (2017)
Paederia foetida L. Family: Rubiaceae	Bakuchi	Leaves	Diarrhoea, dysentery, skin sores and tooth infections	Rath and Padhy (2015)
Hypoxis olchicifolia Baker. Family: Hypoxidaceae	African potato	Tuber	Headaches, dizziness, mental disorders, to treat cancers, inflammation, HIV, diarrhoea	Bisi-Johnson et al. (2017)
Pelargonium sidoides DC. Family: Geraniaceae	Rose-scented Pelargonium	Root	Gonorrhoea, diarrhoea, dysentery, root decoction severe diarrhoea, stomach ailment in children	Bisi-Johnson et al. (2017)
Psidium guajava L. Family: Myrtaceae	Guava	Leaves	Leaves used for diarrhoea, infusion of leaves for bloody diarrhoea andenema for severe diarrhoea	Bisi-Johnson et al. (2017)
Schizocarphus nervosus (Burch.) van der Merwe Family: Asparagaceae	White Scilla	Corms	Rheumatic fever, dysentery. All-purpose herb.	Bisi-Johnson et al. (2017)

glycosides, steroids, essential oils, saponins, anthraquinones, polyketones, peptides (Omojate Godstime et al. 2014). These compounds have a broad range of pharmaceutical applications like anti-microbial, anti-fungal, anti-biofilm, anti-cancer, cardiovascular, anti-epidemic, anti-hypertensive, anti-diabetic, anti-thrombotic, anti-atherogenic, anti-glycaemic etc. (Singh et al. 2017). Nowadays, phytochemical compounds are more preferred as it has very few side effects.

The largest group of secondary metabolites produced by plants are alkaloids. These nitrogenous compounds have anti-microbial activity against enteropathogenic organisms. The condensation products of sugars are glycosides and many glycosidic compounds have been isolated like prunasin from *Prunus*, trophanthidin from *Strophanthus*, salicin from *Salix*, digitoxin from *Digitalis*, cantharidin from *Cantharides*, barbaloin from *Aloes*. Glycosides are bitter and thus act on gustatory nerves, thereby increasing the gastric juice and saliva flow thereby aid indigestion. The bitterness is due to the presence of the lactone group. Over 4000 flavonoids exist in many plants like quercetin, kaempferol etc. Other flavonoid groups include catechin, flavones, flavans, chalcones, dihydroflavons, anthocyanidins, flavonols and leucoanthocyanidins. Phenolics are omnipresent as colour pigments giving colour to fruits of plants. Their most important role is in plant defence against pathogens. Saponins are soaplike and cause cattle poisoning and hemolysis of blood. Tannins have phenolic groups, and this gives its antiseptic property. Formulations rich in tannin-based plants are used in herbal medicines to treat diarrhoea-like diseases. Chemically diverse groups are terpenes. Many sesquiterpene lactones have been isolated, proven to show antibacterial properties (Omojate Godstime et al. 2014).

20.7 Phytochemicals Against Enteropathogens

20.7.1 Curcumin

EAEC and EPEC are known to produce and secrete auto transporter (AT) proteins that act as toxins. These kinds of AT proteins include the plasmid-encoded toxin (Pet) and EPEC-secreted protein C (EspC). Pet and EspC come under the subfamily of Serine Protease Auto transporters of *Enterobacteriaceae* and both toxins result in cytotoxic effects upon entry into the cytoplasm of epithelial cells. Natural compounds including that of phytochemicals can act as therapeutics by acting on those specific molecular targets on the pathogen, thus leaving the pathogen harmless to the host. Curcumin isolated from *Curcuma longa* is a polyphenolic compound that showcases a wide range of biological activities that includes antimicrobial property. Curcumin has demonstrated to inhibit biofilm formation that has a pivotal role in disease establishment. In an animal model study, curcumin was reported to give a protective effect against the infection by enterotoxigenic *E. coli* (ETEC). Curcumin does not have an inhibitory effect on the growth of the two pathotypes, EAEC and EPEC, while at the same time, it showed to decrease the production of toxic proteins Pet and EspC (Sanchez-Villamil et al. 2019). Two hours after the treatment with curcumin, it was observed that the AT proteins were evidently decreased in production by the pathogen. The study confirmed that both Pet and EspC secretion was inhibited in concentration and time-dependant manner. Gene expression studies confirmed that curcumin did not affect the expression of *pet* and *espC* genes, while it blocked the delivery of mature Pet and EspC proteins from the outer membrane region to the extracellular space. The mechanistic action of curcumin in halting the release of AT proteins is carried out via inhibiting the proteolytic cleavage of mature Pet and EspC, thus allowing the passenger domain to remain associated with the outer membrane. This inhibition process by curcumin leads to the accumulation of AT proteins on the bacterial surface. Molecular docking studies showed that curcumin interacted with the translocation domain of AT proteins. In the case of Pet, curcumin associated with residues N_{1018}–N_{1019}, and for EspC, the residues were N_{1029}–N_{1030}. These interaction results suggest that curcumin covers the cleavage site, thereby blocking the proteolytic process. This study brings curcumin into the forefront of antimicrobial therapeutics and can be employed in combination with other antibiotics for synergistic effect (Sanchez-Villamil et al. 2019).

20.7.2 Quassinoids

Antiparasitic compounds that can act on enteropathogens like *Entamoeba histolyica* have been isolated from plant sources. Simaroubaceous plants like *Brucea antidysenterica*, *B. sumatrana* and *Castela nicholsoni* were traditionally used in different parts of the globe for treating dysentery (Geissman 1964; Gillin et al. 1982). The active compound in these plants that act against *E. histolyica* is quassinoids. Ailanthone and glaucarubin are quassinoid compounds that showed effective anti-amoebial activity against *E. histolyica,* in both in vitro and animal model experiments (Gillin et al. 1982).

20.7.3 Glaucarubin

Simarouba amara and *S. glauca* are also known to be effective against *E. histolyica*. Glaucarubin is the active compound in these plants responsible for anti-amoebic activity and the compound is a crystalline glycoside (Del Pozo and Alcaraz 1956). Glaucarubin was earlier tested in many animals like dogs, mice, rats and guinea pigs. In a preclinical study done in dogs, it was observed that the animals severely infected with *E. histolyica,* oral doses of 0.5 and 2.5 mg/kg of glaucarubin were able to remove the pathogen on the fifth and eighth day of treatment. The animals showed no clinical symptoms of amoebic infection for nearly 1 month, microscopic and culture studies did not show any sign of pathogen (Del Pozo and Alcaraz 1956). In a clinical study that included 78 clinical cases with chronic amoebiasis and nine cases with amoebic dysentery, the daily dose of glaucarubin was in the range of 10–280 mg with the duration of treatment varying from 5 to 32 consecutive days. The patients showed clinically relevant improvements from the first day to the sixth day, toxic side effects were not reported and the drug was tolerated well. Clinical pathogenic examination showed the presence of pathogens only in two patients, thus confirming the effective removal of the infectious agent in the majority of the patients in the study. The liver function test and blood count test did not report any abnormalities. Fifty-four patients were followed for a duration on 13 months, and in eight patients who had chronic amoebiasis showed recurrence of the pathogen (Del Pozo and Alcaraz 1956).

20.7.4 Galacto-Glycerolipid (GGL)

Oxalis corniculata is a weed endowed with medicinal properties which can cure dysentery and diarrhoea, owing to these properties the plant has traditional importance in India. Bioactivity-guided study against *E. histolyica* led to the isolation of few compounds from this plant that showed evident antiamoebic activity (Manna et al. 2010). The active compounds present were a mixture of long-chain saturated fatty acids (C24–C28), a mixture of long chain primary alcohols (C18–C28), and a newly isolated single pure compound, galacto-glycerolipid (GGL). The first two mixtures showed moderate antiamoebic and antigiardial activity, while GGL effectively killed the studied enteropathogens even at lower concentrations. The three above components showcased activity in concentration-dependent manner. In vitro experiments revealed that the growth of *E. histolytica* was completely inhibited by GGL within 48 h, leading to lysis of the pathogen after aggregation. The activity by GGL was in comparable range with that of the standard antiamoebic therapeutic metronidazole. In the case of the two mixtures, significant antiamoebic activity was observed in the initial 24 h, with the cell viability ranging from 33 to 38%. After 24 h death of the organism was not observed, and the proliferation of organisms resulted in hiking the viability to 57% at 72 h. Microscopic observations showed that treatment with GGL has resulted in detachment and lysis of *E. histolytica* cells. The Oxalis compounds were tested against another enteropathogen known to cause diarrhoea, *Giardia lamblia* and among the compounds tested GGL showed the most potential antigiardial activity with an IC_{50} of approximately 3.7 µg/mL. The mixture of saturated fatty acid and long chain alcohols did not show a potential antigiardial activity with the IC_{50} registering at 184 µg/mL and 206 µg/mL, respectively. GGL treatment resulted in the death of nearly 87% *Giardia* cells after 24 h of treatment, and complete cell death was attained at 72 h. Microscopic studies again evidently showed detachment and lysis of *G. lamblia* cells. Thus the study showed that GGL is the most potential in killing the studied enteropathogens (Manna et al. 2010).

The study by Manna and team (2010) showed that GGL was able to induce apoptotic cell death in *Entamoeba* and *Giardia* trophozoites. The apoptotic analysis was conducted with the help of membrane selective amphiphilic styryl dye, FM-64. The amphiphilic property of FM-64 does not allow the dye to cross an intact cell membrane, thereby leading to the association of the dye to the outer leaflet of the bilayer. In the event of treatment with GGL, membrane blebs are formed in both amoeba and giardia cells, and increased permeability of the cell membrane allows the dye to move into intracellular regions resulting in accumulation. Microscopic examination revealed intracellular accumulation of FM-64 thereby concluding apoptosis occurrence upon treatment with GGL. GGL was able to inhibit the activity of both pathogens at low concentrations. The inhibitory concentration of GGL was checked for the presence of toxic effect in HEK 293 and no significant side effects were observed. Concentrations that were eight or tenfold higher than the achieved IC_{50} values showed only 15% proliferation inhibition in HEK-293 cells. This shows that the compound is safe for normal cells and also indicates the possible therapeutic potential of the compound as a clinical candidate. A promising therapeutic clinical candidate for treatment against enteropathogens, like *E. histolytica* and *G. lamblia*, should not affect the normal gut microflora. The gut flora at the same time should not restrict the activity of the compound and render it ineffective against gut pathogens. The antiamoebic activity of GGL was unaffected by the presence of gut flora, GGL was able to induce apoptosis in *E. histolytica* cells even in the presence of gut flora. GGL also did not induce apoptosis death in *E. coli* or gut flora thereby proving that the compound is selective for amoeba and giardia.

The culturing of enteropathogens like *E. histolytica* and *G. lamblia* requires a medium that has bovine serum for the proper growth of these pathogens. The bovine serum has the presence of lipid-binding proteins and factors. Bovine serum albumin (BSA) is present in higher quantities in bovine serum, and it is well known for its capacity to bind free fatty acids. These factors can be responsible for reducing the availability of GGL to enteropathogens in culture, thus reducing the efficiency of GGL. Manna et al. (2010) observed the efficiency of GGL in killing enteropathogens in TYI medium containing 1% bovine serum rather than 10%. Results showed that TYI medium supplemented with 1% bovine serum showed much faster death of amoeba and giardia cells when compared with media containing 10% bovine serum. Thus the study suggested that the presence of BSA will sequester GGL and thus decrease the availability of GGL to enteropathogens, so thereby provides a protective effect. The addition of a mixture of saturated fatty acid and a mixture of long-chain alcohols to GGL enhanced the antiamoebic effect of GGL, thus validating that formulations may work better in treating infections by *E. histolytica*. The same combination did not give any enhancement for antigiardial activity, thus confirming that the combination was more specific for amoebic infections.

20.7.5 Carvacrol

The effect of Carvacrol on the growth and toxin production of *B. cereus* and *Clostridium* spp. was studied. It was found to reduce toxin production by downregulating the gene of toxin production and altering lipid composition in the cell membrane of pathogen, thus modulating fluidity (Upadhyay et al. 2015).

20.7.6 Others

For centuries, plant extracts and their active components are used to treat diarrhoea caused by *Vibrio* spp. Tea catechins and dihydroisosteviol were found to reduce fluid secretion without altering fluid absorption and altering cell viability, caused by

V. cholera toxin. It reversibly targets the cystic fibrosis transmembrane regulator (Upadhyay et al. 2015). RG-tannin and apple phenols reduce fluid accumulation caused by toxin by inhibiting ADP-ribosyltransferase activity, which is critical for the action of cholera toxin.

Capsaicin, the major component in chili pepper and other spices, downregulated the critical virulence genes that code for toxin production in *V. cholerae* like ctxA, tcpA and toxT. A similar effect is seen in plant compounds such as eugenol, thymol, red bayberry and carvacrol. Gallic acid, one of the major components present in *Galla chinensis* reduced the effect of diarrhea caused by ETEC toxin. It acts by inhibiting the binding of enterotoxins of ETEC to GM1 receptor thereby reducing toxin-mediated diarrhea. A similar mode of action is shown by zingerone (vanillylacetone), the main component of ginger, thereby preventing the accumulation of fluid. Berberine (benzyltetrahydroxyquinoline), a plant alkaloid, showed a 70% reduction in ETEC enterotoxin mediated diarrhea. A similar effect was seen with other plant-derived compounds like betulinic acid, λ-carragenan, oleanolic acid, ursolic acid, eugenol, catechin, epigallocatechin, carvacrol, thymol and beta-resorcylic acid (Upadhyay et al. 2015) (Table 20.3).

20.8 Ancient Systems of Medicine: Traditional Medicines

One of the ancient forms of natural therapies and perhaps the oldest form of healthcare used for the prevention and treatment of diseases, physical and mental illnesses are traditional medicines also known as ethnic medicines. Most of the traditional medicines are also derived from natural products only. Some forms of traditional medicines practised all around the world for thousands of years are Kampo (Japan), Acupuncture (China), Ayurveda (India), Unani (Arabic countries), Traditional Chinese medicine (TCM) and Traditional Korean medicine. Considerable developments have been achieved in the case of traditional Chinese medicines with respect to the selection of herb, time for acquiring different plants, identification of medicinal materials and methods of preparation (Yuan et al. 2016). Traditional medicines make use of natural compounds, thus it proffer much merits like isolation of lead bioactive molecules as a drug, examining drug activity and investigate the pharmacokinetic and toxicological parameters. If traditional medicines could be applied successfully, it would serve in the development of new drugs thereby reducing the cost.

Ayurveda means knowledge of life. Ayurveda believes in positive health, which includes a well-balanced metabolism with a healthy state of being. Ayurveda is the most ancient of all traditional medicines, older than traditional Chinese medicine. Earliest evidence of diseases, illness and drugs can be found in the Rigveda and Atharvaveda, dating back to 2000 BCE. For many diseases like degenerative and metabolic disorders, there are no adequate drugs for the treatment in modern medicine, wherein ayurvedic drugs are alluring (Ramawat et al. 2009). Ayurvedic drugs are often multicomponent, so they act on a number of targets simultaneously, so possibly to be more effective than drugs acting on a single target. More

Table 20.3 Efficacy of plant-derived compounds and plant extracts against enteropathogens

Plant compounds	Plant	Target microbe	Pathogenic agent	Potential mechanism of action/target site	References
Carvacrol, trans-cinnamaldehyde	*Thymus vulgaris*	*Clostridium* spp.	Toxin	Down-regulate toxin-producing genes, modulation of transcriptional repressor	Upadhyay et al. (2015)
Diallyl sulphides	*Allium ampeloprasum* (elephant garlic)	*V. cholerae*	Toxin	Growth inhibition	Dubreuil (2013)
Galactomannans	*Trigonella foenumgraecum* (fenugreek)	ETEC and *V. cholerae*	Toxin	Inhibits binding of LT and CT to GM1	Dubreuil (2013)
Flavonoids	*Theobroma cacao* (cocoa)	ETEC	Toxin	Inhibits CFTR	Dubreuil (2013)
Applephenon	*Malus* spp. (apple)	*V. cholerae*	Toxin	Inhibits CT ADP-ribosylation	Upadhyay et al. (2015)
Toosendanin	Citrus fruits	*Clostridium* spp.	Toxin	Inhibits entry of toxin into cell cytoplasm	Upadhyay et al. (2015)
Polygallate (rhubarb galloyl tannin)	*Rhei rhizoma* (Daio)	*V. cholerae*	Toxin	Inhibits CT ADP-ribosylation activity	Dubreuil (2013)
Berberine (alkaloid)	*Berberis aristata*	ETEC and *V. cholerae*	Toxin	Effects tight junctions, NHE3 and AQP4 inhibits secretory response of STa	Upadhyay et al. (2015)
Procyanidins	*Humulus lupulus*	*V. cholerae*	Toxin	Inhibits CTADP-ribosylation activity	Dubreuil (2013)
Carvacrol	Essential oil from *Origanum vulgare*	*Bacillus* spp.	Toxin	Modification of bacterial membranes	Upadhyay et al. (2015)
Catechins (EGCG)	*Camellia sinensis* (black tea)	ETEC	Toxin	Modulation of transmembrane regulators	Upadhyay et al. (2015)

(continued)

Table 20.3 (continued)

Plant compounds	Plant	Target microbe	Pathogenic agent	Potential mechanism of action/target site	References
Epicatechin	*Chiranthodendron pentadactylon* (Flor de manita)	ETEC and *V. cholerae*	Toxin	Interacts with CTA subunit	Dubreuil (2013)
Betulinic acid, Oleanic acid and ursolic acid.	*Chaenomeles speciosa*	*V. cholera* and ETEC	Toxin	Inhibits LTB binding to GM1	Dubreuil (2013)
Essential oils, eugenol, 4-hydroxytyrosol	*Thymus vulgaris, Zataria multiflora*	*Staphylococcus aureus*	Toxin	Reduced expression of toxin-producing genes, *sea, seb, tst, hla*	Upadhyay et al. (2015)
Tea catechins, Dihydroisosteviol	Tea	*Vibrio cholerae*	Toxin	Modulation of transmembrane regulators	Upadhyay et al. (2015)
RG-tannin	Apples	*Vibrio cholerae*	Toxin	ADP-ribosyltransferase activity inhibited	Upadhyay et al. (2015)
Eugenol, Thymol, Carvacrol	Red chilli, white pepper, sweet fennel, red bayberry	*Vibrio cholerae*	Toxin	Toxin-producing genes modulated like toxT, ctxA, tcpA	Upadhyay et al. (2015)
Leanolic acid, ursolic acid, and betulinic acid	*Galla Chinensis, Berberis aristata, Cymbopogon martini, C. winterianus* and *Psidium guajava*	ETEC	Toxin	Inhibiting intestinal secretion of enterotoxins, blocks the binding of enterotoxin to GM1, and toxin mediated cellular pathology	Upadhyay et al. (2015)
Eugenol, Catechin, Cinnamon bark oil, Cinnamaldehyde, Epigallocatechin, Carvacrol, Thymol, beta-resorcylic acid	*Limonium californicum* (Boiss.) A. Heller, *Jusiaea peruviana* L., *Cupressus lustianica* Miller, *Salvia urica* Epling	EHEC	Verotoxin	Reducing the transcription of genes stx1 and stx2, decreasing toxin production, reducing expression of globotriaosylceramide (Gb3) receptor by mimicking toxin receptors.	Upadhyay et al. (2015)

investigation into the pharmacodynamics and pharmacokinetic activity of these Ayurveda drugs will bring Ayurveda into the prevailing system of medicines.

Traditional Chinese medicine has been in use for more than 200 years. It includes massage (tuina), breathing exercise (qi gong), acupuncture and dietary therapy. Chinese herbal prescriptions have approximately 6000 drugs of which 480 are of plant origin. The aim of TCM is to restore health by removing the cause and correcting abnormal functioning, *Artemesia annua, Paeonia lactiflora, Ephedra sinica, Rheum palmatum, Angelica polymorpha* var. *sinensis, Peuraria lobata* and *Panax ginseng* are the important medicinal plants of TCM (Ramawat et al. 2009). Extracts from *G. chinensis* and *Berberis aristata* are used in TCM for the treatment of diarrhoea (Upadhyay et al. 2015).

TCM was introduced to Japan between the fifth and sixth centuries, which was altered by Japanese specialists suitable for a particular condition, and this evolved as Kampo, the traditional medicine of Japan. Japanese physicians are using Kampo as preferred medication and also Kampo, together with radiotherapy or chemotherapy is used for the treatment of cancer. This proves how well modern medicines can be integrated with Traditional medicines (Yuan et al. 2016).

Unani is the ancient Greek medical system, nearly 2500 years old and it has been integrated into the Indian national health care system (Yuan et al. 2016). The current scenario of using traditional medicines is different in different countries. Eighty percent of African folk use traditional medicines. But in Australia, traditional medicine is not much preferred as the majority of people prefer conventional medicine. The same case in many other countries also. Thus by connecting modern medicines with ethnomedicine, great reforms in developing new drugs can be achieved.

20.8.1 Ayurvedic View of Diarrhoea

In Ayurveda diarrhoea is described as "Atisara" meaning excessive passage of watery stool. In Ayurveda many aetiological (Nidan) factors have been mentioned like 'ahara janya' (dietary) factors, 'atiruksha' (dry diet), 'atimatra' (heavy quantity), 'atidrava' (excessive liquids), 'virudha-ashna' (incompatible diets), 'krisha-shushka mamsa' (unhygienic meat), 'dushita jala' (contaminated water), 'atijalakrida' (too much swimming), 'vishaprayoga' (ingestion of poisonous substances), 'vega vighata' (suppression of natural urges). It may also be caused by secondary infections like 'krimi doshas' (intestinal infestation of worms), 'shosa' (tuberculosis) and improper 'sodhana therapy' (body purification treatment) (Mishra et al. 2016).

20.8.2 Single Drug Remedies for Treating Diarrhoea in Ayurveda

Many single drug remedies and compound formulations are described in Ayurveda for the treatment of different types of diarrhoea which are enlisted below (Mishra et al. 2016).

Citraka (*Plumbago zeylanica* Linn.) with buttermilk or only tender fruits of Bilva (*Aegle marmelos* Correa.), Pippali (*Piper longum* Linn.) with honey, is effective in controlling diarrhoea.

Ahiphen (*Papaver somniferum* Linn.) mixed with the bark of Kupilu (*Strychnos nux-vomica* Linn. f.) with honey controls all types of diarrhoea.

Ativisha (*Aconitum heterophyllum* Wall.) along with Bilva (*Aegle marmelos* Linn.), Amra (*Mangifera indica* Linn.), Mocarasa (oleo gum resin obtained from *S. malabarica* Schott and Endl.), Dhataki (*Woodfordia fruticosa* Kurz.), Lodhra (*Symplocos racemosa* Roxb.), seed decoction is very effective in treatment of severe diarrhoea.

Ankot (*A. salviufolium* Wang.) root powder or Patha (*Cissampelos pareira* Linn.) leaf paste along with buffalo's buttermilk is used to cure diarrhoea.

Diarrhoea can be checked locally by application of Amalaki paste (*Emblica officinalis* Linn.) filled with Adaraka (*Zingiber officinale* Rosc.) juice in periumblical region.

Decoction of Badara fruits (*Zizyphus jujube* Mill.) mixed with jaggery and oil is taken orally in case of diarrhoea.

Equal amount of Bhanga (*Cannabis sativa* Linn.) and Jatiphala (*Myristica fragrans* Houtt.) is mixed with double amount of Indrayava (seed of *Holarrhena antidysentrica* Linn.) and made into linctus. It cures all types of diarrhoea.

Bark of Tinduka (*Diospyros peregrina* (Gaertn.) Gurke.) with leaves of Kasmari (*Gmelina arborea* Roxb.) with paste of earth and cooked in mild fire. The extract is mixed with honey and taken. It cures all types of diarrhoea.

Bark of Aralu (*Ailanthus excelsa* Roxb.) with ghee is steamed followed by smashing and mixing with honey controls severe diarrhoea.

Bark of Sallaki (*Boswellia serrata* Roxb. Ex Coleb), Amra (*Mangifera indica* Linn.) Jambu (*Syzygium cuminii* (Linn.) Skeels), Arjuna (*T. arjuna* (Roxb.) W. & A.), Badri (*Z. jujube*) and Priyala (*Buchanania lanzan* Spreng), mixed with honey milk checks haemorrhage.

Latex of Udumber (*Ficus racemosa* Linn.) is mixed with Bhanga (*Cannabis sativa* L.) and made into pills. It cures all types of diarrhoea.

Chandan (*Santalum album* Linn.) combined with sugar and rice water on consumption, gives relief from burning sensation and hemorrhage diarrhoea.

A non-unctuous enema is prepared with decoction of Dashmula with honey and milk for managing diarrhoea with pain and tenesmus.

Ghee mixed with oil should be given to patients with diarrhoea, which should be followed by warm milk or with cooked Eranda (*Ricinus communis* L.) root or tender fruits of Bilva (*A. marmelos* L.).

Powder of Pippali (*Piper longum* Linn.) or Marica (*P. nigrum* Linn.) is used to cure chronic diarrhoea.

Satavari paste *(Asperagus racemosus* Wild.) is taken with milk followed by milk diet helps to control diarrhoea with blood.

Ghee cooked with leaf-buds of Kshirivriksa Ashvattha (*Ficus religiosa* Linn.), Parisha (*Thespesia papulnea* Soland. Ex. Correa), Udumbara (*F. racemosa* Linn.),

Vata (*F. benghalensis* Linn.) and Plaksha (*F. lacor* Buch.-Ham.) is taken with honey and sugar to cure diarrhoea.

Cold extract of Shalmali (*Salmalia malabarica* DC.) petioles, kept overnight is taken after adding Madhuka (*Madhuca indica* J.F. Gmel.) and honey for managing diarrhoea.

Gajapippali (*Scindapsus officinalis* Schott.) mixed with honey and sugar cures diarrhoea with blood and mucus.

Covering of Amlika seed (*Tamarindus indica* Linn.), Yavani (*Trachyspermum ammi* Linn. Sprague) and Sunthi (*Z. officinale* Rosc.) rock salt are mixed together and taken with fresh buttermilk for managing diarrhoea.

In diarrhoea caused by pitta, powder of Bilva (*A. marmelos* L.), Durlabha (*Fagonia arabica* L.), Hribera (*Coleus vettiveroides* K.C. Joseph) and Daruharidra (*Berberis aristata* DC.) bark mixed with honey is taken along with rice water.

Goat's milk processed with kamala (*Nelumbo nucifera* Gaertn.), Utpala and Lajjalu (*Mimosa pudica* Linn.), Madhuka (*M. indica* J.F. Gmel.) and Lodhra (*Symplocos racemosa* Roxb.) is mixed with honey and sugar and is used for drinking, eating and sprinkling around anus in case of diarrhoea.

Decoction made of the crushed tender leaves of Simsapa (*Dalbergia sissoo* Roxb. ex DC.) and Kovidara (*Bauhinia purpurea* Linn.) with barley, ghee and milk should be given as slimy enema in discharge of mucus and tenesmus.

Khada made of Cangeri (*Oxalis corniculata* L.), Cukrika and Dugdhika (*Euphorbia thymifolia* Linn.) along with fatty layer of curd, ghee and pomegranate (*Punica granatum* Linn.) seeds is given for diarrhoea.

Decoction of Sunthi (*Z. officinale* Rosc.), Dhanyaka (*Coriandrum sativum* Linn.), Balaka and Bilva (*A. marmelos* L.) reduce pain and helps in digestion and is a good appetizer for diarrhoea patients.

Hanging roots of Nygrodha (*F. benghalensis* L.) powdered with cow's butter milk is taken to cure acute diarrhoea.

Decoction of Indrayava (seed of *H. antidysentrica* (L.) Wall. ex A. DC) and Patol (*Trichosanthes dioica* Roxb.) mixed with honey and sugar cures all types of diarrhoea.

20.8.3 Compound Formulation for Diarrhoea in Ayurveda

In Ayurveda some traditional formulations are prescribed for treating diarrhoea in children as diarrhoea is the leading cause of childhood mortality in India. Few are mentioned below (Niraj and Varsha 2018)

1. Gangadhar Churna: The main ingredients are Ativish, Sugandhbala, Mustaka, Kutaj seed, Araluka, Patha, Shunthi, Mocha rasa, Lodhra and Bilva. It stops diarrhoea from profressing and gives primary relief.
2. Balchaturbhadra Churna: It consists of Karkatsringi, Mustha, Pippali and Ativisha. It helps to cure diarrhoea as well as cold and cough.

3. Dhanya Panchak Kwatha: It constitutes Netrabala, Dhanyaka, Bilva, Shunthi, and Mustaka. This combination has pachana, deepan, and grahi properties also. It cures diarrhoea by reducing and improving the consistency of stool.

20.9 Conclusions

In developed countries, preference is now given for scientific research on the development of new drugs from plants. This research has been carried out in two stages: first, from the prior literature knowledge and survey, already known bioactive compounds from plants with healing properties are selected. The second phase has led to the discovery of new drugs from new bioactive compounds of unexplored medicinal plants from remote areas of the world. Scientific investigation of all traditional medicines from Ayurveda, Unani and Siddha should be carried out and results validated with clinical trials. Many private and government agencies have welcomed the trend and have already validated many formulations for new drug discovery. This is an indispensable approach to correlate the traditional ethnobotanical knowledge with the modern practice for the betterment of human health. Using modern techniques, checking the quality of herbs and herbal formulations is the need of the hour. Herbal pharmacopeias are maintained by many countries and updated from time to time with new procedures and monographs to improve the quality of herbal products.

Advanced technology like high throughput screening, combinatorial chemistry, genomics and proteomics has paved the way to identify novel drug leads in less time. The success chances are limited but in the near future, these significant investments will come out to be fruition.

References

Akah PA, Aguwa CN, Agu RU (1999) Studies on the antidiarrhoeal properties of *Pentaclethra macrophylla* leaf extracts. Phytother Res 13(4):292–295

Atta AH, Mouneir SM (2005) Evaluation of some medicinal plant extracts for antidiarrhoeal activity. Phytother Res 19(6):481–485

Bernstein DI (2009) Rotavirus overview. Paediatr Infect Dis J 28(3):S50–S53

Bisi-Johnson MA, Obi CL, Samuel BB, Eloff JN, Okoh AI (2017) Antibacterial activity of crude extracts of some South African medicinal plants against multidrug resistant etiological agents of diarrhoea. BMC Comple Altern Med 17(1):321

Dubreuil JD (2013) Antibacterial and antidiarrheal activities of plant products against enterotoxinogenic Escherichia coli. Toxins 5(11):2009–2041

Finkelstein RA (1996) Cholera, Vibrio cholerae O1 and O139, and other pathogenic vibrios. In: Medical microbiology, 4th edn. University of Texas Medical Branch at Galveston

Geissman TA (1964) New substances of plant origin. Annu Rev Pharmacol 4(1):305–316

Gillin FD, Reiner DS, Suffness M (1982) Bruceantin, a potent amoebicide from a plant, *Brucea antidysenterica*. Antimicrob Agents Chemother 22(2):342–345

Jamadagni PS, Pawar SD, Jamadagni SB, Chougule S, Gaidhani SN, Murthy SN (2017) Review of Holarrhena antidysenterica (L.) wall. Ex a. DC.: Pharmacognostic, pharmacological, and toxicological perspective. Pharmacogn Rev 11, 141(22)

Leitch GJ, He Q (2011) Cryptosporidiosis—an overview. J Biomed Res 25(1):1–16
Lelario F, Scrano L, De Franchi S, Bonomo MG, Salzano G, Milan S, Milella L, Bufo SA (2018) Identification and antimicrobial activity of most representative secondary metabolites from different plant species. Chem Biol Technol Agri 5(1):13
Manna D, Dutta PK, Achari B, Lohia A (2010) A novel galacto-glycerolipid from *Oxalis corniculata* kills *Entamoeba histolytica* and *Giardia lamblia*. Antimicrob Agents Chemother 54(11):4825–4832
Mekonnen B, Asrie AB, Wubneh ZB (2018) Antidiarrheal activity of 80% Methanolic leaf extract of *Justicia schimperiana*. Evid Based Complement Alternat Med 2018:3037120. https://doi.org/10.1155/2018/3037120
Mishra A, Seth A, Maurya SK (2016) Therapeutic significance and pharmacological activities of antidiarrheal medicinal plants mention in Ayurveda: a review. J Intercult Ethnopharmacol 5(3):290
Niraj S, Varsha S (2016) Antibacterial activity of Kutaj (*Holarrhena antidysenterica* Linn.) in childhood diarrhoea: -in vitro study. Pharma Innov 4(4, Part B)
Niraj S, Varsha S (2018) Chronic and persistent diarrhoea in children and its treatment in ayurveda. J Pharmacogn Phytochem 7(6):43–45
Omojate Godstime C, Enwa Felix O, Jewo Augustina O, Eze Christopher O (2014) Mechanisms of antimicrobial actions of phytochemicals against enteric pathogens—a review. J Pharm Chem Biol Sci 2(2):77–85
del Pozo EC, Alcaraz M (1956) Clinical trial of glaucarubin in treatment of amebiasis. Am J Med 20(3):412–417
Rahman MK, Barua S, Islam MF, Islam MR, Sayeed MA, Parvin MS, Islam ME (2013) Studies on the anti–diarrheal properties of leaf extract of *Desmodium puchellum*. Asian Pac J Trop Biomed 3(8):639–643
Ramawat KG, Dass S, Mathur M (2009) The chemical diversity of bioactive molecules and therapeutic potential of medicinal plants. In: Herbal drugs: ethnomedicine to modern medicine. Springer, Berlin, pp 7–32
Rath S, Padhy RN (2015) Antibacterial efficacy of five medicinal plants against multidrug-resistant enteropathogenic bacteria infecting under-5 hospitalized children. J Integr Med 13(1):45–57
Reygaert WC (2017) Antimicrobial mechanisms of Escherichia coli. In: Escherichia coli—recent advances on physiology, pathogenesis and biotechnological applications, IntechOpen
Robilotti E, Deresinski S, Pinsky BA (2015) Norovirus. Clin Microbiol Rev 28(1):134–164
Ross AG, Olds GR, Cripps AW, Farrar JJ, McManus DP (2013) Enteropathogens and chronic illness in returning travelers. N Engl J Med 368(19):1817–1825
Sanchez-Villamil JI, Navarro-Garcia F, Castillo-Romero A, Gutierrez-Gutierrez F, Tapia D, Tapia-Pastrana G (2019) Curcumin blocks cytotoxicity of enteroaggregative and enteropathogenic Escherichia coli by blocking Pet and EspC proteolytic release from bacterial outer membrane. Front Cell Infect Microbiol 9:334
Shrivastava AK, Kumar S, Mohakud NK, Suar M, Sahu PS (2017) Multiple etiologies of infectious diarrhoea and concurrent infections in a paediatric outpatient-based screening study in Odisha, India. Gut Pathogens 9(1):16
Singh M, Kumar A, Singh R, Pandey KD (2017) Endophytic bacteria: a new source of bioactive compounds. 3 Biotech 7(5):315
Subramani R, Narayanasamy M, Feussner KD (2017) Plant-derived antimicrobials to fight against multi-drug-resistant human pathogens. 3 Biotech 7(3):172
Tomás JM (2012) The main Aeromonas pathogenic factors. ISRN Microbiol 2012
Troeger C, Blacker BF, Khalil IA, Rao PC, Cao S, Zimsen SR, Albertson SB, Stanaway JD, Deshpande A, Abebe Z, Alvis-Guzman N (2018) Estimates of the global, regional, and national morbidity, mortality, and aetiologies of diarrhoea in 195 countries: a systematic analysis for the Global Burden of Disease Study 2016. Lancet Infect Dis 18(11):1211–1228

Upadhyay A, Mooyottu S, Yin H, Nair MS, Bhattaram V, Venkitanarayanan K (2015) Inhibiting microbial toxins using plant-derived compounds and plant extracts. Medicines (Basel) 2 (3):186–211

Valle DL Jr, Andrade JI, Puzon JJM, Cabrera EC, Rivera WL (2015) Antibacterial activities of ethanol extracts of Philippine medicinal plants against multidrug-resistant bacteria. Asian Pac J Trop Biomed 5(7):532–540

Velmurugan G, Anand SP, Doss A (2014) *Phyllodium pulchellum*: a potential medicinal plant-a review. Int J Pharm Rev Res 4(4):203–206

Waldor MK, Mekalanos JJ (1994) ToxR regulates virulence gene expression in non-O1 strains of Vibrio cholerae that cause epidemic cholera. Infect Immun 62(1):72–78

de Wet H, Nkwanyana MN, van Vuuren SF (2010) Medicinal plants used for the treatment of diarrhoea in northern Maputaland, Kwa Zulu-Natal Province, South Africa. J Ethnopharmacol 130(2):284–289

Yuan H, Ma Q, Ye L, Piao G (2016) The traditional medicine and modern medicine from natural products. Molecules 21(5):559

Molecular Chaperones and Their Applications

21

Gayathri Valsala, Shiburaj Sugathan, Hari Bharathan, and Tom H. MacRae

Abstract

Chaperone proteins play a vital role in maintaining cellular protein homeostasis. They assist in folding of newly synthesised nascent peptides and also in protecting proteins from denaturing when exposed to stress. Different classes of chaperone proteins and their functions are discussed with special reference to their importance in plant stress tolerance. Various applications of chaperone proteins in agriculture, medical and industrial fields are also explored.

Keywords

Chaperone proteins · Stress tolerance · Heat shock proteins · Biotech applications

21.1 Introduction

All organisms encounter a range of internal and external stresses on a daily basis. They have developed different coping mechanisms to protect against these stresses. The internal environment of biological systems exists in a state of dynamic equilibrium called homeostasis that is necessary for its normal functioning. Any external or internal factor that interferes with this equilibrium is termed a stressor, and stress can be defined as a condition in which homeostasis is under threat. Stresses include external factors such as, foreign toxins, oxidants, change in temperature, pH and

G. Valsala · S. Sugathan
Division of Microbiology, KSCSTE—Jawaharlal Nehru Tropical Botanic Garden and Research Institute, Thiruvananthapuram, Kerala, India

H. Bharathan (✉)
Department of Zoology, Sree Narayana College, Kollam, India

T. H. MacRae
Department of Biology, Dalhousie University, Halifax, NS, Canada

radiations, and internal factors such a,s diseases, ageing, etc. Depending on the extent of damage caused by the stress, cells can either carryout repair to re-establish homeostasis or induce cell death (Poljšak and Milisav 2012).

The sessile nature of plants makes them continuously exposed to biotic and abiotic stresses. To overcome such detrimental factors, plants have evolved a number of defence mechanisms. Exposure to external stresses like drought, temperature extremes, high salinity and pathogen attack can activate various signalling pathways, which in turn regulate the expression of stress proteins that help in the protection of cellular components (Gilroy et al. 2016). A major portion of these proteins are comprised of chaperone proteins or foldases, which binds to misfolded proteins and assist their proper folding. In this chapter, we explore the different chaperone proteins that are involved in survival under extreme conditions. The possible applications of various plant stress proteins are also explored.

21.2 Different Types of Chaperone Proteins

Cellular stress response is dependent on the damage to a macromolecule and is often not specific to the stressor responsible for the damage. Depending on the type and threshold of stress, the cell can induce cell repair to regain homeostasis, activate responses involved in temporary adaptation to the stressor, induce autophagy or in the case of repairable damage and activate one of the many cell death pathways (Fulda et al. 2010).

Molecular chaperones are a group of unrelated protein families that assist correct folding of unfolded proteins, thereby stabilising them. The functioning of a protein molecule is dependent on its three-dimensional structure, which it attains through specific folding process. Newly synthesised peptides that have not yet been folded are at risk of being misfolded leading to loss of activity. This happens because the hydrophobic regions in the protein chain that are normally hidden in the interior of native proteins are exposed in unfolded, misfolded or partially folded proteins that tend to cluster together causing protein aggregation (He et al. 2016). Such aggregation is prevented by molecular chaperones, which bind to unfolded or partially folded (non-native) protein molecules and stabilising them. Chaperone binding motifs in proteins are usually short stretches of hydrophobic amino acids with basic residues flanking them. These regions lack acidic amino acid residues. The presence of such motifs in many proteins are the reason for the ability of chaperone proteins to act on multiple targets. These proteins then undergo correct folding with the help of chaperone proteins, or transported across subcellular membranes, or in case of irreversible damage, directed to degradation pathways. The chaperone proteins do not interact with native proteins, which are active proteins with correct folding, and they neither form a part of them.

Chaperone proteins can act as 'holdases' that bind to misfolded proteins, thus preventing their irreversible loss of activity, 'foldases' that actively fold a protein in non-native state to its active form or 'unfoldases' that unfold misfolded protein aggregates. sHSPs are involved in the holding of substrates, which does not require energy, while folding performed by HSPs like HSP70, HSP90 or HSP60 requires ATP binding and hydrolysis. Chaperone proteins like ClpB promotes disaggregation of misfolded proteins and increase their solubility with the help of ATP (Barnett et al. 2000).

Many of the chaperone proteins are expressed constitutively in cells, but when under stress like heat shock or oxidative stress that can result in protein denaturation, the expression of some chaperones is upregulated. The activity of most chaperone proteins are non-specific with the ability to bind and assist folding of a number of cellular proteins, but some of them are highly specific in their targets (Saibil 2013).

The dynamic nature of protein synthesis and functioning makes it necessary for cells to have a round-the-clock chaperone surveillance system to safeguard proteostasis or protein homeostasis. Chaperones are involved in the maintenance of proteome through de novo folding of nascent peptides, refolding of denatured proteins, protein trafficking, oligomer assembly and proteolysis (Hartl et al. 2011). Table 21.1 enlists the different classes of chaperone proteins and their functions.

A number of structurally different chaperones are involved in a multiple cellular pathway like protein folding, protein transport and stress response. Some of the chaperone proteins are named heat shock proteins (HSPs) because of upregulation in their expression during stress. The HSPs are classified based on their molecular weight into HSP40 (DnaJ), HSP60 (chaperonins), HSP70 (DnaK), HSP90, HSP100 (Clp proteins) and small heat shock proteins (sHSP). In addition to the HSPs, chaperones also include calnexin, calreticulin, protein disulphide isomerase (PDI) and peptidyl prolyl cis–trans-isomerase (PPI).

21.2.1 Small Heat Shock Proteins (sHSPs)

As the name suggests, sHSPs are small molecular weight proteins, normally in the range of 12–43 kDa. Some of them are related to α-crystallin present in the eye lens of vertebrates. sHSPs are a diverse group of proteins with little sequence conservation across species, found in association with cytoskeleton, nuclei and membranes. Although sHSPs vary at sequence level, they show conservation in their secondary and tertiary structures. They act as a first line of defence against cellular stress by binding to denaturing proteins and prevent them from reaching a state of irreversible damage and aggregation. These ATP-independent chaperones are also known as holdases because they bind to misfolded proteins preventing further denaturation and hold them in a partially unfolded form on which other ATP-dependent HSPs can act and actively refold. The expression of sHSPs are upregulated when a cell encounters some form of stress, and they play a crucial role in apoptosis, cellular signal transduction, host–pathogen interactions and prevention of protein aggregation (Marklund et al. 2018).

A common feature of members of this family is the presence of a conserved α-crystalline domain composed of ~90 amino acid residues. Although the amino acid sequence in this domain varies with the exception of a few positions, the secondary structure of this domain consists of a number of beta-strands that are

Table 21.1 Different classes of chaperone proteins and their functions

Chaperone family	Structural features	Functions
Small HSP (sHSP)	α-Crystallin domain	Prevent aggregation of unfolded protein, maintenance of membrane fluidity, cellular signal transduction
HSP40 (DnaJ)	N-terminal J-domain	Co-chaperones of HSP70 involved in complex formation and regulation of ATPase activity
HSP60 (chaperonins)	Double heptameric ring structure	Protein folding and translocation in association with other chaperone proteins like GroES and HSP70
HSP70 (DnaK)	N-terminal ATP binding domain and C-terminal substrate binding domain	Folding, translocation and degradation of proteins, regulation of stress responses, prevent protein aggregation
HSP90 (HtpG)	Homodimers with N-terminal ATP binding domain, C-terminal domain involved in dimerization and co-chaperone binding	Regulation of cellular proteostasis during normal physiological conditions and under stress
HSP100 (Clp)	AAA+ module involved in ATP binding and hydrolysis	Solubilisation and reactivation of protein aggregates, degradation of misfolded proteins
Calnexin and calreticulin	Proline-rich P domain and a calcium-binding C-domain	Assist folding, subunit assembly and maturation of glycoproteins transported through the endoplasmic reticulum (ER)
Protein disulphide isomerase (PDI)	Thioredoxin domains with βαβαβαββα fold	Catalyse formation and breakage of disulphide bonds between cysteine residues
Peptidyl prolyl cis–trans-isomerase (PPI)	Catalytic domain made up of β-folds	Regulate protein folding at proline residues
Universal stress proteins (USPs)	α/β subdomain	Prevent protein denaturation, cellular growth regulation, protein transportation, hypoxia responses, ion scavenging

assembled into two beta-sheets that are involved in the formation of dimers, the basic building block of most sHSPs. The amino-terminal region of sHSPs is concerned with oligomerisation and substrate attachment, while the carboxy-terminal region takes care of chaperone activity, solubility and also oligomerisation through linkage between subunits (Sun and MacRae 2005). Two motifs specific for sHSPs are carboxy-terminal conserved I/V/L-X-I/V/L motif and amino-terminal WDPF motif, both of which are involved in oligomerisation. A fragment extending from the α-crystallin domain end to the end of I/V/L-X-I/V/L motif helps in anchoring of

Fig. 21.1 Organisation of different domains and motifs in sHSPs

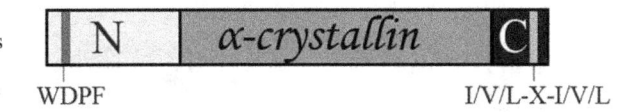

| N | α-crystallin | C |

WDPF   I/V/L-X-I/V/L

sHSPs during oligomerisation and is hence named C-terminal Anchoring Module (CAM) (Poulain et al. 2010).

X-ray crystallography and electron microscopy studies of sHSPs have revealed that the outer diameters of sHSP oligomers are of the range 100–180 Å. Different sHSPs form oligomers of varying number of subunits and shapes. HSP26 from *Saccharomyces cerevisiae* with 24 subunits and HSP16.3 from *Mycobacterium tuberculosis* with 12 subunits form spherical hollow structures, while HSP16.9 from *Triticum aestivum* (wheat) has 12 subunits that forms a barrel-shaped oligomer made up of two hexameric double discs. sHSP oligomer formation is highly dynamic, which makes it difficult to study the structural basis to chaperone activity and oligomerisation (Haslbeck et al. 2005). Unlike most sHSPs that form heterogeneous oligomeric assemblies, HSP16.5 from *Methanococcus jannaschii*, a hyperthermophilic methanogenic archaeon, assembles into homogeneous oligomers of 400 kDa made up of 24 monomers (Kim et al. 1998). Figure 21.1 shows the organisation of different domains and motifs in sHSPs.

Most chaperone proteins switch between two forms with low and high affinity to misfolded proteins depending on the cellular conditions. In sHSPs, this switching is made possible by different oligomeric assemblies, and an oligomeric equilibrium is essential for the regulation of its chaperone activity (Shashidharamurthy et al. 2005). sHSP oligomeric complexes can bind to multiple non-native proteins simultaneously. These structures are highly dynamic and may contain multiple types of sHSPs forming hetero-oligomeric assemblies. Varying binding ability of different homo-oligomeric and hetero-oligomeric structures could be responsible for the difference in binding affinities to different substrates (Basha et al. 2012).

21.2.2 HSP40 (DnaJ)

HSP40 (DnaJ) family comprises of proteins of 40 kDa molecular weight that act as co-chaperones involved in regulating complex formation between HSP70 (DnaK) and target proteins. These proteins also control the ATPase activity of HSP70. They are a large family that are characterised by the presence of a conserved signature J-domain made up of ~70 amino acids that help in the interaction with HSP70. HSP40 is also known as J-proteins because of its N-terminal J-domain. J-domain is made up of four α-helices with helix II and III tightly packed in opposite direction connected by a flexible loop structure containing the signature HPD motif that is involved in the regulation of ATPase activity of HSP70 (Walsh et al. 2004).

In addition to the J domain, many HSP40 proteins contain a cysteine-rich zinc-finger central domain that has four repeats of the motif CXXCXGXG, a G/F region, and a variable C-terminal domain. The zinc finger domain is made up of two distinct

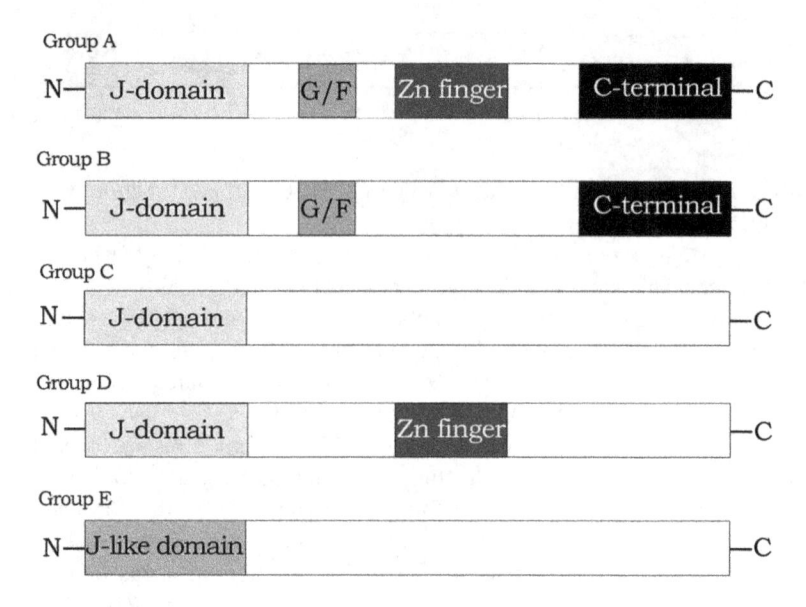

```
Group A
N—[ J-domain ][G/F][Zn finger][   C-terminal   ]—C

Group B
N—[ J-domain ][G/F][            C-terminal    ]—C

Group C
N—[ J-domain ][                                ]—C

Group D
N—[ J-domain ][       Zn finger                ]—C

Group E
N—[J-like domain][                              ]—C
```

Fig. 21.2 Domain organisation in different types of HSP40

Table 21.2 HSP40 proteins and its classification based on domains

Sl. no	Group	Description
1	Group A	HSP40 proteins with J-domain, G/F region, zinc finger domain and the C-terminal domain with low conservation. E.g. DnaJ of *E. coli*.
2	Group B	HSP40 proteins lacking zinc finger domain. The rest of the domains are present in them.
3	Group C	HSP40 proteins having only J-domain.
4	Group D	HSP40 proteins containing the J-domain and zinc finger domain, but lacking C-terminal domain.
5	Group E	They are considered as J-like proteins that have a J-domain that lack the HPD motif. They usually have conserved DKE motif.

clusters of CXXCXGXG repeats with each cluster binding to one zinc unit. This domain exhibits disulphide isomerase activity. The G/F region is rich in glycine and phenylalanine, which acts as a flexible linker that connects the J-domain to the rest of the protein (Fig. 21.2). It helps in stabilising and positioning of the J domain for proper interaction with HSP70. The C-terminal domain is less conserved and is responsible for HSP40 dimerisation, substrate binding and assisting HSP70 during protein folding (Craig et al. 2006). HSP40 proteins are classified into five groups based on the domains present (Pulido and Leister 2018) (Table 21.2).

Members of HSP40 family usually coexist with HSP70 family members in same cellular compartments and function in association with each other. HSP40 chaperones function by regulating the ATP-dependent binding of HSP70 proteins to non-native polypeptides. At first, the HSP40 proteins bind to target peptides through these polypeptide-binding domains (PPDs) and take the bound polypeptides to HSP70. Next, the ATPase-regulating property of HSP40 stabilises the polypeptide-HSP70 complex by mediating conversion of HSP70 from its ATP form to ADP form. Various HSP40 proteins present at different locations in a cell organelle can interact with an HSP70 protein resulting in varying polypeptide binding abilities.

21.2.3 HSP60 (Chaperonins)

Chaperonins or HSP60 family comprises of a group of highly conserved chaperone proteins of ~60 kDa molecular weight that occur in mitochondria, chloroplast, plastids and cytosol of all eukaryotic organisms and eubacteria. They can also function on cellular surfaces and extracellular region. Chaperonins cooperate with other chaperone proteins like HSP10 (GroES) and HSP70 in carrying out functions such as folding, assembly and translocation of newly synthesised or denatured proteins (Bukau et al. 2006). These proteins are classified into two groups based on distinct evolutionary history which have similar structure but no sequence similarity.

In eukaryotes, group I chaperonins are present in organelles like mitochondria and chloroplast, plasma membrane, cytosol and intercellular spaces and also in biological fluids like blood, lymph and secretions like saliva. In prokaryotes, they occur in the cytoplasm. Group I chaperonins form homo-tetradecamers (14 subunits) of ~800 kDa with two heptameric rings that function in association with HSP10 proteins, which form a detachable lid that covers the central cavity. Folding of unfolded or partially folded peptides occur inside the central chamber with the help of ATP. In addition to the functional tetradecamer structure, group I chaperonins can also occur as monomer and oligomers with fewer number of subunits in cells. GroEL in bacteria and Rubisco binding protein (RuBisCoBP) in chloroplasts belong to this group (Rowland and Robb 2017). Each subunit of group I chaperonins contains three domains: the apical domain, the equatorial domain and the intermediate domain. Apical domain is responsible for substrate binding, while equatorial domain contains the ATP-binding site (Spiess et al. 2004). Equatorial domain is more conserved, while high level of sequence variation can be seen in apical domain that has the substrate binding sites. Apical domain forms the opening of the double ring structure with few hydrophobic amino acids exposed to the inner surface of the cavity for binding to substrate proteins (Spiess et al. 2004). The intermediate domain connects apical and equatorial domains, and they undergo a conformational change when bound to ATP mediating a switch between hydrophilic and hydrophobic substrate binding sites. When inactive the cavity is in hydrophobic state, but on activation by binding of seven ATP molecules, the intermediate domain shifts its conformation leading to exposure of hydrophilic region. In addition, HSP10 lid closes the opening of the active HSP60 leading to release of substrate into the enlarged hydrophilic inner cavity. Substrate is folded in the cavity along with the

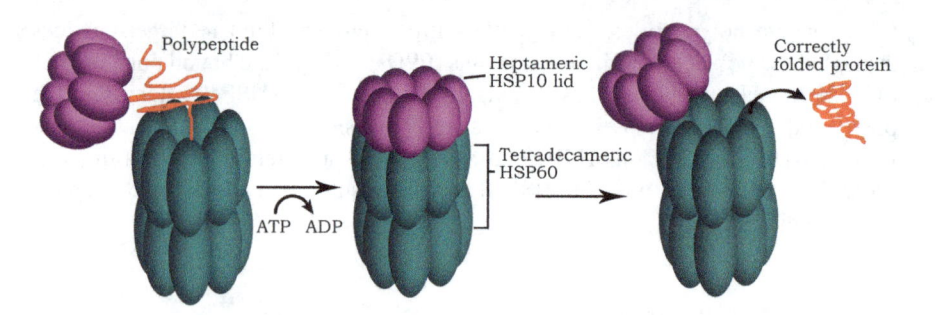

Fig. 21.3 Simplified structure of group I HSP60 tetradecamer with HSP10 lid and the protein folding process

hydrolysis of ATP to ADP. The release of ADP and HSP10 lid occurs along with the properly folded protein, when a new substrate and ATPs bind to the opposite ring (Hartl and Hayer-Hartl 2002) (Fig. 21.3).

Group II chaperonins are found in eukaryotic cytosol and archaea. They also have a double ring structure similar to that of group I chaperonins composed of eight subunits in eukaryotes and 2–3 subunits in archaea. These proteins do not require HSP10 as co-chaperone for their activity. In eukaryotes, group II chaperonins occur in the cytosol and are named CCT/TRiC while in archaea they are called thermosome/TF55. Group II chaperonins have a structure similar to the group I chaperonins, but they differ in having a conserved helical protrusion that acts as an in-built lid. These extensions contain hydrophobic residues that can bind to substrate proteins (Horwich et al. 2007). The protein folding mechanism of CCT is similar to that of group I chaperonins. Unfolded protein substrates bind to the hydrophobic residues in the helical protrusion, when the heterooligomeric CCT ring is in open conformation. The lid structure closes when ATP binds to equatorial ATP-binding domains, and it changes to a folding active state. A difference in the folding process when compared to group I system is that the binding of ATP and release of substrates from the subunits occurs sequentially rather than simultaneously as in GroEL system (Horovitz and Willison 2005).

Chaperones belonging to HSP60 family carried out a number of crucial functions at various cellular locations. Like most HSPs, a number of HSP60 proteins have been observed to show an upregulation of expression under conditions of stress. In mitochondria, they are involved in correct folding of proteins bound for the matrix and maintenance of unfolded state of proteins that are destined to be transported across the mitochondrial inner membrane. Proteins that are to be transported outside mitochondria has an amphiphilic alpha-helix, which is absent in proteins that should remain inside the mitochondria. HSP60 identifies the destination of proteins with the help of this amphiphilic alpha helix. In addition to protein folding, HSP60 also helps with mitochondrial DNA replication and transmission (Kaufman et al. 2003). Another role of HSP60 in mitochondria is in the synthesis and assembly of new HSP60 proteins. HSP60 proteins are coded by nuclear genes and are translated in the cytoplasm, from where the subunits are transported to mitochondria. The subunits are assembled in the mitochondria to their functional structure by existing HSP60 proteins there.

Studies have shown that cytoplasmic HSP60 are involved in regulation of apoptosis, immune response and cancer (Zhou et al. 2018). The detection of HSP60 proteins on cellular surfaces and extracellular secretions in prokaryotes and eukaryotes has led to the discovery of their role in immune response signalling. It has been found that some immune cells like peripheral blood mononuclear cells releases HSP60 proteins, and these proteins are capable of activating monocytes, macrophages and dendritic cells in addition to inducing cytokine release (Marker et al. 2012).

21.2.4 HSP70 (DnaK)

HSP70 or DnaK belong to a family of HSPs of ~70 kDa molecular weight that have been extensively studied and characterised. These monomeric proteins contain two domains and are involved in de novo protein folding, correct folding of misfolded proteins, translocation of proteins across membrane and regulation of stress responses (Craig 2018).

Functioning of HSP70 proteins require the assistance of two co-chaperones, HSP40 (DnaJ) and GrpE. HSP40 interacts with HSP70 through its J-domain and is involved in substrate binding and ATP hydrolysis. It can bind independently to unfolded proteins with low affinity and then direct them to the high-affinity binding sites in HSP70. Multiple HSP40 co-chaperones can associate with each HSP70, and each complex has a distinct substrate binding ability (Mayer and Bukau 2005). The co-chaperone GrpE is a homodimer made of 20-kDa subunits that act as a nucleotide exchange factor and triggers the dissociation of ADP from HSP70 and subsequent release of bound substrate accompanied by the attachment of ATP to HSP70 (Harrison 2003).

HSP70 proteins have an N-terminal nucleotide binding domain that is in charge of the ATPase activity and a C-terminal substrate binding domain connected through a highly conserved linker region. The substrate binding domain is further subdivided into a β-sandwich substrate binding subdomain and a C-terminal α-helical subdomain. A substrate-binding pocket present in the substrate binding subdomain can bind to hydrophobic and neutral amino acids with high affinity. The C-terminal α-helical sub-domain acts as a lid covering the substrate binding pocket. This lid structure is made up of five helices that are named A, B, C, D and E (Mayer and Bukau 2005). HSP70 preferentially bind to a heptameric stretch of amino acids containing a core of 4–5 hydrophobic residues rich in leucine and flanked by basic amino acids present in substrate proteins. Such motifs occur very frequently in protein molecules, and this could be the reason for the promiscuous binding ability of HSP70 proteins (Rudiger et al. 1997).

The protein folding activity by HSP70 occurs in two steps: peptide binding and ATP hydrolysis. The binding affinity of HSP70 to non-native protein targets is dependent on whether it is bound to ATP or ADP. HSP70 bound to ATP has low affinity for target proteins while ADP bound form exhibits high affinity. This is because of the open conformation of the C-terminal α-helical sub-domain lid, which releases the bound peptides quickly when bound to ATP. When bound to ADP, the lid structure is closed, and the substrate protein remains tightly attached to the substrate binding domain (Young 2010). HSP40 co-chaperones catalyses the conversion of the HSP70 bound ATP to ADP, which results in the stabilisation of HSP70–substrate complex. The release of bound ADP is triggered by the nucleotide exchange factor GrpE which in turn initiated the release of the bound substrate protein (Laufen et al. 1999).

21.2.5 HSP90 (HtpG)

HSP90 chaperone proteins helps in the regulation of cellular proteostasis during normal physiological conditions and under stress. They are of ~90 kDa in molecular weight and are one among the most abundant proteins in eukaryotes. These proteins are conserved across species from bacteria to eukaryotes and are involved in cellular processes like cell cycle control, telomere maintenance, apoptosis, protein degradation, vesicle-mediated transport, innate immunity and in signalling pathways (Jackson 2013). These proteins assists the folding of a number of target proteins including those involved in signal transduction like protein kinases, transcription factors, steroid hormone receptors and E3 ubiquitin ligases (Schopf et al. 2017).

Different chaperone systems in a cell work in co-ordination for the proper folding of nascent peptides or misfolded proteins. A protein molecule that has been folded by HSP70 can be transferred to chaperonins, where the folding process gets completed. HSP90 does not bind to nascent peptides, but instead they act on proteins that have already been folded to a near native state (Young et al. 2001).

HSP90 proteins have three structural domains: a highly conserved N-terminal ATP binding domain, a middle domain and a C-terminal domain involved in dimerisation and co-chaperone binding. The N-terminal domain is connected to the middle domain through a variable charged linker region. This linker region was first believed to be just a spacer that mediated conformational modifications to assist binding of co-chaperones, ATP or target proteins; but recent studies have revealed that the sequence of the linker region can also modify the activity of HSP90. The middle domain contains an arginine residue that is essential for the ATPase activity and is also the binding site of target proteins and the co-chaperone AHA1 (Jackson 2013). The C-terminal domain also contains an alternative ATP-binding site. At the C-terminus, there is a highly conserved MEEVD motif which binds to tetratricopeptide repeat (TPR) containing co-chaperones.

HSP90 forms constitutive homodimers with the C-terminus of subunits bound together when they are in open conformation. ATP binding leads to a closed conformation, the N-termini of the subunits also come in contact and form a circular molecular clamp structure that assists the folding of client proteins to their native forms (Meyer et al. 2003). Once the bound ATP gets hydrolysed, the binding at the N-termini dissociates. A number of co-chaperones are involved in the regulation of

HSP90 activity and binding of client proteins (Jackson 2013). The co-chaperone CDC37 mediates binding of certain kinase substrates and is an inhibitor of ATPase activity while AHA1stimulates ATP hydrolysis. HOP is an inhibitor N-terminal binding of subunits, whereas p23 assists the stabilisation of the circular dimerised HSP90 structure (Hartl et al. 2011).

In recent years, HSP90 proteins have been used as a therapeutic target for cancer treatment (Whitesell and Lindquist 2005). This is because of the role of HSP90 in stabilizing many proteins like kinases that are essential for the growth of tumour cells. Thus inhibitors of HSP90 such as geldanamycin that interferes with the ATPase activity of HSP90 are being used for the treatment of certain types of cancers (Barrott and Haystead 2013). In addition to this, studies are underway for the use of HSP90 as target for neurodegenerative disorders and also against viral and protozoan infections (Wang et al. 2017).

21.2.6 HSP100 (Clp)

HSP100 family contains ATP-dependent chaperone proteins with molecular weight between 100 and 104 kDa that mediate solubilisation and reactivation of protein aggregates and degradation of misfolded proteins. The disaggregation by HSP100 is carried out with the help of HSP70 and HSP40 proteins. HSP100 proteins belong to the AAA+ superfamily of proteases. Protein aggregates formed under stress are solubilised by the ATP-dependent protein unfoldase activity of HSP100. These proteins are not required under normal physiological conditions, but their expression is induced under stresses like temperature extremes. They are more abundant in prokaryotes in comparison to eukaryotes were only one or two copies occur in the cytoplasm or mitochondria (Zolkiewski et al. 2012).

Proteins belonging to HSP100 family contains a nucleotide binding domain, the AAA+ module, made up of ~200–250 amino acid residues that comprises of conserved motifs like Walker A and B motifs that are involved in ATP binding and hydrolysis, sensor-1 and sensor-2 motifs, which contribute to nucleotide hydrolysis and in distinguishing between ATP and ADP present in the binding pocket, and the arginine finger motif, which also participate in ATP hydrolysis (Wendler et al. 2012). HSP100 chaperone proteins are divided into two classes based on the number of AAA+ module in them. Class I HSP100 proteins contain two AAA+ modules, named D1 and D2, while class II has only one module. Class I includes HSP104, ClpB, ClpA and ClpC, whereas class II contains ClpX and HslU (Doyle and Wickner 2009). The AAA+ module also takes part in oligomerisation of the protein into two-tier hexameric ring-like structures containing a central pore. In addition to the nucleotide binding domain, HSP100 proteins also contain some less-conserved domains that are concerned with substrate preference and localisation to different cellular compartments (Mogk et al. 2003). HSP104 contains an N-terminal domain of about 150 residues that helps in its binding to aggregated client proteins and a middle domain (M-domain) that mediates binding to HSP70 (Zolkiewski et al.

2012). It also has a small C-terminal domain composed of 38 amino acids that is involved in oligomerisation and thermotolerance (Doyle and Wickner 2009).

Biologically active form of HSP100 proteins is a ring-shaped oligomer with six subunits, stabilised by the binding of ATP to N-terminal domain. They function synergistically with HSP70 in mediating the disaggregation of proteins or target the aggregates for proteosomal degradation with the help of cofactor ClpP (Zolkiewski et al. 2012).

Solubilisation of aggregated proteins occurs in a step-by-step manner, with HSP70 chaperones first binding to the surface of the aggregates and then they recruit HSP100 to the site through direct physical contact. This interaction occurs between the M-domain of HSP100 and ATPase domain of HSP70. The M-domains of HSP100 hexamers interact with neighbouring ATPase ring, when in repressed state with low ATP turnover. When M-domain interacts with HSP70, it dissociates from the ATPase ring leading to the activation of the complex. The activated state is stabilised by the binding of HSP70, which also mediates the initial disaggregation of the target protein aggregate and facilitate interaction between single protein molecules in the aggregate and HSP100 complex. The target molecule then passes through the central pore of the HSP100 hexamer, where it interacts with aromatic residues present in the mobile pore loops. These loops undergo a conformational change during ATP binding and hydrolysis, which generates a pulling force that moves the loop segments downwards and mediate unfolding and threading of substrate molecule. The unfolded target protein undergoes spontaneous refolding or HSP70 or HSP60 chaperone system assisted folding on exiting the HSP100 hexamer channel. Protein aggregates that cannot be reactivated are degraded by HSP100–peptidase proteolytic complexes (Hayashi et al. 2017).

21.2.7 Calnexin and Calreticulin

Calnexins and calreticulins are calcium-binding chaperone proteins located in the endoplasmic reticulum (ER). They are components of ER quality control system, where they assist folding, subunit assembly and maturation of glycoproteins that are transported through the ER. Calnexin is a membrane protein, whereas calreticulin is its soluble analogue (Leach and Williams 2003). They bind to exposed hydrophobic residues in non-native proteins preventing their aggregation and mediating proper folding or initiate degradation of misfolded proteins. In addition to protein folding activity, calnexin and calreticulin blocks the export of non-native proteins from ER (Danilczyk et al. 2000).

21.2.8 Protein Disulphide Isomerase (PDI) and Peptidyl Prolyl *Cis–Trans*-Isomerase (PPI)

Protein disulphide isomerase (PDI) and peptidyl prolyl *cis–trans*-isomerase (PPI) are enzymes that act as foldases by catalysing the rate limiting steps in protein folding.

PDI catalyse the formation and breakage of disulphide bonds between cysteine residues and thereby assist folding of protein molecules into their tertiary and quaternary structure (Wilkinson and Gilbert 2004). In eukaryotes, PDI occurs mainly in the lumen of ER, but it has also been detected in cytosol and cell surfaces. In bacteria, it exists in the periplasm. This enzyme belongs to the thioredoxin superfamily of redox proteins and is capable of three different catalytic activities: thiol-disulphide oxireductase, disulphide isomerase and redox-dependent chaperone (Ali Khan and Mutus 2014). Proteins of PDI family have thioredoxin domains with characteristic βαβαβαββα fold (Kemmink et al. 1997). They have the ability to refold denatured proteins. Recent nuclear magnetic resonance (NMR) studies have revealed that PDI can distinguish between non-native and native proteins (Irvine et al. 2014).

PPIs are ubiquitous proteins found in eukaryotic and prokaryotic cells. They regulate folding of proteins at proline residues by catalysing the *cis–trans* isomerisation of peptidyl bonds present in N-terminal to proline residues (Shaw 2002). PPI proteins are involved in folding of nascent peptides, and they have been found to show chaperone-like activity. They are classified into three different families, namely, cyclophilins (Cyp), FK506 binding proteins (FKBPs) and parvulins (Fischer et al. 1998). Although the amino acid sequences of members of these three families differ, they show similarity in their 3D structures. All PPI proteins contain a single catalytic domain made up of β-folds, in addition to the domains involved in interaction with substrate proteins (Nath and Isakov 2015).

21.2.9 Universal Stress Proteins (USP)

The Universal Stress Proteins (USPs) are a class of conserved proteins whose expression is upregulated on exposure to external stressors like extreme temperature, starvation and osmotic stress. This first USP to be identified was in *Escherichia coli*, after which a number of proteins belonging to this class were identified in other bacteria, fungi, archaea, protozoa, metazoans and plants (Siegele 2005). The overexpression of USP helps the organism to cope up with the stresses by mechanisms that are not fully understood. USPs have been reported to provide protection against mutagens and respiratory uncouplers (Kvint et al. 2003). They are involved in cellular growth regulation, prevention of globular protein denaturation, protein transportation, hypoxia responses and ion scavenging (Chi et al. 2019; Vollmer and Bark 2018). The structure and functions of USP in *Arabidopsis thaliana* (AtUSP) has been extensively studied, and it was found to exist in different forms such as monomers, dimers and as higher oligomeric structures. The inactive smaller USP complexes associate to form large active oligomeric structures when the plant is exposed to stress, which is triggered by a shift in redox status inside the cell. A study by Jung et al. (2015) demonstrated that the chaperone activity of USPs are similar to sHSPs in that they function as holdases preserving cellular proteins from denaturation.

USP proteins have an α/β subdomain that is involved in cellular defence signalling. They exhibit high diversity of structure and contain a variety of motifs. USPs are classified into two groups based on their structural similarity to two USPs. Included in the first category are those USPs similar to MJ0577 protein from *M. jannaschii*, which contains a C- terminal ATP-binding motif and an α/β-core made up of five β-strands and four α-helixes. The second group contains USPs with structural similarity to USPA from *Haemophilus influenza*, which lack ATP binding ability (Sousa and McKay 2001).

21.3 Stress Response in Plants

The sessile nature of plants makes them particularly vulnerable to environmental stresses. Extreme temperatures, salinity, drought, nutrient deficiency and toxic chemicals are some of the abiotic stresses to which plants are exposed regularly. In addition to these, plants also encounter biotic stress such as herbivore grazing and pathogen infection. In order to overcome these factors, plants have developed a number of defence mechanisms that are activated by signal pathways triggered on exposure to external stress. The regulatory process behind plant stress responses have been widely investigated using *A. thaliana* as a model (Chi et al. 2019). The different signalling pathways that are activated in plants on exposure to stress include variation in cellular Ca^{2+} ion concentration, alteration of redox potential, change in membrane fluidity and generation of reactive oxygen species (ROS) (Choudhury et al. 2017). The activation of these multi-level signal transduction processes are the result of a number of mechanisms including regulation of transcription of stress proteins, alternative splicing and rapid production of regulatory proteins through post-translational modifications like ubiquitination (Haak et al. 2017).

The abiotic stress signals can be distinguished into primary and secondary. The primary stress signals are those which immediately arise on exposure to stresses like salinity, drought, temperature extremes or chemicals, while the secondary stress signals arise in response to oxidative stress, impaired metabolism and damage to cellular membrane or macromolecules, caused by primary stress. Both primary and secondary stress signals activate downstream signalling cascades and transcription regulation leading to protection of proteins, membranes and re-establishment of cellular homeostasis. Inability of stress signals in initiating an adequate stress response can cause irreversible damage to cellular homeostasis through denaturation of proteins and disruption of cell membrane, ultimately resulting in cell death (Zhu 2016).

Many of the crop cultivars used in agriculture are sensitive to various environmental stresses like hot/cold temperature, salinity and drought, limiting the geographical area that can be used for their cultivation. There is a dire need to increase the productivity of available land to provide adequate food for world human population that is increasing at a rapid rate. One strategy to achieve this is by developing stress-tolerant crops through genetic modification by the insertion of specific stress tolerant genes from other organisms. Such crops can be cultivated in land areas that were previously left barren due to factors like high salinity or low water availability. Thus, a deeper understanding of the stress signalling pathways and functioning of stress proteins like HSPs are essential to create stress tolerant plant varieties (JK Zhu 2016). An interesting observation made regarding stress response in plants is that the exposure of a particular stress can make the plant resistant to other stresses. This cross tolerance has been observed even between abiotic and biotic stresses, and it helps plants adapt to changing environmental conditions. Achuo et al. (2006) reported twofold higher abscisic acid production and higher resistance to infection by *Botrytis cinerea* and *Oidium neolycopersici* in tomatoes, following exposure to drought stress. The opposite also occurs, with biotic stress conferring resistance against biotic stress. Stomatal closure is triggered in plants during infections in a bid to prevent pathogen entry, which can also result in reduced water loss and better resistance to abiotic stress. There are reports on plants acquiring resistance to drought in response to viral infections (Xu et al. 2008).

21.4 Chaperone Proteins Involved in Plant Stress Response

All living organisms encounter different forms of stress throughout its life cycle. They overcome these conditions either through avoidance or through tolerance. Being sessile, plants are incapable of avoidance and overcome stress through morphological and physiological adaptations. A high percentage of the molecules activated or overexpressed during exposure to stress are chaperone proteins like HSPs.

HSPs belonging to the families HSP100, HSP90, HSP70, HSP60 and sHSP have been identified to be a part of plant stress response, with their expression regulated by heat stress transcription factors (HSFs). Around 20–40 sHSPs are usually found in a single plant localised in the cytosol and organelles like chloroplast, mitochondria, nucleus and endoplasmic reticulum. They all play different roles in the maintenance of cellular proteostasis (Al-Whaibi 2011).

The regulation of HSP expression is controlled by a group of proteins called heat stress transcription factors (HSFs). In plants, HSFs are actively involved in the activation of stress proteins like HSPs on exposure to abiotic stress like high temperature and drought. In addition to their regulatory role, HSFs also take part in the maintenance and recovery phases of heat stress responses (Al-Whaibi 2011). Both HSPs and HSFs are potential proteins for the development of genetically engineering plants with resistance to abiotic stress like drought, high temperature or exposure to chemicals like herbicide and pesticides (Guo et al. 2016)

Recent studies have shown that HSPs play an active role in the innate immunity of plants (Saijo 2010). The plant innate immunity is composed of PAMP-triggered immunity (PTI) and effector-triggered immunity (ETI). In PTI, the two types of pattern recognition receptors (PRRs) in cell surface, namely, surface-localised receptor kinases (RKs) and receptor like proteins (RLPs), recognise PAMPs (pathogen-associated molecular patterns) or damage-associated molecular patterns

(DAMPs) and bind to them triggering intracellular signalling (Zipfel 2014). Since PAMPs are conserved across different classes of microbes, PTI can provide protection against multiple pathogens. The synthesis PRRs occurs in the endoplasmic reticulum, a process in which a number of HSPs like HSP70 luminal-binding protein (BiP) and HSP40 ERdj3B assist in correct folding and quality control (Saijo 2010).

Another plant innate immunity component is ETI in which intracellular immune receptors recognise virulence factors (effectors) produced by pathogens on infected cells and triggers immune response signalling pathways. These receptors are called R proteins belonging to nucleotide-binding domain and leucine-rich repeat (NB-LRR)-containing protein family and HSP90 is essential for their stability (Rajamuthiah and Mylonakis 2014).

21.5 Role of Chaperone Proteins in Plants

21.5.1 HSP100

Proteins belonging to HSP100 family assist the resolubilisation of aggregated cellular proteins and degradation of proteins that have been irreversibly denatured. The protein disaggregase action by HSP100 is mediated in co-operation with sHSPs and HSP70 and is essential for the survival of cells during heat stress.

A number of studies have shown that HSP100 proteins play an important role in heat tolerance of plants. Hong and Vierling (2001) found that cytosolic HSP100 was essential for heat tolerance in Arabidopsis. High sensitivity to heat was observed in Arabidopsis, maize and rice plants with inactive ClpB/Hsp100 (Mishra and Grover 2016). When plants are subjected to stress, stress granules composed of mRNAs and RNA-binding proteins are formed which protects incompletely translated mRNAs from digestion. Once the conditions improve, HSP100 and HSP70 are directed to the granules and mediate their resolubilisation to reinstate mRNA translation (McLoughlin et al. 2019).

HSP100 have been identified in a number of plants like wheat, tepary bean, pea, maize and kidney bean. In most of the plants studied, heat stress has been found to be the key inducer for the expression of HSP100, but a study by Pareek et al. (1995) showed that HSP100 can also be expressed on exposure to high salinity, desiccation, abscisic acid (ABA) or cold stress. Expression of wheat clpB protein, belonging to HSP100 family, was also found to be regulated by dehydration and ABA in addition to heat stress (Campbell et al. 2001). In addition to the role of HSP100 in the recovery phase, there are also reports on their accumulation during exposure to stress suggesting that they play an important role in protecting cellular proteins even while the cells are under stress (Agarwal et al. 2003).

Transgenic plants that constitutively expressed HSP101 were found to have better heat tolerance, while those in which heat stress mediated HSP101 induction was impaired showed reduced seed germination and seedling growth when grown at high temperature. A notable finding is that the growth and development of plants overexpressing or underexpressing HSP101 remained unaffected, making HSP100 genes attractive targets for developing stress tolerant plants (Queitsch et al. 2000).

21.5.2 HSP90

HSP90 functions in cooperation with HSP70, cofactors and co-chaperones to assist refolding of damaged proteins or mediating their degradation. They also regulate the expression of heat shock genes. A unique characteristic of HSP90 is its substrate specificity (Kozeko 2019).

HSP90 proteins are essential for embryo development, seed germination, hypocotyl elongation, pest resistance, stress tolerance and disease resistance in plants (Sangster et al. 2008). They also play an important role in the regulation of auxin signalling during exposure to high temperature. There are reports on HSP90 inhibition leading to aberrations in plant phenotype, such as symmetry of cotyledons, abnormal root hairs and epinastic cotyledon (Xu et al. 2012).

A study by Wang et al. (2016b) revealed that HSP90 and its co-chaperone SGT1 (suppressor of G2 allele kinetochore protein) regulates plant growth in accordance with environmental conditions through their control over auxin and temperature signalling pathways. HSP90 gene silencing experiments showed various defects like retarded growth due to damage to meristem and loss of disease resistance against potato virus X and tobacco mosaic virus in *Nicotiana benthamiana* (Liu et al. 2004; Lu et al. 2003). Phenotypic variations like change in flowering time and morphological characteristics have also been observed in plants with inactive HSP90 (Sangster et al. 2007).

HSP90s are involved in plant immunity, and they play a major role in disease resistance and insect resistance. They activate and regulate the functioning of cytosolic R proteins with the help of co-chaperones RAR1 (required for Mla12 resistance) and SGT1. The R proteins recognise effector molecules produced by pathogens and activate plant immune response. (Shirasu 2009). Disease resistance against *Pseudomonas cloves* in tomatoes, tobacco improved mosaic virus (TMV) in tobacco, powdery mildew in barley and leaf rust in wheat have been linked to HSP90 (Hein et al. 2005; Liu et al. 2004). Chen et al. (2010) reported chitin signalling and anti-fungal immune response in rice mediated by a chaperone complex which had cytosolic HSP90 and co-chaperone Hop/Sti1 as components. Chitin response involves the release of chitinase enzymes by plant cells in response to chitin present in the cell wall of fungi, during a fungal infection and the resultant chitin fragments serve as effectors that induce innate immunity against the pathogen.

HSP90 proteins are also involved in pest resistance in plants through their control over signal transduction pathways. A study by Bhattarai et al. (2007) found that and HSP90 and SGT1 are essential for *Mi-1*-mediated resistance of tomato plants to nematodes and aphids. Mi-1 is a type of R protein containing CC–NBS–LRR (coiled coil–nucleotide-binding site–Leu-rich repeat) motifs, which confer resistance against root-knot nematodes, tomato–potato aphid, potato psyllid and sweet potato whitefly, in addition bacteria, viruses and fungi (Casteel et al. 2006). There exists a common signal transduction pathway that mediates resistance in plants towards multiple organisms, and HSP90 is an essential component required for its activation (Bhattarai et al. 2007).

Abiotic stress response in plants is also influenced by HSP90, but the exact mechanism is unknown. Overexpression of organellar and cytoplasmic HSP90 in *Arabidopsis* resulted in higher sensitivity to drought, salt and oxidative stresses, while inhibition of HSP90 trigged expression of heat-inducible genes and better stress tolerance (Song et al. 2009). The reason for such observations might be because of the role of HSP90 as an active suppressor of stress response during normal conditions, mediated by the binding and inhibiting constitutively expressed HSFs like HsfA1d. On exposure to environmental stress, the available HSP90 gets depleted, as they are recruited to protect stress sensitive proteins, leading to termination of their inhibition on HSFs and thus inducing stress response (di Donato and Geisler 2019; Song et al. 2009). A conflicting observation made in an experiment, where a decrease in abiotic stress linked damages was seen in *Arabidopsis* plants cloned and overexpressing soybean HSP90 (Xu et al. 2013).

21.5.3 HSP70

HSP70 family comprises of highly conserved chaperone proteins that require ATP for their functioning. They are ubiquitously present in all organisms and are involved in both housekeeping activities of cells like folding of nascent peptides and also in stress response. HSP70 proteins also assist in the transportation of synthesised proteins to specific organelles like mitochondrion and chloroplast and in the degradation of unwanted proteins (Usman et al. 2017).

In plants, HSP70 plays an important role in resistance against biotic and abiotic stresses. A number of studies have revealed the role of HSP70 in plant immune response. Heat Shock Cognate 70 (HSC70) protein, a homologue of HSP70, was found to control immunity in *Arabidopsis* in association with SGT1 (Noel et al. 2007). *Arabidopsis* deficient in AtHsp70-15 were more susceptible to infection by turnip mosaic virus (Jungkunz et al. 2011). There are also reports on the involvement of cytoplasmic HSP70 in non-host resistance to *Pseudomonas cichorii* and INF1-mediated hypersensitive response (HR) against *Phytophthora infestans* in *Nicotiana benthamiana* (Kanzaki et al. 2003). A study by NH Kim and Hwang (2015) found that cytoplasmic HSP70a accumulates in leaves of pepper during *Xanthomonas campestris* pv *vesicatoria infection and activates* hypersensitive cell death response resulting in the induction of defence and cell death-related genes.

HSP70 proteins also play a significant role in abiotic stress resistance and impart drought and heat tolerance in plants (Alvim et al. 2001). Drought resistant varieties of plants like pepper, maize and barley were found to express higher quantity of HSP70 when compared to sensitive varieties (Landi et al. 2019; Usman et al. 2017).

Transgenic plants developed with higher expression of HSP70 had higher tolerance to heat stress. Overexpression of cytosolic Hsc70-1 and Hikeshi-like protein have been shown to increase stress tolerance in *A. thaliana* (Koizumi et al. 2014).

Zhao et al. (2019) reported higher thermotolerance and lower accumulation of hydrogen peroxide (H_2O_2), superoxide anion free radical ($O_2{}^{\cdot-}$) and relative electric conductivity (REC) in *Arabidopsis* constitutively overexpressing *Paeonia lactiflora* HSP70. Increased tolerance to heat stress has also been observed in tobacco plants overexpressing BiP, an endoplasmic reticulum Hsp70 and HSP70 from *Brassica campestris* (Wang et al. 2016a).

The role of HSP70 in stress tolerance in plants has been further demonstrated by an increase in mortality in *Arabidopsis* plants deficient in AtHsp70–15 when exposed to high temperature. They also showed retarded growth and higher sensitivity to abscisic acid (ABA) treatment (Jungkunz et al. 2011). A study by Hu et al. (2010) found that HSP70 is essential for ABA-induced upregulation of antioxidant enzymes during heat and drought stress in maize that results in lowering of cellular reactive oxygen species (ROS) levels. In addition to stress response, HSP70 proteins are crucial for many other plant processes. HSP70 and HSP90 were identified to be essential for the initiation and elongation stages of fibre development in cotton (Sable et al. 2018).

21.5.4 HSP40

HSP40/DnaJ, a co-chaperone of HSP70, is essential in plant immunity against viral pathogens and in stress resistance. HSP40 and HSP70 are involved in microbial pathogenesis, especially in movement of viruses from cytosol to membrane or between plant cells through plasmodesmata. This is mediated by the interaction between the HSPs to viral capsid protein (CP) or movement protein (MP) that are bound to viral nucleic acids (Hofius et al. 2007). HSP40 and HSP70 also enhance plant disease resistance by stabilising R proteins, which detect effector molecules of pathogens and activate plant immune response (Hafren et al. 2010). Gene silencing of HSP40 of soybean resulted in an increased susceptibility to Soybean mosaic virus which demonstrates their role in defence against viral infections (Liu and Whitham 2013).

A number of studies have shown the role of HSP40 in environmental stress tolerance. Zhichang et al. (2010) reported higher tolerance to NaCl-stress in transgenic *Arabidopsis* constitutively expressing high levels of HSP40. The roots of these transgenic plants were longer than wild-type plants. Whole-genome analysis of *Capsicum annuum* L. identified 76 putative DnaJ genes, of which the expression of 54 genes was found to be induced when exposed to heat stress (Fan et al. 2017). This indicates the role of HSP40 in protecting plants against abiotic stress.

21.5.5 Small HSPs (sHSP)

Higher plants usually contain around 20 sHSPs in them, but the number can go as high as 40. Expression of sHSPs is induced on exposure to stress such as temperature extremes, high salinity and drought. Presence of diverse sHSPs are believed to be a reason for better adaptability of certain plants to heat stress (Al-Whaibi 2011).

In plants, sHSPs have been reported to assist immune response, enhance stress tolerance and also in their development. RSI2 protein (Required for Stability of I-2) of tomato (*Solanum lycopersicum*), belonging to sHSP family, stabilises the R protein I-2 that confers disease resistance against *Fusarium oxysporum* and other pathogens (Van Ooijen et al. 2010). It is also suggested that RSI2 is essential for the initiation of hypersensitive response, which results in rapid cell death of infected regions.

In *Nicotiana tabacum*, the sHSP Ntshsp17 was determined to be involved in HR-independent disease resistance against *Ralstonia solanacearum*. A decrease in the expression of defence genes and fast growth of *R. solanacearum* were observed in plants lacking active Ntshsp17 gene (Maimbo et al. 2007).

Xanthomonas oryzae pv. oryzae is a bacteria responsible for bacterial blight in rice. A study by Kuang et al. (2017) found that overexpression of the sHSP OsHsp18.0 in susceptible rice variety resulted in disease resistance against *X. oryzae*, while silencing of OsHsp18.0 gene in resistance varieties made them susceptible to infection. In addition, OsHsp18.0 also acted as a positive regulator for abiotic stress. As with HSP40 and HSP70, sHSPs are also involved in viral infection (Verchot 2012).

The role of sHSPs in abiotic stress tolerance of plants have been extensively studied. Some sHSPs in plants are specific to a particular stress like heat shock, while others are expressed on exposure to a number of stressors. For example, the expression of Hsp18.1CI in *A. thaliana* is upregulated on exposure to high temperature but not during any other type of stress, while Hsp17.4CI, Hsp17.6ACI, Hsp17.6BCI and Hsp17.6CCI are expressed during oxidative stress, osmotic stress and exposure to UV-B in addition to heat stress (Waters 2013). Twenty-three sHSP genes were identified in rice by Sarkar et al. (2009), among which 19 were upregulated during heat stress. Many of the sHSPs were found to be differentially expressed on exposure to various biotic and abiotic stresses.

Various sHSPs take part in different developmental stages of plant growth. High-level expression of sHSPs has been observed during embryogenesis, seed germination, pollen development and fruit maturation (Sarkar et al. 2009; Waters 2013). A study by Wehmeyer and Vierling (2000) provided evidence for the role of sHSPs in desiccation tolerance of seeds.

21.5.6 Universal Stress Proteins (USP)

USPs have a number of cellular functions like preventing protein denaturation, DNA repair, and cellular protein transport. They act as holdases that protect proteins from irreversible protein damage. In plants, USPs provide protection against abiotic stresses like high and low temperature, drought stress, oxidative stress and salt stress. They are also involved in defence response against pathogens (Chi et al. 2019).

Expression of USPs in tobacco, tomato and cotton are induced by salt stress and they play a crucial role in salt stress tolerance (Loukehaich et al. 2012). Transgenic tobacco plants expressing SpUSP from tomato had higher salt tolerance and resistance against osmotic stress. In addition to salt stress, SpUSP gene is also induced on exposure to extreme temperatures, wounding, drought, and phytohormone (Chi et al. 2019).

Expression of USPs is upregulated during pathogen attack. Treatment of *Arabidopsis* plants with *Phytophthora infestans* zoospores resulted in phosphorylated of two USPs, AtPHOS32 and AtPHOS34 among which the former is a component of plant immune response (Merkouropoulos et al. 2008). Thus, USPs are essential for both biotic and abiotic stress resistance in plants.

21.6 Applications of Chaperone Proteins

Molecular chaperones like HSPs have several applications in biotechnology and medicine. Transgenic plants expressing high levels of chaperone proteins can be developed that can grow under extreme climatic conditions, thereby expanding the land area used for agriculture. HSPs can be used to prevent inclusion body formation during recombinant protein expression in *E. coli*. Pathogenesis of diseases like Parkinson's disease and Alzheimer's have been linked to accumulation of protein aggregates. Chaperone proteins can be used as a therapeutic agent for the treatment of such diseases. HSPs also play an important role in tumorigenesis and can hence be used as a biomarker for the diagnosis of cancer. There are also reports on the possible application of HSPs as therapeutic targets for cancer treatment (Boshoff et al. 2004; Wu et al. 2017).

21.6.1 Crop Improvement

An important challenge currently faced by humanity is the lack of cultivable land to sustain a rapidly increasing population. Even the land area that is currently under agriculture is turning non-cultivable due to clanging climatic conditions, leading to desertification and an increase in soil salinity caused by over irrigation. Development of stress-tolerant plants that can tolerate high temperature, drought or salinity is essential for ensuring food supply for future generations. A number of naturally stress resistant plants have been identified, which can be directly used for cultivation, or used for the production of hybrids or transgenic plants transformed with the genes responsible for stress resistance (Zhang et al. 2018).

Stress-tolerant plants can be developed by overexpression of either chaperone proteins, or transcription factors like HSFs. Enhanced heat tolerance has been reported in transgenic *Arabidopsis* and tobacco expressing high levels of HSP70 (Lee and Schoffl 1996; Ono et al. 2001). Stress tolerance in tobacco was improved by the expression of a sHSP from Chinese cabbage and in carrot through the transfer of HSP17.7 (Malik et al. 1999; Sanmiya et al. 2004). The holdase activity of USP proteins can also be employed in developing plants resistant to abiotic stress. Improved salt and osmotic stress tolerance have been observed in transgenic tobacco plants expressing SpUSP from tomato (Chi et al. 2019).

Constitutive expression of HSFs in plants generally leads to increased HSP synthesis and better stress tolerance, but in some cases, unaltered or decreased HSP expression were also observed (Gurley 2000). This could be because a single transcription factor could be responsible for inducing expression of multiple genes, and altering their expression can lead to unexpected results. A good understanding of plant stress responses and the components involved is essential for developing resistant crop varieties to ensure food security by improved utilisation of farmland.

In addition to abiotic stress resistance, high expression of chaperone proteins can also confer protection against pathogens. This strategy can be used for the development of disease resistant varieties of transgenic plants that requires only minimal pesticide use.

21.6.2 Industrial Applications

Enzymes and other proteins have become an essential component of many products and in manufacturing processes. Recombinant proteins are being widely used in research and medical fields. Chaperone proteins can be used in stabilising these proteins under conditions that can cause denaturation. They can also help in improving yield during expression of recombinant proteins.

21.6.3 Prevention of Inclusion Body Formation

Recombinant proteins expressed in bacterial cells can form insoluble aggregates called inclusion bodies that can affect the protein yield. The misfolding can result from an overproduction of recombinant protein, causing an overload for the protein folding chaperone networks (Carrio and Villaverde 2002). Modification of culture conditions can reduce the formation of inclusion bodies. To recover active protein from inclusion bodies, the sample is first denatured using chemical denaturants and then refolded in a refolding buffer. Inclusion body formation can reduce the yield of active protein recovered. Co-expression of HSPs along with the recombinant protein can reduce inclusion body formation by inhibiting inappropriate interactions and promoting correct protein folding (Boshoff et al. 2004). HSP70 and HSP40 were found to improve refolding of recombinant proteins. Another reason for misfolding is the incapability of *E. coli* to perform post-translational modifications in expressed eukaryotic proteins. In an experiment by Ahn and Yun (2004), co-expression of human HSP40 along with human cytochrome P450 3A4 in *E. coli* resulted in 15-fold increase in cytochrome P450 3A4 compared to the enzyme expressed alone. There are also reports on concomitant expression of sHSPs improving the solubility of recombinant proteins (de Marco et al. 2007; Guzzo 2012; Han et al. 2004).

21.6.3.1 In Vitro Stabilisation of Proteins Under Denaturing Conditions

HSPs can protect commercially used proteins under in vitro conditions. Many of the reactions involving enzymes used in industries occur at hostile conditions, which can adversely affect the activity of the enzymes. Incorporation of chaperone proteins to such reaction mixtures can improve the activity and durability of enzymes. Laksanalamai et al. (2006) reported better stability of Taq DNA polymerase at high temperatures when used in combination with sHSP and HSP60 from *Pyrococcus furiosus*. In another study, sHSP from *P. furiosus* was found to protect Taq DNA polymerase, *Hind*III restriction endonuclease and lysozyme when exposed to high temperature (H Chen et al. 2006). Lin et al. (2010) reported that the ability of sHSP from *Xanthomonas campestris* in protecting *Eco*RI from thermal inactivation. Ability of the sHSP, AgsA in preventing heat denaturation of luciferase and β-galactosidase was demonstrated by Tomoyasu et al. (2013). These reports show that HSPs have the potential to be used at stabilisers of proteins used in in vitro reactions involving high temperature.

Several enzymes are used in food manufacturing processes. sHSPs can be added to these reactions to enhance stability of enzymes and products. Among the different classes of HSPs, sHSPs are most suited for this application because of their ATP and co-factor-independent activity. Milk coagulation for curd and cheese manufacturing use 'rennet', a mix of enzymes isolated from stomachs of ruminants. It was reported that the addition of sHSP Lo18 during the milk coagulation process reduced the coagulation time and produced firmer dairy gel (Guzzo 2012).

21.6.3.2 Stabilisation of Immobilised Proteins

In protein-biochip techniques, protein molecules are immobilised on surfaces like glass or silicon slide, nitrocellulose membrane or microtitre plate. The stable attachment and maintenance of these proteins is essential for the functioning of these techniques. Binding of protein molecules to the surfaces can result in variation in their active three-dimensional structure. The attached protein molecules can also undergo denaturation. Use of sHSPs is stabilising the immobilised proteins in biochips is a strategy that is being explored (Guzzo 2012). Use of HSP25 in stabilising the antigenicity of proteins used in immune detection was studied by Ehrnsperger et al. (1998). HSP25 was found to enhance binding and stabilisation of troponin T, an indicator of cardiac muscle damage and peptides used for detection viral infection. Immobilised chaperone proteins like HSP70 and α-crystalline were found to preserve firefly luciferase activity after prolonged storage at room temperature (Yang et al. 2003). The immobilisation procedure used should be carefully selected for each protein because it was reported by Shridas et al. (2001) that α-crystalline cross-linking resulted in a decrease in chaperone activity. These data indicate that properly immobilised HSPs can improve the stability and shelf life of biosensors and biochips containing proteins (Han et al. 2008).

21.6.3.3 Inhibition of Proteolysis

Proteolytic degradation of protein samples during different analytical techniques like two-dimensional gel electrophoresis (2-DE) and isoelectric focusing is a major problem faced by researchers. Protease inhibitors are usually used to prevent this, but many proteases still remain active (Finnie and Svensson 2002). Proteolytic cleavage can impair the quality of data because protein degradation can result in missing spots on 2-DE gels, leading to wrong conclusions. The sHSPs IbpA and IbpB from *E. coli* and HSP26 from *Saccharomyces cerevisiae* were found to show protective action against proteolytic cleavage by trypsin and proteinase K under denaturing conditions (Han et al. 2005). The addition of sHSPs resulted in a 50% increase in protein spots in 2-DE gels when compared to those performed with routinely used protease inhibitors. Thus incorporation of sHSPs to samples can improve the proteome profiling data obtained in techniques like 2-DE and liquid chromatography–mass spectrometry (LC-MS) based proteomic studies (Guzzo 2012).

21.7 Conclusions

Each and every living cell is susceptible to external stress factors. Well-defined cellular/molecular mechanisms are available to respond to stress in a variety of ways. Plants are more sessile to different stress factors. Drought, salinity, temperature (cold or hot) and chemicals are the primary stresses affecting plants. These often lead to secondary stresses like osmotic and oxidative stresses. Exposure to stress factors induce the production of a group of special proteins called heat stress proteins (Hsps) or stress-induced proteins within the cells. These proteins are of ubiquitous in nature. Based on molecular weight, these proteins are grouped into five classes: (1) Hsp100, (2) Hsp90, (3) Hsp70, (4) Hsp60 and (5) small heat-shock proteins (sHsps). Most of these proteins have molecular chaperone functions either ATP dependent or independent manner. They have wide-range applications in modern science, namely, in crop improvement, as diagnostic biomarkers, as therapeutic proteins, in recombinant technologies, enzyme stabiliser, etc. In conclusion, major heat-shock proteins have roles in preventing misfolding and aggregation of native proteins at stress conditions and their molecular chaperone functions can be used for various applications in modern science.

References

Achuo EA, Prinsen E, Höfte M (2006) Influence of drought, salt stress and abscisic acid on the resistance of tomato to *Botrytis cinerea* and *Oidium neolycopersici*. Plant Pathol 55(2):178–186. https://doi.org/10.1111/j.1365-3059.2006.01340.x

Agarwal M, Sahi C, Katiyar-Agarwal S, Agarwal S, Young T, Gallie DR, Sharma VM, Ganesan K, Grover A (2003) Molecular characterization of rice hsp101: complementation of yeast hsp104 mutation by disaggregation of protein granules and differential expression in indica and japonica rice types. Plant Mol Biol 51(4):543–553. https://doi.org/10.1023/a:1022324920316

Ahn T, Yun CH (2004) High-level expression of human cytochrome P450 3A4 by co-expression with human molecular chaperone HDJ-1 (Hsp40). Arch Pharm Res 27(3):319–323

Ali Khan H, Mutus B (2014) Protein disulfide isomerase a multifunctional protein with multiple physiological roles. Front Chem 2:70. https://doi.org/10.3389/fchem.2014.00070

Alvim FC, Carolino SM, Cascardo JC, Nunes CC, Martinez CA, Otoni WC, Fontes EP (2001) Enhanced accumulation of BiP in transgenic plants confers tolerance to water stress. Plant Physiol 126(3):1042–1054. https://doi.org/10.1104/pp.126.3.1042

Al-Whaibi MH (2011) Plant heat-shock proteins: a mini review. J King Saud Univ Sci 23(2):139–150. https://doi.org/10.1016/j.jksus.2010.06.022

Barnett ME, Zolkiewska A, Zolkiewski M (2000) Structure and activity of ClpB from *Escherichia coli*: role of the amino- and carboxy-terminal domains. J Biol Chem 275(48):37565–37571. https://doi.org/10.1074/jbc.M005211200

Barrott JJ, Haystead TA (2013) Hsp90, an unlikely ally in the war on cancer. FEBS J 280(6):1381–1396. https://doi.org/10.1111/febs.12147

Basha E, O'Neill H, Vierling E (2012) Small heat shock proteins and alpha-crystallins: dynamic proteins with flexible functions. Trends Biochem Sci 37(3):106–117. https://doi.org/10.1016/j.tibs.2011.11.005

Bhattarai KK, Li Q, Liu Y, Dinesh-Kumar SP, Kaloshian I (2007) The MI-1-mediated pest resistance requires Hsp90 and Sgt1. Plant Physiol 144(1):312–323. https://doi.org/10.1104/pp.107.097246

Boshoff A, Nicoll WS, Hennessy F, Ludewig M, Daniel S, Modisakeng KW, Shonhai A, McNamara C, Bradley G, Blatch GL (2004) Molecular chaperones in biology, medicine and protein biotechnology. S Afr J Sci 100(11-12):665–677

Bukau B, Weissman J, Horwich A (2006) Molecular chaperones and protein quality control. Cell 125(3):443–451. https://doi.org/10.1016/j.cell.2006.04.014

Campbell JL, Klueva NY, Zheng HG, Nieto-Sotelo J, Ho TD, Nguyen HT (2001) Cloning of new members of heat shock protein HSP101 gene family in wheat (*Triticum aestivum* (L.) Moench) inducible by heat, dehydration, and ABA(1). Biochim Biophys Acta 1517(2):270–277. https://doi.org/10.1016/s0167-4781(00)00292-x

Carrio MM, Villaverde A (2002) Construction and deconstruction of bacterial inclusion bodies. J Biotechnol 96(1):3–12

Casteel CL, Walling LL, Paine TD (2006) Behavior and biology of the tomato psyllid, *Bactericerca cockerelli*, in response to the mi-1.2 gene. Entomol Exp Appl 121(1):67–72. https://doi.org/10.1111/j.1570-8703.2006.00458.x

Chen H, Chu Z, Zhang Y, Yang S (2006) Over-expression and characterization of the recombinant small heat shock protein from *Pyrococcus furiosus*. Biotechnol Lett 28(14):1089–1094. https://doi.org/10.1007/s10529-006-9058-y

Chen L, Hamada S, Fujiwara M, Zhu T, Thao NP, Wong HL, Krishna P, Ueda T, Kaku H, Shibuya N, Kawasaki T, Shimamoto K (2010) The Hop/Sti1-Hsp90 chaperone complex facilitates the maturation and transport of a PAMP receptor in rice innate immunity. Cell Host Microbe 7(3):185–196. https://doi.org/10.1016/j.chom.2010.02.008

Chi YH, Koo SS, Oh HT, Lee ES, Park JH, Phan KAT, Wi SD, Bae SB, Paeng SK, Chae HB, Kang CH, Kim MG, Kim W-Y, Yun D-J, Lee SY (2019) The physiological functions of universal stress proteins and their molecular mechanism to protect plants from environmental stresses. Front Plant Sci 10:750–750. https://doi.org/10.3389/fpls.2019.00750

Choudhury FK, Rivero RM, Blumwald E, Mittler R (2017) Reactive oxygen species, abiotic stress and stress combination. Plant J 90(5):856–867. https://doi.org/10.1111/tpj.13299

Craig EA (2018) Hsp70 at the membrane: driving protein translocation. BMC Biol 16(1):11. https://doi.org/10.1186/s12915-017-0474-3

Craig EA, Huang P, Aron R, Andrew A (2006) The diverse roles of J-proteins, the obligate Hsp70 co-chaperone. Rev Physiol Biochem Pharmacol 156:1–21

Danilczyk UG, Cohen-Doyle MF, Williams DB (2000) Functional relationship between calreticulin, calnexin, and the endoplasmic reticulum luminal domain of calnexin. J Biol Chem 275(17):13089–13097

di Donato M, Geisler M (2019) HSP90 and co-chaperones: a multitaskers' view on plant hormone biology. FEBS Lett 593(13):1415–1430. https://doi.org/10.1002/1873-3468.13499

Doyle SM, Wickner S (2009) Hsp104 and ClpB: protein disaggregating machines. Trends Biochem Sci 34(1):40–48. https://doi.org/10.1016/j.tibs.2008.09.010

Ehrnsperger M, Hergersberg C, Wienhues U, Nichtl A, Buchner J (1998) Stabilization of proteins and peptides in diagnostic immunological assays by the molecular chaperone Hsp25. Anal Biochem 259(2):218–225. https://doi.org/10.1006/abio.1998.2630

Fan F, Yang X, Cheng Y, Kang Y, Chai X (2017) The DnaJ gene family in pepper (*Capsicum annuum* L.): comprehensive identification, characterization and expression profiles. Front Plant Sci 8:689. https://doi.org/10.3389/fpls.2017.00689

Finnie C, Svensson B (2002) Proteolysis during the isoelectric focusing step of two-dimensional gel electrophoresis may be a common problem. Anal Biochem 311(2):182–186

Fischer G, Tradler T, Zarnt T (1998) The mode of action of peptidyl prolyl cis/trans isomerases *in vivo*: binding vs. catalysis. FEBS Lett 426(1):17–20

Fulda S, Gorman AM, Hori O, Samali A (2010) Cellular stress responses: cell survival and cell death. Int J Cell Biol 2010:23. https://doi.org/10.1155/2010/214074

Gilroy S, Białasek M, Suzuki N, Górecka M, Devireddy AR, Karpiński S, Mittler R (2016) ROS, calcium, and electric signals: key mediators of rapid systemic Signaling in plants. Plant Physiol 171(3):1606–1615. https://doi.org/10.1104/pp.16.00434

Guo M, Liu J-H, Ma X, Luo D-X, Gong Z-H, Lu M-H (2016) The plant heat stress transcription factors (HSFs): structure, regulation, and function in response to abiotic stresses. Front Plant Sci 7:114–114. https://doi.org/10.3389/fpls.2016.00114

Gurley WB (2000) HSP101: a key component for the acquisition of thermotolerance in plants. Plant Cell 12(4):457–460. https://doi.org/10.1105/tpc.12.4.457

Guzzo J (2012) Biotechnical applications of small heat shock proteins from bacteria. Int J Biochem Cell Biol 44(10):1698–1705. https://doi.org/10.1016/j.biocel.2012.06.007

Haak DC, Fukao T, Grene R, Hua Z, Ivanov R, Perrella G, Li S (2017) Multilevel regulation of abiotic stress responses in plants. Front Plant Sci 8:1564. https://doi.org/10.3389/fpls.2017.01564

Hafren A, Hofius D, Ronnholm G, Sonnewald U, Makinen K (2010) HSP70 and its cochaperone CPIP promote potyvirus infection in *Nicotiana benthamiana* by regulating viral coat protein functions. Plant Cell 22(2):523–535. https://doi.org/10.1105/tpc.109.072413

Han MJ, Park SJ, Park TJ, Lee SY (2004) Roles and applications of small heat shock proteins in the production of recombinant proteins in *Escherichia coli*. Biotechnol Bioeng 88(4):426–436. https://doi.org/10.1002/bit.20227

Han MJ, Lee JW, Lee SY (2005) Enhanced proteome profiling by inhibiting proteolysis with small heat shock proteins. J Proteome Res 4(6):2429–2434. https://doi.org/10.1021/pr050259m

Han MJ, Yun H, Lee SY (2008) Microbial small heat shock proteins and their use in biotechnology. Biotechnol Adv 26(6):591–609. https://doi.org/10.1016/j.biotechadv.2008.08.004

Harrison C (2003) GrpE, a nucleotide exchange factor for DnaK. Cell Stress Chaperones 8(3):218–224

Hartl FU, Hayer-Hartl M (2002) Molecular chaperones in the cytosol: from nascent chain to folded protein. Science 295(5561):1852–1858. https://doi.org/10.1126/science.1068408

Hartl FU, Bracher A, Hayer-Hartl M (2011) Molecular chaperones in protein folding and proteostasis. Nature 475:324. https://doi.org/10.1038/nature10317

Haslbeck M, Franzmann T, Weinfurtner D, Buchner J (2005) Some like it hot: the structure and function of small heat-shock proteins. Nat Struct Mol Biol 12(10):842–846. https://doi.org/10.1038/nsmb993

Hayashi S, Nakazaki Y, Kagii K, Imamura H, Watanabe Y (2017) Fusion protein analysis reveals the precise regulation between Hsp70 and Hsp100 during protein disaggregation. Sci Rep 7(1):8648. https://doi.org/10.1038/s41598-017-08917-8

He L, Sharpe T, Mazur A, Hiller S (2016) A molecular mechanism of chaperone-client recognition. Sci Adv 2(11). https://doi.org/10.1126/sciadv.1601625

Hein I, Barciszewska-Pacak M, Hrubikova K, Williamson S, Dinesen M, Soenderby IE, Sundar S, Jarmolowski A, Shirasu K, Lacomme C (2005) Virus-induced gene silencing-based functional characterization of genes associated with powdery mildew resistance in barley. Plant Physiol 138(4):2155–2164. https://doi.org/10.1104/pp.105.062810

Hofius D, Maier AT, Dietrich C, Jungkunz I, Bornke F, Maiss E, Sonnewald U (2007) Capsid protein-mediated recruitment of host DnaJ-like proteins is required for potato virus Y infection in tobacco plants. J Virol 81(21):11870–11880. https://doi.org/10.1128/jvi.01525-07

Hong SW, Vierling E (2001) Hsp101 is necessary for heat tolerance but dispensable for development and germination in the absence of stress. Plant J 27(1):25–35. https://doi.org/10.1046/j.1365-313x.2001.01066.x

Horovitz A, Willison KR (2005) Allosteric regulation of chaperonins. Curr Opin Struct Biol 15 (6):646–651. https://doi.org/10.1016/j.sbi.2005.10.001

Horwich AL, Fenton WA, Chapman E, Farr GW (2007) Two families of chaperonin: physiology and mechanism. Annu Rev Cell Dev Biol 23:115–145. https://doi.org/10.1146/annurev.cellbio.23.090506.123555

Hu X, Liu R, Li Y, Wang W, Tai F, Xue R, Li C (2010) Heat shock protein 70 regulates the abscisic acid-induced antioxidant response of maize to combined drought and heat stress. Plant Growth Regul 60(3):225–235. https://doi.org/10.1007/s10725-009-9436-2

Irvine AG, Wallis AK, Sanghera N, Rowe ML, Ruddock LW, Howard MJ, Williamson RA, Blindauer CA, Freedman RB (2014) Protein disulfide-isomerase interacts with a substrate protein at all stages along its folding pathway. PLoS One 9(1):e82511. https://doi.org/10.1371/journal.pone.0082511

Jackson SE (2013) Hsp90: structure and function. Top Curr Chem 328:155–240. https://doi.org/10.1007/128_2012_356

Jung YJ, Melencion SM, Lee ES, Park JH, Alinapon CV, Oh HT, Yun DJ, Chi YH, Lee SY (2015) Universal stress protein exhibits a redox-dependent chaperone function in Arabidopsis and enhances plant tolerance to heat shock and oxidative stress. Front Plant Sci 6:1141. https://doi.org/10.3389/fpls.2015.01141

Jungkunz I, Link K, Vogel F, Voll LM, Sonnewald S, Sonnewald U (2011) AtHsp70-15-deficient *Arabidopsis* plants are characterized by reduced growth, a constitutive cytosolic protein response and enhanced resistance to TuMV. Plant J 66(6):983–995. https://doi.org/10.1111/j.1365-313X.2011.04558.x

Kanzaki H, Saitoh H, Ito A, Fujisawa S, Kamoun S, Katou S, Yoshioka H, Terauchi R (2003) Cytosolic HSP90 and HSP70 are essential components of INF1-mediated hypersensitive response and non-host resistance to *Pseudomonas cichorii* in *Nicotiana benthamiana*. Mol Plant Pathol 4(5):383–391. https://doi.org/10.1046/j.1364-3703.2003.00186.x

Kaufman BA, Kolesar JE, Perlman PS, Butow RA (2003) A function for the mitochondrial chaperonin Hsp60 in the structure and transmission of mitochondrial DNA nucleoids in *Saccharomyces cerevisiae*. J Cell Biol 163(3):457–461. https://doi.org/10.1083/jcb.200306132

Kemmink J, Darby NJ, Dijkstra K, Nilges M, Creighton TE (1997) The folding catalyst protein disulfide isomerase is constructed of active and inactive thioredoxin modules. Curr Biol 7 (4):239–245

Kim NH, Hwang BK (2015) Pepper heat shock protein 70a interacts with the type III effector AvrBsT and triggers plant cell death and immunity. Plant Physiol 167(2):307–322. https://doi.org/10.1104/pp.114.253898

Kim R, Kim KK, Yokota H, Kim SH (1998) Small heat shock protein of *Methanococcus jannaschii*, a hyperthermophile. Proc Natl Acad Sci U S A 95(16):9129–9133

Koizumi S, Ohama N, Mizoi J, Shinozaki K, Yamaguchi-Shinozaki K (2014) Functional analysis of the Hikeshi-like protein and its interaction with HSP70 in *Arabidopsis*. Biochem Biophys Res Commun 450(1):396–400. https://doi.org/10.1016/j.bbrc.2014.05.128

Kozeko LY (2019) The role of HSP90 chaperones in stability and plasticity of ontogenesis of plants under normal and stressful conditions (*Arabidopsis thaliana*). Cytol Genet 53(2):143–161. https://doi.org/10.3103/S0095452719020063

Kuang J, Liu J, Mei J, Wang C, Hu H, Zhang Y, Sun M, Ning X, Xiao L, Yang L (2017) A class II small heat shock protein OsHsp18.0 plays positive roles in both biotic and abiotic defense responses in rice. Sci Rep 7(1):11333. https://doi.org/10.1038/s41598-017-11882-x

Kvint K, Nachin L, Diez A, Nystrom T (2003) The bacterial universal stress protein: function and regulation. Curr Opin Microbiol 6(2):140–145

Laksanalamai P, Pavlov AR, Slesarev AI, Robb FT (2006) Stabilization of Taq DNA polymerase at high temperature by protein folding pathways from a hyperthermophilic archaeon, *Pyrococcus furiosus*. Biotechnol Bioeng 93(1):1–5. https://doi.org/10.1002/bit.20781

Landi S, Capasso G, Ben Azaiez FE, Jallouli S, Ayadi S, Trifa Y, Esposito S (2019) Different roles of heat shock proteins (70 kDa) during abiotic stresses in barley (*Hordeum vulgare*) genotypes. Plants (Basel) 8(8). https://doi.org/10.3390/plants8080248

Laufen T, Mayer MP, Beisel C, Klostermeier D, Mogk A, Reinstein J, Bukau B (1999) Mechanism of regulation of hsp70 chaperones by DnaJ cochaperones. Proc Natl Acad Sci U S A 96(10):5452–5457

Leach MR, Williams DB (2003) Calnexin and calreticulin, molecular chaperones of the endoplasmic reticulum. In: Eggleton P, Michalak M (eds) Calreticulin, 2nd edn. Landes Bioscience, Georgetown, pp 49–62

Lee JH, Schoffl F (1996) An Hsp70 antisense gene affects the expression of HSP70/HSC70, the regulation of HSF, and the acquisition of thermotolerance in transgenic *Arabidopsis thaliana*. Mol Gen Genet 252(1-2):11–19. https://doi.org/10.1007/s004389670002

Lin CH, Lee CN, Lin JW, Tsai WJ, Wang SW, Weng SF, Tseng YH (2010) Characterization of *Xanthomonas campestris* pv. campestris heat shock protein A (HspA), which possesses an intrinsic ability to reactivate inactivated proteins. Appl Microbiol Biotechnol 88(3):699–709. https://doi.org/10.1007/s00253-010-2776-z

Liu JZ, Whitham SA (2013) Overexpression of a soybean nuclear localized type-III DnaJ domain-containing HSP40 reveals its roles in cell death and disease resistance. Plant J 74(1):110–121. https://doi.org/10.1111/tpj.12108

Liu Y, Burch-Smith T, Schiff M, Feng S, Dinesh-Kumar SP (2004) Molecular chaperone Hsp90 associates with resistance protein N and its signaling proteins SGT1 and Rar1 to modulate an innate immune response in plants. J Biol Chem 279(3):2101–2108. https://doi.org/10.1074/jbc.M310029200

Loukehaich R, Wang T, Ouyang B, Ziaf K, Li H, Zhang J, Lu Y, Ye Z (2012) SpUSP, an annexin-interacting universal stress protein, enhances drought tolerance in tomato. J Exp Bot 63(15):5593–5606. https://doi.org/10.1093/jxb/ers220

Lu R, Malcuit I, Moffett P, Ruiz MT, Peart J, Wu A-J, Rathjen JP, Bendahmane A, Day L, Baulcombe DC (2003) High throughput virus-induced gene silencing implicates heat shock protein 90 in plant disease resistance. EMBO J 22(21):5690–5699. https://doi.org/10.1093/emboj/cdg546

Maimbo M, Ohnishi K, Hikichi Y, Yoshioka H, Kiba A (2007) Induction of a small heat shock protein and its functional roles in *Nicotiana* plants in the defense response against *Ralstonia solanacearum*. Plant Physiol 145(4):1588–1599. https://doi.org/10.1104/pp.107.105353

Malik MK, Slovin JP, Hwang CH, Zimmerman JL (1999) Modified expression of a carrot small heat shock protein gene, hsp17. 7, results in increased or decreased thermotolerancedouble dagger. Plant J 20(1):89–99. https://doi.org/10.1046/j.1365-313x.1999.00581.x

de Marco A, Deuerling E, Mogk A, Tomoyasu T, Bukau B (2007) Chaperone-based procedure to increase yields of soluble recombinant proteins produced in *E. coli*. BMC Biotechnol 7(32). https://doi.org/10.1186/1472-6750-7-32

Marker T, Sell H, Zillessen P, Glode A, Kriebel J, Ouwens DM, Pattyn P, Ruige J, Famulla S, Roden M, Eckel J, Habich C (2012) Heat shock protein 60 as a mediator of adipose tissue inflammation and insulin resistance. Diabetes 61(3):615–625. https://doi.org/10.2337/db10-1574

Marklund EG, Zhang Y, Basha E, Benesch JLP, Vierling E (2018) Structural and functional aspects of the interaction partners of the small heat-shock protein in. Synechocystis 23(4):723–732. https://doi.org/10.1007/s12192-018-0884-3

Mayer MP, Bukau B (2005) Hsp70 chaperones: cellular functions and molecular mechanism. Cell Mol Life Sci 62(6):670–684. https://doi.org/10.1007/s00018-004-4464-6

McLoughlin F, Kim M, Marshall RS, Vierstra RD, Vierling E (2019) HSP101 interacts with the proteasome and promotes the clearance of ubiquitylated protein aggregates. Plant Physiol 180 (4):1829–1847. https://doi.org/10.1104/pp.19.00263

Merkouropoulos G, Andreasson E, Hess D, Boller T, Peck SC (2008) An *Arabidopsis* protein phosphorylated in response to microbial elicitation, AtPHOS32, is a substrate of MAP kinases 3 and 6. J Biol Chem 283(16):10493–10499. https://doi.org/10.1074/jbc.M800735200

Meyer P, Prodromou C, Hu B, Vaughan C, Roe SM, Panaretou B, Piper PW, Pearl LH (2003) Structural and functional analysis of the middle segment of HSP90: implications for ATP hydrolysis and client protein and cochaperone interactions. Mol Cell 11(3):647–658

Mishra RC, Grover A (2016) ClpB/Hsp100 proteins and heat stress tolerance in plants. Crit Rev Biotechnol 36(5):862–874. https://doi.org/10.3109/07388551.2015.1051942

Mogk A, Schlieker C, Friedrich KL, Schonfeld HJ, Vierling E, Bukau B (2003) Refolding of substrates bound to small Hsps relies on a disaggregation reaction mediated most efficiently by ClpB/DnaK. J Biol Chem 278(33):31033–31042. https://doi.org/10.1074/jbc.M303587200

Nath PR, Isakov N (2015) Insights into peptidyl-prolyl cis-trans isomerase structure and function in immunocytes. Immunol Lett 163(1):120–131. https://doi.org/10.1016/j.imlet.2014.11.002

Noel LD, Cagna G, Stuttmann J, Wirthmuller L, Betsuyaku S, Witte CP, Bhat R, Pochon N, Colby T, Parker JE (2007) Interaction between SGT1 and cytosolic/nuclear HSC70 chaperones regulates *Arabidopsis* immune responses. Plant Cell 19(12):4061–4076. https://doi.org/10.1105/tpc.107.051896

Ono K, Hibino T, Kohinata T, Suzuki S, Tanaka Y, Nakamura T, Takabe T, Takabe T (2001) Overexpression of DnaK from a halotolerant cyanobacterium *Aphanothece halophytica* enhances the high-temperatue tolerance of tobacco during germination and early growth. Plant Sci 160(3):455–461. https://doi.org/10.1016/s0168-9452(00)00412-x

Pareek A, Singla SL, Grover A (1995) Immunological evidence for accumulation of two high-molecular-weight (104 and 90 kDa) HSPs in response to different stresses in rice and in response to high temperature stress in diverse plant genera. Plant Mol Biol 29(2):293–301. https://doi.org/10.1007/bf00043653

Poljšak B, Milisav I (2012) Clinical implications of cellular stress responses. Bosn J Basic Med Sci 12(2):122–126

Poulain P, Gelly J-C, Flatters D (2010) Detection and architecture of small heat shock protein monomers. PLoS One 5(4):e9990. https://doi.org/10.1371/journal.pone.0009990

Pulido P, Leister D (2018) Novel DNAJ-related proteins in *Arabidopsis thaliana*. New Phytol 217 (2):480–490. https://doi.org/10.1111/nph.14827

Queitsch C, Hong SW, Vierling E, Lindquist S (2000) Heat shock protein 101 plays a crucial role in thermotolerance in *Arabidopsis*. Plant Cell 12(4):479–492. https://doi.org/10.1105/tpc.12.4.479

Rajamuthiah R, Mylonakis E (2014) Effector triggered immunity. Virulence 5(7):697–702. https://doi.org/10.4161/viru.29091

Rowland SE, Robb FT (2017) Structure, function and evolution of the Hsp60 chaperonins. In: Kumar CMS, Mande SC (eds) Prokaryotic chaperonins: multiple copies and multitude functions. Springer, Singapore, pp 3–20

Rudiger S, Buchberger A, Bukau B (1997) Interaction of Hsp70 chaperones with substrates. Nat Struct Biol 4(5):342–349

Sable A, Rai KM, Choudhary A, Yadav VK, Agarwal SK, Sawant SV (2018) Inhibition of heat shock proteins HSP90 and HSP70 induce oxidative stress, suppressing cotton fiber development. Sci Rep 8(1):3620. https://doi.org/10.1038/s41598-018-21866-0

Saibil H (2013) Chaperone machines for protein folding, unfolding and disaggregation. Nat Rev Mol Cell Biol 14(10):630–642. https://doi.org/10.1038/nrm3658

Saijo Y (2010) ER quality control of immune receptors and regulators in plants. Cell Microbiol 12 (6):716–724. https://doi.org/10.1111/j.1462-5822.2010.01472.x

Sangster TA, Bahrami A, Wilczek A, Watanabe E, Schellenberg K, McLellan C, Kelley A, Kong SW, Queitsch C, Lindquist S (2007) Phenotypic diversity and altered environmental plasticity in *Arabidopsis thaliana* with reduced Hsp90 levels. PLoS One 2(7):e648. https://doi.org/10.1371/journal.pone.0000648

Sangster TA, Salathia N, Undurraga S, Milo R, Schellenberg K, Lindquist S, Queitsch C (2008) HSP90 affects the expression of genetic variation and developmental stability in quantitative traits. Proc Natl Acad Sci U S A 105(8):2963–2968. https://doi.org/10.1073/pnas.0712200105

Sanmiya K, Suzuki K, Egawa Y, Shono M (2004) Mitochondrial small heat-shock protein enhances thermotolerance in tobacco plants. FEBS Lett 557(1-3):265–268. https://doi.org/10.1016/s0014-5793(03)01494-7

Sarkar NK, Kim YK, Grover A (2009) Rice sHsp genes: genomic organization and expression profiling under stress and development. BMC Genomics 10:393. https://doi.org/10.1186/1471-2164-10-393

Schopf FH, Biebl MM, Buchner J (2017) The HSP90 chaperone machinery. Nat Rev Mol Cell Biol 18(6):345–360. https://doi.org/10.1038/nrm.2017.20

Shashidharamurthy R, Koteiche HA, Dong J, McHaourab HS (2005) Mechanism of chaperone function in small heat shock proteins: dissociation of the HSP27 oligomer is required for recognition and binding of destabilized T4 lysozyme. J Biol Chem 280(7):5281–5289. https://doi.org/10.1074/jbc.M407236200

Shaw PE (2002) Peptidyl-prolyl isomerases: a new twist to transcription. EMBO Rep 3 (6):521–526. https://doi.org/10.1093/embo-reports/kvf118

Shirasu K (2009) The HSP90-SGT1 chaperone complex for NLR immune sensors. Annu Rev Plant Biol 60:139–164. https://doi.org/10.1146/annurev.arplant.59.032607.092906

Shridas P, Sharma Y, Balasubramanian D (2001) Transglutaminase-mediated cross-linking of alpha-crystallin: structural and functional consequences. FEBS Lett 499(3):245–250

Siegele DA (2005) Universal stress proteins in *Escherichia coli*. J Bacteriol 187(18):6253–6254. https://doi.org/10.1128/JB.187.18.6253-6254.2005

Song H, Fan P, Li Y (2009) Overexpression of organellar and cytosolic AtHSP90 in *Arabidopsis thaliana* impairs plant tolerance to oxidative stress. Plant Mol Biol Rep 27:342–349. https://doi.org/10.1007/s11105-009-0091-6

Sousa MC, McKay DB (2001) Structure of the universal stress protein of Haemophilus influenzae. Structure 9(12):1135–1141. https://doi.org/10.1016/s0969-2126(01)00680-3

Spiess C, Meyer AS, Reissmann S, Frydman J (2004) Mechanism of the eukaryotic chaperonin: protein folding in the chamber of secrets. Trends Cell Biol 14(11):598–604. https://doi.org/10.1016/j.tcb.2004.09.015

Sun Y, MacRae TH (2005) Small heat shock proteins: molecular structure and chaperone function. Cell Mol Life Sci 62(21):2460–2476. https://doi.org/10.1007/s00018-005-5190-4

Tomoyasu T, Tabata A, Ishikawa Y, Whiley RA, Nagamune H (2013) Small heat shock protein AgsA: an effective stabilizer of enzyme activities. J Biosci Bioeng 115(1):15–19. https://doi.org/10.1016/j.jbiosc.2012.08.001

Usman MG, Rafii MY, Martini MY, Yusuff OA, Ismail MR, Miah G (2017) Molecular analysis of Hsp70 mechanisms in plants and their function in response to stress. Biotechnol Genet Eng Rev 33(1):26–39. https://doi.org/10.1080/02648725.2017.1340546

Van Ooijen G, Lukasik E, Van Den Burg HA, Vossen JH, Cornelissen BJ, Takken FL (2010) The small heat shock protein 20 RSI2 interacts with and is required for stability and function of tomato resistance protein I-2. Plant J 63(4):563–572. https://doi.org/10.1111/j.1365-313X.2010.04260.x

Verchot J (2012) Cellular chaperones and folding enzymes are vital contributors to membrane bound replication and movement complexes during plant RNA virus infection. Front Plant Sci 3:275. https://doi.org/10.3389/fpls.2012.00275

Vollmer AC, Bark SJ (2018) Twenty-five years of investigating the universal stress protein: function, structure, and applications. Adv Appl Microbiol 102:1–36. https://doi.org/10.1016/bs.aambs.2017.10.001

Walsh P, Bursac D, Law YC, Cyr D, Lithgow T (2004) The J-protein family: modulating protein assembly, disassembly and translocation. EMBO Rep 5(6):567–571. https://doi.org/10.1038/sj.embor.7400172

Wang X, Yan B, Shi M, Zhou W, Zekria D, Wang H, Kai G (2016a) Overexpression of a *Brassica campestris* HSP70 in tobacco confers enhanced tolerance to heat stress. Protoplasma 253 (3):637–645. https://doi.org/10.1007/s00709-015-0867-5

Wang R, Zhang Y, Kieffer M, Yu H, Kepinski S, Estelle M (2016b) HSP90 regulates temperature-dependent seedling growth in *Arabidopsis* by stabilizing the auxin co-receptor F-box protein TIR1. Nat Commun 7:10269. https://doi.org/10.1038/ncomms10269

Wang Y, Jin F, Wang R, Li F, Wu Y, Kitazato K, Wang Y (2017) HSP90: a promising broad-spectrum antiviral drug target. Arch Virol 162(11):3269–3282. https://doi.org/10.1007/s00705-017-3511-1

Waters ER (2013) The evolution, function, structure, and expression of the plant sHSPs. J Exp Bot 64(2):391–403. https://doi.org/10.1093/jxb/ers355

Wehmeyer N, Vierling E (2000) The expression of small heat shock proteins in seeds responds to discrete developmental signals and suggests a general protective role in desiccation tolerance. Plant Physiol 122(4):1099–1108. https://doi.org/10.1104/pp.122.4.1099

Wendler P, Ciniawsky S, Kock M, Kube S (2012) Structure and function of the AAA+ nucleotide binding pocket. Biochim Biophys Acta 1823(1):2–14. https://doi.org/10.1016/j.bbamcr.2011.06.014

Whitesell L, Lindquist SL (2005) HSP90 and the chaperoning of cancer. Nat Rev Cancer 5 (10):761–772. https://doi.org/10.1038/nrc1716

Wilkinson B, Gilbert HF (2004) Protein disulfide isomerase. Biochim Biophys Acta 1699 (1-2):35–44. https://doi.org/10.1016/j.bbapap.2004.02.017

Wu J, Liu T, Rios Z, Mei Q, Lin X, Cao S (2017) Heat shock proteins and cancer. Trends Pharmacol Sci 38(3):226–256. https://doi.org/10.1016/j.tips.2016.11.009

Xu P, Chen F, Mannas JP, Feldman T, Sumner LW, Roossinck MJ (2008) Virus infection improves drought tolerance. New Phytol 180(4):911–921. https://doi.org/10.1111/j.1469-8137.2008.02627.x

Xu Z-S, Li Z-Y, Chen Y, Chen M, Li L-C, Ma Y-Z (2012) Heat shock protein 90 in plants: molecular mechanisms and roles in stress responses. Int J Mol Sci 13(12):15706–15723. https://doi.org/10.3390/ijms131215706

Xu J, Xue C, Xue D, Zhao J, Gai J, Guo N, Xing H (2013) Overexpression of GmHsp90s, a heat shock protein 90 (Hsp90) gene family cloning from soybean, decrease damage of abiotic stresses in *Arabidopsis thaliana*. PLoS One 8(7):e69810. https://doi.org/10.1371/journal.pone.0069810

Yang Y, Zeng J, Gao C, Krull UJ (2003) Stabilization and re-activation of trapped enzyme by immobilized heat shock protein and molecular chaperones. Biosens Bioelectron 18 (2-3):311–317

Young JC (2010) Mechanisms of the Hsp70 chaperone system. Biochem Cell Biol 88(2):291–300. https://doi.org/10.1139/o09-175

Young JC, Moarefi I, Hartl FU (2001) Hsp90: a specialized but essential protein-folding tool. J Cell Biol 154(2):267–273

Zhang H, Li Y, Zhu JK (2018) Developing naturally stress-resistant crops for a sustainable agriculture. Nat Plants 4(12):989–996. https://doi.org/10.1038/s41477-018-0309-4

Zhao D, Xia X, Su J, Wei M, Wu Y, Tao J (2019) Overexpression of herbaceous peony HSP70 confers high temperature tolerance. BMC Genomics 20(1):70. https://doi.org/10.1186/s12864-019-5448-0

Zhichang Z, Wanrong Z, Jinping Y, Jianjun Z, Zhen L, Xufeng L, Yang Y (2010) Over-expression of *Arabidopsis* DnaJ (Hsp40) contributes to NaCl-stress tolerance. Afr J Biotechnol 9:972–978. https://doi.org/10.5897/AJB09.1450

Zhou C, Sun H, Zheng C, Gao J, Fu Q, Hu N, Shao X, Zhou Y, Xiong J, Nie K, Zhou H, Shen L, Fang H, Lyu J (2018) Oncogenic HSP60 regulates mitochondrial oxidative phosphorylation to support Erk1/2 activation during pancreatic cancer cell growth. Cell Death Dis 9(2):161. https://doi.org/10.1038/s41419-017-0196-z

Zhu JK (2016) Abiotic stress signaling and responses in plants. Cell 167(2):313–324. https://doi.org/10.1016/j.cell.2016.08.029

Zipfel C (2014) Plant pattern-recognition receptors. Trends Immunol 35(7):345–351. https://doi.org/10.1016/j.it.2014.05.004

Zolkiewski M, Zhang T, Nagy M (2012) Aggregate reactivation mediated by the Hsp100 chaperones. Arch Biochem Biophys 520(1):1–6. https://doi.org/10.1016/j.abb.2012.01.012

Bioprospecting of Ethno-Medicinal Plants for Wound Healing

22

S. R. Suja, A. L. Aneeshkumar, and R. Prakashkumar

Abstract

Wound healing is an intricate anabolic progression via restoration of tissue or skin integrity after damage, which is due to the involvement of cell and matrix signaling. Chronic wounds have high incidence and economic burden in spite of the new advances in the field of medicine and their management is still challenging. Wounds are the physical injuries caused by the disruption of anatomical, cellular, and functional continuity of a living tissue. A potent natural wound care product should be capable of cellular reformation at the wounded site and must exclude secondary infections and minimize pain, discomfort, and scarring in order to promote healing in the shortest time possible. Plants on the planet earth are the largest healers in the world. Ethno-medico-botanical surveys and ethnopharmacological research leads to the way of natural products exploration which continues to discover a variety of lead structures, which may practice as templates for the progress of new cost-effective drugs for the well-being of humans. There are several phytocompounds which provide significant wound-healing efficacy even though the commercialization of these compound-based drugs is very less. A paradigm shift is needed to enhance the effective translation of new phytoconstituents into clinical therapies as well as understand the pathophysiology of signaling pattern. The combination of traditional knowledge and validation with modern instrumentation and advanced nanomaterials like nanocellulose scaffolds coupled with drug delivery for skin tissue engineering lead to the development of safe and effective therapeutics for cuts and wounds.

S. R. Suja (✉) · A. L. Aneeshkumar · R. Prakashkumar
Ethnomedicine & Ethnopharmacology Division, KSCSTE—Jawaharlal Nehru Tropical Botanic Garden and Research Institute, Palode, Thiruvananthapuram, Kerala, India

Keywords

Herbal drugs · Wound care · Phytoconstituents · Nano scaffold · Secondary metabolites · Ethnopharmacology · Bioprospecting

22.1 Introduction

Nature is a creative architecture that hides many wonders inside the magnitude. The extreme power which hides within the nature was used for the well-being of living organisms in the world. India is one of the richest biological diversity countries in the world and has massive resources of medicinal herbs. Plants on the planet earth are the largest healers throughout the world. Among 52,885 medicinal plant species used worldwide, 3000 are used in India (Schippmann et al. 2002).

India ranked third in the herbal medicine category, with less than 2% of global market share, while China occupies nearly 30% of the market (Patwardhan et al. 2005). Ninety percent of India's medicinal plants diversity is spread in the forests of Western Ghats and Eastern Himalayas. More than 1.5 million traditional and folk medicine practitioners use ~25,000 plant-based formulations for healthcare in India itself. This pharmacological validation on Indian medicinal plants is very limited and a large number of plants used in tribal and folklore medicine with enormous potential have not been validated for their wound-healing activity.

Healthcare sector around the world is live through increased levels of chronic illness and is becoming more and more expensive to common men. Noncommunicable diseases kill around 15 million people between the ages of 30 and 69 each year globally of these over 80% of deaths occur in low- and middle-income countries. Plants have been used as curative mediators since prehistoric times by people globally. Nowadays a resurgence of interest in herbal medicine and related products makes it important to understand the potential interactions between herbs and prescribed drugs (Fugh-berman and Ernst 2001). Ethno-medicobotanical research leads to the way of natural products research, which continues to discover a variety of lead structures, which may be used as templates for the development of new drugs by the pharmaceutical industry (Patwardhan et al. 2004).

According to World Health Organization (WHO) estimates, by the year 2050, the demand for medicinal plants would be US$ ~5 trillion (Aneesh et al. 2009). The annual growth rate in India averaged 6.12% from 1951 until 2017 (India GDP 2018) and is the positive indication that India could gain a momentous competitive edge in the global marketplace, specifically in the pharmaceutical, beauty care and healthcare segments. To improve India's contribution in the escalating global market, it is the need of the hour for a renewed development of India's medicinal plants sector.

A major therapeutic practice is directly or indirectly depending on the herbals which are prescribed by traditional medical practitioners/tribal healers. Government had formulated national policies to encourage the use of herbal medicine and to promote their safety and quality (WHO 1993). A WHO/IUCN/WWF International

Consultation on Conservation of Medicinal Plants pointed that the loss of indigenous cultures directly affects to bring out new medicinal species to benefit the wider community.

22.2 Herbalism in Drug Discovery

Botanicals can alter the effect of drugs, resulting in an effect that is additive, synergistic or one that has an antagonistic action. There are times that naturopathic doctors, or trained practitioners, will choose specific herbs with the aim of decreasing the dosage or use of specific drugs. The likelihood of herb–drug interactions could be higher than drug–drug interactions, since drugs usually contain single chemical entities, while almost all herbal medicinal products (even single-herb products) contain mixtures of pharmacologically active constituents (Fugh-berman and Ernst 2001). Traditional and complementary medicine is used not only to treat diseases but also used in disease prevention, health promotion and health maintenance, and also has proved to be cost-effective, devoid of side effects (WHO 2004). Plant-based drugs such as morphine, codeine, reserpine, digoxin, vinblastine, taxol, quinine, and artemisinin are used for the treatment of different ailments (Sharma et al. 2014; Mohanraj et al. 2017).

Plants produce immense metabolites of which secondary metabolites are used for their self-protection against parasites, pathogens, and predators. Beyond that most of these compounds possess a wide array of therapeutic actions that are utilized by humans from prehistoric times. The information regarding the utilization of such chemical constituents as various ailments for different diseased conditions has helped a long way for developing uncountable drugs. Still the knowledge regarding the use of unexplored plants as medicine is on the way which could serve as newer leads and clues for modern drug designing. The most important of these bioactive chemical constituents of plants include alkaloids, tannins, flavonoids, and phenolic compounds. Correlation between these phytoconstituents and bioactivity of a plant is prudent to know for the synthesis of compounds with precise activities to treat various health ailments and chronic diseased conditions as well. Plants also contain inactive substances, which become active when chemically manipulated during the process of manufacturing chemical medicines regarded as medicinal plants (Mafimisebi and Oguntade 2010). Owing to the significance in the above contexts, such phytochemical activity screening of plants is needed in order to discover and develop novel therapeutic agents with improved efficacy.

22.3 Traditional Medicine

India is bestowed with rich biological heritage. Nearly 70% of tribal and rural inhabitants of India are to a large extent depended on medicinal plants for their primary healthcare management and the utility of wild plants for various purposes has got recognized with generations due to either inaccessible or less availability of modern healthcare system. But the knowledge of traditional medicine in India came down in the recent past due to increased pace of urbanization and vanishing ethnic communities from their aboriginal life style pattern.

Traditional systems of medicine provide valuable information on natural remedies. They have played a significant role in discovering novel drug leads. Today, traditional and alternative remedies are used to treat many chronic conditions in developed countries. This could be due to the lack of effective modern remedies for chronic conditions. Traditional systems of medicine usually have a holistic approach and provide us with many remedies for various chronic medical conditions. These systems of medicine provide remedies for treating wounds as well. Over the past few decades, we have observed increasing interest in researches based on traditional medicines of the world.

22.4 Global Relevance

Knowledge of useful plants must have been first acquired by man to satisfy his hunger, heal his wounds, and treat various ailments (Kshirsagar and Singh 2001, Schultes 1984). Traditional healers employ methods based on the ecological, sociocultural, and religious background of their people to provide healthcare (Anyinam 1995; Gesler 1992). Therefore, practice of ethnomedicine is an important vehicle for understanding indigenous societies and their relationships with nature (Anyinam 1995; Rai and Lalramnghinglova 2011).

Globally traditional medicine has been used before the era of modern medicine and people urged the knowledge over generations with trial and error method. Many countries have developed their own traditional or indigenous forms of therapeutic formulas since the beginning of civilization, which was deeply rooted in their antiquity and ethos. Herbalism or botanical medicine is the traditional medicinal practice via the treatment or prevention of disease using plants and plant extracts (Ernst 2000). Such treatment or prevention of diseases using plant-based medicines; various medical cultures have produced their own type of herbalism, e.g., Ayurveda (India), Kampo (Japan), traditional Chinese medicine, etc. There are over 1000 herbs used in the practice of naturopathic medicine. Naturopathic medicine uses traditional healing system and natural remedies with modern medical knowledge in a distinct and comprehensive way of healthcare in order to support and stimulate the patient body's self-healing process to achieve health and wellness. Even Western medicine has its roots in the use of herbs. Until the 1950s, herbs were the basis of all pharmaceutical drugs. Even today, about 25% of pharmaceuticals are derived from herbs, such as opium, aspirin, digitalis, quinine, etc. Other medicines are also available which are extracted from plants such as corticosteroids and oral contraceptives.

Medicinal systems in India date back at least 5000 BC, coinciding with the emergence of Indus Valley civilization. Traditional systems of medicine were followed in India since the British era. Folk/Traditional refers to those medicinal practices which are generally transmitted orally from generation to generation and whose use is restricted to a specific geographical area or group of people such as a tribe, caste, or community (Mazid et al. 2012). It is estimated that there are 66.5 million tribals in India. With few exceptions, the majority of them continue to depend on forests for their livelihood. Large-scale urbanization, changing socio-ethnic moral values, overexploitation of the raw materials especially biodiversity resources and tribal exploiters mainly by civilized people are the major threats for folk and tribal medicine because of that these traditions are slowly dying out. Traditional medical practices are an important part of the primary healthcare system in the developing world (Fairbairn 1980).

Many pharmaceutical companies of international and national repute are continuously attempting to get benefit from traditional knowledge since countless pharmaceuticals products have been derived from plants, of which 75% of these were discovered by scrutinizing the use of plants in traditional medicine (King and Tempesta 1994).

From the history of Traditional Medicine (TM), it has been noted that many plants and herbs are being used to treat skin disorders, including wound injuries (Lodhi et al. 2006). Natural as well as synthetic bioactive materials with wound-healing potential, have the propensity for antioxidant, chelation, and antimicrobial activities; and its action is related by one or more of these mechanisms (Mallefet and Dweck 2008; Builders et al. 2013). The use of traditional medicinal information has contributed to drug discovery worldwide. Currently, the available wound-healing remedies are suboptimal, but using traditional medical knowledge in this area might provide a good direction for future medical and pharmaceutical research.

22.5 Ethno-Medico-Botanic Approach

Ethnopharmacology is a wide area of interdisciplinary scientific investigation of biologically active herbs traditionally employed or observed by man (Holmstedt and Bruhn 1995). The major focus in this area is the scientific assessment and verification for the validation of traditional usage of plants either via., the separation of active ingredients, or via., pharmacological experiments on indigenous drug preparations that can contribute to new herbal formulations for the development of various effective novel therapeutic drugs as part of integrated healthcare system, which opens new frontiers in the combination of traditional and western healing practices (Cotton and Wilkie 1996). Ethnopharmacological experiments encompass with the documentation and testing of the effectiveness and toxicity analysis of plant extracts, and the identification of the active components. This would allow TK of plants used in medicine practice to be gathered and well-kept-up and recommendations for practice to be based on an experimental thought. Scientific validation of the ethnopharmacological claims about traditional medicine is the present scenario for its globalization and fortification.

Various ethnomedicinal plant products have been used in treating wounds over the years. Herbal products used for wound healing promote blood clotting, fight against infections, and accelerate the healing process. There is an increasing interest in finding herbal medicines with wound-healing efficacy even though the use of such herbals for treating cuts and wounds are common in traditional medicinal practices. The phytoconstituents derived from plants that are responsible for management of wounds need to be identified and screened for antimicrobial activity. Topical antimicrobial therapy is one of the essential components of wound care (Esimone et al. 2008). Various in vitro assays are experimentally useful as they are quick and relatively inexpensive. In case of in vivo studies, small animals provide a multitude of model choices for various human wound conditions. A good herbal product must effectively arrest bleeding from fresh wounds, inhibit microbial growth, and accelerate wound healing (Okoli et al. 2007). Hence, enhanced wound-healing potency of potent herbal extract must be attributed to free radical-scavenging action and the antimicrobial property of the phytoconstituents present in the extract. The quicker process of healing of wounds could be a function of either an individual or the synergistic effects of bioactive molecules.

22.6 Wound Healing

Wounds are the physical injuries caused by the disruption of anatomical, cellular, and functional continuity of a living tissue. It may be produced by physical, thermal, chemical, immunological, or microbial impact to the tissue. When skin is disrupted by cut, torn, or puncture it is an open wound and when blunt force trauma causes a contusion, it is a closed wound, whereas wounds caused by fire, chemicals, electricity, heat, radiation, or sunlight come under burn wounds (Jalalpure et al. 2008). In developing countries, wounds and burn trauma are a major problem often leading to severe complications and involves significant period of hospitalization, multiple cure procedures, long-term rehabilitation, and expensive medication.

Chronic wounds are a challenging clinical condition which affects 6 million patients annually. Today there are a variety of treatment options for chronic wounds and many of them are ineffective. There is increasing number of products in this area at considerably high costs. Various herbal constituents have proven wound-healing properties. As an example, tannins could promote wound healing through free radical removal, increasing the contraction of the affected area and increasing the formation of blood vessels and fibroblasts. Other active principles such as triterpenes, alkaloids, and flavonoids have proven to be effective in this process.

Burns injuries are one of the most common and severe forms of trauma imparting major healthcare problems. Burn wound needs at most care according to the severity of burn. Healing impairment in burns was characterized by increased delayed granulation tissue formation, free-radicals-mediated damage, decreased angiogenesis, and reduced collagen reorganization ending to chronic wound healing. Burn management is entailed with long-term disability, prolonged hospitalization, with expensive medication, multi-operative procedures, and prolonged period of rehabilitation (Arturson 1996). The development of new therapeutic agent or improving the efficacy of existing medicine is a time in need to the benefit of society in general (Shalom et al. 2011).

Healing of skin tissue is fundamentally a connective tissue response which involves a well-organized cascade of dynamic process involving orderly progression of complex interactions of cellular, biochemical, and physiological cascades leading to the restoration of disrupted anatomical continuity that promotes tissue repair and regeneration at the wound site (Singer and Clark 1999; Aarabi et al. 2007; Gurtner et al. 2008; Tanaka and Galliot 2009; Reinke and Sorg 2012). The activity of wound healing is an intricate network of blood cells, cytokines, and growth factors, which aids in the restoration of the injured skin or tissue to its normal condition (Arturson 1996; Sultana and Anwar 2008; Sánchez et al. 2009). Wound care is important to shorten the difficulties faced during the duration of treatment for wound patients and to help regenerate tissues. A potent natural wound care capable of cellular reformation at the wounded site must exclude secondary infections and minimize pain; discomfort and scarring in order to promote healing in the shortest time possible (Percival 2002).

After a wound occurs, various categories of cells are engaged to participate in the repairing process (Werner and Grose 2003). Fibroblasts are one of the connective tissue components, which play a significant role in wound healing. During the injury period, inflammatory cells were proliferated and migrated to the wound environment, which stimulates the fibroblasts. These fibroblasts are sticking on wound margins and start to grow (Clark 1989, 1993). The matured fibroblasts were migrated into provisional matrix formation in the wounded area and developed to form granulation tissue where it synthesizes new extracellular matrix (Spyrou et al. 1998; Liu et al. 2012; Desmoulière et al. 2005; Guo and DiPietro 2010; Clark 2013).

HPR is an imino acid released by a posttranslational product of proline. It is primarily found in the collagen fibers of the connective tissue for the stabilization of helical structure and also for the assembly of collagen molecules into micro fibrils. By the evaluation of HPR content, the metabolism and regulation of collagen can be studied (Nemethy and Scheraga 1986; Reddy and Enwemeka 1996). Also, the HPR value can be used as marker to determine the collagen synthesis (Reddy et al. 2008). Hexosamine and hexuronic acids are matrix molecules that form the ground substratum for the synthesis of new ECM. The glycosaminoglycans stabilize the collagen fibers by strengthening electrostatic and ionic interactions with it and by controlling their resulting alignment and characteristic size (Thakur et al. 2011). They are also identified as important determinants of cellular responsiveness in development, homeostasis, and disease due to their ability to bind and alter protein–protein interactions (Trowbridge and Gallo 2002).

Growth factors are a cluster of hormone-like polypeptides that played a vital role in various phases of wound healing (Yu et al. 1994). The peptide growth factors such as TGF-β1, PDGFα, and VEGF are released from injured as well as inflammatory cells at the onset of occurrence of tissue damage (Cromack et al. 1990; Rappolee et al. 1988; Wahl et al. 1989). During the proliferation phase of tissue repair system, various secreted chemotactic molecules and growth factors, including transforming growth factor-β (TGF-β) as well as platelet-derived growth factor (PDGF), control motion of macrophages/fibroblasts (Kondo and Ishida 2010; Schiller et al. 2004). Immediately after wounding, TGF-β 1 is released in large amounts from platelets

(Assoian et al. 1983). TGF-β is a prominent pro-fibrogenic growth factor which plays a critical role in regeneration of damaged tissue and regulates different stages of wound healing (Amento and Beck 1991). Most of the cells involved in wound healing express TGF-β1, which strongly serves as a chemoattractant for neutrophils, macrophages, and fibroblasts to the site of injury and helps in fibroblast proliferation and ECM production (Chen and Davidson 2005) and these cell types further enhance TGF-β1 amount in various cell types. This growth factor induces the expression of smooth muscle actin and promotes the synthesis and maturity of collagen-I in addition to its action on reducing collagen degradation by inhibiting the expression of MMP-1 and increasing the production of tissue inhibitors of metalloproteinases (TIMP-1) (Huang et al. 2012; Simon et al. 2012). It also maintains a balanced state between synthesis and degradation of ECM via increased synthesis and decreased proteolytic activity of matrix components (Wells 2000). When wound repair is completed, the activity of TGF-β1 is normally turned off.

PDGF is released from platelets/monocytes/macrophages and is responsible for stimulating the proliferation and migration of mesenchymal cell (Bauer et al. 1985; Shimokado et al. 1985). PDGFs comprise of a family of homo- or heterodimeric growth factors (Kischer 1992). After binding to three different transmembrane tyrosine kinase receptors, which are homo- or heterodimers of an α- and a β–chain, they start their functions (Heldin et al. 2002; Heldin and Westermark 1999). PDGF is the first growth factor that acts as a chemoattractant for cells migrating into the healing skin wound like monocytes, neutrophils, and fibroblasts. In these cells it also amplifies fibroblasts proliferation and production of ECM. Furthermore, based on its expression pattern in the healing wound, PDGF has been suggested to have two major but distinct roles in wound repair: an early function to stimulate fibroblast proliferation for contraction of collagen matrices and a later activity is to induce the myofibroblast phenotype in these cells (Clark 1993; Heldin and Westermark 1999). A series of experimental and clinical studies have demonstrated PDGFs to be a major player in wound healing and have proved the beneficial effect of PDGF for the treatment of wound-healing disorders (Heldin and Westermark 1999). In addition, it was the first growth factor to be approved for the treatment of human ulcers (Mandracchia et al. 2001; Embil and Nagai 2002).

VEGF is an important pro-angiogenic cytokine that is capable of enhancing angiogenesis during wound healing through proliferation of endothelial cell, migration, and adhesion of leukocytes, thereby indorse reepithelialization as well as collagen deposition in the wound area. Delayed or aberrant revascularization at the wound area enhances the etiology of chronic wounds (Gupta et al. 2008). In order to support wound repair after cutaneous injury expression of VEGF gene was found to be strongly induced with keratinocytes and macrophages being the major producers (Brown et al. 1992; Frank et al. 1995). Furthermore, its receptors were seen on the blood vessels of the granulation tissue (Lauer et al. 2000; Peters et al. 1993). This pattern of expression of VEGF suggested that it is efficient in wound angiogenesis in a paracrine manner.

The healing process consists of three different phases such as inflammatory phase, proliferative phase, and the remodeling phase that determines the strength and appearance of the healed tissue (Sumitra et al. 2005; Stroncek and Reichert 2008; de Oliveira Gonzalez et al. 2016). These processes of wound healing must progress in a predictable, timely manner, any delay influenced by key factors such as infections, nutrition, hormones and drugs, sites and type of wound, and some disease conditions may cause inappropriate healing, it can lead to either a chronic wound such as a venous ulcer or pathological scarring such as a keloid scar (Martin 1997; Guo and DiPietro 2010; Anderson and Hamm 2014). Nonhealing wounds cause little discomfort to persons and without proper treatment they get prone to microbial attack and become chronic (Dash et al. 2018). Therefore, wound care and maintenance have to be given proper attention that involves a number of measures including dressing and administration of pain relievers, use of anti-inflammatory products, application of systemic topical antimicrobial agents, and healing imparting drugs.

The inflammatory responses are necessary for the initiation of creating appropriate environment in the wounded area for initiation of process of repairing. If the inflammatory responses are severe, it adversely delays the wound repairing cascade (dos Santos Gramma et al. 2016). At the time of injury, the release of these inflammatory mediators plays a significant role in the wound site such as trigger the proliferation of keratinocyte and fibroblast cells, proper direction of the immune response, and synthesis and degradation of extracellular matrix proteins (Bootun 2012).

The main inflammatory response of post-wound, mainly the leukocyte migration is essential to restore homeostasis. So, the quantification of inflammatory cells such as neutrophils and macrophages from the healed skin tissue was analyzed by estimating their specific activity enzymes like MPO, seen in the azurophilic granules of neutrophils and N-acetyl-ß-D-glucosaminidase (NAG), found in lysosomes of activated macrophages, respectively.

The proliferative phase consists of flow of events that decides wound healing power of a drug or natural product. It is characterized by angiogenesis, granulation tissue formation, collagen deposition, epithelialization, and later leading to wound contraction. Angiogenesis is the formation of new blood vessel growth from endothelial cells. During fibroplasia and granulation tissue formation, the fibroblasts secrete collagen and fibronectin that helps to form extracellular matrix. Further, epithelial cells move across the wounded area to seal it which is later on contracted by myofibroblasts, imparting grip for the wound edges and undergo wound contraction using a similar mechanism as that in smooth muscle cells (Nayak et al. 2007a, b).

In normal conditions, wound healing is a multistep process that is guided through the interconnected multiple signaling pathways. Various cytokines and chemokines from endothelial cells, macrophages, platelets, keratinocytes, and fibroblasts put forth in the proper healing of wounds. The vascular tissues in the injured area lead to impaired oxygen deficiency. This hypoxial condition aggravates a rapid influx of inflammatory cells and plays a crucial role in granulation and reepithelialization (Bosco et al. 2008). Therefore, acute hypoxia induces a momentary rise in cellular replication and plays to start healing process (Davidson and Mustoe 2001; Hong et al. 2014; Tandara and Mustoe 2004). This hypoxic gradient in the stromal cell trigger proliferation of fibroblasts, macrophages, and epithelial cells, which release VEGF (Falanga and Kirsner 1993). This growth factor migrates to bone marrow, where it undergoes phosphorylation and activates eNOS. The increased eNOS produces elevated NOS activity, thereby produce NO, which stimulates bone marrow–derived circulating endothelial progenitor cells (EPCs) and passes to stroma (Yoder and Ingram 2009; Morris et al. 2010). Stromal cell-derived factor-1 alpha (SDF 1α), a chemokine which promotes the homing of EPCs to the injury site, where they have neovascularization, formation of new blood vessels at the site of injury (Guo et al. 2015).

Diabetics are the results of abnormal insulin levels, which is required for the breakdown of glucose into energy molecule. The maintenance of proper balance of insulin is very vital for lowering the blood glucose. The high amount of glucose in the blood interferes with the proper healing of impaired tissues. Various factors such as hyperglycemia (Goodson and Hunt Goodson III and Hunt 1979), neuropathy (Frykberg and Mendeszoon 2000), depletion of growth factors (Greenhalgh 1996), microbial infections (Greenhalgh 2003), uremia (Yue et al. 1987), edema, and microvascular diseases (Shimomura and Spiro 1987) all these impair in proper healing of injuries in diabetic situations. Thus, proper management of blood glucose level is essential to cure wound especially on the diabetic conditions. Low blood glucose also impairs the function of WBC, resulting in an inability to fight microbes.

In diabetic conditions, the endothelial nitric oxide synthase (eNOS) activation is impaired, which limits the release of EPCs mobilization from the bone marrow, which directly affects the angiogenesis process. The SDF 1α expression is also decreased in fibroblast cells and epithelial cells in diabetic wound, which prevents the migration of EPCs homing to injured sites, as a result the healing is delayed (Brem and Tomic-canic 2007; Gallagher et al. 2007; Lerman et al. 2003).

22.7 Botanicals in Wound Care Management

Wound healing (cicatrization) is an intricate anabolic progression via., restoration of tissue or skin integrity after damage, which is due to the involvement of cell and matrix signaling, which involves four overlapping phases such as hemostatics, inflammation, proliferation, and remodeling (Davydov et al. 1991; Kirsner and Eaglstein 1993). Vasoconstriction and platelet aggregation are the initial steps to induce blood clotting followed by the vasodilatation and phagocytosis making the starting point of wound healing. If any delay occurs, the infected/injured tissue undergoes necrosis, which interferes with the proper healing mechanisms via inadequate supply of blood, lymphatic blockage, bacterial infection, and diabetes (Puratchikody et al. 2006; Sahu et al. 2010. The inflammatory reactions start immediately after the injury, followed by the triggering of hemostatic mechanisms. The sequential, overlapping biochemical cascade of acute and chronic wound healing is orchestrated by signature events and cells, and their molecular regulators.

In the proliferative phase, fibroblasts produce glycosaminoglycans and collagen which are necessary for wound healing. Shenoy et al. (2009) reported that both alcoholic and aqueous extracts of leaves of *Hyptis suaveolens* (L.) Poit, a traditional medicinal herb significantly increased collagen content in the proliferative phase of wound repair in rats. Petroleum ether extracts showed faster wound contraction and period of epithelization, high hydroxyproline content, and tensile strength. The tensile strength of the healed tissue was accelerated via the cross linking of collagen by vitamin C-dependent hydroxylation.

The aqueous leaf extract of *Hippophae rhamnoides* L. (Elaeagnaceae) was studied for its action on burn wounds in rats. The 5.0% w/w leaf extract ointment showed faster reduction in wound area, increased the level of hydroxyproline and hexosamine content. The regulation of collagen type-III, matrix metalloproteinases (MMP)-2 and 9 expressions further established the healing efficacy. In this study, significant increase in both enzymatic and nonenzymatic antioxidants and decreased malondialdehyde (MDA) levels were observed. The extract also increased angiogenesis and upregulated the vascular endothelial growth factor (VEGF) expression (Upadhyay et al. 2011).

There is immense evidence reported to illustrate various phytoconstituents that are available at affordable, safe, and have significant effect for treating infections and wounds (Grierson and Afolayan 1999; Aiyegoro et al. 2009). Seventy percent of the commercialized products contain active chemical compounds from herbals, rest 20% are from mineral base, and 10% are of animal based and more than 13,000 have been proved to support wound healing (Patel et al. 2019) (Table 22.1).

22.8 Essential Oils

Essential oils are the volatile, low-molecular weight secondary metabolites produced from aromatic plants (Aharoni et al. 2005). The major pathways of origin of these oils are terpenic pathway and non-terpenic compounds by the phenylpropanoids pathway (eugenol, cinnamaldehyde, and safrole). The essential oils are localized in cytoplasm of unicellular epidermal secretory hairs and internal secretory cells of plants. These essential oils are complex mixtures of more than 300 different compounds. Various chemical classes of essential oils are alcohols (Geraniol, α-bisabolol), ethers or oxides, aldehydes (citronellal, sinensal), ketones (menthone, p-vetivone), esters (γ-tepinyl acetate, Cedryl acetate), amines, amides, phenols (thymol), heterocycles, and terpenes. Essential oils were collected from flowers (orange, pink, and lavender), flower bud (clove), bracts (Ylang-ylang), leaves (eucalyptus, mint, thyme, bay leaf, savoy, sage) needles (pine), roots (vetiver), rhizomes (ginger and sweet flag), seeds (carvi and coriander), fruits (fennel and anise), epicarps (citrus), wood (sandalwood and rosewood), and bark (Cinnamon).

Table 22.1 List of ethnomedicinally important plant species and phytoconstituents in wound management

Plant species (Family)	Phytoconstituents	Treatment	References
Aloe vera (L.) Burm.f. (Xanthorrhoeaceae)	Pyrocatechol, anthraquinone, oleic acid, phytol, glucomannan Mannose-6-phosphate, tannic acid	Microbial infections, skin irritations, burn healing	Davis et al. (1994), Pugh et al. (2001), Yusuf et al. (2004), Tanaka et al. (2006), Pandey and Mishra (2010)
Lantana camara L. (Verbenaceae)	22β-acetoxylantic acid, tetradecane, isocaryophyllene, lantadene-A, bicyclogermacrene	Bacterial infection, tissue repair processes	Barre et al. (1997), Dash et al. (2001), Nayak et al. (2009a, b), Karakaş et al. (2012), Passos et al. (2012)
Equisetum arvense L. (Equisetaceae)	Onitin-9-O-glucoside, apigenin, luteolin, rosmarinic acid, hexahydrofarnesyl acetone	Free radical scavenging, skin injuries	Holzhuter et al. (2003), Milovanović et al. (2007)
Mussaenda frondosa L. (Rubiaceae)	Isovanillin, β sitosterol-glucoside, hyperin, ferulic and sinapic acid	Microbial infections, burn healing	Basavaraja et al. (2011), Patil et al. (2012)
Camellia sinensis (L.) Kuntze (Theaceae)	Epigallocatechin3-gallate, cetophenone, afzelechin, arbutin, camellianin, thymol, theophylline	Inflammation, acceleration of keratinocyte cell differentiation for wound-healing process	Donà et al. (2003), Gordon and Wareham (2010), Shi et al. (2011)
Alternanthera sessilis (L.) R.Br. ex DC. (Amaranthaceae)	β-sitosterol, β-carotene, palmitic acid, and linoleic acids	Immunomodulation, wound-healing activity	Guerra et al. (2003), Jalalpure et al. (2008)
Calendula officinalis L. (Compositae)	Umbelliferone, esculetin, zeaxanthin, sitosterols, Stigmasterols, Faradiol 3-O-laurate, palmitate	Microbial infections, inflammation, lipid peroxidation, proliferation, and migration of fibroblasts	Zitterl-eglseer et al. (1997), Chandran and Kuttan (2008), Fronza et al. (2009)
Chromolaena odorata (L.) R.M. King and H. Rob. (Compositae)	Rutin, isoquercitrin, a-amyrin, geijerene, β-cubebene, trans-caryophyllene, delta-humulene	Protection of skin cells against oxidative damage, skin infections, wounds, burns	Phan et al. (2001), Thang et al. (2001)
Carica papaya L. (Caricaceae)	Vitamins B and C, chymopapain, papain, danielone, carpaine protocatechuic acid, oleic acid,	Microbial infections, burn healing	Canini et al. (2007), Anuar et al. (2008)

Plant (Family)	Phytoconstituents	Activity	References
Azadirachta indica A. Juss. (Meliaceae)	Margosic acid, azadirachtin, nimbin, salannin, α-terpinene, vitamin E, trace valeric acid	Microbial infections, tissue swellings, hepatoprotection, wound healing	Biswas et al. (2002), Chattopadhyay (2003), Rajasekaran et al. (2008)
Matricaria chamomilla L. (Compositae)	Luteolin, alpha-bisabolol, herniarin, chamazulene, Chamazulene	Inflammation, wound healing	Avallone et al. (2000), Martins et al. (2009)
Ginkgo biloba L. (Ginkgoaceae)	Ginkgolides, aglycones, isorhamnetin, bilobol	Inhibition of free radicals, efficient in wound breaking strength and hydroxyproline content of granulation tissue	Hasler et al. (1992), Bairy and Rao (2001), Trompezinski et al. (2010)
Cinnamomum verum J. Presl (Lauraceae)	Cinnamaldehyde, 2-hydroxycinnamaldehyde	Diabetic wound healing, repairment of osteoporotic bone defect.	Yuan et al. (2018), Seyed et al. (2019), Weng et al. (2019)
Anethumgraveolens L. (Apiaceae)	Cis-carvone, limonene, α-phellandrene, and anethofuran α-Phellandrene	Infectious wound healing.	Manzuoerh et al. (2019)
Eucalyptus globulus Labill. (Myrtaceae)	Cineole, camphene, and phellandrene, citronellal	Essential oil loaded lipid nanoparticles in wound healing.	Hukkeri et al. (2002), Saporito et al. (2018), Akkol et al. (2017): Alam et al. (2018)
Trigonellafoenum-graecum L. (Leguminosae)	Diosgenin, Yamogenin, Gitogenin, Tigogenin, and Neotigogens 4-Hydroxyisoleucine	Wound healing	Taranalli and Kuppast (1996), Muhammed and Salih (2012)
Nelumbo nucifera Gaertn. (Nelumbonaceae)	Neferine	Osteolytic bone conditions	Mukherjee et al. (2000), Chen et al. (2019)
Matricaria chamomilla L. (Compositae)	Chamazulene, α bisabolol, Bisabolol oxides, Spiroethers, and flavonoids	Linear incisional wound healing, wound dressing mats based on electrospun nanofibrous	Jarrahi et al. (2010), Duarte et al. (2011), Motealleh et al. (2014)
Aegle marmelos (L.) Correa (Rutaceae)	Omethylhalfordional and Isopentylhalfordinol	Wound healing	Azmi et al. (2019), Gautam et al. (2014), Jaswanth et al. (1994)
Linum usitatissimum L. (Linaceae)	α-Linolenic acid	Burn wound healing	Trabelsi et al. (2019), Beroual et al. (2017), Paul-Victor et al. (2017)

(continued)

Table 22.1 (continued)

Plant species (Family)	Phytoconstituents	Treatment	References
Rosmarinus officinalis L. (Lamiaceae)	Terpenoids, limonene, 1, 8-cineol, Carnosic acid, Rosemarinic acid, and α-pinene	Excision cutaneous wounds in diabetic conditions, infected wound healing.	Abu-Al-Basal (2010), Nejati et al. (2015), Khezri et al. (2019)
Allium sativum L. (Amaryllidaceae)	Alliin, Allyl cysteine, Allyl disulfide, and Allicin	Wound healing	Farahpour et al. (2017)
Vitis vinifera L. (Vitaceae)	Vitamin E, linoleic acid phenolic acids, Stilbenes, and Anthocyanins	Burn wound healing in rabbits	Hemmati et al. (2011)
Curcuma longa L. (Zingiberaceae)	Curcumin, Curcuminoids Turmerone, Atlantone and Zingiberene	Woundgauzes augmented the granule Curcumin cross-linked with chitosan-PVA membranes and fibrous connective tissues formation, muscle healing	Abbas et al. (2019), Kazemi-Darabadi et al. (2019)
Aspilia africana (Pers.) C.D. Adams (Asteraceae)	α-pinene, carene, Phytol, and Linolenic acid	Reduce wound bleeding	Komakech et al. (2019)
Pistacia atlantica Desf. (Anacardiaceae)	Resin oil	Burn healing	Shahouzehi et al. (2019)
Astragalus propinquus Schischkin (Leguminosae)	Polysaccharide	Wound healing via inhibiting inflammation, accelerating cell cycle, and promoting the secretion of repair factors.	Zhao et al. (2017)
Morinda citrifolia L. (Rubiaceae)	Epicatechin Gallate, Scopoletin	Full thickness incisional wound healing	Kapoor et al. (2004), Nayak et al. (2007a, b)
Lindera erythrocarpa Makino (Lauraceae)	Lucidone	Wound healing through keratinocyte/fibroblast/endothelial cell growth and migration and macrophage inflammation via PI3K/AKT, Wnt/β-catenin and NF-κB signaling cascade activation.	Yang et al. (2017)

Lithospermum erythrorhizon Siebold and Zucc. (Boraginaceae)	Shikonin	Periodontal tissue wound healing	Imai et al. (2019)
Trifolium canescens Willd. (Leguminosae)	Isoflavones	Excision and incision type wound healing	Renda et al. (2013)
Genista tinctoria L. (Leguminosae)	Genistein aglycone	Skin wound healing in postmenopausal women	Marini et al. (2010)
Medicago sativa L. (Leguminosae)	Phytoestrogens and chlorophylls	Developing nanofibers bioscaffolds for regenerative wound dressings	Ahn et al. (2019)
Centella asiatica (L.) Urb. (Apiaceae)	Madecassoside, Asiaticoside	Burn wound healing	Wu et al. (2012), Bylka et al. (2013)

22.9 Bioprospecting in Wound Healing

The whole process of systematic search, development, and commercialization of a novel product or chemical compounds from natural resources is called biodiversity prospecting or bioprospecting. This field gives major significance in the diverse interdisciplinary sectors such as agriculture, food and medicine, bioremediation, and pharmaceuticals. Documentation and scientific validation of traditional knowledge existing in diverse cultures if exposed to prospecting can produce prime leads in evolving novel drugs/molecules of great trade value. Plants in medicinal biodiversity have served as one of the moneyed sectors of bioprospecting which leads to the novel drug discoveries that have greatly contributed to human health (Cragg and Newman 1999). Diverse list of medicinal plants was documented from tribal areas which showed potent wound healing property; however, a few of them are commercially successful in the market (Rajith et al. 2016; Shedoeva et al. 2019). A traditional plant species *Passiflora edulis* Sims (Passifloraceae) was extracted, characterized, and incorporated in to a chitosan membranes for potential application as wound dressing (Melo et al. 2019). In addition to promoting economic development of the host country, bioprospecting is often associated with sustainability and the preservation of local biodiversity. A good number of drugs based on traditional knowledge had been developed through bioprospecting (Table 22.2).

22.10 Future Scenario

The future of herbal-based phytoconstituents is very challenging, particularly in the field of extraction of phytoconstituents, biocompatibility with the cells of the targeted tissue, lack of toxicity, and clinical studies. The shortage of bioresources due to its unscientific harvesting practices, climatic changes etc., are worldwide problems faced by natural chemists which directly affect the new scientific explorations in synthesis, production, and marketing of herbal-based drugs. There are immense phytocompounds which provide significant wound healing efficacy, even though, the commercialization of these compound-based drugs especially either in the form of an ointment or cellulosic wound beds are very less. The herbal based Nano scaffolds in skin tissue engineering and wounds repair as a material for advanced wound dressings coupled with drug delivery, transparency and sensorics. (Adeli-Sardou et al. 2019; Bacakova et al. 2019; Ezhilarasu et al. 2019; Ghaseminezhad et al. 2019; Moura et al. 2013; Shahbazi and Bahrami 2019; Tiwari et al. 2019). Keratinocyte growth factor, transforming growth factor, beta, and vascular endothelial growth factors are the most intriguing line of target sites for reepithelialization, matrix deposition, and angiogenesis respectively and they stand out as future candidates for development (Appleton 2003). The bioavailability of these marker growth factors can enhance newly available perioperative drug delivery systems (Demidova-Rice et al. 2012), which have a very short in vivo half-life after topical administration (Yamakawa and Hayashida 2019). The delivery of growth

Table 22.2 Bioprospecting of drugs obtained from plants (market name is indicated in parenthesis)

Drug molecule	Plant species	Family	Uses	References
Morphine (Roxanol)	*Papaver somniferum* L.	Papaveraceae	Narcotic analgesics	Clark et al. (2007)
Acetyl salicylic acid (Anacin, Acupri)	*Spiraea altaica* Pall.	Rosaceae	Analgesic, anti-inflammatory and antipyretic	Sudhof (2001)
Quinine (Qualaquin)	*Cinchona officinalis* L.	Rubiaceae	Anti-malarial	Ballestero et al. (2005)
Artmisinin (coartem)	*Artemisia annua* L.	Asteraceae	Anti-malarial	White (1997)
Vinblastin (Alkaban–AQ)	*Catharanthus roseus* (L.) G. Don	Apocynaceae	Anti-cancer	Jain and Kumar (2012).
Paclitaxel (Abraxane, Taxol)	*Taxus brevifolia* Nutt.	Taxaceae	Anti-cancer, antiviral	Koehn and Carter (2005)
Reserpine (Diupres-2)	*Rauvolfiaserpentina* (L.) Benth. ex Kurz	Apocynaceae	Anti-hypertensive activity	Baumeister et al. (2003)
Digitoxin	*Digitalis purpurea* L.	Plantaginaceae	Antiarrhythmic agent	Srivastava et al. (2004)
Warfarin	*Melilotus officinalissubsp. alba* (Medik.) H. Ohashi and Tateishi	Leguminosae	Anticoagulant effect	Hirsh et al. (2007)
Prostratin	*Homalanthus nutans* (G. Forst.) Guill.	Euphorbiaceae	Anti-viral	Wender et al. (2008)

factors via gene therapy is an emerging treatment option for wound repair (Desmet et al. 2018; Shi et al. 2018). A paradigm shift is needed to enhance the effective translation of new phytoconstituents into clinical therapies as well as to understand the pathophysiology of signaling pattern. A modern clinical translation research in the postoperative surgery is the development of genetically modified microalgae-based bioactivated surgical sutures (Centeno-Cerdas et al. 2018), which help to release both oxygen and recombinant growth factors directly into the wound area. This new generation of surgical sutures warranted further clinical studies to elucidate the effectiveness and well-being.

22.11 Conclusions

Secondary metabolites from botanicals provide immense opportunity to develop new type of natural moieties, and the topical ointments developed by these are helpful in enhancing repair mechanism of wounds. The documentation of traditional knowledge about plants, drug development process, and method of application for healing purpose are vital to find out the new natural resource for wound care. Concurrently, the scientific validation of raw materials and usages were mandatory for marketing new herbal drugs in wound treatment, which has got enormous scope for commercialization. Future studies aimed to elucidate the active chemical molecules from herbal preparations help incorporation of the same into suitable cellulosic materials or nanocomposites as a bedding for covering wounded surfaces. This ensures rapid therapeutic action in wound healing. The traditional knowledge system is fast eroding and there is an urgent need to preserve and conserve the rare and endangered species. Further, adequate caution should be taken to categorize the threatened status of plants at regional scale (Aravind et al. 2004). Ethnomedicinal plants should be cultivated in herbal gardens, agroforestry systems, and home gardens to encourage their sustainable utilization and hence conservation. Assessment of the populations of threatened species, development of an appropriate strategy, action plan for the conservation, and sustainable utilization of such components of plant diversity are highly recommended (Samant and Pant 2006).

References

Aarabi S, Longaker MT, Gurtner GC (2007) Hypertrophic scar formation following burns and trauma: new approaches to treatment. PLoS Med 4(9):e234

Abbas M, Hussain T, Arshad M, Ansari AR, Irshad A, Nisar J, Hussain F, Masood N, Nazir A, Iqbal M (2019) Wound healing potential of curcumin cross-linked chitosan/polyvinyl alcohol. Int J Biol Macromol 140:871–876. https://doi.org/10.1016/j.ijbiomac.2019.08.153

Abu-Al-Basal MA (2010) Healing potential of *Rosmarinus officinalis* L. on full-thickness excision cutaneous wounds in alloxan-induced-diabetic BALB/c mice. J Ethnopharmacol 131 (2):443–450. https://doi.org/10.1016/j.jep.2010.07.007

Adeli-Sardou M, Yaghoobi MM, Torkzadeh-Mahani M, Dodel M (2019) Controlled release of lawsone from polycaprolactone/gelatin electrospun nano fibers for skin tissue regeneration. Int J Biol Macromol 124:478–491. https://doi.org/10.1016/j.ijbiomac.2018.11.237

Aharoni A, Jongsma MA, Bouwmeester HJ (2005) Volatile science? Metabolic engineering of terpenoids in plants. Trends Plant Sci 10(12):594–602

Ahn S, Ardoña HAM, Campbell PH, Gonzalez GM, Parker KK (2019) Alfalfa nanofibers for dermal wound healing. ACS Appl Mater Interfaces 11(37):33535–33547. https://doi.org/10.1021/acsami.9b07626

Aiyegoro OA, Afolayan AJ, Okoh AI (2009) Synergistic interaction of *Helichrysum pedunculatum* leaf extracts with antibiotics against wound infection associated bacteria. Biol Res 42 (3):327–338. https://doi.org/10.4067/S0716-97602009000300007

Akkol EK, Tumen I, Guragac FT, Keles H, Reunanen M (2017) Characterization and wound repair potential of essential oil *Eucalyptus globulus* Labill. 9th Annual European Pharma Congress Fresenius Environmental Bulletin 26(11):6390–6399. https://doi.org/10.4172/2167-7689-C1-025

Alam P, Shakeel F, Anwer MK, Foudah AI, Alqarni MH (2018) Wound healing study of Eucalyptus essential oil containing nanoemulsion in rat model. J Oleo Sci:ess18005. https://doi.org/10.5650/jos.ess18005

Amento EP, Beck LS (1991) TGF-beta and wound healing. Ciba Found Symp 157:115–129

Anderson K, Hamm RL (2014) Factors that impair wound healing. J Am Coll Clin Wound Spec 4(4):84–91

Aneesh TP, Hisham M, Sekhar S, Madhu M, Deepa TV (2009) International market scenario of traditional Indian herbal drugs–India declining. Int J Green Pharm (IJGP) 3(3)

Anuar NS, Zahari SS, Taib IA, Rahman MT (2008) Effect of green and ripe *Carica papaya* epicarp extracts on wound healing and during pregnancy. Food Chem Toxicol 46(7):2384–2389

Anyinam C (1995) Ecology and ethnomedicine: exploring links between current environmental crisis and indigenous medical practices. Soc Sci Med 40:321–329

Aravind NA, Manjunath J, Rao D, Ganeshaiah KN, Shaanker RU, Vanaraj G (2004) Are red-listed species threatened? A comparative analysis of red-listed and non-red-listed plant species in the Western Ghats, India. Curr Sci 88:258–265

Appleton I (2003) Wound healing: future directions. IDrugs 6(11):1067–1072

Arturson G (1996) Pathophysiology of the burn wound and pharmacological treatment. The Rudi Hermans lecture, 1995. Burns 22(4):255–274

Assoian RK, Komoriya A, Meyers CA, Miller DM, Sporn MB (1983) Transforming growth factor-beta in human platelets. Identification of a major storage site, purification, and characterization. J Biol Chem 258(11):7155–7160

Avallone R, Zanoli P, Puia G, Kleinschnitz M, Schreier P, Baraldi M (2000) Pharmacological profile of apigenin, a flavonoid isolated from *Matricaria chamomilla*. Biochem Pharmacol 59(11):1387–1394

Azmi L, Shukla I, Goutam A, Rao CV, Jawaid T, Kamal M, Awaad AS, Alqasoumi SI, AlKhamees OA (2019) In vitro wound healing activity of 1-hydroxy-5, 7-dimethoxy-2-naphthalene-carboxaldehyde (HDNC) and other isolates of Aegle marmelos L.: enhances keratinocytes motility via Wnt/β-catenin and RAS-ERK pathways. Saudi Pharm J 27(4):532–539. https://doi.org/10.1016/j.jsps.2019.01.017

Bacakova L, Pajorova J, Bacakova M, Skogberg A, Kallio P, Kolarova K, Svorcik V (2019) Versatile application of nanocellulose: from industry to skin tissue engineering and wound healing. Nano 9(2):164. https://doi.org/10.3390/nano9020164

Bairy KL, Rao CM (2001) Wound healing profiles of *Ginkgo biloba*. J Nat Remed 1(1):25–27

Ballestero JA, Plazas PV, Kracun S, Gómez-Casati ME, Taranda J, Rothlin CV, Katz E, Elgoyhen AB (2005) Effects of quinine, quinidine, and chloroquine on alpha9alpha10 nicotinic cholinergic receptors. Mol Pharmacol 68(3):822–829. https://doi.org/10.1124/mol.105.014431

Barre JT, Bowden BF, Coll JC, De Jesus J, Victoria E, Janairo GC, Ragasa CY (1997) A bioactive triterpene from *Lantana camara*. Phytochemistry 45(2):321–324

Basavaraja BM, Vagdevi HM, Srikrishna LP, Hanumanthapa BC, Joshi SD, Vaidya VP (2011) Antimicrobial and analgesic activities of various extracts of *Mussaenda frondosa* L. bark. J Glob Pharma Technol 3(2):14–17

Bauer EA, Cooper TW, Huang JS, Altman J, Deuel TF (1985) Stimulation of in vitro human skin collagenase expression by platelet-derived growth factor. Proc Natl Acad Sci 82(12):4132–4136. https://doi.org/10.1073/pnas.82.12.4132

Baumeister AA, Hawkins MF, Uzelac SM (2003) The myth of reserpine-induced depression: role in the historical development of the monoamine hypothesis. J Hist Neurosci 12(2):207–220. https://doi.org/10.1076/jhin.12.2.207.15535

Beroual, K., Agabou, A., Abdeldjelil, M.C., Boutaghane, N., Haouam, S. and Hamdi-Pacha, Y., 2017. Evaluation of crude flaxseed (*Linum usitatissimum* L.) oil in burn wound healing in New Zealand rabbits. Afr J Trad Complement Altern Med. 14(3):280-286. Doi: https://doi.org/10.21010/ajtcam.v14i3.29

Biswas K, Chattopadhyay I, Banerjee RK, Bandyopadhyay U (2002) Biological activities and medicinal properties of neem (*Azadirachta indica*). Curr Sci 82(11):1336–1345

Bootun R (2012) Effects of immunosuppressive therapy on wound healing. Int Wound J 10 (1):98–104

Bosco MC, Puppo M, Blengio F, Fraone T, Cappello P, Giovarelli M, Varesio L (2008) Monocytes and dendritic cells in a hypoxic environment: spotlights on chemotaxis and migration. Immunobiology 213(9–10):733–749. https://doi.org/10.1016/j.imbio.2008.07.031

Brem H, Tomic-Canic M (2007) Cellular and molecular basis of wound healing in diabetes. J Clin Invest 117(5):1219–1222. https://doi.org/10.1172/JCI32169

Brown LF, Yeo K, Berse B, Yeo TK, Senger DR, Dvorak HF, Van De Water L (1992) Expression of vascular permeability factor (vascular endothelial growth factor) by epidermal keratinocytes during wound healing. J Exp Med 176(5):1375–1379

Builders PF, Kabele-Toge B, Builders M, Chindo BA, Anwunobi PA, Isimi YC (2013) Wound healing potential of formulated extract from Hibiscus sabdariffa calyx. Indian J Pharm Sci 75 (1):45

Bylka W, Znajdek-Awiżeń P, Studzińska-Sroka E, Brzezińska M (2013) *Centellaasiatica* in cosmetology. Postepy Dermatol Alergol 30(1):46. https://doi.org/10.5114/pdia.2013.33378

Canini A, Alesiani D, D'Arcangelo G, Tagliatesta P (2007) Gas chromatography—mass spectrometry analysis of phenolic compounds from *Carica papaya* L. leaf. J Food Compos Anal 20 (7):584–590

Centeno-Cerdas C, Jarquín-Cordero M, Chávez MN, Hopfner U, Holmes C, Schmauss D, Machens HG, Nickelsen J, Egaña JT (2018) Development of photosynthetic sutures for the local delivery of oxygen and recombinant growth factors in wounds. Acta Biomater 81:184–194. https://doi.org/10.1016/j.actbio.2018.09.060

Chandran PK, Kuttan R (2008) Effect of *Calendula officinalis* flower extract on acute phase proteins, antioxidant defense mechanism and granuloma formation during thermal burns. J Clin Biochem Nutr 43(2):58–64

Chattopadhyay R (2003) Possible mechanism of hepatoprotective activity of *Azadirachta indica* leaf extract: Part II. J Ethnopharmacol 89(2–3):217–219

Chen MA, Davidson TM (2005) Scar management: prevention and treatment strategies. Curr Opin Otolaryngol Head Neck Surg 13(4):242–247

Chen S, Chu B, Chen Y, Cheng X, Guo D, Chen L, Wang J, Li Z, Hong Z, Hong D (2019) Neferine suppresses osteoclast differentiation through suppressing NF-κB signal pathway but not MAPKs and promote osteogenesis. J Cell Physiol. https://doi.org/10.1002/jcp.28857

Clark RAF (1989) Wound repair. Curr Opin Cell Biol 1:1000–1008

Clark RAF (1993) Regulation of fibroplasia in cutaneous wound repair. Am J Med Sci 306 (1):42–48. https://doi.org/10.1097/00000441-199307000-00011

Clark, R.A.F, 2013. The molecular and cellular biology of wound repair. Springer Science & Business Media

Clark JD, Shi X, Li X, Qiao Y, Liang D, Angst MS, Yeomans DC (2007) Morphine reduces local cytokine expression and neutrophil infiltration after incision. Mol Pain 3:28. https://doi.org/10.1186/1744-8069-3-28

Cotton CM, Wilkie P (1996) Ethnobotany: principles and applications. Wiley, Chichester

Cragg GM, Newman DJ (1999) Discovery and development of antineoplastic agents from natural sources. Cancer Investig 17(2):153–163

Cromack DT, Porras-Reyes BEATRIZ, Mustoe TA (1990) Current concepts in wound healing: growth factor and macrophage interaction. J Trauma 30(12 Suppl):S129–S133. https://doi.org/10.1097/00005373-199012001-00026

Dash GK, Suresh P, Ganapaty S (2001) Studies on hypoglycaemic and wound healing activities of *Lantana camara* Linn. J Nat Remed 1(2):105–110

Dash BC, Xu Z, Lin L, Koo A, Ndon S, Berthiaume F, Dardik A, Hsia H (2018) Stem cells and engineered scaffolds for regenerative wound healing. Bioengineering 5(1):e.23. https://doi.org/10.3390/bioengineering5010023

Davidson JD, Mustoe TA (2001) Oxygen in wound healing: more than a nutrient. Wound Repair Regen 9(3):175–177. https://doi.org/10.1046/j.1524-475x.2001.00175.x

Davis RH, DiDonato JJ, Johnson RW, Stewart CB (1994) Aloe vera, hydrocortisone, and sterol influence on wound tensile strength and anti-inflammation. J Am Podiatr Med Assoc 84(12):614–621

Davydov I, Larichev AB, Abramov A, Men'kov KG (1991) Concept of clinico-biological control of the wound process in the treatment of suppurative wounds using vacuum therapy. Vestnikkhirurgiiimeni II Grekova 146(2):132–136

Demidova-Rice TN, Hamblin MR, Herman IM (2012) Acute and impaired wound healing: pathophysiology and current methods for drug delivery, Part 1: normal and chronic wounds: biology, causes, and approaches to care. Adv Skin Wound Care 25(7):304. https://doi.org/10.1097/01.ASW.0000416006.55218.d0

Desmet CM, Preat V, Gallez B (2018) Nanomedicines and gene therapy for the delivery of growth factors to improve perfusion and oxygenation in wound healing. Adv Drug Deliv Rev 129:262–284. https://doi.org/10.1016/j.addr.2018.02.001

Desmoulière A, Chaponnier C, Gabbiani G (2005) Tissue repair, contraction, and the myofibroblast. Wound Repair Regen 13(1):7–12. https://doi.org/10.1111/j.1067-1927.2005.130102.x

Donà M, Dell'Aica I, Calabrese F, Benelli R, Morini M, Albini A, Garbisa S (2003) Neutrophil restraint by green tea: inhibition of inflammation, associated angiogenesis, and pulmonary fibrosis. J Immunol 170(8):4335–4341

dos Santos Gramma LS, Marques FM, Vittorazzi C, de Andrade TAM, Frade MAC, de Andrade TU, Endringer DC, Scherer R, Fronza M (2016) *Struthanthus vulgaris* ointment prevents an over expression of inflammatory response and accelerates the cutaneous wound healing. J Ethnopharmacol 190:319–327

Duarte CM, Quirino MR, Patrocínio MC, Anbinder AL (2011) 2011. Effects of *Chamomilla recutita* (L.) on oral wound healing in rats. Med Oral Patol Oral Cir Bucal 16(6):e716–e721. https://doi.org/10.4317/medoral.17029

Embil JM, Nagai MK (2002) Becaplermin: recombinant platelet derived growth factor, a new treatment for healing diabetic foot ulcers. Expert Opin Biol Ther 2(2):211–218. https://doi.org/10.1517/14712598.2.2.211

Ernst E (2000) The usage of complementary therapies by dermatological patients: a systematic review. Br J Dermatol 142:857–861

Esimone CO, Nworu CS, Jackson CL (2008) Cutaneous wound healing activity of a herbal ointment containing the leaf extract of *Jatropha curcas* L.(Euphorbiaceae). Int J Appl Res Nat Prod 1(4):1–4

Ezhilarasu H, Ramalingam R, Dhand C, Lakshminarayanan R, Sadiq A, Gandhimathi C, Ramakrishna S, Bay BH, Venugopal JR, Srinivasan DK (2019) Biocompatible *Aloe vera* and tetracycline hydrochloride loaded hybrid nanofibrous scaffolds for skin tissue engineering. Int J Mol Sci 20(20):5174. https://doi.org/10.3390/ijms20205174

Fairbairn JW (1980) Perspectives in research on the active principles of traditional herbal medicine. A botanical approach: identification and supply of herbs. J Ethnopharmacol 2:99–104

Falanga V, Kirsner RS (1993) Low oxygen stimulates proliferation of fibroblasts seeded as single cells. J Cell Physiol 154(3):506–510. https://doi.org/10.1002/jcp.1041540308

Farahpour MR, Hesaraki S, Faraji D, Zeinalpour R, Aghaei M (2017) Hydroethanolic Allium sativum extract accelerates excision wound healing: evidence for roles of mast-cell infiltration and intracytoplasmic carbohydrate ratio. Braz J Pharm Sci 53(1). https://doi.org/10.1590/s2175-97902017000115079

Frank S, Hübner G, Breier G, Longaker MT, Greenhalgh DG, Werner S (1995) Regulation of vascular endothelial growth factor expression in cultured keratinocytes. Implications for normal and impaired wound healing. J Biol Chem 270(21):12607–12613. https://doi.org/10.1074/jbc.270.21.12607

Fronza M, Heinzmann B, Hamburger M, Laufer S, Merfort I (2009) Determination of the wound healing effect of Calendula extracts using the scratch assay with 3T3 fibroblasts. J Ethnopharmacol 126(3):463–467

Frykberg RG, Mendeszoon ER (2000) Charcot arthropathy: pathogenesis and management. Wounds 12(6; SUPP/B):35B–42B

Fugh-Berman A, Ernst E (2001) Herb–drug interactions: review and assessment of report reliability. Br J Clin Pharmacol 52(5):587–595

Gallagher KA, Liu ZJ, Xiao M, Chen H, Goldstein LJ, Buerk DG, Nedeau A, Thom SR, Velazquez OC (2007) Diabetic impairments in NO-mediated endothelial progenitor cell mobilization and homing are reversed by hyperoxia and SDF-1α. J Clin Invest 117(5):1249–1259. https://doi.org/10.1172/JCI29710

Gautam MK, Purohit V, Agarwal M, Singh A, Goel RK (2014) In vivo healing potential of *Aegle marmelos* in excision, incision, and dead space wound models. Sci World J 2014:740107. https://doi.org/10.1155/2014/740107

Gesler WM (1992) Therapeutic landscapes: medical issues in light of the new cultural geography. Soc Sci Med 34:735–746

Ghaseminezhad K, Zare M, Lashkarara S, Yousefzadeh M, Aghazadeh Mohandesi J (2019) Fabrication of *Althea officinalis* loaded electrospun nanofibrous scaffold for potential application of skin tissue engineering. J Appl Polym Sci, p 48587. https://doi.org/10.1002/app.48587

Goodson WH III, Hunt TK (1979) Wound healing and the diabetic patient. Surg Gynecol Obstet 149:600–608

Gordon NC, Wareham DW (2010) Antimicrobial activity of the green tea polyphenol (−)-epigallocatechin-3-gallate (EGCG) against clinical isolates of Stenotrophomonas maltophilia. Int J Antimicrob Agents 36(2):129–131

Greenhalgh DG (1996) The role of growth factors in wound healing. J Trauma Acute Care Surg 41(1):159–167

Greenhalgh DG (2003) Wound healing and diabetes mellitus. Clin Plast Surg 30(1):37–45. https://doi.org/10.1016/S0094-1298(02)00066-4

Grierson DS, Afolayan AJ (1999) Antibacterial activity of some indigenous plants used for the treatment of wounds in the Eastern Cape, South Africa. J Ethnopharmacol 66(1):103–106. https://doi.org/10.1016/S0378-8741(98)00202-5

Guerra RNM, Pereira HA, Silveira LMS, Olea RSG (2003) Immunomodulatory properties of *Alternanthera tenella* Colla aqueous extracts in mice. Braz J Med Biol Res 36(9):1215–1219

Guo SA, DiPietro LA (2010) Factors affecting wound healing. J Dent Res 89(3):219–229. https://doi.org/10.1177/0022034509359125

Guo R, Chai L, Chen L, Chen W, Ge L, Li X, Li H, Li S, Cao C (2015) Stromal cell-derived factor 1 (SDF-1) accelerated skin wound healing by promoting the migration and proliferation of epidermal stem cells. In Vitro Cell Dev Biol-Animal 51(6):578–585. https://doi.org/10.1007/s11626-014-9862-y

Gupta A, Upadhyay NK, Sawhney RC, Kumar R (2008) A poly-herbal formulation accelerates normal and impaired diabetic wound healing. Wound Repair Regen 16(6):784–790. https://doi.org/10.1111/j.1524-475X.2008.00431.x

Gurtner GC, Werner S, Barrandon Y, Longaker MT (2008) Wound repair and regeneration. Nature 453(7193):314

Hasler A, Sticher O, Meier B (1992) Identification and determination of the flavonoids from Ginkgo biloba by high-performance liquid chromatography. J Chromatogr A 605(1):41–48

Heldin CH, Westermark B (1999) Mechanism of action and in vivo role of platelet-derived growth factor. Physiol Rev 79(4):1283–1316. https://doi.org/10.1152/physrev.1999.79.4.1283

Heldin CH, Eriksson U, Östman A (2002) New members of the platelet-derived growth factor family of mitogens. Arch Biochem Biophys 398(2):284–290. https://doi.org/10.1006/abbi.2001.2707

Hemmati AA, Aghel N, Rashidi I, Gholampur-Aghdami A (2011) Topical grape (*Vitis vinifera*) seed extract promotes repair of full thickness wound in rabbit. Int Wound J 8(5):514–520. https://doi.org/10.1111/j.1742-481X.2011.00833.x

Hirsh J, O'Donnell M, Eikelboom JW (2007) Beyond unfractionated heparin and warfarin: current and future advances. Circulation 116(5):552–560. https://doi.org/10.1161/CIRCULATIONAHA.106.685974

Holmstedt BR, Bruhn JG (1995) Ethnopharmacology – a challenge. In: Schultes RE, von Reis S (eds) Ethnobotany. Evolution of a Discipline. Dioscorides Press, Portland, pp 338–343

Holzhuter G, Narayanan K, Gerber T (2003) Structure of silica in *Equisetum arvense*. Anal Bioanal Chem 376(4):512–517

Hong WX, Hu MS, Esquivel M, Liang GY, Rennert RC, McArdle A, Paik KJ, Duscher D, Gurtner GC, Lorenz HP, Longaker MT (2014) The role of hypoxia-inducible factor in wound healing. Adv Wound Care 3(5):390–399. https://doi.org/10.1089/wound.2013.0520

Huang C, Akaishi S, Ogawa R (2012) Mechanosignaling pathways in cutaneous scarring. Arch Dermatol Res 304(8):589–597. https://doi.org/10.1007/s00403-012-1278-5

Hukkeri VI, Karadi RV, Akki KS, Savadi RV, Jaiprakash B, Kuppast IJ, Patil MB (2002) Wound healing property of Eucalyptus globulus L. leaf extract. Indian Drugs 39(9):481–483. ISSN: 0019-462X

Imai K, Kato H, Taguchi Y, Umeda M (2019) Biological effects of Shikonin in human gingival fibroblasts via ERK 1/2 signaling pathway. Molecules 24(19):3542. https://doi.org/10.3390/molecules24193542

India GDP Annual Growth Rate 2018. Tradingeconomics.com (cited 22 October 2018). Available from:. https://tradingeconomics.com/india/gdp-growth-annual

Jain D, Kumar S (2012) Snake venom: a potent anticancer agent. Asian Pacific J Cancer Prev 13(10):4855–4860. https://doi.org/10.7314/apjcp.2012.13.10.4855

Jalalpure SS, Agrawal N, Patil MB, Chimkode R, Tripathi A (2008) Antimicrobial and wound healing activities of leaves of *Alternanthera sessilis* Linn. Int J Green Pharm (IJGP) 2(3)

Jarrahi M, Vafaei AA, Taherian AA, Miladi H, Rashidi Pour A (2010) Evaluation of topical *Matricaria chamomilla* extract activity on linear incisional wound healing in albino rats. Nat Prod Res 24(8):697–702. https://doi.org/10.1080/14786410701654875

Jaswanth A, Eswari Akilan D, Loganathan V, Manimaran S (1994) Wound healing activity of *Aegle marmelos*. Indian J Pharm Sci 56(1)

Kapoor M, Howard R, Hall I, Appleton I (2004) Effects of epicatechin gallate on wound healing and scar formation in a full thickness incisional wound healing model in rats. Am J Pathol 165(1):299–307. https://doi.org/10.1016/S0002-9440(10)63297-X

Karakaş FP, Karakaş A, Boran Ç, Türker AU, Yalçin FN, Bilensoy E (2012) The evaluation of topical administration of *Bellis perennis* fraction on circular excision wound healing in Wistar albino rats. Pharm Biol 50(8):1031–1037

Kazemi-Darabadi S, Nayebzadeh R, Shahbazfar AA, Kazemi-Darabadi F, Fathi E (2019) Curcumin and Nanocurcumin oral supplementation improve muscle healing in a rat model of surgical muscle laceration. Bull Emerg Trauma 7(3):292–299. https://doi.org/10.29252/beat-0703013

Khezri K, Farahpour MR, Mounesi Rad S (2019) Accelerated infected wound healing by topical application of encapsulated rosemary essential oil into nanostructured lipid carriers. Artif Cells Nanomed Biotechnol 47(1):980–988. https://doi.org/10.1080/21691401.2019.1582539

King SR, Tempesta MS (1994) From shaman to human clinical trials: the role of industry in ethnobotany, conservation and community reciprocity. CIBA Found Symp 185:197–206. discussion 206-13

Kirsner RS, Eaglstein WH (1993) The wound healing process. Dermatol Clin 11(4):629–640. https://doi.org/10.1016/S0733-8635(18)30216-X

Kischer CW (1992) The microvessels in hypertrophic scars, keloids and related lesions: a review. J Submicrosc Cytol Pathol 24(2):281–296

Koehn FE, Carter GT (2005) The evolving role of natural products in drug discovery. Nat Rev Drug Discov 4(3):206–220. https://doi.org/10.1038/nrd1657

Komakech R, Matsabisa MG, Kang Y (2019) The wound healing potential of Aspiliaafricana (Pers.) CD Adams (Asteraceae). Evid Based Complement Alternat Med 2019:7957860. https://doi.org/10.1155/2019/7957860

Kondo T, Ishida Y (2010) Molecular pathology of wound healing. Forensic Sci Int 203(1–3):93–98. https://doi.org/10.1016/j.forsciint.2010.07.004

Kshirsagar RD, Singh NP (2001) Some less known ethnomedicinal uses from Mysore and Coorg districts, Karnataka state, India. J Ethnopharmacol 75:231–238

Lauer G, Sollberg S, Cole M, Krieg T, Eming SA, Flamme I, Stürzebecher J, Mann K (2000) Expression and proteolysis of vascular endothelial growth factor is increased in chronic wounds. J Investig Dermatol 115(1):12–18. https://doi.org/10.1046/j.1523-1747.2000.00036.x

Lerman OZ, Galiano RD, Armour M, Levine JP, Gurtner GC (2003) Cellular dysfunction in the diabetic fibroblast: impairment in migration, vascular endothelial growth factor production, and response to hypoxia. Am J Pathol 162(1):303–312. https://doi.org/10.1016/S0002-9440(10)63821-7

Liu LN, Guo ZW, Zhang Y, Qin H, Han Y (2012) Polysaccharide extracted from *Rheum tanguticum* prevents irradiation-induced immune damage in mice. Asian Pac J Cancer Prev 13(4):1401–1405. https://doi.org/10.7314/APJCP.2012.13.4.1401

Lodhi S, Pawar RS, Jain AP, Singhai AK (2006) Wound healing potential of *Tephrosia purpurea* (Linn.) Pers. in rats. J Ethnopharmacol 108(2):204–210

Mafimisebi TE, Oguntade AE (2010) Preparation and use of plant medicines for farmers' health in Southwest Nigeria: socio-cultural, magico-religious and economic aspects. J Ethnobiol Ethnomed 6(1):1

Mallefet P, Dweck AC (2008) Mechanisms involved in wound healing. Biomed Sci 52(7):609–615

Mandracchia VJ, Sanders SM, Frerichs JA (2001) The use of becaplermin (rhPDGF-BB) gel for chronic nonhealing ulcers. A retrospective analysis. Clin Podiatr Med Surg 18(1):189–209

Manzuoerh R, Farahpour MR, Oryan A, Sonboli A (2019) Effectiveness of topical administration of anethumgraveolens essential oil on MRSA-infected wounds. Biomed Pharmacother 109:1650–1658. https://doi.org/10.1016/j.biopha.2018.10.117

Marini H, Polito F, Altavilla D, Irrera N, Minutoli L, Calo M, Adamo EB, Vaccaro M, Squadrito F, Bitto A (2010) Genistein aglycone improves skin repair in an incisional model of wound healing: a comparison with raloxifene and oestradiol in ovariectomized rats. Br J Pharmacol 160(5):1185–1194. https://doi.org/10.1111/j.1476-5381.2010.00758.x

Martin P (1997) Wound healing--aiming for perfect skin regeneration. Science 276(5309):75–81

Martins MD, Marques MM, Bussadori SK, Martins MAT, Pavesi VCS, Mesquita-Ferrari RA, Fernandes KPS (2009) Comparative analysis between *Chamomilla recutita* and corticosteroids on wound healing. An in vitro and in vivo study. Phytother Res 23(2):274–278

Mazid M, Khan TA, Mohammad F (2012) Medicinal plants of rural India: a review of use by Indian Folks. Indo Glob J Pharm Sci 2:286–304

Melo MDSF, Almeida LAD, Castro KCD (2019) Chitosan membrane incorporated with *Passiflora edulis* Sims extract for potential application as wound dressing. Adv Tissue Eng Regen Med Open Access 5(4):103–108

Milovanović V, Radulović N, Todorović Z, Stanković M, Stojanović G (2007) Antioxidant, antimicrobial and genotoxicity screening of hydro-alcoholic extracts of five Serbian Equisetum species. Plant Foods Hum Nutr 62(3):113–119

Mohanraj K, Karthikeyan BS, Vivek-Ananth RP, Chand RB, Aparna SR, Mangalapandi P, Samal A (2017) IMPPAT: a curated database of Indian medicinal plants, phytochemistry and therapeutics. Sci Rep 8:4329. https://doi.org/10.1038/s41598-018-22631-z

Morris LM, Klanke CA, Lang SA, Pokall S, Maldonado AR, Vuletin JF, Alaee D, Keswani SG, Lim FY, Crombleholme TM (2010) Characterization of endothelial progenitor cells mobilization following cutaneous wounding. Wound Repair Regen 18(4):383–390. https://doi.org/10.1111/j.1524-475X.2010.00596.x

Moteallehe B, Zahedi P, Rezaeian I, Moghimi M, Abdolghaffari AH, Zarandi MA (2014) Morphology, drug release, antibacterial, cell proliferation, and histology studies of chamomile-loaded wound dressing mats based on electrospun nanofibrous poly (ε-caprolactone)/polystyrene blends. J Biomed Mater Res B Appl Biomater 102(5):977–987. https://doi.org/10.1002/jbm.b.33078

Moura LI, Dias AM, Carvalho E, de Sousa HC (2013) Recent advances on the development of wound dressings for diabetic foot ulcer treatment—a review. Acta Biomater 9(7):7093–7114. https://doi.org/10.1016/j.actbio.2013.03.033

Muhammed DO, Salih NA (2012) Effect of application of fenugreek (Trigonellafoenum-graecum) on skin wound healing in rabbits. AL-Qadisiyah J Vet Med Sci 11(2):86–93

Mukherjee PK, Mukherjee K, Pal M, Saha BP (2000) Wound healing potential of *Nelumbo nucifera* (Nymphaceae) rhizome extract. Phytomedicine 7(2):66–74

Nayak BS, Anderson M, Pereira LP (2007a) Evaluation of wound-healing potential of *Catharanthus roseus* leaf extract in rats. Fitoterapia 78(7–8):540–544

Nayak BS, Isitor G, Davis EM, Pillai GK (2007b) The evidence based wound healing activity of *Lawsonia inermis* Linn. Phytother Res 21(9):827–831

Nayak BS, Raju SS, Eversley M, Ramsubhag A (2009a) Evaluation of wound healing activity of *Lantana camara* L.–a preclinical study. Phytother Res 23(2):241–245

Nayak BS, Sandiford S, Maxwell A (2009b) Evaluation of the wound-healing activity of ethanolic extract of *Morinda citrifolia* L. leaf. Evid Based Complement Alternat Med 6(3):351–356. https://doi.org/10.1093/ecam/nem127

Nejati H, Farahpour MR, Nagadehi MN (2015) Topical rosemary officinalis essential oil improves wound healing against disseminated *Candida albicans* infection in rat model. Comp Clin Pathol 24(6):1377–1383. https://doi.org/10.1007/s00580-015-2086-z

Nemethy G, Scheraga HA (1986) Stabilization of collagen fibrils by hydroxyproline. Biochemistry 25(11):3184–3188. https://doi.org/10.1021/bi00359a016

Okoli CO, Akah PA, Nwafor SV, Anisiobi AI, Ibegbunam IN, Erojikwe O (2007) Anti-inflammatory activity of hexane leaf extract of *Aspilia africana* CD Adams. J Ethnopharmacol 109(2):219–225

de Oliveira Gonzalez AC, Costa TF, de Araujo Andrade Z, Medrado ARAP (2016) Wound healing—a literature review. An Bras Dermatologia 91(5):614–620

Pandey R, Mishra A (2010) Antibacterial activities of crude extract of *Aloe barbadensis* to clinically isolated bacterial pathogens. Appl Biochem Biotechnol 160(5):1356–1361

Passos JL, Barbosa LCA, Demuner AJ, Alvarenga ES, da Silva CM, Barreto RW (2012) Chemical characterization of volatile compounds of Lantana camara L. and L. radula Sw. and their antifungal activity. Molecules 17(10):11447–11455

Patel S, Srivastava S, Singh MR, Singh D (2019) Mechanistic insight into diabetic wounds: pathogenesis, molecular targets and treatment strategies to pace wound healing. Biomed Pharmacother 112:108615. https://doi.org/10.1016/j.biopha.2019.108615

Patil MVK, Kandhare AD, Bhise SD (2012) Anti-arthritic and anti-inflammatory activity of *Xanthium srtumarium* L. ethanolic extract in Freund's complete adjuvant induced arthritis. Biomed Aging Pathol 2(1):6–15

Patwardhan B, Vaidya AD, Chorghade M (2004) Ayurveda and natural products drug discovery. Curr Sci:789–799

Patwardhan B, Warude D, Pushpangadan P, Bhatt N (2005) Ayurveda and traditional Chinese medicine: a comparative overview. Evid Based Complement Alternat Med 2(4):465–473

Paul-Victor C, DalleVacche S, Sordo F, Fink S, Speck T, Michaud V, Speck O (2017) Effect of mechanical damage and wound healing on the viscoelastic properties of stems of flax cultivars (*Linum usitatissimum* L. cv. Eden and cv. Drakkar). PLoS One 12(10):e0185958. https://doi.org/10.1371/journal.pone.0185958

Percival NJ (2002) Classification of wounds and their management. Surgery (Oxford) 20 (5):114–117

Peters KG, De Vries C, Williams LT (1993) Vascular endothelial growth factor receptor expression during embryogenesis and tissue repair suggests a role in endothelial differentiation and blood vessel growth. Proc Natl Acad Sci 90(19):8915–8919. https://doi.org/10.1073/pnas.90.19.8915

Phan TT, Wang L, See P, Grayer RJ, Chan SY, Lee ST (2001) Phenolic compounds of *Chromolaena odorata* protect cultured skin cells from oxidative damage: implication for cutaneous wound healing. Biol Pharm Bull 24(12):1373–1379

Pugh N, Ross SA, ElSohly MA, Pasco DS (2001) Characterization of Aloeride, a new high-molecular-weight polysaccharide from Aloe vera with potent immunostimulatory activity. J Agric Food Chem 49(2):1030–1034

Puratchikody A, Devi CN, Nagalakshmi G (2006) Wound healing activity of Cyperus rotundus Linn. Indian J Pharm Sci 68(1):97. https://doi.org/10.4103/0250-474X.22976

Rai PK, Lalramnghinglova H (2011) Ethnomedicinal plants of india with special reference to an Indo-Burma hotspot region: an overview. Ethnobot Res Appl 9:379–420

Rajasekaran C, Meignanam E, Vijayakumar V, Kalaivani T, Ramya S, Premkumar N, Siva R, Jayakumararaj R (2008) Investigations on antibacterial activity of leaf extracts of *Azadirachta indica* A. Juss (Meliaceae): a traditional medicinal plant of India. Ethnobot Leaflets 2008(1):161

Rajith NP, Navas M, Swapna MM, Mohanan N (2016) Traditional wound healing plants of Kasaragod district, Kerala state, India. J Tradit Folk Pract:2–4

Rappolee DA, Mark D, Banda MJ, Werb Z (1988) Wound macrophags express TGF-a and other growth factors in vivo: analysis by mRNA phenotyping. Science 241(4866):708–712. https://doi.org/10.1126/science.3041594

Reddy GK, Enwemeka CS (1996) A simplified method for the analysis of hydroxyproline in biological tissues. Clin Biochem 29(3):225–229. https://doi.org/10.1016/0009-9120(96)00003-6

Reddy BS, Reddy RKK, Naidu VGM, Madhusudhana K, Agwane SB, Ramakrishna S, Diwan PV (2008) Evaluation of antimicrobial, antioxidant and wound-healing potentials of *Holoptelea integrifolia*. J Ethnopharmacol 115(2):249–256. https://doi.org/10.1016/j.jep.2007.09.031

Reinke JM, Sorg H (2012) Wound repair and regeneration. Eur Surg Res 49(1):35–43

Renda G, Yalcın FN, Nemutlu E, Akkol EK, Suntar I, Keles H, Ina H, Calıs I, Ersoz T (2013) Comparative assessment of dermal wound healing potentials of various Trifolium L. extracts and determination of their isoflavone contents as potential active ingredients. J Ethnopharmacol 148(2):423–432. https://doi.org/10.1016/j.jep.2013.04.031

Sahu K, Verma Y, Sharma M, Rao KD, Gupta PK (2010) Non-invasive assessment of healing of bacteria infected and uninfected wounds using optical coherence tomography. Skin Res Technol 16(4):428–437. https://doi.org/10.1111/j.1600-0846.2010.00451.x

Samant SS, Pant S (2006) Diversity, distribution pattern and conservation status of the plants used in liver diseases/ailments in Indian Himalayan Region. J Mt Sci 3:28–47

Sánchez M, Anitua E, Orive G, Mujika I, Andia I (2009) Platelet-rich therapies in the treatment of orthopaedic sport injuries. Sports Med 39(5):345–354. https://doi.org/10.2165/00007256-200939050-00002

Saporito F, Sandri G, Bonferoni MC, Rossi S, Boselli C, Cornaglia AI, Mannucci B, Grisoli P, Vigani B, Ferrari F (2018) Essential oil-loaded lipid nanoparticles for wound healing. Int J Nanomedicine 13:175. https://doi.org/10.2147/IJN.S152529

Schiller M, Javelaud D, Mauviel A (2004) TGF-β-induced SMAD signaling and gene regulation: consequences for extracellular matrix remodelling and wound healing. J Dermatol Sci 35(2):83–92. https://doi.org/10.1016/j.jdermsci.2003.12.006

Schippmann U, Leaman DJ, Cunningham AB (2002) Impact of cultivation and gathering of medicinal plants on biodiversity: global trends and issues. Biodivers. Ecosyst. Approach Agric. For. Fish, Rome

Schultes RE (1984) Fifteen years of study of psychoactive snuffs of South America: 1967-1982-a review. J Ethnopharmacol 11:17–32

Seyed SA, Farahpour MR, Hamishehkar H (2019) Topical application of cinnamon verum essential oil accelerates infected wound healing process by increasing tissue antioxidant capacity and keratin biosynthesis. Kaohsiung J Med Sci 35(11):686–694. https://doi.org/10.1002/kjm2.12120

Shahbazi E, Bahrami K (2019) Production and properties analysis of honey Nanofibers enriched with antibacterial herbal extracts for repair and regeneration of skin and bone tissues. J Pharm Pharmacol 7:37–50. https://doi.org/10.17265/2328-2150/2019.02.001

Shahouzehi B, Sepehri G, Sadeghiyan S, Masoumi-Ardakani Y (2019) Ameliorative effects of *Pistacia atlantica* resin oil on experimentally-induced skin burn in rat. Res J Pharmacogn 6(1):29–34. https://doi.org/10.22127/rjp.2018.80368

Shalom A, Kramer E, Westreich M (2011) Protective effect of human recombinant copper–zinc superoxide dismutase on zone of stasis survival in burns in rats. Ann Plast Surg 66(6):607–609. https://doi.org/10.1097/SAP.0b013e3181fc04e1

Sharma A, Dutta P, Sharma M, Rajput NK, Dodiya B, Georrge JJ, Kholia T, Bhardwaj A (2014) BioPhytMol: a drug discovery community resource on anti-mycobacterial phytomolecules and plant extracts. J Chem 6(1):46. https://doi.org/10.1186/s13321-014-0046-2

Shedoeva A, Leavesley D, Upton Z, Fan C (2019) Wound healing and the use of medicinal plants, Evid Based Complement Altern Med:2019. https://doi.org/10.1155/2019/2684108

Shenoy C, Patil MB, Kumar R (2009) Wound healing activity of *Hyptis suaveolens* (L.) Poit (Lamiaceae). Int J Pharm Tech Res 1(3):737–744

Shi CY, Yang H, Wei CL, Yu O, Zhang ZZ, Jiang CJ, Sun J, Li YY, Chen Q, Xia T, Wan XC (2011) Deep sequencing of the *Camellia sinensis* transcriptome revealed candidate genes for major metabolic pathways of tea-specific compounds. BMC Genomics 12(1):131

Shi R, Lian W, Han S, Cao C, Jin Y, Yuan Y, Zhao H, Li M (2018) Nanosphere-mediated co-delivery of VEGF-A and PDGF-B genes for accelerating diabetic foot ulcers healing in rats. Gene Ther:1. https://doi.org/10.1038/s41434-018-0027-6

Shimokado K, Raines EW, Madtes DK, Barrett TB, Benditt EP, Ross R (1985) A significant part of macrophage-derived growth factor consists of at least two forms of PDGF. Cell 43(1):277–286. https://doi.org/10.1016/0092-8674(85)90033-9

Shimomura H, Spiro RG (1987) Studies on macromolecular components of human glomerular basement membrane and alterations in diabetes: decreased levels of heparan sulfate proteoglycan and laminin. Diabetes 36(3):374–381. https://doi.org/10.2337/diab.36.3.374

Simon F, Bergeron D, Larochelle S, Lopez-Vallé CA, Genest H, Armour A, Moulin VJ (2012) Enhanced secretion of TIMP-1 by human hypertrophic scar keratinocytes could contribute to fibrosis. Burns 38(3):421–427. https://doi.org/10.1016/j.burns.2011.09.001

Singer AJ, Clark RA (1999) Cutaneous wound healing. N Engl J Med 341(10):738–746

Spyrou GE, Watt DA, Naylor IL (1998) The origin and mode of fibroblast migration and proliferation in granulation tissue. Br J Plast Surg 51(6):455–461. https://doi.org/10.1054/bjps.1997.0277

Srivastava M, Eidelman O, Zhang J, Paweletz C, Caohuy H, Yang Q, Jacobson KA, Heldman E, Huang W, Jozwik C, Pollard BS, Pollard HB (2004) Digitoxin mimics gene therapy with CFTR and suppresses hypersecretion of IL-8 from cystic fibrosis lung epithelial cells. Proc Natl Acad Sci U S A 101(20):7693–7698. https://doi.org/10.1073/pnas.0402030101

Stroncek JD, Reichert WM (2008) Overview of wound healing in different tissue types. In: Indwelling neural implants: strategies for contending with the in vivo environment. CRC Press, Boca Raton, pp 3–40

Sudhof TC (2001) Alpha-Latrotoxin and its receptors: neurexins and CIRL/latrophilins. Annu Rev Neurosci 24:933–962. https://doi.org/10.1146/annurev.neuro.24.1.933

Sultana B, Anwar F (2008) Flavonols (kaempeferol, quercetin, myricetin) contents of selected fruits, vegetables and medicinal plants. Food Chem 108(3):879–884

Sumitra M, Manikandan P, Suguna L (2005) Efficacy of Butea monosperma on dermal wound healing in rats. Int J Biochem Cell Biol 37(3):566–573

Tanaka E, Galliot B (2009) Triggering the regeneration and tissue repair programs. Development 136:349–353

Tanaka D, Kagari T, Doi H, Shimozato T (2006) Essential role of neutrophils in anti-type II collagen antibody and lipopolysaccharide-induced arthritis. Immunology 119(2):195–202

Tandara AA, Mustoe TA (2004) Oxygen in wound healing—more than a nutrient. World J Surg 28(3):294–300. https://doi.org/10.1007/s00268-003-7400-2

Taranalli AD, Kuppast IJ (1996) Study of wound healing activity of seeds of *Trigonellafoenum graecum* in rats. Indian J Pharm Sci 58(3):117

Thakur R, Jain N, Pathak R, Sandhu SS (2011) Practices in wound healing studies of plants. Evid Based Complement Alternat Med. https://doi.org/10.1155/2011/438056

Thang PT, Teik LS, Yung CS (2001) Anti-oxidant effects of the extracts from the leaves of *Chromolaena odorata* on human dermal fibroblasts and epidermal keratinocytes against hydrogen peroxide and hypoxanthine–xanthine oxidase induced damage. Burns 27(4):319–327

Tiwari S, Patil R, Bahadur P (2019) Polysaccharide based scaffolds for soft tissue engineering applications. Polymers 11(1):1. https://doi.org/10.3390/polym11010001

Trabelsi I, Slima SB, Ktari N, Bardaa S, Elkaroui K, Abdeslam A, Salah RB (2019) Purification, composition and biological activities of a novel heteropolysaccharide extracted from *Linum usitatissimum* L. seeds on laser burn wound. Int J Biol Macromol S0141-8130 (19):35157–35158. https://doi.org/10.1016/j.ijbiomac.2019.10.077

Trompezinski S, Bonneville M, Pernet I, Denis A, Schmitt D, Viac J (2010) Gingko biloba extract reduces VEGF and CXCL-8/IL-8 levels in keratinocytes with cumulative effect with epigallocatechin-3-gallate. Arch Dermatol Res 302(3):183–189

Trowbridge JM, Gallo RL (2002) Dermatan sulfate: new functions from an old glycosaminoglycan. Glycobiology 12(9):117R–125R. https://doi.org/10.1093/glycob/cwf066

Upadhyay NK, Kumar R, Siddiqui MS, Gupta A (2011) Mechanism of wound-healing activity of *Hippophae rhamnoides* L. leaf extract in experimental burns. Evid Based Complement Altern Med. https://doi.org/10.1093/ecam/nep189

Wahl SM, Wong H, McCartney-Francis N (1989) Role of growth factors in inflammation and repair. J Cell Biochem 40(2):193–199. https://doi.org/10.1002/jcb.240400208

Wells RG (2000) V. TGF-β signaling pathways. Am J Physiol Gastrointest Liver Physiol 279(5): G845–G850. https://doi.org/10.1152/ajpgi.2000.279.5.G845

Wender PA, Kee JM, Warrington JM (2008) Practical synthesis of prostratin, DPP, and their analogs, adjuvant leads against latent HIV. Science 320(5876):649–652. https://doi.org/10.1126/science.1154690

Weng SJ, Yan DY, Tang JH, Shen ZJ, Wu ZY, Xie ZJ, Yang JY, Bai BL, Chen L, Boodhun V, Yang L (2019) Combined treatment with Cinnamaldehyde and β-TCP had an additive effect on bone formation and angiogenesis in critical size calvarial defect in ovariectomized rats. Biomed Pharmacother 109:573–581. https://doi.org/10.1016/j.biopha.2018.10.085

Werner S, Grose R (2003) Regulation of wound healing by growth factors and cytokines. Physiol Rev 83(3):835–870. https://doi.org/10.1152/physrev.2003.83.3.835

White NJ (1997) Assessment of the pharmacodynamic properties of antimalarial drugs in vivo. Antimicrob Agents Chemother 41(7):1413–1422

WHO (1993) World Health Organization research guidelines for evaluating the safety and efficacy of herbal medicines. Regional Office for the Western Pacific, Manila

WHO (2004) World Health Organization guidelines on safety monitoring of herbal medicines in pharmacovigilance systems, Geneva

Wu F, Bian D, Xia Y, Gong Z, Tan Q, Chen J, Dai Y (2012) Identification of major active ingredients responsible for burn wound healing of *Centella asiatica* herbs. Evid Based Complement Altern Med:2012. https://doi.org/10.1155/2012/848093

Yamakawa S, Hayashida K (2019) Advances in surgical applications of growth factors for wound healing. Burns Trauma 7(1):10. https://doi.org/10.1186/s41038-019-0148-1

Yang HL, Tsai YC, Korivi M, Chang CT, Hseu YC (2017) Lucidone promotes the cutaneous wound healing process via activation of the PI3K/AKT, Wnt/β-catenin and NF-κB signalling pathways. Biochim Biophys Acta Mol Cell Res 1864(1):151–168. https://doi.org/10.1016/j.bbamcr.2016.10.021

Yoder MC, Ingram DA (2009) The definition of EPCs and other bone marrow cells contributing to neoangiogenesis and tumor growth: is there common ground for understanding the roles of numerous marrow-derived cells in the neoangiogenic process? Biochim Biophys Acta 1796 (1):50–54. https://doi.org/10.1016/j.bbcan.2009.04.002

Yu W, Nairn JO, Lanzafame RJ (1994) Expression of growth factors in early wound healing in rat skin. Lasers Surg Med 15(3):281–289. https://doi.org/10.1002/lsm.1900150308

Yuan X, Han L, Fu P, Zeng H, Lv C, Chang W, Runyon RS, Ishii M, Han L, Liu K, Fan T (2018) Cinnamaldehyde accelerates wound healing by promoting angiogenesis via up-regulation of PI3K and MAPK signaling pathways. Lab Investig 98(6):783. https://doi.org/10.1038/s41374-018-0025-8

Yue DK, McLennan S, Marsh M, Mai YW, Spaliviero J, Delbridge L, Reeve T, Turtle JR (1987) Effects of experimental diabetes, uremia, and malnutrition on wound healing. Diabetes 36(3):295–299. https://doi.org/10.2337/diab.36.3.295

Yusuf S, Agunu A, Diana M (2004) The effect of *Aloe vera* A. Berger (Liliaceae) on gastric acid secretion and acute gastric mucosal injury in rats. J Ethnopharmacol 93(1):33–37

Zhao B, Zhang X, Han W, Cheng J, Qin Y (2017) Wound healing effect of an Astragalus membranaceus polysaccharide and its mechanism. Mol Med Rep 15(6):4077–4083. https://doi.org/10.3892/mmr.2017.6488

Zitterl-Eglseer K, Sosa S, Jurenitsch J, Schubert-Zsilavecz M, Della Loggia R, Tubaro A, Bertoldi M, Franz C (1997) Anti-oedematous activities of the main triterpendiol esters of marigold (*Calendula officinalis* L.). J Ethnopharmacol 57(2):139–144

CPSIA information can be obtained
at www.ICGtesting.com
Printed in the USA
BVHW012353081220
595246BV00001B/2

9 789811 551352